José Galizia Tundisi

Takako Matsumura-Tundisi

Limnology

CRC Press
Taylor & Francis Group
Boca Raton   London   New York   Leiden

CRC Press is an imprint of the
Taylor & Francis Group, an **informa** business

A BALKEMA BOOK

CRC Press
Taylor & Francis Group
6000 Broken Sound Parkway NW, Suite 300
Boca Raton, FL 33487-2742

First issued in paperback 2017

Originally published in Portuguese as: "Limnologia",
© 2008 Oficina de Textos, Editora Signer Ltda, São Paulo, Brazil

English edition 'Limnology', CRC Press Balkema, Taylor & Francis Group, an informa Business
© 2012 Taylor & Francis Group, London, UK

Translation to English: Lyle Prescott
English version proofreading: Dr Jack Talling

© 2012 Taylor & Francis Group, London, UK
CRC Press/Balkema is an imprint of the Taylor & Francis Group, an informa business

Typeset by MPS Limited, a Macmillan Company, Chennai, India

British Library Cataloguing in Publication Data
A catalogue record for this book is available from the British Library

Library of Congress Cataloging-in-Publication Data

Tundisi, J. G.
    Limnology / José Galizia Tundisi, Takako Matsumura Tundisi.
        p. cm.
    Includes bibliographical references and index.
    ISBN 978-0-415-58835-5 (hbk. : alk. paper) 1. Limnology—Latin America.
2. Limnology—Brazil. 3. Reservoirs—Latin America. 4. Reservoirs—Brazil.
I. Tundisi, Takako Matsumura. II. Title.

    QH106.5.T86   2012
    551.48—dc23
                                                2011042600

Visit the Taylor & Francis Web site at
http://www.taylorandfrancis.com

and the CRC Press Web site at
http://www.crcpress.com

ISBN 13: 978-0-415-58835-5 (hbk)
ISBN 13: 978-1-138-07204-6 (pbk)

Limnology

This book is dedicated to: Professor John E.G. Raymont; Professor A.P.M. Lokwood from the Department of Oceanography, University of Southampton (England); Professor Marta Vannucci from the Oceanographic Institute of University of São Paulo, Brazil. To Dr. José Eduardo Matsumura Tundisi for his permanent dedication to International Institute of Ecology, São Carlos, Brazil and to the support to our research work.

The authors

Acknowledgements

The translation to English from the Portuguese edition of this book, was financially supported by the São Paulo State Foundation for Research (FAPESP). The authors would like to thank CNPq, CAPES, FINEP and FAPESP for their support in countless limnological research projects, participation in scientific meetings and publications, and for their ongoing support of our Masters and Doctoral fellows; and to FAPESP for their assistance in three thematic projects: typology of dams, comparison of Barra Bonita and Jurumirim, and Biota/FAPESP.

We would also like to thank the Organization of American States; the National Science Foundation; Japan's Ministry of Education, Science and Culture; the *Instituto Estadual de Florestas* (MG) ["State Forestry Institute"] of Brazil; and Eletronorte, Eletrobrás and Elektro for supporting limnological research projects on Brazilian dams, and the CESP (SP) for encouraging research on the lakes in Parque Florestal do Rio Doce and on Amazonian lakes. The authors also extend their thanks to Furnas, which recently supported a large-scale research project (carbon balance in the reservoirs of Furnas power plants), the United Nations Environmental Program (UNEP), Investco, University of São Paulo, and the Federal University of São Carlos.

We also thank the Conrado Wessel Foundation for presenting its 2005 award of the Applied Water Science (2005) to José Galizia Tundisi, an important stimulus for the continuation of this work; as well as the Brazilian Academy of Sciences, and the United Nations University.

Special thanks to Dr. José Eduardo Matsumura Tundisi for his steady support and encouragement of our work, Dr. Aristides Pacheco Leão and Dr. José Israel Vargas for their decisive support of the study of lakes in Parque Florestal do Rio Doce; Dr. Paulo Emilio Vanzolini, for his support and encouragement of our limnological research work, to he former presidents of the Federal University of São Carlos and Professors Luiz Magalhaes and Edmundo de Souza Heitor Gurgulino for supporting limnological research and its consolidation in UFSCar; to Dr. Odette Rocha for reviewing parts of the work and the bibliographic compilation and Vera Huszar for revising some of the figures; to Dr. Naércio Aquino Menezes for reviewing and compiling the table on the Orders of freshwater fish. We also thank Dr. Milan Straškraba, Dr. Colin Reynolds, Dr. Ramón Margalef, Dr. Henry Dumont, Dr. Clóvis Teixeira; Dr. Yatsuka Saijo, Dr. Francisco A. Barbosa, Dr. Ernesto Gonzalez, Dr. Guilhermo Chala, Dr. Marcos Gomes Nogueira, Dr. Adriana Jorcin, Dr. Arnola Rietzler, Dr. Raoul Henry, Dr. Evaldo Espindola, Dr. Sven Jørgensen, Dr. Joan Armengol, and Dr. Abilio Lopes de Oliveira Neto for the opportunity to exchange information, publications and sharing of studies that resulted in the publication of this work; and Dr. Sydney and Dr. Magela Thomaz Luiz Mauricio Bini for permission to publish figures from "Aquatic Macrophytes" (Eduem). Many thanks to photographers Mario Pinedo Panduro and Luiz Marigo and the New York Botanical Gardens, for permission to use some of their photos.

Thanks to researchers from the International Institute of Ecology who collaborated with photos, reviews of parts of the work, suggestions and criticism; to Dr. Donato Seiji Abe and Dr. Corina Sidagis Galli, Daniela Cambeses Pareschi, Anna Paula Lucia, Guilherme Ruas Medeiros, Thaís Helena Prado, Fernando de Paula Blanco, Nestor Freitas Manzini, Paulo Henrique Von Haelin, Eduardo Henrique Frollini, José Augusto Fragale Baio, Juan Carlos Torres Fernández, Heliana Rosely Neves Oliveira, Rogério Flávio Pessa, and Valeria Teixeira da Silva. Thanks also to secretarial assistance from Miriam Aparecida Meira, Denise Helena Araujo, Luciana Zanon, Natalia Andricioli Periotto, Suelen Botelho, and José Jesuel da Silva who typed, corrected and formatted the first versions of the work. And to Mr. João Gomes da Silva, for continued support for our field work (40 years), and to Marta Vanucci.

Last but not least, our thanks go to our tireless, professional and competent team at the publisher *Oficina de Textos*: to our editor Shoshana Signer for her decisive and determined support, editorial manager Ana Paula Ribeiro, art director Malu Vallim, and typesetter Gerson Silva.

Foreword by Dr. Jack Talling

This book provides an expansive and detailed account of limnology – the science of inland waters – from a tropical and Brazilian viewpoint. Although it draws extensively upon the international literature, and is richly illustrated with imported examples, it gives special emphasis to tropical conditions and the Brazilian experience. Here both Amazonia and the authors' distinguished record and enterprises are prominent. It is primarily not a descriptive text, but aims at generalised classification of examples and a dynamic and functional approach. Issues of applied limnology – especially with reservoirs – are treated extensively. There is also opinion on past trends and future prospects in the science. These features, with the combination of length and exceptionally rich illustration, mark a unique contribution to the environmental literature, and especially apt for developing countries.

Dr. Jack Talling
FRS, Freshwater Biological Association,
Cumbria,
United Kingdom

Foreword by Dr. Joan Armengol Bachero

The authors have asked me to write an introduction to this book, an honor that I accept with pleasure as it gives me an opportunity to express the admiration I have for them, based on knowledge of their lifetime's work studying inland waters, and on the deep friendship we share, fruit of our extensive collaboration over the years.

José G. Tundisi and Takako Matsumura Tundisi wrote this book after many years of teaching limnology and for this reason the book follows a pattern typical of courses in this field. But the book is also the result of many years of field and laboratory research, basic and applied research, and exploration of aquatic ecosystems throughout Brazil. The work includes the search for solutions to reduce human impact, establish criteria for water resource management, and restore altered or contaminated systems. In short, the versatility of concepts and systems studied make this a multifaceted book.

The book has a clear geographic component, Brazil, and thus highlights tropical and subtropical limnology.

The book's structure follows a modern approach, with the first 10 chapters devoted to processes that we might call physical, chemical and biological limnology. The second part, Chapters 11 to 17, corresponds to the limnology of systems. Finally, the last three chapters focus on applied limnology.

I would like to highlight aspects of the content that, independent of the book's structure, I find particularly innovative. In order of appearance, the treatment given to hydrodynamics seems most appropriate, with an agile yet rigorous presentation of physical processes that govern the movement of water masses. The authors' experience is evident in the study of reservoirs where hydraulic stratification and meteorological effects are particularly relevant to the chemical and biological processes. A strong naturalist component is also present, logically to be expected given the authors' background. Reservoirs are treated intensively since the researchers have devoted much of their research to these ecosystems. I am pleased to see that the book has a section on estuaries and coastal lagoons, not just because of their intrinsic importance, but because transitional ecosystems tend to be neglected in many of the treatises on inland or marine waters. I especially appreciate the chapters on regional limnology not just because it is a pioneering issue in limnology, but also because it forms the basis of many modern theories of inland aquatic ecology. In the second chapter the authors discuss many such studies conducted on a global scale. I have always believed that the speciality that best describes the authors is that of regional limnologists. Their study on the typology of reservoirs in São Paulo is not only the first in its field in South America, but it also marks the beginning of a school of limnology under their tutorship. It is therefore no exaggeration to say that under the auspices of this project, a generation of Brazilian limnologists was formed who are currently scattered throughout the country working in the discipline. Finally, applied limnology is covered in the last three chapters.

To be honest, the chapters on reservoirs and regional and applied limnology give the book its personality. Since these are the fields most impacted by the authors' lifework, these chapters clearly reflect their vision of limnology.

I want to point out that a book of this calibre is not a casual undertaking. It is the result of an educational process first, followed by study; the education of specialists, teaching and mastery, of study in function of the needs of a country. It is important to remember that Brazil has 14% of the freshwater reserves of the biosphere, and that the country has clearly opted to develop hydropower. Among its many attributes, Brazil has both the largest river on the planet and serious water shortages in a large part of the country. This range of factors has generated a need for knowledge, basic information and application of results. These are the fibers of which this basket is made. For this reason while the book has two authors – José Tundisi and Takako Matsumura Tundisi – it also has many companions who, far from being anonymous, have joined forces to make it possible for a level of knowledge to be attained in Brazil so that a book such as this can be written.

Dr. Joan Armengol Bachero
Ecology Professor
Department of Ecology
University of Barcelona

Foreword by Dr. Blanca Elena Jiménez Cisneros

Limnology, first considered to be a science in the 19th century, has undergone major advances in recent years. A better understanding of the bodies of water located in tropical and semi-tropical climates is one notable recent contribution. The authors of this book, Dr. José Galizia Tundisi and Dr. Takako Matsumura Tundisi, have made many significant contributions to this the field. Consequently, this text allows the reader not only to learn about the basic principles of global limnology, but also to understand previously little-known features of tropical and Neotropical bodies of water common in many developing countries. In spite of the importance of these bodies of water, they have rarely been addressed in international literature.

Many of the book's examples are taken from bodies of water in Brazil, which is appropriate because of the country's abundance of water (nearly 14% of the world's water resources are found there) and the wide variety of climates due to the country's size and location. These examples represent a wide diversity of geographic regions and illustrate conditions also found in many other countries. Moreover, to make their coverage truly universal, the authors describe examples from other regions, such as Africa, Asia and Europe. The text covers nearly all types of inland bodies of water, including some not often seen in books on limnology, such as estuaries, which are of particular interest to most Latin America and Caribbean countries.

Finally, the book does not limit itself to an academic focus, as each chapter balances a mixture of scientific and technical aspects with practical aspects to help explain, analyse and make better use of water in every way. In fact, several of the final chapters discuss how to conserve and restore aquatic ecosystems with a focus on the drainage basin. This mix makes the book an excellent tool for undergraduates and postgraduates, as well as for professionals from other disciplines concerned with the management and reasonable use of water resources. In particular, by highlighting the importance of combining sustainable development with economic development, the book provides a much needed perspective for readers in developing countries.

With this in mind, I leave the reader to enjoy the book and would like to invite the authors to translate their work into Spanish in order to share their knowledge more extensively.

Dr. Blanca Elena Jiménez Cisneros
Senior Professor and Researcher
Department of Environmental Engineering
National Autonomous University of Mexico (UNAM)
Recipient of the National Award in Ecology (2006)

Foreword by Dr. Odete Rocha

This book is very important in the context of aquatic ecology and Brazilian limnology, as it will fill a great gap in textbooks and popular books on the science of limnology, making the science more available to the Portuguese-speaking population. It also represents a framework for tropical limnology by inserting widely and carefully results from research in the Neotropics, which is not very common in similar textbooks, written and published by authors on other continents.

The vast experience in limnology of the authors of this book is refected in the excellent quality of this work. They have dedicated an enormous effort and a large part of their professional careers.

This book can be widely used by many people at different academic levels to contemplate the basic themes of this science, written in accessible language, with examples and illustrations that facilitate the understanding of the topics addressed. Because research results on tropical ecosystems were included, in addition to other regions, the book will even serve as a source of information for many researchers in different areas. It will also be a significant work for administrators, managers and decision-makers and activities related to water resources in the Brazil.

Brazil is a country rich in inland aquatic resources, from the Amazonian mega-basin to an endless network of micro-basins. The rational use and preservation of these resources, however, depend on education in limnology, and the present book makes a signficant contribution toward that end.

Dr. Odete Rocha
Professor in the Department of Ecology and Evolutionary Biology
Federal University of São Carlos

Contents

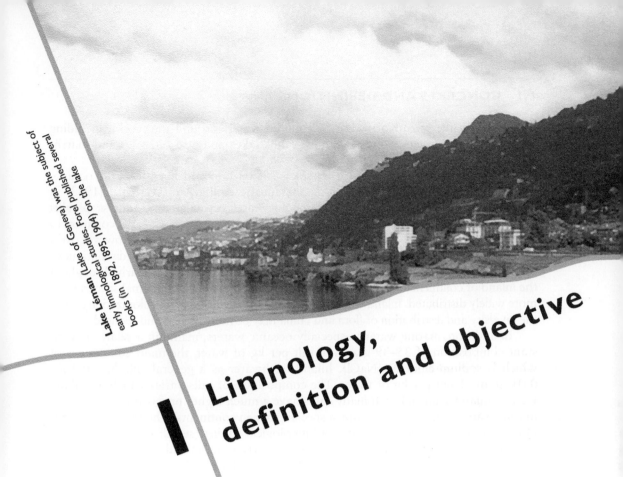

Lake Léman (Lake of Geneva) was the subject of early limnological studies. Forel published several books (in 1892, 1895, 1904) on the lake

Limnology, definition and objective

SUMMARY

The history of limnology has steadily evolved over the last 120 years, both conceptually and technically. Beginning with Forel's classic work on Lac Léman, published in three volumes (1892, 1895 and 1904), and Forbes' classic work on lakes as microcosms (1887), scientific interest in limnology – which encompasses the physics, chemistry and biology of lakes – has continued to grow. Since the early 20th century, research laboratories in many countries in the Northern Hemisphere have continuously promoted limnological research and training of human resources. Initially regarded as the science of lakes, limnology now encompasses **freshwater** and **saline lakes** in the inland of continents, rivers, estuaries, **reservoirs**, wetlands, marshes and all physical, chemical and biological interactions in these ecosystems.

Limnology has contributed significantly to the grounding and expansion of **theoretical ecology**, and modern management of aquatic systems cannot ignore the need for a limnological knowledge base for effective long-term management.

Tropical limnology has advanced through studies undertaken by geographical expeditions and consolidated by research in laboratories in the vicinity of several lake and river systems in Central and South America, Africa, Southeast Asia, and Australia. In Brazil, early limnological studies focused on **fishing**, aquaculture and applied studies in the area of health. In the last 30 years in Brazil, limnology has advanced significantly as a result of studies on several artificial and natural ecosystems, the founding of the Brazilian Limnology Society, the Congress of the International Association of Limnology held in São Paulo (1995), and the relevance of basic research in the management of **water basins**, fishing and lakes, reservoirs and **wetlands**.

1.1   CONCEPTS AND DEFINITIONS

Limnology is the scientific study of all **inland waters** around the world, including lakes, reservoirs, rivers, ponds, swamps, saline lakes and also estuaries and marshlands in coastal regions.

There are several definitions of limnology: Forel (1892) defined it as the oceanography of lakes, Lind (1979) as non-marine aquatic ecology, and Margalef (1983) as the ecology of non-oceanic waters. Limnology and oceanography address parallel problems and processes, as the liquid medium (i.e., the **substrate** water) is common to **oceans**, lakes, and rivers and has certain fundamental properties. Oceans, however, are a *continuum* in space and are much older than inland waters. Inland waters are discontinuous in space, relatively ephemeral (in geological time), and distributed irregularly throughout the inland of continents. The continuity of oceans enables plant and animal species to be more widely distributed. Inland waters rely on the diverse processes of **colonization** and so the diversity and distribution of flora and fauna may be more limited and reduced.

In addition, marine waters, especially oceanic waters, maintain a relatively constant composition of 35–39 grams of salt per kg of water, the main component of which is sodium chloride (NaCl). Inland freshwater as a general rule has at least 0.01 gram of salt per kg of water. The composition of salts varies widely in inland waters. Inland saline lakes in many cases have a much higher proportion of salt than marine waters. Saline lakes occupy a special place in continents. Their ecosystems are unusual and thus are also the focus of limnological studies (see Chapter 15).

Chemical processes and mechanisms occurring in inland waters are highly dependent on the geochemistry of soils in drainage basins. Aquatic systems interact with their drainage basin in various sub-systems and components. That concept and comprehensive studies on drainage basins in relation to lakes, rivers, reservoirs and wetlands are more recent (Borman and Likens, 1967, 1979). The features of water basins determine, for example, the origins of the matter contributing to the formation and functioning of lakes, rivers and reservoirs (see Chapter 11).

Another definition of limnology is from Baldi (1949). He defined limnology as the scientific study of the interrelated processes and methods by which matter and energy are transformed in a lake. He considered the essence of limnology to be the study of movement of matter in a body of water.

In all these definitions, two main aspects must be considered: the descriptive and the functional, along with the necessary synthesis.

The following definition was made as a broad summary:

Limnology is the science of inland waters studied as ecosystems:

▸ An ecosystem is a natural unit consisting of living components (biotic) and non-living components (abiotic) that belong to a system of **energy flow** and cycling of matter.

▸ In the structural analysis, two basic aspects are included: first, description of the abiotic components and their properties (**physical** and chemical **factors**, concentrationss and intensities); second, assessment of biotic communities (composition of species, abundance, **biomass** and life cycles).

▸ Analysis of the interrelated functions in an ecosystem includes research on the elements responsible for the cycling of matter, **dynamic processes** in abiotic systems, relationships of the organisms to environmental factors and relationships between organisms.

▸ Limnological research includes analytical field and laboratory research, the results of which may achieve limnological synthesis. Even in studies with limited objectives, realistic connections to the system as a whole should be established. Such studies should contribute to the body of knowledge on the entire system's structure and function.

▸ Assessment of the practical merits of research in limnology should be made based on detailed analysis of scientific teachings and other knowledge such as basic limnology. A limnologist who acquires experience through his/her own research (individually or in team work) will be better prepared to conduct research and provide interdisciplinary training in limnology. Development and consolidation of research on one of the central topics in limnology should be valued, instead of research on peripheral problems. (Summary based on a presentation made to the International Society of Limnology, 1989, Munich, Germany.)

1.1.1 Contributions of limnology to theoretical ecology

Throughout its history as a science, limnology has contributed significantly to the development of theoretical ecology. Contributions include:

▸ **Community succession** and factors that control it (studies of **phytoplankton succession**, development of the benthic community in different types on substrata, **periphyton succession**, and succession of **fish communities**);

▸ **Evolution of communities** (studies on eutrophication in lakes, **restoration** of eutrophic lakes and reservoirs);

▸ **Community diversity** and **spatial heterogeneity** (studies of periphyton and phytoplankton in different ecosystems, **aquatic insects**, comparative studies on lakes, reservoirs and floodplains. Theory and studies of **ecotones**);

▸ **Primary production** and energy flow (studies on the primary productivity of phytoplankton, **aquatic macrophytes** and periphyton, **feeding** habits of zooplankton and fishes. Physiological responses of phytoplankton to light intensity and **concentration of nutrients**);

▸ **Distribution of organisms** and factors that control dispersal and colonization mechanisms (studies on the vertical migration of zooplankton, vertical distribution of phytoplankton, colonization in reservoirs and **temporary waters**, distribution of **aquatic organisms** in lakes, rivers and reservoirs);

▸ Evolution of ecosystems (studies on **eutrophication**, reservoirs, monitoring reservoirs and alterations resulting from **human activities**).

The contribution of limnology to theoretical ecology took place over a period of approximately 100 years, from early limnological studies and their organization as a science in the final decades of the 19th century. Because of the relative ease of studying organisms, **populations** and communities, the contribution of limnology to theoretical ecology has been invaluable. The paradigm of this contribution can be seen in the summary published by Reynolds (1997) and in the various summaries and hypotheses by Margalef (1998).

1.2   LIMNOLOGY: HISTORY AND DEVELOPMENT

The scope and definitions of any science cannot be considered without looking at the history of its development and the individuals, institutions and groups that have contributed to its progress. This historical overview highlights some pivotal concepts and theoretical trends, as well as theoretical tendencies and basic lines of reasoning employed in many countries and regions. The early history of limnology was described by Elster (1974) and Ueno (1976), with a more recent extension by Talling (2008).

Aquatic organisms attracted the attention of scientists and naturalists in the 17th, 18th and 19th centuries, as can be seen in the works of Leeuwenhoek, Müller, Schaffer, Trembley, Eichhorn, Bonnet, and Goetze. The studies focused on aquatic organisms and their behavior and propagation in water.

With the discovery and early descriptions of marine **plankton** by Müller in 1845, interest in freshwater organisms grew, especially **plankton** in lakes (Schwoerbel, 1987).

The description and measurement of internal **waves** by F. Duvillier and the first descriptions of thermal structure, **wind action** and light penetration in deep lakes by J. Leslie (1838) were important milestones in the progress of limnology (Goldman and Horne, 1983). Also, **diurnal fluctuations** in photosynthetic activities were described by Morren and Morren (1841).

Junge (1885) and Forbes (1887) were the first to treat a lake as a microcosm. In particular, Forbes' work, *The lake as a microcosm* (see Figure 1.1), made a big impact because it highlighted the fact that lakes formed a microcosm in which all elemental forces are at play and the life forms constitute an interrelated complex (Forbes, 1887, p. 537).

Forbes' work had important consequences that stimulated limnology, but the work of F. A. Forel (1901) was the first synthesis and first book on this science.

In his extensive monograph on Lake Léman (see Chart 1.1), Forel studied the lake's biology, physics, and chemistry, and also formulated original concepts on different types of lakes.

The development of limnology as an organized science began to grow at the end of the 19th century, and already by the early 20th century many limnological field stations and working laboratories had been established near lakes. In 1901, for example, Otto Zacharias founded a limnology research institute in Plön (1981), Germany, which has played an important role ever since (it is now the Max-Planck Institute for Evolutionary Biology).

Subsequent developments in limnology, still in the early 20th century, can be seen in the work of Thienemann (1882–1960) of Germany, and Naumann (1891–1934) of Sweden, who worked independently at first and then together established the first comparative studies on the European continent. The work established an orderly classification, taking into account regional characteristics and **biogeochemical cycles**. The oligotrophic-eutrophic system introduced by these studies, using concepts from Weber (1907), constituted a very important base for advancing the development of limnology. The classification of lakes by trophic status was a first step in the development of this science.

The **typology of lakes** proposed by Birge and Juday (1911) considered the relationships between productivity of organic matter, lake depth, lake morphology and **dissolved oxygen** content.

STATE OF ILLINOIS
DEPARTMENT OF REGISTRATION AND EDUCATION
DIVISION OF THE
NATURAL HISTORY SURVEY
STEPHEN A. FORBES, Chief

Vol. XV. BULLETIN Article IX.

The Lake as a Microcosm

BY
STEPHEN A. FORBES

PRINTED BY AUTHORITY OF THE STATE OF ILLINOIS

URBANA, ILLINOIS
November, 1925

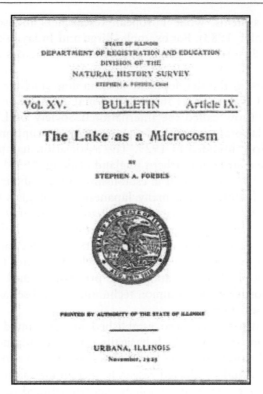

Figure 1.1 Reproduction of the cover of Forbe's study published in Illinois (USA) in 1925 (Vol. XV, Bulletin, Article IX).

Chart 1.1 Books published by F.A. Forel and their contents.

Volume 1 (1892)	*Volume 2 (1895)*	*Volume 3 (1904)*
1. Geography 2. Hydrography 3. Geology 4. Climatology 5. Hydrology	6. Hydraulics 7. Techniques 8. Optics 9. Acoustics 10. Chemistry	11. Biology 12. History 13. Navigation 14. Fishing

Source: Le Leman. *Monographie limnologique*.

In North America, L. Agassiz (1850) was an early pioneer; essential contributions were later made by Birge (1851–1950) and Juday (1872–1944), who studied the effect of **thermal** and chemical **stratification** on the composition of plankton. They also conducted comparative studies on North American lakes and studied quantities such as water transparency, organic matter and **phosphorus**, developing graphic correlations with frequency distributions and trends (Juday and Birge, 1933). Juday (1916) also conducted comparative studies on several lakes in Central America.

An important difference between the development of limnology in Europe and in the United States is that from very early on, American researchers focused on chemical cycles in systems, while European researchers focused on the study of communities (Margalef, 1983). For example, Birge and Juday used the concentration and distribution of dissolved oxygen to express a set of factors describing the functioning of lakes.

Important events in the development of limnology include the 1922 founding by Thienemann and Naumann of the International Association of Theoretical and Applied Limnology (now called the International Society of Limnology), and the establishment of a laboratory at Windermere (1931) in support of the Freshwater Biological Association, founded in 1929. The association has produced significant works from the lake district in northern **England** (Talling, 1999). In **Japan**, the work of Yoshimura (1938) was important in establishing a scientific information base. An important scientific feature is that many Japanese limnologists have also produced scientific works in oceanography. Japan may be one of the only countries where the boundaries blur between limnology and oceanography, mainly because of the common focus of application in terms of eutrophication studies on inland and coastal waters. Inland, a major focus is on **aquaculture**, which requires thorough knowledge of the major limnological and oceanographic processes – and even a comparative understanding – in order to use common techniques to utilise lacustrine and marine systems for food production.

Research laboratories were also established in the United States and Europe near lakes or regional systems, which basically served as adjuncts to large internationally renowned institutions. The laboratories were active centres that gathered scientific information and conducted research on regional **aquatic ecosystems**. The ever-growing body of research in limnology has contributed to the development of theories, evolution of the science, and greater understanding of regional systems.

Another important factor in this robust and growing science has been the continuing education and training of qualified researchers who across several decades have contributed significantly to scientific advancement in several areas.

Tropical limnology largely advanced through research by groups from temperate regions. As Margalef (1983) points out, an understanding of basic limnological processes should certainly take into account systems in **tropical regions**. In the **Sunda Expedition** in 1928–1929, Thienemann, Ruttner and Feuerborn (Talling, 1996) produced important comparative data, as did the early work of Worthington and of Beadle (1932) in Africa. The work of Thienemann (1931) highlighted the absence of hypolimnetic oxygen in lakes in Java, Sumatra and Bali, and revealed problems in the traditional **oligotrophic/eutrophic** classification used for temperate-region lakes.

The classifications described by Thienemann and Naumann had an important catalytic effect on the scientific development of limnology, and by the 1950s, the classification of lakes had become a fundamental factor. Thienemann (1925) added the term *dystrophy* to the eutrophic-oligotrophic terms to describe lakes with high concentrations of **humic substances**.

In Central and South America, North American and European influences alternated. In South America, major rivers and **deltas** were studied extensively by researchers from the Max Planck Institute (Sioli, 1975) and the National Institute for Amazonian Research (INPA). The Paraná, Uruguay and Bermejo Rivers were studied extensively by Bonetto (1975, 1986a, b), Neiff (1986), and Di Persia and Olazarri (1986).

In Central America, researchers conducted field work during various expeditions to different countries. An early study, already mentioned, was published by Juday (1908). Guatemalan and Nicaraguan lakes were studied by Brezonik and Fox (1975), Brinson and Nordlie (1975), Cole (1963), Covich (1976), Cowgill and Hutchinson (1966). Studies on **Lake Amatitlan** in Guatemala (Basterrechea, 1986) and **Lake Managua** (Lake Xolotlán) in Nicaragua (Montenegro, 1983, 2003) awoke interest in limnology in these two countries. In Chile, limnological studies were conducted at the **Rapel Dam** (Bahamonde and Cabrea, 1984) and in **Venezuela**, limnological studies were conducted by Infante (1978, 1982), Infante and Riehl (1984), and more recently by Gonzales (1998, 2000).

Comparative tropical limnology also developed in Africa, based on numerous expeditions to deep and shallow lakes (Beadle, 1981). Key contributions to tropical limnology were made by studies on **African lakes**, such as **Lake Victoria** (Talling, 1965, 1966) and other lakes (Talling, 1969). Talling and Lemoalle (1998) presented an extremely relevant summary of tropical limnology (see Chapter 16).

More recently, the **International Biological Programme** (IBP) has extensively studied Lake Chad (Carmouze *et al.*, 1983) and Lake George (Ganf, 1974; Viner, 1975, 1977).

IBP was very important for limnology because it allowed the comparison of lakes at different latitudes, the standardization of methods and the quantification of processes. In particular, it encouraged the **study of ecological processes**, enabling a more dynamic and comparative approach. It also established a scientific basis for a more comprehensive quantitative approach in the study of lakes and a comparative study of processes, such as **primary production of phytoplankton** and its **limiting factors** (Worthington, 1975).

IBP summaries promoted innumerable modifications in research methods used to study **primary productivity** (Vollenweider, 1974) and biogeochemical cycles (Golterman *et al.*, 1978), and also stimulated more advanced interpretation of data and correlations between seasonal cycles, hydrological cycles, biogeochemical cycles and primary productivity of lakes, rivers and reservoirs (see Chapter 11).

The evolution of limnology has also been influenced by the construction of large dams (Van der Heide, 1982) in South America and Africa (Balon and Coche, 1974). In Spain, research comparing 100 reservoirs opened huge prospects for the process of classification and **typology of reservoirs** (Margalef, 1975, 1976), mainly focusing on the concept that the study of artificial reservoirs can elucidate the processes occurring in water basins. Margalef's work established an important theoretical and conceptual angle in limnology.

The study of reservoirs differs from the study of lakes, as reservoirs are much younger. They present unique characteristics, with a continuous flow and, in many cases, widely fluctuating levels, reflecting the system's ecological structure. Reservoirs provide the opportunity for important qualitative and quantitative theoretical comparisons with natural lakes (see Chapters 3 and 12).

Figure 1.2 shows the interrelationships between several areas of ecology, limnology, oceanography, hydrobiology and fisheries management, according to Uhlmann (1983).

The wide range of studies in the last 30 years clearly demonstrates the robust and diverse ideas in limnology. The great classic work on limnology in the 20th century

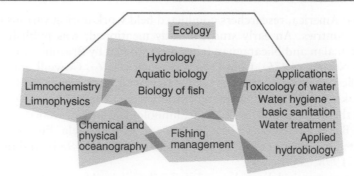

Figure 1.2   Reproduction of Uhlmann's concept (1983) on ecology, limnology, oceanography, hydrobiology, and fishing management, and their interrelationships and applications.

was by Hutchinson (1957, 1967, 1975, 1993). Other classic works influencing scientific research and the training of new researchers include those by Whipple, Fair and Whipple (1927), Welch (1935, 1948), Ruttner (1954), Dussart (1966), Hynes (1972), Golterman(1975),Wetzel(1975),Whitton(1975),Cole(1983),Uhlmann(1983),Stumm (1985) and more recently, Mitsch and Gosselink (1986), Burgis and Morris (1987), Schwoerbel (1987), Moss (1988), Margalef (1983, 1991, 1994), Patten (1992a, b), Goldman and Horne (1994), Hakanson and Peters (1995), Schiemer and Boland (1996), Lampert and Sommer (1997), Margalef (1997), Talling and Lemoalle (1998), Kalff (2002), and Carpenter (2003).

Many works on individual lakes were also published (see Chapter 15).

Chart 1.2 shows the main stages in the development of limnology and the conceptual progress starting with the work of Forel. This is not an exhaustive evaluation but rather it seeks to put into context the major milestones and scientists who introduced the main concepts in this branch of science.

Significant scientific progress in limnology was made during the 20th century. After the descriptive and comparative phase initiated with the development of Thienemann and Naumann's works up until the early 1960s, there was a major breakthrough in knowledge on the processes occurring in lakes, reservoirs, rivers and wetlands. These processes can be summarized as follows:

▶ Phytoplankton succession and the primary **forcing functions** that define and influence it. **Temporal** and spatial **phases** in phytoplankton succession (Harris, 1978, 1984, 1986; Reynolds, 1994, 1995, 1996, 1997; Bo Ping Han *et al.*, 1999);

▶ **Transfer of energy**, phyto-zooplankton integration mechanisms and the composition and structure of the **food network** (Porter, 1973; Lampert, 1997; Lampert and Wolf, 1986);

▶ **Hydrogeochemical** and **geochemical studies** on sediment, **sediment-water** interactions and chemical processes in lakes (Stumm and Morgan, 1981);

▶ Greater understanding of species distribution, **biogeography, biodiversity** and regulating factors (Lamotte and Boulière, 1983);

▶ Greater scientific knowledge on interactions between aspects of geography, climate, and hydrology and their effects on primary production and biogeochemical

Chart 1.2 Principal developments in limnology and conceptual advances based on Forel's studies.

1901	F.A. Forel	Physical classification based on the **thermal characteristics** of lakes
1911	E.A. Birge and C. Juday	**Chemical classification** based on **stratification** and dissolved oxygen
1915	A. Thienemann	Chemical and **zoological** classification based on balance of oxygen and **colonization of sediments**
1917	E. Naumann	Biological photoautotrophic production in the water column, linked to the concentration of organic material in sediment and balance of oxygen
1932	A. Thienemann and F. Ruttner	The Sunda expedition in Indonesia
1938	S. Yoshimura	Oxygen and **vertical distribution of temperature** in lakes in Japan, comparative analyses
1941	C.H. Mortimer	**Sediment-water interactions.** Circulation in lakes
1942	R. Lindeman	**Dynamic trophic theory** applied to lakes. Introduction of the concept of lakes as functional systems
1952	E. Steemann Nielsen	Introduction of technique to measure primary productivity with radioisotopes (^{14}C)
1956	E.P. Odum	Development of technique to measure **metabolism** in rivers
1956	G.E. Hutchinson and H. Löffler	**Thermal classification** of lakes
1958	R. Margalef	Introduction of the theory of information in the processes of phytoplanktonic succession
1964	R. Margalef	Beginning of studies on the theory of information in the processes of phytoplanktonic succession
1964	IBP	International Biological Programme founded
1968	R.A. Vollenweider	Concept of load from water basins and the effects on eutrophication in lakes
1974	C.H. Mortimer	**Hydrodynamics of lakes**
1974	J. Overbeck	Aquatic and biochemical microbiology
1975	G.E. Likens and Borman	Introduction of the study of the hydrographic basin as a unit
1990	R.G. Wetzel	Interactions between the littoral systems and the **pelagic zone** in lakes
1994	J. Imberger	Hydrodynamics of lakes. New methodologies for calculations in real time
1997	C.S. Reynolds	Summary of the temporal and spatial scales in phytoplanktonic cycles
2004	Goldman, Sakamoto and Kumagai	**Impacts of global changes** on lakes and reservoirs

Source: Various sources.

cycles (Straškraba, 1973; Le Cren and Lowe-McConnell, 1980; and Talling and Lemoalle, 1998).

Two important conceptual frames of reference were presented by Kalff (2002):

▸ 1960–1970 – Improved measurements of rates and processes, in addition to calculations of abundance and quantity of organisms. Development of **modelling** systems to simulate scenarios and evaluate impact on lakes. Establishment and expansion of the concept of phosphorus loading by Vollenweider (1968). Studies on the transport of nutrients in aquatic systems.
▸ 1970–2000 – Emphasis on the role of wetlands as functional systems. Re-integration of studies on ichthyology and limnology. Relevant research on **toxic substances**, **acidification** and eutrophication. Surging interest in studying the interactions of structure and function in lakes, rivers, reservoirs and wetland areas.

There was a surge in studies on **aquatic microbiology** and interactions between phyto- and bacterioplankton and phyto- and zooplankton. Concentration, distribution, and fluctuations in the stock of dissolved carbon were extensively studied, as well as temporal variations in DOC and POC (Sondergaard, 1997). The **microbial loop**, already studied by Krogh (1934), was later studied by Pomeroy (1974). Azam *et al.* (1983), in their model of trophic relationships and interactions in the pelagic zone, included bacterial production as an important quantitative process, based on DOC released by phytoplankton and zooplankton. Cladocerans partly play an important role in the food chain because of their ability to graze efficiently on **bacteria**. Key qualitative and quantitative relationships were identified from the dissolved organic carbon, bacterial succession and interactions with zooplankton (Wetzel and Richard, 1996).

During the last several decades, there has been a growing number of studies on distribution, **population dynamics** and biogeography of fish related to chemical composition, **trophic status** and pollution, including studies on **morphometry** and the organizational structure of lakes and reservoirs and the ichthyofauna (Barthem and Goulding, 1997).

As previously mentioned, these advances have had a decisive impact on the formulation of concepts in theoretical ecology and its application.

At the ecosystem level, significant progress has been made in understanding the hydrology of rivers and interactions with floodplain lakes (Neiff, 1996; Junk, 1997); operating mechanisms in wetlands (Mitsch, 1996); comparative studies on reservoirs (Margalef *et al.*, 1976; Straškraba *et al.*, 1993; Thorton *et al.*, 1990), saline lakes (Williams, 1996), interactions between terrestrial systems and aquatic systems (Decamps, 1996), and the ecology of large and small rivers (Bonetto, 1994; Walker, 1995).

Greater understanding of the processes and operating mechanisms at the level of interactions between components of the system or at the ecosystem level gave rise to a large number of contributions on the restoration and management of lakes and reservoirs. Outstanding works include those of Henderson-Sellers (1984) and Cooke *et al.* (1993), and application of models in management and **prediction** of scenarios and impacts (Jorgensen, 1980).

Over the last decade, **biomanipulation** studies on populations have applied scientific knowledge and advances in experimental limnology to the management of lakes and restoration of aquatic systems (De Bernardi and Giussani, 2000; Starling, 1998).

Much progress has been made by studies on shallow lakes (average depth <3 meters), **polymictic** lakes colonized to a great extent by macrophytes and subject to constant oscillations due to fluctuating levels and climatic effects such as wind and solar radiation, with significant interactions between the sediments and water column (Scheffer, 1998).

A discussion on theory and applications, philosophy of the science of limnology and its role in resolving problems was presented by Rigler and Peters (1995).

1.3 TROPICAL LIMNOLOGY

The mapping of the main rivers and watersheds of the continents of South America and Africa was completed in the 19th century.

The major rivers (Nile, Amazon, Zambezi, Niger and **Congo**) were explored and their size and **hydrographic networks** identified. Already by 1910, Wesemberg-Lund had drawn attention to the need for comparative studies on tropical lakes in order to expand the conceptual base. A number of expeditions were undertaken after 1900 (see Chart 1.3) to study biogeography, hydrology and geography. The Sunda Expedition

Chart 1.3   Important ecological expeditions in tropical limnology, in chronological order, 1894–1940.

	Neotropics	Africa	Asia
1940		Brunelli[16] Cannicci Morandini	
–	Gilson et al.[4]	Beauchamp[15]	
1930	Omer-Cooper[11]	Woltereck[18] Damas[14] Cambridge[13]	
–	Carter[3] Carter and Beadle[2]	Jenkin[12] Graham and Worthington[10]	Sunda Exp.
1920			
–	Juday[1]	Stappers[8] Cunnington[7]	Bogert[17] (Apstein)
1910		Fulleborn[6]	
		Moore[5]	
1900			

The numbers identify the localities and the lakes:
1. Guatemala, El Salvador; 2. Paraguay, Brazil; 3. British Guyana, Belize; 4. Lake Titicaca, the Andes; 5. **Lake Tanganyika**; 6. **Lake Nyasa**, Malawi; 7. Lakes Tanganyika, Nyasa and Victory; 8. Lakes Tanganyika and Moero; 9. Lake Victoria; 10. Lakes Kioga and **Albert**; 11. Ethiopia; 12. Kenyan rift lakes; 13. Kenya, Uganda; 14. Lakes **Kivu**, Edward, and Ndalaga; 15. Lakes Tanganyika and Nyasa; 16. Ethiopia: **Lake Tana**, rift lakes; 17. Ceilão (Sri Lanka); 18. The Philippines, Celebes Islands.
Source: Talling (1996).

(1928–1929) was a major event that brought Thienemann, Ruttner, Feuerborn and Herrmann together for a joint limnological project (Talling, 1996).

A series of expeditions gathered valuable information on tropical lakes and provided a foundation for future long-term research, mainly on African lakes Talling and Talling (1965), Talling (2005) and in the early 1950s on the Amazon (Sioli, 1984; Junk, 1997).

Integrative works on tropical limnology placed more emphasis on the limnology of **large African lakes** and less attention on the tropical limnology of South and Central America. This trend has shifted in recent years, especially due to the growing number of international journals published in South America, putting the contribution of limnologists from tropical regions in proper perspective.

Progress in understanding the functioning of tropical lakes comes from various sources and extensive studies by scientists who have conducted on-going research on many tropical lakes and reservoirs over the last 50 years (Beadle, 1981).

1.4   LIMNOLOGY IN THE 21ST CENTURY

In the last decade of the 20th century, conceptual advances in limnology and scientific investment in the process of discovery led to a significant increase in management and restoration programs for inland aquatic systems. Basic science can decisively support this application. Restoration systems have become more sophisticated, and **ecological models** play an important role in **scenario planning** and preparation (Jørgensen, 1992).

That decade also saw a major breakthrough in technology and development of more precise methods, including automated measurements and real-time data compilation.

The main conclusions that can be reached by analysing the progress of limnology include:

▸ Knowledge of processes led to the conclusion that lakes are not "freshwater islands" isolated in continents, but depend on interactions with the water basin (Likens, 1992);
▸ The response of lakes to human activities in water basins varies widely, depending on the morphology and the activities' intensity (Borman and Likens, 1979);
▸ Lakes respond to the most varied human activities in nearby areas in the drainage basin and also to alterations that occur in distant areas of the basin;
▸ The accumulation of information by lakes results in responses ranging from physical and chemical processes to community responses reflected in productivity, biodiversity, composition and genetic changes (Kajak and Hillbricht-Ilkowska, 1972; Reynolds, 1997a, b; Talling and Lemoalle, 1998);
▸ The recognized **complexity of inland aquatic ecosystems** should be the source of ongoing analytical study, integration, and comparison. Each aquatic system in its water basin is unique (Margalef, 1997);
▸ **Response times** to **climatic factors** and **anthropogenic activities** in water basins vary with the activities' intensity, the characteristics of the ecosystem and the stage of organization (Falkenmark, 1999). Progress has been made in understanding **spatial and temporal stages** in lakes and in response time to forcing functions.

The Sunda Expedition and tropical limnology

The Sunda Expedition took place in 1928–1929 and lasted for 11 months (7 September 1928 to 31 July 1929). The expedition travelled around the lakes, rivers and wetland areas of Java, Bali and Sumatra (now Indonesia). It was led by four researchers: Thienemann, Ruttner, Feuerborn and Herrmann, who traveled extensively throughout the region to make observations, collect biological samples and take meteorological measurements. The limnologists participating in the expedition, particularly Thienemann and Ruttner, were, respectively, specialists in typology of lakes (influenced by the Swedish limnologist Naumann) and in hydrochemical circulation processes and planktonic distribution. The expedition's main areas of focus were the taxonomy of the flora and fauna, stratification and circulation in lakes, and comparison and typology of lakes. A summary of expedition's early results was published by Thienemann (1931).

The Sunda Expedition set a standard for future limnological expeditions in the tropics, especially in terms of biogeography and the ecological characteristics of aquatic flora and fauna in tropical regions in Africa. Other expeditions in Africa followed (Damas, 1937) with the encouragement of the works published by the Sunda Expedition.

Source: Schiemer and Boland, 1996.

For example, spatial and temporal phases in the vertical and horizontal circulation in lakes, reservoirs and rivers seemed to be influenced by several factors, such as basic forces in physical and chemical processes (Imberger, 1994);

▸ Any progress in managing inland aquatic systems depends and will depend on understanding the operating principles of the systems, and sustainable management will only be possible with **comprehensive management** of the water basin (Murdoch, 1999). The **self-designing capacity** of aquatic ecosystems depends on feedback mechanisms and constant adjustments to structurally dynamic changes (Jorgensen and De Bernardi, 1998).

1.5 LIMNOLOGY IN BRAZIL

Limnology in Brazil developed based on scientific studies that began in the latter part of the 19th century. Oswaldo Cruz (1893) was an important pioneer. Many other observations and early studies of inland and estuarine systems took place in the 18th and 19th centuries (Esteves, 1988). Other contributions to the development of limnology in Brazil arose from medical applications, **microbiology** and the study of fish communities to expand production capacity (fisheries and aquaculture). To lay a solid foundation for aquaculture, it was necessary to better understand the aquatic system, especially natural lakes and reservoirs, and maintain solid conceptual bases for implementation. In this stage of limnology in Brazil, the works of Spandl (1926), Wright (1927, 1935, 1937), Lowndes (1934) and Dahl (1894) played a relevant role.

Esteves (1988) published a detailed description of the evolution of limnology in Brazil, outlining the pivotal points in this science's progress through the 20th century.

Esteves, Barbosa and Bicudo (1995) published a comprehensive summary of the development of limnology in Brazil from its inception up to 1995.

Another significant contribution was made by Branco (1999), who conducted a series of **hydrobiological studies** applying the lessons of **aquatic biology** to sanitation and to the promotion of new comprehensive technologies to coordinate the work of sanitation engineers, biologists and limnologists.

An important milestone in Brazil beginning in 1971 was a set of systematic studies at the UHE Carlos Botelho Dam (Lobo-Broa) and water basin, which introduced many innovative methods for the study of Brazilian aquatic ecosystems. The first publications addressed issues of spatial heterogeneity, horizontal and vertical temperature gradients, distribution of planktonic organisms and phyto-zooplanktonic interactions. Studies were also conducted on fish communities, especially on growth, **reproduction** and feeding, and lake benthos (Tundisi *et al.*, 1971, 1972).

These last studies also focused on **seasonal processes** and established new perspectives to understand interactions between **climatic cycles**, hydrological cycles and the primary productivity of plankton and biochemical cycles (Tundisi, 1977a, b).

The project in this artificial, shallow, **turbulent** ecosystem was also significant because it coincided with the establishment of a human resources training program that led to the inauguration of the graduate program in Ecology and Natural Resources at the Federal University of São Carlos (UFSCar) in 1976. The program enabled the "São Carlos School" to be established in Brazil to train limnologists and ecologists, and it has expanded to a large number of centres of excellence in many regions of the country. Today, limnologists trained at the São Carlos School work in 20 universities and ten research institutes in Brazil and in 15 countries in Latin America and three countries in Africa.

For the education and training of qualified human resouces, not only are there masters and doctoral graduate programs, but 12 international courses on specialized subjects were also offered at São Carlos from 1985–2003, which enabled the training of specialists from Latin American countries, Africa and Brazil. The four initial graduate programs in Ecology in Brazil (INPA – **Manaus;** Unicamp – Campinas; UNB – Brasilia; and UFSCar – São Carlos) had different approaches; UFSCar and Inpa were more focused on aquatic ecology and limnology. Also, the First National Conference on Limnology, Aquaculture and Inland Fishing was held in Belo Horizonte in 1975 (Vargas, Loureiro and Milward de Andrade, 1976).

Over the past 25 years, with the founding of the Brazilian Society of Limnology (1982), the consolidation of limnology conferences and publication of the journal *Acta Limnologia Brasiliensa* (Brazilian Limnology Review), limnology has definitively become established as a science in Brazil. It should also be noted that another milestone was the Congress of the International Association of Theoretical and Applied Limnology held in 1995 in Sao Paulo. That congress was attended by 1065 scientists from 65 countries, with 470 presentations by Brazilian researchers and graduate students, leading to widespread international recognition of limnology in Brazil and numerous interactions in several lines of scientific research that bore fruit after the congress.

The capacity of scientific production in limnology in Brazil can be measured by the growing number of publications, especially in the last 20 years, which has helped consolidate trends, programs and approaches and promote significant progress in the science.

Many summaries and articles have been published in Brazil over the last 30 years. The early and conceptually sound work of Kleerekoper (1944) was followed by the work of Schafer (1985) and Esteves (1988), and summaries produced by Tundisi (1988), Pinto Coelho, Giani and Von Sperling (1994), Barbosa (1994), Tundisi, Bicudo and Matsumura Tundisi (1995), Agostinho and Gomes (1997), Junk (1997), Tundisi and Saijo (1997), Henry (1999a, b, 2003), Nakatami *et al.* (1999), Tundisi and Straŝkraba (1999), Junk *et al.* (2000), Santos and Pires Salatiel (2000), Tundisi and Straŝkraba (2000), Medri *et al.* (2002), Bicudo, Forti and Bicudo (2002), Brigante and Espindola (2003), Thomaz and Bini (2003), and Bicudo and Bicudo (2004).

These recent studies show that limnology in Brazil is undergoing a rapid transformation, shifting away from the simple description of systems and organisms to the interpretation of ecological processes, mathematical modelling, prediction and quantification.

Esteves (1998) published a volume on the ecology of coastal lagoons in Restinga de Jurubatiba National Park and the municipality of Macae (Rio de Janeiro). The work examines the historical, ecological, structural and dynamic aspects of the ecosystems located in the northern part of the state of Rio de Janiero.

1.6   IMPORTANCE OF LIMNOLOGY AS A SCIENCE

The study of limnology is, like other sciences, basically a search for principles. Those principles that are involved in certain processes and operating mechanisms can be used to predict and compare. In particular, the comparative aspect of limnology should be emphasized. For example, when comparing the **hydrodynamics** of rivers, lakes and reservoirs, certain basic functional aspects are immediately understood that affect the **life cycle** and distribution and biomass of aquatic organisms.

This approach, taken by Legendre and Demers (1984), examines the early work on vertical **stability** and instability, phytoplanktonic succession (Gran and Braarud, 1935; Bigelow, 1940) and productivity (Riley, 1942), and in light of recent data on biological variability, sampling and **microscale** techniques, *in vivo* fluorescence (Lorenzen, 1966) and quantifying techniques such as spectral analysis. Other important factors analysed relate to the study of the **physiological behaviour** of phytoplankton and its response to shifting light intensities due to **turbulence**. This new approach views hydrodynamics and its effects on vertical structure as principal controlling factors in phytoplanktonic succession in the system. The approach opens immense theoretical and practical perspectives in the study of limnology, and is one of the advanced points in current limnology, closely aligned with oceanography from the conceptual point of view.

The ability to make predictions and forecasts also qualifies limnology as a science, and is important in applied limnology. In recent years, inland water systems have been increasingly degraded by various types of waste, through deforestation of water basins, and air pollution leading to acid rain. Halting these processes of deterioration and correcting and preventing alterations that damage inland waters can only be achieved if a solid base of scientific knowledge exists.

On the other hand, human interference in aquatic life (overexploitation of aquatic plants and animals, **introduction of exotic species**) has produced many changes in the structure of aquatic ecosystems.

In addition to the problems of pollution, eutrophication, and deterioration that inland waters have suffered, it should be kept in mind that proper management of these ecosystems is also important for better use of existing resources in lakes, rivers and reservoirs. For example, in many countries, construction of dams has significantly modified the structure of natural terrestrial and aquatic ecosystems and introduced new ecosystems with particular characteristics. Management of these systems for diverse purposes represents a considerable investment that should be encouraged, using basic knowledge.

In addition to scientific interest and deepening basic knowledge, limnology can provide important applications.

Another important aspect of the interface between basic and applied limnology is the study of the evolution of lakes and reservoirs. These systems evolved under different types of pressure and the progressive introduction of **ecological filters** that resulted in characteristic mechanisms and the resulting community. This study – which includes aspects of geomorphology, hydrodynamics, composition of the sediment, the ratio of **allochthonous** (externally derived) to **autochthonous** (internally derived) material, and composition of the community – is key to understanding the effects of human activities on inland water systems. Comparisons between lakes of different origins and reservoirs can provide much scientific information needed for understanding ecosystems, as well as for enabling important diagnoses, such as an integrated process of watershed hydrology, extending the concept of limnology to a more global view that not only includes the liquid medium in which organisms live but also the complex system of interactions developed in the terrestrial system around the lake or inland **aquatic ecosystem**.

Figure 1.3 is modified from an original concept presented by Rawson (1939), in which interactions were identified and key objectives in limnology were summarized. The diagram provides an important framework for knowledge on inland aquatic systems. Although other interpretations and processes resulting from further studies will be presented in later chapters, the concepts in the diagram show that a comprehensive and systematic viewpoint already existed in 1939.

The application of basic limnology to the various aspects of **regional planning** is discussed in Chapter 19.

The most important progress in limnology as a science over the last ten years has been the growing understanding of the dynamic ecology of aquatic systems and applications in solving problems of protection, conservation and restoration of lakes.

Another important development in limnology has been the ability to predict the trends and characteristics of lakes and reservoirs over time, especially in terms of controlling influences such as eutrophication and fish stock. The **prognosis** is usually calculated with multiple variables and relatively simple models, based on extensive collection of data.

Utilization of these techniques enables the extensive application of limnology in planning and solving problems, such as those previously cited (see Chapter 19).

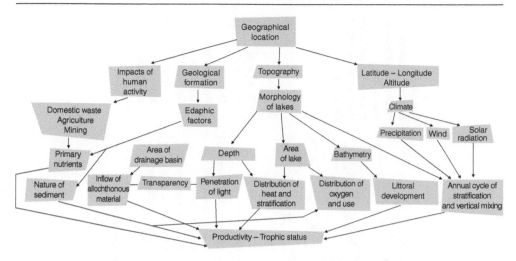

Figure 1.3   Important frameworks in the understanding of inland aquatic systems.
Source: Modified from Rawson (1939).

Chart 1.4   Principal attributes and the hierarchy of factors that act on inland aquatic ecosystems, according to Rawson.

Regional properties	Climate	Geology	Topography	Retention time of sedimentation
Characteristics of the drainage basin	Vegetation	Soil	Hydrology	
Attributes and characteristics of the systems	Morphometry	Circulation-stratification		
Physical and chemical properties	Light penetration, Water temperature	Turbidity and **conductivity**	Humic substances	Nutrients Toxins
Biological and ecological properties	Biomass	Productivity	Trophic Structure; Biodiversity	
Impacts of human activities	Destruction of systems and habitats Reduction of biodiversity	Inflow of nutrients and sediments	Climatic alterations	Toxic substances

Source: Based on Horne and Goldman (1994) and Kalff (2002).

Chart 1.4 summarizes **Rawson's concept** and sets out the key attributes and hierarchy of factors involved in inland aquatic ecosystems, including the impact of human activities. The table also consolidates the conception and organization of this work.

Figure 1.4 (Likens, 1992) shows this author's concept of the ecosystem, energy matrix and levels of organization and study based on individuals and communities. The matrix summarizes the current major developments in the approach to and study of aquatic systems, their activities and communities.

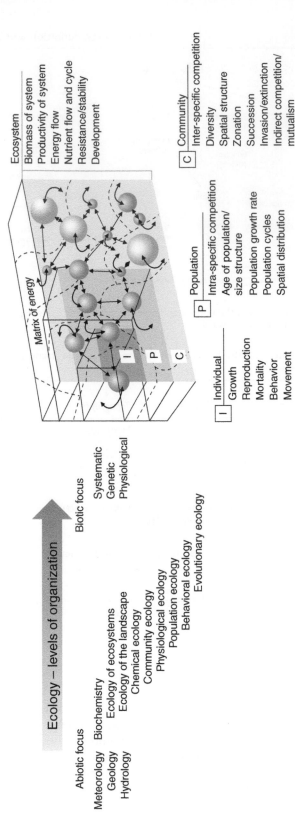

Figure 1.4 Energy matrix of an ecosystem and levels of organization and study based on individual and communities.
Source: Modified from Likens (1992).

Principal scientific publications on limnological studies.

Algological Studies
Amazoniana – Limnologia Et Oecologia Regionalis
   Systemae Fluminis Amazonas
Ambio
American Scientist
Annals of the Entomological Society of America
Applied Geochemistry
Aquaculture
Aquatic Botany
Aquatic Ecology
Aquatic Ecosystem Health & Management
Aquatic Insects
Aquatic Microbial Ecology
Aquatic Toxicology
Archiv für Hydrobiologie
Archive of Fishery and Marine Research
Australian Journal of Freshwater and Marine
   Science
Biodiversity and Conservation
Biological Conservation
Biological Invasions
Bioscience
Biotropica
British Antarctic Survey Journal
British Journal of Phycology
Bulletin Ecological Society of America
Canadian Journal of Fisheries and Aquatic Sciences
Conservation Biology
Ecohydrology & Hydrobiology
Ecological Modelling
Ecological Monographs
Ecology
Ecology of Freshwater Fish
Ecotoxicology
Environmental Biology of Fishes
Environmental Conservation
Estuaries
Fisheries Management and Ecology
Fisheries Research
Freshwater Reviews
Freshwater Biology

Hydrobiologia
Hydroecology and Hydrobiology
Inland Waters
Interciência
International Journal of Ecology and Environmental
   Sciences
International Review of Hydrobiology
Journal of Applied Microbiology
Journal of Coastal Research
Journal of Ecology
Journal of Fish Biology
Journal of Freshwater Biology
Journal of Freshwater Ecology
Journal of Great Lakes Research
Journal of Hydrology
Journal of Lake and Reservoir Management
Journal of Phycology (US)
Journal of Plankton Ecology
Journal of Plankton Research
Journal of Tropical Ecology
Lakes & Reservoirs Research and Management
Limnetica
Limnologica
Limnology and Oceanography
Marine and Freshwater Behaviour and Physiology
Memorie dell' Istituto Italiano di Idrobiologia
Microbial Ecology
Nature
Naturwissenschaften
New Zealand Journal of Freshwater and Marine
   Science
Oikos
Phykos
Polar Research
Proceedings of the International Association of
   Theoretical and Applied Limnology
Proceedings of the Royal Society (UK) Series B.
Restoration Ecology
Swiss Journal of Hydrobiology
Water Research
Water Resources Research

In Brazil, the most relevant publications in this area are:

Acta Amazonica
Acta Botanica Brasilica
Acta Limnologica Brasiliensia
Anais da Academia Brasileira de Ciências
Atlântica
Biota Neotropica

Brazilian Journal of Oceanography
Boletim do Laboratório de Hidrobiologia
Brazilian Archives of Biology and Technology
Brazilian Journal of Biology
Brazilian Journal of Ecology

2 Water as a medium

SUMMARY

Water is an extremely unusual substance. It exists in three states: solid, liquid and gas. The transition from one phase to another depends on a rearrangement of the molecules and the configuration of its aggregates.

The **physical properties of water,** especially its temperature-related anomalies in density, play a key role in the circulation and stratification processes in lakes and reservoirs, and in the vertical organization of the system in in temperate lakes in winter, when surfaces freeze.

The physical and chemical properties of water, in particular its anomalies in density, **surface tension** and thermal features, are important to aquatic organisms inhabiting the liquid medium. Another significant biological property is that surface tension enables the existence of special forms of aquatic life. Viscosity is another important property, as the mobility of aquatic organisms in the liquid medium depends on it.

The **hydrologic cycle** on the planet includes the components of **evaporation,** transport by wind, **precipitation,** and **drainage.** Driven by solar radiation and wind energy, the cycle depends on the perpetual transition from the liquid phase in the oceans to the gaseous phase in the atmosphere to precipitation over the continents.

The global distribution of water is irregular. Some regions have plenty and others have shortages. The availability of liquid water depends on a reserve of inland waters, characterized by waters in lakes, rivers, reservoirs, wetlands and groundwater.

The volume and **quality of water** in underground **aquifers** depend on the **vegetation cover,** which promotes re-charge and helps maintain water quality.

Distribution of fresh water is also irregular in Brazil. Some regions have abundant resources of surface and underground waters and relatively sparse population. Other regions have relatively sparse water resources and are densely populated, such as the highly urbanized regions in the south-east. The *per capita* distribution of water in the country is uneven.

2.1   WATER'S PHYSICAL AND CHEMICAL PROPERTIES

The essential life processes of all organisms depend on water. Water is the **universal solvent** carrying the dissolved gases, elements, substances, and organic compounds that form the basis of all plant and animal life on the planet.

Hydrogen in water functions as a source of electrons in **photosynthesis**. Water's unique properties are related to its atomic structure, intermolecular hydrogen bonds and the molecular associations in the solid, liquid and gas phases. Oxygen is highly electronegative; in water, the oxygen atom binds with two hydrogen atoms that retain a positive charge.

The asymmetric charge in water molecules enables the oxygen in a molecule to form a weak bond with the hydrogen of two adjacent molecules. Such covalent O-H bonds result in strong intermolecular attraction.

Water in its gaseous phase has no structure, and the gas is essentially monomeric. The solid phase, ice, is extremely orderly, and its structure is well known. Each atom of oxygen in the structure is connected, through hydrogen, to four more oxygen atoms, forming a tetrahedron at a distance of 2.76 Å from the central oxygen atom. The structural organization can be represented by a network of hexagonal rings with spaces between the molecules, enabling ice to float on liquid water. At temperatures of approximately 0°C, molecules of water in the liquid phase begin to oscillate between 10^{11}–10^{12} times per second. Already, the molecules of ice at temperatures close to 0°C present movements, oscillating from 10^5–10^6 times per second. It is the hydrogen bridges that hold the water molecules together. Hydrogen bridges are also present in other molecules, as shown in the examples of Table 2.1.

Temperature variations affect **intermolecular distances**. When ice melts, the empty spaces in the molecular structure disappear, increasing the water's density, which reaches its maximum at 4°C. As the temperature increases, the liquid expands due to increased intermolecular distance and lower density.

Because water is a strong **dipole** with two hydrogen atoms (positive) and one oxygen atom (negative), in addition to the distance (as charge × distance), there are significant effects on water's physical properties. Without this strong dipolar feature, water would not be liquid.

Water molecules are strongly attracted to each other, forming spherical, linear or area aggregates. The electrostatic action of the molecules causes a redistribution

Table 2.1   Other substances (besides water) with hydrogen bonds.

Bond	Substance	Bond Energy (kcal/Mol)	Bond Length (Angstroms)
F – H – F	H_6F_6	6.7	2.26
O – H --- O	H_2O_2 (ice) H_2O_2	4.5 4.5	2.76
N – H – N	NH_3	1.3	3.38
N – H – N	NH_4F	5.0	2.63

of their charges, resulting in the formation of hydrogen bonds (see Figure 2.1). Each water molecule is connected to four other molecules through the hydrogen bonds, leading to a tetrahedral arrangement of electrons around the oxygen atom.

The physical characteristics of water present many anomalies. Chart 2.1 presents these properties and their importance in physical and biological environments.

Figure 2.2 presents the relationship between the density of water and its corresponding temperature. Table 2.2 presents the physical properties of water, and Table 2.3 presents the values for various temperatures, densities and specific volumes of water.

The **anomaly of density** can be demonstrated by the following properties:

▸ Water in lakes, in the deepest parts, cannot be colder than water at its **maximum density**, which is approximately 4°C;
▸ Masses of water freeze from the surface toward the bottom, so the layer of ice protects the deeper water from freezing. This property has important consequences for the distribution and survival of aquatic organisms.

The anomalies of density are important not only in the process of circulation of water and during the winter period in temperate or arctic lakes, they are also fundamental in the stability/**circulation of lakes** in the tropics and in periods of

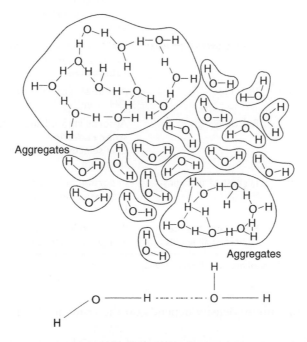

Figure 2.1   Because of its tendency to form hydrogen bonds, water plays an extremely active role in the metabolism of organisms.
Source: Schwoerbel (1987).

Chart 2.1   Physical features of liquid water.

Property	Comparison with other substances	Physical and biological significance
Fusion point and latent heat of fusion	High, with exception of NH_3	Thermostatic effect at the freezing point due to **absorption** or release of latent heat
Specific heat (amount of heat in calories needed to raise the unit weight of substance by 1°C)	Highest of the solids and liquids (except NH_3)	Impedes extreme changes in temperature Transfer of very high heat due to movements of water Maintains uniform body temperature
Evaporation point and latent heat of evaporation	Highest of all substances	Extremely important in the **transfer of heat** and water in the atmosphere
Thermal expansion	The temperature of maximum density for pure water is 4°C This temperature decreases with increased salinity	Maximum density of fresh water and diluted sea water is above the freezing point
Surface tension	The highest of all liquids	Important in cellular **physiology** Controls surface phenomena Decreases with increasing temperature There are organisms adapted to this layer of surface tension Organic compounds reduce surface tension
Power of solution	High (universal solvent)	Highly important due to the dissociation capacity of **dissolved organic substances**
Electrolytic dissociation	Very low	A neutral substance that contains H^+ and OH^- ions
Transparency	Relatively high	Absorbs infrared and ultraviolet rays of **solar radiation** Low selective absorption in visible spectrum
Conduction of heat	Highest of all liquids	Important in live cells, the molecular processes can be affected by the condition for diffusion
Dielectric constant	Pure water has the highest of all liquids	Results in high dissociation of dissolved inorganic substances

Source: Modified from Sverdrup, Johnson and Fleming (1942).

circulation. The maximum density depends on the content of salt in the water and on pressure.

For each 1% increase in the salt content of water, the temperature of maximum density decreases by about 0.2°C (Schwoerbel, 1987). Therefore, the maximum density of sea water is approximately −3.5°C and it freezes at −1.91°C.

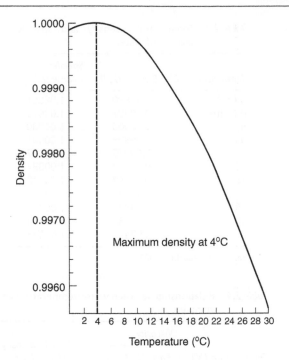

Figure 2.2   Relationship between water density and temperature.
Source: Beadle (1981).

Table 2.2   Physical properties of water.

Density	(25°C) kg/m³	997.075
Maximum density	kg/m³	1,000.000
Temperature of maximum density	°C	3.840
Viscosity (Pascal/sec)	25°C	$0.890 \cdot 10^{-3}$
Kinetic viscosity (m²/s)	25°C	$0.89 \cdot 10^{-6}$
Freezing point (°C)	(101,325 Pa) Pat n	0.0000
Boiling point (°C)	(101,325 Pa) Pat n	100.00
Latent heat of ice	(kg/mol)	6.0104
Latent heat of evaporation	(kg/mol)	40.66
Specific heat	(15°C in J/kg · °C)	4.186
Thermal conductivity	(25°C) J/cm · °C	0.00569
Surface tension	W/m	$71.97 \cdot 10^{-3}$
Dielectric constant	25°C	78.54

Source: Schwoerbel (1987); Wetzel (2001).

2.1.1   The importance of water's physical and chemical properties for aquatic organisms

The entire life cycle and behaviour of aquatic organisms are influenced by the physical and chemical properties of water, especially its density, anomalies of density, thermal

Table 2.3   Temperature, density and specific volume of water.

Temperature (°C)	Density (kg/L)	Specific volume (L/kg)
0 (ice)	0.91860	1.08861
0 (water)	0.99987	1.00013
4	1.00000	1.00000
5	0.99999	1.00001
10	0.99973	1.00027
15	0.99913	1.00087
18	0.99862	1.00138
20	0.99823	1.00177
25	0.99707	1.00293
30	0.99568	1.00434
35	0.99406	1.00598

Source: Schwoerbel (1987).

Table 2.4   Relationship between water temperature and viscosity.

Temperature (°C)	Viscosity		Kinetic viscosity $(M^2/s) \cdot 10^{-6}$
	$Pa\,s \cdot 10^{-3}$	%	
0	1.787	100.0	1.771
5	1.561	84.8	1.561
10	1.306	78.7	1.304
15	1.138	63.7	1.139
18	1.053	58.9	1.054
20	1.002	56.0	1.004
25	0.890	49.8	0.892
30	0.798	44.7	0.801
35	0.719	40.3	0.723

Source: Schwoerbel (1987).

properties and capacity as a universal solvent. The surface tension of water, which also has great biological significance, varies with temperature and the concentration of dissolved solids.

Water's surface tension enables a set of organisms – **neuston** or **pleuston** – to utilize the interface between the water and the atmosphere for support as well as for movement.

Another biologically important property of water is its **dynamic viscosity**, which is the force required to move 1 kg a distance of 1 m in 1 second through a mass of water. Viscosity depends on the temperature of the water and its salt content. Therefore, the movement of aquatic organisms depends on the water's viscosity, temperature, and the shape of the body, and the energy field for the organisms in movement is related to these factors. The unit of viscosity is the Pascal-second ($Pa\,s = 1\,kg\,m^{-1}\,s^{-1}$).

Table 2.4 shows the relationship between water temperature and viscosity. Kinematic viscosity is the ratio of viscosity to density, and is approximately the same as the classic viscosity of water. The dimensional unit is m^2/s (1 Stokes $= 10^{-4}\,m^2/s$).

2.2   THE WATER CYCLE AND DISTRIBUTION OF WATER ON THE PLANET

The water cycle is the basic **unifying principle** of everything referring to water on the planet. The cycle is the model representing the interdependence and continuous movement of water in the solid, liquid and gas phases. All water on the planet is in continual movement in a cycle between the solid, liquid and gas reserves. Clearly, the phase of greatest interest is liquid, which is essential for use and to satisfy the needs of humans and all other organisms, both animal and plant. The components of the water cycle (Speidel *et al.*, 1988) include:

▶ Precipitation: water added to the Earth's surface from the atmosphere. Can be liquid (rain) or solid (snow or ice);
▶ Evaporation: the transformation process of liquid water to gas (water vapor). Most evaporation occurs over oceans, followed by lakes, rivers and reservoirs;
▶ Transpiration: the process by which plants release water vapor, which is dispersed into the atmosphere.
▶ **Filtration:** the process by which water is absorbed by the soil.
▶ **Percolation:** the process by which water moves through the soil and rock formations down to the phreatic level.
▶ Drainage: the movement of displacement of the water on surfaces during precipitation.

Water that reaches the surface of a water basin can then be drained, held in lakes and reservoirs, and from there evaporate into the atmosphere or filter and percolate into the soil.

Figure 2.3 illustrates the steps of the water cycle and its principal processes.

Until the late 1980s, it was believed that Earth's water cycle was a closed system, i.e., the total amount of water on the planet has always remained the same since the beginning of the Earth. No water on the planet was thought to enter from or leave into outer space. Recent discoveries, however, suggest that 20–40-ton snowballs (called "small comets" by scientists) can reach Earth's atmosphere from other regions of the solar system. The snowballs vaporize when nearing the Earth's atmosphere and may add 3 trillion tons of water every 10,000 years (Frank, 1990; Pielou, 1998).

The velocity of the water cycle varies from one geological period to another, as well as the total proportions of fresh water and sea water. During periods of **glaciation**, for example, the proportion of liquid fresh water was reduced. In warmer periods, the liquid form was more common.

According to Pielou (1998), the water cycle can be considered a life cycle, and the natural history of water on the planet is related to life cycles and the history of life.

Table 2.5 shows the distribution of water and the principal reservoirs of water on Earth. Figure 2.4 shows the distribution of water on the planet and percentages of salt water and fresh water.

The distribution of water on the planet is not homogeneous. Table 2.6 shows the distribution of the renewable water supplies by continent and percentages of the world's population.

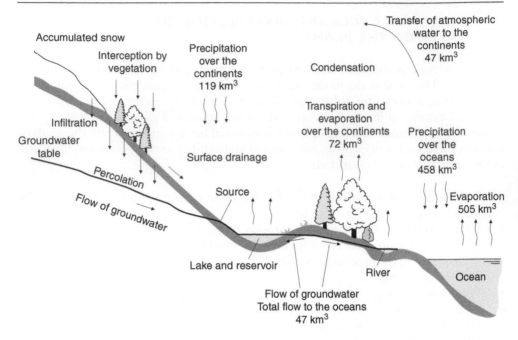

Figure 2.3 The water cycle. Numbers in km³ (× 10³) indicate yearly flows of evaporation, precipitation and drainage to the oceans.
Source: Adapted from various sources.

Table 2.5 Total and relative areas and volumes of water in the world's main bodies of water.

Reserve of water	Area (10³ km²)	Volume (10³ km³)	% of total volume	% of volume of fresh water
Oceans	361,300	1,338.000	96.5	–
Ground water	134,800	23.400	1.7	–
Fresh water	–	10.530	0.76	30.1
Humidity in soil	–	16.5	0.001	0.05
Polar caps	16,227	24.064	1.74	68.7
Antarctic	13,980	21.600	1.56	61.7
Greenland	1,802	2.340	0.17	6.68
Arctic	226	83.5	0.006	0.24
Glaciers	224	40.6	0.003	0.12
Frozen ground	21,000	300	0.022	0.86
Lakes	2,058.7	176.4	0.013	–
Fresh water	1,236.4	91	0.007	0.26
Salt water	822.3	85.4	0.006	–
Swamps	2,682.6	11.47	0.0008	0.03
Flow of rivers	148,800	2.12	0.0002	0.006
Water in the biomass	510,000	1.12	0.0001	0.003
Water in the atmosphere	510,000	12.9	0.001	0.04
Total	510,000	1,358.984	100	–
Total reserve of fresh water	148,800	35.029	2.53	100

Source: Shiklomanov (1998).

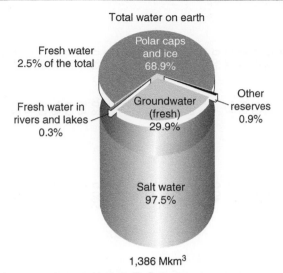

Figure 2.4   Distribution of water on Earth at any given moment.
Source: Shiklamanov (1998).

Table 2.6   Distribution of renewable water supply by continent.

Region	Average annual drainage (km³)	% of global drainage	% of global population	% stable
Africa	4,225	11	11	45
Asia	9,865	26	58	30
Europe	2,129	5	10	43
North America	5,960	15	8	40
South America	10,380	27	6	38
Oceania	1,965	5	1	25
Ex-Soviet Union	4,350	11	6	30
World	38,874	100	100	36

Source: Adapted from L'vovich (1979).

The major rivers and lakes on Earth are important reserves of fresh water. Located in the interior of continents and draining large areas, these vast reserves are critical to the survival of plant and animal species, and the very survival of *Homo sapiens*. Table 2.7 shows the main rivers on the planet. Figure 2.5 presents the major water basins. Artificial man-made reservoirs (created by dams) constitute another important freshwater reserve, holding about 9,000 km³ of water (Straškraba *et al.*, 1993a, b).

Table 2.8 shows that most of the Earth's large lakes are located in temperate regions in the Northern Hemisphere. The **African continent** also has a significant number of lakes with sizable areas and volume. Large lakes are rare in South America, especially in the tropics; small floodplain lakes and extensive swampy areas with many lakes predominate. In the southern part of South America, there are larger lakes, in surface area and volume, in Patagonia and the Andean region.

Table 2.7 Main features of the world's important rivers.

River	Length (km)	Area of basin (km²)	Discharge (km³/year)	Intensity mm/year (D/C)	Transport of dissolved substances t/km²/year (Td)	Transport of suspended solids t/km²/year (Ta)	Ta/Td	Total amount transported (T × 10⁶/year)
Amazon	7,047	7,049,980	3,767.8	534	46.4	79.0	1.7	290.0
Congo	4,888.8	3,690,750	1,255.9	340	11.7	13.2	1.1	47.0
Yangtze	6,181.2	1,959,375	690.8	353	N/A	490.0	N/A	N/A
Mississippi-Missouri	6,948	3,221,183	556.2	173	40.0	94.0	2.3	131.0
Yenisei	5.58	2,597,700	550.8	212	28.0	5.1	0.2	73.0
Mekong	4.68	810,670	538.3	664	75.0	425.0	5.8	59.0
Orinoco	2,309.4	906,500	538.2	594	52.0	91.0	1.7	50.0
Paraná	4,330.8	3,102,820	493.3	159	20.0	40.0	2	56.0
Lena	6,544.8	2,424,017	475.5	196	36.0	6.3	0.15	85.0
Brahmaputra	1.8	934,990	475.5	509	130.0	1,370.0	10.5	75.0
Irrawaddy		431,000	443.3	1,029	N/A	700.0	N/A	N/A
Ganges	1.8	488,992	439.6	899	78.0	537.0	6.9	76.0
Mackenzie	3.663	1,766,380	403.7	229	39.0	65.0	1.7	
Ob	6.1578	3,706,290	395.5	107	20.0	6.3	0.3	50.0
Amur	4.86	1,843,044	349.9	190	10.9	13.6	1.1	20.0
Saint Lawrence	2,808	1,010,100	322.9	320	51.0	5.0	0.1	54.0
Indus	3.24	963,480	269.1	279	65.0	500.0	8.0	68.0
Zambezi	3.06	1,329,965	269.1	202	11.5	75.0	6.5	15.4
Volga	4,123.8	1,379,952	256.6	186	57.0	19.0	0.3	77.0
Niger	4.68	1,502,200	224.3	149	9.0	60.0	6.7	10.0
Columbia	2,185.2	668,220	210.8	316	52.0	43.0	0.8	34.0
Danube	3,198.6	816,990	197.4	242	75.0	84.0	1.1	60.0
Yukon	3,562.2	865,060	193.8	224	44.0	103.0	2.3	34.8
Fraser	1,530	219,632	112.4	512	N/A	N/A	N/A	N/A
São Francisco	3,576.6	652,680	107.7	165	N/A	N/A	N/A	N/A
Hwang-Ho (Yellow River)	5,221.8	1,258,740	104.1	83	N/A	2,150.0	N/A	N/A
Nile	7,482.6	2,849,000	80.7	28	5.8	37.0	6.4	10.0
Nelson	2.88	1,072,260	76.2	71	27.0	N/A	N/A	31.0
Murray-Darling	6,067.8	1,072,808	12.6	12	8.2	30.0	13.6	2.3

Source: Based on various sources.

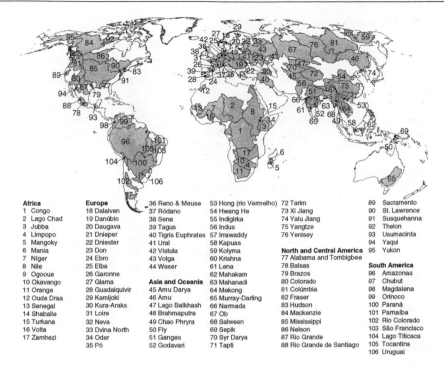

Africa	Europe		53 Hong (rio Vermelho)	72 Tarim	89 Sacramento
1 Congo	18 Dalalven	36 Reno & Meuse	54 Hwang He	73 Xi Jiang	90 St. Lawrence
2 Lago Chad	19 Danúbio	37 Ródano	55 Indigirka	74 Yalu Jiang	91 Susquehanna
3 Jubba	20 Daugava	38 Sena	56 Indus	75 Yangtze	92 Thelon
4 Limpopo	21 Dnieper	39 Tagus	57 Irrawaddy	76 Yenisey	93 Usumacinta
5 Mangoky	22 Dniester	40 Tigris Euphrates	58 Kapuas		94 Yaqui
6 Mania	23 Don	41 Ural	59 Kolyma	North and Central America	95 Yukon
7 Níger	24 Ebro	42 Vístula	60 Krishna	77 Alabama and Tombigbee	
8 Nile	25 Elba	43 Volga	61 Lena	78 Balsas	South America
9 Ogooue	26 Garonne	44 Weser	62 Mahakam	79 Brazos	96 Amazonas
10 Okavango	27 Glama		63 Mahanadi	80 Colorado	97 Chubut
11 Orange	28 Guadalquivir	Asia and Oceania	64 Mekong	81 Colúmbia	98 Magdalena
12 Oude Draa	29 Kemijoki	45 Amu Darya	65 Murray-Darling	82 Fraser	99 Orinoco
13 Senegal	30 Kura-Araks	46 Amu	66 Narmada	83 Hudson	100 Paraná
14 Shaballe	31 Loire	47 Lago Balkhash	67 Ob	84 Mackenzie	101 Pamaíba
15 Turkana	32 Neva	48 Brahmaputra	68 Salween	85 Mississippi	102 Rio Colorado
16 Volta	33 Dvina North	49 Chao Phryra	69 Sepik	86 Nelson	103 São Francisco
17 Zambezi	34 Oder	50 Fly	70 Syr Darya	87 Rio Grande	104 Lago Titicaca
	35 Pó	51 Ganges	71 Tapti	88 Rio Grande de Santiago	105 Tocantins
		52 Godavari			106 Uruguai

Figure 2.5   World's principal drainage basins illustrated by continent.
Source: Revenga et al. (1998).

The North American Great Lakes are important freshwater reserves and are an international water resource shared by Canada and the United States. In South America, **Lake Titicaca** is shared by Bolivia and Peru, and there is a major international project underway to restore the Lake Titicaca watershed and lake. All the large lakes presented in Table 2.8 have important **multiple uses**, with enormous and significant impact on the economies of many countries and regions. These lakes have significant **aquatic biota** and are important systems for maintaining **aquatic biodiversity**. **Ancient lakes**, such as **Baikal** in Russia, or Tanganyika in Africa, are rich in diversity.

The volume of **suspended material** transported by rivers depends on the uses of the water basin, and on the degree of deforestation or vegetation cover. The suspended material is either deposited in the deltas and estuaries or is transported by sea currents to accumulate in gulfs or bays. Dams built for various different reasons alter the flows and **sediment transport** of rivers, often impacting coastal regions and deltas. Recent scientific data show that reservoirs in the Tietê, in the State of São Paulo, may retain up to 80% of suspended material (Tundisi, 1999).

The draining by rivers, which represents the renewal of water resources, is the most important component of the water cycle. The Amazon, the world's largest river, produces 16% of global drainage, and accounts for 27% of the total drainage of the rivers represented by the Amazon, Ganges-Brahmaputra, Congo (Zaire), **Lantz** and Orinoco.

Table 2.8   The world's major freshwater lakes.

Lake	Area (km²)	Volume (km³)	Maximum depth (m)	Continent
Superior	82,680	11,600	406	North America
Victoria	69,000	2,700	92	Africa
Huron	59,800	3,580	299	North America
Michigan	58,100	4,680	281	North America
Tanganyika	32,900	18,900	1,435	Africa
Baikal	31,500	23,000	1,741	Asia
Nyasa	30,900	7,725	706	Africa
Great Bear Lake	30,200	1,010	137	North America
Great Slave Lake	27,200	1,070	156	North America
Erie	25,700	545	64	North America
Winnipeg	24,600	127	19	North America
Ontario	19,000	1,710	236	North America
Ladoga	17,700	908	230	Europe
Chad	16,600	44.4	12	Africa
Maracaibo	13,300	–	35	South America
Tonlé Sap	10,000	40	12	Asia
Onega	9,630	295	127	Europe
Turkana	8,660	–	73	Africa
Nicaragua (Cocibolca)	8,430	108	70	Central America
Titicaca	8,110	710	230	South America
Athabasca	7,900	110	60	North America
Reindeer	6,300	–	–	North America
Tung Ting	6,000	–	10	Asia
Vanerm	5,550	180	100	Europe
Zaysan	5,510	53	8.5	Asia
Winnipegosis	5,470	16	12	North America
Albert	5,300	64	57	Africa
Mweru	5,100	32	15	Africa

Source: Adapted from Shiklomanov, in Gleick (1998).

Not all water basins empty into the oceans. Those that do not drain into oceans are called endorheic regions (internal drainage), and account for an area of 30 million km² (20% of the Earth's total land area). In contrast, regions that drain into oceans are exorheic.

2.2.1 Groundwater

Water found in the subsoil of the Earth's surface is called groundwater. It occurs in two zones. The upper zone extends from the surface to depths ranging from less than 1 meter to several hundred meters in semi-arid regions. This region is called the unsaturated zone, since it contains water and air. The saturated zone, which occurs just below it, contains only water. Figure 2.6 presents the main features of groundwater and the general terminology used. A portion of the precipitation that reaches the Earth's surface percolates down through the unsaturated zone to the saturated zone. These areas are called recharge areas, as they replenish underground aquifers.

The movement of groundwater includes lateral dislocations in which hydraulic gradients occur in the direction of the discharge of the aquifers.

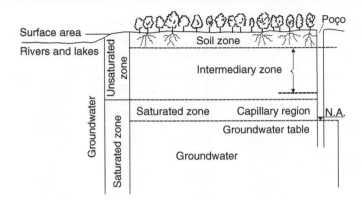

Figure 2.6   Groundwater. The upper portion of the saturated zone is occupied by water in the capillary region held by surface tension.
Source: Modified from Speidel *et al.* (1988).

Table 2.9   Contribution of groundwater flow to the discharge of rivers (km³/year).

Continents/Resources	Europe	Asia	Africa	North America	South America	Australia/ Oceania	ex-USSR	World total
Surface drainage	1,476	7,606	2,720	4,723	6,641	1,528	3,330	27,984
Underground contribution	845	2,879	1,464	2,222	3,736	483	1,020	12,689
Total average discharge of rivers	2,321	10,485	3,808	6,945	10,377	2,011	4,350	40,673

Source: Tundisi (2003); Rebouças *et al.* (2006).

Water that percolates through the soil surface forms non-confined aquifers, in contrast to confined aquifers, in which water is retained by less permeable soil. All types of rocks (igneous, sedimentary and metamorphic) confine water in different regions. Important sources of groundwater deposits include limestone, dolomite, basalt and sandstone. It is important to emphasize that water present in the soil supports biomass from various natural or cultivated sources. Groundwater can be found in all regions around the world and is an important natural resource. It is often used for domestic supply, **irrigation** in rural areas, and industry.

The generalized uses of groundwater are also due to its availability near the site of use and also to its quality, because it can be free of pathogens and contaminants. The perpetual of long-continued availability of groundwater is another reason for its extensive use. The contribution of the flows of groundwater to the discharge from rivers by continent is shown in Table 2.9.

2.2.2 Distribution of continental waters in Brazil

Brazil is estimated to contain between 12–16% of all the fresh water on the planet, unevenly distributed, as shown in Table 2.10 and Table 2.11. Figure 2.7 shows annual average distribution.

Table 2.10   Hydric balance in the principal water basins of Brazil.

Drainage basin	Area (km²)	Average precipitation (m³/s)	Average discharge (m³/s)	Evapotranspiration (m³/s)	Discharge/ precipitation (%)
Amazon	6,112,000	493,191	202,000	291,491	41
Tocantins	757,000	42,387	11,300	31,087	27
North Atlantic	242,000	16,388	6,000	10,388	37
Northeast Atlantic	787,000	27,981	3,130	24,851	11
São Francisco	634,000	19,829	3,040	16,789	15
Northwest Atlantic	242,000	7,784	670	7,114	9
Southwest Atlantic	303,000	11,791	3,710	8,081	31
Paraná	877,000	39,935	11,200	28,735	28
Paraguay	368,000	16,326	1,340	14,986	8
Uruguay	178,000	9,589	4,040	5,549	42
South Atlantic	224,000	10,515	4,570	5,949	43
Brazil, including Amazon basin	10,724,000	696,020	251,000	445,020	36

Source: Braga *et al.* (1998).

Table 2.11   Availability of water for society and demands by state in Brazil.

State	Water potential (km³/year)	Population**	Availability of water for society (m³/ individual/year)	Population density (inhabitants/ km²)	Total use*** (m³/individual/ year)	Level of use in 1991
Rondônia	150.2	1,229,306	115,538	5.81	44	0.03
Acre	154.0	483,593	351,123	3.02	95	0.02
Amazonas	1,848.3	2,389,279	773,000	1.50	80	0.00
Roraima	372.31	247,131	1,506,488	1.21	92	0.00
Pará	1,124.7	5,510,849	204,491	4.43	46	0.02
Amapá	196.0	379,459	516,525	2.33	69	0.01
Tocantins[1]	122.8	1,048,642	116,952	3.66	–	–
Maranhão	84.7	5,022,183	16,226	15.89	61	0.35
Piauí	24.8	2,673,085	9,185	10.92	101	1.05
Ceará	15.5	6,809,290	2,279	46.42	259	10.63
R. G. do Norte	4.3	2,558,660	1,654	49.15	207	11.62
Paraíba	4.6	3,305,616	1,394	59.58	172	12.00
Pernambuco	9.4	7,399,071	1,270	75.98	268	20.30
Alagoas	4.4	2,633,251	1,692	97.53	159	9.10
Sergipe	2.6	1,624,020	1,625	73.97	161	5.70
Bahia	35.9	12,541,675	2,872	22.60	173	5.71
M. Gerais	193.9	16,672,613	11,611	28.34	262	2.12
E. Santo	18.8	2,802,707	6,714	61.25	223	3.10
R. Janeiro	29.6	13,406,308	2,189	305.35	224	9.68
São Paulo	91.9	34,119,110	2,209	137.38	373	12.00
Paraná	113.4	9,003,804	12,600	43.92	189	1.41
Sta. Catarina	62.0	4,875,244	12,653	51.38	366	2.68
R. G. do Sul	190.0	9,634,688	19,792	34.31	1,015	4.90
M. G. do Sul	69.7	1,927,834	36,684	5.42	174	0.44
M. Grosso	522.3	2,235,832	237,409	2.62	89	0.03
Goiás	283.9	4,514,967	63,089	12.8	177	0.25
D. Federal	2.8	1,821,946	1,555	303.85	150	8.56
Brazil	5,610.0	157,070,163	35,732	18.37	273	0.71

Source: *DNAEE, 1985, [1]Srhimma, **IBGE Census, 1996, ***Rebouças, 1994.

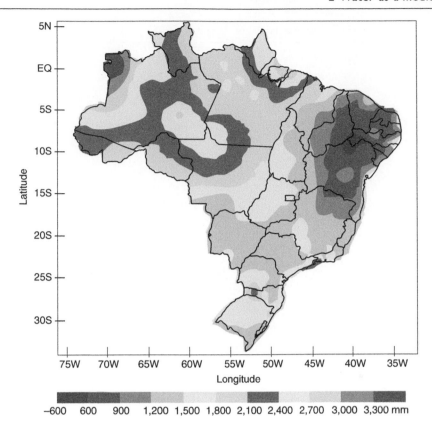

Figure 2.7   Characteristics of **average annual precipitation** (in mm) in Brazil (CPTEC/Inpe)
(See also color plate section, plate 1).
Source: Rebouças *et al.* (2002).

Lake Balaton in Hungary. This site, which has been extensively studied, which lake has been extensively studied, limnological lake has been one of the lakes (average depth: 3 m). makes it one of shallow lakes (average depth: 3 m). makes on shallow lakes. studies on shallow lakes. Source: ILEC.

3 | The origin of lakes

SUMMARY

This chapter examines the **origin of natural lakes**, beginning with specific geomorphological events and **patterns of drainage** and the respective types and characteristics.

Lakes' origins and morphometry play an important role in their physical, chemical and biological processes, since, along with regional climatic processes, these factors contribute to the functioning of the lakes.

In addition to classifying lakes according to their origin and different configurations, also presented are morphometric variables and parameters, morphological types, bathymetric profiles, and zonation of lakes and reservoirs. This chapter also includes the global distribution of lakes by origin, examples of the age of lakes as a function of their origin, and the **global distribution of fluvial lakes**.

3.1   GENERAL FEATURES OF LAKES AND DRAINAGE BASINS

The **study of geomorphology** contributes significantly to understanding the origin of lakes and the dynamics of the formative processes of lake ecosystems (Swanson, 1980). **Morphology,** the study of lake shapes, is related to the origins of each system. **Morphometry** deals with the quantification of these forms and elements. Lake morphology and morphometry basically depend on the processes from which lakes originated.

Natural lakes, rivers, streams and reservoirs have a short lifespan from the geological point of view. The disappearance of lakes can be the object of prognosis. Some are very old, and the history of the events that occurred in the water basin and the lake itself is recorded in the sediments. These sediments can be dated with techniques that determine naturally occurring ^{14}C activity. Countless fragments of undecomposed aquatic organisms and vegetation can be used to determine the sequence of events occurring in lakes: the remains of **diatoms** and zooplankton, the remains of vertebrates and pollen in the sediments. These fragments can also provide precise information on the changes of vegetation in the drainage basin over geological time.

Lakes that were formed from specific geomorphological events in certain geographical areas have similar characteristics and are therefore grouped into **lake districts.**

Although these characteristics are similar, there are differences in the **morphometry, productivity** and **chemical composition** of the water. Comparative studies of lakes in the same district and between different lake districts allow for regional classification. For example, in the **lakes of the Médio Rio Doce** in eastern Brazil, the process that originally formed the lakes is probably the same (De Meis and Tundisi, 1986), but considerable differences exist among the lakes in terms of morphometry, productivity, and chemical composition of water. These differences are due to the presence of relatively ancient rivers and streams in the aquatic system that gave rise to the lakes.

Lakes are also referred to as **lentic systems** (from the Latin *lentus,* meaning *lens*).

The **geomorphology** (basin structure) of lakes plays a significant role in establishing the physical, chemical and biological conditions, whose series of events, given the limits of the climatic conditions in the lacustrine basin, largely depend on the basic operating mechanisms provided by the initial morphometric and morphological conditions established by the geomorphological pattern. The influx of nutrients, stratification, **thermal de-stratification** and **retention time** all depend on the lake's geomorpology.

The rivers that form a hierarchical hydrographic system comprise the **hydrographic network** and occupy a **drainage basin,** the territorial expanse of land bathed or traversed by a hydrographic network (Ab'Saber, 1975). The drainage basin may cover tens of thousands or millions of km² in area. The localization of lakes in the original fluvial system leads to a discussion of the most common types of flow that occur.

The principal forms of hydrographic networks are categorized in several **drainage patterns** important in characterizing the types of regional evolution of the networks of rivers and the interrelationships between climatic factors, rock type, and the nature of terrains. Drainage patterns also provide basic information on the processes of formation of lakes in the regional context (see Figure 3.1).

Drainage basins, which are regional domains of hydrographic networks, can be classified based on the **destination** of the running water. Chart 3.1 shows various types of **drainage basins.**

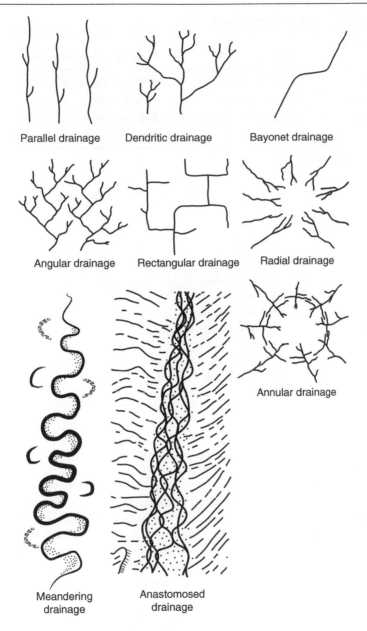

Figure 3.1   Principal types of drainage in water basins.
         Source: Ab'Saber (1975).

3.2 ORIGIN OF LAKES

According to the original definition by Forel (1892), a lake is a body of stationary water, occupying a given basin and not confluent with the ocean. Several authors make the distinction between lakes, ponds, wetland areas, coastal lagoons and fluvial

Chart 3.1   Types of drainage basins.

Types of drainage	Features	Destination of running water
Exorheic	Open drainage	Perennial and periodic rivers
Endorheic	Closed drainage	Periodic rivers
Dry	Diffused desert drainage	Sporadic rivers
Cryptoreic	Karst drainage	Underground rivers and labyrinths

Source: Ab'Saber (1975).

Chart 3.2   Classification of lakes by origin.

Origin

1. Tectonic
2. Volcanic
3. Movements of earth
4. Glaciation
5. Solution lakes
6. Fluvial action
7. Wind action
8. On the coast
9. Organic accumulation
10. Constructed by organisms
11. Impact of meteorites

Source: Hutchinson (1957).

ponds adjacent to large, medium or small rivers (from a few square meters to tens or hundreds of thousands or millions of square kilometers).

All inland water systems originated from a variety of natural processes and diverse formative mechanisms that vary from region to region and from one geologic era to another. Hutchinson (1957) identified 76 types of lakes grouped into 11 types of formation (see Chart 3.2). Other authors, such as Bayly and Williams (1973), listed morphogenetic classifications of lakes based on regional experiences. Bayly and Williams' classification, for example, is based on Australian lakes.

In addition classifying lakes based on **geomorphology**, it is important to consider the continuity or **intermittence** of systems, lakes with wet seasons and extremely irregular flooding, wetland areas, flood-plains and channels of rivers, lakes, and coastal lagoons, perennial or intermittent lagoons that are unconnected to coastal waters or permanently connected through channels.

All inland water systems are subject to a continuous process of change produced by the contribution of the respective drainage basins, and therefore are transitory features in the landscape. Inland water systems contain:

▶ Sediments from tributaries of the water basin (linear and **laminar**, or layered, erosion) or from diffuse drainage;
▶ Material accumulated through transport by wind;

▶ **Deposits** from wind and tide activity;
▶ Structured biological material deposited in lakes (remains containing carbonates, phosphorus).

The average lifespan of natural lakes and artificial reservoirs vary according to their volume, area (lake) or area of drainage (reservoir), maximum and average depth, retention time of water, the morphometry and morphology of the drainage basin and the lake, and also of the reservoir.

Below we describe some of the most common mechanisms that lead to the formation of lakes (see Figure 3.2).

3.2.1 Tectonic

The lake is formed by movements of the Earth's crust, such as faults that result in **depressions.** They are often formed in **rift valleys** (graben). Well-known examples are Lake Baikal (Siberia), Lake Tanganyika (Africa) and Lake Victoria (Africa), which was facilitated by the obstruction of the Kagera and Katonga Rivers, creating a 68,422-km^2 basin.

Tectonic movements can occur through the **emergence** or **subsiding** (lifting or sinking) of areas with shifts in sea level. The formation of lakes then begins with isolation from the ocean. Some lakes, which were ancient fjords, formed when their connection with the sea was closed. Many lakes of this type can be found in Norway, British Columbia, New Zealand and Scotland (northern British Isles).

3.2.2 Volcanic

The formation of depressions, or hollows that do not drain naturally, produces a series of volcanic lakes. Volcanoes are common in areas where tectonic movements occur. Lava discharged by active volcanoes can obstruct a river and form lakes. Examples include some usually small lakes in Africa, Asia, Japan and New Zealand. Lake Kivu, in Central Africa, is an example of a lake formed when a valley was obstructed by volcanic lava (Horne and Goldman, 1994).

3.2.3 Glaciation

Many present-day lakes were formed by glacial action. These movements, which can be catastrophic, cause deposits or corrosion of masses of ice and their depositions, with subsequent thawing. Massive glaciation in the **Pleistocene,** followed by subsequent retreat, for example, led to the formation of many lakes in the Northern Hemisphere. Examples can be found in the Lake District of England, lakes in Finland and Scandinavia, and alpine lakes. In some cases, the transport of rocks and material obstructed valleys and depressions, producing moraine lakes. When this glacial activity formed lakes near the coast, where sea water remained on the bottom, **meromictic lakes** could result (see Chapter 4). In North America, the Great Lakes (Superior, Michigan, Huron, Erie and Ontario) are examples of lakes formed by glaciation.

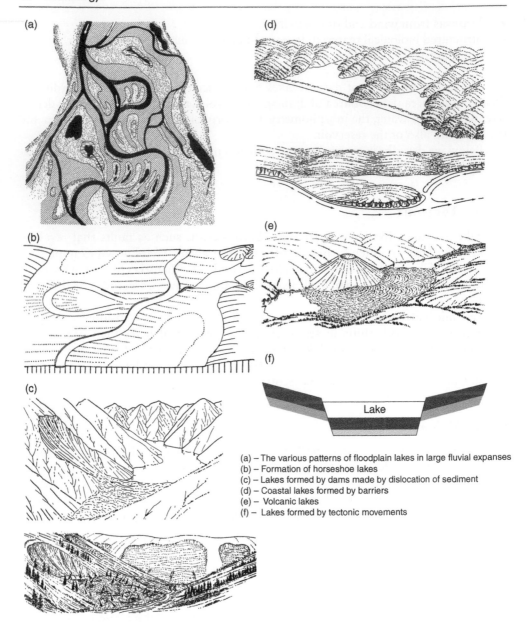

(a) – The various patterns of floodplain lakes in large fluvial expanses
(b) – Formation of horseshoe lakes
(c) – Lakes formed by dams made by dislocation of sediment
(d) – Coastal lakes formed by barriers
(e) – Volcanic lakes
(f) – Lakes formed by tectonic movements

Figure 3.2   Formation mechanisms of lakes and morphological characteristics.
Sources: Modified from Welcomme (1979); Horne and Goldman (1994); and Wetzel (2001).

3.2.4   Solution lakes

Many lakes are formed when deposits of soluble rock are gradually dissolved by **percolating water**. For example, solution lakes are formed by the **dissolution** of $CaCO_3$ from slightly acidic water containing CO_2. These lakes can be found in karst regions of the Balkan Peninsula, Florida, the Brazilian state of Minas Gerais, the Yucatan Peninsula of Mexico, and northern Guatemala.

3.2.5 Lakes formed by fluvial activity

Rivers, as they flow, have an obstructive capacity (due to **deposition of sediments**) and an erosive capacity (due to transport of sediment). Many lateral lakes are formed by the activity of rivers depositing sediments that obstruct tributaries. In regions with intensive sedimentation, water tends to flow around the sediments in a U-shaped pattern. In many cases, delta-like lakes may develop when obstruction occurs near alluvial deposits in coastal regions or even in large internal deltas of rivers such as the Amazon or the Paraná. Horseshoe-shaped lakes are formed by this type of system (*oxbow lakes*, or 'billabongs' in Australia). Fluvial lakes are therefore related to erosion and sedimentation in rivers.

3.2.6 Lakes formed by wind action

Depressions formed by wind action, or blocked by accumulation of dunes, can also lead to the formation of lakes. Such lakes are ephemeral, since hollows thus formed retain water only during rainy periods, and become increasingly saline due to evaporation in the dry season, and, finally, they dry up. Many of these lakes occur in endorheic regions of South America, Asia and Australia, and in **arid regions** of the United States.

3.2.7 Lakes formed by organic deposits

The growth of plants and associated **detritus** can cause obstructions in small rivers and depressions. This is another important factor in the formation of lakes in certain regions, such as the Arctic tundra. These depressions form a series of small shallow lakes, with a pronounced tendency to have extensive coverage of macrophytes. Artificial dams produced by beavers and humans can also cause the accumulation of organic matter and plants and produce small lakes.

3.2.8 Landslides

Large-scale movements of rock or soil, resulting from abnormal weather events such as excessive rainfall or the action of earthquakes, can produce lakes by blocking valleys. Such lakes are usually temporary, due to rapid erosion that occurs in the unconsolidated dam.

3.2.9 Coastal lagoons

Deposition of material on the coast, produced in regions where there are bays or inlets, can lead to the creation of coastal lakes. In many cases, incomplete separation occurs, and so periods of fresh water and brackish water alternate in the lakes. Small coastal lakes can also form adjacent to large inland lakes (see Chapter 14).

3.2.10   Lakes of meteoric origin

On rare occasions a meteorite striking the Earth's surface can lead to a depression that later accumulates water, forming a lake (for example, Lake Bosumtwi, in West Aftica).

3.2.11   Multi-origin lakes

Several of the processes described – such as glaciation, heavy precipitation, and tectonic movement – can interact and lead to the formation of a lake or a complex of lakes.

3.3   LAKE MORPHOLOGY AND MORPHOMETRY

A lake's origin establishes some basic morphological and morphometric conditions. These conditions change over time, depending on a number of factors, especially human activity and events occurring in the catchment area. The interaction of such events with the lake or reservoir is very important, since there is a relationship between the area of the reservoir or lake, the area of the water basin, and the **retention time** of the water, which is the time needed for all the water in the lake to be replaced.

All human activities occurring in the drainage basin affect the lake system, especially if the drainage network is dendritic, since there are close interconnections between the various rivers and streams. Consequently, knowledge about the retention time is essential for control, **monitoring** pollution, and calculating nutrient balance. Retention time also has important biological consequences and affects eutrophication processes (see Chapter 18).

To understand the origins and formation of lakes, it is first necessary to determine their bathymetry and varying depths. Today bathymetry is performed with echo-sounding equipment with GPS (geographical positioning systems) to determine each point's exact position. Typically, bathymetry is conducted in transects that enable reconstruction of the lake bottom's main contour lines and calculations of total volume of the lake or reservoir.

The principal measurements and indices used to describe morphometric features are shown in Chart 3.3.

Chart 3.3   Morphometric parameters.

Area (km^2)	A
Volume	V
Maximum length	L_{max}
Maximum width	La_{max}
Maximum depth	Z_{max}
Average depth	Z
Relative depth	Z_r
Perimeter	M
Index of development of margin	D_s
Development of volume	D_v
Average slope	(d)

Identifying these variables and a lake's bathymetry play a key role in quantifying morphometry and morphological structures.

3.3.1   Area (A) of lake or reservoir (km²)

A distinction is generally made between total area (A1), which includes islands, and the liquid surface. This area can be calcualted based on measurements made with planimeters or maps, aerial photographs or global-positioning satellite imagery.

3.3.2   Volume (V)

Volume is determined by measuring the area of each contour, finding the volume between the planes of each successive contour, and totaling these volumes.

3.3.3   Maximum length (L_{max}) and maximum width (La_{max})

The maximum length of a lake is determined based on maps or aerial photographs or satellite images. It is the distance between the two furthest points of the lake in a straight line. The **maximum effective length** has an important hydrographic and limnological application, because it corresponds to the maximum distance between two uninterrupted points on the lake or reservoir. The distance between these two points is called the *fetch* (Von Sperling, 1999). The longer the fetch, the greater the effects of wind on the lake or reservoir's surface.

The **maximum width** (La_{max}) is the maximum distance between the banks at right angles to the maximum length (L_{max}).

3.3.4   Maximum depth (Z_{max})

The maximum depth is determined directly from the bathymetric map and has an important value for calculations and future measurements of vertical circulation in lakes and reservoirs.

3.3.5   Average depth (Z)

The average depth is determined by dividing the lake's volume (V) by its area (A). It is an important parameter because the lake's biological productivity is generally related to its average depth, according to Rawson (1955).

3.3.6   Relative Depth (Z_r)

The relative depth is given as a percentage and is defined by the ratio of the lake's maximum depth (Z_{max}) and average diameter. The greater the depth, the more probable it is that the lake presents more stable thermal stratification.

3.3.7   Perimeter (m)

The perimeter corresponds to measurements of the lake's contour, and can be obtained using maps or aerial photographs and measuring, by various techniques, the values of the **perimeter** in meters. Häkanson (1981) detailed these techniques, also described by Von Sperling (1999).

3.3.8   Shoreline development index (D_L)

The shoreline development index is a measure of the degree of irregularity of the shoreline. It is the relationship between the length of the shoreline and the length of a circumference of a circle with an area equal to that of the lake. It measures the degree of deviation of the lake or reservoir from a circular pattern.

The shoreline development index (D_L) is given by:

$$D = \frac{S}{Z\sqrt{A/\pi}}$$

▸ In a perfectly round lake $D_L = 1.0$, while in lakes with shapes that deviate moderately from a circle, $D_L = 1.5$ to 2.5.
▸ Very irregular dendritic lakes have values from 3 to 5.

3.3.9   Volume development index (D_v)

This indicator is used to express the type of lake basin and is defined as the ratio of the lake's volume to the volume of a cone with a base area equal to the lake's area and a height equal to the lake's maximum depth. The D_v is generally about three times the ratio $Z:Z_{max}$.

3.3.10   Average slope (α)

The average slope (α) is determined by the formula:

$$\alpha = \frac{(l_{0/2} + l_1 + l_2 + \cdots + l_{n-2} = l_{n/2})Z_m}{10An}$$

where:
α = average slope as a percentage
$l_0, l_1, \ldots, l_n$ = perimeter of various contours in km
Z_m = maximum depth in meters
n = number of contour lines
A = area of lake in km^2

Timms (1993) presented another method to calculate the average slope, which is:

$$\alpha = \frac{100Z_m}{\sqrt{A/x}}$$

Hypsographic curves allow visualization between depth and area, and are constructed by locating the depth and the cumulative area (see Figure 3.3).

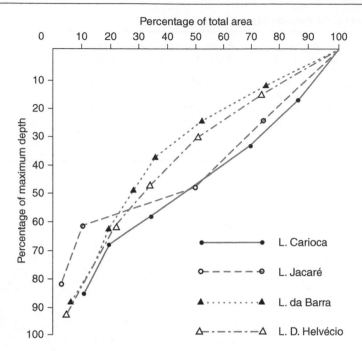

Figure 3.3   Hypsographic curves of four lakes in the Parque Florestal do Rio Doce (MG).
Source: Tundisi and Musarra (1986).

Hutchinson (1957) presents a number of different types of morphology for lakes, including:

▶ Circular: volcanic lakes, especially lakes in old craters;
▶ Sub-circular: slightly modified from the circular, due to wind action and transport of material – commonly found in glacial lakes;
▶ Elliptic: arctic lakes;
▶ Elongated sub-rectangular: lakes of glacial origin in glacier-carved valleys, approximately rectangular in shape;
▶ Dendritic lakes: originated from submerged valleys obstructed by sedimentation, with many arms and bays;
▶ Triangular form;
▶ Irregular: in regions where basins fused, resulting in irregular shapes;
▶ Crescent shaped: some lakes with half-moon shapes in flooded valleys or volcanic regions.

Knowledge of the shape of the lake is essential, because there is a relationship between the shape and circulation of waters and mechanisms in the functioning of lakes.

A large number of lakes have the ratio $Z:Z_{max} > 0.33$, i.e., the development of the volume is greater than the standard. In a study using photographs and maps of the Radam Brazil project, Melack (1984) showed the typology (see Table 3.1) for the morphology of Amazonian lakes. The table shows the number of lakes in each shape category in the **Amazon River**: the total for the entire water basin; upstream water

Table 3.1   Lakes in the Amazon Basin.

Form	Total	Basins upstream Amazonia	Basins upstream Middle Amazonia	Basins upstream lower-middle Amazonia	Basins downstream lower Amazonia
Round/oval	5,010	600	1,450	2,080	890
Dammed	1,530	480	860	170	10
Dendritic	830	50	170	570	40
Crescent	140	20	60	50	10
Horseshoe	270	220	50	10	0
Compound	270	80	80	100	10

Source: Melack (1984).

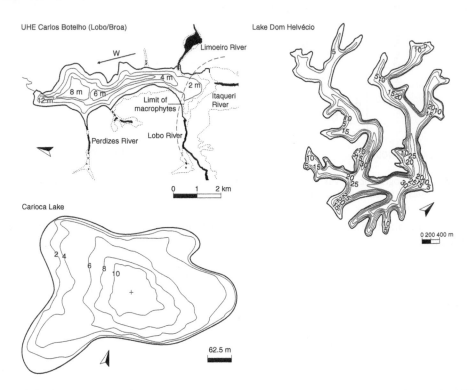

Figure 3.4   Bathymetric profiles of two lakes in the Parque Florestal do Rio Doce (MG) and the **UHE Carlos Botelho reservoir (Lobo/Broa)**.
Source: Tundisi (1994), Tundisi and Musarra (1986).

basins from Peru to the edge of the **Jutaí River;** upstream water basins in the **Middle Amazon,** from the **Japurá River** to Manaus; the **Middle-Lower Amazon,** Manaus to the **Trombetas River;** the **Lower Amazon,** from **Tapajos River** to **Xingu River.**

Morphological and morphometric studies of lakes in the Medio Rio Doce showed that a dendritic pattern occurs in many lakes, but there are a certain number of lakes with circular or elliptical shapes, including many in the Mato-Grosso **swamp,** and marginal lakes in the **flood valleys** of the Paraná River and the Bermejo River.

Identifying a lake's **bathymetric profile** is also an important factor because of the relationship between irregularities and depressions and circulation. These depressions

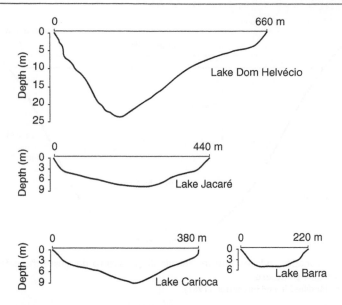

Figure 3.5   Bathymetric profiles of four lakes in the Parque Florestal do Rio Doce (MG) and the UHE
Carlos Botelho reservoir (Lobo/Broa).
Source: Tundisi (1994), Tundisi and Musarra (1986).

can alter thermal and chemical conditions during the **period of stratification**. Welch
(1935) called this process 'submerged depression individuality.'

Figure 3.4 and Figure 3.5 show the bathymetric profiles of natural lakes and artificial reservoirs in Brazil.

Significant morphological modifications can occur when a lake is subject to many
impacts, especially from human activities. For example, deforestation in water basins
can significantly alter a lake's morphometry and morphology, due to suspended sediment. In **Jacare Lake,** of the Medio Rio Doce (Brazil), changes in **bottom sediment**
were noted, resulting from the clearing of forests and cultivation of *Eucalyptus* species
(Saijo and Tundisi, 1985).

Changes in the flow of rivers in sub-lacustrine channels can also alter the morphology of the bottom of a lake.

The maximum depth of some lakes is below current sea level. Such lakes are
called **cryptodepressions**. Well-known cases of tectonic lakes include Lake Baikal and
the **Caspian Sea**. Some glacial lakes in Norway, Scotland and England are also cryptodepressions. Of the sub-alpine lakes in Central Europe, large lakes in northern Italy
have depths below the current sea level.

Hutchinson (1957) estimates that there are a total of at least 1,000 lakes in
cryptodepressions.

3.4   ZONATION IN LAKES

The morphology and morphometry of lakes are particular to each **lake district,** and
different kinds of lakes can occur in a single district. However, certain structures of
lakes are common and should always be determined in initial studies (see Figure 3.6).

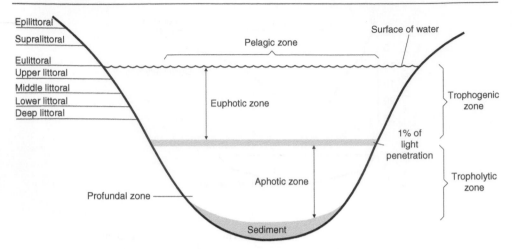

Figure 3.6   Classic zonation in lakes with the terminology used to designate the different regions in the vertical structure.
Source: Modified from Hutchinson (1967).

The **littoral zone** of a lake extends from the lake's edge to the depth where the non-stratified upper waters reach the bottom when the lake is stratified. It can also be viewed as the region into which sufficient light penetrates for aquatic macrophytes to grow. Below the littoral zone is the dimly lit **sublittoral zone**, where few macrophyte species can survive at low light intensities.

The **deep zone** consists of sediments of very fine particles, transported from the banks and the continuous sedimentation of suspended particulates, plankton and the remains of other dead organisms. The **deep benthos** is ecologically important and can provide valuable information on the functioning and characteristics of the lake. Light generally does not reach the deep zone, and in stratified lakes, gases such as methane and hydrogen sulphide accumulate.

In addition, the lake also presents the pelagic or **limnetic** zone, a common region in deep lakes where there is little influence of the bottom.

The illuminated region in the lake extends to the depth where 1% of surface light reaches. This region is called the **euphotic zone**. The **aphotic zone** refers to the non-illuminated regions. In many clear lakes, the aphotic zone is extremely small and the euphotic zone extends to the bottom.

The zonation in depth and the zonation of light penetration are key elements in a lake's vertical structure. With the thermal characteristics in the **vertical profile** (see Chapter 4), these elements in a lake's structure determine basic operational processes.

It should also be mentioned that an aquatic ecosystem presents three important interfaces that regulate countless mechanisms: the air-water, sediment-water, and organism-water interfaces (see Figure 3.7).

The vertical structure is dynamic, it becomes modified over time, and each interface plays an important role in the balance of substances in the lake, including **vertical transport,** horizontal transport, diffusion, precipitation, and sedimentation.

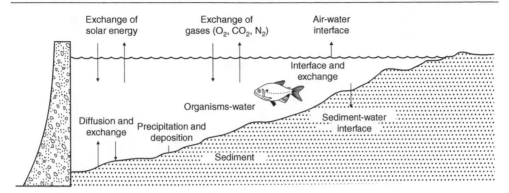

Figure 3.7   The principal interfaces in aquatic ecosystems: air-water interface; sediment-water interface; organisms-water interface.

Some important relationships between the structural components of lakes are:

Z_{eu}/Z_{max}: relationship between the depth of the euphotic zone and maximum depth;
Z_{eu}/Z_{af}: relationship between the depths of the euphotic zone and the aphotic zone.

There is a large complex set of other relationships in the vertical axis of lakes, rivers and reservoirs, which will be described in Chapter 4.

The morphometry of lakes, as well as some other biological and chemical parameters, can be used to express a lake's trophic state. Schindler (1971) used morphometric data to calculate differences in trophic state. Whereas in many lakes the only source of nutrients comes from the drainage area of the water basin and precipitation on the surface of the lake, Schindler proposes:

$$\frac{Ad+Ao}{v}$$

where:
Ad = drainage of the water basin
Ao = precipitation on the lake's surface
v = volume (**dilution** factor)

This ratio should therefore be proportional to the nutrient levels and biological productivity of the lake.

Rawson (1951) proposed the term morpho-edaphic index, empirically relating **total dissolved solids** (in mg/liter) and average depth (in meters).

This calculation was originally applied to determine the production of commercial fishing in lakes in Canada and Finland, and it shows a good correlation (Cole, 1983). However, certain problems with high salinity can complicate use of the index. It is relatively difficult to apply for turbulent shallow lakes in tropical and equatorial regions. A major problem in studying the morphometry of lakes is understanding the relationship between lake area and the total area of the basin in which it lies. For lakes in the Medio Rio Doce, in Minas Gerais, Pflug (1969) found that 14 out of 21 lakes presented an inter-relationship of the area of the water basin to the area of the lake of 3.0–5.0,

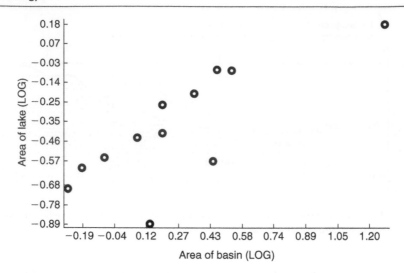

Figure 3.8 Relationship between the area (in km²) of the drainage basin and the area of the lakes in the lake system of the Medio Rio Doce.
Source: De Meis and Tundisi (1997).

while for seven other lakes, values were much higher, around 13 (see Figure 3.8). In a later work, Moura *et al.* (1978) documented the existence of two separate groups of lakes: one group in which lake area is extremely large compared to the total watershed area (65–75%), and another in which the respective lake area is very small in comparison to total watershed area (around 25%).

Shallow lakes are extremely important limnologically and ecologically, and some examples can be seen in Table 3.2. Löffler (1982) distinguishes shallow lakes with little accumulation of sediment and shallow lakes with deep sediment, resulting from **siltation**, which differ considerably in limnological features. Many **shallow saline lakes** show **horizontal gradients** of salinity. The basic feature of these lakes is that their vertical structure is always under the influence of the wind, and turbulence affects the entire water-mass as well as the upper part of the sediment. These lakes generally have an average depth of 10 meters or less, and a flat or slightly concave bottom.

The following general features are common in shallow lakes:

▶ Irregular sedimentation interrupted by periods of erosion;
▶ The **horizontal zonation** of different abiotic and biotic parameters is always greater than the vertical zonation;
▶ Effective vertical and horizontal circulation and absence of prolonged stratification periods; lakes generally polymictic (see Chapter 4);
▶ Tendency toward eutrophication;
▶ Lack of internal waves and exposure of submerged areas during periods of **fluctuation in levels**;
▶ Large numbers of **aquatic birds**, causing additional impact on eutrophication and biogeochemical cycles.

Shallow lakes resulting from the process of siltation in old deep lakes can be found in mountainous regions and highlands, such as in South America and Mexico, and also in

Table 3.2   Some well-known shallow lakes.

Name	Area (km²)	Z (m)	DS (m)	Type	Chemical characteristics	Area covered by emergent vegetation
Parakrama Samudra (reservoir/Sri Lanka)	18.2	3.9	0.3–1.3	P	Fresh water	10%
Nakuru (Africa)	40.0	2.3	–	–	Alkaline	10%
George (Africa)	250			S	Fresh water	30%
Neusiedlersee (Austria)	300	0.5	0.05–1.40	P	Fresh water/alkaline water	50%
Balaton (Hungary)	600	3.3	0.20–4.00	S	Fresh water/alkaline water	10%
Niriz (Iran)	1,240	~0.5	1.0		Fresh water/alkaline water	10%
Chad (Africa)	20,900	~3.4	0.08–0.8		Fresh water/alkaline water	10%

P – Primary shallow lake; S – Secondary shallow lake; Z – Average depth; DS – **Secchi disk** reading.
Source: Modified from Löffler (1982).

Asia. These lakes are called secondary shallow lakes. Primary shallow lakes accumulate little sediment and are located in marginal areas of tropical rivers and shallow tectonic basins (also lakes in permafrost areas). Many primary shallow lakes are in arid or semiarid regions, and are highly saline.

An illustrative example of interactions between the origin of lakes, geomorphological processes and the effects on the **limnological functioning** of lacustrine ecosystems is the process occurring in the lake system of the Medio Rio Doce, located in eastern Brazil.

In these lake systems (De Meis and Tundisi, 1997), the irregular distribution of the processes of erosion and sedimentation results in a large set of lakes, wetlands and swamps, with varying morphometries, sizes, depths and differentiated limnological features.

The dynamics of the Upper Quaternary in the region was studied by Pflug (1969a, 1969b). According to this author, the Rio Doce depression (see Figure 3.9) results from the process of pediplanization during the Quaternary's semi-humid climate. Later, in periods varying from 3,000–10,000 years ago, there was a process of incision of valleys and a tributary system was formed. These tributaries with dendritic morphology were then subjected to the process of obstruction, and lakes and wetlands were formed.

The maximum depths of lakes are related to their location in the regions of tributaries of higher or lower altitudes. Lakes originating from higher-altitude tributaries are shallower than those from lower altitudes, resulting in systems with different shoreline development indices, average depths and volumes (see Figure 3.10 and Figure 3.11).

The various sedimentation processes with different rates over time led to the creation of the lake system. The different sedimentation rates are considered the main factor controlling the spatial and temporal development of this system.

As a result of these basic processes, many limnological features are unique to the lake system:

▶   processes of thermal stratification and vertical circulation;
▶   processes of **biological stratification** in summer and winter;

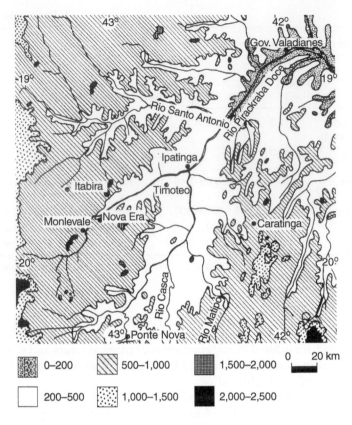

Figure 3.9 The interplateau depression of the Rio Doce.
Source: De Meis and Tundisi (1997).

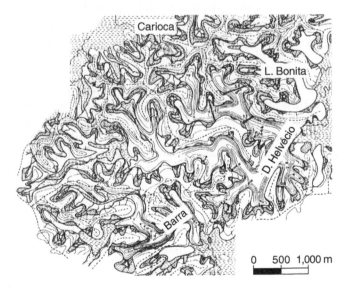

Figure 3.10 Topographical and geomorphic elements of three lakes of the Rio Doce (D. Helvécio, Carioca, and Barra), illustrating the morphological complexity of the region and the relationship between the system's origin and function.
Source: De Meis and Tundisi (1997).

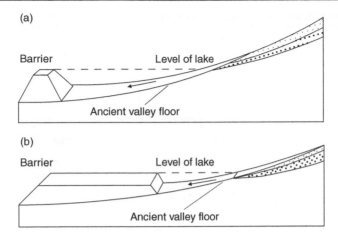

Figure 3.11   Profile of (a) a deep lake and (b) a shallow lake in the lake system of the Médio Rio Doce, illustrating the effect of the dam on the lakes' average depth and maximum depth.
Source: De Meis and Tundisi (1997).

▶   processes of **chemical stratification** resulting from thermal stratification;
▶   fluctuating diurnal patterns in lakes.

A summary of the mechanisms operating in lakes in the Medio Rio Doce and their origins is presented in Chapter 16.

3.5   ARTIFICIAL DAMS

Humans have been damming rivers for thousands of years. However, at the end of the 19th century and throughout the 20th century, these artificial systems, constructed around the globe, became very large (more than $1 \, km^3$ of volume in the majority of reservoirs and flooded areas often of several hundreds or thousands of square kilometers).

Artificial reservoirs present very different characteristics from lakes and affect drainage basins and hydrological cycles. In this chapter, the main differences between artificial reservoirs and lakes are presented (see Chart 3.4). Details of the operating mechanisms in reservoirs as ecosystems are presented in Chapter 11, and a comparison of lakes, rivers and reservoirs, from the hydrodynamic point of view, is presented in Chapter 4.

Another important difference between lakes and reservoirs is that reservoirs have very characteristic and accentuated longitudinal gradients, in which three regions can be distinguished:

▶   region under the influence of **tributary rivers**;
▶   transitional region functioning as an intermediate zone between river and lake;
▶   region that is more lacustrine, subject to the opening/closing actions in the spillways and turbines.

Chart 3.4   Differences between lakes and reservoirs.

Characteristics	Lakes	Reservoirs
Qualitative differences (absolute)		
Nature	Natural	Artificial
Geological age	Pleistocene or earlier	Young (<100 years)
"Aging"	Slow	Fast
Formation	Various origins	More frequent flooding in river valleys
Morphometry	Generally regular, oval, round	General dendritic
Index of development of the margin	Low	High
Maximum depth	Generally near the centre	Generally near the dam
Sediments	Generally autochthonous	Generally allochthonous
Longitudinal gradients	Directed by the wind	Directed by the flow
Height of the discharge	On the surface	Deep
Quantitative differences (relative)		
Area of the water basin/area of the lake-reservoir	Low	High
Retention time	High	Low and variable
Connection with the hydrographic basin	Small	Large
Fluctuations in level	Generally small	Generally higher
Hydrodynamics	More regular	More varied and less regular
Cause of the pulses	Generally natural	Generally artificial, produced by man

Source: Straškraba *et al.* (1933a, b).

The expansion and contraction of these three areas depend on the flow of water, the entry of tributaries, retention time, and the construction features of the dam.

One basic difference between lakes and reservoirs is in each system's hydrodynamics. The circulation system in reservoirs is principally driven by the operating process. The spatial and temporal succession of planktonic populations, horizontally and vertically, depends on vertical circulation and horizontal hydrodynamic patterns.

Hydrodynamic flow and retention times affect the distribution of phytoplankton and zooplankton in reservoirs, and also affect the life cycles and reproduction of species. The transport of toxic substances, carbon, **nitrogen** and phosphorus can

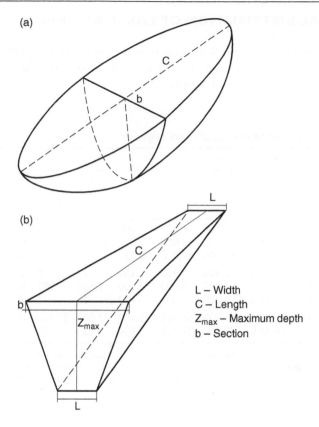

Figure 3.12   Typical morphometry of (a) lakes and (b) reservoirs.
Source: Straškraba *et al.* (1993a, b).

also be greatly influenced by vertical circulation and lateral and horizontal transport (Tundisi *et al.*, 1998).

A cascade of reservoirs (as in the case of many systems on the main rivers in southeastern Brazil) causes a series of effects on the distribution and reproduction of organisms, biogeochemical cycles, horizontal downstream circulation, and produces new hydrodynamic patterns as well as chemical and biological patterns (Barbosa *et al.*, 1999; Straškraba and Tundisi, 1999). Variations in retention time in reservoirs establish much more dynamic and differentiated horizontal patterns than in lakes.

Reservoirs and lakes can function under the direct action of external and internal forces. External forces include climate, solar radiation, precipitation and wind. Such forces determine the intensity of the different processes, the **vertical and horizontal gradients, resuspension** and sedimentation.

Natural lakes occupy **natural depressions** in the local or regional topography. Consequently, they are located in the centre of symmetric and contiguous drainage basins. Reservoirs are generally built downstream from a drainage basin. Horizontal gradients are therefore more evident and more common in reservoirs, as shown in Figure 3.12 (see Chapter 12).

3.6   GLOBAL DISTRIBUTION OF LAKES BY ORIGIN

Table 3.3, Table 3.4 and Chart 3.5 summarize, respectively, the global distribution of fluvial lakes, examples of the age of lakes in mode of origin, and global distribution of lakes in accordance with their origin.

Table 3.3   Global distribution of lakes of fluvial origin.

		Area in each class (km²)								Total area of lakes	Limnic ratio RL (%)
		0.01	0.1	1	10	10^2	10^3	10^4	10^5		
Floodplain lakes (1)	d_L	100,000	44,000	6,000	6,000	30	2	0		200,000	5.9
	A_0	8,800	39,000	52,000	52,000	26,500	22,560	0			
	n	340,000	150,000	20,000	2,000	102	8	0			
Lakes in deltas (2)	d_L	9,000	7,400	900	140	23	2	0		18,200	1.7
	A	260	2,000	2,600	3,900	6,500	2,980	0			
	n	10,000	8,000	1,000	150	25	2	0			
All lakes of fluvial origin	A_0	9,000	41,000	54,600	55,900	33,000	24,600	0		218,000	
	n	350,000	158,000	21,000	2,150	127	10				

A_0 – Total area of lakes; n – total number of lakes; d_L – density of lakes (number per million of km²); LR – Limnic Ratio (expressed in percentage) is defined as the ratio between the total area of lakes to the total documented area in which the census was made; (1) For a total area of floodplains of $3.4 \cdot 10^6$ km²; (2) For a total area of deltas of $1.08 \cdot 10^6$ km²
Source: Meybeck (1995).

Table 3.4   Global distribution of lakes based on their origin.

Origin	Class of lakes by area (km²)	0.01	0.1	1	10	100	1,000[a]	10,000[a]	100,000[a]	Total area of lake (km²)
Tectonic	A_0	5,000	10,000	20,000	30,000	52,000	134,900	267,300	347,000	893,000
	n	200,000	40,000	8,000	1,100	200	40	8	1	
Glacial	A_0	85,000	144,000	165,000	175,000	197,000	136,000	345,000	0	1,247,000
	n	3,250,000	554,000	63,000	6,800	710	52	9		
Fluvial	A_0	9,000	41,000	54,600	55,900	33,000	24,600	0	0	218,000
	n	350,000	158,000	21,000	2,150	127	10	0		
Crater	A_0	130	130	390	800	610	1,100	0	0	3,150
	n	500	500	150	30	4	1			
Coastal lakes	A_0	700[b]	3,400[b]	5,700[b]	9,400[b]	15,600[b]	15,060[b]	10,140	0	60,000
	n	25,000	13,000	2,200	360	600	10	1		
Miscellaneous	A_0	13,000	15,000	15,000[b]	15,000[b]	15,000[b]	15,000	0	0	88,000
	n	500,000	60,000	6,000	600	60	11	0	0	
Total	d_L	32,000	6,200	750	80	8.6	0.93	0.13	0.0075	
	A_0	113,000	213,000	261,000	286,000	313,000	327,000	623,000	374,000	2,510,000
	n	4,300,000	825,000	100,000	10,600	1,150	124	18	1	4,400,000

A_0 – Total area of lakes; n – total number of lakes; d_L – density of lakes (number per millions of km²); [a] – data from Herdendorf census (1984, 1990) for lakes larger than 500 km²; [b] – estimated values.
Source: Meybeck (1995).

Chart 3.5   Examples of age of lakes depending on their origin.

Origin	Lake	Age	Reference
Tectonic	Baikal	$\approx$20 M	
	Issyk-Kul	$\approx$25 M	Tiercelin and Mondeguer (1991)
	Tanganyika	20 M	Timms (1993)
	George	4–8 M	
Diverse tectonic origins	Caspian[a]	>5 M	Stanley & Wetzel (1985)
	Aral[a]	>5 M	Stanley & Wetzel (1985)
	Ohrid[a]	>5 M	Stankovic (1960)
	Prespa[a]	>5 M	Stankovic (1960)
	Maracaibo	>36 M	Fairbridge (1968)
	Biwa	2 M	Timms (1993)
	Eyre	20–50 M	Adamson/Williams (1980)
	Victoria	$\approx$20,000 [B.P.]	Imboden *et al.* (1977)
	Tahoe	2 M	
Tectonic volcano	Kivu	1 M[b]	Degens *et al.* (1973)
	Toba	10,000 [B.P.] [c]	Tiercelin/Mondeguer (1991)
	Lanao (Phillippines)	75,000 [B.P.]	Dawson (1992)
		>2 M	Frey (1969)
Glacial	**Great Lakes (USA-Canada)**	8,000 [B.P.] [d]	Dawson (1992)
Fluvial	**Lakes of the Mississippi River**	9,000 [B.P.]	Dawson (1992)
Sedimentation	**Vallon**	1943	Dussart (1992)
	Sarez (Pamir)	Feb. 1911	Hutchinson (1957)
Crater impact	**Crater** (Quebec)	1.3 M	Oullet and Page (1990)
	Bosumtwi (Ghana)	1.3 M	Livingstone and Melack (1984)
Volcanic crater	Wisdom	300 [B.P.]	Ball/Glucksman (1978)
	Crater (Oregon)	6,500 [B.P.]	Fairbridge (1968)
	Atitlan (Guatemala)	84,000 [B.P.]	Dawson (1992)
	Le Bouchet (France)	>250,000 [B.P.]	Williams *et al.* (1993)
	Viti (Iceland)	17 May 1724	Hutchinson (1957)

B.P. – Before the present
a – Relicts of the Thetis Sea
b – Age of tectonic depression
c – Date of discharge to the Nile and connection with the Zaire through lake Tanganyika
d – In its present configuration
M – Millions of years
Source: Meybeck (1995).

Classic turbulence. The breaking wave off Kanagawa. Hokusai

4 Physical processes and circulation in lakes

SUMMARY

This chapter describes the factors involved in the vertical and horizontal transport mechanisms in lakes, the major forces acting on turbulence, and the physical processes that affect the distribution of elements, substances and organisms in lakes, reservoirs and rivers.

The characteristics and differences between the circulation in lakes and reservoirs are presented, as well as the distinction between **turbulant flow** and **laminar flow**. The scales of horizontal and vertical circulation are defined and described. Thermal, chemical and biological stratification are discussed, and also how turbulent kinetic energy (TKE) operates in the distribution and dispersion of suspended particles (plankton, suspended inorganic and organic material – detritus).

The chapter also presents dimensionless numbers, which are important tools in defining circulation mechanisms in lakes, reservoirs and rivers.

4.1   PENETRATION OF SOLAR ENERGY IN WATER

The intensity and spectral distribution of the solar radiation received by the Earth depend on the characteristics of emission and distance from the sun. Total solar energy received is approximately 5.46×10^{24} J per year.

Electromagnetic radiation exists as discrete units of energy. They are known as **quanta** or **photons**. Each photon is characterized by a specific energy (in units of joules), a wavelength frequency U (units of cycles s^{-1}), a specific wavelength (ζ) (in units of meters), the speed of light in a vacuum (c) ($3 \cdot 10^{8}$ m·s^{-1}) and Planck's constant h (h = $6625 \cdot 10^{-34}$ joule·s).

However, in some regions of the **electromagnetic spectrum,** it may be more convenient to express wavelength in **nanometers** (1 nanometer is equivalent to 10^{-9} meters).

Therefore:

$$\varepsilon = h \cdot v \cdot \frac{c}{\zeta}$$

where:
ε = energy of a photon (or quantum of radiation)
v = frequency of electromagnetic waves
ζ = electromagnetic wavelength
c = speed of light ($3 \cdot 10^{8}$ m·s^{-1})
h = planck's constant (h = $6625 \cdot 10^{-34}$ J·s).

The term *light* is generally used to refer to the portion of the electromagnetic spectrum to which the human eye is sensitive (i.e., the region of the spectrum considered visible – in the range of 390–740 nm). The integration of this radiation in the spectral band with the sensitivity and response of the human eye results in the neurophysiologic sensation of colour (Bukata *et al.*, 1995). The aesthetic value of a body of water is related to its colour and hence to the quality of the water.

Life essentially depends on the amount and quality of solar energy that is available on the surface and distributed in the water column. In any medium, light is related to its colour, and this in turn, to water quality.

The solar radiation reaching the top of the atmosphere (in unit area and unit time) is called the **solar constant.** The intensity and quality of solar radiation changes, and its quantity is substantially reduced during its passage through the atmosphere.

This reduction in intensity occurs because of **scattering** by air molecules and particles of dust, and also because of the absorption of water vapor, oxygen, ozone and **carbon dioxide** in the atmosphere.

Part of the radiation in the scattering process is lost in space and the other part reaches the Earth's surface.

Only a portion of the solar radiation that reaches the surface of **terrestrial** and aquatic **ecosystems** can be used in the **process of photosynthesis**. This photosynthetically active radiation (PAR) is equivalent to approximately 46% of the total PAR, corresponding to the virtual spectrum of radiation. The intensity of the radiation

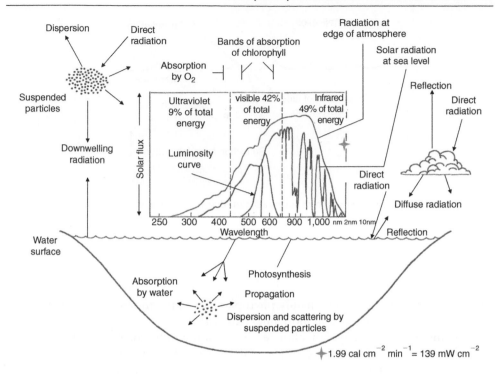

Figure 4.1 Spectrum of solar radiation and its penetration in the aquatic environment.

reaching the surface of lakes, oceans and continents varies with **latitude**, season, time of day, and condition of cloud cover. Solar radiation on Earth has two components: direct radiation coming directly from the sun, and indirect radiation, which is the result of **reflection** and **refraction** by clouds.

Figure 4.1 shows the various characteristics of the spectrum of solar radiation that reaches the surface of ecosystems, and the various processes that accompany penetration of solar energy into the aquatic environment.

The penetration of solar energy into the aquatic environment basically depends on processes of absorption – it has specific properties in terms of variations in the **coefficient of attenuation K** (m^{-1}) by absorption and scattering – and its underwater intensity varies with wavelength.

The solar radiation that reaches the surface of the water is modified by reflection and refraction at the air/water interface. The effect of refraction is to modify the underwater angular distribution in the water. Reflection and refraction depend not only on the angle of incidence of the light, but also on the status of the surface of the bodies of water (calm or with waves under the **effect of the wind**).

Figure 4.2 shows the reflection and refraction of light on the air/water surface.

All the absorption of solar radiation that reaches a body of water is attributed to four components of the aquatic environment: the aquatic medium, dissolved compounds,

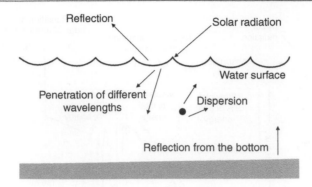

Figure 4.2   Reflection and refraction of solar radiation that reaches the water surface.

photosynthesizing organisms (mainly phytoplankton, and aquatic macrophytes) and particulate matter (organic and inorganic) (Kirk, 1980).

A complete description of the climate of **underwater radiation** requires that the flow of radiation be described for each angle and wavelength. This climate of underwater radiation requires that all optical properties of the medium be known: the scattering in various directions and depths (Sathyendranath and Platt, 1990). Light intensity decreases exponentially with depth; the loss is expressed by the coefficient of extinction (or attenuation), which depends on the fraction of light absorbed per meter of water. The larger the coefficient of extinction, the less the light is transmitted or the less transparent is the body of water.

For a beam of monochromatic light, light intensity at depth z, when the sun is in a vertical position, is:

$$I_z = I_o \cdot e^{-kt}$$

where:
I_z = irradiation at depth z
I_o = irradiation on the surface
e = base of natural logarithms
kt = coefficient of total attenuation of underwater irradiation
The total attenuation coefficient (unit, m^{-1}) is given in terms of its components:

$$Kt = Kc + Kw + Kx$$

where:
Kt = total light attenuation coefficient;
Kw = attenuation coefficient due to water and dissolved substances;
Kx = attenuation coefficient due to suspended matter (organic or inorganic);
Kc = attenuation due to suspended **chlorophyll**.

Table 4.1 Coefficients of absorption due to "yellow substances" (gilvin).

Coefficient of absorption at wavelength *440 nm due to the gilvin and* **particulate matter.**

	440 nm (m^{-1})	440 nm (m^{-1})
1. Oceanic waters		
Sargasso Sea	0	0.01
Pacific Ocean (coast of Peru)	0.05	–
Oligotrophic waters	0.02	–
Mesotrophic waters	0.03	–
Eutrophic waters	0.09	–
2. Coastal waters		
Estuary of the Reno River	0.86	0.572
Baltic Sea	0.24	
3. Lakes and reservoirs		
Lake Victoria (Africa)	0.65	0.22
Lake Neusidlersee (Austria)	2.0	
Guri Reservoir (Venezuela)	4.84	

Source: Modified from Kirk (1986).

The attenuation coefficient Kc is given by C. 0.016, where C is the concentration of chlorophyll and 0.016 is an average representative of the specific attenuation coefficient of chlorophyll (Yentsch, 1980).

Possibilities exist to determine the contribution of each component in the attenuation process.

Kirk (1980) considers, for photosynthetically active radiation, the equation:

$$Kt = Kw + Kg + Ktr + Kf$$

where:

Kw = coefficient of partial attenuation of the water;

Kg = coefficient of partial attenuation of the "gilvin" (humic substances = dissolved yellow substances; from the Latin: gilvus – yellow);

Ktr = partial coefficient of trypton (non-living particulate matter);

Kf = partial attenuation coefficient of phytoplankton.

There is significant variation in these coefficients of attenuation, depending on the suspended **organic material**, alive or dead, on the dissolved substances in the water and on the concentration of phytoplankton (see Table 4.1).

In general, lakes or dams with concentrations of chlorophyll between 10–20 µg/L or <10 µg/L present a predominance of vertical attenuation due to non-living components including trypton. Lakes or reservoirs in which the concentration of chlorophyll is >20 µg/L present predominantly vertical attenuation coefficients due to phytoplankton.

Coastal waters naturally have a higher absorption at 440 nm than that of ocean waters. In inland waters, this absorption is extremely high in waters that drain tropical forests or in swampy areas.

The terms of *absorption coefficient, attenuation coefficient* and *extinction coefficient* are often used synonymously. They refer to calculations that use natural logarithms.

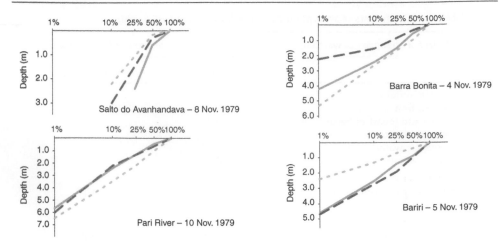

Figure 4.3   Relative penetration of solar radiation of various wavelengths (logarithmic scale) in various reservoirs in the state of São Paulo (see also color plate section, plate 2).
Source: Reservoir Typology project of the state of São Paulo – Fapesp).

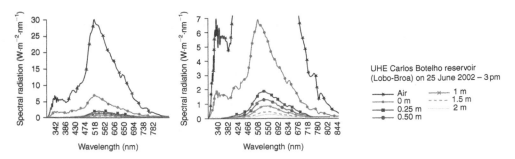

Figure 4.4   Spectral distribution of underwater solar radiation in the UHE Carlos Botelho (Lobo/Broa) reservoir 25 June 2002 (3 pm) (see also color plate section, plate 3).
Source: Rodrigues (2003).

In addition to attenuation and absorption of light, **transmittance** (T) can be considered, which is the percentage of light passing through 1 meter. It is $100 e^{-k}$, where k is the coefficient of vertical absorption.

Transmittance varies for different wavelengths. For example, when λ (transmittance of a given wavelength) is in the red wavelength 680 mm, $T = 100^{-k\lambda_{680}}$. Therefore, when a beam of incident solar radiation, which consists of many wavelengths, reaches the surface of the water, the total radiation is extinguished only approximately exponentially, but the transmittance and attenuation/absorption coefficient vary for different wavelengths, depending on the suspended material, phytoplankton and dissolved organic substances. Figure 4.3 shows the transmittance, on a logarithmic scale, of various wavelengths for lakes and reservoirs.

Solar radiation in the **infrared zone** is rapidly absorbed in the first few meters of the water column. The spectrum of the immediate subsurface also presents ultraviolet radiation, which is rapidly absorbed in the vertical column.

The spectral variations of radiation in the photosynthetically active band also vary daily and seasonally, depending on latitude, nature of the body of water, concentration

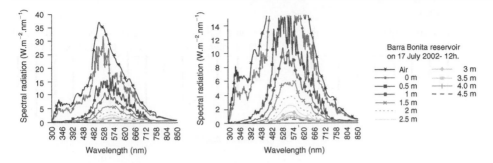

Figure 4.5 Spectral distribution of underwater solar radiation in the Barra Bonita (SP) reservoir of São Paulo, on 17 July 2002 (12 pm) (see also color plate section, plate 4).
Source: Rodrígues (2003).

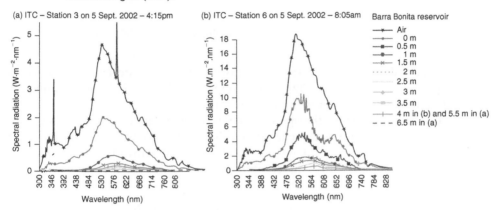

Figure 4.6 Spectral distribution of underwater solar radiation in the Barra Bonita (SP) reservoir (see also color plate section, plate 5).
Source: Rodrigues (2003).

of dissolved substances, suspended particulate matter (trypton or phytoplankton), the effects of circulation on **vertical mixing** of the water column, the patterns of wave formation and stability on the surface of lakes, rivers, reservoirs and estuaries.

Figure 4.4 and Figure 4.5 show the spectral distribution of the **underwater solar radiation** measured with spectro-radiometer in the Carlos Botelho (Lobo/Broa) and Barra Bonita (Medio Tietê) reservoirs.

One of the techniques used to determine an **optical signature** for the water column is to determine the absorption spectrum of the filtered and centrifuged water and compare it with a sample that is distilled (see Figure 4.6).

Variations in the spectral quality of solar energy penetrating estuarine waters during different tidal periods were discussed by Tundisi (1970). The climate of underwater radiation can also be modified by the effect of the wind on the surface of the water, which amplifies the effects of reflection produced by small waves on the surface. The reflection of solar energy on the water surface is also higher near sunrise and sunset. The percentage of light that is reflected on the surface of the water is called the albedo.

Figure 4.7 shows the seasonal variation of Secchi disc readings (maximum depth for visibility of a white disc) in reservoir of the **Barra Bonita dam** (SP).

When the lake's surface is covered with ice and snow, the albedo reaches 90%. Differences in reflection caused by the accumulation of **pollutants**, suspended

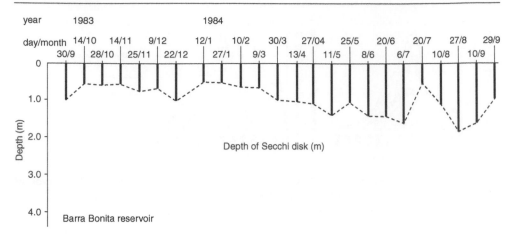

Figure 4.7   Seasonal variations in depth of Secchi disk in the Barra Bonita (SP) reservoir.
      Source: Tundisi and Matsumura Tundisi (1990).

particulates, or inflow of rivers with a lot of dissolved humic material can be detected by aerial photographs or by satellite, which helps considerably in understanding the distribution of masses of water in large lakes or reservoirs.

One method to compare the spectral quality of underwater radiation is to calculate the amounts of underwater energy and the coefficients of vertical attenuation (Kd) for wavelengths of 450, 550, 650 nm, compared to the absorption coefficient in the red portion of the spectrum. This compound value, called the spectral attenuation ratio, is given by:

$$SAR = \frac{Kd\ (blue)}{Kd\ (red)} : \frac{Kd\ (blue)}{Kd\ (red)} : \frac{Kd\ (red)}{Kd\ (red)}$$

4.1.1   The euphotic zone

The euphotic zone is the illuminated layer of a mass of water. Its depth depends on the absorption by the water, of the transmittance, concentration of suspended particulate matter (trypton), phytoplankton and zooplankton, and the amount of dissolved organic substances. It can vary daily, seasonally, or from important climatic events such as **cold fronts**, periods of circulation and stability of the water column.

The relationships z_{eu}/z_{af} (euphotic zone/aphotic zone) and z_{eu}/z_{max} (euphotic zone depth/maximum depth of the lake, river or reservoir) are important for understanding the vertical functioning of aquatic ecosystems.

Table 4.2 shows data from reading a Secchi disc for several ecosystems.

Table 4.3 and Table 4.4 show the total attenuation coefficient of solar radiation in water due to dissolved particulate components.

4.1.2   Underwater radiation and aquatic organisms

The climate of underwater radiation is very important to aquatic organisms. Certain plants present specific pigments that use the solar energy available in the depths at which they live. This chromatic (colour) adaptation can also exist in organisms living

Table 4.2   Comparative data of Secchi disk readings for various lakes and reservoirs (in meters).

Lake	Secchi disc reading
Crater Lake (Oregon, USA)	38.00 m
Crystal Lake (Wisconsin, USA)	14.00 m
Carlos Botelho UHE reservoir – Lobo/Broa (São Paulo, Brazil)	1.20 m (winter)
Carlos Botelho UHE reservoir – Lobo/Broa (São Paulo, Brazil)	1.70 m (summer)
Lake Dom Helvécio (Minas Gerais, Brazil)	6.0 m
Tururuí reservoir (Amazonas, Brazil)	5.00 m
Barra Bonita reservoir (São Paulo, Brazil)	1.15 m (winter)
Lake Cocibolca (Nicaragua)	0.50 m
Lake Amatitlan (Guatemala)	2.40 m
Lake Atitlan (Guatemala)	6.10 m
Lake Gatún (Panamá)	7.00 m

Source: Various sources.

Table 4.3   Coefficient of total attenuation of the solar radiation, Kt, and its components (Kw, Kc and Kx) for the Barra Bonita reservoir, at a point in the central body of the reservoir, P5B, in December 1999 (rainy period).

Site of study	PAR (%)	Depth (m)	Kt (m^{-1})	Kw (m^{-1})	%	Kc (m^{-1})	%	Kx (m^{-1})	%
P5B	100	0.00	2.903	0.046	1.6	1.396	48.1	1.461	50.3
8 Dec 99	30	0.50	2.379	0.048	2.0	1.569	66.0	0.762	32.0
1:45 pm	10	1.00	2.347	0.041	1.7	1.354	57.7	0.952	40.6
	0.9	2.25	2.072	0.041	2.0	0.536	25.9	1.495	72.1
	Aphotic zone	5.00	1.825	0.041	2.3	0.351	19.2	1.433	

Source: Rodrigues (2003).

Table 4.4   Coefficient of total attenuation of the solar radiation, Kt, and its components (Kw, Kc and Kx) for the UHE Carlos Botelho reservoir (Lobo/Broa), in December 1999 (rainy season).

Site of study	PAR (%)	Depth (m)	Kt (m^{-1})	Kw (m^{-1})	%	Kc (m^{-1})	%	Kx (m^{-1})	%
P1	100	0.00	2.401	0.039	1.6	0.079	3.3	2.283	95.1
10 Dec 99	20	0.50	3.219	0.039	1.2	0.000	0.0	3.180	98.8
9:40 am	11	1.50	1.465	0.037	2.5	0.061	4.2	1.367	93.3
	1	4.25	1.092	0.041	3.8	–	–	–	–
	Aphotic zone	6.00	1.201	0.044	3.7	0.084	7.0	1.073	89.3

P1 – located in the central body of the reservoir
Source: Rodrigues (2003).

in regions with high constant light intensity, with high levels of protective carotenoid pigments, such as occurs in certain Antarctic lakes (Horne and Goldman, 1994).

The behaviour of many aquatic organisms is greatly influenced by the intensity of underwater solar energy and also the spectral quality of light present at different depths. For example, the vertical migration of zooplankton, and benthic forms from the surface of the sediment in shallow lakes is strongly influenced by solar energy.

Even the low intensity radiation that exists during periods of the full moon can influence the migration of lacustrine zooplankton (Gliwicz, 1986). Problems of photosynthetic responses of phytoplankton to the varying light intensities produced by vertical circulation will be discussed in Chapter 9.

Besides providing energy, which is the basic source of life for aquatic organisms, underwater radiation is widely used by organisms for orientation and information on the environment in which they live. Many aquatic organisms also play an important role in controlling this underwater radiation, not only from the qualitative but also quantitative point of view. For example, Tundisi (unpublished results) showed that banks of *Eichhornia crassipes* in lakes of the Mato Grosso Swamp can reduce solar radiation that reaches them by up to 90% of the level on the surface of the water. Extensive blooms of **cyanobacteria** (blue-green algae) can reduce solar radiation reaching the surface of lakes, reservoirs and rivers to just a few centimeters. Therefore the euphotic zone is extremely influenced by the concentration of these organisms.

Ultraviolet radiation can affect aquatic organisms, as will be discussed in Chapter 7.

4.1.3   Remote sensing of aquatic systems and solar radiation

Measurements of solar radiation penetration in water and studies on the optics of aquatic systems traditionally are mostly related to the evaluation of the energy available for photosynthesis, the responses of phytoplankton to different light intensities, and also to the use of the light emitted by fluorescence to measure the concentration of chlorophyll.

In remote sensing, the technology uses the light reflected by the surface of a body of water to measure and classify the reflectance, due to its composition. Sensors located on satellites measure this reflectance after corrections (for example, correcting for the reflectance produced by aerosols). The sensors produce images of optical properties that are very useful for classifying lakes, reservoirs or rivers and make it possible to detect impacts of uses of the water basin in these systems. Measurements of this reflectance should also be accompanied by ground-based measurements of physical, chemical and biological variables at the same time as the passing of the satellite.

4.2   HEAT BALANCE IN AQUATICS SYSTEMS

Figure 4.8 shows the principal mechanisms for the transfer of heat through the surface of the water. The main processes consist of radiation, heat conduction and **evaporation**. The **liquid heat flow** is described as a border condition, directly on the free surface. The absorption of radiation occurs not only on the air/water interface, but also in a layer of water near the surface, and the thickness of this layer depends on the characteristics of absorption, such as, for example, turbidity. Determining the **heat balance** of inland aquatic ecosystems, especially for lakes and reservoirs, is of fundamental importance for understanding the physical effects on the systems and their repercussions on chemical and biological processes. Diurnal variations that occur in heat transfers are responsible for thermal heating and cooling, and for the behaviour of masses of water at different times of day and night.

Method to measure penetration of solar energy in water

The tool most commonly used to measure the total incidence of solar radiation on the surface of lakes, rivers, coastal waters, estuaries, oceans and terrestrial systems is the pyrheliometer, which measures total direct or indirect solar radiation during the complete period of one day. The measurements of underwater solar radiation and, more specifically, of the climate of underwater radiation, can be done with waterproof measurement equipment located in the water at various depths, including estimates of the underwater radiation at wavelengths not visible to the human eye. These tools, hydrophotometers, can be used with transmittance filters at various wavelengths, which enables the passage of certain bands of non-visible wavelengths (infrared, ultraviolet), plus visible (purple, blue, yellow, orange). Qualitative data can therefore be collected on the spectral quality of light.

Other more sophisticated instruments are underwater spectroradiometers, which can read every 2 nanometers of different spectral bands and conduct a complete scan of the underwater spectrum.

An old robust instrument, widely used more than 100 years ago, is the Secchi disc. In 1865, Father Pietro Angelo Secchi (Tyler, 1968) was asked to make measure the transparency of the coastal waters of the Mediterranean. In April 1865, he placed a white disc, 43 cm in diameter, in the water, and for the first time measured the depth at which the disc disappeared from view. Since then, this depth has been called the **Secchi depth**. He concluded that the critical factors in estimating this depth were the diameter of the disk, its spectral reflectance, the presence of waves, reflections of the sun and sky in the water and the levels of plankton – variables that even today are used to describe the climate of underwater radiation.

Several studies were conducted to determine the empirical relation between the Secchi depth (z_{DS}) and the euphotic zone (z_{eu}). It is generally accepted, according to Margalef (1983), that $z_{eu} = 3.7z_{DS}$. Poole and Atkins (1929) obtained a different calculation $Kt = 1.7/z_{DS}$, which estimates the total attenuation coefficient of the water mass. Kirk (1982) considered that the value $Kt = 1.44/z_{DS}$ is more appropriate to determine the attenuation coefficient.

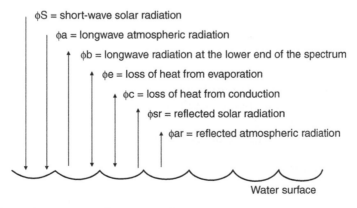

φS = short-wave solar radiation

φa = longwave atmospheric radiation

φb = longwave radiation at the lower end of the spectrum

φe = loss of heat from evaporation

φc = loss of heat from conduction

φsr = reflected solar radiation

φar = reflected atmospheric radiation

Water surface

Figure 4.8   Fluxes of radiation on the water surface.

Figure 4.9   Heat-content for the UHE Carlos Botelho reservoir (Lobo/Broa) (1988).
Source: Henri and Tundisi (1988).

In general, the flux through the surface of the water can be determined by the sum of the individual terms:

$$\phi n = \phi s - \phi sr + \phi a - \phi ar - \phi br - \phi e - \phi c$$

ϕn depends on **weather conditions** (variations in air and water temperature, solar radiation, relative humidity and wind speed), which depend on the latitude, longitude and altitude where the body of water is located.

Figure 4.9 shows variation in the heat balance for the reservoir of the UHE Carlos Botelho (Lobo/Broa) dam in southeastern Brazil (Henry and Tundisi, 1988).

4.3   PHYSICAL PROCESSES IN LAKES, RESERVOIRS AND RIVERS

The principal mechanisms and functions of physical force that affect the vertical and horizontal structure of lakes and reservoirs include the following:

External mechanisms

▶ Wind
▶ Barometric pressure
▶ Heat transfer
▶ **Intrusion** (natural or artificial)
▶ Downstream flow (natural or artificial)
▶ **Coriolis effect** (from the rotation of the Earth)
▶ Discharges on the surface
Plumes and jets on the surface of lakes and reservoirs

Internal mechanisms

▶ Stratification
▶ Vertical mixing

▶ Selective removal or selective loss downstream (natural or artificial)
▶ **Density currents**
  Formation of internal waves

These mechanisms drive the processes of vertical organization in lakes and reservoirs, and have fundamental chemical and biological consequences for the functioning of these ecosystems. Both internal and external mechanisms are influenced by climatic and hydrological factors, which are functions of force acting on systems.

The physical processes of stratification and vertical mixing are of fundamental importance for the structure and organization of chemical and biological processes in lakes, reservoirs, rivers and estuaries. In freshwater ecosystems, the processes of stratification and **mixing** result from the cumulative effects of heat exchange and energy intake; absorption of solar radiation with depth (which depends on the optical conditions of the surface water), the direction and strength of the wind, the kinetic energy of water intakes, and the direction and force of water outlets. Mixing and **vertical stratification** are dynamic processes. Morphometric characteristics are important in vertical and horizontal mixing: volume, maximum and average depth, and location (latitude, longitude and altitude). The basic mechanisms of generating and dissipating turbulent kinetic energy are the same in lakes and oceans. Differences are caused by density (due to salinity of sea water and effects of the Earth's rotation on oceans or large lakes).

Wind exerts an action of **turbulence stress** on the water surface. As a consequence, the following phenomena occur:

▶ Surface currents;
▶ Accumulation of water on the surface, in the direction of wind, and an oscillation in the stratified interface;
  Turbulence in the surface layers, which can increase during breaking waves.

Figure 4.10 presents a photograph of the phenomenon of turbulence on a small scale (cm) along with a drawing illustrating the physical process, which has frequent chemical and biological consequences.

The amplitude and vertical dimension of these events depend on wind speed; location in relation to the major axis of the lake, reservoir, or river (fetch); and the site's topography. Waves are **periodic** and rhythmic **oscillations** of masses of water, with intense vertical movement, and currents are unidirectional and often non-periodic flows of masses of water. Part of the kinetic energy of wind produces waves on the surface, which dissipate and lose energy; part of the energy is transferred to the currents. In addition, wind can induce internal waves in the thermocline and **hypolimnion**.

The effect of the turbulent stress of wind, Tw, which occurs on the water surface, is usually represented as:

$$Tw = Cd \cdot pa \cdot V^2$$

where:
V = velocity of wind measured at a certain height, usually 10 m;
pa = air density;
Cd = coefficient of wind stress, which depends on the status of the surface (smooth or turbulent) and quantity of waves on the surface in height, shape and velocity.

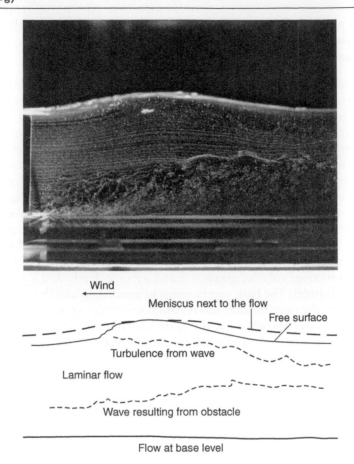

Figure 4.10   Small-scale waves. The depth of unperturbed water is 15 cm.
Source: Banner and Phillips (1974).

Wind-produced kinetic energy, therefore, generates currents, waves, turbulence and transient situations that promote mixing and dissipation.

4.4 TYPES OF FLOW

In laminar flow, the movement of water is regular and horizontal, without effective mixing of the various layers. Mixing that occurs is at the molecular level. Laminar flow, not surprisingly, is relatively rare in natural conditions.

Turbulent flow is irregular, with **water velocity** varying vertically and horizontally, characterized by erratic movements. Apparently chaotic, it can be calculated statistically. Its properties can be analyzed by determining the average values of its properties, such as fluctuations in velocity.

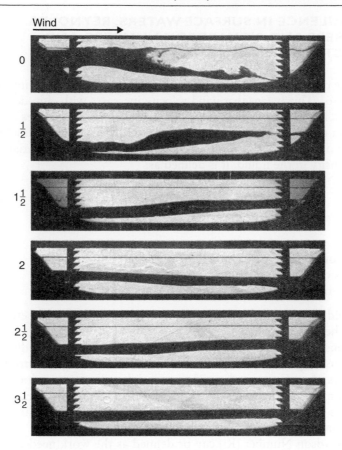

Wind

0

$\frac{1}{2}$

$1\frac{1}{2}$

2

$2\frac{1}{2}$

$3\frac{1}{2}$

Figure 4.11   Model lake with internal waves after wind stress. The initial condition, under severe stress from wind, is presented in the first illustration. The following illustrations, with no wind, show successive stages in the oscillations after the wind stops.
Source: Adapted from Mortimer (1951).

A **transition zone** exists between turbulent flow and laminar flow that essentially depends on water temperature. Above the sediment in a river or lake, there is a very fine layer of water with a rapid flow, called an interface layer.

In rivers, this layer is turbulent and may extend throughout the entire water column, with gradients of velocity from the surface to the bottom. The type of bottom and its irregularities can alter this interface layer. The sediment/water interaction, which is of fundamental importance in chemical and biological processes, depends greatly on the characteristics of this layer. The processes of erosion, deposition and re-suspension of sediment depend on variations in the state of turbulence and viscosity and irregularities on the bottom.

Figure 4.11 shows the process of generating waves and currents, as well as the production of turbulence in the water created by wind action.

4.5 TURBULENCE IN SURFACE WATERS, REYNOLDS NUMBERS AND RICHARDSON NUMBERS, AND EFFECTS OF DENSITY AND STRATIFICATION

In general, the profile of the velocity of wind over the water is logarithmic (Phillips, 1966 in Harris, 1986). This means that wind velocity increases with altitude so that a graph correlating the logarithm of wind velocity (U) with height (z) produces a straight line.

The effect of wind on the surface T_o is approximately proportional to U_2.

The **Reynolds Number** (Re) is the relationship between the **inertial forces** and the **forces of viscosity** in the liquid. It is defined as:

$$Re = \frac{U \cdot d}{V}$$

where:
U = **velocity of current**
d = thickness of layer
V = viscosity of water

A Reynolds Number below 500 indicates a laminar flow, and between 500–2000 (for water), the flow is turbulent.

When **density gradients** in a mass of water occur as a result of different temperatures and/or accumulation of suspended material in several layers, turbulence between adjacent layers of water is very low. The vertical distribution of velocity and the transition from laminar to turbulent flow are also affected and become reduced. Because of these layers of different density, there is resistance to vertical mixing indicated by the **Richardson Number**, which describes the stability of flow.

The Richardson Number (Ri) can be defined as the work needed to produce turbulence in layers of water with different densities.

$$Ri = \frac{g}{f} \cdot \frac{df}{dz} \frac{(du)^{2-}}{(dz)}$$

where:
f = density of the liquid
u = average velocity of water
g = acceleration of gravity
z = depth

When the liquid is homogeneous, $Ri = 0$; if $Ri > 0$, the stratification is considered stable; if $Ri < 0$, the stratification is unstable; if $Ri < 0.25$, the stratification is destroyed and turbulence increases, and when $Ri > 0.25$, the layers of water will present a flow without turbulence, remaining superimposed.

The density-currents formed in lakes or reservoirs are important for the process of the flow of water, over, under or through a mass of water. These currents are ecologically important because of the isolation of masses of water in lakes and reservoirs, the effect on horizontal and vertical circulation, and the **transport of nutrients**, suspended matter and organisms.

As was seen, the two important parameters in the mechanism of turbulence and circulation of the mass of water are wind and heat balance, which together produce a series of vertical gradients and a spectrum of varying degrees of stability and turbulence. One of the important measures normally used to indicate the frequency of oscillation, without the component of kinetic energy induced by wind, is that given by the Brunt-Väisälä equation (Brunt-Väisälä Frequency):

$$N^2 = -\frac{g}{\rho} \cdot \frac{dp}{dz}$$

where:
g = gravity acceleration
ρ = density
z = depth

This frequency, measured on the basis of differences in density of the water, enables the determination of periodic oscillations without wind action. It is a useful parameter in the study of the spectra of the diversity of phytoplankton as a function of the vertical oscillation in the system.

4.5.1   Internal waves

When turbulence increases in the epilimnion, displacement also occurs in the **metalimnion** (i.e., region of the thermocline), forming an internal wave, or seiche, which is basically an oscillating motion in the **metalimnetic layer** produced in stratified lakes.

Small instabilities in the metalimnion, based on this epilimnetic turbulence, can also be formed and then immediately collapse (Kelvin-Helmholtz instability).

4.5.2   The Coriolis effect and Langmuir circulation

The Earth's rotational effects produce intense horizontal movements in masses of water, causing the appearance of currents and circular movements that are common in oceans and large masses of inland water. These effects are significant in lakes or reservoirs with a width of approximately >5 km (Mortimer, 1974, in Harris, 1986). The Coriolis effect can be calculated:

$$f = \phi\Omega\sin\phi$$

where:
Ω – angular velocity of Earth's rotation
ϕ – latitude

The Coriolis effect acts in a counterclockwise direction in the Northern Hemisphere (to the right) and in a clockwise direction in the Southern Hemisphere (to the left). These rotational effects also affect internal waves, causing successive dislocations, such as occurs in the Great Lakes of North America. The Coriolis effect, the effect of winds, the functions of force and pressure act on the large masses of inland water, producing horizontal and vertical oscillations of great ecological importance.

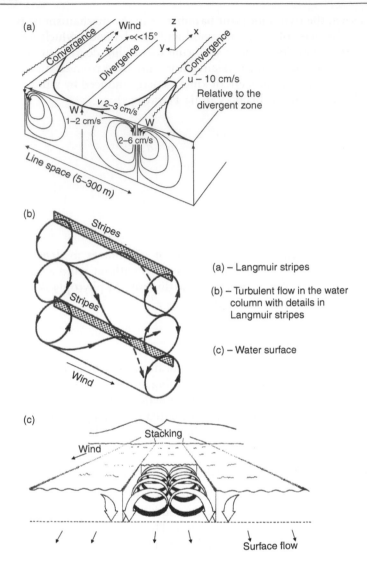

Figure 4.12   Wind action and the production of turbulent kinetic energy in water.
Source: Adapted from Reynolds (1984).

The **Langmuir rotation** or circulation (Langmuir, 1938) consists of the formation of convection currents by wind action over $3\,\text{m}\cdot\text{sec}^{-1}$ (approximately $11\,\text{km}\cdot\text{h}^{-1}$) and there is a direct correlation between the length of these Langmuir cells and their depth. Wind action produces a downward movement on masses of water that then form vertical cells that appear on the surface as grooves, which concentrate dissolved particles and substances in the water. For example, Horne and Goldman (1994) noted that the concentrations of *Daphnia* spp. and other cladocerans in these grooves have the functions of enabling escape from predators and providing food to the species, given the concentration of phytoplankton and particulate material in these grooves.

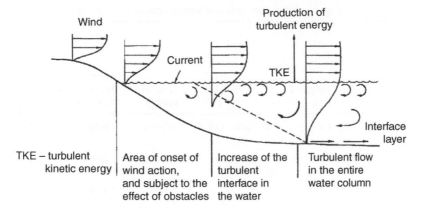

Figure 4.13   The effect of wind on spatial variation of turbulent kinetic energy (TKE) in a lake or reservoir.
Source: Bloss & Harleman (1979).

4.5.3   Ekman spirals

In 1905, V. W. Ekman presented a theory that wind, constantly blowing over an extremely large water surface of infinite depth and with uniform viscosity, shifts this surface at a 45-degree angle to the right of its direction in the Northern Hemisphere and to the left in the Southern Hemisphere. Water masses in lower layers are moved successively to the right (or left) until at a certain depth the direction of water displacement is opposite to that of the action of the wind on the surface, and also the velocity is greatly reduced. This deflection of currents near the surface is more common in deep large lakes and oceans. The **horizontal displacement** of a mass of water can also cause a circular movement with a stagnant center. Such a type of circulation, described by oceanographers, is called a gyre (Von Arx, 1962). It can be important in the distribution of planktonic organisms and pollutants.

Figure 4.12 and Figure 4.13 show the action of wind and its effects on the production of turbulent energy and turbulent flow in the water column, as well as the effects of turbulent kinetic energy on the water surface.

Table 4.5 and Figure 4.14 show the different types of circulation and movements in the water column, their scale and ecological effects.

4.6   THERMAL STRATIFICATION AND VERTICAL AND HORIZONTAL CIRCULATION IN INLAND AQUATIC ECOSYSTEMS

The principal heat-generating process is solar radiation reaching the water surface. The long-wavelength infrared radiation is absorbed in the first centimeters. The thermal heating establishes a less-dense layer of water with a higher temperature near the surface. Such thermal and density stratification is an important phenomenon in inland

Table 4.5 Time and space scales, velocities and kinetic energy and their ecological effects.

Processes and types of circulation	Temporal scale	Spatial scale H	V	Scale of velocity	Importance for the spectrum of kinetic energy	Ecological importance – effects on the phytoplankton and on the recycling of nutrients
Waves on the surface	I s	I to 10 m	I m	10 m·s⁻¹	Little	Little
Turbulent vertical mixing	s to min	I to 100 m		2 cm·s⁻¹	Little	Little
Oscillations of water masses in stratified columns	s to min	100 m	2 to 3 m	I to 30 cm·s⁻¹	Large	Large
Langmuir circulation (Langmuir spirals)	5 min	5 to 100 m	2 to 20 m	0 to 8 cm·s⁻¹	Moderate	Moderate
Effects of windshear on water column	hrs	100 m to I km	2 m	2 cm·s⁻¹	Large	Moderate
Internal waves (short)	2 to 10 min	100 m	2 to 10 m	2 cm·s⁻¹	Moderate	Moderate
Internal waves (long)	I day	10 km	2 to 20 m	50 cm·s⁻¹	Large	Large
Free circulation in stratified systems	I min-hr	I cm to I m	I to 10 cm	I cm·s⁻¹	Large vertical movement	Moderate
Lateral convection due to cooling and heating	hrs	I to 5 m	2 to 5 cm	2 cm·s⁻¹	Moderate horizontal and vertical movement	Moderate
Turbidity currents	min-hrs	I to 10 m	I to 5 m	I cm·s⁻¹	Moderate horizontal and vertical movement	Moderate
Circulation in the hypolimnion	long periods	>I km	2 m	0.5 cm·s⁻¹	Small	Moderate
Annual cycles of stratification and destratification	weeks and months	10 m to km	m to km	0.1 cm·s⁻¹	Small	Long-term recycling effects
Horizontal gradients from the water basin	min to hrs	m to kms	Cm to m	1–10 m·s⁻¹	Large	Large direct and indirect effects
Meromixis	years	km	km	<I cm·s⁻¹	Small	Moderate or highly relevant

Source: Adapted from Mortimer (1951), Thorpe (1977), Dillon (1982), Mortimer (1974), Spiegel and Imberger (1980), Tundisi *et al.* (1977), Barbosa and Tundisi (1980), Lombardi and Gregg (1989), Horne and Goldman (1994), Imberger (1994), Tundisi (1997), Tundisi and Saijo (1997), Tundisi (1999), Romero and Imberger (1999), Kennedy (1999), Tundisi and Straškraba (1999).

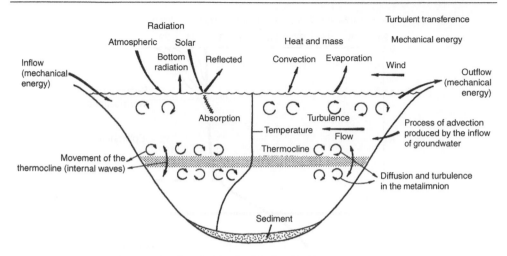

Figure 4.14   The different mechanisms of inflow and outflow of mechanical energy, flows, movement of water masses and absorption of solar radiation in lakes.

aquatic systems, and a large part of the operating processes and mechanisms result from the vertical gradient thus formed.

The upper layer of warmer, less-dense water – the epilimnion – is also quite homogeneous due to wind action and diurnal thermal heating and nocturnal **thermal cooling**, forming a **temporary thermocline** (during the diurnal period).

The lower layer of water, denser and cooler, is called the hypolimnion. The depths of the hypolimnion and epilimnion depend on the geography of the lake, its average and maximum depth, regional characteristics in terms of wind (position of the aquatic system, direction and strength) and the lake's position in the water basin.

The *metalimnion* (Wesenberg-Lund, 1910) or *mesolimnion* is an intermediate layer between the hypolimnion and epilimnion, which shows a gradual decline in temperature in relation to the epilimnion, and it is difficult to identify its limits. In the metalimnion, the region in which the temperature drops at least 1°C per meter was called the *thermocline* by Birge (1897). Hutchinson (1957), modifying this idea, defined it as **planar thermocline**, i.e., the region that includes an imaginary plane in the lake at an intermediate level between the two depths where the thermal difference is greater (see Figure 4.15). The concept of an imaginary plane in the lake, dividing two layers – one penetrated by light and with total productive circulation, the other dark with reduced circulation and where the predominant process is **decomposition** – is very useful for understanding the operating processes of lakes and reservoirs.

More important than the concept of **absolute temperature**, it is essential to understand that the thermal differences in the vertical gradient and the respective densities are the principal differentials that determine the basic characteristics of the functioning of lakes. The interrelationships between water temperature and density were already discussed in Chapter 2. At normal pressure, water reaches its maximum density at 4°C when 1 cm³ of water has the mass of 1 gram. Water becomes less dense below 4°C, so ice floats to the surface, covering the lakes permanently or for certain periods. Differences in density tend to increase with higher temperatures, so then the process

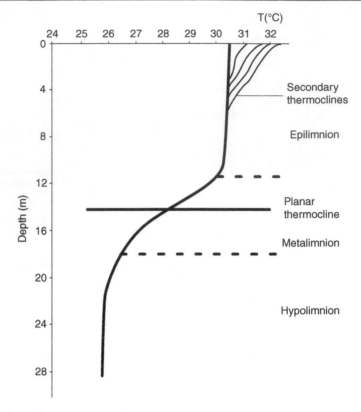

Figure 4.15   Thermal stratification in lakes.

of stratification and stabilization of the lake is greater, even when the thermal differences between the surface temperature and depth temperature are relatively small. Table 4.6 shows this process.

Between 4–5°C, the change in density is $81 \cdot 10^{-7}$. Between 23–24°C, the change in density is $2.418 \cdot 10^{-7}$; the ratio is $2.418/81 = 29.08$. This means that the change in density from 23–24°C is 29 times greater than the change in density between 4–5°C. This is example illustrates the importance of thermal differences and stability in tropical lakes, even when these differences are relatively small.

Factors that can alter the density of water and vertical gradients include pressure (which lowers the temperature of maximum density) and presence of dissolved substances (dissolved salts increase density). Stratification in saline lakes, in this case, can present anomalies such as high temperatures in the deepest layer of the lake (Cole, 1983).

Suspended materials that increase the density of water can produce well-defined density-currents in lakes and reservoirs. Surface-bound currents produced by rivers with higher density are more common in lakes and reservoirs. In regions where there is dense vegetative cover in rivers because of **riparian vegetation**, water temperature can decrease by 2–3°C in comparison to the temperature of the lake or reservoir. **Advection** (externally introduced) currents with denser water were recorded at the Lobo reservoir (State of São Paulo), for example, in several of its tributaries. Tundisi

Table 4.6   Relationship between water and density.

Water temperature (°C)	Density $10^{-7}g/cm^3$	Change in density $10^{-7}g/cm^3$
4–5	9999919 (at 5°C)	81
23–24	9973256 (at 24°C)	2,418

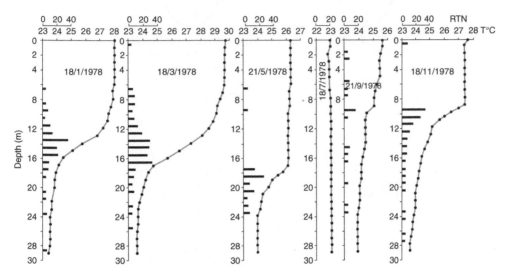

Figure 4.16   Seasonal pattern of vertical stratification and circulation in Lake Dom Helvécio, eastern Brazil.
Source: Tundisi and Saijo (1997).

(unpublished results) recorded advection currents in **Lake Cristalino** (**Negro River**), in tributaries from cooler forest streams. Lakes and reservoirs can be grouped into categories related to their vertical **thermal pattern** and their evolution during the seasonal cycle of stratification and circulation.

Thus, the systems can be classified as:

▶ Monomictic Lakes – have a regular period of total circulation occuring at some time during the year. There are two basic types of **monomictic lakes**: warm monomictic lakes, which circulate during winter and are not covered with ice; and cold monomictic lakes, which present a reverse stratification in winter with temperatures of 0°C and ice cover on the surface, and 4°C below the ice. Circulation occurs during spring and summer.

The thermal profiles shown in Figure 4.16 show the evolution of the process of stratification and de-stratification in a warm monomictic lake.

The stratification process in warm monomictic lakes occurs because of thermal heating on the surface. Also, in certain lakes during the summer, there is a process of shifts in density in the lower layers, due to the contribution of denser cooler water from precipitation. The effect of rainwater helps establish the thermocline and metalimnion,

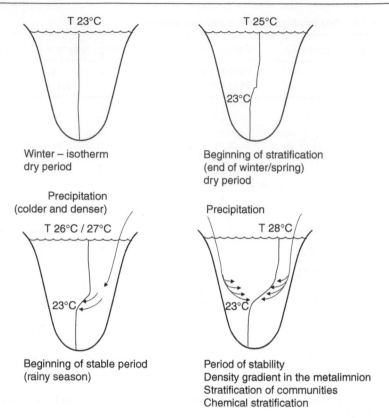

Winter – isotherm
dry period

Beginning of stratification
(end of winter/spring)
dry period

Beginning of stable period
(rainy season)

Period of stability
Density gradient in the metalimnion
Stratification of communities
Chemical stratification

Figure 4.17   Effect of precipitation on density stratification in the deepest lakes of the Parque Florestal
do Rio Doce.
Source: Tundisi (1997) in Tundisi and Saijo (1997).

quickly stabilizing the lake. Such a phenomenon was described for the lakes of the
Rio Doce Valley, in Minas Gerais (see Figure 4.17) (Tundisi, 1997), especially for
Lake Dom Helvécio.

There is also a difference in stratification between the deepest parts of the lake
and the shallower regions that can present complete **isothermy** during certain periods,
and then return to sharp stratification again. **Pulses of circulation** and diverse stratifications can occur in these shallower regions.

▶ **Dimictic Lakes** – present two periods of circulation annually. Stratification occurs
   in summer and remains until autumn, when a complete circulation occurs. A
   period of thermal cooling follows along with the presence of an inverse thermocline with ice coverage on the surface. In spring, thermal heating produces new
   circulation as the ice sheet disappears. Dimictic lakes tend also to occur in the
   subtropics at high altitudes.

▶ **Polymictic lakes** – present many periods of circulation annually. Diurnal temperature variations and formation of a thermocline occur during the diurnal period,
   which can be more extreme in temperature differences than **seasonal variations**. In

general, shallow lakes that experience year-round wind action present this type of circulation. Stratification can occur for a few hours or even days, but then quickly disappears. Examples include Lake George (Uganda) and **Clear Lake** (California). The reservoir behind the Lobo dam (Broa), in the State of São Paulo, is an example of a classic polymictic reservoir.

▶ **Meromictic Lakes** – These lakes never have complete circulation and present a lower layer called the **monimolimnion** that never has circulation. The term *meromictic* was introduced by Findenegg (1935). These lakes have high concentrations of dissolved substances in the lower layer. In such lakes, added density is the principal stratifying factor and not temperature. The **meromixis** can be biogenic when biological material accumulates in the deepest part of the lake, with dissolved organic substances from bacterial decomposition and salts, for example. **Anoxia at the bottom** occurs in these lakes and establishes a chemical concentration gradient or **chemocline** between the top layer (named **mixolimnion** by Hutchinson, 1937) and the hypolimnion. Meromictic lakes with **biogenic meromixis** and substantial contributions of biological material from internal or external (autochthonous or allochthonous) sources – such as **litter** from forests surrounding the lake – are found in many regions, and can sometimes present complete circulation during a certain period. An example occurs in some lakes of the Médio Rio Doce, which may remain stratified for long periods beyond the annual cycle. This layering is mainly due to differences in density caused by the accumulation of biological matter.

Another type of meromixis can be of external origin or ectogenic, which occurs in water with higher salinity in deeper layers. This accumulation of water is the result of contributions from outside the lake. A classic example was reported by Matsuyama (1978), who studied the **Mikata** group of lakes, located near the seacoast of Japan, which receive water from the sea in their deepest layer, causing a sharp **saline stratification** and also a characteristic chemocline. Lakes of this type can also be found in some coastal areas in fjords in Norway.

Crenogenic meromixis results from the **intrusion** of saltier water from subsurface sources, establishing steep vertical gradients of salinity. A classic example is Lake Kivu in Africa. In contrast, Lake Malawi and Lake Tanganyika show a possibly biogenic meromixis.

Most existing lakes can be classified in the four categories mentioned. Another less-common category, called amictic, has permanent ice on the surface and can occur at elevated altitudes in low latitudes, as demonstrated by Löffler (1964) in Andean lakes in Peru. Hutchinson and Löffler (1956) proposed a minimum of 6,000 meters altitude for **amictic lakes** in equatorial regions.

Holomictic lakes present complete periodic circulation and do not have occasional stratification (unlike polymictic lakes, which can experience occasional stratification processes).

Hutchinson and Löffler (1956) used the term **oligomictic** to describe shallow lakes that circulate at irregular intervals and are rapidly stratified, with reduced short-term stability.

Later, the term **atelomixis** was used by Lewis (1974) to define prevalence of incomplete vertical mixing.

Imberger (1994) studied the dynamics of the thermal structure in greater detail. He compared stratification patterns in Southern Hemisphere lakes with lakes in the Northern Hemisphere, and concluded that the year-round occurrence of temperature gradients is predominant in the Southern Hemisphere. Due to thermal heating in summer and the inflow of colder water in winter, the temperature gradients can reach 10–15°C during periods of intense stratification. Thus, the stability of stratification can only be altered during periods of thermal cooling on the surface and with wind action combined with the formation of internal waves, which enables deepening of the diurnal mixed layer.

The average temperature of the water column and differences in density are therefore crucial in Southern Hemisphere lakes. The vertical and horizontal temperature **distribution patterns**, associated with the effects of wind on the system and the distribution of phytoplankton, gases and other dissolved substances, can be altered by the presence of vegetation, or in higher-altitude lakes. The process of stratification on a small scale, studied by Imberger (1985), shows great complexity in the vertical gradient, and the differences that occur in the **horizontality of the isotherms** (lines of the same temperature) are caused by **advection processes** in the transport of laminar flows on a small scale, several centimeters deep.

The 'buoyancy frequency' can be calculated by the formula (Imberger, 1994):

$$N^2 = -\frac{g}{\rho_o} \cdot \frac{d\rho_e}{dz}$$

where:
g = acceleration due to gravity
ρ_o = average density
ρ_e = average density of the layer introduced by stabilization
z = vertical coordinate (depth)

When this formula is applied to a vertical stratification profile, zones of greater stability in a seasonal thermocline can be identified, given by a peak of N^2, and a region of descending variation of N^2, which is the metalimnion, composed of relatively small vertical gradients. Below the metalimnion is a constant region, with stable N^2, which is the hypolimnion. Between these two regions is a poorly defined transitional layer important both ecologically and biologically.

Analysis of this **microstructure** shows variable effects of the wind and the thickness of the water/air contact layer as a significant factor in the processes of shifts in the vertical thermal structure at relatively shallow depths (1–2 meters). Phytoplankton is quite sensitive to these rapid modifications in the density and turbulence of the superficial layers. Due to rapid changes, precise measurements of the weather conditions on the lake's surface and near the **air-water interface** need to be taken frequently.

The processes that influence the **dynamics of thermal heating** and cooling of lakes and reservoirs, during short periods or in **seasonal cycles**, can be summarized as follows: differential heating of the lake's surface, which causes horizontal gradients and advection; differential absorption in the water column, due to particulate material and colored substances; diurnal emergence of cooler waters; dynamics of inflows, based on the variability of **horizontal transport** in rivers (Fisher and Smith, 1983) and mixing mechanisms in the water column resulting in complex vertical and horizontal structures (Imberger and Hamblin, 1982) (see Figure 4.18).

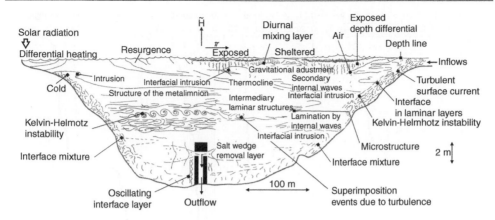

Figure 4.18   Scales of mobility in a stratified lake.
Source: Imberger and Hamblin (1982).

4.7   THERMAL STRATIFICATION AND DE-STRATIFICATION IN RESERVOIRS

Thermally, the physical processes occurring in reservoirs are the same as in lakes. However, in reservoirs, which are almost always subject to unidirectional flow and variations in this flow, additional processes can occur.

One such process is **hydraulic stratification**, caused by the height of the water outlet at different depths. This produces extremely sharp thermal stratification and density stratification, similar to the natural process. In this case, **accumulation of reductive substances** and deoxygenation of the hypolimnion also occur. This hydraulic stratification (Tundisi, 1984) is characteristic of deep reservoirs, where it is necessary to create a steep artificial vertical gradient (to generate power), which may be partial and occur only in the portion of the reservoir subjected to flow.

Other stratification processes that can occur in reservoirs are steep vertical gradients resulting from advection produced by the entry of rivers or cooler sources (which is relatively common), and stratification caused by the height of the outflow of water from the spillway, which can also lead to a horizontal gradient over a certain extension. Reservoirs with a dendritic pattern may have more pronounced stratification in the compartments than in the main channel. In reservoirs with excessive macrophytes or uncut vegetation, the effect of the wind is greatly reduced and, thus, the processes of small-scale turbulence, which can reduce stratification, are curtailed.

There are many differences in the thermal behavior and vertical stratification in large and small reservoirs (between volumes of 10,000 km^3 and 5 km^3) and reservoirs with different **morphometric patterns**.

Tundisi (1984) classified the reservoirs in the State of São Paulo in terms of stratification and de-stratification as follows:

▶   Reservoirs with prolonged stratification periods: 8–10 months, and de-stratification in winter (monomictic heating);

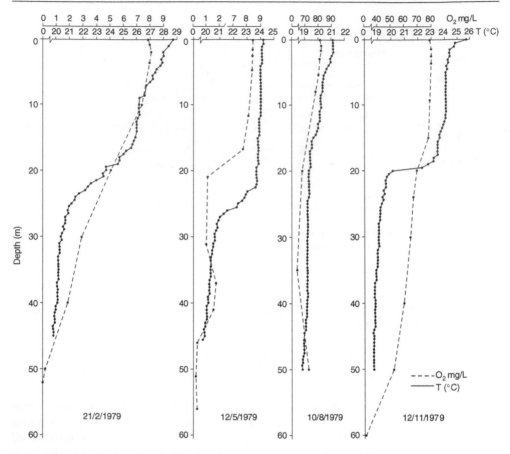

Figure 4.19   Thermal profile and profile of dissolved oxygen in the Promissão reservoir (SP).
Source: Reservoir typology project in the state of Sao Paulo – Fapesp.

▶   **Polymictic reservoirs** with occasional stratification periods;
Reservoirs with "hydraulic stratification" resulting from the dam's operating characteristics (water outlet).

Figure 4.19 shows thermal profiles and dissolved oxygen in the **Promissão reservoir** (São Paulo).

Figure 4.20 shows several thermal profiles in the Barra Bonita reservoir (São Paulo) from various sampling sites, showing the reservoir's degree of vertical and horizontal heterogeneity.

4.8   DIEL VARIATIONS OF TEMPERATURE

Studies on **diel** (day-night) thermal variations in temperature and de-stratification during the nocturnal period have been conducted on many lakes, especially in tropical regions where annual temperature cycles may be less extreme than diurnal cycles

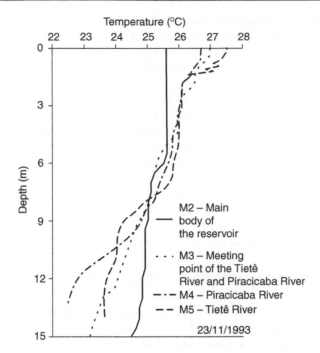

Figure 4.20 Thermal profiles in the Barra Bonita reservoir (SP) at various collection points on the same day and changes in the mixing layer (0 to 2 meters) during the day.

(Hare and Carter, 1984). These diurnal variations in thermal structure are accompanied by shifts in the concentration of dissolved gases (O_2 and CO_2) at various depths, and also in the vertical distribution of planktonic organisms.

The diurnal thermal stratification process and the formation of secondary thermoclines have many other physical, chemical and biological consequences. Tundisi and Barbosa (1980) and Barbosa (1981) observed this process in **Lake Carioca**, a small lake in the Medio Rio Doce, and showed that the degree of variation in the thermocline in the epilimnion, during the day, depended on the time of year. Even in periods of total isothermy, stratification occurs with the formation of secondary thermoclines measuring up to 2 m depth (for a lake with a maximum depth of 12 m). The thermal **micro-stratification** process on the surface is important due to the formation of gradients with different densities, with the compartmentalisation of masses of surface water into microstructures.

Haze and Carter (1984) described a diurnal stratification and nocturnal cooling cycle in **Opi Lake** in Nigeria (Africa). During the dry season, with stronger winds and shallower lake, nocturnal de-stratification resulted in complete vertical circulation. During the rainy season, with greater lake depth, deep vertical circulation did not occur, despite partial de-stratification of the lake. These results also show the effects of lateral circulation, due to the entry of rainwater into the lake and the stratification that it causes.

Diurnal chemical variations associated with the photosynthetic process are known since the work of Morren and Morren (1841).

In studies conducted in three different lake systems (a small lake near swamps and papyri, a reservoir, and a bay in Lake Victoria, in Africa), Talling (1957) described diurnal variations, the process of nocturnal isothermy and thermal stratification during the day. The periods of isothermy coincided with the thermal nocturnal cooling.

Diurnal and nocturnal variations occurring at the same time as thermal variations in **concentrations of dissolved oxygen, photosynthetic activity,** and the vertical distribution of *Anabaena flos-aquae* var. *intermedia* forma *spiroides* were shown by Talling (1957).

The work of Ganf and Horne (1975) on **Lake George,** Uganda (Africa), identified three distinct phases in the thermal cycle: isothermy, intense stratification, and complete total mixing. During the stratification period, the surface temperature could reach 36°C, while the bottom remained at 25°C. In a rather shallow lake (2.5 m) with a surface area of 250 km², intense maxima of dissolved oxygen, pH and chlorophyll caused marked diurnal oscillations.

In Lake **Jacaretinga** in the Amazons, Tundisi *et al.* (1984) described pulses of stratification and de-stratification, mainly during periods of low lake depth (2.5 m). These pulses are important for the process of liberating nutrients, resulting from **anoxia,** which generally accompanies diurnal thermal stratification.

4.9 STABILITY OF LAKES AND RESERVOIRS

The term and concept of stability (S), introduced by Schmidt (1915, 1928, in Cole, 1983), represents the **quantity of work** per unit area necessary for the lake or reservoir to present a uniform density without adding or subtracting heat. The value of S is equal to zero in the case where the lake's density is uniform, from surface to bottom. The heating of the surface layer and subsequent decrease in density establish a vertical gradient and the calculation of stability, the units of which can be given in g cm/cm² (representing the work per unit of lake area), which is necessary to reduce the stability to zero.

The most well-known and utilized formula is that of Idso (1973):

$$S = \frac{1}{A_0} \frac{Z_m}{Z_0} (\rho z - \overline{\rho})(A_z)(z - z\overline{\rho})dz$$

where:
A_0 = surface area (in centimeters)
A_z = area at the depth of any z (in cm²)
ρ = average density of the water column that would result from the mixing process
ρ_z = density at depth z
$z\overline{\rho}$ = depth (in cm) where the final or average ($\overline{\rho}$) density occurs before the mixing
Z_m = maximum depth (in cm)
Z_0 = zero depth at the surface
z = depth (in cm)

4.10 IMPORTANCE OF THE PROCESS OF THERMAL STRATIFICATION AND DE-STRATIFICATION IN THE DIURNAL AND NOCTURAL TEMPERATURE CYCLES OF WATER

Thermal stratification and de-stratification are accompanied by a series of other physical and chemical alterations in the water. The most significant include the vertical

Atelomixis

The term atelomixis was proposed and defined by Lewis (1973) in his study on Lake Lanao in the Philippines.

Atelomixis – vertically incomplete mixing – is related to the phenomenon of diurnal thermal heating and nocturnal thermal cooling, producing diel temperature variations that can exceed seasonal variations (annual). This process was also reported and described by Barbosa and Tundisi (1980) for Lake Carioca, a small lake in Parque Florestal do Rio Doce, in the lacustrine system of the Medio Rio Doce in southeastern Brazil.

The diel phenomenon is not exclusive to tropical lakes. It is caused by the effect of thermal heating and heat loss with increasing density of water in the nocturnal periods (Barbosa and Padisák, 2002).

Such thermal variations of temperature and density occur throughout the water column, as described by Lewis (1973) or only in the epilimnion, when stable stratification occurs with well-established **primary thermoclines**. In this case, the atelomixis can be called partial atelomixis and considered a common phenomenon in tropical lakes (Lemoalle and Talling, 1998; Barbosa and Padisak, 2002).

The phenomenon is critically important in the vertical reorganization of the thermal structure and density of lakes and their interactions with the vertical distribution of phytoplankton and nutrients, as can be seen in Chapter 7.

distribution of dissolved gases, the vertical distribution of nutrients (with accumulation of substances and **chemical elements** in the hypolimnion during stratification), and more homogeneous vertical concentrations in the water column, or precipitation and total re-circulation. Changes also occur in the distribution of phytoplanktonic and zooplanktonic organisms and accumulation of certain components of the community, such as bacteria in the temperature and density gradients. The presence of gases at higher concentrations in the hypolimnion is another important consequence of stratification, with profound ecological effects in the case of the interruption of this stratification by strong winds or other occasional climatic phenomena (extremely rapid thermal cooling) (see Chapter 7).

The periodic thermal stratification and de-stratification process in monomictic lakes, and the permanent stratification process in meromictic lakes, are important ecological factors in vertical organization and seasonal re-organization of biological communities. Reynolds *et al.* (1983) demonstrated the **stratification of populations** of limnetic *Lyngbya* in Carioca pond (Medio Rio Doce) when thermal stratification and density stratification occur. Such stratification is periodic and is repeated during the phase of stability in the lake. Hino *et al.* (1986) reported a clear example of marked biological stratification of the phytoplankton community.

Thermal variations presented by inland aquatic ecosystems, whether during the seasonal or day-night (diel) cycles, and vertical and horizontal gradients are also physiologically important for aquatic organisms. Upper and lower limits of **tolerance** to temperature for adults and larvae can result in specific distributions in the vertical profile or horizontal gradient.

Water temperature controls patterns of physiological behaviour (**respiration**, for example). It can limit or accelerate the growth of organisms, and is a factor in **reproductive** processes. On the other hand, rapid variations in temperature in a 24-hour cycle or less, with large thermal differences, imply mechanisms of tolerance and adaptation to these differences. For example, Tundisi (1984) observed variations of 11°C over a 24-hour period on the surface of a small lake in the Mato Grosso Swamp. These variations are accompanied by variations in the density of water, **solubility of gases** (see Chapter 5) and result in the formation of thermal microstructures and extremely dynamic density. Measuring these variations and determining their ecological and physiological significance are still important tasks to be done in many shallow tropical lakes and small reservoirs.

In inland saline lakes, diel temperature variation is also important in terms of the organisms found in these lakes and their tolerance to salinity. In general, polymixis in saline lakes is common, since the lakes are shallow and subject to constant wind action.

Meromixis in saline lakes has been described by many authors. Hutchinson (1937b) described of meromixis process in Big Soda Lake, Nevada, USA (**ectogenic** meromixis). Saline lakes have greater gradients of thermal variation than do freshwater lakes.

The classification of lakes by thermal patterns includes some unusual types, studied by Yoshimura (1938). Included among these types are **dichothermal lakes**, where an increase in temperature occurs in the monimolimnion of meromixis lakes, in some cases due to biological activity (metabolism of bacteria) or geothermal sources. **Mesothermal lakes** have increased temperatures at intermediate depths, with a higher-temperature layer between two cooler layers.

Poikilothermic lakes are saline and present complex thermal structures due to the existence of several layers. The layers are the result of different densities with circulation limited to a very thin layer (from a few centimeters to a meter). In lakes with winter temperatures below 4°C, a 'thermal bar' forms in spring, because of the heating of the shallow waters and continual circulation of the deepest waters. The thermal barrier consists of a denser layer of water at 4°C that separates the more heated and superficial masses of water in the shallower regions. In this layer, blooms of phytoplankton can occur, caused by the accumulation of nutrients.

4.11 ECOLOGICAL SIGNIFICANCE OF THE METALIMNION AND IMPORTANCE OF THE MEROMIXIS

The thermal and chemical gradient established in the metalimnion of monomictic or meromictic lakes determines a series of important ecological conditions in the vertical gradient. These conditions result in part from a decrease in the **rate of sedimentation** of organic particles (due to increased viscosity), which is responsible for the following processes:

▶ Increased concentration of nutrients in the euphotic zone;
▶ Increased retention time of nutrients in the metalimnion and epilimnion;
  Increased concentration of phytoplankton in the epilimnion and metalimnion.

Often, in monomictic or meromictic lakes in which the metalimnion is still within the euphotic zone, populations of cyanobacteria and **photosynthetic bacteria** develop, at a level of about 1–2% of the intensity of light reaching the surface. King and Tyler (1983) described meromictic lakes in southwest Tasmania, noting the dense zone of sulphurous photosynthetic bacteria in the upper part of the monimolimnion.

A maximum of zooplankton also occurs in the metalimnion, usually associated with high concentrations of photosynthetic bacteria or cyanobacteria (Gliwicz, 1979). In Lake Dom Helvécio, Rio Doce, thermal stratification leads to a process of biological and chemical stratification (see Chapter 7).

The term meromixis applies not only to lakes with permanent incomplete circulation, but also to periods of intense stratification and isolation of water masses. Aberg and Rodhe (1942) used the expression **spring meromixis** to refer to the duration of the meromixis that can extend to from brief periods of marked stratification to thousands of years. When the process of meromixis refers to circulation, it should therefore consider incomplete vertical circulation.

The terms used to characterize meromictic lakes refer to circulation processes in the upper mixolimnion, in the deep monolimnion and the **chemocline** – a transitional zone between the two layers.

Meromictic lakes occur in many regions. Walter and Likens (1975) identified 117 meromictic lakes in North America, Africa, Europe and Asia.

4.12 PRINCIPAL INTERACTIONS OF THE PROCESSES OF CIRCULATION, DIFFUSION, CHEMICAL COMPOSITION OF WATER AND OF COMMUNITIES IN LAKES, RESERVOIRS, AND RIVERS

The perpetual mobility of water masses has considerable influence on chemical and biological processes. All vertical or horizontal transport is performed by these movements, which depend on kinetic energy from wind action or factors of diffusion caused by turbulence.

Scales of mobility vary. There are diel movements associated with climatic variations in short time periods, and movement with seasonal scale, related to periodic effects of wind and formation of a thermocline. Seasonal interactions between energy flows, thermal heating, and wind action produce seasonal patterns of circulation and horizontal and vertical dislocation of water masses. The prognosis of thermal structures for lakes and reservoirs can be made using data on solar radiation and wind-produced kinetic energy, and plays an important role in management of these inland systems.

Because of the close interrelationships of the atmosphere and lakes, it is of value to monitor the relationship between climatic factors (such as solar radiation, winds and precipitation) and events in the lake (thermal structure, vertical and horizontal circulation). The use of climate data and the study of climatic/hydrological interactions is essential for understanding many of the processes in lakes and reservoirs. Chart 4.1 defines a set of formulae and **dimensionless numbers** that enable the calculation of various processes in lakes and reservoirs.

4.13 CIRCULATION IN LAKES, RESERVOIRS AND RIVERS

There are several differences in the circulation in lakes, reservoirs and rivers. For example, the **selective removal** of water from different depths in the reservoir produces some specific circulation mechanisms, mainly advection currents.

Chart 4.1 Main formulae and definitions in the circulation and mixing processes in different lakes and reservoirs and the application of adimensional numbers.

Thermal resistance to circulation (TRC)	$RTC = \dfrac{dt_2 dt_1}{dH_2O(4)dH_2O(5)}$	**TRC – Thermal resistance to circulation** dt_1 – density of water and temperature t_1 dt_2 – density of water and temperature t_2 $dH_2O(4)$ – density of water and temperature of $4°C$ $dH_2O(5)$ – density of water and temperature of $5°C$
Work of the wind (B)	$B = \dfrac{1}{A_0} \, z_m \, f_{Z_0} Z(\phi_1 - \phi_2) A_Z d_2$	A_o – surface area of lake or reservoir A_z – area in cm² at depth Z Z – depth considered positive (in cm) Z_0 – Depth zero Z_m – Maximum depth ϕ_1 – initial density ϕ_2 – density observed at depth Z
Stability of the system (S)	$S = \dfrac{1}{A_0} \dfrac{Z_m}{Z_0}(\rho_z - \overline{\rho})(A_z)(z - z_{\overline{\rho}})dz$	A_0 – surface area (in cm) A_z – area at any depth z (in cm) ρ – average density of the column of water that would result from the process of mixture ρ_z – density at depth z $z\overline{\rho}$ – depth (in cm) where the final density of the average (ρ) occurs before the mixture Z_m – maximum depth (in cm) Z_0 – zero depth at the surface z – depth (in cm)
Wedderburn number (W)	$W = \dfrac{\Delta \rho g' h^2}{\rho \mu^2 L} \quad \mu^2 = v\dfrac{\rho_{ar} \cdot Cd}{\rho H_2O} \quad g' = \dfrac{g\Delta\rho}{\rho_0}$	$\Delta\rho$ – difference in density of base of surface layer (which varies) ρ_0 – density of hypolimnetic water
Lake Number (LN)	$LN = \dfrac{f[Z - L(H)A_z\phi_z d_z(1 - [H - L]/H)}{\phi_0 u^2 \cdot A_0^{3/2}(1 - [L - h]/H)}$ The formula in the sequence is characterized by the center of the water mass and can be written as follows: $L(H) = \dfrac{{}_0f^H ZA(Z)d_z}{{}_0f^H A(Z)d_z}$	A_z – surface of lake at depth z A_0 – (A/0) area of surface of lake H – total depth given for the center of the water mass L(H) – center of volume H – distance from the middle of the metalimnion to the surface of lake
Richardson Number (R)	$Ri = g\dfrac{\dfrac{d\rho/\rho}{d_z}}{\dfrac{\rho(du)^2}{d_z}}$ Low Ri values indicate turbulence; high values (Ri > 0.25) indicate stability.	g – gravitational acceleration d_ρ/d_z – vertical gradient of density d_u/d_z – vertical gradient of horizontal velocities
Froude Number Influence of intrusions and outflows of water in lakes and reservoirs	*Intrusions* (F_1): $Fi = \dfrac{Ue}{\sqrt{gi'H}}$	Ue – velocity of entrance of water Z (flow of entrance) gi' – modified acceleration of flux due to the differences of acceleration of gravity between the water of the lake and the water of intrusion H – Hydraulic depth of flow
Froude Number	(F_o): $Fo = \dfrac{Qs}{H_2\sqrt{g_0' H}}$	Qs – velocity of discharge of water (discharge flow) H – depth at point of discharge g'_0 – difference in acceleration of gravity between the surface of the lake or reservoir and the point of discharge of the water

The nature of turbulence in a lake or reservoir can be determined by the relationship of the Froude Number and the Reynolds Number. Active turbulence is found in the region of FR > 1 and Re > 1.

Chart 4.2 Comparative data on the hydrodynamics of rivers, lakes and reservoirs.

	Rivers	*Reservoirs*	*Natural lakes*
Fluctuations in level	Large Rapid Irregular	Large Irregular	Small and stable
Intrusions	Highly irregular surface and underground braking	Intrusion via tributaries Intrusion of water in various layers in superficial or deep flux	Intrusion via tributaries and strong diffuse Intrusions on surface or deep
Discharges	Irregular, depending on precipitation and surface drainage	Irregular, depending on uses of water; Discharge from surface or hypolimnion	Relatively stable, frequently on the surface
Flows	Rapid, unidirectional, horizontal	Variable depending on the use of the water. At various depths depending on construction and operation	Constant, little variability, at various depths

Source: Wetzel (1990b, 2001).

The very characteristic of the use of the dam with through-flow to produce electricity can cause short-term differences in horizontal and vertical circulation.

Pollutants can be distributed horizontally by advection or accumulate in the hypolimnion, in sediments and their **interstitial water**. Circulation (both horizontal and vertical) in lakes contributes greatly to the **dispersion** and concentration of **heavy metals** and toxic substances in the system's various spatial compartments (horizontal and vertical). In the case of stratified lakes, accumulation of pollutants or toxic substances can occur in the metalimnion along with the accumulation of suspended material. Horizontal movements or vertical instabilities of water masses in certain periods can increase the dispersion in this layer and in the deep part of the epilimnion.

Chart 4.2 presents comparative hydrodynamic characteristics among rivers, reservoirs and lakes. These differentiated hydrodynamics have repercussions on biogeochemical cycles, distribution of organisms, and aquatic biodiversity.

4.14 DIFFUSION

The processes of diffusion correspond to chaotic and random movements. These processes are related to concentration gradients between a particular substance and that concentration which already exists in the surrounding water. Diffusion is the in-liquid movement of an element or substance, down a gradient of its concentration.

Molecular diffusion of ionic solutions in porous media (sediments) refers to the diffusion, within a single phase of atomic constituents, i.e., atoms, ions or molecules. Molecular diffusion is an important process, for example, in the sediment/water interaction.

Vertical and horizontal turbulent diffusions occur on the surface and in the thermocline of lakes. In general, the horizontal turbulent diffusion on the surface

Table 4.7   Coefficients of vertical and horizontal turbulent molecular diffusion.

Molecular diffusion	Coefficent of diffusion $cm^2 \cdot s^{-1}$
Ionic solutions in porous medium	approx. $10^{-8} - 10^{-3}$
Turbulent vertical diffusion	approx. $10^{-2} - 10$
Turbulent horizontal diffusion	approx. $10^2 - 10^6$

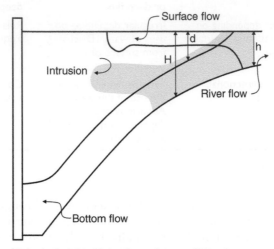

H, h, d – height of intruding columns of intrusion

Figure 4.21   Intrusion of waters on the surface or the bottom of lakes and rivers.
Source: Modified from Imberger and Patterson (1990).

accompanies the **advection process**, which involves distances greater than 1,000 meters. Small-scale diffusion or turbulence occurs at distances of less than 100 m. When the wind's effect diminishes on the surface, horizontal turbulent diffusion predominates. Coefficents of vertical and horizontal turbulent diffusion and their tanges of variation are presented in Table 4.7.

4.15   INTRUSION IN LAKES AND RESERVOIRS

Lakes and reservoirs receive their water supply from rivers. When the river meets the most static waters of the lake or reservoir, it generally encounters water masses of different temperature, salinity and turbidity. The intruding water can therefore be denser or less dense than the surface water of the lentic ecosystem. There are different points of intrusion – on the surface, below the surface, or on the bottom (see Figure 4.21). Such intrusion signifies transport of suspended material, loading of nutrients or organisms that are transported to the different depths. In estuaries, an intrusion generally occurs on the surface from less-dense tributary waters, producing a salinity gradient and promoting the differentiated distribution in space of organisms based on varying tolerance to salinity (Tundisi and Matsumura Tundisi, 1968).

The intrusion of water on the surface or bottom can lead to accumulation of suspended material or pollutants and contaminants from the water basin (see Figure 4.22).

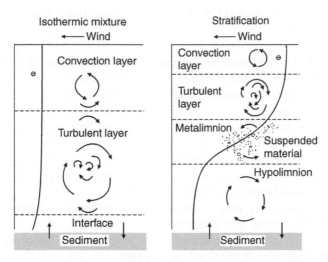

Figure 4.22   Principal systems of vertical mixing in periods of circulation and stratification.
Source: Modified from Reynolds (1984).

Definitions

Thermal resistance to circulation: expresses the resistance of the water (based on different densities) and circulation promoted by the wind.

Work of the wind: expresses the work needed by the wind to promote complete mixing in the water column.

Stability of the system: expresses the stability of stratification of the water column when the winds are not sufficiently strong to create homogeneity of density.

Wedderburn number: determines the response of surface layer to wind action. It is based on relationships between layers of water with different densities.

For $W \gg 1$, oscillations of surface isotherms due to wind are small and horizontal variations are negligible.

For $W \ll 1$, oscillations will be high and there is a general resurgence in the region of the lake or reservoir located in the path of the wind. For intermediate values of $W \sim 1$, upwelling and **horizontal mixing** are equally important.

Lake Number: expresses the response of the entire lake to the kinetic energy promoted by the wind.

   LN between 0–2 indicates weak stratification and strong effect of kinetic energy from the wind.

   The Lake Number allows a more comprehensive characterization of the water mass, taking into account the arbitrary vertical stratification Δ^9.

Transport mechanisms in lakes and reservoirs: Overview

Many processes related to vertical and horizontal movements of water masses have qualitative and quantitative effects with significance in bio-geochemical and biological phenomena.

According to Fisher *et al.* (1979), Ford and Johnson (1986), Thornton *et al.* (1990), and Reynolds (1997):

- Advection – forced transport by a system of currents produced by inflow of rivers, discharges downstream, and effects of wind on the surface.
- Convection – vertical transport induced instabilities in density, when, for example, there is surface cooling.
- Turbulence – described as a set of turbulence that have varying ranges down to molecular motion. Can be generated by winds, convection, inflow.
- Diffusion – the mechanism by which there is transfer of certain properties of fluids through a concentration gradient. In aquatic systems, it occurs at the molecular level and the level of turbulence (turbulent diffusion).
- Shearing – generated by wind at the air/water interface for advection currents on the bottom and internal density currents.
- Dispersion – the combined effect of shear and diffusion. Generally prevalent in regions with high velocity of the inflowing water.
- Intrusion – type of advective transport in which waters of different densities are added in layers with defined gradients, producing inflow with various effects on the transport of nutrients and organisms.
- Mixing – vertical or horizontal, is any of the processes that produce mixing of water masses, including diffusion, shear, dispersion and intrusion.
- Sedimentation – the settling of particles with higher density than the surrounding fluids and another important transport process in lakes, reservoirs and rivers in all inland systems.

These transport mechanisms occur in all aquatic systems, either simultaneously or at different periods in spatial and temporal scales depending on the morphometric characteristics of the system, its latitude, longitude and altitude. The magnitude of these variations and of these phenomena and their duration depend on the external forcing functions that occur in the various systems and on their equilibria with the internal forces.

A proposal by Lewis

Lewis (1983) proposed a revision of the original classification by Hutchinson-Löffler (1956). His proposal includes:

Amictic lakes – lakes that are permanently covered with ice.
Cold monomictic lakes – lakes with ice covering the surface most of the year, without ice during the summer, but with temperatures that never exceed 4°C.
Cold polymictic lakes – lakes covered with ice for part of the year, without ice during the summer, with temperatures above 4°C and stratified at least during the day.

Discontinuous cold polymictic lakes – lakes covered with ice for part of the year, free of ice above 4°C, stratified during the summer, but with interruption of stratification and total circulation at irregular intervals.

Dimictic lakes – lakes covered with ice for part of year, stratified during the summer, with circulation during periods of transition between these two states of vertical organization.

Warm monomictic lakes – no ice cover at any time of the year, stable stratification during part of the year and vertical mixing during one period of the year.

Discontinuous warm polymictic lakes – no ice coverage at any time of the year, stratified for days or weeks, but with vertical circulation several times a year.

Continuous warm polymictic lakes – Permanent circulation, with no ice cover throughout the year, and a few hours of stratification at certain periods.

The relationship between morphometry of lakes (in which maximum and average depths are critical), their geographic location (latitude, longitude, altitude), and the effects of climatic factors (such as solar radiation and wind action) are fundamental factors in the thermal behavior lakes and the processes of stratification and vertical mixing.

Lewis' classification categorizes lakes according to the following criteria: ice cover on the surface, vertical circulation, and stratification.

São Francisco River.
Photo: José G. Tundisi

5 | The chemical composition of water

SUMMARY

The chemical composition of natural inland waters is complex because of the large number of **dissolved ions** and organic substances resulting from the natural conditions of drainage basins and human activities. Another contributing source of substances and elements is the atmosphere. The chemistry of natural waters varies greatly due to the **geochemistry** of soil and rocks that constitute the substratum of water basins. Activities of organisms (excretion, respiration, **bioperturbation**) also play a role in the **balance of materials** in aquatic systems.

Dissolved ions and organic substances play diverse biological roles such as regulation of the **physiological processes** in organisms, including activities of membranes and activation of enzyme systems. Of the gases dissolved in water, oxygen and carbon dioxide are key because they are involved in the processes of production of organic matter by primary producers (photosynthesis) and respiration by all organisms. Diurnal variations in the levels of these gases are caused by changes in the processes of photosynthesis, respiration, and circulation of water masses. The vertical distribution of dissolved ions, organic substances, and gases depends on horizontal and vertical circulation processes, stratification mechanisms, and interactions of tributaries with aquatic ecosystems.

5.1 INTRODUCTION

Natural water is chemically complex because it contains a large number of dissolved substances. Dissolved chemicals and elements found in inland aquatic ecosystems originate from the geochemistry of the soil and rocks underlying water basins, through which water drains into rivers and lakes. Another source is the atmosphere, which varies considerably. In many industrial regions with high levels of atmospheric sulphur, rain can become acidic. Especially over deserts, rain water contains dust particles. Rain water varies in concentration from region to region, which affects the chemical composition of water draining through the soil. In coastal regions, the composition of rain water can also be affected by the presence of salt from the sea. Small air bubbles formed by wind action on the sea's surface can carry particles of water and be blown over the continents, thus contributing to the chemistry of inland waters.

Interactions between the main **ions in solution** also affect the chemical composition of inland waters. The **equilibrium theory** can be used to describe the chemistry of these waters, based on the distributions of equilibria between metallic ions and complex ions. Through application of this theory, a prediction can be made whether an ion will be present as a free ion or in complexes of various types.

Air pollution is another component that affects the chemical composition of rain water, contributing various different ions, including HSO_4^- and nitrogen oxides (general formula: NOx), such as from oil drilling and extraction. The pH of rainwater can drop to 2.1–2.8 (generally below 4.0), as occurs in some industrial regions in England, Scandinavia and the United States. Rainwater with acidic pH affects the chemical composition of the water flowing through drainage basins into rivers and lakes. As water drains through soils of different origins and chemical compositions based on local geology, complex chemical interactions take place that are specific to each water basin and, within each basin, in each sub-basin. Alterations made by human activities also lead to changes in the composition of natural waters. Removal of plant cover, different treatments of soil, and industrial and agricultural effluents are all factors. The chemical composition of natural waters that drain basins on all continents is the result of a complex set of chemical processes and interactions occurring between land and water, and between water and atmospheric systems.

Emissions of **ammonia**, nitrogen and sulphur chemically alter the atmosphere. Oxidation from these chemical substances affects the quality of rainwater and increases N levels in terrestrial and aquatic ecosystems. In northern Europe and northeastern United States, such effects have been recorded for more than 20 years. Recent evidence, however, shows that the same phenomenon is occurring in Southeast Asia (Lara *et al.*, 2001). Martinelli *et al.* (1999) estimated that the total amount of leaves burned by the sugarcane industry in the Piracicaba River drainage basin is 20 tons per hectare annually, which releases approximately 100,000 tons of organic matter and 50 tons of carbon into the atmosphere. This biomass also releases organic acids, **nitrate** and sulphate into the atmosphere, which can substantially alter the chemical composition of rain water and negatively influence surface waters and groundwater.

Lara *et al.* (2001) measured the chemical composition of precipitation in the Piracicaba River basin over one entire year. The study's results are presented in Table 5.1. According to the authors, the three factors determining the composition of

Table 5.1   Average concentrations of ions in rain water at four collection points in the state of São Paulo (values in $\mu eq \cdot L^{-1}$; DOC and DIC in $\mu mol \cdot L^{-1}$)

	Bragança			Campinas			Piracicaba			Santa Maria		
	Dry	Wet	Annual	Dry	Wet	Annual	Dry	Wet	Annual	Dry	Wet	Annual
pH	4.6	4.4	4.4	4.6	4.5	4.5	4.8	4.5	4.5	4.4	4.3	4.4
H^+	22.5	39.2	36.3	26.1	31.1	29.7	17.4	34.0	33.0	35.9	40.6	39.7
Na^+	2.2	2.3	2.3	3.1	2.7	2.7	4.2	2.1	2.7	5.6	4.2	4.5
NH_4^+	23.8	17.2	18.6	19.9	14.3	15.4	26.0	11.6	17.1	21.3	12.7	14.5
K^+	2.6	2.5	2.5	2.7	1.6	3.4	4.5	2.1	2.9	3.2	2.5	3.5
Mg^{2+}	1.3	1.2	1.2	1.6	1.2	1.3	3.1	1.9	2.3	2.7	2.3	2.3
Ca^{2+}	3.6	3.4	2.3	3.9	3.6	3.7	7.8	4.3	5.3	10.3	7.0	7.7
Cl^-	3.5	5.2	4.9	5.1	6.3	6.0	8.4	6.0	7.0	11.1	8.0	8.8
NO_3^-	17.0	14.5	15.0	18.6	17.9	18.0	20.6	13.8	16.6	18.3	12.3	13.5
SO_4^{2-}	15.1	17.3	17.0	19.6	19.9	19.7	27.4	14.8	18.7	15.3	11.5	12.3
HCO_3^-	0.4	0.3	0.3	0.3	0.3	0.3	0.9	0.3	0.4	0.9	0.4	1.2
DOC	84.9	57.1	58.8	80.4	47.1	50.8	134.5	78.8	94.4	100.5	43.9	7.66
DIC	55.7	48.4	50.0	23.7	33.5	30.7	34.2	43.5	43.9	81.8	62.4	67.8
Total cations	56.0	65.7	69.3	79.0	63.2	56.1	62.9	56.1	54.5	57.3	64.2	72.3
Total anions	36.6	38.0	37.7	45.3	34.0	45.1	44.7	36.4	60.2	46.6	47.5	38.0
TOTAL	92.5	103.7	101.9	102.6	102.0	102.7	123.1	92.5	107.7	24.1	109.2	113.5
DEF	19.4	27.7	26.6	12.1	7.0	9.5	2.7	19.7	18.3	33.9	35.4	34.3

DEF – Deficit of anions (DEF $-$ E cations $-$ E anions) in $\mu eq \cdot L^{-1}$
DOC – Dissolved organic carbon
DIC – Dissolved inorganic carbon
Source: Lara *et al.* (2001).

rain water in the Piracicaba River basin are ions coming from the dust of soils (Ca^{2+} and Mg^{2+}), the burning of sugarcane, and industrial emissions from the Campinas and Piracicaba regions. These activities caused significant levels of acid rain as well as high rates of nitrogen deposition. Changes in the atmosphere's chemistry and precipitation of chemicals from the rain produce changes in soil and in the chemical composition of waters in the region.

The 'material balance' of a lake, river or reservoir is also the result of activities of organisms that affect the chemical cycles and the chemical composition of the water. Figure 5.1 illustrates these processes (Schwoerbel, 1987).

5.2   DISSOLVED SUBSTANCES IN WATER

As discussed in Chapter 2, water is the universal solvent. Chart 5.1 presents a list of substances commonly dissolved in natural waters.

The most common ions are called 'conservative ions' because their concentrations vary little as a function of activities of organisms. The main nutrient ions are not

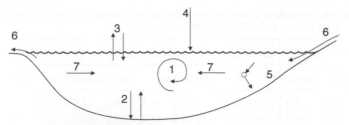

1 Recycling of biogenic materials
2 Sedimentation and processes of exchange at sediment-water interface
3 Exchange between water and atmosphere
4 Influx from precipitation
5 Adsorption and desorption of dissolved substances on the surface of
   particulates in suspension
6 Inflows and outflows
7 Lateral transportation

Figure 5.1   Diagram of the principal processes that determine the balance of materials in a lake.

Chart 5.1   Dissolved substances in water and the analytic complexity for their determination.

Principal ions	$Na^+, K^+, Mg^{2+}, Ca^{2+} SO_4^{2-}, Cl^-$ and HCO_3^- Dissolved in $mg \cdot L^{-1}$ (parts per million)	
Atmospheric gases	Nitrogen (N_2) Oxygen (O_2) Carbon dioxide (CO_2)	
Principal nutrient ions	$PO_4^{3-}, HPO_4^{2-}, H_2PO_4^-, NO_3^-, NH_4^+, Si(OH)_4, Fe^{3+}, Mn^{3+},$ CO_2 and HCO_3^- dissolved with concentrations between $\mu g \cdot L^{-1}$ and $mg \cdot L^{-1}$	
Trace ions	$Cu^{2+}, V^{5+}, Zn^{2+}, B^{2+}, F^-, Br^-, Co^{2+}, Mo^{6+}$ and $Hg^{2+}, Cd^{2+},$ $Ag^+, As_1, Sb_2, Sn^{4+}$ Dissolved in concentrations of $mg \cdot L^{-1}$ and $\mu g \cdot L^{-1}$	
Organic refractory substances (difficult to decompose)	Dissolved in various concentrations $ng \cdot L^{-1}$ and $\mu g \cdot L^{-1}$	
Labile organic substances (very reactive)	Dissolved in various concentrations from $\mu g \cdot L^{-1}$ to $mg \cdot L^{-1}$	

Increase in analytic complexity (vertical label, arrow pointing down)

[1] As^{3+} found in water as $H_2AsO_4^-$
[2] Sb^{3+} found in water as SbO_2^-
Fe and Mn can be found in water in reduced form (Fe^{2+} or Mn^{2+}) or in an oxidized form (Fe^{3+} or Mn^{3+})
Source: Modified from Moss (1988).

conservative, i.e., their concentrations, which are less than those of the most common ions, vary considerably depending on activities of organisms.

N_2, a dissolved gas, is essential in the nitrogen cycle and important for a group of organisms that can fix it from the atmosphere. O_2 is essential in respiratory processes and CO_2 can be a limiting factor for primary producers under certain conditions.

Many organisms require trace ions. The processes of reduction and oxygenation are important for some elements, such as Fe and Mn.

Chart 5.2   Origins and nature of organic substances dissolved in water.

Origin from living organisms in drainage basins	Dissolved organic derivatives in drainage water
Proteins	Methane, peptides, amino acids, urea, phenols, mercaptans, fatty acids, melanin, 'yellow substances' (Gelbstoff)
Lipids (fats, oils) and hydrocarbons	Methane, aliphatic acids (acetic, glycolic, lactic, citric, palmitic, oleic) carbohydrates, hydrocarbons
Carbohydrates (cellulose, starch, hemicellulose, lignin)	Methane, glucose, fructose, arabinose, ribose, xylose, humic acids, fulvic acids, tannins
Porphyrins and pigments, chlorophylls of plants (carotenoids)	Phytane, pristane, alcohols, cetane acids, porphyrins, isoprenoids

Source: Moss (1988).

Some of these elements are toxic to aquatic organisms when the levels rise due to industrial discharges, human activities, or natural processes such as occur in volcanic areas or in natural waters draining through soils where high levels of these elements occur naturally. Such is the case with mercury and arsenic in certain areas.

Organic substances occurring in natural waters have a complex origin (see Chart 5.2) and countless and varied reactions in the water, depending on photo-reductive and photo-oxidative processes. These **dissolved organic substances** include various stages of decomposition of natural vegetation and they play an essential role in inland aquatic systems.

In general, **dissolved organic matter** (DOM) in water is classified into two groups:

▶ Humic substances: defined as a 'general category of naturally occurring highly heterogeneous biogenic organic substances, characteristically yellow and black in color, with high molecular weight and refractory' (Aiken *et al.,* 1985). The authors defined as 'humic acids' those that are not soluble in water with acidic pH (below 2), but may be soluble at higher pH levels; and

▶ Non-humic substances, such as amino acids, carbohydrates, oils and resins.

Fulvic acids are the 'portion of humic substances soluble in all pH conditions,' and humic acids are the 'portion not soluble in water under any pH condition' (Aiken *et al.,* 1985).

Figure 5.2 presents levels of dissolved organic carbon and **particulate organic carbon** in various different natural waters. Figure 5.3 presents the distribution of total organic carbon in Finnish waters, and Figure 5.4 presents total carbon levels in natural waters of Brazil.

Humic substances thus constitute an essential component of dissolved organic matter in natural waters.

Water light brown in color is one of the special characteristics of temperate-region lakes with dissolved humic substances, as described by Naumann (1921, 1931, 1932). Lakes with large concentrations of these waters were called **dystrophic**. Later, Aberg and Rodhe (1942) discovered the predominance of light penetration in the infra-red portion of the spectrum (>800 nanometers). The same observation was made by

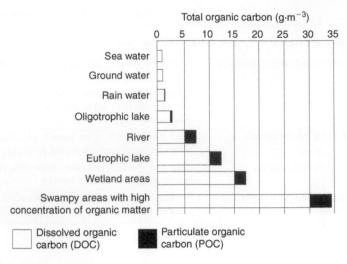

Figure 5.2 Concentration of dissolved organic carbon and particulate organic carbon in several different types of natural waters.
Source: Thurman (1985).

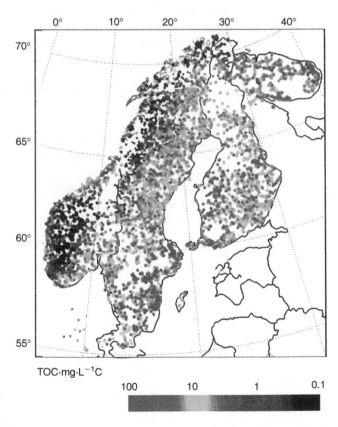

Figure 5.3 Distribution of total organic carbon in waters in Scandinavia (see also color plate section, plate 6).
Source: Skjelvale *et al.* (2001) in Eloranta (2004).

Figure 5.4 Concentration of total carbon in natural surface water of Brazil (Brazil Water Project) (see color plate section, plate 7).

Tundisi (1970), who measured increased penetration of infra-red light in the dark mangrove waters of the **lagoon area of Cananéia,** in the State of São Paulo. High levels of dissolved organic substances with complex molecules increase oxygen consumption in all layers of the water, from the surface to the bottom, and affect the 'underwater radiation climate' of lake systems, reservoirs and rivers.

Humic substances are present in all natural waters as dissolved molecules, **colloidal suspensions** or particulate matter. The dissolved component is always the most significant in terms of impact on the water's biology and chemistry.

Concentrations of these humic substances can range from 100 to $500\,\text{mg}\cdot\text{m}^{-3}$ in seawater, and between $1–2\,\text{mg}\cdot\text{m}^{-3}$ up to $15\,\text{mg}\cdot\text{m}^{-3}$ in groundwater. In lakes with large amounts of decomposing macrophytes and peat on the shoreline, concentration levels can reach $60\,\text{mg}\cdot\text{m}^{-3}$.

The concentration of dissolved organic carbon (DOC) in water filtered through 0.45-μm-pore size Millipore filters can be determined by carbon catalysers. In unfiltered waters, the concentration of carbon is called **total organic carbon** (TOC). The level of carbon calculated is given in $g\cdot m^{-3}$.

In many natural bodies of water, humic substances account for about 50% of the dissolved organic carbon.

Important elements in humic substances include oxygen (35–40% by weight), hydrogen (4–5% by weight), and nitrogen (2%). The carbon present in the material retained by fine filters (i.e., 0.45-μm-pore Millipore filters) is the **particulate organic carbon** (POC).

The terms *total organic material* (TOM), **dissolved organic material** (DOM) and **particulate organic material** (POM) are similar to carbon-based TOC, DOC and POC. However, these terms (TOM, DOM and POM) refer to all material present, including oxygen, hydrogen and nitrogen. In general, the values are twice as high as TOC, DOC and POC alone.

Methods for estimating organic substances in water include biochemical oxygen demand (BOD) and chemical oxygen demand (COD). The latter method uses a powerful chemical oxidant such as permanganate to measure the amount of oxygen consumed, the former measures that consumed by bacteria. The methods do not, however, directly determine carbon concentration in water.

The concentration of humic substances in water can be indirectly determined through the spectrum of water filtered through 0.45-μm Millipore filters and centrifuged. Waters from different sources can be compared to determine which has greater or lesser concentration of humic substances.

The reading of UV absorbance of water samples filtered through 0.45-μm–245-nanometer Millipore filters is a simple and quick way to estimate the levels of organic matter. This method is based on the linear relationship between carbon content and absorption of ultraviolet light.

The variability of dissolved organic carbon in natural waters is large and depends on internal (autochthonous) and external (allochthonous) inputs, periods of drought and precipitation, and internal processes in lakes and reservoirs (decomposition, bacterial action, water temperature, turbulence, and stratification).

Dissolved organic substances, especially humic substances, play an important role in the availability of organic and inorganic nutrients for bacteria, fungi, phytoplankton and aquatic macrophytes.

Dissolved organic matter plays an important role in the complexation, absorption and immobilization of many contaminating organic substances and heavy metals. Such absorption can also make these contaminants available for organisms and increase their bio-availability.

One source of **dissolved salts** in lakes is the steady contribution from drainage through igneous or sedimentary rocks, and because of this, the salt levels vary in inland waters. The composition of waters that drain through these rocks reflects the relative contribution of soluble ions that make up these rocks – in general, $Mg^{2+} > Ca^{2+} > Na^+ > K^+$ – but depending on the region, the sequence may be $Na^+ > Mg^{2+} > Ca^{2+} > K^+$.

The capacity of water to dissolve these ions increases with temperature, acidity, water flow and levels of dissolved oxygen. Acid hydrolysis, for example, dissolves aluminum below pH 4.5 and silicic acid, $Si(OH)_4$, is also released. In clay soils of volcanic origin, iron is released. Sedimentary rocks can release sulphate, carbonate and phosphate, or bicarbonate.

Concentrations of dissolved salts vary greatly in inland waters as a result of the particular features of each **region's hydro-geochemistry** and drainage through igneous

Table 5.2   Global distribution of dissolved salts (DS) in lakes.

Type of lake		Area ($10^3 km^2$)	Volume ($10^3 km^2$)	DS (kg/m^3)	Tsal ($10^{15}g$)
Saline[a] (Vertical)	Caspian	374	78.2	13.0	1,016
	Other endorheic saline lakes	204	4.16	32.0	133
	Coastal saline lakes	40	0.128	5.0[c]	0.64
	Total	618	82.5	13.9	1,150
Fresh Water[b] (Vertical)	Tectonic	424	54.6	0.29	16.1
	Glacial	1,247	38.4	0.10	3.8
	Fluvial	218	0.58	0.10[c]	0.058
	Volcanic	3.1	0.58	0.080	0.046
	Miscellaneous	88	0.98	0.30[c]	0.33
	Total	1,980	95.14	0.213	20.3

Tsal – Total mass of salt
[a]Terminal endorheic lakes and coastal lakes
[b]Exorheic lakes, except coastal lakes and endorheic lakes without terminal outlet
[c]Estimated
Source: Meybeck *et al.* (1989).

or sedimentary rocks. Table 5.2 presents the total distribution of dissolved salts (**total dissolved salts – TDS**) in aquatic ecosystems in saline lakes and lakes with different origins.

Climatic and lithological differences explain the different compositions of dissolved salts. For example, in arid or semi-arid regions, high evapotranspiration rates lead to increased salt levels. In other regions, impacts of drainage can occur in areas with high hydrothermal discharge. Volcanic lakes have concentrations similar to rivers that drain through volcanic rocks.

Table 5.3 shows the geographical variability of common dissolved elements (in $mg \cdot L^{-1}$) in pristine natural fresh waters draining through the most common types of rocks.

Table 5.4 compares the **chemical composition of fresh waters** (of rivers) with seawater in percentages of total (weight/weight). Table 5.5 shows the average ionic composition in river waters in two temperate regions (North America and Europe) and two tropical regions (South America and Africa). Ionic levels and composition vary significantly. Tropical rivers tend to have much lower concentrations than those in temperate-zone rivers, which have much higher levels of **calcium** and **bicarbonate**. In many tropical regions, there is a predominance of sodium, chlorine, silicate and iron. The calcium carbonate present in the headwaters of the Amazon River is an exception, where 85% of dissolved salts come from recent Andes rocks (Gibbs, 1972).

Figure 5.5a presents the characteristics and effects of the main factors determining the composition of inland surface water, including precipitation and evaporation and their influence in the process. The Figure reaffirms that the principal processes affecting total dissolved salt composition in inland waters are **dissolution**

Table 5.3   Geographic variability of the principal dissolved elements in natural pristine fresh water that drains through the most common types of rocks.

	Conductivity $\mu S \cdot cm^{-1}$	pH	ΣCations	Ca^{2+}	Mg^{2+}	Na^+	K^+	Cl^-	SO_4^-	HCO_3^-	SiO_2
Granite	35	6.6	3.5 (166)	0.8 (39)	0.4 (31)	2.0 (88)	0.3 (8.8)	0	1.5	7.8	9
Rocks of various origins; schists, quartz, feldspathic	35	6.6	4.1 (207)	1.2 (60)	0.7 (57)	1.8 (80)	0.4 (10)	0	2.7 (56)	8.3 (136)	7.8 (130)
Volcanic rocks	50	7.2	8.0 (435)	3.1 (154)	2.0 (161)	2.4 (105)	0.5 (14)	0	0.5 (10)	25.9 (425)	12.0 (200)
Sandstone	60	6.8	4.6 (223)	1.8 (88)	0.8 (63)	1.2 (51)	0.8 (21)	0	4.6 (95)	7.6 (125)	9.0 (150)
Clayey rocks	ND	ND	14.2 (770)	8.1 (404)	2.9 (240)	2.4 (105)	0.8 (20)	0.7 (20)	6.9 (143)	35.4 (580)	9.0 (150)
Carbonated rocks	400	7.9	60.4 (3.247)	51.3 (2.560)	7.8 (640)	0.8 (34)	0.5 (13)	0	4.1 (85)	194.9 (3,195)	6.0 (100)

The values are in $mg \cdot L^{-1}$; the values between parentheses are in $\mu eq \cdot L^{-1}$; the values of silica are in $mg \cdot L^{-1}$, with $\mu mol \cdot L^{-1}$ in parenthesis; ND- Not determined.
Source: Meybeck *et al.* (1989).

Table 5.4   Comparison between the chemical composition of river waters and sea water in total % (weight/weight).

	Ocean water	River water
CO_3^{2-}	0.41 (HCO_3^-)	35.15
SO_4^-	7.68	12.14
Cl^-	55.04	5.68
NO_3^-	–	0.90
Ca^{2+}	1.15	20.39
Mg^{2+}	3.69	3.41
Na^+	30.62	5.79
K^+	1.10	2.12
$(Fe, Al)_2 O_3$	–	2.75
SiO_2	–	11.67
Sr^{2+}, H_3BO_3, Br^-	0.31	–

Source: Schwoerbel (1987).

and **draining** of salts from rocks, **atmospheric precipitation** (rainfall) and **processes of evaporation** (crystallization). Precipitation, a primary influencing factor in the tropics, is not only a direct source of ions, but it also provides a means to dissolve rocks and soils (Payne, 1986).

Table 5.5   Average ionic composition of river water in different continents (in mg·L^{-1}).

	HCO_3^-	SO_4^-	Cl^-	SiO_2^-	NO_3^-	Ca^{2+}	Mg^{2+}	Na^+	K^+
North America	67.7	40.3	8.1	4.2	0.23	42.0	10.2	9.0	1.6
South America	31.1	9.6	4.9	5.6	0.16	14.4	3.6	3.9	0.0
Europe	95.2	48.0	6.7	3.5	0.84	62.4	11.4	5.3	1.6
Africa	68.9	9.3	20.2	22.2	0.17	7.9	7.8	21.5	–

Source: Payne (1986).

Figure 5.5b shows the relationship (in weight) of Na/Na + Ca with total dissolved salts in surface waters, and Figure 5.5c shows the ratio (in weight) of HCO_3^- ($HCO_3^- + CO_3^{2-}$) with total dissolved salts in waters on all continents.

Table 5.6 presents the ionic composition of lakes and rain water in tropical Africa. The table also presents electrical conductivity (K_{20}, in µS cm^{-1}), approximate salinity, and pH level.

Table 5.7 presents the ionic composition of Lake Carioca, in Parque Florestal do Rio Doce in East Minas Gerais. In the vertical profile there are increased levels of Ca, Fe, Mn, Na and SO_4, probably also due to the increase in dissolved salts resulting from the stratification period, water-sediment interactions, the organisms, and the chemistry of the water.

Table 5.8 shows the ionic composition of **Lagoon 33** in the same park. This is a wetland area where aquatic macrophytes predominate, especially *Typha dominguensis*.

The **ionic composition of reservoir waters** varies depending on the characteristics of the terrain, flooded areas and the presence of vegetation, retention time and land use over time, which affect the characteristics of the drainage water. Table 5.9 shows the ionic composition of the Barra Bonita reservoir and Promissao reservoir, both on the **Tietê River**, in February/March 1979.

Poorly mineralized waters, with low dissolved ionic composition, occur in many regions of Brazil.

Ionic levels in Amazonian waters illustrate well the regional differences as well as the chemical characteristics of natural waters, including rivers and lakes that are hydrologically and chemically dependent.

Table 5.10 (Sioli, 1984) presents the average levels of different ions in the waters of the Solimões, Negro and Tarumã-Mirim Rivers; Lakes Jacaretinga, Calado and Castanho; from rivers in the forest and rain water. According to Furch (1984), comparing the natural waters of the Amazonian aquatic systems with the world average for surface waters, Amazonian waters may be considered '**chemically poor.**' However, there are major differences in the way the various different chemical components express this shortage. For example, alkaline-ferrous metals (Ca + Mg + Sr + Ca) are less than 0.5% of the world average. Amazonian waters are 50 times richer in trace elements than the world average.

Furch (1984) classified the waters in the western region of the Amazon – including the **Solimões River** and associated floodplain lakes – as carbonate-rich waters. They

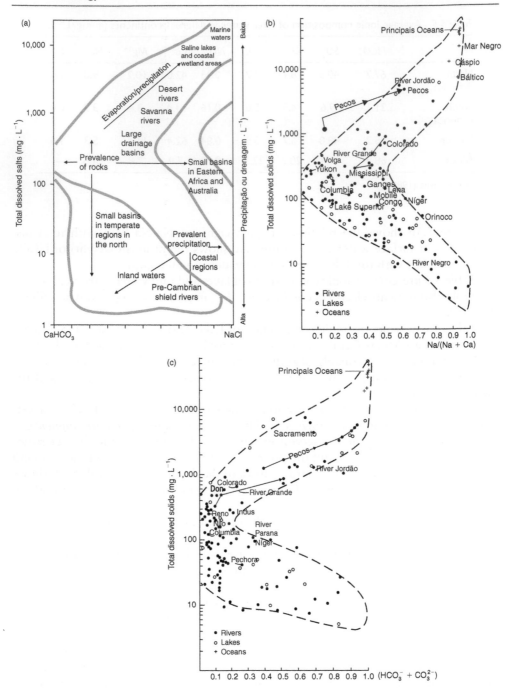

Figure 5.5   a) Schematic representation of mechanisms that control the chemistry of surface water;
b) Ratio (by weight) of Na/Na + Ca with total dissolved salt in surface waters of rivers,
lakes and oceans; c) Ratio (by weight) of $(HCO_3^- + CO_3^{2-})$ with the total dissolved salt in
inland waters.
Source: Modified by Gibbs (1970).

Table 5.6 Ionic composition (units meq · L⁻¹) of some lakes in tropical Africa and rain waters.

Lakes	Data sampling	K_{20} Conductivity ($\mu mho \cdot cm^{-1}$)	Approx. salinity % (g/L)	pH interval	Na^+	K^+	Ca^{2+}	Mg^{2+}	$CO_3^- + HCO_3$	Cl^-	SO_4^{2-}	Cations	Anions	Reference
Lungwe	1953	15–17	0.010	6.5–6.7			0.07	0.030						Dubois (1955)
Tumba	1955	24–32	0.016	4.5–5.0			0.03	0.020	0					Dubois (1959)
Nabugabo	June 1967	25	0.015	7.0–8.2	0.090	0.028	0.060	0.020	0.140	0.04	0.019	0.198	0.199	Beadle and Heron, in Talling and Talling (1965)
Bangweulu	1960	35	0.023	7.0–8.3	0.114	0.033	0.075	0.066	0.260	0.009	0.021	0.288	0.290	Harding and Heron, in Talling and Talling (1965)
Victoria	May 1961	96	0.093	7.1–8.5	0.430	0.095	0.280	0.211	0.900	0.112	0.037	1.02	1.05	Talling and Talling (1965)
George	June 1961	200	0.139	8.5–9.8	0.59	0.90	1.00	0.67	1.91	0.25	0.23	2.35	2.39	Talling and Talling (1965)
Chad (Baga Sola)	July 1967	180	0.165	8.0–8.5	0.5	0.2	0.8	0.3	1.8	0	0.1	1.8	1.9	Maglione (1969)
Malawi	Sept. 1961	210	0.192	8.2–8.9	0.91	0.16	0.99	0.39	2.36	0.12	0.11	2.46	2.56	Talling and Talling (1965)
Tanganyika	Jan. 1961	610	0.530	8.0–9.0	2.47	0.90	0.49	3.60	6.71	0.76	0.15	7.46	7.62	Talling and Talling (1965)
Albert	Feb. 1961	735	0.597	8.9–9.5	3.96	1.67	0.49	2.69	7.33	0.94	0.76	8.81	9.03	Talling and Talling (1965)
Edward	June 1961	925	0.789	8.8–9.1	4.78	2.23	0.57	3.98	9.85	1.03	0.89	11.65	11.77	Talling and Talling (1965)
Kivu	Feb. 1954	1,240	1.115	9.1–9.5	5.7	2.17	1.06	7.00	16.40	0.89	0.33	15.93	17.62	Van der Ben (1959)
Turkana	Jan. 1961	3,300	2.482	9.5–9.7	35.30	0.54	0.28	0.25	24.50	13.50	1.40	36.37	39.40	Talling and Talling (1965)
Rain waters														
Kampala (Uganda)	1960			7.7–8.1	0.28	0.10	0.005			0.05	0.05			Visser (1961)
Gambia	1963 (9 km from coast)				0.026	0.01	0–1.10							Thornton (1965)

Table 5.7   Ionic composition of water in Lake Carioca (Parque Florestal do Rio Doce – MG) in 7 September 1978 (in mg·L^{-1}).

Depth (m)	SO_4^-	Ca^{2+}	Fe^{2+}	Ions K^+	Mg^{2+}	Mn^{2+}	Na^+	Si
0.0	<1.00	2.06	0.12	0.90	0.80	<0.01	2.16	1.99
1.0	<1.00	2.12	<0.10	0.95	0.80	<0.01	1.80	2.00
2.0	<1.00	2.09	<0.10	0.95	0.80	<0.01	2.09	2.00
3.0	<1.00	2.11	0.10	0.95	0.82	0.05	2.28	2.01
4.0	<1.00	2.42	0.41	0.95	0.86	0.22	2.56	2.07
5.0	1.62	2.41	2.41	0.95	0.84	0.18	1.82	2.03
6.0	1.62	2.50	1.61	1.00	0.86	0.86	2.52	2.06
7.0	3.60	2.45	3.74	1.00	0.86	0.86	2.36	2.10
9.0	4.59	2.43	3.94	0.95	0.86	0.86	2.35	2.12

Table 5.8   Ionic composition of surface water in Lagoon 33 (Parque Florestal do Rio Doce – MG) in different seasons of the year (in mg·L^{-1}).

Depth(m)	Data	SO_4^{2-}	Ca^{2+}	Fe^{2+}	Ions K^+	Mg^{2+}	Mn^{2+}	Na^+	Si
0.00	19/3	2.61	2.43	0.26	0.58	0.92	0.05	2.93	0.36
0.00	19/5	2.61	1.70	0.88	0.32	0.68	0.05	2.07	0.03
0.00	21/7	3.93	2.35	2.20	0.58	0.91	0.05	2.29	<1.00
0.00	23/9	4.59	2.31	1.94	0.43	0.92	0.03	2.19	<1.00
0.00	21/11	4.92	2.16	1.81	0.43	0.87	0.04	2.29	<1.00

Table 5.9   Ionic composition of water in Barra Bonita reservoir and Promissao reservoir (Tiete River – SP), in February/March 1979 (in mg·L^{-1}).

Sta	Dep (m)	pH	Conductiv $\mu S \cdot cm^{-1}$	SiO_3^{2-}	SO_4^{2-}	Ca^{2+}	Mg^{2+}	Fe^{2+}	Na^+	K^+	Cl^-	Total CO_2	Temp °C
Barra bonita reservoir	0.0	8.70	112	2.44	0.00	4.10	2.15	1.07	7.35	2.45	6.39	24.3	29.7
	0.2	8.50	112	0.00	0.00	0.00	0.00	0.00	0.00	0.00	0.00	24.3	
	0.5	8.40	111	0.00	0.00	0.00	0.00	0.00	0.00	0.00	0.00	24.7	29.7
	0.7	8.50	114	0.00	0.00	0.00	0.00	0.00	0.00	0.00	0.00	25.1	
	1.3	8.60	111	0.00	0.00	0.00	0.00	0.00	0.00	0.00	0.00	24.6	29.1
	5.0	7.80	112	3.87	0.00	4.37	2.14	1.07	7.40	6.39	6.39	24.7	28.0
	10.0	7.60	113	3.94	0.00	4.37	2.14	1.14	7.19	6.39	6.39	25.5	26.0
	15.0	7.60	113	3.90	0.00	4.37	2.14	1.17	7.50	6.39	6.39	25.8	25.7
Promissao reservoir	0.0	8.10	90	2.81	0.00	4.19	1.94	0.10	6.09	1.77	4.20	0.0	28.6
	5.0	7.80	89	3.06	0.00	4.19	1.94	0.10	6.24	1.64	4.20	0.0	27.8
	10.0	7.60	90	2.74	0.00	4.19	1.94	0.10	6.14	1.64	4.20	0.0	27.6
	15.0	7.50	100	3.02	0.00	4.19	1.94	0.10	6.14	1.64	4.20	0.0	27.2
	20.0	7.50	98	3.16	0.00	4.19	1.97	0.10	5.94	1.64	4.20	0.0	27.1

Table 5.10 Average concentration ($\bar{x}$) for different chemical variables in waters of Amazonia.

	Solimões River			Lake Jacaretinga			Lake Calado (1)			Lake Castanho			Lake Calado (2)			Negro River			Taruma-mirim			Forest Rivers			Rain Water		
	n	x	s	n	x	s	n	x	s	n	x	s	n	x	s	n	x	s	n	x	s	n	x	s	n	x	s
Na (mg·L⁻¹)	29	2.3	0.8	25	2.5	0.7	23	1.6	0.6	30	1.6	0.4	27	1.3	0.5	24	380	124	23	335	88	20	216	58	25	119	97
K (mg·L⁻¹)	29	0.9	0.2	25	1.4	0.4	23	0.9	0.7	30	0.9	0.2	27	0.6	0.3	24	327	107	23	312	98	20	150	108	25	100	104
Mg (mg·L⁻¹)	29	1.1	0.2	25	1.4	0.4	23	0.9	0.5	30	0.9	0.2	27	0.7	0.4	24	114	35	23	99	44	20	37	15	25	21	17
Ca (mg·L⁻¹)	29	7.2	1.6	25	8.5	1.8	23	6.2	3.1	30	5.0	1.2	27	4.3	2.5	24	212	66	23	186	83	20	38	34	25	72	78
Na+K+Mg+Ca (mg·L⁻¹)	29	11.5	2.6	25	13.8	3.1	23	9.6	4.6	30	8.4	1.7	27	6.9	3.6	24	1020	312	23	926	285	20	441	182	25	312	275
Conductivity (µS·cm⁻¹)	27	57	2.6	23	60	18	23	47	19	27	42	9	24	38	12	22	9	2	21	9	2	20	10	3		N.D.	
pH	27	6.9	0.4	23	6.9	0.3	23	6.6	0.4	27	6.7	0.3	24	6.5	0.4	22	5.1	0.6	21	5.0	0.5	20	4.5	0.2		N.D.	
total P (µg·L⁻¹)	28	105	58	25	57	26	21	62	38	25	40	14	26	50	33	24	25	17	23	22	21	20	10	7		N.D.	
total C (mg·L⁻¹)	28	13.5	3.1	25	16.2	5.8	22	12.8	4.2	28	12.4	1.8	26	10.8	2.7	24	10.5	1.3	23	9.9	1.6	20	8.7	3.8		N.D.	
HCO₃–C (mg·L⁻¹)	26	6.7	0.8	24	8.5	1.7	22	5.6	2.2	28	5.0	1.1	26	4.3	1.9	24	1.7	0.5	23	1.6	0.3	20	1.1	0.4		N.D.	
Cl (mg·L⁻¹)	26	3.1	2.1	24	2.9	1.7	22	2.5	1.2	28	2.0	1.0	26	2.1	1.0	24	1.7	0.7	23	1.8	0.7	20	2.2	0.4		N.D.	
Si (mg·L⁻¹)	28	4.0	0.9	25	4.3	1.1	22	3.6	1.1	28	3.8	1.3	26	3.0	0.9	24	2.0	0.5	23	1.7	0.4	20	2.1	0.5		N.D.	
Sr (µg·L⁻¹)	29	37.8	8.8	25	39.7	11.0	23	27.5	11.2	30	24.4	8.0	27	23.0	13.3	24	3.6	1.0	23	2.8	1.0	20	1.4	0.6	23	0.7	0.5
Ba (µg·L⁻¹)	29	22.7	5.9	25	21.7	6.6	23	16.1	6.2	30	16.9	6.1	27	15.0	7.0	24	8.1	2.7	23	7.1	3.2	20	6.9	2.9	23	4.4	3.0
Al (µg·L⁻¹)	29	44	37	25	20	14	23	26	18	30	23	16	27	21	14	24	112	29	23	119	40	20	90	36	23	10	8
Fe (µg·L⁻¹)	29	109	76	25	123	79	23	111	68	30	83	38	27	85	49	24	178	58	23	136	59	20	98	47	23	26	31
Mn (µg·L⁻¹)	29	5.9	5.1	25	3.0	2.3	23	4.4	3.2	30	2.8	2.5	27	3.5	2.7	24	9.0	2.4	23	7.9	2.9	20	3.2	1.2	23	1.4	0.7
Cu (µg·L⁻¹)	29	2.4	0.6	25	1.6	0.9	23	2.1	0.9	29	2.2	1.1	27	1.7	0.6	24	1.8	0.5	23	1.6	0.6	20	1.5	0.8	23	3.3	2.1
Zn (µg·L⁻¹)	29	3.2	1.5	25	2.2	1.1	23	3.4	1.7	30	2.9	1.6	27	3.0	1.6	24	4.1	1.8	23	4.0	1.8	20	4.0	3.3	23	4.6	3.5

n – number of samples; s – standard deviation; N.D. – not determined
Source: Furch (1984).

have the highest **ionic concentration** of HCO_3^-, Ca^{2+}, Mg^{2+}, Na^+, K^+, Ba^{2+}, and Sr^{2+}, approximating the global averages for natural waters.

In the waters of the Negro River and adjacent lakes, there is a shortage of carbonates and a high level of trace elements such as Fe and Al (ten times more than in the Solimões River). These waters are acidic (pH circa 5.1) and also have high levels of humic substances and dissolved organic matter.

The most unusual chemical characteristics are in small streams in the forests of the central Amazon, where carbonate levels are absent or much lower than in the Negro River and its tributaries, the pH is acidic (4.5), and there is a high percentage of trace elements and predominance of **alkaline metals**, with low Ca^{2+} and Mg^{2+} levels.

Amazonian rain water, also analyzed by Furch (1984), contains trace metals, and proportions of alkaline-ferrous metals and alkaline metals similar to the waters of the Negro River.

Table 5.11 shows the conductivity values for different aquatic ecosystems. Table 5.12 shows statistical correlations between water conductivity and the levels of several different elements and ions.

Salinity of natural waters

The salinity of natural waters, in milligrams per liter ($mg \cdot L^{-1}$) or milli-equivalents per litre ($meq \cdot L^{-1}$), is the sum of the dissolved salts in the water.

The electrical conductivity or specific conductance, is an indicator of the salinity resulting from concentration of salts, acids and bases in the natural waters. It is measured by the electrolyte content of water, through the flow of current between two platinum electrodes: the higher the measurement the greater the concentration. Conductivity also increases with temperature.

The units that express conductivity (at 25°C) are: micro-Siemens ($\mu S \cdot cm^{-1}$), milli-Siemens ($mS \cdot cm^{-1}$) ($1 \, mS \cdot m^{-1} = 1000 \, \mu S \cdot cm^{-1}$), or micromho ($cm^{-1}$) (1 micromho $cm^{-1} = 1 \, \mu S \cdot cm^{-1}$).

The conductivity of different waters reflect a large number of complex phenomena: depending on the ionic concentration, there is a correlation between conductivity and nutrients of phytoplankton and macrophytes. In certain lakes and reservoirs, conductivity also depends on the alkalinity or acidity (pH) of the water.

The total dissolved solids (TDS) include all salts present in the water and non-ionic components. Dissolved organic compounds contribute to total dissolved solids and can be measured by the total content of dissolved carbon (TDC), as already explained.

The content of TDS is obtained by filtering a sample of water, evaporating the filtrate and measuring the dry weight from the principal solutes remaining. Total content of TDS is used by geomorphologists interested in determining the effects of chemical erosion in different regions.

Golterman (1988) verified that the salinity (S) of fresh water can be estimated as: $S = {\sim}0.75C$ (where $C = \mu S \cdot cm^{-1}$) and $S = mg \cdot L^{-1}$, or $S = {\sim}0.01C$ (where $C = \mu S \cdot cm^{-1}$ and $S = meq \cdot L^{-1}$).

Williams (1986) examined the inter-relationships between salinity and conductivity. In a gradient of conductivity (C) from $5{,}500–100{,}000 \, \mu S \cdot cm^{-1}$, the salinity in $mg \cdot L^{-1}$ is $= {\sim}0.6–0.7C$.

Table 5.11   Values of conductivity for different aquatic ecosystems.

Carlos Botelho (Lobo/Broa) reservoir	$10–20\,\mu S\cdot cm^{-1}$
Natural waters in the Negro River	$9–10\,\mu S\cdot cm^{-1}$
Barra Bonita (SP) reservoir	$100\,\mu S\cdot cm^{-1}$ (1974)
Barra Bonita (SP) reservoir	$370\,\mu S\cdot cm^{-1}$ (2002)
Saline lakes of Africa (Turkana Lake)	$2.482\,\mu S\cdot cm^{-1}$
Rain water in regions not impacted by human activity	$10–15\,\mu S\cdot cm^{-1}$
Solimões River	$57\,\mu S\cdot cm^{-1}$
Jacaretinga River (Amazonia)	$60\,\mu S\cdot cm^{-1}$
Salto de Avanhandava reservoir (Tiete River, SP)	$74\,\mu S\cdot cm^{-1}$
Capivara reservoir (River Paranapanema)	$54\,\mu S\cdot cm^{-1}$
Atlantic Ocean	$43.000\,\mu S\cdot cm^{-1}$

Table 5.12   Statistical correlations (example-set) between the conductivity of water and the concentration of different elements and ions.

Component	Correlation with Conductivity
Ca	0.973
HCO_3	0.961
Cl	0.928
Na	0.090
Sr	0.898
Mg	0.862
SO_4	0.730

Source: Margalef (1993).

5.3   IONIC COMPOSITION IN SALINE LAKES AND INLAND WETLANDS

The ionic composition of inland saline lakes varies considerably and differs greatly from the ionic composition of rivers with low conductivity. Ionic composition depends on the evaporation rate, which produces precipitation of ions as a function of **solubility**. As evaporation occurs and the volume of water **decreases**, there is **differential precipitation** of Na^+, Mg^{2+}, Cl^- and SO_4^{2-} ions. In waters with high levels of Ca^{2+} and Mg^{2+}, $MgCO_3$ precipitates as crystals of dolomite [$CaMg(CO_3)_2$]. Salts deposited in dry river-beds may be transported by wind, causing **impacts on human health** and on the agriculture in surrounding regions (see **Aral Sea**, Chapter 18).

The relationship of **water temperature to evaporation and precipitation** is critical for the precipitation of ions and solubility in lakes in arid and semi-arid regions.

5.4   THE ROLES OF CATIONS AND ANIONS IN BIOLOGICAL SYSTEMS

The roles of cations and anions in biological systems are many and varied. Important functions include:

▸   activation of enzyme systems;
▸   stabilization of proteins in solution;
▸   the development of electrical excitability;
▸   regulation of membrane permeability;

▶ maintenance of a state of dynamic isotonic equilibrium between cells and extracellular fluids.

High levels of salt in solutions absorb much free water in solution and therefore
tend to precipitate proteins. An ion that activates enzymes can be an integral part of
an enzyme, function as a bond between the enzyme and the substrate, alter the balance of an enzymatic reaction, or inactivate the enzyme system (Lokwood, 1963).

Sodium – is the main cation in extracellular fluids in many animals. High concentrations can inhibit **enzyme** systems.

Potassium – is the principal cation in cells. It plays a role in establishing membrane potential, and is also a component in the activation of certain enzymes.

Calcium – decreases the permeability of cell membranes and ions. Some enzyme
systems are also inhibited by high calcium levels. As a divalent ion, calcium is important in stabilising colloids.

Magnesium – forms the nucleus of the chlorophyll molecule. It is an ion that activates
many enzymes involved in energy transfer. Large quantities of magnesium and calcium
can decrease membrane permeability, and reduce the oxygen consumption by cells.

pH – chemical properties of proteins shift with pH. Changes in pH can play an
important role in enzymatic activity, the osmotic pressure of colloids, and shifts in the
acidity or basicity of extracellular fluids.

Anions – phosphate and bicarbonate have buffering effects on cells and extracellular fluids. High phosphate levels tend to inhibit activities that depend on calcium.
Bicarbonate correlates with the retention of potassium in muscles.

Osmotic regulation in freshwater organisms (plants, vertebrates and invertebrates) is an essential physiological feature. As the levels of cations and anions in
inland waters vary, these organisms also regulate their cation and anion levels to
enable the functioning of enzymes. Such osmotic regulation is performed by the active
absorption of water, elimination of the water in urine and the subsequent absorption
of cations and anions through surfaces and gills. Freshwater organisms tend to maintain higher internal concentrations of salt than is found in the medium.

Ionic levels in aquatic ecosystems constitute a key factor in the distribution of
aquatic organisms and the **colonisation of environments** (with different conductivi-
ties), which trigger regulatory processes and tolerance that vary for different groups of
aquatic animals and plants. For example, changes in the **diversity** of calanoid species
in zooplankton in reservoirs in the Medio Tietê can be related to successive increases
in the electrical conductivity of the water and its shifts in ionic levels (Matsumura
Tundisi and Tundisi, 2003).

Colonisation by **invasive species** largely depends on tolerance to ionic levels,
availability of cations and anions, and their total amounts in water.

Of course, the anions PO_4^{3-}, NO_3^-, NO_2^- and the cation NH_4^+ are fundamentally
important as nutrients for **photosynthetic phytoplankton**, aquatic macrophytes, and
photosynthetic bacteria.

5.5   DISSOLVED GAS: AIR-WATER INTERACTIONS AND
THE SOLUBILITY OF GASES IN WATER

The air-water interface is important in aquatic ecosystems because of the **exchange of
energy** and gases that takes place there. Of the gases dissolved in water, oxygen and

Table 5.13   Principal gases in the atmosphere.

Gas	%
Nitrogen (N_2)	78.084
Oxygen (O_2)	20.946
Argon (Ar)	0.934
Carbon dioxide (CO_2)	0.038

carbon dioxide are chemically and biologically important. Table 5.13 shows the main gases found the atmosphere.

The solubility of gases in water depends on the physical and chemical characteristics of the water mass, as well as on pressure, temperature and salinity. According to Henry's law, 'the quantity of gas absorbed by a determined volume of liquid is proportional to the atmospheric pressure exercised by the gas.' According to this law (Cole, 1983):

$$C = K \times p$$

where C is the concentration of gas, p is the partial pressure exercised by the gas, and K is a solubility factor that differs for each gas.

Levels of dissolved gases in water can be expressed in mg/liter ($mg \cdot L^{-1}$) or millimoles/liter ($mmol \cdot L^{-1}$). Most gases follow Henry's law. The solubility of gases in water depends, then, on altitude, water temperature (the solubility of gas decreases with increased temperature) and salinity. The solubility of dissolved oxygen drops, for example, in waters with high saline levels, as in the case of inland saline lakes. Copeland (1967) observed that in the **Tamaulipas Lagoon** (Mexico), levels of dissolved oxygen decreased from $6.6\,mg \cdot L^{-1}$ (for 25°C ocean water) to $3\,mg \cdot L^{-1}$ (for water with salinity of 22%).

5.5.1   Dissolved oxygen

The level of dissolved oxygen in water is one of the most important variables in limnology. Oxygen is extremely important biologically, and in water it participates in many chemical reactions. It quickly dissolves in water, depending on air/water interactions, i.e., on the water temperature and atmospheric pressure.

Dissolved oxygen levels can be measured by $mg \cdot L^{-1}$, $mL \cdot L^{-1}$ or millimoles $\cdot L^{-}$, and also by the percentage of **saturation** of dissolved oxygen in the water. Thus, 100% saturation theoretically means the maximum dissolved oxygen possible for a given temperature and pressure. This percentage, in a sea-level water mass, can represent higher saturation values than at 1000 meters of altitude, for example. Figure 5.6 shows a nomogram in which the percentage of **saturation of dissolved oxygen** in water can be determined as a function of temperature, levels measured in $mg \cdot L^{-1}$ or $mL \cdot L^{-1}$ and atmospheric pressure. Corrections in altitude, atmospheric pressure and solubility factors for gases in general can be obtained by using tables (Wetzel, 1975).

It is worth noting that solubility is always considered to be a relationship between the levels of O_2 in solution and the levels above the solution – in the air – hence, saturation is presented in relative terms.

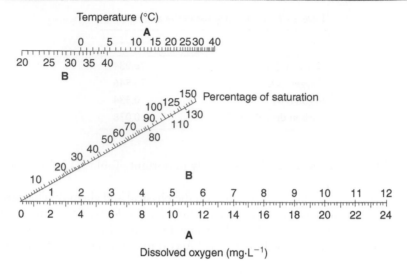

Figure 5.6 Nomogram for determining the percentage of saturation of dissolved oxygen in water. *Source:* Hutchinson (1957).

Table 5.14 presents the solubility of oxygen in pure water in equilibrium with the air saturated with water vapour at 1 atmosphere.

Dissolution of oxygen through the air-water interface generally occurs under conditions of intense vertical movement or in a process of slow diffusion and transport by convection. **Dissolution of oxygen** by molecular diffusion through a calm surface is extremely slow and insignificant (Hutchinson, 1957).

The levels of dissolved oxygen in water depend on the coefficients of oxygen exchange between the atmosphere and the water surface. Movement of oxygen through the air-water interface is given by the Bohr equation (Hutchinson, 1957):

$$\frac{dO}{dt} = a \cdot \alpha(P - p_t)$$

where:
a = area of interface;
P = partial atmospheric pressure of the gas;
p_t = pressure at which the level of gas in the water at a determined time is in equilibrium.
α = coefficient of entry

An exit coefficient (B) is also considered.

Another important dissolution mechanism results from thermal cooling, which occurs during evaporation and results in increased salinity with subsequent vertical movement by convection currents. The process results in exchanges of gases.

Major sources of dissolved oxygen in water

The atmospheric source of dissolved oxygen and its dissolution in water depends, as described, on the specific conditions of the liquid mass. **Processes of vertical transport**

Table 5.14   Solubility of dissolved oxygen in pure water and in equilibrium with the air saturated at 1 atmosphere.

Temperature (0°C)	Concentration mg·L^{-1}	Temperature (0°C)	Concentration mg·L^{-1}
0	14.62	21	8.91
1	14.22	22	8.74
2	13.83	23	8.58
3	13.46	24	8.42
4	13.11	25	8.26
5	12.77	26	8.11
6	12.45	27	7.97
7	12.14	28	7.83
8	11.84	29	7.69
9	11.56	30	7.56
10	11.29	31	7.43
11	11.03	32	7.30
12	10.78	33	7.18
13	10.54	34	7.06
14	10.31	35	6.95
15	10.08	36	6.84
16	9.87	37	6.73
17	9.66	38	6.62
18	9.47	39	6.51
19	9.28	40	6.41
20	9.09		

To convert to ml·L^{-1}, multiply by 0.70.
Source: Cole (1983).

of oxygen resulting from the effects of wind-caused turbulence play a decisive role in this dissolution. The turbulent flow promotes oxygenation in the upper layers. In some cases, dissolution from turbulence can lead to supersaturation. This is the case, for example, in outflows of water from dams, usually in volume-regulating floodgates. Cases of up to 150% saturation have been reported. This mechanism of increasing dissolved oxygen levels has generally been explored in reservoirs. In turbulent rivers, the saturation levels of dissolved oxygen increase. As a result, rivers can provide an effective aeration system and replenishment of dissolved oxygen in water, resulting in **auto-purification**.

Photosynthetic activity is an important source of dissolved oxygen in water. Such production of oxygen is restricted to the euphotic zone and occurs during the day. Therefore, the vertical distribution of dissolved oxygen is strongly related to the vertical distribution of phytoplankton in the euphotic zone. In lakes with high levels of chlorophyll due to phytoplankton in the epilimnion in eutrophic conditions, **supersaturation of dissolved oxygen** can occur with values up to 130–150% during the day. Tundisi (unpublished results) measured levels of 120% saturation on the surface in Barra Bonita reservoir, São Paulo, with chlorophyll levels of approximately 200 µg/liter.

Supersaturation can also occur in transparent shallow waters, with high biomass levels in the form of immersed macrophytes, phytobenthos and **periphyton**. Tundisi *et al.* (1984) showed that, in general, the relatively high diurnal levels of dissolved oxygen in Lake Jacaretinga (Central Amazon) were the result of photosynthesis by **submerged macrophytes** near the surface and by **periphyton**.

The presence of high biomass levels from primary producers leads to intense diel fluctuations in the levels of dissolved oxygen, which will be described below.

Losses of dissolved oxygen

Respiration by aquatic plants and animals and decomposition through bacterial activity are important sources of loss of dissolved oxygen. At the **sediment-water interface**, substantial loss of oxygen can occur in the water because of bacterial activity and **chemical oxidation**. The effects of oxidation on oxygen consumption can be measured by various techniques. One method is to place sediment in closed flasks with water and periodically determine the dissolved oxygen in the **supernatant** water.

Wind agitation in shallow lakes also leads to decreased oxygen levels in the water, due to the resuspension of sediments and organic materials. Oxygen loss in water can also result from high levels of mortality of aquatic organisms. In the eutrophication process, with extensive cyanobacterial blooms, a mass kill occurs after a **period of senescence,** producing increased oxygen consumption. Such episodic processes of decreased dissolved oxygen may also result from periods of intense stratification followed by circulation, in which the anoxic hypolimnion undergoes a mechanical action from the effects of wind and the layer of epilimnetic water comes into contact with the anoxic water. On these occasions, fish kills and mass deaths of other organisms can occur.

The levels of dissolved oxygen drop drastically when the levels of suspended matter markedly increases after heavy precipitation and drainage into lakes, ponds or rivers. Tundisi (1995, unpublished results) recorded a drastic drop in dissolved oxygen in the Barra Bonita reservoir, São Paulo, with levels ranging from $0.00 \, mg \cdot L^{-1}$ to a maximum of $5.00 \, mg \cdot L^{-1}$ at the surface and anoxic environment at depths below $12 \, m$ (to a maximum depth of $25 \, m$) with high levels of NH_4^+. This process caused a fish kill. The low levels of dissolved O_2 were the result of high levels of suspended material draining into the reservoir after heavy precipitation.

Vertical distribution of dissolved oxygen

Varying oxygen levels at different depths in lakes result from the processes of stratification, de-stratification, vertical circulation and its efficiency, and the vertical distribution and activity of organisms.

Surface waters generally present oxygen levels close to saturation. Supersaturation on the surface or subsurface can occur in cases of high levels of phytoplankton. Higher temperatures in the epilimnion cause loss of oxygen through the air-water interface. When complete circulation occurs in an unproductive lake, dissolved oxygen is distributed uniformly all the way to the bottom. This type of vertical distribution is called **orthograde**. The slight increase in dissolved oxygen that occurs with depth is due to higher solubility at lower temperatures.

In productive lakes, summer stratification is characterized by a vertical distribution with an anoxic hypolimnion and oxygen levels close to saturation or supersaturation in the epilimnion. This curve, **clinograde**, is typical of a stratified, eutrophic lake in summer. Decaying material accumulates in the hypolimnion and consumption of oxygen there is high.

There is an increase in dissolved oxygen in some stratified lakes due to the accumulation of phytoplankton in the upper metalimnion. Accumulation of cyanobacteria in the metalimnion was described for several lakes in temperate and tropical regions. For example, dense layers of *Oscillatoria agardhii* were observed in many temperate lakes, and the presence of *Lyngbya limnetica* was described for Carioca Lake (Parque Florestal do Rio Doce – MG) by Reynolds *et al.* (1983).

This type of curve is called positive **heterograde**, with saturations of up to 300%. Circulation and horizontal transport resulting from the production of oxygen in photosynthesis by macrophytes, in the shallowest parts of the lakes, can also cause an increase in oxygen, producing this **positive heterograde curve**.

On the other hand, high consumption of oxygen, resulting from the levels of organisms or decomposing organic material, can occur in the lower portion of the metalimnion, resulting in a **negative heterograde curve**. This minimum in the metalimnion, associated with density gradients and increased respiration, is common in monomictic and meromictic lakes.

Another less-common type of vertical distribution occurs when there is a maximum in the hypolimnion. This results from transport mechanisms and horizontal circulation caused by the inflow of denser and cooler waters. Rivers commonly cause horizontal stratification in temperature and dissolved oxygen.

Figure 5.7 shows the various types of curves and vertical profiles of dissolved oxygen in lakes.

In polymictic lakes, dissolved oxygen is generally distributed homogeneously in the vertical profile. In reservoirs, the processes of circulation and vertical distribution of dissolved oxygen can be more complex. For example, in flooded areas of reservoirs with no deforestation after filling, there is almost permanent anoxia resulting from decomposing plant material. The process of re-oxygenation on the surface is difficult, since the presence of vegetation prevents horizontal and vertical circulation (see Chapter 12).

In meromictic lakes, there is permanent anoxia in the hypolimnion.

The vertical distribution of oxygen in lakes is also important in relation to the chemical processes of precipitation, re-dissolution, and biogeochemical cycles of elements (see Chapter 9).

Dissolved oxygen deficit

A lake's oxygen deficit is defined as 'the difference between the saturation level of dissolved oxygen and water temperature, the lake's surface pressure and the observed value' (Hutchinson, 1957). The real oxygen deficit is the difference between the amount of O_2 measured at a certain depth and the amount presented if the water were saturated in the same conditions of pressure and temperature (Cole, 1975).

An assessment of the real deficit of oxygen (RDO) in a lake at a certain time of the year can be calculated by the formula:

$$RDO = \sum_{i=1}^{n} G_i \cdot V_i$$

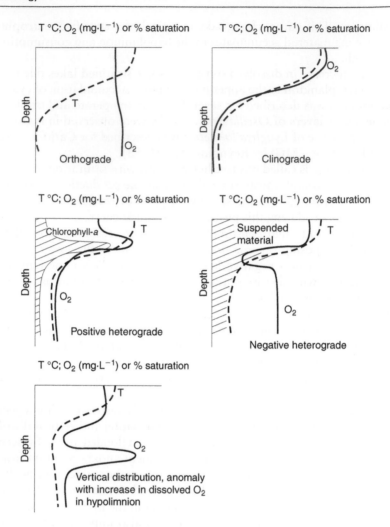

Figure 5.7   Vertical curves and profiles of dissolved oxygen in lakes.
Source: Modified from Cole (1983).

where:

G_1 – measures the real oxygen deficit (mg·cm^{-3}) found in the upper and lower limits of each of the lake's strata (or that which can be considered as a layer in cm);

V_1 – volume (cm^3) of each stratum in the lake, which can be calculated by the formula:

$$V_1 = \frac{h}{3}\left(S_1 + S_2 + \sqrt{S_1 \cdot S_2}\right)$$

where:

h = height between the upper and lower planes of each strata;

S_1 and S_2 = areas (in cm^2) in each layer between two strata.

The **relative deficit** of dissolved oxygen is the difference between two calculations: one made during the period of maximum stratification and the other during the period of maximum circulation. The relative deficit makes it possible to calculate the depletion of oxygen during a given period by unit area in the hypolimnion. Indications of biological productivity in lakes can also be estimated though this calculation.

The relative oxygen deficit is calculated for the entire hypolimnion and expressed in unit of area (Deficit of Hypolimnion Oxygen by Area – DOHA, mg $O_2 \cdot cm^{-2} \cdot day^{-1}$) applying the the formula:

$$DOHA = \frac{M_1 - M_2}{\Delta t}$$

where:

M_1 = content of oxygen in the period of maximum circulation in a volume equal to the volume of the hypolimnion at M_2;

M_2 = content of oxygen observed in the hypolimnion at a specific time of year;

(Δt) = time interval (days) between M_1 and M_2.

$$(M_1 - M_2) = \Sigma^i\, G_i - V_i/H$$

where:

G_1 = average concentration of oxygen in the upper and lower limits of each stratum of the hypolimnion ($mg \cdot cm^{-3}$)

V_1 = volume (cm^3) between the metalimnion and the hypolimnion

H = border plane (cm^2) between the metalimnion and hypolinmnion, calculated using the lake's thermal profile corresponding to the inflection point of the curve.

In long series of sequential data, it is important to determine the oxygen deficit in the hypolimnion, as an indication of progression of eutrophication. Eberly (1975) described a method for determining the level of eutrophication in a dimictic lake in the **temperate region** from the deficit of dissolved oxygen in the hypolimnion. This methodology uses the volume of various different strata in the hypolimnion and saturation values at the hypolimnetic temperature. Table 5.15 presents these values for **Lake Mendota** (Wisconsin, United States) over a period of 50 years.

Table 5.15   Deficits of dissolved oxygen and temperatures of the hypolimnion in Lake Mendota (USA).

Date	Temperature of hypolimnion (°C)	Deficit of oxygen (%)	Deficit by area (g/m²)
18 Aug. 1912	8.3	74.9	54.78
13 July 1927	9.4	82.8	56.55
16 July 1931	8.3	96.3	70.01
21 July 1953	8.3	99.1	72.87
12 Sept. 1962	7.3	99.9	75.97

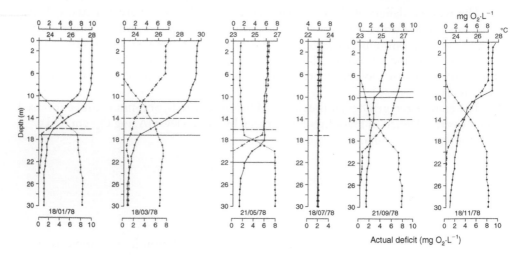

Figure 5.8   Results of oxygen deficit for Lake Dom Helvécio (Parque Florestal do Rio Doce – MG). *Source:* Henry *et al.* (1989).

In this case, the eutrophication process resulting from the increase in the deficit per area was caused by allochthonous material, mainly domestic sewage. Hutchinson (1957) suggests that an oxygen deficit of $0.05\,mg \cdot cm^{-2} \cdot day^{-1}$ already indicates eutrophication.

Henry *et al.* (1989) calculated the oxygen deficit in Lake Dom Helvécio (Parque Florestal do Rio Doce – MG) from vertical profiles of water temperature and dissolved oxygen levels. The results are presented in Figure 5.8. In July (winter period with no stratification), the real deficit in the water column was $2.0\,mg \cdot L^{-1}$ The deficit rises gradually, reaching $8.0\,mg \cdot L^{-1}$ of oxygen during thermal stratification of the hypolimnion. The relative deficits of dissolved oxygen in the water column do not exceed $6.5\,mg \cdot L^{-1}$. The real total oxygen deficit, expressed per unit of the lake's surface area, ranges from 119.07 tons of O_2 and $1.73\,mg\,O_2 \cdot cm^{-2}$ (September, period of stratification) to 163.28 of O_2 and $2.37\,mg\,O_2 \cdot cm^{-2}$ (May, period of limited circulation). The hypolimnetic oxygen deficit, expressed per unit area, varies from 0.56 to $1.30\,mg\,O_2 \cdot cm^{-2}$.

Henry *et al.* (1989) attribute the lake's oxygen deficit to several factors, such as decomposition of organic matter in the water column and oxygen consumption in the sediment produced by accumulation of non-decomposed organic matter in the water column. Such consumption can vary in different lake regions (Lasemby, 1975).

5.6 THE CO_2 SYSTEM

Carbon dioxide is another biologically important gas. It dissolves in water to form soluble carbon dioxide, which reacts with water to produce non-dissociated

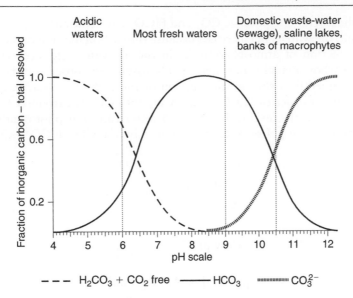

Figure 5.9   Interrelations of pH, free dissolved CO_2, bicarbonate ion (HCO_3^-) and carbonate ion (CO_3^-).

carbonic acid. The following equation describes the main phases of the CO_2 system in water:

$$CO_2 \underset{gás}{} \xrightarrow[\text{rapid}]{\text{rapid}} \frac{\text{dissolved}\ CO_2 + H_2O}{H_2CO_3} \xrightarrow[\text{slow}]{\text{slow}} HCO_3^- + H^+ \rightleftharpoons CO_3^{2-} + H^+$$

Figure 5.9 lists the various different forms of carbon occurring in water at different pH levels. The levels in each phase depend on the water's temperature and ionic level. The level of free CO_2 to maintain HCO_3^- in solution is called the equilibrium CO_2.

Aquatic plants can use CO_2, HCO_3^-, and more rarely, CO_3^{2-} as sources of carbon. Some aquatic macrophytes use HCO_3^-, after converting it to CO_2 by the action of carbonic anhydrase enzyme. The majority of lakes have sufficient HCO_3^- levels for photosynthesis, in a pH range that varies from approximately 6.0 to 8.5. CO_2 is the dominant form at low pH and CO_3^{2-} is the dominant form for pH >10.5.

5.6.1 CO_2 levels and pH

The term pH is defined as the negative logarithm of the concentration of hydrogen ions. On a scale from acidity to alkalinity, 0–7 pH indicates acidity and 7–14 indicates alkaline status. Most lakes have a pH value between 6.0 and 9.0; lakes with high concentrations of acids have a pH of between 1.0–2.0; extremely eutrophic lakes with extreme CO_2 depletion or high levels of carbonate (soda lakes) can have pH values >10.

In the photosynthetic process, CO_2 and HCO_3^- are removed by primary producers. As a result, the water's pH increases because the carbon fixation capacity is greater than the dissolution of atmospheric CO_2 in the air-water interface. Therefore, by reducing the carbon available in water, the photosynthetic process leads to increased pH and, consequently, a shift in the reaction of the carbonate system. As shown in the equation, the transfer of H_2CO_3 and HCO_3^- into CO_2 is extremely fast, but continuous reduction in the level of free CO_2 in water can limit photosynthesis. Under normal conditions, the pH of water is regulated by the CO_2, $HCO_3^- - CO_3^{2-}$ system. The formula:

$$\log \frac{[HCO_3]}{[CO_2 \text{total}]} = pH - pK_1$$

applies between 15–20°C and below pH 7:

$$[HCO_3^-] = 4\,[CO_2\text{total}]$$

The CO_2 system is the principle source of inorganic dissolved carbon for aquatic plants, and the three forms tend to equilibrium with each other and the atmosphere. CO_2 is more abundant in water than in air and nearly 200 times more soluble than oxygen. The photosynthetic process plays an important role in the reduction of CO_2 in water. Diel fluctuations in the CO_2 and O_2 in water, produced by photosynthetic cycles and respiration, will be discussed below.

The terms alkalinity, alkaline carbonate, and alkaline reserves are used to designate the total quantity of base that can be measured with a strong acid titration. In general, alkalinity can be expressed in $mg \cdot L^{-1}$ or $meq \cdot L^{-1}$.

Alkalinity can be expressed as:

▸   alkalinity in $mg \cdot L^{-1}$ of HCO_3^-;
▸   alkalinity in $mg \cdot L^{-1}$ of CO_3^-;
▸   alkalinity in $mg \cdot L^{-1}$ of HCO_3^- and CO_3^-;

According to Hutchinson (1957), alkalinity of bicarbonates is the most useful, since the pH of many natural waters is in the range associated with HCO_3^-.

By titrating a water sample with strong acid and calculating the water's alkalinity, we can start to estimate the total amount of CO_2 present $(CO_2 + HCO_3^- + CO_3^{2-})$. With a complete titration curve and potentiometric determination of pH, total CO_2 can be calculated. CO_2 can also be measured by means of gas chromatography or infra-red detectors.

Calculation of the total **inorganic carbon** present in a body of water is also important because of the relationship of CO_2 to the processes of photosynthesis and respiration in the epilimnion and hypolimnion. In the case of slightly buffered water, the release of CO_2 in the hypolimnion significantly reduces pH. In general, the pH of water is related to its chemical properties, the **geochemistry of the water basin,** in addition to the effects of **biological processes** such as photosynthesis, respiration and the decomposition of organisms. Hutchinson (1957) presented examples of **heterograde pH distribution** in lakes with slightly buffered water, in which ferrous and manganous ions balanced by bicarbonates accumulated in the hypolimnion within increase of bicarbonate alkalinity. Also, vertical distribution of **heterograde pH** can occur with an increase of pH in the metalimnion, as a result of an increased photosynthesis and removal of CO_2 in the layer.

Total alkalinity corresponds to the excess of cations compared with **strong** anions.

$$Total\ alkalinity = HCO_3^- + CO_3^- + OH^- - H^+$$

Alkalinity resulting from boric acid, important in seawater, may be insignificant in fresh water. The interrelationship between total inorganic carbonate and alkalinity depends on the water's pH. Alkalinity – and thus total inorganic carbon – can be determined by the titration and displacement of weak acids (for example: HCO_3^-; $H_2BO_3^-$; $H_3SiO_4^-$) by strong acid (sulphuric or hydrochloric) up to a pH in which all inorganic carbon present has definitely been dislocated (usually in the pH range 2–3). Inorganic carbon levels can be determined, therefore, from the water's alkalinity. Table 5.16 presents a factor that can be used (multiplied by alkalinity, in $meq \cdot L^{-1}$) to determine the level of inorganic carbon (Margalef, 1983).

The term **hardness** or **degree of hardness** expresses the amount of carbonates, bicarbonates or sulphates present in water. Scales can be used to calculate hardness. For example, in the French scale, hardness is given in parts of $CaCO_3$ per 100,000 parts of water. Table 5.17 presents ratios (in percentage) between the main chemical forms of calcium and inorganic carbon.

5.6.2 Vertical distribution of CO_2

In lakes with a vertical temperature gradient and a clinograde curve of dissolved oxygen, CO_2 accumulates in the hypolimnion, which can result from metabolic activity. In part, bicarbonate increases due to the presence of $(NH_4)HCO_3$. Transport of bicarbonate by solution, from the sediment in the bottom, occurs more easily in **anaerobic sediment** than in aerobic sediments.

Vertical distribution of CO_2 in general accompanies the distribution of dissolved oxygen and the oxygen deficit. Ohle (1952) concluded that accumulation of CO_2 in the hypolimnion is a more accurate measurement of a lake's metabolism than the oxygen deficit.

Sedimentation of $CaCO_3$, which occurs in some lakes with high rates of photosynthesis and elevated calcium levels, increases the CO_3^{2-} precipitating to the bottom of lakes, thus increasing the carbon available for photosynthesis. Introduction of calcium ions and bicarbonate in the lake (there is excess bicarbonate) produces the following reaction:

$$Ca + 2HCO_3 \rightleftharpoons \underset{\text{soluble}}{Ca(HCO_3)_2} \longrightarrow \underset{\text{insoluble}}{CaCO_3} + H_2O + CO_2$$

Table 5.16   Factor by which the alkalinity (in meq·L^{-1}) multiplies to obtain the total concentration of inorganic carbon (in mg·L^{-1}). Temperature 15°C. For lower temperatures, increasing 1% per degree, for higher temperatures, decrease at the same proportion.

pH	Factor	pH	Factor
6.0	44.16	8.0	12.60
6.5	22.08	8.2	12.12
7.0	17.16	8.5	12.00
7.5	14.04	9.0	11.64
7.8	12.96		

Source: Margalef (1983).

Table 5.17   Relationship (in %) between the principal species of calcium and inorganic carbon in sea water and in fresh water.

	Sea water	Fresh water
Free HCO_3^-	63–81	99.23
$NaHCO_3^0$	8–20	0.04
$MgHCO_3^+$	6–19	0.21
$CaCO_3^+$	1–4	0.52
CaO_3^{2-}	8–10	31.03
$NaCO_3^-$	3–19	0.03
$MgCO_3^0$	44–67	6.50
$CaCO_3$	21–38	62.44
Free Ca^{2+}	85–92	96.89
$CaSO_4^0$	8–13	1.45
$CaHCO_3^+$	0.1–1	1.32
$CaCO_3$	0.1–0.9	0.33

Sources: Atkinson et al. (1973); Hanor (1969); Pytkowicz and Hawley (1974); Garrels and Thompson (1962); Millero (1975a, b); Dyrssen and Wedborg (1974); Kester and Pytkowicz (1969).

Increased photosynthetic activity always produces increased insoluable CO_3^{2-}, which precipitates to the bottom or remains in suspension.

When the dissolved oxygen curve is orthograde, so is the CO_2 curve. When vertical distribution of dissolved oxygen occurs with a clinograde curve, an inverse CO_2 clinograde curve also occurs. When a heterograde dissolved oxygen curve occurs, there is also an inverse vertical heterograde distribution of CO_2.

5.6.3   Respiratory quotient in lakes

The total **respiratory quotient** of lakes can be estimated by calculating the balance between dissolved oxygen produced by photosynthesis, the consumption of oxygen

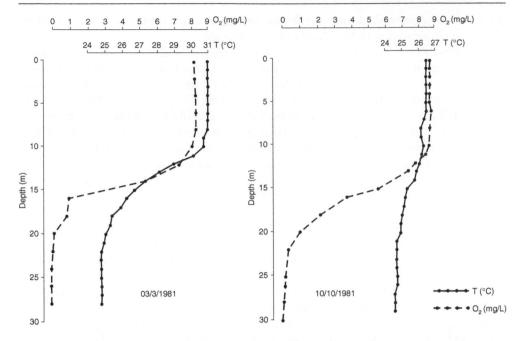

Figure 5.10 Vertical profile of dissolved oxygen in periods of thermal stratification in Lake Dom Hevécio (Parque Florestal do Rio Doce – MG).

for total respiration, and the amount of CO_2 produced by respiration. According to Hutchinson (1957), the metabolic CO_2:O_2 quotient is 0.85.

Calculating these relationships in a lake's metabolism is essential for studying the balance of gases, the interrelationship between photosynthesis and respiration, and interactions in the air-water interface.

Figure 5.10 presents the typical distribution of dissolved oxygen and water temperature in a stratified lake (warm monomictic lake) in summer.

5.7 SEASONAL AND DIURNAL VARIATIONS IN O_2 AND CO_2

Patterns of diurnal variation of dissolved O_2 and CO_2 gases, and their interrelationship with the processes of thermal stratification, nocturnal cooling and vertical circulation were described for several lakes, particularly focusing on the studies in tropical lakes by Talling (1957, 1969) Barbosa (1981), Ganf (1974), Ganf and Horne (1975), Hare and Carter (1984), and Melack and Fisher (1983).

Variations occurring in the epilimnion of lakes during the diurnal and nocturnal periods are related to the distribution of CO_2 through photosynthetic activity, increased pH and, in some cases, high levels of chlorophyll, an increase in dissolved O_2 that can reach high saturation values. Measurements of the processes of daily variations related to dissolved gases and modifications in thermal structure show that in some cases these changes are more significant than those occurring in the seasonal cycle.

In many shallow tropical lakes, increases in pH and supersaturation of O_2, as well as depletion of O_2 and low pH during the nocturnal period, depends on the vertical distribution of phytoplankton, periphyton and aquatic macrophytes. Wide variations also occur in regions with high levels of solar energy.

Studies by Ganf and Horne (1975) in Lake George (Uganda), a shallow equatorial lake with high levels of cyanophytes, illustrate these variations. Thermal stratification was present by day, with values of 10°C difference from a lake's surface temperature to its bottom temperature (maximum depth 2.50 m), along with the stratification of chlorophyll a, which, in the afternoon period (16 h), could be $100 \, mg \cdot m^{-3}$ at the surface and $400 \, mg \cdot m^{-3}$ on the bottom. During the nocturnal period, vertical distribution of chlorophyll remains homogeneous. In periods of intense thermal stratification (35°C), photosynthesis drops on the surface, along with increased respiration.

The **inhibition of photosynthetic activity** in conditions of intense solar radiation and high temperatures is a fairly common phenomenon in tropical lakes with reduced diurnal circulation.

Levels of dissolved O_2 and pH also fluctuated widely in Lake George. During the day, values of up to 250% saturation were found, with a pH up to 9.7, the result of intense photosynthesis. This shift in pH was accompanied by a reduction in total CO_2 from 70 to $49.5 \, mg \cdot L^{-1}$.

The study also identified fluctuations in the vertical distribution of photosynthesis and the point of optimum photosynthesis in the vertical profile. Diurnal fluctuations in photosynthetic activity and vertical distribution of high photosynthesis were also observed by Tundisi (1977) in the Lobo reservoir. Such fluctuations reflect alterations in the vertical distribution of CO_2 and O_2.

In a study on diurnal variations in water temperature, dissolved gases, chlorophyll and photosynthesis in the Carioca lakes (Parque Florestal do Rio Doce -MG), Barbosa (1981) found that during the months of greatest thermal stability (summer), chlorophyll is sharply stratified, with high levels in the hypolimnion, following thermal stratification. Small oscillations of dissolved oxygen in the epilimnion resulted from photosynthetic and respiratory activity. Diel variations in 24-hour cycles of oxygen and CO_2 were also described by Tundisi et al. (1984) for Lake Jacaretinga (Amazons) and Tundisi et al. (results unpublished) for Lake Dom Helvécio (Parque Florestal do Rio Doce – MG).

The diurnal O_2 and CO_2 cycles enable the calculation, based on variations in dissolved gases, of the community's photosynthetic production and respiration. Hourly measurements of dissolved gases coupled with thermal measurements allow calculations that provide approximate and valuable information on the cycles of O_2/CO_2, respiration/photosynthesis and metabolism in lakes (see Figure 5.11).

The diurnal balance between production of dissolved O_2 and consumption and production of CO_2 (i.e., photosynthesis and respiration) is also important for maintaining ecological stability in shallow lakes (Ganf and Viner, 1973). As Ganf and Horne pointed out in the case of Lake George, the seasonal variations occurring in temperate-zone lakes appear to have been compressed into a 24-hour cycle.

Seasonal variations in levels of dissolved oxygen and CO_2 depend on biomass levels, stratification and thermal de-stratification cycles, the seasonal inflow of precipitation, and the processes of advection. In warm monomictic lakes, a hypolimnion develops during the period of stratification, and a more uniform distribution of oxygen occurs during the circulation period. CO_2 accumulates in the hypolimnion during periods of intense stratification. High levels of other gases such as methane can also

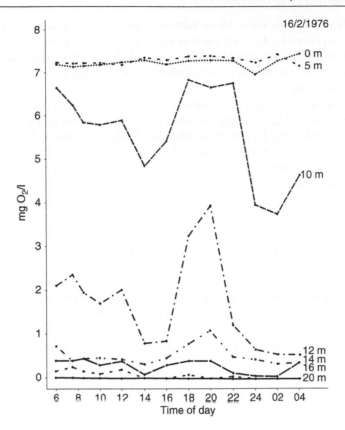

Figure 5.11   Diel variations in levels of dissolved oxygen in summer at various depths in Lake Dom
Helvécio (Parque Florestal do Rio Doce – MG).

occur. Meromictic lakes have an anoxic hypolimnion, as well as high levels of methane and H_2S.

Polymictic lakes have high levels of dissolved oxygen during most of the seasonal cycle, with short periods of anoxia or reduced levels.

Seasonal variations of CO_2 and O_2 are thus related to the seasonal cycle of stratification, de-stratification and circulation, the vertical activity and distribution of organisms in different periods, and the lake's interactions with climatic factors such as precipitation and wind activity.

In a study of 80 diurnal cycles of dissolved oxygen in an **Amazon floodplain lake** (Lake Calado), Melack and Fisher (1983) concluded that the diurnal variation of dissolved oxygen in the lake showed the prevalence of respiration of organisms over phytoplankton photosynthesis. **Aerobic and anaerobic respiration** are important, producing oxygen depletion and significant quantities of reduced substances. After vertical mixing, the reduced substances cause oxygen levels to drop in the epilimnion. The main implication of this cycle is that the planktonic organisms of Lake Calado consume more oxygen than they produce by photosynthesis.

Floodplain lakes in the Amazon produce abundant amounts of organic matter. The predominance of respiration over production of oxygen by photosynthetic activity explains why waters in these lakes are generally subsaturated with dissolved oxygen and supersaturated with carbon dioxide. According to Melack and Fisher (1983), plankton organisms in these lakes most likely use **allochthonous** organic matter as a source of energy. Inflow of atmospheric oxygen is the main source of dissolved oxygen in Lake Calado.

Scientific studies have shown evidence for sub-saturation of dissolved oxygen and anoxic hypolimnion in **lakes of the Amazon floodplain** (Schmidt, 1973; Santos, 1973; Marlier, 1967) and in the Negro River (Rai and Hill, 1981; Reiss, 1977), **Purus River** (Marlier, 1967) Tapajós River (Braun, 1952) and Trombetas River (Braun, 1952). Gessner (1961) also identified sub-saturation in the Negro River. In areas with high levels of organic matter, in floating aquatic forests, sub-saturation of oxygen is common (Junk, 1973).

5.8 OTHER GASES DISSOLVED IN WATER

Despite high atmospheric levels of N_2, N_2 in water is almost chemically inert, and its gaseous cycle is much less significant compared with that of O_2 and CO_2. In water, nitrogen is present in high amounts as a gas but is virtually inert. It is used only by organisms that can fix nitrogen, such as some cyanobacteria and bacteria. In some lakes and reservoirs, **biological fixation** of nitrogen is an important source of nutrient (in this case, nitrogen).

Methane (CH_4) gas is extremely common in the hypolimnion of permanently stratified lakes or lakes or with reduced circulation (see Chapter 10).

Levels of dissolved oxygen in reservoirs with submerged forests are extremely low (see Chapter 12).

Principles for measuring dissolved oxygen and CO_2 in the water

Calculations of the levels of O_2 and CO_2 in water are important ecologically and chemically and also from the experimental viewpoint as well, since the two gases are involved with the process of photosynthesis. These calculations can thus be used to quantify photosynthesis, primary production and respiration of aquatic plants and organisms in general.

Dissolved oxygen in water is generally estimated using the traditional Winkler method (which has undergone many technical modifications). Manganese sulphate and potassium iodide are added to the water in an alkaline medium. The oxygen in the water oxidizes the Mn^{2+}, which becomes Mn^{3+}, forming a complex. The amount of Mn^{3+} is proportional to the concentration of O_2 present. The manganese oxide complex is formed and settles to the bottom of the 250–300-ml amber bottle and is dissolved by the action of H_2SO_4. The iodine released is equivalent to the concentration of dissolved O_2 and titration is done with thiosulphate sodium NaS_2O_3, using starch as indicator.

Total CO_2 and inorganic carbon can be calculated from the level of alkalinity of the water, which consists of adding a strong acid (usually H_2SO_4 or HCl) to the water sample and titrating until the pH reaches ~4–5.

6 Organisms and communities in inland aquatic ecosystems and estuaries

SUMMARY

This chapter presents the groups of organisms (plants and animals) that form the major components of inland aquatic ecosystems. Dispersal mechanisms are discussed, as well as isolation, geographical distribution and factors that limit or control the **diversity of aquatic biota**. Information is presented on the overall diversity of the genera and species of the different groups.

Factors affecting the **spatial distribution of aquatic communities** are also presented, including the main characteristics of these communities, their relative importance and their composition.

Emphasis is given to the description and composition of **communities in the Neotropical region**. Sampling methods in these communities, assessment of aquatic biodiversity and different approaches to their study and spatial and temporal complexities are also presented.

Lists of species from different regions in Brazil are presented in the appendix and in several tables in the chapter, as a reference and example of **Neotropical aquatic biota**.

6.1 COLONIZATION OF AQUATIC ENVIRONMENTS

The totality of organisms living in different inland aquatic systems – lakes, rivers, reservoirs, artificial tanks, small natural pools, wetlands and estuaries – is a complex with great botanical, zoological, ecological and economic significance. A large number of plant and animal groups are present in these different ecosystems, as illustrated in Figure 6.1.

It is generally accepted that in the early stages of evolution, very simple **eukaryotic cells** (with true nucleus) captured and ate **prokaryotic cells**, which then became organelles. Prokaryotes are organisms with no nucleus, Golgi apparatus, endoplasmic

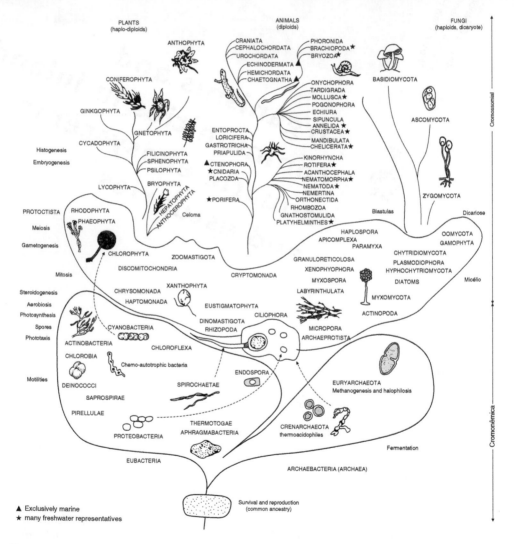

Figure 6.1 Principal phyla on planet Earth showing some common freshwater animals and microorganisms.
Source: Modified from Margulis and Schwartz (1998).

reticulum, mitochondria or plastids. The various different organisms classified as **eukaryotes** have all these structures. Prokaryotic organisms include bacteria and cyanobacteria; all other living organisms are eukaryotes. For a long time the prevailing theory was that eukaryotes evolved through further differentiation of prokaryotes. But more recently, through the use of biochemical techniques, electron microscopy and molecular biology studies, the theory of **symbiosis** put forward by Mereschkowsky (1905) has become more accepted. This theory holds that organelles such as chloroplasts and mitochondria were independent prokaryotes, which were later incorporated into eukaryotes. Chloroplasts originated from cyanobacteria, and mitochondria from bacteria. Initially, bacteria and cyanobacteria lived as guests in cells; gradually they transformed into organelles.

Further evidence supports the theory of symbiosis: chloroplasts and mitochondria are, in a way, independent. Chloroplasts of certain **algae** have retained more cyanobacterial features than typical chloroplasts and can therefore be considered intermediaries between chloroplasts and cyanobacteria. For a long time some chloroplasts living inside cells were considered to be cyanobacteria living symbiotically inside a heterotrophic host.

Recent comparisons of ribosome RNA sequences in mitochondria, chloroplasts and prokaryotes confirmed a genetic relationship between chloroplasts and **photoautotrophic cyanobacteria**, and between mitochondria and **heterotrophic bacteria**.

A large body of evidence currently supports the theory that all *phyla* of photoautotrophic and heterotrophic eukaryotes evolved directly from a single **ancestral eukaryote** (Hoek, Mannard and Jahnsl, 1998).

Colonization of inland aquatic environments can be challenging due to the physiological limitations of plants and animals and the challenges of competition and interactions in food webs. Conditions vary more widely in inland aquatic systems than in oceans, where physical and chemical variables tend to be relatively constant. The **average mineral concentration** in inland aquatic systems is from 1/1000 to 1/100 times less than in seawater systems and ionic levels and composition vary widely. The pH level varies greatly and water temperatures fluctuate seasonally and daily. Dissolved oxygen levels also vary and fluctuate, as shown in Chapter 5.

Meanwhile, in the evolution of invertebrates and many vertebrates, displacements occurred between aquatic systems and terrestrial systems. Many organisms in the marine invertebrate phyla have been confined to marine environments (Moss, 1988). Others adapted to the variations in osmotic concentrations in estuaries (15 to 900 m osmolarity $\cdot$ L^{-1}) and others adapted to freshwater conditions ($<$15 m osmolarity $\cdot$ L^{-1}).

Osmolarity measures the quantity of ions or molecules that are not dissociated in a kilogram of water.

Aquatic invertebrates in fresh water need to maintain a concentration between 30 to 300 osmolarity $\cdot$ L^{-1} for the effective functioning of enzymes, which means constantly maintaining osmotic equilibrium.

The concentration and composition of marine waters are extremely similar to the composition of **cellular fluids** in many marine invertebrates. Inland waters, freshwater bodies of water, have low ionic levels, very dilute for the functioning of protoplasm and enzymes. Animals inhabiting fresh water must therefore maintain higher concentrations in their intracellular fluids than those found in the aquatic environment in which they live. These organisms must eliminate water that enters by osmosis and replace ions that are eliminated by movement down the concentration gradient between the organisms and the waters where they live.

A range of specific physiological conditions must therefore exist to enable the colonization of different inland aquatic ecosystems. The wide variety of ionic concentrations in inland systems should also be considered. Saline lakes may have even higher saline levels than those of marine systems (Lockwood, 1963).

Osmotic concentrations in freshwater fish are also similar to those in freshwater invertebrates. Like the invertebrates, freshwater fish need to maintain the same processes that regulate osmotic pressure and physiological adjustment.

Movements between inland aquatic systems, marine systems and terrestrial systems have occurred. Figure 6.2 illustrates some of these movements. It is currently accepted that the earliest forms of life originated in oceans and from there colonized freshwater systems and terrestrial habitats (Barnes and Mann, 1991).

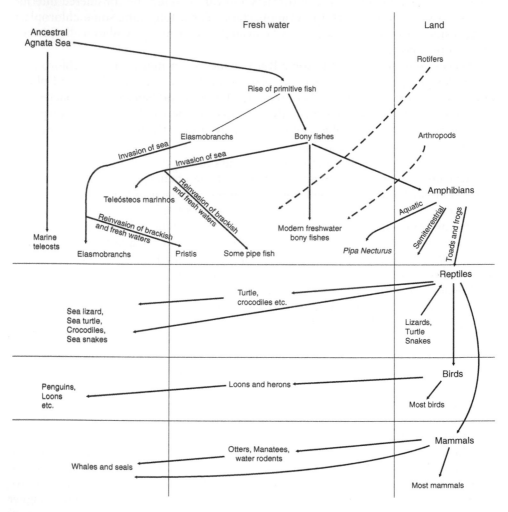

Figure 6.2   Movement of organisms from marine systems to inland aquatic systems and land systems.
    Source: Modified from Lockwood (1963).

Colonization of inland aquatic systems from terrestrial systems also appears to have taken place, based on evidence of aquatic insects, **lunged molluscs** and **vascular plants**. This type of colonization often implies more limited access to oxygen, and therefore depends on the capacity to utilize greater quantities of energy. However, water is more viscous than air and has the capacity to support structures (bones, for example), and the temperature is more stable.

Many insects whose larval stages develop in water have systems for breathing oxygen from the air. Some adult insects that live in water use an air sac that allows them to extract oxygen from the air. These examples illustrate the physiological challenges of life in inland aquatic systems and, as explained by Moss (1988), the lower diversity of freshwater biota compared with that in marine waters and terrestrial systems. The lower diversity must result from the physiological challenges of colonizing **aquatic habitats**.

The lower diversity in freshwater systems may be mainly due to the wide **physical climatic variability** and chemical variability in inland aquatic systems and the changes that have occurred over thousands of years. In addition to fluctuations occurring over geological time that have disrupted biological systems and led to extinction of flora and fauna, there is also **physical discontinuity** in freshwater ecosystems, making it difficult, for example, to recolonize and maintain a set of species and populations after a catastrophe occurs. Lakes are also a relatively recent phenomenon – typically around 10,000 years old – a short geological time in terms of evolutionary processes and speciation. The oldest lakes (such as Malawi and Tanganyika in Africa and Baikal in Russia), due to total age and steady environmental conditions over thousands of years, have high diversity and **endemic species** of fish resulting from evolutionary change and specialism, such as the exploration of many **feeding niches**.

In addition to the constantly fluctuating natural processes in freshwater systems, human exploration and use of the water in these systems has generated additional chemical, physical and even biological variability, making colonization and maintenance of diversity even more difficult. Over the last hundred years, human activity has had a huge impact on fresh water. For example, the construction of dams has significantly modified river systems around the world, and the contributions of drainage basins and terrestrial systems near rivers and lakes have greatly increased (see Chapter 12 and Chapter 18).

An important conclusion about inland waters and their chemical composition can be drawn from the fact that ionic levels, although low, are crucial for the survival of plants and animals colonizing these systems. The presence and distribution of organisms in an aquatic ecosystem depend in part on the water's ionic composition and hence its conductivity. Ecological evidence shows that in highly saline inland waters, above a certain conductivity value the composition of the aquatic flora and fauna shifts significantly. For example, when total salinity reaches 5–10% $(5–10\,g\cdot L^{-1})$, many species disappear from 'fresh water.' Alterations in a water's ionic composition and relative proportion of ions can also lead to shifts in diversity and distribution of aquatic organisms.

Evidence that ionic levels play an extremely important role in the distribution of organisms is shown by studies on the distribution of aquatic plants and animals and laboratory-controlled experiments where they are exposed to different ionic levels to test their survival and reproductive capacity (Tundisi and Matsumura Tundisi, 1968).

Lévêque *et al.* (2005) defined 'freshwater species' as follows:

▶ Some species depend on fresh water for all stages of their life cycles; for example, freshwater fish, **crustaceans** and **rotifers**. In fish, the exception should be made for diadromous species, i.e., those species that migrate between marine and freshwater systems.
▶ Some species, such as **amphibians** and insects, need fresh water to complete their life cycle.
▶ Some species need only **moist habitats**, such as certain species of Collembola.
▶ Some species are dependent on fresh water for food or habitat, for example, '**aquatic birds**', **mammals** or **parasites** that use a freshwater organism as a host. Thus, there are 'truly freshwater' species as well as 'freshwater dependent' species.

Two groups of freshwater animals can be identified:

▶ One group of marine origin, which includes **primary aquatic animals** – without terrestrial ancestors – that invaded fresh waters directly from the oceans: **metazoans** located on the lowest phylogenetic levels, brachiated **molluscs** (mussels and prosobranchs), crustaceans, lampreys and fish. Some of these species can live in fresh or salt water. Many species of fish and some crustaceans are diadromous, with a life cycle that includes fresh water and salt water. Some groups, such as Echinodermata, Ctenophora and Chaetognata, are exclusively marine (they do not colonize freshwater habitats).
▶ A second group, of terrestrial orgin, evolved in terrestrial systems and then colonized fresh water. This group includes lunged molluscs, which have a primitive lung enabling them to breathe air.

Many insects spend part of their life cycle in fresh water and are in that second group of freshwater-dependent organisms. Species that depend on fresh water occupy a wide range of habitats.

6.2   DIVERSITY AND DISTRIBUTION OF ORGANISMS: LIMITING AND CONTROLLING FACTORS

The diversity of species in inland aquatic ecosystems depends on various evolutionary factors and processes in which interactions occur between species, colonization mechanisms and duration, as well as the response of juveniles and adults to environmental conditions such as ionic levels in water, temperature, effects of parasites, predators, and other physical, chemical and biological factors.

A species' **distribution area** may be very wide, or limited to a few sites at certain latitudes and altitudes. Climatic factors (such as precipitation and drought, with effects on water chemistry, temperature and dissolved oxygen) may function as barriers for species to expand and colonize inland waters.

Some aquatic species, however, have succeeded in becoming widely distributed around the world (i.e., cosmopolitan) due to evolutionary and physiological factors that help reduce the effects of ecological barriers that restrict distribution. Aquatic organisms that use different modes of dispersion can also overcome such barriers. Some of the most diverse dispersal mechanisms include spores and **resistant eggs** carried by the wind, birds or other aquatic organisms (vertebrates, for example); and dispersal by currents or surface or ground drainage.

Species diversity can present regional patterns, as shown in the case of zooplankton in Alto Tietê reservoirs (Matsumura Tundisi *et al.*, 2003, 2005), or spatial patterns within one aquatic ecosystem (horizontal and vertical) (Matsumura Tundisi *et al.*, 2005).

Endemic organisms occurring in specific regions are limited by different barriers, and the degree of endemism in an ecosystem can be a reflection of the period of isolation.

Physical, chemical and biological factors influencing the distribution of a given organism present vertical or horizontal gradients, which to a greater or lesser degree affect all species in an aquatic ecosystem (see Figure 6.3).

Factors influencing the distribution of aquatic organisms include variations in salinity of estuaries, alterations in the substratum in rivers, gradients of conductivity in rivers and reservoirs (both horizontal and vertical), periods of drought and temporary **desiccation** of rivers and lakes, intensity of underwater radiation, and horizontal and vertical gradients of dissolved oxygen. These factors can also limit or expand diversity.

The **environment** of each species includes a complex set of interacting physical, chemical and biological factors (Hutchinson, 1957).

The **biotic factors** responsible for the diversity of species and their distribution include a wide variety of processes: **exclusive competition**, effects of **predation**, **parasitism**, production of **inhibitory substances** and chemical interactions between species, populations and communities (Lampert, 1997).

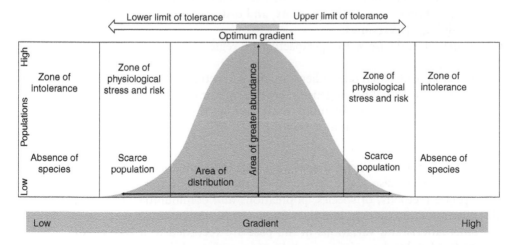

Figure 6.3   Gradient of organisms' tolerance to several environmental factors and interactions with other species.
Source: Modified from Cox and Moore (1993).

Biodiversity

Biodiversity is defined as the variability of living organisms (marine, freshwater and land) and the ecological complexes to which they belong. It includes diversity within a species, between species, and in ecosystems.

Biodiversity is the basis of a wide range of **ecosystem services** that contribute to human well-being in a relevant and indispensable way, and is important in both natural systems and systems modified by man.

Human-produced alterations in biodiversity directly and indirectly affect the well-being of human populations.

The measure of biodiversity is complex: **species richness** may be one of the measures (number of species per area or volume), but should be integrated with other measures. These measures (including taxonomic, functional and genetic bases) should also be accompanied by other key attributes, such as variability, quantity, distribution and abundance.

Knowledge of the taxonomic diversity in biodiversity is still incomplete and has many flaws, especially in the tropics and subtropics. Estimates of all species on Earth range from 5 to 30 million species. From 1.7 to 2.0 million species have been formally identified (Millennium Ecosystem Assessment, 2005).

This brief introduction to the principal factors responsible for the diversity and distribution of freshwater organisms outlines the complexity of the problem, and considers the following scales:

▶ Long-range time scale, in which evolutionary processes have progressed, influenced by geomorphological, climatic and biological changes;
▶ A shorter-range time scale, in which seasonal and diurnal variations affect the composition of species, biodiversity, and distribution of organisms, with varying physiological responses;
▶ A spatial scale from the perspective of regions, continents or large drainage basins;
▶ A spatial scale based on gradients in microstructures (vertical and horizontal) and micro-scale spatial heterogeneity (centimeters or meters);
▶ Spatial scales in **microhabitats** with various different physical, chemical and biological gradients and distribution.

Methods for sampling and analysing species, populations, and aquatic communities must take these scales into consideration. Continuous observation, through appropriate methods, of species, populations and communities helps measure variables in aquatic ecosystems and their fluctuations in space and time.

Fluctuations and variations in the past are recorded in a lake's sediment, and the study of **palaeolimnology** can help measure variability based on climatic conditions, uses of drainage basins, and the succession of species, populations and communities.

Diversity of fauna in fresh water

Estimates of animal diversity in fresh water vary widely. One statistical source, probably very low, puts identified species at about 100,000, half of which are freshwater insects. About 20,000 vertebrate species (35–40% of total) live in or are dependent on fresh water.

Most information was obtained from studies of inland waters in the Northern Hemisphere (North America and Europe). In South America and Africa, there is still a huge area of research yet to be developed.

Vertebrates are the most-studied group of animals. Studies on invertebrates have focused on those organisms that are **vectors** of diseases.

6.3   COMMUNITIES IN INLAND AQUATIC ECOSYSTEMS

Organisms and communities play an essential role in the operational processes of rivers, lakes, reservoirs and wetlands. Since the basic unit of reference in limnology and ecology is the ecosystem, it is important to study the principal interactions between community components and physical/chemical environments. These interactions can be described through a continuous process of measuring, sampling and experimentation. At the community level, biomass (the amount of living matter existing at a given point of time in unit area or volume), species diversity, the co-existence of various species, horizontal and vertical distribution, fluctuations, and cycles should all be studied.

According to Margalef (1983), each species, sub-species or individual can be viewed as a filter that can provide invaluable information on the community's composition and structure, since the significance of these various filters multiplies. A classified list of all the organisms found in a lake thus has enormous ecological value, once other information on structures can be determined. Such information also has historical value, since the presence of particular organisms in a body of water at a given time is significant in the history of evolution.

According to Margalef (1978), it is important to classify communities utilising tertiary (not binary) systems. A community's organization is reflected in the number of species present, their interactions, fluctuations, cycles, and distribution.

As already mentioned, the interdependent relationships among a community's various components are fundamental and the dynamic study of these relationships can help identify the main functions of the system's components and structure.

6.4   DISPERSAL, EXTINCTION, SPECIATION AND ISOLATION OF AQUATIC BIOTA

Dispersal of aquatic plants and animals is generally accomplished by the movement of particles of wind-blown dust, insects, birds, aquatic mammals, **reptiles**, amphibians and fish that carry algae, larvae and fish eggs, **protozoa**, eggs and resistant forms of aquatic organisms. It is thus understandable why certain species of cyanobacteria (of the genera *Microcystis, Oscillatoria, Planktolyngbya* and *Anabaena*, for example) can rapidly colonize certain aquatic ecosystems. Ostracod eggs and resistant eggs of rotifers and copepods can also get transported – and thus dispersed – by a variety of aquatic plant and animal organisms.

Dispersal can also result from human activities. The ballast water of ships and mobile man-made structures (for example, vehicles that are moved from one region to another) can be responsible for introducing invasive species across wide continents.

Extinctions of species in aquatic ecosystems can result from human pressure and the introduction of exotic species. Biogeographically, lakes and reservoirs can be viewed as islands, and island biogeography models can be applied to lake ecosystems to study biotic equilibrium and ratios of number of species that illustrate immigration, extinction and colonization.

According to Margalef (1983), the distribution of taxonomic units in space results in the equilibrium between **extinction, dispersion** and **speciation**. In some ancient lakes, there are many endemics; for example, in Lake Malawi, there are more than 200 species of **cichlids,** of which at least 196 are endemic. In Lake Victoria, there were (before recent predation) about 170 species of *Haplochromis,* a genus of cichlids. Lakes Tanganyika and Baikal are so old that both have many endemic species, of major importance for the study of aquatic biodiversity.

In Lake Baikal, at least 240 species of amphipods, mostly endemic, have been identified.

Isolation in many aquatic ecosystems has resulted in speciation in mollusc, ostracod, Trichoptera and fish species. Lakes, temporary waters, inland shallow wetlands, saline lakes and water in bromeliads are all isolation-supporting environments. For small freshwater invertebrate species, this isolation is less common, but for some vertebrates, such as **lungfish,** isolation also occurs. Isolation occurs insymbiont or commensal parasite species that live with isolated single-species hosts, symbionts or commensal organisms.

Given that flows in inland aquatic systems vary (and can sometimes be quite intermittent), **processes of extinction, isolation, speciation** or **dispersion** can be quite significant. The presence of non-contiguous liquid water propels organisms to remain in certain regions, resisting temporary desiccation and the rigour of the factors **of desiccation** (high temperatures, high salinity and ionic levels).

According to Banarescu (1995), evolution and dispersion of **aquatic fauna** are related to changes in drainage basins, formation and disappearance of lakes, and the sustained presence of biodiversity and endemic fauna over long periods of time in deep ancient lakes. Lake Titicaca in the Andes, Lake Baikal in Siberia, and **Lake Ohrid** in southeastern Europe (Macedonia) are much older that many drainage basins and have remained intact for long periods. Most endemic animals in these lakes can be classified as secondary freshwater species, such as cichlids in African lakes and cyprinodonts in Lake Titicaca. Some fish and invertebrates in these lakes are not related to the fauna in other inland lakes. Brachiate molluscs and amphipods in the latter are primarily descendents of freshwater organisms.

6.5   PRINCIPAL GROUPS OF ORGANISMS IN AQUATIC COMMUNITIES

6.5.1   Viruses

Viruses are generally small agents of organisms (*circa* 0.02 μm) visible only with the use of special techniques and electron microscopes. Their role in aquatic ecosystems is still unclear. However, they can cause diseases such as hepatitis if they survive in natural waters. Recently, the *influenza virus* was detected in aquatic birds in China and it is possible that eutrophic systems contain a virus population that affects human health. In some cases, viruses are present in the decomposition of the cyanobacteria of plankton or periphyton.

6.5.2 Bacteria and fungi

Bacteria and fungi play a significant role in the ecosystem, recycling organic and inorganic matter. They are the intermediaries in a large number of chemical transformations in nature. Heterotrophic bacteria decompose organic matter in rivers and lakes and provide food for **scavengers**. Bacteria play a role in the nitrogen cycle (**nitrification** and **denitrification**) and in the mineralisation of sulphur and carbon. **Chemolithotrophic bacteria** found in sediment or particulate matter are responsible for the oxidation of Fe^{2+} to Fe^{3+}, the oxidation of ammonia to **nitrite** and nitrate, and H_2S to SO_4^{2-} Photosynthetic bacteria (photo-autotrophic or **photo-lithotrophic**) use H_2S and CO_2 as substrates and are found in anoxic and illuminated regions in the metalimnion of permanently stratified (of meromictic) lakesforming layers several centimeters thick at depths with low light intensity (from 0.1–1% of the light intensity reaching the surface) (see Chapter 8).

Matsuyama (1980, 1984, 1985) has extensively studied the vertical distribution of bacteria in meromictic lakes in Japan.

Heterotrophic bacteria that do not harvest underwater solar energy can grow in a wide variety of organic and inorganic substrates such as cellulose, chitin, CO_2, SO_4, N_2, CH_4 and H_2S. These bacteria actively participate in **recycling** organic matter and produce gases and dissolved substances (Abe *et al.*, 2000).

It is relatively difficult to collect and study bacteria in aquatic systems. **Bacterial metabolism** is complex and diverse. Bacterial abundance can be determined with existing techniques; metabolic activities can be studied by growing bacteria in solutions with varying concentrations of nutrients. Also, adding substrates to a water sample taken from a lake can alter the metabolic rate of bacterial populations.

In addition to total bacterial numbers (measured by using special filters, or by growing bacteria on plates in an appropriate medium) identification is desirable, as well as studies on specific activities and vertical distribution of the quantity of bacteria at different depths. Bacteria can be identified by morphology, type of nutrition, and response to specific substrates, which, when added to the samples, help to identify which types promote growth and the products released by bacterial activities (Jones, 1979).

Free-living bacteria that fix nitrogen have been identified in various inland aquatic environments. Such bacteria, found around the roots of aquatic plants or in anaerobic sediments unassociated with plant roots, were described by several authors, including Brezonik and Harper (1969) and Santos (1987) in association with macrophytes. The nitrogen provided by these bacteria can facilitate and accelerate plant growth.

Another group of bacteria, coliforms, can impact the health of other species, since they develop in the digestive tract of animals. High levels of these bacteria are an indication of contamination by **organic detritus** from decomposing animals.

Aquatic fungi participate in the decomposition and recycling of organic matter in aquatic ecosystems. They act as parasites or saprophytes that use organic matter for growth.

Bacteria and fungi are an important source of food for other aquatic organisms, partly since they form a layer of organic material on the surface, which can be used as a food source by scavengers.

Bacteria and fungi play an important role in reducing organic and inorganic pollution in aquatic ecosystems. For example, after bacteria were identified that can remove oils from the surfaces of lakes and rivers, there has been intensive technological development in order to grow and utilize these bacteria. The role of bacteria and fungi in recycling organic and inorganic matter is extremely important in aquatic ecosystems, and these organisms form an essential part of the **trophic network** in any ecosystem. Their concentration and density (total numbers or biomass) depend on the levels and types of organic and inorganic material and the availability of such material in natural or artificial systems (Walker, 1978).

6.5.3   Algae

According to Reynolds (1984), the term 'algae' is generally used to describe **photo-autotrophic organisms** (other than certain bacteria) and has no taxonomic significance.

'Algae' can be part of phytoplankton or may be supported on a substratum. Algae inhabit a wide range of freshwater and marine ecosystems. Algae are important primary producers of organic matter; however, in shallow illuminated regions, immersed or **submerged macrophytes** can be the most important primary producers. Algae are a diverse group, and may be unicellular or colonial or multicellular; many are filamentous. They can reproduce vegetatively or develop special reproductive cells.

The cell walls of algae are composed of silica, protein, fat, cellulose and other polysaccharides, which together produce characteristic and diverse cell walls.

Silica is an important component in the frustules of algae known as diatoms. Through 'core' analysis of sediment, different periods in the ecological history of a lake can be identified, since frustules remain intact after death.

The main pigments of these organisms are chlorophylls and caratenoids. In the photosynthetic process, Chlorophyll-*a* produces **chemical energy**, since it can 'donate' electrons as an effect of the excitation produced by solar energy between 360 and 700 nm. The other pigments – chlorophyll-*b* and -*c*, carotenoids, xanthophylls, phycocyanins, and phycoerythrins – play accessory roles (see Figure 6.4).

Chart 6.1 presents the main groups of algae commonly found in oceans, lakes, rivers, reservoirs, wetlands and planktonic areas, or supported on a substratum. The chart presents examples of the most common genera and species in the Neotropical region and in inland systems in Brazil.

The term phytoplankton was first used in 1867 to describe a diverse, polyphyletic group of photosynthesizing unicellular and colonial organisms found in oceans and in a wide range of inland waters and estuaries. These organisms are estimated to be responsible for more than 45% of the global **primary production**.

Photosynthesis by photo-autotrophic organisms produces the oxygen that contributes to Earth's atmosphere, and fixes the CO_2 in the atmosphere and water, which is considered an important source of the world's sink of carbon. Table 6.1 shows the **phylogenetic distribution** of terrestrial and aquatic photo-autotrophs based on morphological features. Photo-autotrophs are widely diverse in aquatic environments,

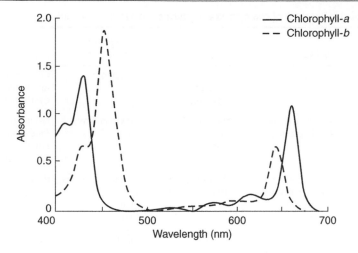

Figure 6.4 Spectrum of chlorophyll-*a* and chlorophyll-*b* in ether solution.
Source: Reynolds (1997) and various sources.

Chart 6.1 Grouping (by class) of common algae in aquatic ecosystems.

Class	Habitat	Morphology	Composition of cellular wall	Examples
Bacillariophyceae	Oceans, lakes, estuaries; planktonic or living on substrata	Unicellular or colonial; microscopic	Silica	*Aulacoseira italica* *Aulacoseira granulata* *Cyclotella meneghiniana* *Navicula rostellata*
Chlorophyceae	Lakes, rivers, estuaries; planktonic or living on substrata	Microscopic or visible, filamentous; colonial, unicellular; some flagellated	Cellulose	*Tetraedron triangulare* *Chlorella vulgaris* *Kirchneriella lunaris* *Selenastrum gracile*
Dinophyceae	Oceans, lakes, estuaries; planktonic	Microscopic, unicellular or colonies; all with flagella	Cellulose and with silica	*Sphaerodinium cinctum* *Durinskia baltica* *Peridinium gatunense* *Dinococcus bicornis*
Cyanophyceae	Lakes and oceans; planktonic or living on substrata	Microscopic or visible; generally filamentous	Mucopeptide – amino sugars – amino acids	*Coelomoron tropicale* *Microcystis wesenbergii* *Sphaerocavum brasiliense* *Anabaena spiroides*
Chrysophyceae	Lakes, rivers, oceans	Microscopic; unicellular or flagellated colonies	Pectin or in algae; silica or cellulose	*Sphaleromantis ochracea* *Rhipidodendron huxleyi* *Dinobryon bavaricum* *Mallomonas kristianienii*

Chart 6.1   Grouping (by class) of common algae in aquatic ecosystems (continued).

Class	Habitat	Morphology	Composition of cellular wall	Examples
Cryptophyceae	Lakes; planktonic	Microscopic; unicellular; flagellates	Cellulose	*Chroomonas nordstedtii* *Rhodomonas lacustris* *Cyathomonas truncata*
Euglenophyceae	Lakes, shallow pools; planktonic	Microscopic; unicellular; flagellates	Protein film	*Gyropaigne brasiliensis* *Rhabdomonas incurva* *Euglena acus* *Phacus curvicauda*
Florideophyceae	Oceans, estuaries, lakes, streams and rivers; living on substratum	Microscopic or visible	Cellulose + gel	*Paralemanea annulata* *Bostrychia moritziana*
Phaeophyceae	Oceans, estuaries; living on substrata or floating	Visible	Cellulose + gel	*Fucus* sp.

Table 6.1   Phylogenic distribution of photo-autotrophs based on morphological characteristics.

		Species in marine waters	*Species in inland waters*
Bacteria	Cyanobacteria	150	1,350
Discicristata	Euglenophyta	30	1,020
Alveolata	Dinophyta	1,800	200
Plantae	Glaucocystophyta	0	13
	Rhodophyta	5,800	120
	Chlorophyceae	100	2,400
	Prasinophyceae	100	20
	Ulvophyceae	1,000	100
	Charophyceae	5	3,395
Cercozoa	Chlorarachniophyta	4	0
Chromista	Cryptophyta	100	100
	Prymnesiophyceae	480	20
	Bacillariophyceae	5,000	5,000
	Chrysophyceae	800	200
	Dictyochophyceae	2	0
	Eustigmatophyceae	6	6
	Phaeophyceae	1,497	3
	Raphidophyceae	10	17
	Synurophyceae	0	250
	Tribophyceae	50	500
	Xanthophyceae	50	500
Lichens	Approximately 13,000 spp. (99.7% terrestrial)		
Embryophyta	272,000 spp. (99% terrestrial)		

Source: Modified from Falkowski *et al.* (2004).

in contrast with terrestrial photo-autotrophs, which are predominantly higher-plant *Embryophyta* species (Figures 6.5 and 6.6).

Within these groups, the *Cyanophyta*, or blue-green algae, are ecologically, evolutionarily and biochemically important. They share a certain affinity with the prokaryotic organization of bacteria cells, which is why these organisms are also called cyanobacteria.

Planktonic algae present a wide variety of shapes and sizes, creating problems for their collection and quantitative study. The surface-to-volume ratios for these algae are also significant in the mechanisms of fluctuation and absorption of nutrients (Munk and Riley, 1952; Reynolds, 1984; Tundisi *et al.*, 1978).

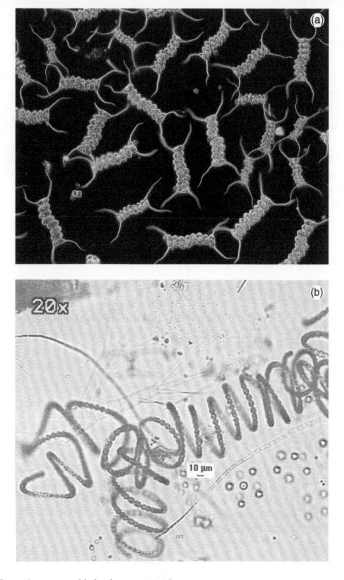

Figure 6.5   a) *Scenedesmus* sp.; b) *Anabaena spiroides*.

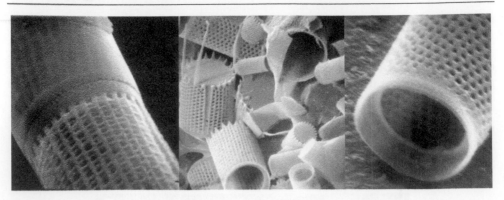

Figure 6.6   Photograph of cell walls of *Aulacoseira italica* (photo taken through electron scanning microscope).

Selective pressures that affect the **process of succession** (and predominance of different species with varying sizes) include **hydrodynamic mechanisms**, vertical circulation, **selective feeding** by herbivores, concentration of nutrients, and effects of wind and precipitation on turbulence, vertical stratification and nutrient distribution (see Chapter 7).

Many planktonic algae can move by means of flagella or by sliding when near a substratum, conferring an obvious advantage in terms of optimizing underwater solar radiation and nutrient concentrations.

Periphytic algae grow on a substratum; common types include Diatomaceae, Cyanophyceae and Chlorophyceae.

6.5.4   Protozoa

Protozoa are found in practically all aquatic systems, and many species are widespread because of the easily dispersed resistant forms. Protozoa feed on detritus, bacteria, and algae, and move with the use of flagella or cilia. Some species of the genus *Stentor* have **stentorin** pigment (quinone radical), and contain live *Chlorella* cells that participate actively in metabolism. Tundisi (1979, unpublished data) observed the presence of high levels of these protozoa in D. Helvecio Lake in Parque Florestal do Rio Doce, causing the formation of a 'red tide' in the lake during the summer of 1978.

Paramecium is one of the most common protozoa, and can be found free-swimming in temporary waters and small ponds. Another fairly common protozoan is *Vorticella*, which is fixed, filters particles and is common in waters with high levels of suspended organic material and detritus.

Protozoa are classified according to their form of locomotion, and so include **flagellates** (example: *Euglena*), **ciliates** (example: *Paramecium*), **amoeboids** (example: *Globigerina*) and **sporozoa** – which are parasites in humans (*Plasmodium*) or fish.

Free-living ciliated protozoa of the genus *Stentor* are found in many lakes and ponds. Due to their extremely efficient forms of resistance, protozoa can also be found in temporary waters in arid and semi-arid regions.

Vertical distribution of protozoa of the family Tracheloceridae in Lake Kaiike was described by Matsuyama (1982), where there is a sharp thermocline with high levels

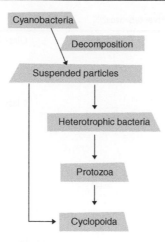

Figure 6.7 Role of Protozoa in recycling organic material.

of bacteria in regions with high H_2S levels. The presence of masses of these ciliates immediately above the bacterial layer suggests feeding relationships between protozoa and bacteria.

Recent data from hypereutrophic lakes in Japan (**Lake Kasumigaura**) show the importance of protozoa in the recycling of organic matter, as illustrated in the process shown in Figure 6.7.

Chart 6.2 lists aquatic invertebrates by major **taxonomic groups**, according to Ismael *et al.* (1999).

6.5.5 Porifera (freshwater sponges)

Freshwater sponges (Porifera) are found in rivers, lakes and reservoirs. The study of Porifera in Brazil was developed and advanced by Vega and Volkmer-Ribeiro (1999). Melao and Rocha (1996) studied *Metania spinata* in **Lake Dourada**, a municipality of Brotas in the state of São Paulo.

According to Volkmer-Ribeiro, in Brazil there are 20 genera of freshwater sponges (one endemic) and 44 species. *Radiospongilla amazonensis* can be found distributed throughout Brazil and probably in Argentina. Sponges are important environmental indicators, inhabiting temporarily flooded areas, deep substrata of rivers in the Amazon region, and even freshwater lagoons or brackish coastal waters of Brazil. The Spongillidae family is represented by 170 species around the planet, of which 27 are found in North America (Lévêque *et al.*, 2005) (see Figure 6.8).

6.5.6 Cnidaria

The Phylum Cnidaria is primarily marine with some freshwater species in the class Hydrozoa. The medusa phase (sexual reproduction) produces the dispersion. There are close to 30 to 45 species of freshwater cnidarians. The *Craspedacusta sowerbii* medusa has colonized all continents except Antarctica.

Chart 6.2   Large groups of aquatic invertebrates.

Phylum Porifera	Class Branchiopoda
Phylum Cnidaria	Order Notostraca
Phylum Platyhelminthes	Order Anostraca
Class Turbellaria	Order Conchostraca
Phylum Nemerta	Order Cladocera
Phylum Gastrotricha	Class Malacostraca
Phylum Nematoda	Subclass Eumalacostraca
Phylum Nematomorpha	Superorder Peracarida
Phylum Rotifera	Order Amphipoda
Phylum Bryozoa	Superorder Syncarida
Phylum Tardigrada	Order Anaspidacea
Phylum Mollusca	Order Bathynellacea
Class Bivalvia	Superorder Pancarida
Class Gastropoda	Order Thermosbaenacea
Phylum Annelida	Superorder Eucarida
Class Polychaeta	Order Decapoda
Class Oligochaeta	Subphylum Uniramia
Class Hirundinea	Class Insecta
Phylum Arthropoda	Subclass Entognatha
Subphylum Chelicerata	Order Collembola
Class Arachnida	Subclass Ectognatha
Subclass Acari	Order Odonata
Order Prostigmata	Order Ephemeroptera
Subphylum Crustacea	Order Plecoptera
Class Copepoda	Order Hemiptera
Order Calanoida	Order Neuroptera
Order Harpacticoida	Order Trichoptera
Order Cyclopoida	Order Lepidoptera
Class Branchiura	Order Diptera
Class Ostracoda	Order Coleoptera

Source: Adapted from Ruppert and Barnes (1996) in Ismael *et al.* (1999).

Figure 6.8   a) *Corvoheteromeyenia heterosclera* (Ezcurra de Drago, 1974). Spongillidae family, common in lakes among the Lençóis Maranhense dunes; b) *Uruguaya corallioides* (Bowerbank, 1863), Potamolepidae family, typical of substrata in large deep rivers, especially in Uruguay. The two species belong to genera that are endemic to South America (Courtesy of Dr. Cecilia Volkmer Ribeiro).

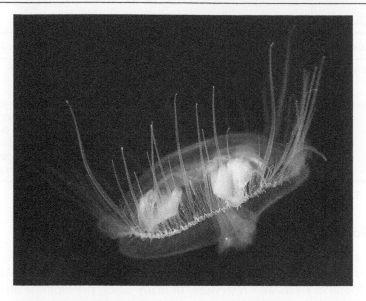

Figure 6.9   Craspedacusta sowerbii (see also color plate section, plate 8).

Limnocnida is another common widespread genus found in Africa (Williams *et al.*, 1991). *Limnocnida tanganicae* is endemic to Lake Tanganyika. *Craspedacusta sowerbii* was identified in two sites in Brazil (new distribution): the reservoir of the **UHE Luis Eduardo Magalhaes dam** (Lajeado) – **Tocantins Rivers** – and a small pond in the Tietê/Jacare basin (state of São Paulo) (Tundisi *et al.*, in press). *Craspedacusta sowerbii* was reported in South America for the first time by Tundisi and Vannucci (1962) (see Figure 6.9).

6.5.7 Platyhelminths

Species of the Platyhelminth Phylum are divided into four classes: Cestoda, Trematoda, Monogenea, and Turbellaria (flatworms). Most species of microturbellaria live in fresh water (400 species). Of the macroturbellaria, 100 species live in fresh water. In Lake Baikal, there are 150 species of free-living flatworms (planarians), 130 of which are endemic.

6.5.8 Rotifers

There are close to 1800 species of rotifers, mostly distributed in inland fresh waters. Rotifers are an important component of zooplankton in lakes and reservoirs with short retention time. They can be attached (sessile). They feed on suspended material, gathered by a crown of cilia that are also used for mobility. Some rotifers are predators. Their structures are called **trophi** and their bodies are protected by a cuticle called **lorica**. Trophi and lorica are used to classify rotifers (see Chart 6.3).

The rotifer fauna in tropical South America and Asia is quite diverse and rich in endemic species. Many different species have been reported in Brazil and in the state of São Paulo (Oliveira Neto and Moreno, 1999).

Chart 6.3 Classification of common rotifers in aquatic ecosystems.

Taxonomic group (class)	Aquatic habitat	Type of food	Examples (genera)
1. Digononta	Fresh waters; planktonic; sessile	Filterers of suspended material	*Phylodina*
2. Monogononta	Fresh water; planktonic; sessile	Filterers of suspended material	*Asplanchna sieboldi* *Brachionus calyciflorus* *Keratella americana* *Keratella cochlearis* *Lecane* spp. *Synchaeta pectinata*

6.5.9 Molluscs

Molluscs include bivalves, lamellibranchs and pulmonates. Many of these organisms feed on detritus, phytobenthons and bacteria. Large molluscs of the Anodontidae family are found in rivers and reservoirs. In many cases, the biomass includes thousands of individuals per square kilometer (see Figure 6.10).

Table 6.2 shows the number of mollusc species in endemic areas (old lakes and some major drainage basins).

Both the Bivalve class and Gastropod class are represented in marine and fresh waters. Some 5,000–6,000 species of bivalves and gastropods have been identified in fresh and marine waters. There is little data on mollusc diversity in South America and Asia.

Among molluscs, species of the *Biomphalaria* genus are medically important because of transmission of schistosomiasis (these molluscs are hosts of the cercariae of *Schistosoma* sp., which occur in the tropics and can cause major **public health problems** in semi-arid regions in Brazil).

6.5.10 Annelids

Annelids include two main classes: Polychaeta and Oligochaeta. Polychaetes are almost exclusively marine with only a few freshwater species. Oligochaetes are well represented in both fresh water and marine environments. There are about 133 Neotropical species. In Lake Baikal, 164 of a total of 194 identified species are endemic.

There are approximately 600 species of oligochaetes in the world. In Brazil, there are nearly 70 species, and 46 species have been reported in the state of São Paulo (Righi, 1984, 1999).

The class Hirudinea includes marine and freshwater fauna. Leeches can prey on macroinvertebrates or are ectoparasites on fish, aquatic birds or mammals.

6.5.11 Decapods

About 10,000 species of **decapods** have been identified in the world, 116 of which are in Brazil. The state of São Paulo has 33 known species.

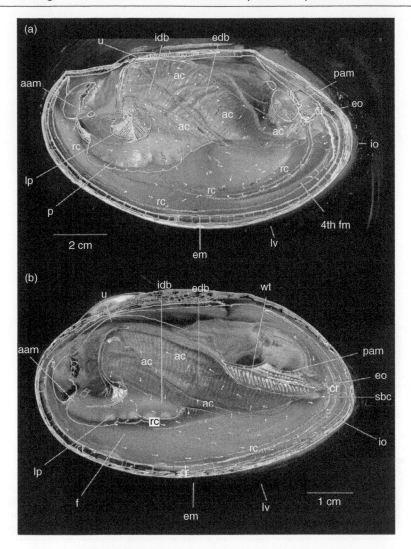

Figure 6.10   a) *Anodontites trapesialis*; b) *A. elongantus*. Schematic representation of the ciliary currents (acceptance and rejection). Abbreviations: eo – exhalant opening, io – inhalant opening, em – edge of mantle, ac – acceptance current, rc – rejection current, sbc – supra branchial chamber, 4th fm – 4th fold of the mantle, edb – external demibranch, idb – internal demibranch, aam – anterior adductor muscle, pam – posterior adductor muscle, f – foot, lp – labial palps, wt – water tubules, lv – left valve, u – umbo (see also color plate section, plate 9).
Sources: a) Lamarck (1819); b) Swaison (1823) in Callil (2003).

Decapods are predominant in tropical and subtropical waters in Central and South America, Europe and Southeast Asia. Crustaceous decapods include economically and ecologically important shrimps and crabs (Branchiura and aeglids). The best-known genera and species in Brazil, especially in the state of São Paulo, are presented in Chart 6.4.

Table 6.2   Number of mollusc species in key areas of endemism: the major,
world's oldest lakes and some of the main drainage basins.

Lakes	Gastropods	Bivalves	Total
Baikal[1]	150 (117)	31 (16)	181 (133)
Biwa	38 (19)	16 (9)	54 (28)
Sulawesi[2]	ca. 50 (ca. 40)	4 (1)	54 (41)
Tanganyika	68 (45)	15 (8)	83 (53)
Malawi	28 (16)	9 (1)	37 (17)
Victoria	28 (13)	18 (9)	46 (22)
Ohrid	72 (55)		
Titicaca	24 (15)		
Drainage basins			
Mobile Bay basin	118 (110)	74 (40)	192 (150)
Lower Uruguay River and Plata River	54 (26)	39 (8)	93 (34)
Mekong River	121 (111)	39 (5)	160 (116)
Lower Congo basin	96 (24)		
Lower Zaire basin	96 (24)		

The numbers of endemic species are in parentheses
[1]Timoshkin (1997)
[2]Lake Poso and Malili lake system
[3]Davis (1982)
Source: Lévêque et al. (2005).

Chart 6.4   Most common decapods in inland aquatic ecosystems in Brazil.

Taxonomic group	Aquatic habitat	Type of food	Examples
Decapods (freshwater crabs, shrimps, and freshwater crayfish)	Fresh water; marine; benthic	Raptorial	Aegla franca Atya scabra Potimirim glabra Procambarus clarkii Macrobrachium brasiliense Palaemon pandaliformis Goyazana castelnaui Trichodactylus fluviatilis

Freshwater crayfish belong to the Parastacidae family. Freshwater crabs belong to the Aeglidae family, whose genus *Aegla* includes 35 species reported in Brazil (Bond-Buckup and Buckup, 1984) (see Figure 6.11).

6.5.12   Crustaceans

Crustaceans are benthic or planktonic organisms and play an important role in the structure and function of lakes, rivers, reservoirs (bodies of fresh water in general), estuaries and marine waters. All crustaceans have a chitinous exoskeleton, which in some cases may also be enriched with calcium carbonate. Crustaceans are classified based on the exoskeleton's features and design, as well as number of segments and appendages. Chart 6.5 presents the taxonomic groups of crustaceans and the most

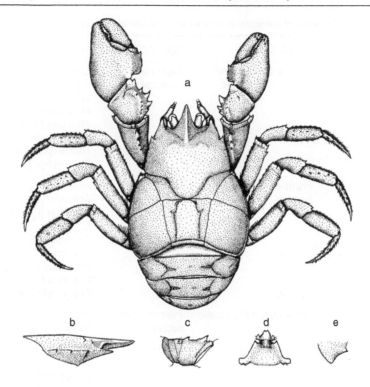

Figure 6.11   *Aegla parva:* a) dorsal view, b) front portion of the carapace (lateral view), c) ischio-base of cheliped (side view), d) third and fourth thoracic sternites (front view), e) epimer 2 (side view).
Source: Melo (2003).

Chart 6.5   Principal groups of common crustaceans in aquatic ecosystems.

Taxonomic group	Aquatic habitat	Type of food	Examples in Brazil
1. Cladocera	Fresh water, planktonic; benthonic or sessile on structure; marine	Predators, filterers	*Bosmina tubicen* *Ceriodaphnia cornuta* *Ceriodaphnia silvestrii* *Daphnia gessneri* *Diaphanosoma spinulosum* *Diaphanosoma brevireme* *Moina minuta* *Sida crystalina*
2. Copepoda; Calanoid	Fresh water; marine, planktonic	Filterers	*Argyrodiaptomus azevedoi* *Argyrodiaptomus furcatus* *Notodiaptomus cearensis* *Notodiaptomus conifer* *Notodiaptomus iheringi* *Notodiaptomus transitans* *Odontodiaptomus paulistanus* *Scolodiaptomus corderoi*

Chart 6.5 Principal groups of common crustaceans in aquatic ecosystems (continued).

Taxonomic group	Aquatic habitat	Type of food	Examples in Brazil
3. Copepoda Cyclopoid	Fresh water; marine planktonic; benthic	Raptorial predators	*Cryptocyclops brevifurca* *Eucyclops encifer* *Ectocyclops rubescens* *Mesocyclops brasilianus* *Microcyclops anceps* *Thermocyclops decipiens* *Thermocyclops minutus* *Tropocyclops prasinus*
4. Harpacticoida	Fresh water; marine; benthic	Filterers and parasites	*Attheyella jureiae* *Attheyella (Canthosella) vera* *Attheyella (Chappuisiella) fuhmanni* *Attheyella (Delachauxiella) broiensis* *Elaphoidella bidens* *Elaphoidella lacinata* *Elaphoidella deitersi*
5. Mysidacea	Fresh water; planktonic	Predators and detritivores	*Brasilomysis castroi* *Mysidopsis tortonesi*
6. Amphipoda	Fresh water; marine benthic	Raptorial	*Ampithoe ramondi* *Cymadusa filosa* *Corophiidae acherusicum* *Hyalella caeca* *Leucothoe spinicarpa* *Sunampithoe pelagica*

Source: Ismael *et al.* (1999); Lévêque *et al.* (2005).

Table 6.3 Global diversity of freshwater crustaceans.

Group	North America	Europe	Asia	Australia	South America	Africa	World
Branchiopoda							
Cladocera	140[1]						500[3]
Phyllopoda	67	72					420
Ostracoda	420[2]	400[5]			500[6]	500[7]	2,000
Copepoda[4]	363	902	927	181	516	524	2,085
Branchiura	3						4,200
Malacostraca							

[1]Pennak (1989); [2]Thorp and Covich (1991); [3]Dumont and Negrea (2001); [4]Dussart and Defaye (2002); [5]Giller and Malmqvist (1998); [6]Martens (1984); [7]Martens and Behen (1994).
Source: Lévêque *et al.* (2005).

common species occuring in Neotropical fresh waters. Table 6.3 presents the overall diversity of freshwater crustaceans.

Figure 6.12 shows the calanoid copepods found in the reservoir of the Carlos Botelho (Lobo/Broa) hydroelectric dam.

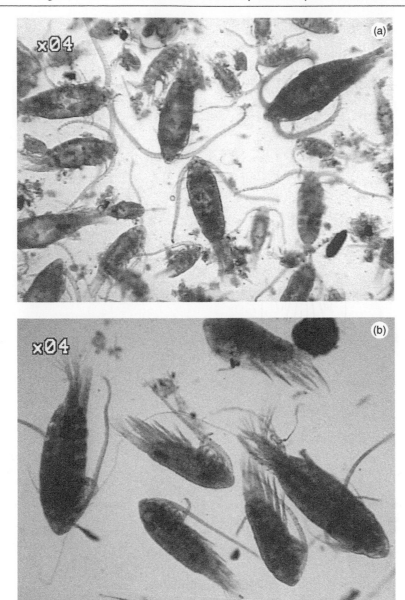

Figure 6.12   Calanoid copepods in UHE Carlos Botlho (Lobo/Broa) reservoir. **Photo:** Thais J. Prado.

Figure 6.13 shows the classical distinction between calanoids, cyclopoids and harpacticoids, which are the main components of copepods in both inland waters and marine waters.

Figure 6.14 presents some zooplankton species commonly found in the Neotropics.

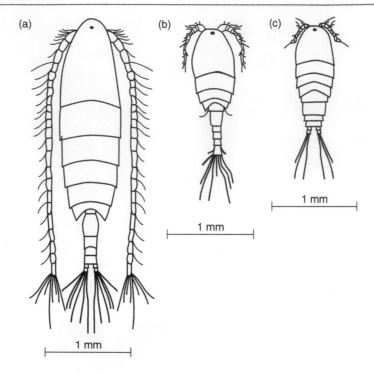

Figure 6.13   a) Calanoid; b) Cyclopoid; c) Harpacticoid.

Cladocerans, copepods, rotifers and protozoa are the principal components of zooplankton in freshwater ecosystems. Freshwater shrimps and crabs are found in rivers and the littoral zone of lakes and estuaries.

These organisms may be strainers (calanoids, for example) or may feed raptorially by seizing (cyclopoids, for example).

6.5.13   Aquatic insects

In rivers, lakes, reservoirs, wetlands and surrounding areas, aquatic insects and their larvae are found in abundance. Many larvae are able to survive in fast-moving currents, with special morphological features to resist moving water. The adult forms of most of these aquatic insects are air-borne and short-lived. The insect larvae can be predators. Adults of the Belostomaticea(aquatic cockroach) family are voracious predators that can capture small fish. Chironomid larvae are important components of the **zoobenthos** (see Chart 6.6).

In some tropical regions, **blackfly larvae** are important and can transmit onchocerciasis. *Chaoborus* **larvae** are the only examples of larvae of aquatic insects among plankton. The majority of insect larvae are found in the benthos.

Table 6.4 presents an estimate of the number of aquatic insects by continent and major biogeographical area.

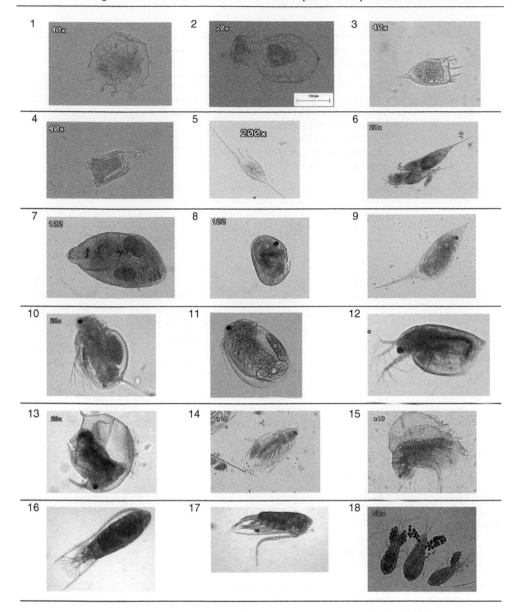

Figure 6.14 **Common planktonic organisms**. Rotifers: *1–6 (1 – Brachionus dolabratus; 2 – Asplanchna sieboldi; 3 – Keratella cochlearis; 4 – Polyarthra vulgaris; 5 – Kellicotia bostoniensis; 6 – Trichocerca cylindrica chattoni); 7 – Turbellaria; 8 – Ostracoda; Cladocerans: 9–15 (9 – Daphnia gessneri; 10 – Moina minuta; 11 – Ceriodaphnia cornuta; 12 – Simocephalus sp.; 13 – Bosmina hagmanni; 14 – Diaphanosoma birge; 15 – Holopedium amazonicum); Copepods: 16–18 (16 – Notodiaptomus iheringi, female; 17 – Notodiaptomus iheringi, female; 18 – Three genera of Cyclopoida: Acanthocyclops, Mesocyclops and Thermocyclops)* (see also color plate section, plate 10).

Chart 6.6   Orders of aquatic insects.

Order	Habitat and diet – geographical distribution	Examples in South America and in Brazil
Coleoptera Holometabolic insects	10% of families are aquatic or semi-aquatic. Aquatic beetles inhabit fresh water, brackish water, and seawater. Most common habitats are areas with abundant **aquatic vegetation**. Predators, filter-feeders or scrapers. Cosmopolitan.	More than 2,000 species identified in South America, distributed in 20 families.
Diptera holometabolic insects	Occur in a wide variety of aquatic systems. At least 30 families have freshwater species. The larvae feed on detritus, can collect particles, or are predators.	Number of medically important species: Tipulidae: 14,000 species; Culicidae: 3,450 species; Anophelidae: 420 species; Simulidae: 1570 species; Ceratopogomidae: 5,000 species; Tabanidae: 4,000 species. Broad geographical distribution, especially in the humid tropics. Neotropical Chironomidae were described in 155 genera and 709 species. 168 species in Brazil, distributed in 32 genera. Pseudochironomus, Malloch (1915). *G. neopictus*, and Trivinho-Strixino and Strixino (1998).
Ephemeroptera hemimetabolic insects; nymphs are aquatic and adults are terrestrial.	The most common habitats are rivers in tropical or subtropical regions. Close to 2,100 species described. Faunas of Ephemeroptera of Australia, New Zealand and South America are similar. The larvae are surface scrapers.	150 valid species in Brazil. *Caenis cuniana*, Froelich (University City, Boraceia Biological Station; Guarapiranga reservoir).
Hemiptera – Heteroptera hemimetablic insects	Semi-aquatic species associated with the interstitial zone or shorelines of lakes with vegetation. Nearly 3,300 species. Wide geographic distribution. Can be predators or filterers of water and detritus.	16 families, 81 genera and 900 species in tropical South America.
Odonata Hemimetabolic insects with aquatic nymphs and terrestrial adults	Some marine species; species of brackish lakes and ponds in semi-arid regions, where the larvae develop in temporary waters. The larvae are predators. Wide geographic distribution, from the tropics to the poles. 5,500 species identified.	
Plecoptera Hemimetabolic insects	The nymphs are aquatic and develop in cold waters, currents, temperature below 25°C, with high content of dissolved oxygen. 2,000 species identified. Many families distributed in temperate zones in the North and South Hemispheres. Larvae are predators.	2,000 species on the Planet, 110 in Brazil and 40 species known in the State of São Paulo.

Chart 6.6   Orders of aquatic insects (continued).

Order	Habitat and diet – geographical distribution	Examples in South America and in Brazil
Trichoptera Holometabolic insects	Larvae and pupae live in temporary waters, cold and warm waters. Wide geographic distribution, except Antartica. The diversity is higher in waters in tropical regions. Some species have predatory larvae. Generally scrapers of surfaces or filterers of suspended material. Larvae live in currents with good water quality, and are good indicators.	There are close to 9,600 species in the world, 330 in Brazil and 45 in the state of São Paulo. Examples: *Simicridea albosigmata*, Ulmer (1907); *S. boraceia*, Flint (1998); *S. froelichi*, Flint (1998), Boracéia Biological Station.
Megaloptera Holometabolic insects	Aquatic larvae in all species; close to 250 to 350 species described. Geographical distribution principally limited to temperate regions. Few species in the tropics.	Number of species in the world: 300. In Brazil: 16.
Lepidoptera Holometabolic insects	Almost all aquatic species are associated with plants.	Little information in Brazil. A descriptive work (Messinian and Dasiluas, 1994).
Neuroptera Holometabolic insects	Most are terrestrial species. There are three families with freshwater species whose larvae are semi-aquatic. There are close to 4,300 species, 100 aquatic. The *Sysyridae* family with 45 species is associated with aquatic sponges of the *Spongillidae* family. Geographic distribution mainly in Australia, with few species in North America and Europe.	A species of the Sysiridae family occurs in Brazil.
Hymenoptera Holometabolic insects	Few aquatic or semi-aquatic species. Many families of parasitic wasps are associated with waters where their hosts occur, especially in the aquatic phase of Collembota, Ephemeroptera and Plecoptera. The hosts live in lotic or lentic waters. There are about 100 species of parasitic wasps.	Little information in Brazil.

Sources: Lévêque et al. (2005); Froehlich (1999); Hubbard and Pescador (1999); Ismael et al; Strixino and Strixino (1999).

6.5.14   Fish

Fish are part of the nektonic (free-swimming) community and are economically, evolutionarily and ecologically important. The interaction of fish with the aquatic ecosystem and biota occurs through inter-relationships in the feeding web and the effects on

Table 6.4   Estimated number of species of aquatic insects for all the continents and large biogeographical areas (adapted and completed by Hutchinson, 1993).

	Afrotropical	Neartic	Eastern Palearctic	Europe	Neotropic	Oriental	Australian	World
Ephemeroptera	295[1]	670[6]		350[6]	170		84[4]	>3,000
Odonata	699[1]	>650[5]		150[7]	800		302[4]	5,500
Plecoptera	49[1]	578[4]		4,234[7]	ND		196[4]	2,000
Megaloptera	8[1]	434		64	63[3]		26[4]	300
Trichoptera	>1,000[1]	1,524[1]	1,228[1]	1,724[1]	2,196[2]	3,522[1]	1,116[1]	>10,000
Hemiptera		404[4]		129[4]	900		236[4]	3,300
Coleoptera		1,655[4]		1,077[4]	2,000		730[4]	>6,000
Diptera		5,547[4]		4,050[4]	709		1,300[4]	>20,000
Orthoptera		Ca 20		0	1			ca 20
Neuroptera		6[4]		94	1		58[4]	ca 100
Lepidoptera		782[8]		54				ca 1,000
Hymenoptera		55[4]		744				>129
Collembola	ND	50	ND	30	ND	ND		

[1]Elouard and Gibon (2001); [2]Flint *et al.* (1999); [3]Contreras-Ramos (1999); [4]Hutchinson (1993); [5]Ward (1992); [6]Resh (2003); [7]European limnofauna (2003); [8]Lange (1996)
Sources: Lévêque *et al.* (2005); Spies and Reiss (1996).

the water's chemical composition (respiration and excretion) and sediment (removal of other organisms, disturbance of sediment). Fish also transport organic material vertically and horizontally, because of their great capacity for movement (including dislocation), and, in some cases, extensive migrations occur between marine and freshwater environments. Many fish species have complex osmotic regulation, which facilitates their ability to colonize inland waters with varying saline concentrations, and migration between inland and marine waters (oceans and estuaries).

In almost all aquatic systems, fish are ecologically and economically important for many human populations, due to intensive fishing and, more recently, inland or marine aquaculture.

Table 6.5 and Table 6.6 present the orders of freshwater fish and the distribution of different fish species on continents, respectively.

Lévêque *et al.* (2005) studied the physiology of fish that migrate between freshwater and marine systems (mullet, for example), of which there are about 500 species. These fish are called **diadromous**. Those that reproduce in fresh waters and live in marine waters, such as salmon, are called **anadromous**. Those that breed in marine waters and live in fresh waters, such as the eel family Anguillidae, are called **catadromous**.

The challenges of dispersion and tolerance to the conditions of desiccation constitute important processes in the distribution of fish in inland waters. The quantity of species is related to the historical processes of dispersion and connection of inland waters of drainage basins.

The Neotropical region, which includes most of Central and South America, has the most diverse fish fauna in the world. The Amazon basin has about 1,300 species described in the Zoological Records (Roberts, 1972). Goulding (1980), however, cites between 2,500–3,000 species of fish in the Amazon Basin, mostly characids and siluroids.

Table 6.5 Orders of fish found in inland waters.

Class	Subdivision	Superorder	Order	Total No. of familes	No. of Species	No. of freshwater species	%
Petromyzonida			Petromyzoniformes	3	38	29	76.3
Chondrichthyes			Charcharhiniformes	8	224	1	0.45
Actinopterygii			Acipenciformes	2	27	14	51.8
			Lepisosteiformes	1			85.7
			Amiiformes	1			100
	Osteoglossomorpha		Osteoglossiformes	4	7	6	83.3
	Elopomorpha		Anguiliformes	15	1	1	83.3
	Ostarioclupeomorpha	Clupemorpha	Clupeiformes	5	218	218	100
		Ostariophysi	Gonoryncliformes	4	37	31	0.75
			Cypriniformes	6	3,268	3,268	21.7
			Characiformes	18	1,674	1,674	83.8
			Siluriformes	35	2,867	2,740	100
			Gymnotiformes	5	134	134	100
		Protacantholpterygll	Salmoniformes	1	66	45	68.2
		Paracanthopterygll	Percopsiformes	3	9	9	100
			Gadiformes	9	555	1	95.5
			Ofidiiformes	5	385	5	100
			Batrachoidiformes	1	78	6	68.2
		Acanthopterygii	Atheriniformes	6	2,312	210	100
			Mugiliformes	1	72	1	1.4
			Gasterosteiformes	11	278	21	7.55
			Synbranchiformes	3	99	96	97
			Scorpaeniformes	26	1,477	60	4
			Perciformes	160	10,033	2,040	20.3
			Pleuronectiformes	14	678	10	1.47
			Tetradontiformes	9	357	14	4
Sarcopterygii	Ceratodontimorpha		Ceratodotiformes	3	6	6	100

Source: Nelson (2006).

Table 6.6   Distribution of the diversity of fish species at the continental level.

Zone	Number of fish species	Fao database
Europe and ex-USSR	360[1]	393 + 448
Africa	3,000[2]	3,042
North America	1,050[1]	1,542
South America	5,000+[1]	3,731
Asia	3,500+[3]	3,443
Australia (New Guinea)	500[1]	
Australasia		616
Total	13,400	13,215

[1]Lundberg et al. (2000); [2]Lévêque (1997); [3]Kottelat and Whitten (1996).

The geographical distribution of fish families, genera and species are related to **geomorphological challenges,** involving connection and isolation of inland waters, the historical basis of the persistence of species, and the relationship between dispersal, extinction and speciation. More recently, the impact of human activities has had a growing impact (fishing, transport of invasive and exotic species, destruction of habitats, and pollution).

Figure 6.15a presents the relative composition of species of freshwater fish from South America, Southeast Asia, and rivers and lakes in Africa. Figure 6.15b presents examples of adaptive radiation in cichlids in **Lake Malawi.**

According to their way of life, fish may be classified as pelagic, living in open waters; or **demersal,** living on or close to the bottom. Each species' shape, body structure, physiology, evolutionary status, feeding ecology and behaviour are all interrelated.

The families of marine fish are obviously more common in oceans than in inland waters. Nelson (1994) presented an estimated total of 28,500 species of fish. Nearly 10,000 species live in inland waters, of which 500 are diadromous. There are many highly endemic regions of fish in inland waters, for example, species of cichlids in large African lakes (Victoria, Malawi, Tanganyika). In Lake Titicaca, 24 species of *Orestias* (Cyprinodontidae) have been identified (Lanzanne, 1982). The endemicity of fish species in rivers is less known.

Most species of fish in inland waters are represented by cyprinids or Cypriniformes (carp), characins (Characiformes) and **catfish** (Siluriformes) from numerous families, according to Lowe-McConnell (1999). Fish species that have evolved in fresh water from marine groups include cichlids (very important in the large African lakes), which can tolerate and survive low-salinity brackish waters.

African and South American fauna include caracoides and cichlids; lungfish and primitive groups are also found there. In South America, the Cypriniformes (cyprinids) are totally absent. In Asia, there are very few cichlids and no characiform fish.

There are an estimated 3000 fish species in Africa (Lévêque, 1997). Estimates of the Neotropical **ichthyofauna** (Central and South America) include 3500–5000 fish species. In tropical Asia, there are an estimated 300 different species.

Annexes 1–3 (pages xxx-xxx) present the composition and descriptions of the fish species found in three major South American drainage basins: the Amazon River basin, **São Francisco River basin** (entirely in Brazilian territory) and the Paraná River

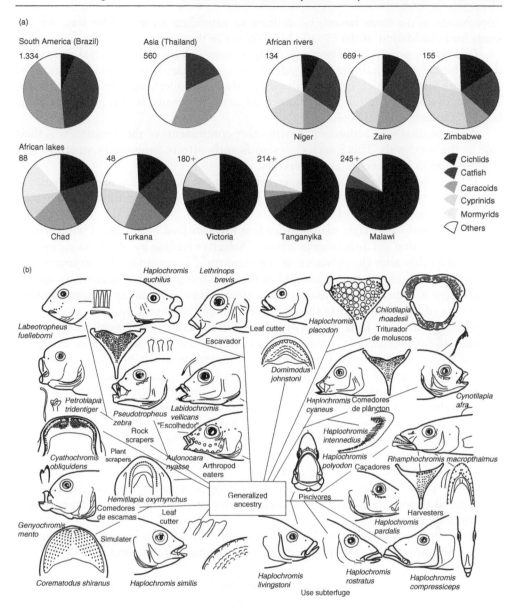

Figure 6.15 a) The relative composition of freshwater fish from South America (Brazil), Southeast Asia (Thailand), and rivers and lakes in Africa, based on the indicated number of species; b) Examples of adaptive radiation in cichlids of Lake Malawi.
Source: Modified from Lowe-McConnell (1999).

basin. The diversity of fish species in these basins, their evolutionary and behavioural characteristics, their physiology and feeding habits are related to the dynamics of each of these basins, i.e., circulation, period of flooding, interactions with vegetation, presence of marginal lakes, competition, **predation** and parasitism. These examples show a widely distributed and diverse fauna with enormous ecological and economic importance in the three basins, in addition to providing an important base for the study and elucidation of the aquatic biodiversity in these drainage basins.

It is worth noting that the genus **Tilapia** (now subdivided in several genera) of the Cichlidae family, abundant in Africa, was introduced to South America. The Mugilidae family (which includes *Mugilcephalus*, or mullet), important in estuaries, feeds on plankton. The Clupeidae family, common in oceans, is also present in some lakes, including tropical lakes.

An important interaction of fish with other components of the community is their role in the food web. This interaction significantly affects community structure and composition in lakes, rivers and reservoirs.

Pelagic fish can be planktophages (plankton-eaters) or piscivores (fish-eaters). Fish species in coastal areas and the deep part of lakes feed on **benthic invertebrates**, detritus, or higher aquatic plants.

The majority of pelagic plankton-feeding fish can feed by filtration with bristles and extremely thin gill-attachments (rakers). Selective feeding by plankton-eating fish can considerably alter the structure of the planktonic community. The digestive system for cyanobacteria requires low pH in the digestive tract and, since many species of fish have a higher pH in their digestive tract, digestive capacity remains limited. In the case of tilapia, the pH of the digestive tract can drop to 1.4.

Most **benthic fish** feed on detritus. In South America, the Prochilodontiae and Curimatiae families include important stocks of fish that, in some regions, comprise 50% of the total fish biomass (ichthyomass). Adaptations of detritivores include alterations in the digestive tract to eliminate suspended inorganic particles, an anterior stomach, a reservoir of ingested material, and an extremely muscular posterior stomach with the ability to knead together food and ingested sand.

Fruit-eating and seed-eating fish, in the Characiformes of the genera *Colossoma* and *Brycon*, are unusual, and only occur in the Amazon region.

Adult tambaquis feed on fruit in flooded forested areas, while the young are found in swampy areas and feed on zooplankton.

Neotropical rivers provide a varied diet for detritus-eating fish. Detritus, as it is transported downstream, typically changes in size and can be utilized compartmentally by different species.

Predators such as piranhas (genus *Serrasalmo*) and peacock bass (*Cicchla occelaris*), when introduced, can produce extensive changes in the food web (Zaret and Paine, 1973).

The importance of fish as food for humans has led to many studies on commercial species and the development of intensive fishing, fish farming activities and aquaculture in many lakes and aquaculture stations, in developed and developing countries alike.

It is worth noting that studies on different fish species require a broad limnological base in order to determine **tolerance limits**, growth rates due to physical and chemical parameters, and inter-relationships with other species.

Problems arising from the introduction of exotic species and their effects on selective predation and transport of other organisms should also be mentioned as important in the basic study of fish as components in the community.

Artificial reservoirs considerably alter the natural conditions of rivers and influence composition of the fish fauna.

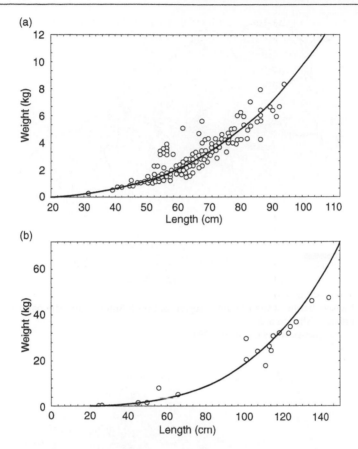

Figure 6.16 Weight/length ratio of catfish (a) surubim (*Pseudoplatystoma fasciatum*), and (b) jaú (*Paulicea lutremi*).
Source: Barthem and Goulding (1997).

Fish of the Amazon basin

The large predatory catfish in the Amazon basin have ecological and commercial importance. Studies of these species of catfish were conducted by Bayley (1981), Bayley and Petrere (1989), and in even further depth by Barthem and Goulding (1997), who produced an important monograph on the ecology, migration and conservation of these fish.

Figure 6.16 shows the weight-length ratios of two important species: the surubim and jau.

In addition to these large catfish from the Amazon basin, other species have enormous ecological and commercial importance: tambaqui (family Characidae, sub-family Serrasalminae, species *Colossoma macropomum*). Araujo-Lima and Goulding (1998) studied this species and produced an excellent and informative monograph describing its distribution, feeding habits, migration, reproduction, nutrition, fishing and intensive fish farming, as well as its evolutionary biology.

The tambaqui is found in the Solimões/Amazon and Orinoco River basins (see Figure 6.17). According to Araujo-Lima and Goulding (1998), the Solimões Rivers

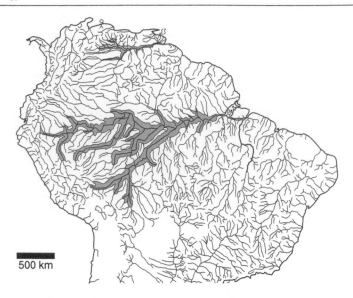

Figure 6.17   Drainage basins of the northern region of South America and the natural distribution of
tambaqui (shaded areas).
Source: Araujo-Lima and Goulding (1998).

Figure 6.18   Tambaqui (*Colossoma macropomus*) (see also color plate section, plate 11).

(Amazon and Madeira) are the main axes of distribution of this species. According
to these authors, the tambaqui depends on muddy water for its survival. However,
it can be found in **blackwater** rivers, and, in this case, along no more than 200–300 km
of muddy rivers.

Young tambaqui are found near the shores of the Solimores/Amazon River, where
there are many lakes. These lakes and floodplains are the main habitat for the species'
reproduction and growth. The **egg-laying** period lasts from two to five months. Adult
tambaqui feed on fruits and seeds. Tambaqui larvae and tambaqui young feed on microcrustaceans, chironomids, insects and grass (see Figure 6.18) (also see Chapter 8).

Fishes of the São Francisco

The São Francisco River basin (631,133 square kilometers) runs through the states of Minas Gerais, Goiás, Federal District, Bahia, Sergipe, Alagoas and Pernambuco. The São Francisco River travels 2,700 kilometers through Brazilian territory and empties into the **Atlantic Ocean** between the states of Sergipe and Alagoas (Sato and Godinho, 1999). Climatic conditions of the basin vary greatly. Rainfall can range from 350 to 1,900 mm. The annual average discharge of the São Francisco River is $3,150\,m^3 \cdot s^{-1}$. There are 36 tributaries and 11 hydroelectric dams along the river (Codevast, 1991).

Britski *et al.* (1984) compiled a list of 133 fish species in the Sao Francisco River basin. The list of species (presented by Sato and Godinho, 1999) shows great diversity, including *Prochilodus margrave* (pacu), *Salminus brasiliensis* (golden dorado), *Schizodon knerii* (white-piau) and *Lophiosilurus alexandri* (pacman catfish). Introduced species include the peacock bass (*Cichla ocellaris*), corvina (*Plagioscion squamosissimus*) and several species of carp, tilapia, tambaqui and African catfish.

Annex 1 (page xxx) lists the fish species of the São Francisco River.

Since 1986, the ichthyofauna of the Paraná River has been studied intensively between the mouth of the Paranapanema River and the Iguaçu River (including the Itaipu reservoir) by the Maringa State University's Limnology, Ichthyology and Aquaculture Center (*Nupelia* in Portuguese) (Agostinho and Julio Ferreira, 1999). Vazzoler and Menezes (1992) classified the species of the Upper Paraná based on **reproductive strategies**. Annex 3 (page xxx) lists the species of fish found in the Paraná River.

6.5.15   Amphibians, reptiles, birds and mammals

Amphibians, reptiles, birds and mammals play an extremely important role in inland aquatic systems, especially shallow lakes, wetlands and estuaries.

From their early stages, amphibians use water; in many species, tadpoles inhabit the shallow waters of rivers and lakes. Amphibians are also important in the food web near the interfaces between terrestrial and aquatic systems.

Reptiles play an important ecological role, especially in the waters of tropical lakes and large internal deltas. Turtles, crocodiles, alligators and some species of snakes inhabit shallow waters and are important predators, with a relevant role in control of fish fauna, small mammals, and birds in wetland areas. There are nearly 200 species of freshwater turtles, mainly in tropical and subtropical regions.

There are 23 species of crocodiles in tropical and subtropical regions. Species of the Alligatoridae family can be found in North America, Central America and South America. Most species of the Crocodylidae family are found in Africa, India and Asia.

The *Cayman crocodilos* inhabits the Pantanal of MatoGrosso in large numbers in swampy areas, shallow lakes and rivers of the Pantanal.

Two of the truly aquatic snake species are adapted to freshwater environments (of the Acrochordidae family). *Eunectes murinus* (anaconda) is a common species in flooded tropical regions and semi-aquatic areas of South America.

In addition to their effects on the food web, these animals play an extremely significant role in recycling nutrients due to excretion. Excretion of ammonia by *Caiman latirostris* was identified by Fitkau *et al.* (1975) in an experiment conducted in the Amazon. Capybaras and nutria (coypu) are large rodents living in aquatic systems in Neotropical regions. They play an important role in recycling nutrients, due to excretion and the permanent removal and alteration of sediment (see Figure 6.19).

Many mammals live in the areas surrounding inland aquatic environments. However, some species live directly in the water, including beavers (better known in South America as capybara – *Hydrochaeris hydrochaeris*), otters, hippopotami (*Hippopotamus amphibious*, only in Africa), buffaloes (*Bubalus bubalis*), as well as species of cetaceans (in Asia), three species of manatee (in South America – genus *Trichechus*) and species of freshwater seals in Europe and Siberia (see Figure 6.20).

Figure 6.19   Capybara (*Hydrochoerus hydrochaeris*), the biggest rodent in the world: 1 m long, 50 cm tall and weighing 60 kg (see also color plate section, plate 12).

Figure 6.20   Amazonian manatee (*Trichechus inunguis*) (see also color plate section, plate 13).

Figure 6.21   White-faced Whistling Duck (*Dendrocygna viduata*) (see also color plate section, plate 14).

Horne and Viner (1971) demonstrated the significance of the hippopotamus' role in Lake George and other African lakes. These animals recycle 30% of the nitrogen in these lakes, through excretion and removal of vegetation that they eat. In South America, the manatee plays an important role in the removal of vegetation and recycling of nutrients.

6.5.16   Aquatic birds

In shallow lakes and wetlands, aquatic birds play a special role, with quantitative and qualitative effects on the food web and nutrient recycling.

These birds can feed on plankton (for example, in **Lake Nakuru** in Africa, where the lesser flamingo *Phoenicopterus minor* feeds on *Spirulina* spp. and microphytobenthos) or fish, significantly altering the biomass of many species. In addition to their role in the food web, aquatic birds also transport organisms over long-distance migrations, and are most likely essential in the colonization of invasive species in many lakes, wetlands and reservoirs around the planet. Aquatic birds transport algae, bacteria, fish eggs and parasites. Recent evidence shows that they played a significant role in transporting the avian flu virus, which may have developed in eutrophic lakes of China (Hahn, 2006).

Approximately 156 species of birds in the Pantanal live in or depend on wetland areas, and 32 species feed on fish (Cintra and Yamashita, 1990) (see Figure 6.21).

Figure 6.22 illustrates some of the interactions between aquatic organisms, amphibians and mammals in their roles of recycling nutrients.

It is clear that all aquatic organisms, from viruses to mammals, play a significant role – quantitatively and qualitatively – in the functioning of aquatic inland water systems. The study of the dynamics of the processes in which they are involved and their biotic and abiotic interactions are essential for understanding evolutionary processes and the capacity to recycle organic matter.

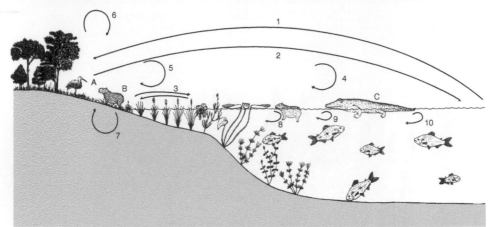

1 – Transport from the aquatic system to land; 2 – Transport from the terrestrial system to the aquatic and influx of nutrients into lake (excretion); 3 – As above; 4,5,8,9 and 10 – Contribution into recycling of nutrients in lake (excretion, waste); 6 and 7 – Contribution into recycling of nutrients in the land system.

Figure 6.22   Interactions of aquatic vertebrates with terrestrial system and aquatic system.

Table 6.7 presents the richness of animal species in inland waters and the number of species in each class/order.

6.6   SPATIAL ORGANIZATION OF AQUATIC COMMUNITIES

Aquatic communities are located in different regions and substrata, in open water or supported by different structures. Each community varies according to its position, which requires, of course, specialized systems of buoyancy or ageing on substrates with different characteristics, roughness and substances of which they are made. This compartmentalization implies a differentiated use of resources (underwater solar radiation, **inorganic nutrients,** and available CO_2 and O_2) that enable colonization and development of different populations and communities. The degree of compartmentalization depends on the volume of the lake, river or reservoir, current velocity, diversity in the bodies of water, and the influences of abiotic factors, such as water temperature and ionic levels (Tundisi *et al.*, 1998).

The totality of aquatic plants and animals appears, as described by Margalef (1983), to be a mosaic or complex of distinct associations.

The '**functional distinction**' of different communities has been deciphered over the course of the hundred-year history of limnology and is one of the products of 'descriptive limnology' and the natural history of aquatic organisms. Despite its limited value – from the point of view of dynamic knowledge of the system, spatial and temporal distribution of different components in the associations – it is possible to verify, along with physical and chemical information, the influence of organisms in aquatic systems and the factors involved in their spatial position. Over the last 20 years, the study of these associations has grown because of their value as indicators of physical and chemical conditions; and also because of the need to better understand the processes of spatial and temporal succession as influenced by the phenomena of

Table 6.7 Richness of animal species in inland waters.

Phyla	Class/order	Number of species
Porifera		197
Cnidaria		30
	Hydrozoa	ca. 20
Nemertea		12
Platyhelminthes		ca. 500
Gastrotricha		ca. 250
Rotifera		1,817
Nematoda		3,000
Annelida	Polychaeta	?
	Oligochaeta	700
	Hirudinae	ca. 300
		70–75
Bryozoa		
Tardigrada	Bivalva	ca. 1,000
Mollusca	Gastropod	ca. 4,000
Arthropoda		
Crustacea		
Branchiopoda	Cladocera	>400
	Anostraca	273
	Notostraca	9
	Conchostraca	130
	Haplopoda	1
Ostracoda		3,000
Copepoda		2,085
Malacostraca		
	Mysidacea	43
	Cumacea	20
	Tamaidacea	2
	Isopoda	ca. 700
	Amphipoda	1,700
	Decapoda	1,700
Arachnida		5,000
Entognatha	Collembola	
Insecta		
	Ephmeroptera	>3,000
	Odonata	5,500
	Plecoptera	2,000
	Megaloptera	300
	Trichoptera	>10,000
	Hemiptera	3,300
	Coleoptera	>6,000
	Diptera	>20,000
	Orthoptera	ca. 20
	Neuroptera	ca. 100
	Lepidoptera	ca. 100
	Hymenoptera	ca. 100
Vertebrata		
	Teleostomi	13,400
	Amphibia	5,540
	Reptilia	ca. 250
	Aves	ca. 1,800
	Mammalia	ca. 100

? Insufficient information.
Source: Lévêque (2005).

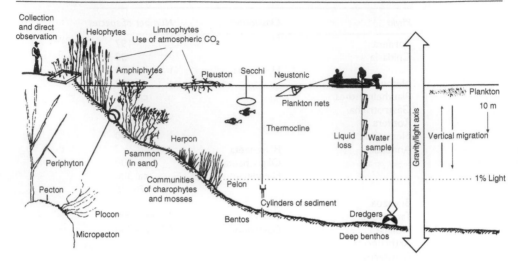

Figure 6.23   Different functional communities in aquatic systems.
        Source: Adapted from Margalef (1983).

circulation, or as a response to the interactions of aquatic ecosystems, such as drainage basins. The study of different communities and their interrelationships is currently receiving growing attention in rivers, lakes, reservoirs, wetlands areas, estuaries and small ponds (Thomaz and Bini, 2003).

Figure 6.23 illustrates the classification of different types of communities in aquatic ecosystems.

Plankton is the community that inhabits open waters with limited capacity of locomotion and with systems that enable permanent or limited buoyancy. **Phytoplankton** and **zooplankton** are respectively **autotrophic** and heterotrophic components of plankton and are interrelated, since the phytoplanktonic community can be used as food by herbivorous zooplankton. Plankton communities typically have a high renewal rate, also related to the retention time of the water and turbulence in the system.

The different common forms of freshwater and marine phytoplanktonic species and groups may be interpreted as **functional adaptations** to unstable and turbulent environments.

The combination of sedimentation and turbulence causing variations in the vertical displacement rate is probably one of most important factors in phytoplankton biology, according to Margalef (1978). The vertical and horizontal transport of phytoplankton cells or colonies, nutrients, and dissolved substances play an essential role in the phytoplankton's production of organic matter in inland waters and oceans. The suspension and survival of these organisms in the euphotic zone is therefore essential for the production of organic material in aquatic ecosystems.

Phytoplankton, according to Margalef (1983) and Reynolds (1984), requires a special condition: it must **remain in suspension**. The main expression of this requirement is the interaction with different currents at varying depths. The requirements to maintain phytoplankton in suspension range from **morphological changes** enabling increased buoyancy to the size of small dimensions and the physiological functioning that decreases the density of the cells or colonies. The size, volume and shape of a single phytoplankton cell range from organisms with a volume of 18 µm³ up to almost spherical colonies of *Microcystis* measuring 0.57 millimeters. Colonies can form filaments, aggregate in forms of sheets or form a pool of cells surrounded by a mucilaginous layer. The ratio of surface area to volume is essential to understanding how natural selection has acted on the ecology of phytoplankton, since the need to remain in suspension and absorb nutrients are decisive factors. The geometrical features of phytoplankton range from spherical cells to cells shaped like cylinders, trapezoids, or cubes, or organisms with a series of spines or other formations protruding from the cells' exterior. All of these alternative variations help preserve the optimum surface/volume ratio of the cells or colonies of cells.

The presence of a layer of **mucilage** surrounding a colony of cells – for example, in several species of *Volvox* spp, – increases the colony's buoyancy, reduces the chance of its ingestion by zooplankton and produces a special environment for the exchange of gases and nutrients. Colonies with mucilage can control, to a certain degree, their distribution and vertical displacement. In its organization, phytoplankton presents the main features occurring in cells of all photosynthesizing eukaryotes: protoplasmic formations, surrounded by a membrane, the **plasmalemma**, which is a set of complex substances that, in some cases, have external mucilage with an outer membrane. Such membranes are made up of different substances and materials composed of cellulose, silica, calcium carbonate or protein. The cells consist of a **nucleus**, containing genetic material in the form of **chromosomes; mitochondria,** in which the respiratory enzymes of the Krebs cycle are located; **chloroplasts,** where the photosynthetic pigments are found; a **Golgi complex,** where extracellular products are produced; **endoplasmic reticulum** with functions of synthesis and transport of substances; **lysosomes,** which contain enzymes and peroxides; **vacuoles** with fluids; **vesicles** or bubbles of oil, starch granules; **microtubules** and **microfibrils** for structural support; and **basal bodies.**

Contractile vacuoles are common in many freshwater flagellates and rare in marine forms (Hutchinson, 1957). Many phytoplanktonic cells synthesize a variety of substances such as glycerol, mannitol, proline, and glycerides, which are important for osmotic regulation in highly saline waters.

The presence of flagellates is important for the movement of organisms and the regulation of depth in relation to light intensity and the flow of nutrients.

The products of photosynthesis are partly reserved in the cytoplasm of phytoplankton and vary according to different groups. Chlorophyta and cryptophytes produce starch. Chrysophytes produce other polysaccharides. Other groups store proteins and lipids. The relative proportions of these cytoplasmic components depend on metabolism, nutritional conditions in the medium, and population dynamics.

Dry weight is a useful measurement of phytoplankton. Dry weight, after removal of water, includes organic and inorganic deposits. After oxidation of the organic matter (heating in air to 500°C), the "ashes" that remain can be used to calculate the ratio of the organic component (free of ash) and the ashes. Organic content, inorganic content and ash can be measured in relation to pigments (chlorophyll). These components in phytoplankton vary greatly. Nalevajko (1966) calculated that on average chlorophyceae ashes account for 10.2% of the dry weight, and diatoms for 44%.

According to Reynolds (1984), the silica content in diatoms varies from 26 to 69% of the dry weight of cells. It is important to determine the silica content in relation to cell volume and surface area in diatoms, as one of the fundamental relationships.

Carbon, nitrogen and phosphorus are elements in phytoplankton cells; carbon concentration is 51 to 56% of the ash-free dry weight (Redfield, 1958). Nitrogen concentration is typically 4 to 9 % of the ash-free dry weight (Lund, 1970), and phosphorus is 0.03 to 0.8% of the ash-free dry weight (Rovard, 1965).

Analysis of phytoplankton therefore presents a reduction to this relationship (called the Redfield ratio) of C:N:P of approximately 106:16:1. According to Lund (1965), most likely this is the ratio of elements required for phytoplankton.

The ratios of chlorophyll to dry weight or volume are also important. Generally chlorophyll has values ranging from 0.9 to 3.9% of ash-free dry weight. The chlorophyll content in cells also varies in relation to cell volume and between different groups.

Figure 6.24 shows the relationships between the dry weight of cells and cell volume for different groups, and the relationship between cell volume and content of chlorophyll in picograms, also for different groups.

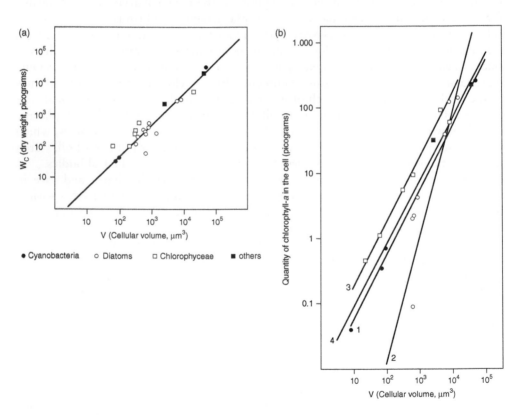

Figure 6.24   a) Relationship between the dry weight (W_c) and the cell volume (V) of phytoplankton. The regression equation is $W_c = 0.47 V^{0.99}$; and b) Relationship between the content of chlorophyll-*a* and the cell volume of phytoplankton. The regression equations are fixed points for cyanobacteria (1: log chl = 1.00 log V − 2.261), diatoms (2: log chl = 1.45 log V − 3.77), chlorophytes (3: log = 0.88 log V − 1.51) and all points (4: log chl = 0.984 log V − 2.072).
Source: Reynolds (1984).

The data on the basic composition of phytoplankton, or the concentration of chlorophyll in relation to dry weight or cell volume, are fundamental for understanding the basic relationships of phytoplankton. The chemical composition and the interrelationships between various elements help to determine estimates of biomass and productivity. These estimates and determinations also help to better establish experiments on the various species in relation to the concentration of elements in the surrounding medium.

Variability in the size of phytoplankton

The various dimensions in phytoplankton size are characterized as follows (Round, 1985):

▶   50–60 µm – microphytoplankton or microplankton
▶   5–50 µm – **nanophytoplankton** or nanoplankton
▶   0.5–5 µm – ultraplankton or ultrananoplankton
▶   0.2–2 µm – **picophytoplankton**

Platt and Li (1986) published a comprehensive volume on photosynthetic picoplankton, focusing on its physiology, vertical distribution in oceans and lakes, optical properties and relationship with detritus. Munk and Riley (1952) discussed phytoplankton's variability in size and differentiated morphology in a key work, which showed that the different size fractions of phytoplankton, compared to their surface/volume ratios and shapes, indicate different possibilities of absorbing nutrients and maintaining buoyancy.

Active movement of cells or cell colonies helps spread and renew the nutrient medium, and enables their regeneration, depending on movements and displacement on the surface of the cells. Phytoplankton that has the ability of displacement, such as flagellates, can quickly move at a rate of 1–2 meters per hour (Taylor, 1980). Flagellates present vertical phototactic migration. Distribution in size of estuary phytoplankton and the importance of nanophytoplankton as a primary producer in marine ecosystems were studied by Teixeira and Tundisi (1967), Tundisi *et al.* (1973) and Tundisi (1977).

Phytoplankton collection, characterization and studies

Phytoplankton is a heterogeneous collection of cells and colonies distributed in the water column at various depths, mixed with dead cells, detritus, colloids and a wide range of components. The study of the characteristics and dimensions of phytoplankton relies on a variety of techniques that utilize different tools and methods of observation. Figure 6.25 presents these various methodologies and their conditions of use.

Zooplankton organisms present a large set of components. However, those of inland waters are much less diverse than marine zooplankton.

Marine plankton includes Foraminifera, radiolarians, Ctenophora, molluscs and Appendicularia, which are not present in freshwater plankton.

According to Margalef (1978, 1983), zooplankton in inland waters is comprised of a set of organisms that went through a rigorous selection, characterized by processes adapting to the fluctuations and variability in freshwater environments. Hutchinson

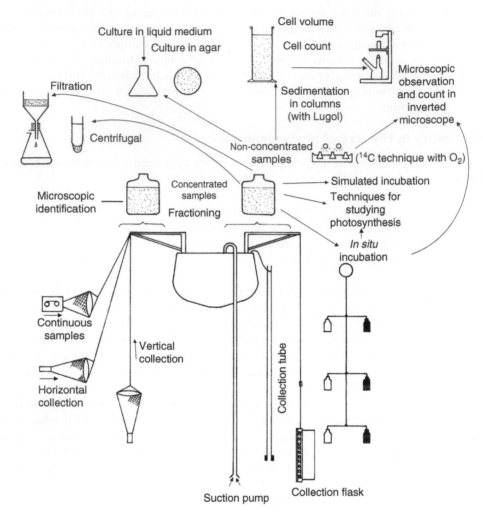

Figure 6.25   Collection and identification methods for phytoplankton.
        Source: Modified from Round (1985).

(1957), Ward and Whipple (1959), and Margalef (1983) objectively present the biology of zooplankton.

Protozoans, rotifers, cladocerans and copepods are basic components of zooplankton (Table 6.8). In some lakes, copepods comprise about 50% of the zooplankton biomass (see Table 6.9 and Chart 6.7 and Chart 6.8).

Other organisms that make up freshwater plankton include the *Craspedacusta sowerbiii* medusa, for example, which has widespread distribution; **ostracods**, with few planktonic species (*Cypria petensis*); and Mysidaceae, such as *Mysis relicta,* found in temperate-region lakes. *Chaoborus* larvae, of the family Chaoboridae, are commonly found in the plankton of many tropical lakes and reservoirs. They spend

Table 6.8 Number of species of Rotifera, Cladocera, Cyclopoida and Calanoida in some bodies of water in Brazil.

Lakes	Drainage basin	Rotifera	Cladocera	Cyclopoida	Calanoida	Researcher
Castanho (WW)	Amazon	21	9	3	1	Hardy (1980)
Cristalino (BW)	Amazon	13	9	3	1	Hardy (1980)
Camaleão	Amazon	175	14	3	4	Hardy et al. (1984)
Batata	Amazon	33	4	1	1	Bozelli (1992)
Açu	Northeast	33	4	1	1	Reid and Turner
Viana	Northeast	6	2	3	–	Reid and Turner
Dom Helvécio	East	4	5	3	2	Matsumura Tundisi (1987)
Aníbal	East	9	2	3	–	Tundisi et al. (1987)
Comprida	Paraná	41	11	4	4	Sendacz (1993)
Reservoirs						
Samuel	Amazon	14	8	1	1	Fallotico (1994)
Vargem das Flores	East	17	7	3	1	Freire and Pinto-Coelho (1988)
Paranoá	Paraná	16	3	1	–	Pinto-Coelho (1987)
Paranoá	Paraná	32	3	1	–	Branco (1991)
Broa	Paraná	15	3	5	1	Matsumura Tundisi and Tundisi (1976)
Billings	Paraná	13	5	3	–	Sendacz et al. (1985)
10 reservoirs in the state of São Paulo (average numbers)	Paraná	7.5	4.5	2.0	–	Arcifa (1985)
Paraibuna	Paraná	44	23	5	1	Cabianca (1991)
Monte Alegre	Paraná	15	9	2	–	Arcifa et al. (1992)
Jacaré Pepira	Paraná	20	16	3	1	Claro (1981)
Dourada lagoon	Paraná	32	8	3	–	Rocha and Sampaio (1991)
Passaúna	Paraná	15	9	3	–	Dias and Schimidt (1990)
Guarapiranga	Paraná	51	20	5	5	Caleffi (1994)

Lakes and rivers	Drainage basin	Rotifera	Cladocera	Copepoda	Researcher
Nhamundá	Amazon	141	5.8	3.2	Brandorff et al. (1982)
Acre (average numbers)	Amazon	23	5.8	4.8	Sendacz and Melo Costa (1991)
Trombetas	Amazon	97	12	6	Bozelli (1992)
São Francisco	East	50	5	2	Neumann-Leitão et al. (1989)
Upper Paraná	Paraná	64	20	11	Sendacz (1993)
Alto Paraná	Paraná	153	11	7	Bonecker (1995)

Relationship of some of the most common species in aquatic systems in brazil

Rotifera	Cladocera	Copepods	
		Cyclopoida	*Calanoida*
Asplanchna sieboldi	Bosmina hagmani	Metacyclops mendocinus	Notodiatomus cearensis
Brachionus zahniseri var guesneri	Ceriodaphna silvestri	Mesocyclops ogunnus	Notodiaptomus conifer
Filinia opoliensis	Daphnia gessneri	Thermocyclops decipiens	Notodiaptomus evaldus
Keratella americana	Diaphanosoma spinulosum	Thermocyclops inversus	Notodiaptomus iheringe
Ptygura libera	Moina micrura	Thermocyclops minutus	
Tricocerca capuccina	Sida crystallina		

Sources: Sendacz and Kubo (1982); Matsumura Tundisi and Rocha (1983); Matsumura Tundisi (1986).

Table 6.9 Global diversity of freshwater copepods.

	North America	Europe	Asia	Australia and New Zealand	South America	Mexico and Central America	Africa	World
Calanoid	111	119	294	63	123	37	113	678
Cyclopoid	105	277	308	52	203	118	228	1,045
Harpacticoid	147	504					183	1,260
Gelyelloida		2						2
Total	363	902[1]	927[2]	181	516	216	524[3]	2,080

[1,2]Including Turkey, Philippines, Indonesia, Malaysia
[3]Including Madagascar
Source: Dussart and Defaye (2002).

Chart 6.7 Occurrence of Cyclopoids in the principal drainage basins of Brazil.

	Amazon basin	Parana basin	Northeast basin	Paraguay basin	East basin	Southeast basin
Macrocyclops albidus (Jurine)	−	−	−	−	−	+
M. ater (Sars)	−	−	−	−	−	+
Paracyclops rubescens (Fischer)	−	+	−	−	−	−
Ectocyclops rubescens (Brady)	+	+	−	−	−	−
Tropocyclops prasinus (Fisher)	−	+	−	−	+	−
T. schubarti (Kiefer)	+	+	−	−	−	−
T. federensis (Reid)	−	+	−	−	−	−
T. nananae (Reid)	−	+	−	−	−	−
T. piscinalis (Dussart)	−	+	−	−	−	−
Eucyclops ensifer (Kiefer)	−	+	−	−	−	+
E. pseudoensifer (Dussart)	−	+	−	−	−	−
Thermocyclops deciplens (Kiefer)	+	+	+	+	+	−
T. inversus (Kiefer)	−	−	+	−	−	−
T. minutus (Lowndes)	+	+	+	+	+	−
T. tenuis (Marsh)	−	−	+	−	−	−
T. parvus (Reid)	+	−	−	−	−	−
Mesocyclops longisetus (Thiebaud)	−	+	−	+	−	+
M. annulatus (Wierzejski)	−	−	−	−	−	+
M. meridianus (Kiefer)	−	+	+	+	−	+
M. meridionalis (Dussart and Frutos)	−	−	−	+	−	−
M. ellipticus (Kiefer)	+	+	−	−	−	−
Metcyclops mendocinus (Wierzo)	−	+	−	+	+	+
M. brauni (Herbst)	+	−	−	−	−	−
Microcyclops ceibaensis (Marsh)	−	−	−	−	−	−

Chart 6.7 Occurrence of Cyclopoids in the principal drainage basins of Brazil (continued).

	Amazon basin	Parana basin	Northeast basin	Paraguay basin	East basin	Southeast basin
M. anceps (Richard)	+	+	−		+	−
M. finitimus (Dussart)	+	−	−	−	−	−
M. varicans (Sars)	+	+	+	+	−	−
Apocyclops procerus (Herbst)	−	−	−	−	+	−
Neutrocyclops bravifurca (Lowndes)	+	+	+	+		
Halicyclops venezuelensis (Lindberg)	+	−	−	−	−	−
Oithona amazônica (Burckhardt)	+	−	−	−	−	−
O. bowmani (Rocha)	+	−	−	−	−	−
O. hebes (Giesbrecht)	−	−	−	−	+	−
O. gessneri	+					
O. oligohallina (Fonseca and Bjornberg)	−	−	−	−	+	+
O. ovalis (Herbst)	−	−	−	−	−	+
O. nana (Wilson)	−	−	−	−	−	+
O. similes	−	−	−	−	−	+
O. plumifera (Wilson)	−	−	−	−	−	+

Source: Rocha et al. (1995).

Chart 6.8 Occurrence of calanoid species in drainage basins of Brazil

	Amazon basin	Parana basin	Northeast Atlantic basin (NE West and NE East)	Paraguay basin	East Atlantic basin	Southeast Atlantic basin
Argyrodiaptomus azevedoi (Wright)	+	+	+	−	−	−
A. furcatus (Sars)	−	−	−	−	+	−
A. robertsonae (Dussart)	+	−	−	−	−	−
Caladiaptomus merillae (Wright)	+	−	−	−	−	−
C. perelegans (Wright)	+	−	−	−	−	−
Dactylodiaptomus persei (Wright)	+	−	−	−	−	−
Notodiaptomus amazonicus (Wright)	+	−	−	−	−	−
N. anisitsi (Daday)	−	−	−	−	−	+
N. brandorfii (Reid)	−	−	+	−	−	−
N. carteri (Lowndes)	−	−	−	−	−	+
N. cearensis (Wright)	−	−	+	−	−	−
N. conifer (Sars)	−	+	−	−	−	−
N. conlferoides (Wright)	+	+	−	−	−	−

Chart 6.8   Occurrence of calanoid species in drainage basins of Brazil (continued).

	Amazon basin	Parana basin	Northeast Atlantic basin (NE West and NE East)	Paraguay basin	East Atlantic basin	Southeast Atlantic basin
N. dahli (Wright)	+	−	−	−	−	−
N. deitersi (Poppe)	−	+	−	+	−	−
N. deeveyorum (Bowman)	−	−	−	+	−	−
N. dubius (Dussart and Matsumura Tundisi)	−	−	−	−	+	−
N. gibber (Poppe)	−	−	−	−	−	+
N. henseni (Dahl)	+	+	+	−	−	−
N. iheringi (Wright)	+	+	+	+	+	−
N. incompositus (Brian)	−	−	−	−	−	+
N. inflatus (Kiefer)	+	−	−	−	−	−
N. isabelae (Wright)	−	+	+	−	+	−
N. jetobensis (Wright)	+	+	+	+	−	−
N. kieferi (Brandorff)	+	−	−	−	−	−
N. nordestinus (Wright)	−	−	+	−	−	−
N. paraensis (Dussart and Robertson)	+	−	−	−	−	−
N. santaremensis (Wright)	+	−	−	−	−	−
N. spinuliferus (Dussart and Matsumura Tundisi)	−	+	−	−	−	−
N. transitans (Kiefer)	−	+	−	−	−	−
Odontodiaptomus paulistanus (Wright)	−	+	−	−	−	−
Rhacodiaptomus calamensis (Wright)	+	−	−	−	−	−
R. calatus (Brandorff)	+	−	−	−	−	−
R. flexipes (Wright)	+	−	−	−	−	−
R. retroflexus (Brandorff)	+	−	−	−	−	−
R. insolitus	+	−	−	−	−	−
Aspinus acicularis (Brandorff)	+	+	−	−	−	−
Trichodiaptomus coronatus (Sars)	+	+	−	−	+	−
Scolodiaptomus corderoi (Wright)	−	+	−	−	+	−
Diaptomus azureus (Reid)	−	−	−	−	+	−
D. fluminensis (Reid)	−	−	−	−	+	−
D. linus (Brandorff)	+	−	−	−	−	−
D. silvaticus (Wright)	+	−	−	−	−	−
D. negrensis (Andrade and Brandorff)	+	−	−	−	−	−
P. gracilis (Dahl)	+	−	−	−	−	−

Source: Rocha *et al.* (1995).

most of the day buried in the sediment. They rise at night and are **carnivorous** predators that can substantially alter the composition of zooplankton due to their intense predation.

Zooplankton organisms have different feeding systems. They filter phytoplankton, bacteria and detritus, and intense **intra-zooplanktonic predation** can lead to significant alterations in trophic networks in certain lakes (Dumont, Tundisi and Roche, 1990) (see Chapter 8).

The collection and characterization of zooplankton can be challenging. The organisms are generally collected vertically, horizontally or obliquely, with planktonic trawl nets with 50 to 68 micrometer apertures, allowing the collection of a variety of the organisms present. Other techniques include pouring water-sampler content through filters in special systems for collecting developed by Patalas (1975), or special pumps that collect at different depths and allow a better quantification of zooplankton (Matsumura Tundisi *et al.*, 1984).

The total zooplankton biomass can be expressed by volume in water volume (cm^3/m^3) or as wet weight (mg/m^3). The chemical composition of zooplankton, as a relationship of carbon per mg of dry weight of zooplankton (or of a specific species), is also used and allows more accurate assessment of zooplankton's role in the flow of energy and the food web.

The terms **euplankton** or **holoplankton** are generally used to designate planktonic organisms that throughout their life cycle remain in the plankton. There are also species of **mesoplankton** that are found in the sediment during part of their life cycle, such as species of the diatom genus *Aulacoseira*.

The terms **limno-**, **heleo-** and **potamo-** are generally used to characterize the plankton in lakes, ponds and rivers, respectively.

Benthos

The communities distributed on the surface of sediment and sediment/water interface are the **benthos**. Benthic organisms live on the substratum or depend on it, spending part or all of their lives in the solid bottom component of aquatic systems.

Benthic organisms include a wide variety of taxonomic groups, as shown in Chart 6.9. There is enormous diversity in feeding habits. The main groups of benthic invertebrates include **insects**, **annelids**, **molluscs** and **crustaceans**. The most common orders of insects, with much greater diversity in lotic water, include Ephemeroptera, Plecoptera, Trichoptera, Diptera and Odonata. The majority of these insects spend their life cycles in the form of larvae, whereas adults are terrestrial and short-lived. The period of emergence of different species depends on the temperature and photoperiod.

The qualitative composition of **benthic fauna** is a good indicator of the trophic conditions and degree of contamination of rivers and lakes, based, for example, on the components of *Chironomus* larvae, because they resist low concentrations of dissolved oxygen.

The distribution of benthic fauna depends of the type of substratum and levels of organic material, the current velocity, and transport of sediment by the current. The erosion of substratum by currents and the subsequent shift in composition of fauna are important factors in the process of spatial succession in the **benthic community**. Considerable differences exist in the qualitative composition of benthic communities

Chart 6.9   Some genera and species of benthic invertebrates in lakes, rivers, ponds and reservoirs with neo-tropical examples.

Taxonomic group	Habitat in which found most often	Food	Examples
1. Turbellarians	Lakes, rivers, reservoirs, ponds	Carnivores	*Catenulidae leuca*
2. Nematodes	Various aquatic ecosystems	Carnivores, herbivores, parasites	*Protoma eilhardi*
3. Annelids Oligochaetes	Various ecosystems	Filter sediment	*Tubifex* sp.
Hirudinea (leeches)	Various ecosystems	Carnivores, detritivores	*Helobdella triserialis lineata*
4. Molluscs gastropods	Various ecosystems, banks of macrophytes, irrigation channels	Grazers	*Planorbis* sp.
Pelecypods (Bivalves)	Rivers	Filterers	*Anodonta* sp.
5. Crustaceans Malacostraca (amphipods crabs)	Various ecosystems	Detritrivores	*Macrobranchium denticulatum*
6. Insects Plecoptera	Well-oxygenated waters	Omnivores	*Tupiperla* sp.
Odonata	Rivers, ponds	Raptorial carnivores	*Libellulla* sp.
Ephemeroptera	Various ecosystems	Raptorial carnivores	*Caenis cuniana*
Hemiptera	Various ecosystems	Filterers	*Belostoma* sp.
Megaloptera	Various ecosystems	Raptorial Carnivores	*Corydalidae* sp.
Tricoptera	Various ecosystems	Filterers	*Dolophilodes sanctipauli*
Coleoptera	Ponds	Raptorial carnivores	*Haliplus* spp.
Diptera	Ponds	Filtrators	*Culex* spp.
	Lakes	Raptorial carnivores	*Chaoborus* spp.
	Fast-moving rivers	Filterers	*Simulium* spp.
	Various ecosystems	Detritivores	*Chironomus* spp.

Source: Adapted from Horne and Goldman (1994).

in lotic (running) and lentic (standing) waters. In general, insects are predominant in **lotic communities**, using various types of substrata, such as the surface of stones and smooth rocks. Mollusca and Turbellaria can establish themselves on smooth rocks and tolerate current speeds of 100–200 cm/s (Macan, 1974). The stability of substrata also allows greater density of organisms (Welch, 1980).

In general, benthic organisms are located in the littoral or sublittoral zone of lakes and in the deep zone, which is relatively more uniform, and, in the case of stratified lakes, often presents low levels of dissolved oxygen (anoxia) and extremely low temperatures. The littoral and sublittoral zones present greater variability and spatial heterogeneity, accumulation of biomass, and greater diversity.

Diel variations are also greater in the littoral zone, particularly of water temperature, dissolved oxygen, pH and CO_2, which implies an environment where there is need for adjustments and bouyancy in short periods of time.

The deep areas of lakes have a simpler composition, due to the special limiting conditions. Common species there include *Chaoborus* larvae, some species of molluscs and oligochaete worms. In the littoral zone there is a wide diversity of organisms and feeding habits, and often, greater biomass. Meanwhile, in lakes with an extensive deep zone and a limited littoral zone, the portion of biomass in the deep zone is high.

Most zoobenthic organisms are detritivores, filtering detritus and suspended organic material or feeding on sediment. Some zoobenthic species are carnivorous and predatory; others are grazers (such as some molluscs). Many benthic animals spend most or all of their life cycle in the sediment. Others, such as the *Chaoborus* larva, which is a predator, migrate to the surface at night and feed on zooplankton, whereas during the day they remain burrowed in the anoxic sediment. This mechanism also avoids predation, since few species are adapted to a totally anoxic environment, as is *Chaoborus*.

Benthic organisms depend in part on the sendimentation of organic material from a lake's upper layers, or transport of material by flow in the case of rivers. In the littoral zone, the contribution of allochthonous material and recycled organic autochthonous material is significant (for example, from decomposition of macrophytes).

The **benthos of the deep zone** is much more dependent on organic material produced in the littoral zone and the epilimnion of stratified lakes. Jónasson (1978) described the inter-relationships between the seasonal cycles of the planktonic and benthic communities of Lake Esrom in Denmark.

In addition to substratum type and current speed, the benthic community can be limited or controlled by temperature and levels of dissolved oxygen in water. Several groups of **benthic macro-invertebrates** are relatively intolerant of low levels of dissolved oxygen. Other organisms with special morphological or physiological adaptations can tolerate low oxygen levels. For example, various insect larvae can breathe oxygen from the air, and larvae of *Chironomus* spp. and *Branchiura sowerbyi* have haemoglobin, which enables them to better tolerate low O_2 levels.

Excess dissolved organic matter, nitrogen and phosphorus, and toxic substances – all resulting from pollution and eutrophication – significantly affect benthic organisms, altering community structure and succession.

The benthic community can be sampled and collected by various techniques. The difficulty of collecting benthic organisms lies in the fact that substrata are different and not all equipment is sufficiently flexible to sample different substrata with the same efficiency. The selectivity of the different methods in the sampling of organisms is another important factor. Generally, biomass or numbers of individuals per area or volume of sediment is used to obtain quantitative data that can be used comparatively.

Samples can be collected from the bottom with the Ekman-Birge grab, which collect a certain volume of sediment, or by dredges that collect organisms from the surface, or the bottom, or even by samplers that collect sediment in cylindrical tubes in which the vertical distribution of fauna can be studied.

The possibility of studying benthos through the use of artificial substrata should also be mentioned, which, though selective, can reveal some important aspects of succession in space and time.

The recovery of benthic fauna in rivers, after the introduction of water treatment and dispersal of the effect of pollutants and toxic substances, can be very fast. Because of the steady flow, after a full hydrological cycle, part of the contamination occurring in the sediment decreases considerably, and recovery can occur after one or two years of treatment. Several cases of full or partial recovery of the benthic fauna have been described for rivers in temperate regions subjected to pollution treatment control (Welch, 1980).

Monitoring the benthic fauna in rivers can therefore be an important indicator of the recovery process.

All material suspended in water is part of the **seston**, in which the **plankton** is the living component and **trypton** (dead organic particles) is the non-living component. The **necton** is made up of organisms with ample mobility (which in freshwaters are mainly fish) and that are distributed by the water column.

Pleuston involves organisms that are found on the water, on the film on the surface of the water.

Neuston involves organisms that stay on the surface due to the surface tension of the water; there is a distinction between **epineuston** and **hyponeuston**.

Research on neustonic organisms has increased in recent years. This microlayer (a few nanometers) of the water's surface – where the neustonic organisms (algae, bacteria, protozoa) are located because of adhesive forces on the water interface – presents a concentration of lipids (fatty acids, phospholipids, glycerides).

Below this layer there is another, consisting of protein or carbohydrate complexes (e.g., polysaccharides), and under that, bacterioneuston, phytoneuston and zooneuston accumulate (Falkowski, 1996). Due to high levels of organic substances in this layer, autotrophic and heterotrophic bacteria have optimal growth rates. Neustonic organisms have high concentrations of **hydrophobic compounds** (for example, mucopolysaccharides, glycoproteins, polymers) on their external cell structures, which enable them to adapt well to the surface layer in aquatic ecosystems.

The **neustonic community** is one link through which organic matter flows from the atmosphere through the water column. Bacterioplankton and bacterioneuston play an important role in the biotransformation of allochthonous and autochthonous organic matter.

Kalwasinska and Donderski (2005) studied a layer of bacterioneuston in Polish lakes and found a high percentage of bacteria with the capacity to decompose lipids. According to the authors, the presence of these bacteria is due to the accumulation of liquids in this neustonic layer (triglycerides, phospholipids, free fatty acids, sterols and oils) in the form of emulsion.

Periphytic algae develop on aquatic macrophytes and use them as a substratum. Organisms that are located on rocks and surfaces on the bottom of rivers and lakes are also called **periphyton**.

The periphyton is an important part of the community, settling on substrata of lentic or lotic waters, contributing significantly to the production of organic matter in shallow and sunny regions of lakes, ponds, rivers or wetlands. In these regions, the periphyton can play an important role in the production of organic matter and metabolism of the lake (Wetzel, 1975).

The development of the **periphytic community** in rivers largely depends on the velocity of the current. The time needed to colonize the substratum depends on the type of substratum and surface roughness. Deep rivers with slow-moving currents limit the development of periphyton (Panitz, 1980).

The periphytic community is composed of diatoms (for example, *Navicula, Synedra, Cymbella*), cyanobacteria (*Oscillatoria* and *Lyngbya*), filamentous green algae (*Cladophora*, for example), filamentous bacteria or fungi, protozoa (*Stentor* or *Vorticella*, for example), rotifers and larvae of some insect species.

Many recent studies have shown that ciliates play an important role in the dynamics of periphytic communities, as well as rotifers, gastropods, Lamellibranchia and insect larvae. This fundamental role of ciliates in the dynamics of periphytic communities plays out especially in trophic networks, because they are important consumers of bacteria and algae, and are also essential components in the diets of rotifers and crustaceans (Mieczan, 2005).

Periphyton can be sampled by removing material from an area of artificial substratum, analyzing the dry weight, wet weight, chlorophyll level, and cell count.

Artificial substrata have been used extensively to determine the growth rate of periphyton, community succession and biomass levels. These substrata can vary from plastic or glass sheets to concrete blocks or panels of wood. Panitz (1980) conducted an extensive study on the growth and succession of periphyton on artificial substrata in the Lobo (Broa) reservoir and concluded that wood substratum enabled rapid growth and stabilization in the chlorophyll level and number of cells after 30 days of colonization.

The artificial substrata can provide basic information on the growth rate of periphyton and biomass, but there are limitations because of the selectivity produced by the type of substratum and the fact that these substrata are originally completely devoid of organisms, which rarely occurs in natural conditions. Nevertheless, the use of these substrata provides basic comparative data on lakes, rivers and reservoirs with different trophic states.

Important primary producers in aquatic systems include **aquatic macrophytes,** higher aquatic plants called **rhizophytes** when roots are accomadate to sustain them; **limnophytes,** when they are found totally submerged; **anphiphytes,** when systems of fluctuation are accommodated (such as the water hyacinth *Eichhornia crassipes*); and **helophytes,** when emergent structures are present.

Definitions of 'aquatic plants' or 'aquatic macrophytes' vary. Pott and Pott (2000) use the term aquatic macrophytes as adopted by the International Biological Programme, which comprises plants inhabiting environments from marshes to truly aquatic ecosystems (Esteves, 1988).

Aquatic macrophytes are a recent evolutionary group in which the general evolutionary tendency in aquatic systems occurred with a partial transition (flowering plants and pollination by wind and insects) or with a complete adaptation with underwater pollination and flowering. Chart 6.10 lists the most common species in South America.

Countless processes occur with aquatic macrophytes, including tolerance to low levels of dissolved oxygen, and unique structures for transporting gases and fixing HCO_3^- in photosynthesis.

In the latter process, the excretion of OH^- by plants, in the measure that HCO_3^- is broken down into CO_2 and OH^-, produces a high pH that leads to precipitation of carbonate ions.

Hydrostatic pressure and the penetration of solar energy limits the distribution of submerged aquatic macrophytes. Light intensity is another basic limiting factor.

Wetzel (1975) summarizes the main features of macrophyte vegetation as follows:

▸ **Emergent macrophytes:** produce aerial reproductive organs, are located in shallow regions (<1.5 m of water). They are usually perennial with developed rhizomes (example: *Typha* spp.).
▸ **Macrophytes with floating leaves:** chiefly angiosperms occurring in regions with depths of 0.5 to 3.0 m. Floating leaves occur at the tips of long petioles or on short petioles. Example: *Nymphaea* spp. Aerial or floating reproductive organs.

Chart 6.10 Most common species of floating, immersed and submerged aquatic macrophytes in the inland aquatic ecosystems of South America.

Eichhornia crassipes (FF)	Typha latifolia (EE)
Eichhornia azurea (FF)	Echinochloa polystachia (EM)
Pistia stratiotes (EM)	Pontederia spp. (EE)
Salvinia herzogii (FF)	Utricularia foliosa (S)
Salvinia auriculata (FF)	Cabomba furcata (S)
Azolla caroliniana	Egeria densa (S)
Egeria najas (S)	Panicum fasciculatum (EM)
Cabomba australis (S)	Paspalum repens (EM)
Ludwigia peploides (EM)	Lusiola spruceana (EM)
Lemma gibba (FF)	Oriza perennis (EM)
Mayaca fluviatilis (S)	Pontedeira cordata (EM)
Nuphar luteum liteum	Pontedeira lanceolata (EM)
Nymphaea ampla (FF)	Cyperus giganteus (EE)
Nymphoides indica (FF)	Cyperus acicularis (EE)
Cabomba pyauhiensis (S)	Ceratophyllum demersum (S)
Scirpus arbensis (EE)	

EM – Emergent
FF – Floating
S – Submerged
Sources: Pott and Pott (2000); Thomaz and Bini (2003).

▶ **Submerged macrophytes:** occur at all depths in the euphotic zone, but as angiosperms are limited to 10 m (1 atm of pressure). Leaf form varies widely; reproductive organs are aerial, floating or submerged. Some **pteridophytes**, charophytes and angiosperms. Example: the genus *Mayaca*.

▶ **Floating macrophytes:** a group without roots in the substratum, which float freely, with various forms. Example: *Eichhornia crassipes*, *Eichhornia azurea*, *Salvinia*, *Azolla*, Lemnaceae. Aerial or floating reproductive organs (example: *Utricularia*).

In Brazil, two volumes have been recently published on the taxonomy, description and management of aquatic plants: Thomaz and Bini (2003) and Pott and Pott (2000). The latter describes the 'life form or biological form', which is the habit (morphology and mode of growth) considered in relation to the water surface, such as:

▶ **Amphibious** or **semi-aquatic:** able to live well both in the flooded areas and out of water, usually morphologically modifying the aquatic phase for land when the water level drops.

▶ **Emergent:** rooted on the bottom, partially submerged and partially out of water.

▶ **Floating fixed:** rooted at the bottom with floating stem and/or branches and/or leaves.

▶ **Floating free:** not rooted in the bottom, can be carried by the current, wind or even by animals.

▶ **Submerged fixed:** rooted in the bottom, with stems and submerged leaves, usually with only the flower out of water.

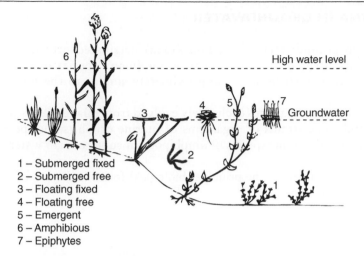

Figure 6.26   Spatial succession of aquatic macrophytes in lakes.
Source: Modified by Thomaz and Bini (2003).

▶ **Submerged free:** not rooted on the bottom, totally submerged, generally with only flowers emerging.
▶ **Epiphytic:** lives on other aquatic plants.

Observations of a macrophyte community in a **transition zone** between river and lake show a quite characteristic spatial succession. Community composition and structure along the horizontal axis depend on current velocity, type of substratum and deposition of sediment, and the velocity (see Figure 6.26).

The term **plocon** is used to describe communities that are located in masses of filamentous algae where an environment forms with respect to dissolved and particulate organic material and where bacteria, protozoa and crustaceans grow. Fixed communities that do not extend to form branches and are located on hard surfaces are called **pecton.** There are also organisms on the bottom that move or slide over the sediment (some species of algae of microphytobenthos, protozoans, ciliates, euglenoids), which are called **herpobenthos** or **herpon.**

Animals that live in the sediment are called **epifaunal** (above the sediment) and **infauna** (within the sediment). Organisms that live between the sand particles are called **psammon** or **pelon** (in this case, those living between finer particles).

6.7   AQUATIC BIODIVERSITY IN THE STATE OF SÃO PAULO

The body of knowledge on aquatic biodiversity of São Paulo was summarized in a series of volumes edited by Joly and Bicudo (1999). In relation to aquatic organisms, it is worth mentioning the text on marine invertebrates, edited by Migotto and Tiago (1999); on freshwater invertebrates, edited by Ismail *et al.* (1999); and vertebrates (Castro, 1998), which provided basic information not only on the study and distribution of aquatic biota in the state of Sao Paulo, but also for all neotropical biota.

6.8   FAUNA IN GROUNDWATER

The fauna in groundwater (**hypogeum fauna**) originated primarily from fresh-water animals adapted to subterranean water (many species of fishes and crustaceans) and their distribution is approximately similar to the fauna of surface inland waters.

Also present are organisms exclusively evolved from inland waters (prosobranch molluscs) and some strains of organisms of marine origin (some species of decapods). There are also some species of amphipods in subterranean waters (Banrescu, 1995).

Chart 6.11 summarizes the principal biological forms and types of communities in inland aquatic systems.

Chart 6.11   Detailed description of principal biological forms and the types of communities that occur in continental systems, especially for primary producers.

Migratory organisms

With little capacity to move and limited transport of organic material.
a) In air-water interface – **Neuston**
b) Organisms located at the bottom or on other organisms – **Tetoplankton**
c) Microscopic organisms with slow sliding movement on the bottom – **Herpon**
d) Migratory organisms with low capacity for locomotion – **Plankton**

Organisms on a fixed surface

a) Organisms that live on a compact substratum, forming fine covers – **Pecton**
b) Filamented organisms that may become detached from their base with their mass at some distance from the base – **Plocon**
c) Organisms that form masses several millimeters in height, generally on macrophytes, rocks and sediment – periphyton or haptobenthos ("Aufwuchs")

Migratory with roots or with roots suspended in water

a) At the air-water interface, using atmospheric CO_2 – **Pleuston**
Floating plants with reduced structure or relatively large structure; with floaters- *Eichhornia, Salvinia, Lemm, Pistia*
b) Plants submerged below the surface with no roots, often resting on the bottom – Megalopleuston or megalopleuston. In some cases, resting on the bottom. *Utricularia* sp.

Fixed on a substratum

a) Plants with roots fixed on substratum, adapted to live in very intense running water – Haptophytes
b) Submerged rooted plants in sediment – Rhizophytes
  b1 – Plants with submerged foliage that use CO_2 dissolved in water.
  b2 – Plants with foliage that is partially in contact with the atmosphere and can use CO_2 from the air.
  b3 – Plants with floating foliage that occurs on the surface – Amphiphytes or Epiphydrophytes. The floating foliage can be round (*Nymphaea, Victoria*) or lanceolate, such as *Polygonum*.
  b4 – Plants with totally immersed foliage, supported on vertical support stems, such as *Typha* spp., *Paspalum* spp., *Phragmites* spp., *Polygonum* – Helophytes or Hyperhydrophytes.

Determination of biomass

Biomass refers to the amount of living matter that exists per unit of volume or surface. It can be expressed in total wet weight, or total dry weight (after drying and discarding the water). It can also be expressed in units of C:N:P per unit dry weight. In most freshwater ecosystems, the maximum total dry, weight is about 1 kg, or $100g\, C \cdot m^{-2}$.

Calculating the biomass of organisms presents major problems. Very different techniques are needed to collect the organisms. The samples must be representative of the communities. Biomass can be expressed as number of individuals per m^2 or m^3 (area or volume), dry weight, or wet weight. It can also expressed in terms of chemical energy. For example, 1 gram of dry organic matter is between 4,000–6,000 calories/gram in the form of stored chemical energy, with 1 gram of organic carbon equivalent to approximately 11 kcal or 45 kJ. Sestonic material with dead particles is $1gC = 8.8\,kcal$.

Collection and categorization of aquatic communities

Collecting and categorizing the components of an aquatic community require the use of methods that allow the determination of quantitative and qualitative measurements of biomass, to characterize the specific composition and type of **association** that occurs in the different spatial compartments. The collection and categorization involve observations in laboratory and field work. Currently a number of more advanced techniques for studies *in situ* (field fluorometers, nets and special collection pumps) allow more consistent quantification of the biomass of organisms.

The use of satellite imagery and aerial photography, together with field collections, also provides a more consistent vision of spatial distribution of biomass and concentrations of organisms, at least in surface waters. Details of collection of organisms and characterization of communities can be obtained from Bicudo and Bicudo (2004).

Of course, for each of the populations and communities found in different spatial extracts, there is special equipment and techniques for collection, observation and experimentation. A combination of these three approaches is fundamental to understanding the relative importance of each component.

How much is the biodiversity of inland waters worth?

The flora and fauna of inland waters play an important role in the functioning of freshwater ecosystems. According to Dumont (2005), there needs to be a way to assign a value to this biodiversity, an economic value, for these undomesticated species. The **assessment of damage to biodiversity** can be made using the contingent valuation method (CVM), which should include the valuation of the loss of functions or 'services' provided by certain aquatic biodiversity. The attitude of **purpose of paying** (PP) or **purpose of accepting** (PA) must also be considered. These concepts refer to species that are aesthetically pleasing to the human eye or of attention-catching size. However, according to Dumont (2005), the **risk of extinction** for any species must be considered as fundamental to the valuation, especially if it is possible to determine the role of the species in the functioning of the ecosystem. The value of individual species and the **resilience** of the species, or that is, the more resistant to extinction, should be considered in this assessment.

And finally, the ability and opportunity to commercialize biodiversity in a sustainable fashion must provide conditions for an **economic valuation** of the biodiversity. Conservation of the biodiversity in general (especially aquatic biodiversity) is critical for maintaining processes in the biosphere and maintaining the natural course of evolution in the systems. In many tropical regions, the studies of aquatic biodiversity are still in an intermediate stage and provide little advanced knowledge. These tropical regions – especially large internal deltas of major rivers in South America, Africa and Southeast Asia – are **active centres of evolution** because of their biodiversity and the processes of generic interaction and flow (Margalef, 1998; Tundisi, 2003).

For tropical regions it is therefore essential to promote and accelerate studies on aquatic biodiversity (structure and function), in order to save it and promote means of economic valuation (Gopal, 2005).

Note: For current information on biodiversity of animal species and vertebrates by zoo-geographical region, see Annex 5.

Mamirauá floodplain
Photo: Luis Marigo

7 | The dynamic ecology of aquatic plant populations and communities

SUMMARY

In this chapter, the main mechanisms and interactions of the components of aquatic plant populations and communities will be examined and discussed, as well as factors that influence their spatial and temporal succession and diversity and distribution in lakes, rivers, reservoirs and wetlands. The factors that limit and control the primary production of phytoplankton, periphyton and macrophytes will be examined, as well as fluctuations in biomass and these organisms' role in biogeochemical cycles and interactions with other aquatic organisms. Examples are presented from case studies of different aquatic plant communities in the Neotropics as a basis for understanding spatial and seasonal succession in different ecosystems typical in this region. Case studies of temperate-zone lakes are presented for comparison. Basic concepts about succession of phytoplankton, periphyton and macrophytes and the effects of disturbances on these communities are discussed as examples in shallow lakes, Amazonian lakes and reservoirs.

An ecological system is composed of **biotic components**, ranging from viruses and bacteria to higher organisms, plants and animals that interact with abiotic physical and chemical components, forming a basic ecological unit – the ecosystem. These organisms, interacting with **abiotic factors**, come from a wide variety of species that individually form the population. A *population*, therefore, is defined as a set of organisms of the same species, and a *community* as a collection of various populations in an ecosystem. Populations and communities present a series of different dynamic processes and attributes. For example, a population has density (i.e., number of organisms/area or volume), a property that cannot be applied to an individual organism. And a community has *diversity of species*, which is not applicable to a population.

7.1 IMPORTANCE OF POPULATION STUDIES IN AQUATIC SYSTEMS

Populations present a series of unique attributes in each group of organisms that, all together, characterize an aquatic ecosystem. It is important to understand the behaviour of species in selection of habitats, interactions with other species, and tolerance of each population to physical and chemical factors in the environment. When studying the biotic components of an aquatic system, the first question about species – composition is: why are certain species present or absent in a given habitat? To answer this question, it is important to consider the approach and analysis presented by Macan (1963, cited in Krebs, 1972) based on the criteria of presence or absence and determining factors (see Figure 7.1).

7.2 MAIN FACTORS IN BIOLOGICAL PROCESSES

Biological processes depend on a number of basic factors that together determine and control the responses of individual organisms, populations and communities. The first factor is **temperature dependence**, which controls and limits the physiological responses of organisms, biochemical activities, growth rates, and reproductive rates.

Another factor is **dependence on available substrate**, i.e., basic nutrition, including **macro-nutrients** such as carbon, nitrogen, phosphorus, silicon, or **micro-nutrients** such as **molybdenum**, zinc, manganese, iron, and copper. Some organisms rely on a single nutrient or on many nutrients simultaneously. The availability and type of nutrients control growth, reproduction and succession in plant and animal communities. Plants are also dependent on the availability, quantity and quality of light.

Organism size is another basic factor, since physiological responses, migration and development are related to distribution in size. Margalef (1978) presented the main ideas and hypotheses on phytoplankton size and vertical distribution in turbulent environments with a high degree of vertical mixing. Also, some aquatic organisms need to survive in an environment that enables them to float and, therefore, there is **dependency on density**, which is also significant in terms of the organisms' vertical distribution, position and capacity to migrate vertically and horizontally.

All these processes act simultaneously. Reproduction, growth, development, migration and physiological behaviour are controlled by these factors, depending on

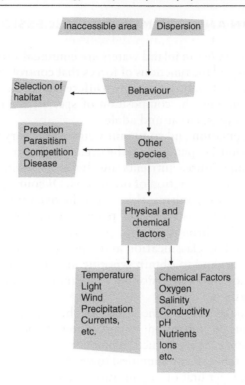

Figure 7.1   Mechanisms for selection of habitat, dispersion, factors that regulate and limit the presence or absence of species in ecosystems.

the aquatic ecosystem, its latitude, longitude, altitude, and physical and chemical conditions that establish regulating and limiting factors.

In plants, physiological processes are limited by light intensity, and light availability and intensity control the distribution, physiology and physiological behaviour of many aquatic plants. Light and water temperature are factors that control and limit growth and photosynthesis in photo-autotrophs. They also act synergistically in many aquatic environments. High light intensities inhibit photosynthesis in phytoplankton and other photo-autotrophs.

Aquatic organisms are also dependent on **oxygen levels** and saturation in water. In water, vertical and horizontal distribution of dissolved oxygen produces changes in the distribution and behaviour of organisms. For many organisms, the anoxic regions of lakes and coastal waters are uninhabitable. The availability of oxygen controls the growth rate and physiological and biochemical responses of many aquatic organisms.

Another important factor on which aquatic organisms depend is **pH**, as it controls many chemical reactions and provides ions (HCO_3^- and CO_3^{2-}) for aquatic plants (pH levels, for example, can limit and control the distribution and growth of aquatic organisms) (Thomaz and Bini, 2005).

These two chemical variables – pH and dissolved O_2 – largely control the growth and physiological responses of aquatic organisms, populations and communities.

7.3 POPULATION AND COMMUNITY SUCCESSION

Populations and communities in inland waters are continually interacting, due to fluctuations in ecosystems and the functions of forces that control and limit the reproduction and development of aquatic organisms. Differences in the seasonal hydrological cycle, for example, can alter the composition of species, community structure, and relative proportions of eggs, larvae and adults.

Time scales in population and community succession vary from extremely short to extremely long periods, depending on the organisms' reproductive capacity, fluctuations in water temperature, nutrients and light, in addition to the controlling factors resulting from the interaction of organisms. Of course, these time scales also vary based on the size of organisms, their reproductive rates and response capacities to abiotic factors, such as effects of temperature, wind speed and direction, and nutrient levels. The chronology of a process of succession in inland aquatic systems is complex and its classification and study depend on the capacity to collect data and specimens, and to obtain synoptic and updated information in order to analyse organization and succession and determine conditions for quantitative and qualitative analysis.

It is difficult to make certain generalizations, but as Margalef (1983) observed, some basic aspects can be recognized: minimization of energy exchange per unit of information, maintaining the reproduction/biomass coefficient an evolutionary trend in ecosystems and, notably, in communities, toward reducing the exchange of energy with 'increased entropy per unit of organization conserved' (p. 127).

Besides the direct effects, it is important to consider the **indirect effects** affecting succession. For example, water temperature, nutrition, parasitism and predation can influence stages of succession in species, populations and communities.

Succession, according to Odum (1969) and Reynolds (1997), is the principal manifestation in the development of ecosystems. Clements (1916) described the basis of the study and classification of ecological succession, indicating the possible biogeophysical control of the environment by the community, which would be able to maintain an internal control process. Theories on internal control of succession and a trajectory toward a **climax** (Tansley, 1939) were challenged by irregularities in the process (disturbances such as fire, drought, **floods,** or hurricanes). The reconciliation of these tendencies – orderly succession as opposed to stochastic (chance) responses resulting from interruptions in the process – is made considering external forces, the functions of force and the respective composition and response of biota.

According to Reynolds (1997), the process of succession cannot be viewed as orderly and predictable. Odum (1969) and Margalef (1991, 1993), however, describe certain singularities in succession, which, according to Reynolds (1984a, 1984b, 1986, 1995), apply to succession in pelagic communities, which respond to different external forces driving different populations and communities in space and especially in time. The general principles of the ecological succession model presented by Odum (1969), still hold up with further information added (see Table 7.1).

Most characteristics and attributes of aquatic communities in the process of succession were determined by multiple studies not only of pelagic communities but also of benthic, periphytic and nektonic communities (Margalef, 1991). Interactions between the components of the different communities and the flow of energy can

Table 7.1   Chemical composition of algae and the relative abundance of the principal components.

	C	H	O	N	P	S	Si	Fe
Redfield atomic ratio (stoichiometric) with respect to phosphorus[a]	106	263	110	16	1	0.7	–	0.05
Ratio for Redfield mass (stoichiometric with respect to phosphorus)[a]	42	8.5	57	7	1	0.7	–	0.1
Ratio for Redfield mass(stoichiometric with respect to sulphur)[a]	60	12	81	10	1.4	1		
Ratio for Redfield mass (stoichiometric with respect to carbon)[a]	100			16.6	2.4			
Chlorella (dry weight related to carbon)[b]	100			15	2.5	1.6	–	
Peridineans (dry weight related to carbon)[c]	100			13.8	1.7		6.6	
Asterionella (dry weight related to carbon)[d]	100			14	1.7			
Lake water (mol·L^{-1})[e]	10^{-3}	10^2	10^2	10^{-4}	10^{-6}	10^{-3}	10^{-2}	$<10^{-5}$

[a] Stumm and Morgan, 1981; [b] Round, 1965; [c] Sverdrup *et al.*, 1942; [d] Lund, 1965; [e] approximations of the author (Reynolds, 1997), omitting gaseous dissolved nitrogen.
Source: Reynolds (1997).

be classified and identified in different stages of succession. Theoretical knowledge and identification of the processes involved in succession also have important practical application. They make it possible to control, to a certain extent, the succession of populations and communities, especially in aquatic ecosystems, where there are opportunities to manipulate functions of external forces, for example, that control retention periods and vertical mixing and stratification (Reynolds, 1997; Tundisi *et al.*, 2004).

7.4   GENERAL FEATURES OF PHYTOPLANKTON

The cellular organization of photosynthetic eukaryotes consists of a cell with organelles including a nucleus, mitochondria, chloroplasts and cell membrane. These cells contain green **chloroplasts** or other chromatophores with photosynthetic pigment, a Golgi complex, an endoplasmatic reticulum, lysosomes containing digestive enzymes, liquid-filled vacuoles, reserve substances containing droplets of oil or starch granules, microtubules and micro-fibrils for structural support, and basal bodies where flagella are fixed. Figure 7.2 presents the general structure of a eukaryote cell of a photosynthetic organism. The protoplasmic formations are enclosed by a complex membrane (plasmalemma) consisting of two or three separate layers. In some cells, the external plasmalemma layer is mucilage. A non-living cell wall can be found in many eukaryotes, composed of carbohydrates, cellulose or inorganic substances, such as carbonates

Chart 7.1   General characteristics of succession processes in ecosystems.

Attributes of ecosystems	Initial phases	Mature phases
Energy of communities		
Gross production/ respiration of community	More or less than 1	Close to 1
Gross production/biomass (P/B ratio)	High	Low
Biomass supported per unit of energy flow (ratio B/Q)	Low	High
Net production (product)	High	Low
Food chains	Linear	Network
Structure of communities		
Total organic material	Low	High
Inorganic nutrients	High levels; outside organisms	Inside organisms
Diversity of species	Low	High
Biochemical diversity	Low	High
Equability of species	Low	High
Structural diversity	Poorly organized	Well organized
Life cycles and history		
Specialization	Wide	Narrow
Size of organisms	Small	Large
Life cycles	Rapid and simple	Slow and complex
Nutrient cycles		
Mineral cycles	Open	Closed and complex with significant controls of biomass
Nutrient exchange between organisms and the environment	Rapid	Slow
Role of detritus in regeneration of nutrients	Not important	Very important
Succession		
Selection by growth	r[1]	K[2]
Production	Aimed at greater quantity	Aimed at greater quality
Homeostasis of community		
Intense symbiosis	Not developed	Developed
Conservation of nutrients	Poor	Good
Resistance to external disturbances	Poor	Good
Entropy	High	Low
Information	Low	High

[1] Quickly reproducing species; [2] Slower-reproducing species.
Source: Odum (1969).

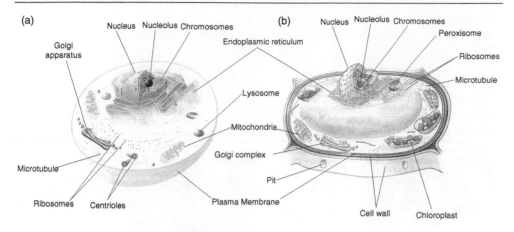

Figure 7.2   Structure of a cell in a photosynthetic organism (b) as compared to an animal or non photosynthetic protist cell (a).
Source: Doodson (2005).

or silica (typical of diatoms) (Taylor, 1980). These cells still have chromatophores that vary greatly in size and number and can take many forms such as platelets or discs. Cyanobacteria are not eukaryotes but prokaryotes, and do not have chromatophores. A detailed description of the characteristics of photosynthetic eukaryotic cells was presented by Oliveira (1996). Photosynthetic eukaryotes have pigments that vary in different groups. All groups of freshwater phytoplankton contain chlorophyll (notably chlorophyll-*a*) and beta-carotene; some contain xanthophylls; phycobilins are limited to cyanobacteria and Rhodophyta (Round, 1981).

Reserve products of photosynthesis and metabolism are found in the cytoplasm of phytoplankton. Chlorophytes and cryptophytes store starch. Chrysophyta produce other polysaccharides such as chrysose and chrysolaminarin. Cyanobacteria store glycogen. Many components of phytoplankton even store proteins and lipids. The rates vary and may be significantly altered by environmental conditions. The levels of these compounds also vary with cell metabolism.

Diatoms have a rigid cell wall, with silica (frustule), consisting of two valves with an epitheca and a hypotheca. The valves are connected by pectin or by protuberances. This 'box' of silica contains cytoplasm, vacuoles and nucleus. Figure 7.3 presents several representative species from the different phytoplankton groups.

7.4.1   Reproduction and life cycles

Phytoplankters normally reproduce by simple division. The rate depends on the cell's physiological conditions, the water temperature, and the nutrient supply.

Cellular division can be synchronised (i.e., **simultaneous division** of all cells in a population), which basically depends on light/dark cycles, nutrient levels and water temperature. Synchronous division can be stimulated in a laboratory through the manipulation of certain nutritional conditions and light/dark cycles. Some studies report examples of synchronous divisions in natural conditions (Nakamoto, Marins and Tundisi, 1976).

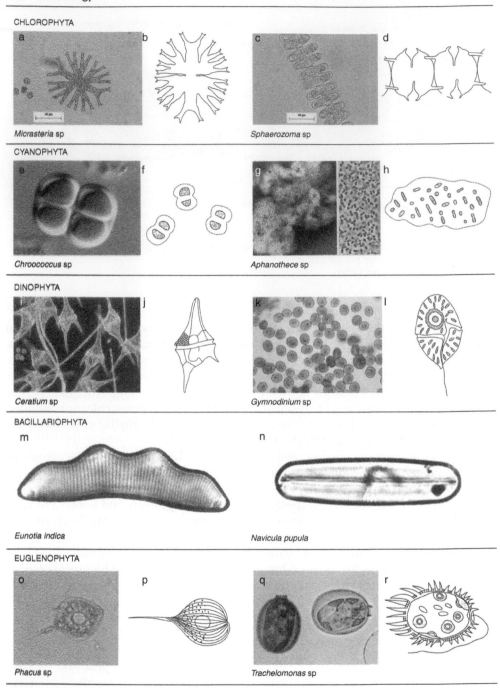

CHLOROPHYTA

a | b

Micrasteria sp

c | d

Sphaerozoma sp

CYANOPHYTA

e | f

Chroococcus sp

g | h

Aphanothece sp

DINOPHYTA

i | j

Ceratium sp

k | l

Gymnodinium sp

BACILLARIOPHYTA

m

n

Eunotia indica

Navicula pupula

EUGLENOPHYTA

o | p

Phacus sp

q | r

Trachelomonas sp

Figure 7.3　Representative examples of the different divisions of phytoplankton (see also color plate section, plates 15).
　　Sources: Prescott, 1978 (b, d), Canter-Lund and Lund, 1995 (e, g, i, k, q); Hino and Tundisi, 1984 (m, n); Mizuno, 1968 (p); Bicudo and Menezes, 2005 (j, l, r); Silva, 1999 (f, h); Thais Ferreira Isabel (a, c); Ana Paula Lucia (o).

Cultures and **natural populations** can divide without synchronisation. In such cases, large numbers of cells in different phases of the cycle are present, sometimes making it difficult to identify and collect organisms.

Prokaryotes can maximally divide roughly every hour, eukaryotes every 8–24 hours. Some species divide during light periods, while others prefer dark periods. In flagellates, reproduction consists of a simple longitudinal division. In species with cell walls, the process is more complicated. In diatoms, a new smaller hypotheca is formed in the daughter cells (with a smaller diameter). The size steadily decreases with each division until the formation of an **auxospore,** which is often related to sexual reproduction.

The formation of **resistant cysts** and **spores** that can survive unfavourable periods is common in many phytoplanktonic organisms. Cysts are formed by **sexual** or **asexual** processes. **Resistant cysts** have reduced chlorophyll levels as well as water loss. In some prokaryotes and cyanobacteria, true spores occur.

Cysts settle rapidly, lodging in sediment until an environmental stimulus (such as temperature, nutrient level, or light intensity) triggers rapid germination. Vegetative cysts occur in many species of phytoplankton.

Sexual reproduction is common in Chlorophyta and diatoms. Many phytoplanktonic organisms have gametes closely resembling the mother cells, causing identification problems. The gametes of some dinoflagellate species, for example, have different morphology and have been mistakenly identified as new species.

The zygote formed by fusion of gametes in flagellates can be mobile (with flagella) or not (hypnozygote). The zygote can be encysted (zygospore), as is common in Chlorophyta such as *Chlamydomonas* or *Volvox*.

The processes that cause phytoplankters to reproduce sexually are not yet fully known. The formation of zygotes has been observed more frequently at the end of extensive blooming or in cultures with excess cells. In conditions of limited nitrogen, a stimulus for sexual reproduction has been observed in cultures of several *Chlamydomonas* species (Lund, 1965; Reynolds, 1984).

External factors, such as thermal perturbations, hydrodynamics, or light intensity, associated with internal conditions, may be responsible for sexual reproduction in phytoplankton.

Lund (1965) reported that three genera of algae (*Asterionella, Fragilaria* and *Tabellaria*) rarely present resistant forms. Another group occasionally produces resistant spores, including some *Aphanizomenon* species. And another group produces resistant spores annually, including Chrysophyta, dinoflagellates and, in some temperate regions, *Anabaena*. The genus *Cyclotella* does not have resistant spores or other resistant forms, nor do *Microcystis* and *Oscillatoria*.

The case of production of resistant spores in several *Aulacoseira* species is classic (Lund, 1965).

7.4.2 Environmental influences on morphology

Polymorphism during the life cycle is common. In general, internal and external factors lead to changes in the shape of some phytoplanktonic species. Influences from light intensity, light quality, nutrient levels in the water, osmotic pressure, and temperature have been described as significant in altering the morphology of cells and colonies in lacustrine and marine phytoplankton. Silica deficiencies in cultures can cause morphological changes in certain diatom species. Traivor *et al.* (1976) showed that iron levels influence the morphology of *Scenedesmus* spp., and in certain species, formation of colonies can also be predicted when organic phosphate is present (Lund, 1965).

7.4.3   Symbiosis and interrelations

Phytoplankters can associate themselves with photosynthetic or non-photosynthetic organisms. For example some non-photosynthetic flagellates can be found on the surface of diatoms. Other **symbiotic associations** have been described, mainly with ciliates.

7.4.4   Characteristics of flagellate forms and sessile forms

In freshwater environments, the main sessile forms include the **diatoms, desmids** and **Chlorococcales**. Flagellate stages occur in the life cycle of **Chlorococcales**.

In the main stage of their life cycle, flagellates present structures that enable locomotion, called flagella. Several groups of flagellates contribute significantly to phytoplankton. Generally there are two flagella, one more developed than the other, which perform homodynamic beats (i.e., even beats) or heterodynamic beats (differing beats). Many flagellates are non-photosynthetic, which makes it difficult to distinguish between the two physiological types unless dyes, or special observational methods such as fluorescence, are applied.

7.4.5   Controlling and limiting factors

According to Reynolds (1997) (also see Chapter 6), phytoplankton is the name given to a community of photo-autotrophic organisms that live most of their life cycle in pelagic zones of oceans, lakes, ponds and reservoirs. **Photo-autotrophic phytoplankton** (described in Chapter 9) plays a key role in producing organic carbon, which supplies food chains in the pelagic zone.

Primary productivity of phytoplankton, biomass, species composition and fluctuations in communities are the main features involved in the dynamics of these organisms. Primary productivity of phytoplankton and how to measure it will be examined in Chapter 9. Measuring primary productivity and biomass, in many cases, is sufficient for the scientific understanding of certain processes, but it is also important to examine species succession and other influencing factors.

Factors that affect the physiology, growth and reproduction of phytoplanktonic organisms include light quality and quantity, which vary based on the climate of solar radiation and the characteristics of the water masses (see Chapter 4). Inhibition of photosynthesis due to high light intensity is another key factor, examined in detail in Chapter 9.

Figure 7.4 illustrates the relationship between maximum depth of penetration of solar energy, 'critical depth' (the point at which **gross photosynthesis** is equal to respiration per unit area), and movement of water masses based on factors such as wind, solar radiation and the consequent vertical transportation of phytoplankton.

The **vertical distribution of productivity**, biomass (chlorophyll-*a*), and their interrelationship are important to identify in order to classify the aquatic ecosystem and its liquid productive capacity of organic matter.

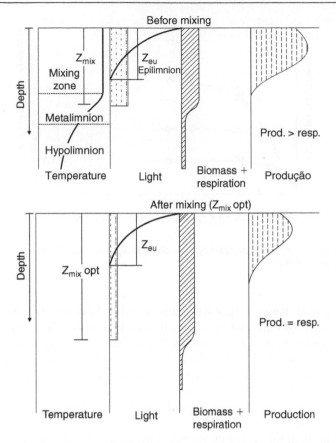

Figure 7.4   Effects of mixing (natural and artificial) on the biomass of phytoplankton and production. The phytoplankton is mixed to depth Z_{mix}, while light penetrates only to depth Z_{eu}. When the mixing is deep, there is low production of phytoplankton and, consequently, reduction in biomass.
Source: Straškrába and Tundisi (2000).

Temperature is another factor influencing phytoplankton growth and response. Within limits the metabolic response of all organisms follows the basic Q_{10} rule, which states that **metabolic processes roughly** double their rate with every 10°C increase in temperature. Temperature limits the saturation rate of **phytoplankton photosynthesis**. At low light intensities, photosynthesis increases in proportion to light intensity, but reaches a maximum that depends on the temperature – with the increase in temperature, the maximum increase in accordance with the Q_{10} rule. Figure 7.5 describes this relationship and Figure 7.6 indicates the growth rates in a group of planktonic algal species related to light intensity.

Interactions between light intensity and temperature influence seasonal succession of phytoplankton species, as reported by Lund (1965), but many subsequent studies indicate greater complexities in the processes of phytoplankton succession.

In addition to the fluidity and transparency common to all aquatic environments, the aforementioned concentration of chemical elements and chemical composition of

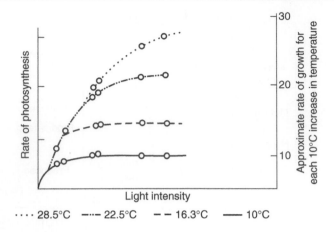

Figure 7.5 Relationship of photosynthesis by *Chlorella* with light and temperature.
   Source: Welch (1980).

the water (resulting from the regional hydro-geochemistry and features of the drain-age basin, geomorphology and activities of the organisms) play a fundamental role in the organization, vertical and horizontal distribution, and succession of the community. Unlike marine waters, the chemical composition of inland waters are varied and complex. They can range from low levels of sodium and potassium to high levels of calcium, magnesium and bio-carbonates. In regions with high evaporation rates and geological deposits, lakes may be highly saline and highly alkaline, as described by Williams (1996).

Many elements in the chemical composition of inland waters are, in the final analysis, vital constituents in the cells of aquatic plants and animals. Approximately 20 elements are required to sustain healthy tissue in plants; many of these are necessary at such low concentrations they are considered to be trace elements or 'micro-nutrients.' Manganese, molybdenum, cobalt, zinc and iron are some of the elements that must be added, if any, to algal cultures in laboratories. Calcium and silicate are also necessary for the growth and development of certain groups of phytoplanktonic algae.

Six elements are considered to be principal nutrients: carbon, hydrogen, oxygen, nitrogen, phosphorus and sulphur. Table 7.1 shows the Redfield ratio, presenting the ratio of **carbon** and **phosphorus** and chemical composition of some typical species of algae with the chemical composition of water in aquatic ecosystems. It should be noted that chemical compositions can vary greatly. They depend on various different relationships among the components in the aquatic system and inflows from the drainage basin, the decomposition rate of dead tissue, and the biogeochemical processes dependent on temperature and dissolved oxygen levels. All existing data and information on **limiting nutrients** (see Chapter 10) indicate that **carbon, phosphorus** and **nitrogen** form the basis for the sustainability and reproduction of **phytoplanktonic populations** and other primary producers (Reynolds, 1997).

Nitrogen – one of the principal nutrients that can limit reproduction, growth and sustainability in aquatic plant populations – is used to synthesize amino acids and proteins. Aquatic plants' main sources of nitrogen include **nitrate, nitrite** and **ammonium**

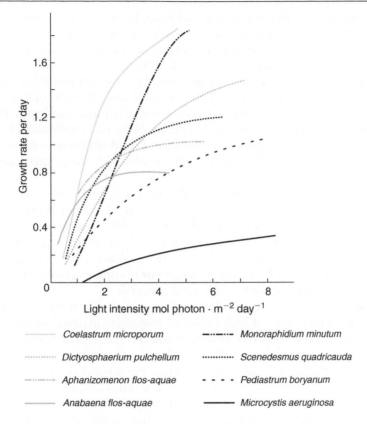

Figure 7.6   Rate of growth of phytoplankton in function of light intensity, at 20°C.
        Source: Modified from Reynolds (1997).

ions, as well as some forms of dissolved nitrogenated organic compounds such as amino acids and urea. Some cyanobacteria with **heterocysts** (or heterocytes) can fix available **atmospheric nitrogen** (*Anabaena, Anabaenopsis, Cylindrospermopsis* and *Gloeotrichia* in inland waters, and *Trichodesmium* in marine waters).

   Phosphorus regulates the productivity of aquatic plants because of its role in intracellular molecular synthesis and transport of ions. It is partly available to aquatic plants in the form of orthophosphate (HPO_4^{2-}, $H_2PO_4^{-}$) ions from organic molecules, all resulting from the decomposition of organisms. Meanwhile, geochemical forms (such as apatites, evaporites and other phosphate minerals) are found at low levels and are not available (Stumm and Morgan, 1981).

   Phosphorus levels are often too low for rapid and sustainable growth of aquatic plants, and the level sets the limit for biological productivity in inland aquatic systems. However, a correlation was found between the **soluble phosphorus level** and phytoplankton biomass (chlorophyll *a*). The mean result (see also Chapter 10) is represented by the equation: $\log[\text{chlor } a]_{max} = 0.585 \log[P]_{max} + 0.801$, where $[\text{chlor } a]_{max}$ is the maximum chlorophyll level in relation to the originally available phosphorus level $[P]_{max}$ (Reynolds, 1978, 1992). Of course, the stoichiometric relationship of N:P is significant in this process.

The relationship between the levels of phosphorus and chlorophyll *a*, (initially described by Sakamoto (1966) and Vollenweider (1968) for log total P and log chlorophyll-*a*) evolved to include nitrogen, transparency, colour, inorganic turbidity and interrelations in the food chain (Huszar *et al.*, 2006).

In the analysis of a database that included 196 low-latitude inland aquatic systems (136 lakes and 56 reservoirs) between 31°N and 30°S latitude, Huszar *et al.* (2006) examined the relationship between chlorophyll and nutrients. The authors compared area, average depth, Secchi disc depth, extinction coefficient, chlorophyll *a*, total phosphorus, total nitrogen, and the ratio of total phosphorus to total nitrogen (see Figure 7.7a). Figure 7.7b compares the results with temperate-zone lakes and reservoirs.

The authors showed that substantial differences existed in the quantitative ratio between chlorophyll *a* and nutrients, and a more variable ratio between log of total phosphorus and log of total chlorophyll *a*, with lower production of chlorophyll *a* per unit of total phosphorus than the regressions calculated for temperate-zone lakes. The differences cited took into consideration the problems of sampling, seasonal differences in the restriction of nutrients (nitrogen or phosphorus) and differences in the reduction of light intensity due to suspended material.

In the specific case of reservoirs, the situation is more complex. Land use, phosphorus and nitrogen loads and high inflow of suspended material during the summer can complicate total N: total P relationship and the ratios total P: chlorophyll *a* and total N: chlorophyll *a*. Similarly, differences between temperate and tropical lakes in terms of the effects of zooplankton predation on phytoplankton influence the results, since cladocerans, rotifers, and copepodites play a significant role in removing phytoplankton from tropical lakes and reservoirs, making the food chain more complex and altering the **prediction capacity**, beginning with data on the ratio of total phosphorus: chlorophyll *a* (Levis, 1990; Arcifa *et al.*, 1995; Fisher *et al.*, 1995; Lazzaro, 1997).

Indices used to quantify eutrophication and the ratio of nutrients to phytoplankton in temperate-zone lakes and reservoirs, for example, must be cautiously applied to tropical and semi-arid lakes and reservoirs. These warmer systems are spatially and temporally more complex, and thus require other indices (Tundisi Matsumura *et al.*, 2006).

Carbon dioxide (CO_2), especially when reduced in aquatic systems with high pH, can limit growth, as demonstrated by Talling (1973, 1976).

Dissolved oxygen and **redox potential** are other essential factors for photoautotrophs in aquatic ecosystems. In anoxic systems, a few highly specialized species can survive, as is the case of some cyanobacteria. A combination of **microbial respiration** with chemical oxidation of organic material reduces the concentration well below the equilibrium levels of dissolved oxygen (between $8–14 \, mg \, O_2 \cdot L^{-1}$ at 0°–25°C temperature). Close to the sediment, the oxygen demand increases.

The relationship between redox potential and available nutrients is key for biochemical cycles and phytoplankton reproduction and growth. With low dissolved oxygen levels, for example, nitrate is reduced to nitrite and nitrogen gas (N_2). At redox potential $<50 \, mV$, Fe^{3+} is reduced to Fe^{2+}, liberating precipitated phosphate as ferrous phosphate at higher redox potential.

The presence of high chlorophyll *a* levels at depths below the thermocline, with low light intensity, high phosphate and anoxia levels can be explained by the availability of these nutrients at low redox potentials (Reynolds *et al.*, 1983).

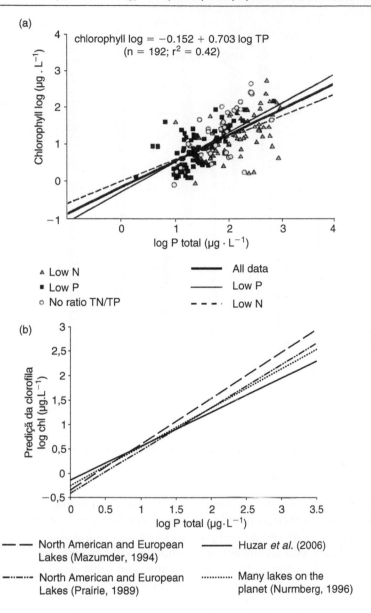

Figure 7.7   a) Relationship between the annual average log of total phosphorus ($\mu g \cdot L^{-1}$) and log chlorophyll ($\mu g \cdot L^{-1}$) for surface waters of 192 tropical and subtropical lakes in Africa, Asia, South America, and North America. Data presented for lakes limited by phosphorus (where the TN:TP ratio is >17 by weight) and lakes limited by nitrogen or nitrogen and phosphorus (where the TN:TP ratio is <17 by weight); b) Comparison of ratio of log of total phosphorus ($\mu g \cdot L^{-1}$) with log chlorophyll ($\mu g \cdot L^{-1}$) between the tropical and subtropical lakes, regressions selected for temperate lakes.
Source: Adapted from Huszar *et al.* (2006).

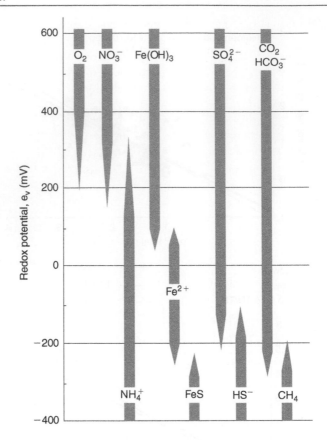

Figure 7.8   Distribution of chemical species of biologically important elements in a spectrum of redox
potentials.
Source: Modified from Reynolds (1997).

Figure 7.8 illustrates the distribution of chemical compounds at various different redox potentials and their availability in water, which affect the reproduction, growth and sustainability of photo-autotrophic communities in aquatic ecosystems. Silicate and iron levels are also significant in phytoplanktonic growth (especially silicate, in the specific case of diatoms). Iron, molybdenum and other elements are essential for growth, especially under chelation by complex chemical components, in some cases, humic substances (Droop, personal communication).

Variable physical and chemical conditions and their interactions in lakes and reservoirs establish a basic hierarchy of factors that control, limit and promote the growth of photo-autotrophic organisms in aquatic ecosystems (i.e., photosynthetic phytoplankton, periphyton, aquatic macrophytes and photosynthetic bacteria).

Another factor that regulates and controls phytoplankton growth is the level of organic material, such as humic substances ('Gelbstoff') and other dissolved organic compounds. Many species may use this material heterotrophically (Droop, 1962; Rodhe, 1962).

In extremely deep lakes, Overbeck and Babenzien (1963) reported assimilation of dissolved organic matter. Further evidence shows that heterotrophic growth utilizing dissolved organic matter occurs in certain types of lakes during specific periods of the year. The release of extracellular substances, which began to be studied in the 1960s (Hellebust, 1965; Tundisi, 1965), led to enormous growth in this field (Fogg, 1962; Stewart, 1963; Veira *et al.*, 1994, 1998) and supported hypotheses of possible re-use of this dissolved material in the growth of phytoplankton, as reflected in the succession of species (Reynolds, 1997).

Parasitism and predation by herbivores are other factors that control productivity, biomass and phytoplankton succession. Many fungi that parasitize species of phytoplankton are generalists. Rotifers, cladocerans and calanoid copepods are predators on algae by the pressure exerted by grazing. Most of these animals are filtrators and ingest relatively small phytoplankters (10–50 μm) (Nauwerck, 1963; Lund, 1965; Reynolds, 1984; Rietzler *et al.*, 2002).

Numerous studies (including Cushing, 1959; Edmondson, 1965; Reynolds, 1997) showed the diverse interactions between the seasonal phytoplankton cycle and the grazing factors, predation and succession. The interrelationship between grazing and phytoplankton succession is greater in marine ecosystems, as shown by Cushing (1963a, 1963b). Pressure on nano-phytoplankton is greater in environments where the <50-μm component is predominant (Tundisi and Teixeira, 1968), but there is evidence of grazing pressure on >50-μm components by marine and freshwater calanoid species (Mullin, 1963; Rocha and Matsumura Tundisi, 1997). Parasitism can affect phytoplankton, reducing populations and altering the pattern of phytoplankton succession (Lund, 1965; Reynolds, 1984).

7.4.6 Buoyancy, sedimentation rates and displacements

Due to differences in light intensity and various vertical gradients (such as in depth of euphotic zone, nutrient levels, and temperature distribution), the phytoplankton photo-autotrophic growth rate is influenced by the organisms' vertical distribution in the column of water.

A central problem in plankton ecology, particularly in the case of phytoplankton, is the fluctuation and vertical distribution produced by turbulence and movement of water masses. Production of organic matter and **phytoplankton biomass** are largely determined by the **sedimentation** rate and the floatation capacity of cells and colonies (see Chapter 9).

Sedimentation of phytoplankton thus has some negative effects on the growth of cells and colonies (such as less available light intensity due to sinking) and positive aspects, such as periodic renovation of nutrient levels near the cells, which occurs as the cells move downward and pass through layers of water with different nutrient levels. In a medium with low nutrient levels and little turbulence or stratification, the permanent sedimentation of phytoplankton is undoubtedly an advantage. In a turbulent medium, such sedimentation is a disadvantage. The characteristics of phytoplankton buoyancy and sedimentation rates are determined by each species' physiological processes and morphology, with obvious selective implications in terms of size, cell volume and factors such as covariance with conditions of turbulence and hydrodynamics (Reynolds, 1973 a, b).

In examining the problem of phytoplankton sedimentation, it is important to consider the physical influences that act on phytoplankton organisms, basically the movement of water masses and the forces that act on inert bodies in viscous fluids. In general, the sedimentation rate of spherical bodies is given by the Stokes equation:

$$V_s = 2/9 \ gr^2 (\rho' - \rho)\eta$$

where:
V_s – rate of sedimentation $(m \cdot s^{-1})$
g – acceleration of gravity $(m \cdot s^{-2})$
η – coefficient of the medium's viscosity $(kg \cdot m^{-1} \cdot s^{-1})$
ρ – density of the medium $(kg \cdot m^{-3})$
ρ' – density of spherical body $(kg \cdot m^{-3})$
r – radius of spherical body (m)

Source: Margalef (1983); Reynolds (1984).

The factors r and ρ' in the Stokes equation are characteristic of individual organisms and mainly determine the sedimentation rate. A large number of phytoplankton species have non-spherical shapes. It is thus necessary to determine the type of shape that causes resistance to sedimentation. Size (which is highly dependent on the movements of the water masses), type of resistance, and cell density are important factors in resistance to sedimentation.

The effect of shape on sedimentation is generally expressed in terms of the drag coefficient ϕ, defined by: $\phi = V_s/V$, where V is terminal velocity of a particle and V_s is the velocity of a sphere of equal density and volume in the same liquid (Walsby and Reynolds, 1981).

Theories of resistant shapes have developed based on ellipsoids; those with elongated shapes are the most resistant to sedimentation, such as certain heavy planktonic diatoms.

The sedimentation rate of a cylinder with a constant diameter increases with increasing length (Hutchinson, 1967). In the case of colonies or chains, despite increased density, the sedimentation rate is slower than that of an equal-volume sphere. Three shapes – spheres, ellipsoids and cylinders – account for a large percentage of the number of shapes present in any phytoplanktonic association. The presence of protuberances, horns and other kinds of formations tends to increase due to cellular surface area/volume.

The cell or colony's orientation during sedimentation is another important factor. The location of the cells' protuberances or horns and the weight differential caused by these locations must also be taken into consideration (Smayda and Boleyn, 1966).

As already described, the main chemical components in the protoplasm of living cells are denser than water. Cells are therefore denser than water. The accumulation of certain substances facilitates buoyancy, as seen in the case of lipids, which can account for up to 40% of the dry weight of phytoplankton. In general, diatoms in a **senescent state** produce excessive lipids. Cultures with high radiation or limited oxygen also produce excess lipids (Fogg, 1965).

The vital regulation of organisms' sedimentation or sinking rate is achieved by regulating the carbohydrate, starch, and glycogen content or by changing the viscosity of the surrounding medium through the release of organic substances. The theory of **structural viscosity** (Margalef, 1983) encapsulates aspects related to the surface produced by the cells in water, as by growing a diatomic 'particle' + water, which, since it is bigger than the diatom, would on average be less dense. The variation of the electrical charge on the surface (zeta-potential) would alter the sinking rate. Likewise, variations in sinking occur during periods of high or low photosynthetic rates. In the case of cyanobacteria with **gas vesicles**, there is a coupling of the respiration/photosynthesis cycle in the process.

7.4.7   Time scales in the dynamic ecology of phytoplankton

The definition of time scales in the ecology of phytoplankton is essential for understanding factors such as horizontal and vertical distribution and phytoplankton succession. Phytoplanktonic aggregations on the horizontal axis of lakes and reservoirs or in backwaters of rivers depend on current speeds in response to forces such as wind and thermal heating and differences in density and/or nutrient levels. Harris (1986) found that vertical mixing in the scale of meters can take 24 hours, while horizontal mixing on the scale of miles occurs in the same time period. Vertical and horizontal dimensions are therefore indispensable in phytoplankton ecology, since they define mechanisms and forces that clearly influence succession and the combination of factors in the process (Tundisi, 1990; Margalef, 1991; Reynolds, 1997).

Figure 7.9 defines the scales in phytoplanktonic ecology and the relationships with the main biogeochemical processes as well as processes of succession.

7.4.8   Succession and the spatial and temporal organization of phytoplankton

Planktonic communities and their organization have been studied extensively. Initially, it was considered that, due to the fluidity of the medium and its wide variability, there was little or no **spatial or temporal structural organization** (reviews of Smayda, 1980; Harris, 1987). Meanwhile, scientific results of many studies showed regularities in the succession of marine phytoplankton species (Margalef, 1967, 1978; Raymont, 1963; Smayda, 1980). These studies describe associations of species common in tropical, temperate and polar regions. Most studies describe the process of succession in shallow neritic ocean waters and the relationship with the pelagic zone and areas of upwelling, estuaries and coastal waters (Teixeira and Tundisi, 1967; Smayda, 1980; Tundisi *et al.*, 1973, 1978).

According to the authors, the most conspicuous and consistent components of marine planktonic communities are diatoms (Bacillariophyceae), dinoflagellates (Pyrrophyta) and coccolithophorids (Haptophyceae – see phytoplanktonic classification in Chapter 6). In ocean regions and along tropical coasts, the cyanobacterium *Trichodesmium* commonly occurs above the thermocline, as well as the dinoflagellates *Gonyaulax* and *Gymnodinium*, which form red tides in tropical and oceanic coastal waters. More recently, Azam *et al.* (1983) described pico-phytoplankton in oceanic areas as significant in the food chains in these oligotrophic regions. The description of regularities in spatial and temporal succession of marine phytoplankton led to a more precise and consistent evaluation of phytoplankton in inland waters (Rodhe, 1948; Rawson, 1956; Lund, 1965; Hutchinson, 1967; Reynolds, 1980).

Reynolds (1997) described a series of associations common in lakes with different vertical mixing processes and nutrient levels, from oligotrophic systems to totally

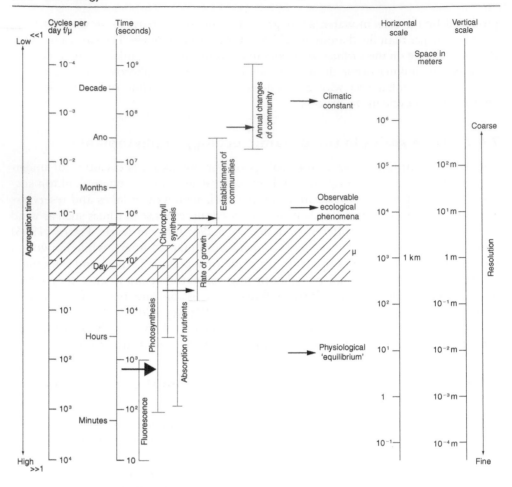

Figure 7.9   Time scales in ecology of phytoplankton.
Source: Modified from Harris (1986).

eutrophic systems. The combination of factors (a matrix of **vertical mixing, intensity of underwater solar radiation,** and **nutrient levels**) need to be studied at the regional level, focusing on individual lakes and reservoirs (Tundisi, 1990).

The existence of these associations depends on a series of covariances among the vertical and horizontal distribution of limiting and controlling factors: light intensity, degree of turbulence, and available inorganic nutrients. However, these associations vary in space and time, and the functions of the physical and chemical forces acting on them are the basic processes driving succession (Harris, 1986). The random occurrence of external disturbances produces variability in ecosystems, which causes alterations in the succession, or continuity, depending on the period of time under consideration.

According to Reynolds (1984, 1997), these associations include:

▶ **Diatom-dominated associations** – occur in turbulent waters; *Cyclotella* is often dominant in oligotrophic waters and *Aulacoseira* in eutrophic waters.

▶ **Chrysophyta-dominated associations** – in high-altitude lakes in the Southern and Northern hemispheres; *Dinobryon* is a significant component in this association.

▶ **Chlorophyta-dominated associations** – order Chlorococcales; *Sphaerocystis* is a common genus. Cell colonies generally held together by mucilage. Other common genera in this group are *Gloeocystis* and *Botryococcus*, which use oil droplets as a product of assimilation. *Botryococcus* is a cosmopolitan genus, common in oligotrophic waters.

▶ **Chlorophyta-dominated associations in eutrophic lakes:** *Scenedesmus, Pediastrum, Ankistrodesmus* and *Tetraedron* are common in this group partly as isolated non-colonial cells; however, the purely colonial and motile genera *Eudorina* and *Pandorina*, found in shallow waters with high nutrient levels, are also representatives of this association.

▶ **Dinoflagellate associations** – *Peridinium* and *Ceratium* spp. often occur at the surface waters of lakes with low nutrient levels. Because of their mobility, they can explore deeper waters, richer in nutrients. In some mesotrophic or eutrophic lakes, *Ceratium* and *Peridinum* spp. develop extensive biomasses and compete with cyanobacteria.

▶ **Cyanobacteria-dominated associations** – Cyanobacteria can be found in a wide range of lakes with various trophic stages. Species that fix atmospheric nitrogen, such as *Anabaena* and *Aphanizomenon* spp., can dominate shallow waters with low nutrient levels. *Cylindrospermopsis* is also common in this association. *Microcystis aeruginosa* is dominant in some tropical eutrophic lakes with high thermal stability and may form an association with *Ceratium* spp. in some lakes. In this group, filamentous forms such as *Planktolyngbya, Lyngbya, Phormidium,* and *Pseudoanabaena* can dominate the plankton of polymictic eutrophic lakes with high turbidity and low light penetration. *Planktothrix agardhii* and *Pseudoanabaena limnetica* occur in these lakes (Post *et al.*, 1985). Other components of this association occur in lakes with high thermal stability and steep chemical gradients. *Planktothrix* spp., and *Lyngbya* were identified in Lake Carioca (Parque Florestal do Rio Doce – MG) (Reynolds *et al.*, 1983) and *Phormidium* spp. (Vincent, 1981) in an Antarctic lake permanently covered with ice.

▶ **Cryptomonas associations** – include species in the *Cryptomonas, Chilomonas* and *Rhodomonas* genera; generally found in mesotrophic or eutrophic lakes. These are biflagellate algae with moderate locomotive capacity that can form 'plates' in stratified lakes.

Nanoplankton and picoplankton (or picophytoplankton) associations include diverse groups of algal species classified by size (picophytoplankton: cells measuring 0.2–2 μm; nano-phytoplankton: cells measuring 2–20 μm;). They dominate superficial waters in stratified lakes and oligotrophic or eutrophic waters dominated by *Chlorella* or *Monoraphidium*.

Picophytoplankton is found in oligotrophic waters dominated by *Synechococcus* spp., *Synechocystis* spp. (cyanobacteria) and by Chlorophyta such as *Chlorella minutissima*. Their play an important role in the food chain in oligotrophic lakes, and can present high productivity during summer months.

Photo-autotrophic bacteria are another association, found in regions of lakes with reducing conditions with low light intensities; a variety of violet-coloured bacteria (*Chromatium* sp., *Thiocapsa*) or green (*Chlorobium, Pelodictyon*) (Vincent and Vincent, 1982; Vicente and Miracle, 1988). Guerrero *et al.* (1987) described these associations in lakes of the Mediterranean region in Spain (see Chapter 9).

An association that can be considered **miscellaneous** occurs in waters with high levels of humic substances and dissolved organic matter dominated by plankton consisting of euglenids (*Euglena* spp.), dinoflagellates (*Peridinium* spp.), or motile diatoms of the genera *Navicula* and *Nitzschia* spp. Such associations were found, for example, during the filling of several dams in Amazonia.

These associations are the result of the responses of phytoplankton to different physical, chemical and biological influences in different aquatic ecosystems and in covariance with the functions of climatological factors (i.e., wind, solar radiation, and rainfall).

7.4.9　Phytoplankton succession and conceptual models

Hydrological, hydraulic and hydrodynamic functions in lakes, rivers and reservoirs, as well as relations between vertical distribution of factors such as light intensity and nutrients, establish the different patterns of phytoplanktonic associations (including macro-phytoplankton and picophytoplankton). The results of **wide-ranging studies** conducted in a group of lakes, reservoirs, rivers and coastal waters established the precise correlation of variables and, to a certain extent, the **predictive capacity** with great theoretical and applied value. Figure 7.10 shows

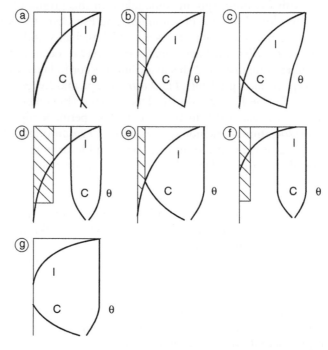

Figure 7.10　Patterns of ideal vertical distribution of underwater solar radiation (I) and the concentration of limiting nutrients (C) in relation to the depth of mixing represented by the temperature distribution (θ) and the ability to support growth of phytoplankton (hatched). a, b, c, d, e, f, g corresponds to co-variances between the different light intensity, nutrient concentration and thermal stability or instability, resulting in different vertical distributions (hatched). *Sources:* Reynolds (1997) and modified from Reynolds (1987).

the vertical patterns postulated by Reynolds (1997) and the relationship between **biomass, nutrients, temperature, water density** (affecting and relative turbulence), and **light intensity**.

Chart 7.2 presents the relationship between spatial and vertical distribution of phytoplankton, its frequency, temporal cycle and seasonal succession (Tundisi, 1990), summarizing the previous discussion. Reynolds (1997) has illustrated the succession of aquatic ecosystems (especially pelagic succession) and the effects of perturbations (see Figure 7.11). The concept of **exergy** introduced by Jørgensen (1992) and Jørgensen *et al.* (1992) shows a progression as communities mature, along with the accumulation of genetic information and the number of genes in the community (Matsumura Tundisi, 2006) (see Figure 7.12).

Case studies in Brazilian lakes and reservoirs and in shallow lakes

The theory and practice of phytoplankton succession
The UHE Carlos Botelho reservoir, also known as the Lobo reservoir (Broa), is a small reservoir with 22,000,000 m³ and an average depth of 3 m. Many ecological, aquatic biological and limnological studies have been conducted there since 1971 (Tundisi *et al.*, 1971a, b; Tundisi *et al.*, 1997; Tundisi and Matsumura Tundisi, 1995).

In this ecosystem, it was determined that the dominant influences (forcing functions) operating on the phytoplanktonic community (biomass and composition), which is

Chart 7.2   Relationship between spatial and vertical distribution of phytoplankton, frequency and seasonal succession.

Spatial distribution of phytoplankton	Vertical thermal structure Advection currents Horizontal flux and effects of wind Temporal compartmentalization
Vertical	Spatial heterogeneity
Horizontal	Cycles of operation in dams + events (natural cyclic climatological – hydrological)
Sequences	Zeu/Zmix; Zeu/Zaf;Zeu/Zmax Force and direction of wind Rate of reproduction Rate of mortality (effect of grazing + sinking + downstream losses)
Temporal cycle	Circulation; wind; stability Precipitation and nutrient flux Pulses and their effects on seasonal cycles Retention time
Seasonal succession	Use of the hydrographic basin and nutrient charges; potential of eutrophication Rate of annual 'aging' of the dam or the lake Development of trophic relationships in the system Grade of organic/inorganic toxicity

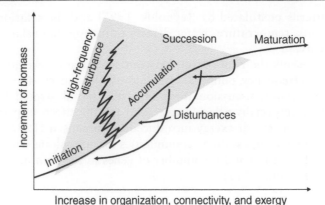

Figure 7.11 Pattern of succession in a pelagic ecosystem, showing the effect of repetitive disturbances in the organization of communities.
Source: Modified from Reynolds (1989).

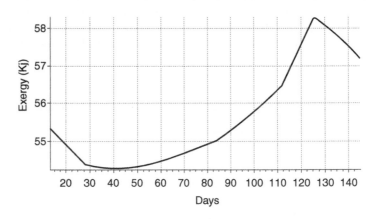

Figure 7.12 Progression of exergy with succession in aquatic phytoplankton communities.
Source: Matsumura Tundisi (2006).

the primary producer of organic matter, are precipitation during summer (November–
March) and wind during the winter (July–September). These two conditions, in terms
of functions of forces, produce the following events: enrichment of nutrients, especially
nitrogen and phosphorus, during summer; and effects of wind-generated turbulence in
the direction of the reservoir's main axis during dry periods in winter.

In the UHE Carlos Botelho reservoir, therefore, nutrient recycling results from
either precipitation, which leads to **'new production'** during summer, or the effects of
turbulence and wind, which promote and stimulate **'regenerated production'** in that
period. Nakamoto *et al.* (1976) described the synchronized growth of *Aulacoseira
italica* filaments in the reservoir immediately after the onset of strong winds
($8-10\,\mathrm{m\cdot s^{-1}}$), which trigger the distribution of filaments in the water and promote

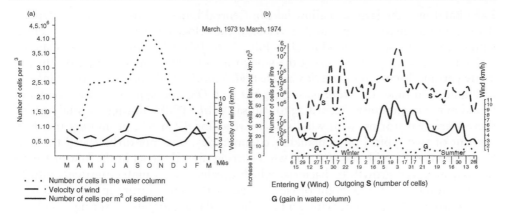

Figure 7.13   a) The seasonal cycle of *Aulacoseira itálica* in the UHE Carlos Botelho (Lobo/Broa) dam, showing the wind-accumulation relationship of colonies in water and in sediment; b) Model generated with information.
Source: Tundisi (1982).

rapid reproduction with an increased number of cells containing cytoplasm in each filament and a more constant number of cells per filament.

In the authors' interpretation, the growth of *Aulacoseira italica* colonies, stimulated by the relocation of filaments from the bottom sediment and their distribution in the water column, promoted rapid multiplication of colonies of diatom cells, dormant in the sediment. The number of same-size cells in the filaments and the approximately equal number of cells in each filament are evidence of synchronized growth, driven by an environmental factor (wind). According to Lund (1965) and other authors, such as Nipkov (1950), *Aulacoseira* spp. can remain for long periods of time in the dark in the sediment deposits of lakes in a strategy of 'physiological rest.' An initial stimulating factor will be needed, in this case the wind and the subsequent movement of filaments into the water column.

The seasonal cycle of *Aulacoseira italica* in the UHE Carlos Botelho reservoir (Lobo/Broa) is a good example of relationships that influence factor-succession in phytoplankton. The cycle is illustrated in Figure 7.13, with the action of wind, precipitation and accumulation of filaments of *Aulacoseira italica* in water and in sediment. This relationship was later studied and a model prepared (Lima *et al.*, 1978) with predictive capacity in relation to wind. The Figure clearly illustrates the function of wind and precipitation as stimulators of 'new production' in summer and 'regenerative production' in winter. The large masses of *Aulacoseira italica* during the winter period are a particular characteristic of this ecosystem and its functioning, a feature that is fully consistent with the diatom-dominated association described by Reynolds (1997) in his model of phytoplankton succession.

Lake Balaton

Another example of phytoplankton succession is presented by Padisák *et al.* (1988). The author and her collaborators conducted extensive studies in **Lake Balaton**, a shallow lake in Hungary where the wind stirs up sediment from the bottom and enriches the water column with nutrients, thus increasing turbidity by turbulence.

Lake Balaton is the largest shallow lake in Central Europe, with an average depth of 3.14 m and a retention time of 3–8 years. Wind velocity reaches 2–12 m·s⁻¹ during summer, and the studies conducted in July 1976, July 1977, and July 1978 showed a fluctuating wind pattern during the study time of approximately 30 days. In each study period, *Aphanizomenon flos-aquae* f. *klebahnii* was dominant, with an increase in duplication time. *Cryptomonas*, *Lyngbya* and *Thischia* spp. were also commonly found in the community. According to Padisák *et al.* (1988), during the study period, there were two storms, with wind velocity reaching 12 m·s⁻¹, promoting the development of two typical phytoplankton communities, one 'pre-turbulence,' the other 'post-turbulence,' with the effect of wind stirring up sediments in the first phase and the synchronized growth of bacteria, which accelerated available nutrients into the water column. From this point of view, consider the temperature in lakes and reservoirs during calm periods, with gentle light winds, analogous to spring circulation in temperate-zone lakes. The succession in the lake of rapidly reproducing species (r-type) and slower reproducers (K-type), according to the author, depends on a 5–7-day period to establish community **K**.

The Lake Balaton studies showed that **physical control**, according to Sommer (1981), can play an essential role in phytoplankton succession in shallow lakes. Growth and loss of species appear to be synchronized with physical forces, especially wind, in this specific case. The established relationships, of r-selection or K-selection, are rapidly destroyed because of the effects of turbulence generated by strong winds and storms, a common phenomenon in shallow polymictic lakes at various latitudes (Branco and Senna, 1996).

Lake Batata (Amazonia)

Another example of theory and practice of phytoplankton succession can be seen in the work by Huszar and Reynolds (1997) in a wetland lake in the Amazonian lowlands (Lake Batata, Para). According to the authors, the lake, connected to the Tronischia River, presents an annual cycle based on hydrology (flood height, flow rates in rivers) and **hydrography** (stability and frequency of vertical mixing in the lake). According to Reynolds (1994), the selection of phytoplankton and the process of succession, particularly in shallow lakes (average depth <5 m), depend on the complex and varied frequencies of turbulence generated by horizontal and vertical movements of water. The gradients of turbulent vertical and horizontal mixing – caused by effects such as wind, precipitation and drainage – overlap and interact with events resulting from the viscosity of the water (residual movements due to dislocations of molecules) and the fluvial dynamic. The gradients of turbulent movement that range from laminar flow to turbulent flow (see Chapter 4) lead to various degrees of suspension and development of **phytoplankton communities**. Figure 7.14 illustrates various stages of succession over time.

Autogenic growth of the population counteracts the impact of allogenic forces such as wind, storms and floods caused by various effects. Autogenic growth leads to structural complexity (as discussed by Odum, 1969; Reynolds, 1997) and lower productivity. Growth subject to allogenic forces implies effects resulting in greater productivity and 'regenerative production' or 'new production' with renewed resources and a simpler structure. The system's perpetual structuring and restructuring (according to Connel and Slayter, 1977; and Margalef, 1991), promotes processes of succession

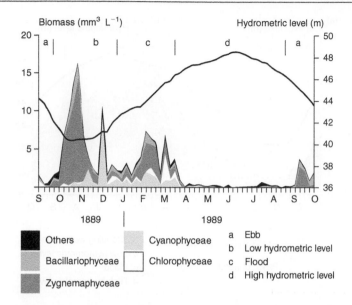

Figure 7.14   Fluctuation of total biomass of phytoplankton (fresh biovolume) of Lake Batata in relation to variation of hydrometric level of the Trombetas River (above sea level). *Source:* Huszar and Reynolds (1997).

that fluctuate between autogenic and allogenic bases. Figure 7.15a shows the flooding sequence in the Trobetas River and the water level in Lake Batata. Figure 7.15b presents the fluctuations in diversity during different flooding periods in Lake Batata. Figure 7.16 illustrates the succession of different phytoplanktonic associations in Lake Batata based on the matrix proposed by Reynolds (1993).

According to Huszar and Reynolds (1997), the frequency of vertical mixing in this lake follows a diel variation pattern, as also shown by Tundisi *et al.* (1984) for a small wetland lake in the Amazons. Seasonal patterns in the volume of river water and the hydrometric level in the lake are superimposed on this variation. Large fluctuations in the river's level, and vertical mixing processes, establish the basic forcing-functions in the lake that regulate and control the essential process of succession.

Other examples of shallow tropical lakes

Examples of shallow tropical lakes can be found in studies on Lake Dom Helvecio (Hino *et al.*, 1986) and Parque Florestal do Rio Doce – MG (Reynolds, 1997). In Lake Dom Helvecio, the vertical distribution of phytoplankton, in a system with great thermal stability, was clearly identified, with a well-established vertical structure of cyanobacteria located in the metalimnion, where half of the **primary phytoplankton production** occurs (total of $377\,\mathrm{mg\,C\cdot m^{-2}\cdot day^{-1}}$). Part of this primary production in the metalimnion was attributed to **microbial biosynthesis,** due to the accumulation of photosynthesizing bacteria (from the **sulphur cycle**) in this region. Phytoplanktonic species in these stratified communities have adapted to low light intensities. In this case, therefore, the availability of inorganic nutrients in the region helps maintain the autotrophic growth of the community.

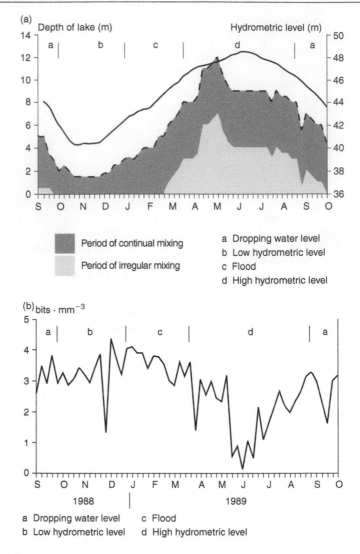

Figure 7.15   a) Sequence of the Trombetas River flood (fluctuation of the water level above sea level) and periods of continuous and uneven mixing. The depth of the lake is represented by the dotted line; b) Changes in species diversity (Shannon-Weaver) according to the hydrological phases of Lake Batata.
Source: Huszar and Reynolds (1997).

In his review on vertical distribution of phytoplankton in monomictic lakes in the Parque Florestal do Rio Doce, Reynolds (1997) concluded that lakes are characterized by communities of algae typical of oligomesotrophic systems. Phytoplanktonic distribution and composition in a lake are more dependent on the lake basin's morphometry than on the water's chemistry. The percentage of **nitrogen fixers** suggests that the **anoxic cycle** can support the biological production and that low representation of diatoms reflects an inadequate and infrequent suspension, due to limited turbulence.

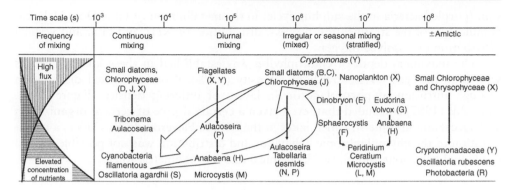

Figure 7.16   Succession of different associations of phytoplankton in Lake Batata (larger arrows) in relation to standard and seasonal vertical mixing, based on the matrix proposed by Reynolds
(1993). In this figure species of the genus *Melosira* were modified to *Aulacoseira*.

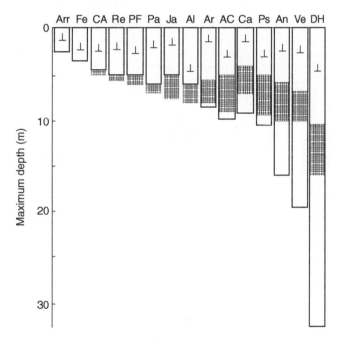

Figure 7.17   Water column in 15 lakes of the Rio Doce Valley with the extension (shaded) of the
metalimnion in December 1985. The depths of the Secchi disc are presented.
Source: Reynolds (1997).

Figure 7.17 shows comparisons of the water column in 15 lakes in the Rio Doce
Valley, the depth of the Secchi disc readings, and the extent of the metalimnion. The
vertical distribution of phytoplankton is regulated by the long period of thermal
stability (such as the case of D. Helvecio Lake and Carioca Lake), the **vertical distribution of nutrients**, and the extension of the euphotic zone. The presence of *Chromatium*

spp. (purple bacteria in the sulphur cycle) in the metalimnion of Lake Dom Helvécio is indicative of the stratification process, distribution of NH_4^+ and H_2S, and vertical distribution of dissolved oxygen.

The maximum depth of chlorophyll-*a* was established in these lakes at depths with high nitrogen levels, low light intensity (<1%), showing typical distribution through intrusion of phytoplankton in layers of differing density (Figure 7.18). Reynolds (1997) concluded that, even given a certain degree of vertical organization and succession in some stratified lakes in the Parque Florestal do Rio Doce, a generalization of the phytoplanktonic behaviour and distribution was not possible, due to individual behaviour and physical, chemical and morphometric forces.

Cyanobacteria

One important aspect of phytoplankton succession in lakes and reservoirs is the occurrence of **cyanobacteria associations** under special conditions, in cases of intense eutrophication. Reynolds (1997) described the conditions under which such

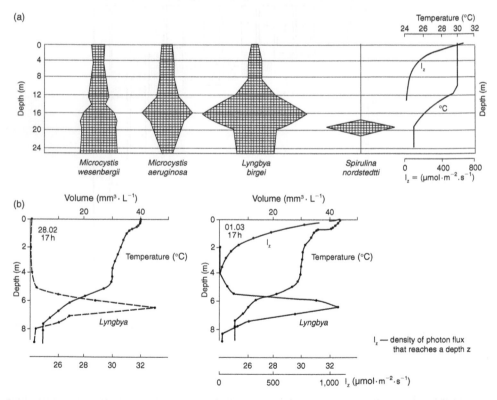

Figure 7.18 Vertical distribution of phytoplankton in Lake Dom Helvécio (Parque Florestal do Rio Doce – MG). a) Distribution of various species of cyanobacteria; b) Vertical distribution of a species of *Lyngbya* in differen Helvécio t times, showing the absence of migration. *Source*: Reynolds, Tundisi and Hino (1983).

associations occur. According to Carmichael (1994), cyanobacteria evolved approximately 3.5 billion years ago, as photo-autotrophic producers that released oxygen into Earth's primitive atmosphere, which at the time was highly reduced. A growing number of studies on cyanobacteria have been conducted in Brazil, especially on reservoirs that supply public water or are used for hydroelectric purposes.

One of the problems produced by cyanobacteria – largely resulting from eutrophication – is the occurrence of toxic species, of which there are 20 in Brazil (in 14 genera). *Microcystis aeruginosa* is the most common. *Anabaena* spp., (*A. circinalis, A. flos-aquae, A. planctonica, A. solitaria,* and *A. spiroides*) are potentially toxic. More recently, *Cylindrospermopsis raciborskii* has been detected and its cycle described in several aquatic ecosystems in Brazil (Branco and Senna, 1994; Sant'Anna and Azevedo, 2000; Huszar, 2000; Conte *et al.*, 2000).

A description of cyanobacteria-produced toxins and their impact on human health can be found Chapter 18. The different toxins have varying effects on human health and aquatic organisms (Chorus and Bartram, 1999). Variations in cyanobacterial toxicity have not been definitively clarified. Chart 7.3 presents the cyanobacterial blooms occurring in Brazilian **springs** up to 2001 (data provided to Dr. Sandra Azevedo by Dr. J.S. Yunes).

The nutrient load is undoubtedly the main cause of cyanobacterial blooms (from **point sources** and diffuse sources), especially if nitrogen deficiencies occur in semi-arid regions, promoting the growth of some species in the genera *Anabaena, Aphanizomenon,* and *Cylindrospermopsis,* capable of fixing atmospheric nitrogen (Reynolds, 1984). Since many cyanobacteria and other phytoplanktonic organisms store phosphorus (Reynolds, 1984; Huni, 1986), their biomass can increase even when phosphorus levels has been depleted. Cyanobacterial blooms occur during periods of high light intensity, high surface temperatures and thermal stratifications that promote stability in the water column.

Gonzalez *et al.* (2004) described the phytoplanktonic composition in an artificial reservoir in Venezuela (Pao-Cachinche) where cyanobacteria from several genera (*Anabaena* spp. *Cylindrospermopsis* raciborskii, *Microcystis* spp., and *Spirulina* spp.) accounted for about 75% of the phytoplanktonic composition during the 18-month study. The reservoir's high temperature ($>28°C$), the permanent thermal stratification, and orthophosphate levels ($>10\,\mu g \cdot L^{-1}$) promoted the growth of cyanobacteria, with primary production values of $1000\,mg\,C \cdot m^{-2} \cdot day^{-1}$, rather high in comparison with other reservoirs in the same region of Venezuela. However, the cyanobacteria level was much lower than in the **hypereutrophic** Lake Valencia ($7,400\,mg\,C \cdot m^{-2} \cdot day^{-1}$) (Infante, 1997).

The presence of mucilagem (mucopolysaccharides that absorb water) in cyanobacteria reduces their density, although it increases cell size. The mucilagem (which also occurs in mobile chlorophyte colonies such as *Gloecystis* sp. and *Oocystis* sp.) can help reduce but not prevent the sinking rate (Reynolds, 1984, 1997). Control of buoyancy by means of the presence of proteinaceous gas cylinders, which are called gas vesicles and regulate buoyancy through production of gas and glycogen resulting from photosynthesis (glycogen used as weight), is one of the most important phenomenon in the physiological regulation of the dislocation of cyanobacteria described by Reynolds and Walsby (1975).

Chart 7.3 Cyanobacterial blooms occurring in Brazilian springs, 1991–2001.

Locale	Year	Predominant cyanobacteria	Toxicity Yes	Toxicity No	Toxicity ND*	Toxins Detected	Method	Source
Barra Marica lagoon (RJ)	1991	Synechocystis aquatilis	X			MCYST	Immuno-assay	Nascimento and Azevedo (1999)
Res. Funil (RJ)	1991/1992	Microcystis aeruginosa	X			MCYST	HPLC-DAD	Bobeda (1993)
Jacarepagua Lake (RS)	1996	Microcystis aeruginosa			X	MCYST	HPLC-DAD	Magalhães and Azevedo (1998)
Itaipu, Iguaçu Park (PR)	1996	Microcystis	X			MCYST	Immuno-assay	Hirooka et al. (1999)
Itaipu reservoir (PR)	1999	Anabaena sp.			X	MCYST	Immuno-assay monoclonal	Kamogae et al. (2000)
Capivara reservoir (PR)	2000	Microcystis sp.			X	MCYST	Immuno-assay monoclonal	Kamogae et al. (2000)
Amparo and Itaquacetuba (SP)	1993/1995	Cylindrospermopsis raciborskii	X		X	SXT, neoSXT, GXT	HPLC-FLD, GXT	Lagos et al. (1999)
Dos Patos lagoon (RS)	1994/1995	Microcystis aeruginosa	X			MCYST-LR, -FR Leul-MCYST	HPLC-DAD HPLC-MS	Matthiensen et al. (2000) Yunes et al. (1996)
Rio Grande (RS)	1995	Anabaena spiroides	X			Anatoxina-a (S)	Inib.AChe	Monserrat et al. (2001)
Rio dos Sinos (RS)	1999	Cylindrospermopsis raciborskii	X			Saxitoxinas equiv.	HPLC-FLD	Conte et al. (2000)
Camaquã (RS)	2000	C. raciborskii, Mocrocystis Pseudoanabaena sp.	X			MCYST; NeoSXt GTX1; GTX2	Immuno-assay HPLC-FLD	Yunes et al. (2000)

Location	Year	Organism			Toxin	Method	Reference
Itapeva (RS)	2000	*Anabaena circinalis, A. spiroides*	X		MCYST; ANTXa; ANTX-a (S)	Immuno-assay HPLC-FLD Inib.AChE	Yunes *et al.* (2000)
Farroupilha, Erechim (RS)	2000	*Microcystis*	X		MCYST	Immuno-assay	Yunes *et al.* (2000)
Peri lagoon, Florianópolis (SC)	2000/2001	*Cylindrospermopsis raciborskii*		X		Bioassay	Relatório: Casan/CNPq/Floran/UFSC
Tapacurá Reservoir (PE)	1998/1999	*Cylindrospermopsis raciborskii*	X		SXT equivalents	Bioassay	Nascimento *et al.* (2000)
Ingazeira Reservoir (PE)	1998	*Cylindrospermopsis raciborskii*	X		SXT equivalents	Bioassay	Bouvy *et al.* (1999)
Itauba (RS)	2000	*Anabaena circinalis*	X		MCYST	Immuno-assay	Werner *et al.* (2000)
Garças lagoon (SP)	1996/1997	*Microcystis aeruginosa; Planktothrix agardhii*		X			Sant'Anna and Azevedo (2000)
Sta. Rita reservoir (SP)	1997	*Microcystis wesenbergii*		X			Sant'Anna and Azevedo (2000)
Juramento reservoir (MG)	2000	*Radiocystis fernandoi* / *Microcystis* spp. ©	X		MCYST	Immuno-assay and HPLC-DAD	Jardim *et al.* (1999, 2000a)
Urban ponds	1998	*Cylindrospermopsis raciborskii* ©	X		GTX	HPLC-FLD	Jardim *et al.* (1999)
Three Marias reservoir	1997	*Microcystis wesenbergii*		X	MCYST	Immuno-assay and HPLC-DAD	Jardim *et al.* (1999, 2000b)
Furnas Dam (Alfenas and Carmo do Rio Claro, MG)	1998	*M. viridis (Radiocystis fernandoi) Microcystis* ©	X	X	MCYST		Jardim (1999); Jardim *et al.* (2000a)

(Continued)

Chart 7.3 Cyanobacterial blooms occurring in Brazilian springs, 1991–2001 (continued).

Locale	Year	Predominant cyanobacteria	Toxicity Yes	Toxicity No	ND*	Toxins Detected	Method	Source
Furnas reservoir (Alfenas and Carmo do Rio Claro, MG)	1998	Cylindrospermopsis raciborskii	X			CYN		Jardim et al. (1999, 2000a)
Vargem das Flores reservoir (MG)	1999	Bloom of Microcystis spp. and Radiocystis fernandoi	X			MCYST	HPLC-DAD	Jardim (1999); Jardim et al. (2000b)
Velhas River (MG)	1999	Aphanizomenon manguinii © Cylindrospermopsis raciborskii©		X	X		HPLC-DAD	Jardim et al. (2000b)
Conselheiro Lafaiete (MG)	1998	Oscillatoria splendida (syn: Geitlerinema splendidum)		X				Jardim et al. (2000b)
Pedra Azul (Medina, Ninheira, MG)	1999/2000	Bloom of Cylindrospermopsis raciborskii	X			Negative p/ CYN SXT	HPLC-DAD HPLC-DAD	Internal Copasa Report
São Simão reservoir	2001	Anabaena circinalis		X		Negative p/MCYST	Immuno-assay	Internal Copasa Report
Ribeirão Ubá (MG)	2000	M. virdis, M. aeruginosa, Anabaena spp., Oscillatoria sp.			X		Immuno-assay	Jardim et al. (2000b)

WWTP: Wastewater Treatment Plant by facultative stabilization ponds; MCYST, microcystin; CYN: cylindrospermopsins; SXT: saxitoxins; ANTX it: anatoxin-a, (S) © crops; HPLC liquid chromatography, high efficiency; DAD: diode photo detector; FLD: fluorescence detector fi; MS: mass spectroscopy. Data transferred by Prof. J. S. Yunes to Sandra Azevedo.

The presence of gas-filled vacuoles was first described in *Gloeotrichia* (Klebahn, 1895). These structures are common in various different species of cyanophytes that develop in extensive water blooms and significantly reduce the density, enabling greater buoyancy. They are also important in the **regulation** of buoyancy in these cyanobacteria. The **gas vesicles** that form the vacuoles are complex structures whose number can be regulated under the effects of pressure (the vesicles support external pressure of up to 4–7 atmospheres) (Grant and Walsby, 1977), thus enabling regulation of the level through increased or decreased density.

Reynolds and Walsby (1975) described in detail the physiological mechanisms of buoyancy regulation in cyanophytes. The relationships of the number of vesicles and vacuoles to the photosynthetic rate and increased or decreased density are essential for buoyancy regulation in *Anabaena flos-aquae* (Figure 7.19). Reynolds (1978) studied the problem further, describing the vertical distributions of *Anabaena circinalis* and *Microcystis aeruginosa*, their control through the production and collapse of gas vacuoles, as well as their growth rate (Krombamp and Mur, 1984; Krombamp *et al.*, 1988; Bitlar *et al.*, 2005).

The presence of bacteria in *Microcystis aeruginosa* blooms was reported by Lial Sandes (1998), who described the dynamics of these blooms and their senescence in a short time period (seven days) in the Barra Bonita reservoir. The development and collapse of cyanobacterial blooms present characteristics that support the theory of catastrophes for the collapse and rejuvenation of populations and communities. Figure 7.20 (Tundisi *et al.*, 2006) illustrates this feature.

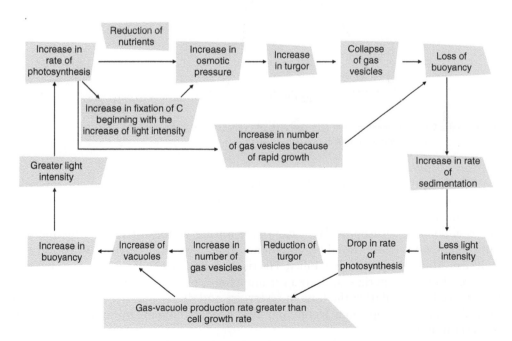

Figure 7.19   Cyanobacterial flotation mechanism with gas vesicles.
Source: Reynolds (1984).

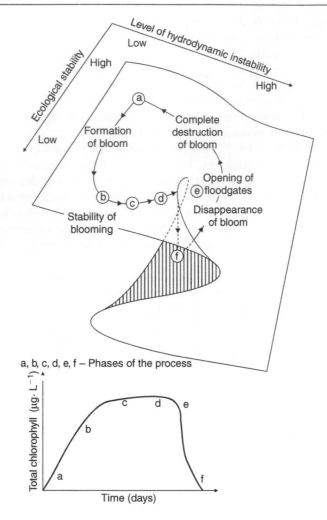

a, b, c, d, e, f – Phases of the process

Figure 7.20   Growth, development and collapse of cyanobacterial blooms.
Source: Tundisi *et al.* (2006).

Harris and Baxter (1996) showed the interannual variability of phytoplankton, especially cyanobacteria, relation to of water volume and nutrient levels in a subtropical reservoir (Figure 7.21). According to the authors, a pattern of diatom predominance in subtropical lakes can occur during periods of vertical mixing and instability, and cyanobacterial predominance in periods of thermal stratification and stability. Along with this variability, Harris and Baxter (1996) suggested that the control of cyanobacteria – and, to a certain extent, phytoplankton succession – can be achieved by regulating the flow in reservoirs and decreasing retention time. Such procedures are an important technology that benefits humans by regulating cyanobacteria blooms.

A recent book on eutrophication in South and Central American lakes and reservoirs (Tundisi *et al.*, 2006) discusses the causes and consequences of eutrophication,

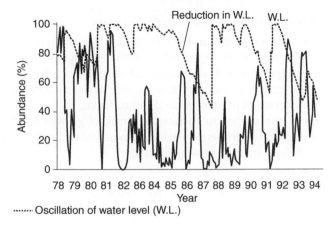

Figure 7.21   Abundance of cyanobacteria (%) in relation to the water storage capacity in a subtropical reservoir. Note that the level of cyanobacteria varies with the water volume of the
reservoir, which increases or decreases the reduction or enhancement, respectively, of
the density of cyanobacteria.
Source: Modified from Harris and Baxter (1986).

as well as technologies for its management and control. The book describes trophic
conditions in reservoirs, the impact of cold fronts on vertical thermal instability, and
the effect of such instability on the process of succession.

7.5 PERIPHYTON

Periphyton algae are another **photo-autotrophic component** of great ecological and
biological importance. The algae are found on the surface of rocks, submerged macrophyte vegetation, boat exteriors, and other natural and artificial surfaces in rivers,
streams, lakes, ponds, wetlands and estuaries. Along with bacteria, fungi, protozoa
and some metazoa, this community – called 'Aufwuchs' – is complex and difficult to
collect and study quantitatively, and therefore periphytonic studies began later than
phytoplanktonic studies.

Substratum heterogeneity and variations in periphytonic communities make classification difficult, particularly qualification of community processes (Wetzel, 1983a).
Roos (1983) used the term **euperiphyton** for a community that is located on and
adheres to a substratum by various mechanisms such as rhizoids, tubules, or other
structures for fixation. Periphyton plays a key role in metabolism in the littoral zone
as well as in biological and biogeochemical processes in wetland areas.

The photo-autotrophic algal component in a periphytonic community can play
an important role in an inland ecosystem's primary productivity, especially in rivers
with large amounts of dissolved allochthonous material. In deep lakes, periphyton's
primary productivity is reduced. In shallow lakes where the euphotic zone reaches the
bottom, the periphytic photo-autotrophic algal component can be significant. In fast
currents, the photo-autotrophic algal component in primary productivity is extremely
high. Wetzel (1964) compared primary productivity in periphyton, phytoplankton
and macrophytes in a shallow saline lake in California and found that, in shallow

Methods for determination of periphyton biomass

Periphyton can be collected by the careful cleaning of surfaces measured in cm^2 or m^2, with the analysis of fresh weight, dry weight, ash-free dry weight, total chlorophyll-a content, number of organisms (cells or colonies of cells) and number of species.

Another method used introduced surfaces of various dimensions and roughness to determine the growth rate of periphytes on these surfaces.

With the use of these artificial surfaces, the rate of growth can be determined after a sequential study of some days or weeks. The seasonal cycle of photo-autotrophic periphytes can be defined by the following equation:

$$\frac{dC}{dt} = \frac{dP}{dt} \cdot C - (G + Pa + D)$$

where: C is the concentration of cell numbers; G, the effect of grazing on algae; Pa, parasitism and diseases affecting the algae; and D, general cell mortality.

Bicudo (1990) discussed the method for counting periphytic algae (1990); Watanabe (1990) compared methods applied to assess pollution and contamination through the study of periphyton.

areas (<2 m), periphytic productivity exceeded macrophytic and phytoplanktonic productivity. In the lake's pelagic zone, **phytoplanktonic productivity** predominates. In shallow lentic waters, **photo-autotrophic periphytic production** may account for more than 60% of the total.

Periphytic productivity rates in rivers can reach up to approximately 1050 mg C·m^{-2}·day^{-1} (compared, for example, with 200 mg C·m^{-2}·day^{-1} for the UHE Carlos Botelho – Lobo/Broa reservoir, or 1500 mg C·m^{-2}·day^{-1} for the Barra Bonita reservoir).

The factors affecting the photo-autotrophic algal community that constitute the primary production component of periphyton are the same that affect photo-autotrophic phytoplankton. Water temperature, light intensity and nutrient availability are key factors in the growth, reproduction and succession of photo-autotrophic periphyton.

7.5.1 Temperature

Changes in water temperature can affect the metabolism of photo-autotrophic periphyton, as reported by McIntyre and Phinney (1965) (see Table 7.2).

The above figures show that the photosynthetic rate varies as shown in Table 7.3 below:

In this case, light intensity remained stable at 20,000 lux.

7.5.2 Effects of light intensity

Many authors – such as Welch (1980), for example – have described the adaptations of periphytic algae to high and low light intensities, as in the case of phytoplankton. McIntyre and Phinney (1965) studied the changes in light intensity in streams, artificial

Table 7.2 Effects of alteration of water temperature on the respiratory rate of periphyton.

Alterations in water temperature	Alterations in respiratory rate
6.5–16.5°C	41–132 mg $O_2 \cdot m^{-2} \cdot h^{-1}$
17.5–9.4°C	105–63 mg $O_2 \cdot m^{-2} \cdot h^{-1}$

Table 7.3 Effects of alteration of temperature of water in rate of photosynthesis of periphyton.

Alterations in water temperature	Alterations in photosynthetic rate
11.9–20°C	335–447 mg $O_2 \cdot m^{-2} \cdot h^{-1}$

rivers and the effects on periphyton growth. The main effect was a higher photosynthetic rate at low light intensities and differences in the responses of diatoms, cyanobacteria and chlorophytes to the same gradient of light intensities. In communities adapted to low light intensities, growth is slower, but the final accumulation of biomass is almost the same as in those adapted to higher light intensities with faster growth.

Periphyton growth is thus controlled by light intensity and temperature, but when temperature and light conditions are stable, nutrients levels can dominate the growth response (Welch, 1980). In certain lakes and reservoirs, higher phytoplanktonic levels limit the availability of light for the periphyton.

Variations in the amounts of suspended matter (which alters light intensity) can rapidly modify the responses of periphyton algae, in particular the growth and photosynthetic rates. Turbidity is a limiting factor in the growth and ecological dynamics of periphytic algae, and can cause reduced biomass.

River current velocity can also alter the composition of periphytic algae and act as a selective factor. Experimental velocities of 38 cm/second, for example, induced the growth of diatoms in the community (McIntyre, 1966). At reduced speeds (9 cm/second), the author found filaments of Chlorophytes (*Oedogonium* and *Tribonema*).

In artificial streams with high nutrient levels in the metropolitan region of São Paulo, Tundisi (2006, unpublished results), found masses of *Scenedesmus* and *Tabellaria* in microbial mats, as well as high levels of bacteria and protozoa.

In the discharge of polluted rivers or rivers with high nitrogen and phosphorus levels, rapid and intense growth of periphyton algae can be observed and the whole complex of organisms that accompany them. This growth, caused by eutrophication, can also lead to deterioration in water quality on the shores of lakes, such as occurred in Lake Erie and Lake Huron (USA/Canada), where masses of *Cladophora* developed very quickly. Phosphorus is usually the most important nutrient in this growth (Welch, 1980).

Periphytonic succession has been studied by many specialists in Brazil. Fernandes (1993), for example, studied the structure of the epiphytic community that develops on the leaves of *Typha dominguensis* in coastal lakes of Jacarepagua in Rio de Janeiro. He concluded that the epiphytic succession was related to the decomposition of the plant's leaves. The system was in an advanced stage of eutrophication, 78 taxa were found, with a predominance of chlorophytes (32%), cyanophytes (23%), Bacillariophyceae (22%), Chrysophyta (6%) and Euglenophyta (5%). The distributions in the community, in this case, were attributed to difference levels of pH, dissolved oxygen, and nutrients.

In the **Lobo River,** in the municipality of Brotas (SP), Chamixaes (1991) conducted a study on **periphytonic colonization** on artificial substrata over a 32-day period. The distribution of the periphyton was slower and more gradual in winter (dry season) and faster in summer (rainy season), with large irregular fluctuations most likely due to discharges of suspended matter or changes in the current velocity, resulting from more intense rainfall in summer. Colonization by epiphytes depends, as shown by Schwarzbold (1992), on the flooding rate. Biomass accumulation also depends on the velocity and magnitude of the flooding pulse. The greater variations in biomass during summer are probably due to the effects of precipitation and the physical consequences on periphytic communities in **lotic systems.**

By studying periphytic succession on artificial substrata in the UHE Carlos Botelho (Lobo/Broa) reservoir, Panitz (1980) found that the type of substratum and its depth were key to the colonization and **succession of periphytic algae.** The greatest biomass was found in summer, on the artificial substrata studied by the author. However, during summers with heavy rainfall, chlorophyll-*a* drastically decreased. Communities near the sediment, on artificial substrata, showed the highest primary production ($55\,mg\,C\cdot m^{-2}\cdot day^{-1}$), probably due to the recycling of nutrients through the decomposition of macrophytes, as well as the higher temperatures.

Studies by Soares (1981) showed that periphyton associated with aquatic macrophytes presented greater biomass and other tendencies in succession. Table 7.4 shows the biomass (chlorophyll *a* or dry weight) of periphyton in various inland ecosystems in Brazil. Experimental studies conducted by Cerrao *et al.* (1991) shows the effects of nitrogen and phosphorus on periphytonic growth. Results of these studies showed that 300 and $30\,mg\cdot L^{-1}$ of these nutrients were more effective in the growth of periphyton than double those concentrations.

Pompeo (1991) studied the primary production of *Utricularia gibba* and periphyton, and showed that periphyton was responsible for 80% of the **gross primary production.** Suzuki (1991) studied the interrelationship between zooplankton, phytoplankton and epiphytes in a marginal lake of the Mogi River (Infernão lagoon, municipality of Luis Antônio, SP). The author used 24 transparent plastic structures enriched with KH_2PO_4 and NH_4NO_3 during studies in winter and summer. The structure of the phytoplanktonic community was altered by the presence of larger organisms. The phytoplankton responded quickly to enrichment, the periphytonic biomass grew slowly and reached higher levels during the rainy season (summer). Primary producers appeared to assimilate nitrogen in the form of ammonium (NH_4) more efficiently. According to Suzuki (1991), phytoplankton is more effective than periphyton in assimilating nutrients from pelagic waters.

Necchi (1992) conducted other periphytonic studies in Brazil, investigating succession of macrophytic communities in rivers. He identified two species of cyanobacteria, one chlorophyte and one rhodophyte, with variations in frequency and biomass. *Klebshormidium subtile* was the common species on the artificial substrata used by the author, who explained the succession based on competition for space on the substratum and reproductive strategies, factors that are determinants in the dynamics of **macro-algal** communities.

In another study, Necchi and Pascoaloto (1993) analyzed macro-algal communities growing on natural substrata in the drainage basin of the Preto River in the state of São Paulo, and concluded that the seasonality is dependent on the substratum. The

Table 7.4 Maximum chlorophyll content (mg m^{-2}) and dry weight (g m^{-2}) of a substratum exposed to colonization of periphyton in aquatic ecosystems in Brazil.

Locale	Substratum	Duration	Frequency of collections (samples)	Chlorophyll-a	Dry weight	Observations	Reference
UHE Carlos Botelho reservoir (Lobo/Broa) – before the observatory (downstream)	Glass plates	31–32 days	Weekly	6.1	–	Summer	Chamixaes (1991)
				9.8	–	Winter	
– after the reservoir (upstream)				2.2	–	Summer	
Itaqueri stream				3.4	–	Summer	
				2.0		Winter	
Perdizes stream				2.0	–	Summer	
				2.0		Winter	
Coastal lagoon of Jacarepaguá (2 sites)	Typha dominguensis	20–28 days	Weekly	–	4.2	Summer of 1990	Fernandes (1993)
Peninsula North Lake	Glass plates (exposed horizontally)	70 days	Weekly	–	74.0	0.10 m	Rocha (1979)
				–	17.7	0.55 m	
				–	87.0	1.00 m	
	Glass plates (exposed vertically)			–	11.8	0.10 m	
				–	15.1	0.55 m	
				–	17.6	1.00 m	
Paranoá Lake	Glass plates (exposed horizontally)	70 days	Weekly	–	85.9	0.34 m	
				–	80.0	1.03 m	
				–	28.0	2.07 m	
	Glass plates (exposed vertically)			–	89.0	0.34 m	
				–	23.0	1.03 m	
				–	12.0	2.07 m	
				–	28.8	6.27 m	
UHE Carlos Botelho Reservoir (Lobo/Broa)	Pontederia cordata	42 days	Weekly	55.5	1.3	—	Soares (1981)

Source: Bicudo et al. (1995).

greatest amounts of macro-algal biomass were measured during lower temperatures, lower current velocities and greater transparency. Five species of green algae were found in the study, as well as three species of cyanobacteria and two of red algae. Lobo *et al.* (1990) conducted other studies on the seasonal cycle and periphytonic colonization in Brazil.

The contribution of epiphytic microorganisms to nitrogen fixation revealed the key role that periphytonic microorganisms play in the nitrogen cycle, especially bacteria that fix this nutrient.

Studies on the **seasonal succession of periphyton** are complex because, along with the factors that influence succession (such as current velocity, nitrogen and phosphorous levels, light intensity and water temperature), other factors not influenced by external aspects are also intrinsic to the complexity of this community. Periphyton is composed of photo-autotrophic components (which are essential for the fixation of CO_2 in the water) and heterotrophic components (which play a role in decomposition processes, oxido-reduction systems, and internal nutrient recycling) (Wetzel, 1983c). The complex metabolic processes that occur between living and non-living components of the periphytic community and the varied nature of organisms in these communities can be studied *in situ* using special techniques with microelectrodes, communities on natural strata, experiments in controlled conditions (artificial streams), as well as studies on artificial substrata.

Microbial mats

The association of photo-autotrophic algae and bacteria that develop in sediment, in some aquatic environments − such as those with high concentrations of nutrients, elevated salinity/conductivity and where there is adequate penetration of photosynthetically active radiation − constitutes a special micro-stratified community, called a microbial mat. This microbial mat is composed of cyanobacteria, microscopic photo-autotrophic algae and bacteria. This set, that presents biological, physical and chemical interactions, occupies shallow lakes with light penetration to the sediment and low impact of predation (McIntyre et al., 1996; Wetzel, 2001; Dodson, 2005).

Photo: Guilherme Ruas Medeiros

The relationship between dissolved organic substances (such as carbohydrates or lipids) and periphytonic growth and succession should also be explored in future studies, because initial experiments showed the important role of these substances in these communities' composition and succession. Changes in the organic and inorganic nutrient levels and composition in contaminated or polluted water caused modifications in the periphytonic biomass and composition, which led to the use of periphyton as an indicator of water quality related to the organic nutrient levels.

Sladecková and Sladeček (1963) proposed terms such as oligosaprobic, mesosaprobic, and polysaprobic (no longer widely used, but they were important at the time they were proposed) as a basis for organizing data on pollution and contamination and organisms' responses. The problem is more complex than simply quantifying and classifying the community's components, as originally proposed. Inorganic nutrients and organic substances in contaminated water stimulate the growth of a **biofilm** of bacteria and fungi upon which microscopic and macroscopic algae thrive, complicating the general definition of communities and saprobes. Periphyton responds to toxic substances and these studies show the effects of heavy metals and organic xenobiotics on the succession and composition of periphytonic communities.

Like phytoplankton, periphyton plays an essential role in the metabolism of lakes, rivers, reservoirs and estuaries. The role of the dense communities in the feeding of organisms and maintenance and development of the food chain will be discussed further.

7.6 AQUATIC MACROPHYTES

Aquatic macrophytes include a large group of organisms, such as thalloid algae, mosses and liverworts, ferns, conifers and flowering plants that grow in inland and brackish waters, estuaries, and coastal waters. Aquatic macrophytes include small (15 mm) floating organisms up to large trees such as the cypress (*Taxodium* spp.) found in swamps in the southern United States. Chapter 6 described aquatic macrophytes as *emergent plants* firmly rooted in submerged soils; **floating macrophytes** with leaves, such as *Nymphaea* spp. and water hyacinth; and **fully submerged macrophytes**. The morphological features of these three types of macrophytes are important, since they show different types of adaptation, such as an **aerenchyma** that helps transport oxygen to the roots of emerging plants, or fine leaves with a very thin superficial film to help fix nutrients and carbon dioxide in the liquid medium, such as occurs in submerged plants. The extremely thin leaves minimize the trajectory of diffusion of nutrients and maximizes the underwater solar energy available for submerged plants.

Floating macrophytes such as *Lemna*, *Eichhornia azurea* or *Eichhornia crassipes* form large mats, which in some cases are connected by roots that absorb all their nutrients directly from the water, and not from the sediment. These floating plants require sheltered locations, are affected by strong waves and wind, and compete directly with phytoplankton and periphyton for nutrients.

Floating, emergent or submerged aquatic macrophytes are extremely active substrata and are important for photo-autotrophic microalgal periphyton and aquatic invertebrates (adults and larvae of aquatic insects, for example). These plants compete with phytoplankton and periphyton for available nutrients and solar radiation, and play a key role in the metabolism of shallow lakes, as seen in the products of their decomposition and their availability as a food source for many aquatic animals, from invertebrates to hippopotami, as is the case in certain African lakes (Carpantes and Lodeje, 1972; Horne and Goldman, 1994). Herbivory on macrophytes can thus play a significant role in the food chain (Lodge, 1991). In the platform formed by aquatic macrophytes, a wide variety of invertebrate animals such as molluscs, Trichoptera and chironomid larvae develop, that are a food source for fish and other invertebrates (Figure 7.22).

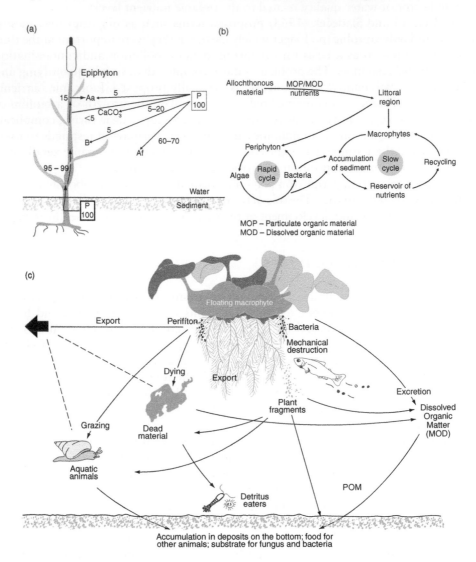

Figure 7.22   a) Illustration of flow of phosphorus (P) among the sediment, macrophytes and epiphytic microflora. Abbreviations: Aa – adnated algae; Af – loosely attached algae; B – bacteria (modified from Wetzel, 1990b); b) Relationship of macrophyte-periphyton complex to the conservation of nutrients; c) (modified from Wetzel, 1990a) The role of macrophytes in biogeochemical cycles and the flow of nutrients and POM between sediment, macrophytes, epiphytic microflora and periphyton.
Source: Modified from Burgis and Morris (1987).

Productivity of macrophytes, especially emergent macrophytes, is extremely high (Moss, 1988), and a large portion of the organic material produced may be consumed directly by herbivores. Aquatic habitats have less available oxygen because of oxygen's low solubility in water. Emergent plants can root into the bed of organic deposits (peat) or they can be exported downstream, settling in remote areas. When the plants thrive, the peat can provoke a series of reactions causing **progressive sedimentation** in lakes, although this process is relatively slow, while the sediment deposited in other locations forms the substratum of future formations. Water that is under dense layers of floating macrophytes, mainly in tropical climates, becomes anaerobic due to the decomposition of this material (Maltchik, Rolon and Groth, 2004).

In emergent or floating swamps, the low oxygen levels favour the evolution of a community of system-dependent animals, such as those that can breathe atmospheric air, necessary for their survival. Nevertheless, animal production can be quite high in these locations. In swamps and other seasonally flooded areas, where grasses eventually sprout up in their vicinity, mammals and birds migrate there seasonally, seeking to take advantage of the wide availability of food provided by macrophytes (Welcomme, 1979). In the tropics, a large number of indigenous species depend on these floodplain systems, since they derive their livelihood from the aquatic and semi-aquatic plant communities that exist there.

Indirect contributions of macrophytes to the food chain include the architectural structure they provide to the lake and their influence on processes involving nutrient cycles (Wetzel, 1990). Macrophyte-dominated zones in a lake generally have higher diversity of animal species than plankton-dominated zones of open water (Macan and Kitching, 1972). Macrophytes provide the physical platforms for niches, resting sites, burrows and hollows for predators to spawn or deposit fertilized eggs. Due to this diversity, there is a great capacity to take advantage of the natural resources represented by light and nutrients, since a differentiated community should have specialization in the diets of its inhabitants; therefore, energy will not be spent on competition between species.

Because of high productivity in macrophyte-dominated habitats, there is a high internal metabolic rate (Mickle and Wetzel, 1978a, b). This is even more true when organic detritus and the communities of microorganisms that colonize the surface of plants fall toward the sediments. The surface of sediments are, in general, **micro-aerophilic** (low in oxygen) or anaerobic, and release a considerable quantity of nutrients, such as phosphorus. The cycle of nutrients within the masses of floating or submerged **emergent macrophytes** is complex, with organic and inorganic transfer between the plants and the open water.

Macrophytes can be found in almost all lakes, except some extremely saline lakes or those damaged by pollution. Macrophytes thus have important functional differences in different types of lakes. In large deep lakes, although there is a littoral strip in the geometry of the lake, the community of open water is much more significant. However, in terms of absolute importance, the littoral zone can be quite significant since the deep water frequently has low fertility, and is thus unproductive.

For riverside dwellers, dependent on fish, littoral waters are much more crucial than the areas of water further away, and this becomes more significant because of the exportation of organic materials and nutrients from the macrophyte-dominated areas to littoral waters, with the subsequent increase in productivity. The communities of fish in the macrophytic zone, captured by simple artisanal methods, can represent a large quantity of fish (Goulding, 1981), although they cannot be exploited on a large scale by mechanical methods, since these are capable of destroying the ecosystem's structure.

Large lakes, although prominent on maps, occupy a much smaller total area than the sum of millions of small shallow lakes, which hold great importance for humanity. It is in these lakes that macrophytes play a more important role, from all points of view. These lakes include naturally occurring shallow basins formed by glacial action, shallow depressions on floodplains, ponds in swamps, marshes and many small man-made reservoirs. They are the main components of extensive wetland systems that cover large areas of the Earth's surface (or did cover at one time, since in some cases they have been severely altered by drainage or flooded to form large reservoirs). Given the importance of these habitats and the great losses suffered by them, it is of paramount importance that, before altering those that remain, careful consideration first be given before implementing any concrete actions.

7.6.1   Studies on aquatic macrophytes in Brazil

In a recent volume, Thomaz and Bini (2003) analyzed the body of research on aquatic macrophytes in Brazil. In the early 1960s, studies were conducted on higher plants in Brazil, and it was generally recognized that aquatic macrophytes play an essential role in the metabolism and functioning of shallow lakes, reservoirs, rivers, coastal areas, estuaries and wetlands (Wetzel, 1990; Esteves, 1998).

In the last twenty years, research on aquatic macrophytes in Brazil has greatly advanced. The main biological forms and taxa studied are presented in Chart 7.4 (Thomaz and Bini, 2003). Hoehme (1948) published *Aquatic Plants*, which became an important reference on ecology and geographical systems and distribution of these plants in Brazil. Pott and Pott's publication (2000) is a particularly important contribution to the knowledge of aquatic plants in the Pantanal, and is also useful as a reference for Brazil.

Chart 7.4   Principal biological forms and taxa studied.

Species/genus	Biological form	Number of studies
Scirpus cubensis	Emerging	13
Eichhornia azurea	Emerging	13
E. crassipes	Live floating	13
Pontederia spp.	Emerging	9
Salvinia spp.	Live floating	8
Nymphoides indica	Floating sheet	7
Echinochloa polystachya	Emerging	7
Typha domingensis	Emerging	6
Cabomba pyahuiensis	Submerged	5
Total		81 (=50% of the studies)

Source: Thomaz and Bini (2003).

Arens (1933, 1936, 1938, 1939, and 1946) focused on aquatic macrophytes' physiological aspects (absorption of bicarbonate) in temperate areas. There also, Steemann-Nielsen (1945) published work on the metabolism of bicarbonate absorption by aquatic macrophytes.

7.6.2   Biomass and succession of aquatic macrophytes

In regions with widely varying hydrological levels, differences occur in the composition of macrophyte communities, because of changing water levels and the shifts from dry or humid conditions to flood conditions. Junk (1986), for example, reported that changes occur in macrophyte succession during periods of low water level and flooding. For example, *Sagittaria sprucei*, in the Alismataceae family, can survive in flooded areas with relatively little water, but cannot adjust to high levels of flooding once it has flowered during the dry season. After the flood and elevated water level, free-floating species are more common, such as *Eichhornia crassipes* and *Salvinia* spp. Macrophytes of the genera *Pistia* or *Eichhornia* can quickly colonize aquatic environments.

Aquatic macrophytes help establish an environment with plenty of organic matter and detritus, as well as providing an important substratum for periphytic algae and aquatic invertebrates. On the other hand, physical and chemical conditions in regions with dense populations of free-floating macrophytes or roots are very particular: there is little light penetration; dissolved oxygen levels vary greatly during 12-hour and 24-hour periods; marsh areas covered by macrophytes have a more pronounced cycle, and this biomass can affect O_2/CO_2 cycles, due to decomposition, the organisms' respiration, and photosynthesis on the part of the associated periphyton. These biomass levels also provide reproductive areas and food for fish and amphibians. The capacity of these plants to affect the biogeochemical cycles and high biomass levels, with resulting effects on the diurnal cycles of physical and **chemical factors,** makes these regions in lakes, rivers and reservoirs very important for studies on temporal and spatial succession as well as on the flora and fauna associated with these communities. Research on the biogeochemical cycles in these regions is particularly important because of the capacity of these plants to concentrate nitrogen and phosphorus or metals, and thus function as a biological concentrator of elements and substances that produce pollution and eutrophication. Results of studies already appear to confirm the possibility of practical application. Macrophytes accelerate the process of sedimentation and clogging in lakes due to the concentration of sediment they produce, mainly in lake regions with gentle **slopes,** which favour the accumulation of sediment.

In the tropics, especially in reservoirs, the rapid growth of free-floating macrophytes such as *Eichhornia*, *Pistia* or *Salvinia* spp. can cause serious problems in system management due to the rapid accumulation of organic material. Control of these plants has been achieved by mechanical removal, biological control, and dredging. In eutrophic reservoirs in the state of São Paulo, *Eichhornia crassipes* is the most common aquatic macrophyte.

The **zonation of aquatic macrophytes** in lakes and reservoirs can serve as an indicator of ecological conditions, of mechanisms of circulation, current velocity, or turbulent regions.

Neiff and Neiff (2003) presented an analysis of **connectivity** as an approach to interpret the interactions between processes and elements in an ecosystem defined by variations in space and time. Connectivity has been considered as given – essential in the large floodplains and internal deltas of rivers, for example, to explain energy transfers between rivers, marginal lakes and the horizontal convection between the different mosaics. Studies on connectivity evaluate the interactions between system components, their complexity, the **indicator species** and the principal functions of factors that have the greatest weight as determinants in the succession of aquatic macrophytes.

Colonization of aquatic macrophytes in lakes and reservoirs – an important process for the future management of these ecosystems – depends on a set of variables, including the diversity of species in nearby areas (in backwaters of rivers or marginal lakes) or the invasion and dispersal rates of exotic species (Thomaz and Bini, 1998, 1999).

The adjustment of macrophyte communities to hydrological variations in the different levels is essential to the process of succession, which involves metabolism, number of species, size, and form of vegetation (Neiff, 1978; Thomaz and Bini, 1999). As described by Neiff and Neiff (2003), seeds from aquatic plants in the Parana River basin do not germinate in flooded soil, only in soil that emerges after flooding. Therefore, a process of germination and inhibition can occur, providing **pulses of biomass** in some species and recession in others. Also, according to the authors, processes associated with succession of macrophytes, in terms of river pulsations, include factors such as:

frequency of changes in level and pulse;
magnitude: intensity of drought or flooding periods;
variability: value of the standard deviation of the average maximum or minimum of a multi-year curve of hydrometric fluctuations;
recurrence – statistical probability that a flood or drought of a given magnitude will occur in a century or millennium;
amplitude – phase of duration of the drought or flood of a certain magnitude in a floodplain;
seasonality – seasonal frequency with which droughts or floods occur.

A series of biogeochemical processes – such as decomposition of organic material, accumulation of litter, nutrient availability, flow, and sediment – depend on the frequency, intensity, duration, seasonality and connectivity between the rivers and marginal lagoons (Neiff *et al.*, 1994).

Knowing the amplitude and gradient of the hydrological variation in which a particular aquatic plant species occurs, it is possible to predict the species' presence or absence. Knowledge of its phenology (seasonal timing) in periods of drought and flooding provides data to evaluate the presence or absence of the species at different stages or periods of flooding. It can also explain the extent of the species' colonization and duration. Neiff and Neiff (2003) propose an equation of fluvial connectivity (QCF):

$$QCF = DI/D_a I_a$$

where:
DI = number of days of flooding (**potamophase**)
$D_a I_a$ = number of days of isolation of marginal ponds (**limnophase**)

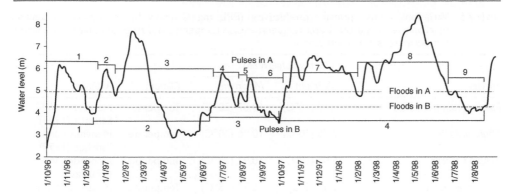

Figure 7.23   Level of water in the Paraná River, with pulses and levels of inundation in different regions of the lowland (the numbers indicate pulses).
Source: Neiff and Pol de Neiff (2003) in Thomaz and Bini (2003).

This quotient could be compared in each pulse with the fluctuations in biomass, density and various sizes of the predominant vegetation (see Figure 7.23).

7.6.3   Limiting factors and primary production of aquatic macrophytes

The quantity and quality of incident solar radiation play a similar role in the physiology of macrophytes, comparable to that in phytoplankton. In general, macrophytes are adapted to higher light intensities and their distribution and abundance depend on the quantity of light. For some species of submerged macrophytes, incident solar radiation at infrared wavelengths is more effective (Welch, 1980). For submerged macrophytes, water turbidity and high phytoplanktonic levels ($>200\,\mu g \cdot L^{-1}$ chlorophyll-*a*, for example) can limit their growth. For example, in the UHE Carlos Botelho (Lobo/Broa) reservoir, **increased turbidity** due to the extraction of sand was responsible, in certain periods, for the disappearance of populations of *Mayaca fluviatilis*, a submerged macrophyte commonly found in this region (the entry of the Itaqueri River into the reservoir).

Temperature acts as a control in the saturation rates of photosynthesis by macrophytes, which, to some extent, behave within Q_{10} regions as do phytoplankton and periphyton.

The question of the effects of nutrient levels on macrophyte growth and productivity has conflicting interpretations in the literature (Welch, 1980). Clearly, higher growth rates occur with high nutrient levels (Finlayson, 1984), as observed in experiments and observations on *Pistia stratiotes*, *Eichhornia crassipes*, *Salvinia molesta* and *Typha dominguensis*, as cited by Campbell *et al.* (2003). However, the nutritional requirements of different macrophytic species vary. For example, Camargo and Esteves (1995) observed extensive banks of *Salvinia* sp. in a marginal pond of **Mogi-Guacu River**, with orthophosphate levels from $514\,\mu g \cdot L^{-1}$.

Underwater incident solar radiation and availability of carbon are the most important factors for the growth and productivity of submerged macrophytes, according to

Table 7.5 Variations in net primary productivity (P.P.L mg O₂/gDW·h) of submerged aquatic macrophytes at various water temperatures, in temperate and tropical climates. The water temperature is in parentheses.

Species	P.P.L (mg O₂/gDW·h)		Climate	Authors
	Minimum	Maximum		
Potamogeton pectinatus	2.19 (10°C)	19.67 (20°C)	Temperate	Menendez and Sanchez (1998)
Chara hispida	2.55 (10°C)	10.86 (20°C)	Temperate	Menendez and Sanchez (1998)
Ruppia cirrhosa	5.00 (10°C)	10.92 (23°C)	Temperate	Menendez and Peñuelas
Ranunculus aquatilis	1.90 (5°C)	5.92 (15°C)	Temperate	Madsen and Brix (1996)
Elodea canadensis	1.12 (5°C)	7.37 (15°C)	Temperate	Madsen and Brix (1996)
Egeria densa (R. Aguapeú)*	5.85 (20°C)	9.23 (21°C)	Tropical	Pessato (1999)
Egeria densa (R. Aguapeú)*	2.76 (21°C)	5.40 (19°C)	Tropical	Pezzato (1999)
Cabomba furcata (R. Mambu)*	5.21 (20°C)	15.62 (23°C)	Tropical	Benassi et al. (2001)
Utricularia foliosa*	3.24 (17°C)	25.55 (24°C)	Tropical	Assumpção (2001)

P.P.L – net primary production; gDW – g of dry weight; h – hour
*experiment in the field.
Source: Thomaz and Bini (2003).

Madsen and Adams (1988). Camargo (1991) reported that in a marginal pond of the Mogi-Guacu River, the maximum biomass growth of *Eichhornia azurea* occurred in the post-filling period, with 171 g/m² of dry live biomass weight (with low turbulence, high temperatures and higher nutrient levels). Table 7.5 shows variations in the primary productivity of submerged aquatic macrophytes at various different water temperatures in both temperate and tropical climates.

Other factors that play a role in the productivity and biomass of floating aquatic macrophytes (immersed or submerged) include current velocity, **interspecific** or **intraspecific competition**, and the role of herbivorous predators, which can drastically dominate macrophyte biomass in a relatively short period of several days or hours (Horne and Goldman, 1994). Table 7.6 shows the maximum density, growth coefficient, and duplication time of some aquatic macrophyte species (Bianchini, 2003).

The decomposition of aquatic macrophytes and the effect on lake metabolism and bio-geochemical cycles

Since aquatic macrophytes occupy many niches in the littoral zone of lakes and reservoirs and in marginal lagoons in large floodplain systems, they play a significant role in biogeochemical cycles. Their intense proliferation and growth produce increased biomass, which, upon decomposition, accelerates the release of phosphorus, nitrogen and particulate matter into the water and sediment, accelerating the biogeochemical cycles of many elements (in particular, carbon, nitrogen and phosphorus), making these elements available to other photo-autotrophic producers such as phytoplankton and periphyton (see Table 7.7).

Table 7.6   Maximum density (K, dry weight), growth coefficient (r_m) and time of duplication (Td) of several species of aquatic macrophytes, in different environments.

Macrophyte	K (gDW/m²)	r_m (day⁻¹)	Td (day)	Reference
Brachiaria arrecta	1,815.0			Moraes (1999)
Cyperus sesquiflorus	1,461.2			Moraes (1999)
Echinochloa polystachya	2,755.9			Pompeo (1996)
Egeria najas	234.0			State University Foundation of Maringá Nupélia/Itaipu Binational (1999)
Egeria najas		0.082	8.5	State University Foundation of Maringá Nupélia/Itaipu Binational (1999)
Egeria najas		0.058	11.9 31.5	State University Foundation of Maringá Nupélia/Itaipu Binational (1999)
Egeria najas		0.022	4.2	State University Foundation of Maringá Nupélia/Itaipu Binational (1999)
Egeria najas-w/ sediment	1,159.3	0.164	4.1	Bitar and Bianchini Junior (in preparation)
Egeria najas-without/ sediment	1,419.5	0.171	4.1	Bitar and Bianchini Junior (in preparation)
E. najas – avg.	1,286.2	0.168	4.1	Bitar and Bianchini Junior (in preparation)
Eichhornia azurea	595.0		4.1	Coutinho (1989)
E. crassipes		0.053	11–15	Penfound and Earle (1948)
E. crassipes	1,638.0			Esteves (1982)
E. crassipes	1,918.8			Moraes (1999)
E. crassipes	294.0			State University Foundation of Maringá Binational Nupélia/Itaipu (1999)
E. crassipes		0.040	17.3	State University Foundation of Maringá Nupélia/Itaipu Binational (1999)
Glyceria maxima	1,507.9	0.050	13.9	Esteves (1979)
Justicia americana	2,385.7	0.092	7.5	Boyd (1969)
Nymphoides indica	322.3			Menezes (1984)
Paspalum repens	1,444.0			Petracco (1995)
P. repens	2,146.2			Meyer (1996)
Pistia stratiotes	881.2			Moraes (1999)
P. stratiotes	372.0			State University Foundation of Maringá Nupélia/Itaipu Binational (1999)
Polygonum spectabile	1,981.2			Petracco (1995)
Pontederia cordata	3,053.3			Menezes (1984)
P. lanceolata	235.9			Penha, Silva e Bianchini Junior (1999)
Salvinia auriculata	102.0			State University Foundation of Maringá Nupélia/Itaipu Binational (1999)
S. auriculata	199.8	0.094	7.2	Saia and Bianchini (1998)
S. auriculata		0.064	10.8	State University Foundation of Maringá Nupélia/Itaipu Binational (1999)
S. molesta		0.036	19.1	Mitchell and Tur (1975)
Scirpus cubensis	1,062.0			Coutinho (1989)
S. cubensis	2,467.0			Carlos (1991)
S. cubensis		0.002	285	Bianchini Junior et al. (in preparation)
Utricularia breviscapa	20.9			Menezes (1984)

gPS/m² – grams of dry weight per square meter.
Source: Bianchini (1998) in Thomaz and Bini (2003).

Table 7.7 **Decay coefficient** (k_3) of some of the organic dissolved compounds, originating from decomposition of aquatic macrophytes, estimated from the environmental conditions. The amounts are calculated from the results in each reference.

Origin	$K_3 (day^{-1})$	References
Carbohydrates leached from *Cabomba piauhyensis* (anaerobic process, neutral medium)	0.043	Campos Junior (1998)
Carbohydrates leached from *Cabomba piauhyensis* (anaerobic process, reduced medium)	0.004	Campos Junior (1998)
Carbohydrates leached from decomposition of *Macaya fluviatilis* (anaerobic process, neutral medium)	0.060	Bianchini Junior (1982)
Carbohydrates leached from the decomposition of *Nymphoides indica*	0.074	Bianchini Junior (1982)
Carbohydrates leached from *Salvinia* sp. (anaerobic process, neutral medium)	0.037	Campos Junior (1998)
Carbohydrates leached from *Salvinia* sp. (anaerobic process, reduced medium)	0.018	Campos Junior (1998)
Carbohydrates leached from *S. cubensis* (anaerobic process, neutral medium)	0.020	Campos Junior (1998)
Carbohydrates leached from *S. cubensis* (anaerobic decomposition, reduced medium)	0.011	Campos Junior (1998)
Labile carbohydrates leached from aerobic decomposition of *Cabomba piauhyensis*	0.22	Cunha and Bianchini Junior (1998)
Refractory carbohydrates leached from aerobic decomposition of *C. piauhyensis*	0.005	Cunha and Bianchini Junior (1998)
Labile carbohydrates leached from the anaerobic decomposition *Scirpus cubensis*	0.20	Cunha and Bianchini Junior (1998)
Carbohydrates leached from the anaerobic decomposition of *C. piauhyensis*	0.030	Cunha and Bianchini Junior (1998)
Carbohydrates leached from the anaerobic decomposition of *Scirpus cubensis*	0.020	Cunha and Bianchini Junior (1998)
DOM leached from the decomposition *Cabomba piauhyensis*	0.196	Cunha (1996)
DOM leached from the decomposition of *Cabomba piauhyensis*	0.025	Cunha (1996)
DOM leached from the decomposition of *Eleocharis mutata* (labile fraction)	0.196	Bianchini Junior and Toledo (1996)
DOM leached from the decomposition of *E. mutata* (resistant fraction)	0.002	Bianchini Junior and Toledo (1996)
DOM leached from the decomposition of *Nymphoides indica*	0.006	Bianchini Junior (1985)
DOM leached from the decomposition of *N. indica* (labile fraction)	0.69	Bianchini Junior and Toledo (1996)
DOM leached from the decomposition of *N. indica* (refractory fraction)	0.009	Bianchini Junior and Toledo (1996)
DOM leached from the decomposition of *Scirpus cubensis* (aerobic process)	0.37	Cunha (1996)
DON (N-kheldahl) leached from the decomposition of *Mayaca fluviatilis*	0.085	Bianchini Junior (1982)
DON (N-kjeldahl) leached from the decomposition of *Nymphoides indica*	0.116	Bianchini Junior (1982)
Polyphenols leached from the decomposition of *Mayaca fluviatilis*	0.081	Bianchini Junior, Toledo and Toledo (1984)
Polyphenols leached from the decomposition of *Nymphoides indica*	0.057	Bianchini Junior, Toledo and Toledo (1984)

Source: Bianchini Junior (1982, 1985) in Thomaz and Bini (2003).

7.6.4   Interactions of aquatic macrophytes with periphyton, zooplankton, and fish and benthic communities

Aquatic macrophyte communities form extensive banks in lakes, reservoirs and rivers, and this accumulation of organic matter favours the development of a community of bacteria, periphyton, zooplankton and **macro-zoobenthos**. Macrophytic banks are known as breeding areas for many fish species (nursery-grounds). In many cases, metabolism in the littoral zone, driven by masses of submerged or immersed macrophytes, controls the metabolism of lakes and reservoirs, since it influences biogeochemical cycles through the export of particulate and dissolved matter.

Photo: Luiz Marigo – Mamirauá

8 The dynamic ecology of aquatic animal populations and communities

SUMMARY

A great number of animal species are distributed throughout inland waters. The various phyla or classes make up different proportions of the inland aquatic fauna.

Many species of the freshwater fauna originated from land animals. Over time, passive and active invasions occurred from land and sea systems into inland aquatic systems. The present chapter describes the dynamics and interactions of the aquatic fauna, including the structure and functioning of food chains, seasonal cycles, migration (horizontal, vertical and latitudinal), and distribution. The composition and abundance of the fauna and the use of aquatic animals as indicators of water pollution are also discussed.

Freshwater animals include a varied and rich population of organisms from many phyla and classes. These organisms can be found in all freshwater ecosystems. They come from different origins: some originated from land-dwelling predecessors that migrated to freshwater ecosystems; others migrated from oceans to inland waters.

8.1   ZOOPLANKTON

Chapter 6 describes the zooplankton in freshwater systems, including a wide variety of **micro-zooplankton** (protozoa and rotifers) and **meso-zooplankton** (crustaceans, cladocerans, and cycloploid and calanoid copepods). *Chaoborus* and Mysidacea larvae sometimes can be found in lakes, reservoirs, and ponds and form part of the **macro-zooplankton**. Most zooplankton organisms measure 0.3 to 0.5 millimetres in length. They represent an important link in the food chain in all freshwater ecosystems, estuaries, oceans, and coastal waters. Most of these organisms feed on phytoplankton or bacterioplankton. Rotifers, cyclopoid copepods and worms may even prey on other types of zooplankton.

The principal metabolic and behavioural features of freshwater zooplankton include seasonal cycles, spatial and temporal succession, vertical migration, reproduction and basic aspects of the life cycle, including development and feeding. The main groups of zooplankton include non-photosynthesizing protozoa, rotifers, many subclasses of crustaceans, and some coelenterates, flatworms, and insect larvae. Very few invertebrate larvae occur in freshwater plankton, as compared with marine plankton. Additional freshwater zooplankton groups include amoebae, ciliates, *Mesostoma* flatworms, and eggs and larvae of several freshwater fish species.

Life cycles and reproduction

The reproductive rates and growth rates of zooplankton depend on environmental factors such as water temperature, food availability, levels of dissolved oxygen, and general conditions of water quality in aquatic ecosystems. Specific reproductive features vary among the different groups of zooplankton. Rotifers are often parthenogenetic, i.e., adults produce eggs with diploid chromosomes and do not require a sexual phase for reproduction. The life cycles of crustacean zooplankton (including both cladocerans and copepods) are more complex.

Cladocerans are usually parthogenetic. In conditions of extreme overcrowding, males develop from diploid eggs; sexual reproduction with eggs and haploid sperm then occurs. Copepods are not parthenogenetic, but they can reproduce rapidly due to the female's sperm reserve, which allows fertilization of multiple clutches from a single copulation (see Figure 8.1).

Cyclomorphosis is an important phenomenon that occurs in zooplankton, particularly in crustaceans. Many zooplankton organisms change their morphology at different phases of the seasonal cycle. In many aquatic ecosystems, summer populations present different characteristics from winter populations. Examples of cyclomorphosis include the formation of spines or "helmets" in some species, as shown in Figure 8.2. Various factors influence and regulate cyclomorphosis, including high temperature, abundance of available food, and turbulence, according to Jacobs (1967). Lampert and Sommer (1997) described how the presence of dissolved chemical substances

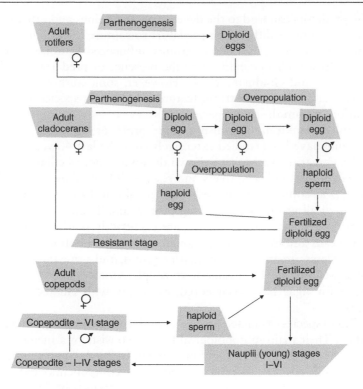

Figure 8.1   Types of reproduction of rotifers, cladocerans and copepods.

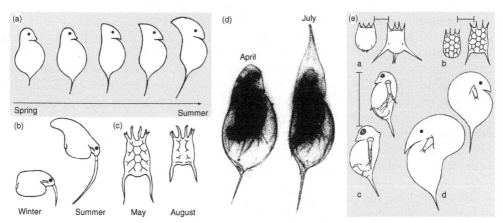

a – Induction of temporary spines in the rotifer *Brachionus calyciflorus* rotifer by the rotifer predator *Asplanchna*; scale 100 μm (*Halbach*, 1969).
b – Induction of permanent spines in rotifer *Keratella testudo* by *Asplanchna*; scale of 100 μm (Stemberger, 1988).
c – Induction of a bulge in the upper part of the carapace in *Daphia pulex* by *Chaoborus* larvae; scale of 1 mm (Havel and Dodson, 1984).
d – Induction of a 'hood' or 'helmet' in *Daphia carnicata* by *Anisops*; size of *Daphnia*, approximately 5 mm (Grant and Dayly, 1981).

Figure 8.2   Cyclomorphosis of zooplankton. a) Seasonal alterations in the morphology of *Daphnia retrocurva*, b) winter and summer forms; c) Morphology of the rotifer *Keratella quadrata* in May and August; d) Cyclomorphosis in *Daphnia cuccullata*; round carapace (April) and with helmets (July); e) The morphs produced in the presence or absence of predators are placed next to each other for comparison.
Source: Modified from Lampert and Sommer (1999).

produced by predators can lead to the development of spines and other formations in *Daphnia*, making it more difficult for predators to capture the organism.

Cyclomorphosis occurs in different groups, influenced by abiotic and biotic factors. A morphological change induced by the presence of predators is called **chemomorphosis** (Horne and Goldman, 1994). However, some authors believe that such morphological variations are, in truth, features of different species and not morphological variations within the same species.

Among microplanktonic organisms, some protozoan species in the protozooplanktonic group have been studied extensively over the last 40 years (Labour-Parry, 1992). Studies in marine waters have shown the abundance of ciliates and flagellates; and subsequent studies in lakes, rivers and reservoirs have shown the importance of these organisms as components of freshwater plankton. Protozooplankton contains representatives of all groups of free-living protozoa: ciliates, flagellates and sarcodines. Some groups, such as Foraminifera, are exclusively marine. Ciliated tintinnids can occur in oceans and fresh water, although few species are found in inland ecosystems. Dinoflagellates are an important marine group, and some important freshwater species exist as well. Among freshwater proto-zooplanktonic organisms, varying degrees of symbiosis with algae have been noted, especially with ciliated protozoa of the *Stentor* genus.

Nearly 7,000 species of ciliated protozoa (both free-swimming and parasitic) have been identified. Their main morphological feature consists of a membrane with cilia and membranelles used for feeding and movement. In harsh conditions of desiccation or salinity, some ciliates or other protozooplanktonic components can form cysts. The cysts, which vary in form, can survive through long periods of desiccation or other unfavourable conditions. The capacity of some *Strombidium* ciliates to retain plastids from various prey is important in the flow of carbon and transfer of energy in aquatic systems (Matsuyama and Moon, 1999) (see Figure 8.3).

Many ciliated proto-zooplankton have a combined autotrophic-heterotrophic mode. Retaining plastids from prey species provides the capacity to fix carbon through photosynthesis. Ciliated plankton in the genera *Vorticella* and *Epistylis* are also common in some lakes. They are sessile ciliates, feeding on small particles and bacteria. In some cases, diatoms serve as a substratum for epibiont protozoa.

Regali Seleghim and Godinho (2004) described the colonization of copepods, cladocerans, and rotifers by *Rhadostyla* spp. and *Scyphidia* spp. These epibionts can take advantage of free transportation to areas where food is more abundant, avoid predation by zooplankton (Henebry and Ridgway, 1979) or be transported to areas with better physical and chemical conditions of survival.

Protozooplanktonic flagellates, common in all aquatic systems, are a morphologically and physiologically heterogeneous group. They use their flagella for movement and feeding. These organisms are generally unicellular, but there are instances of phytoflagellate and zooflagellate colonies. Reproduction is asexual in all flagellate groups.

The **heterotrophic flagellate** group includes a variety of physiologically and morphologically diverse species of fungi (*Phytomastigophora* and *Zoomastigophora*). Among phyto-flagellates, the dino-flagellate group is common in marine and coastal waters (e.g., *Noctiluca* spp.). The most common genera of freshwater dino-flagellates

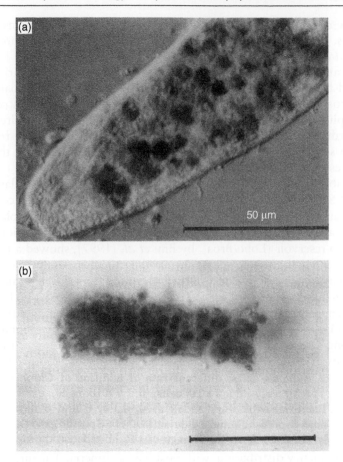

Figure 8.3   Microphotographs of *Tracheloraphis* sp. (a) and *Spirotrichia* (b). The two species include many cells of *Chromatium* sp. and *Macromonas* sp. in their structure (Lake Kaiike).
Source: Matsuyama and Moon (1999).

are *Ceratium, Peridinium, Gymnodinium* and *Cystodinium*. Choano-flagellates are common in saltwater but not in fresh water.

Phagotrophic phytoflagellates ingest suspended particles, bacteria and dissolved organic matter (Porter, 1988). Mixotrophic phytoflagellates, especially dinoflagellates, are common in some freshwater ecosystems. Foraminifera, Radiolaria and Heliozoa are Sarcodines that are found mainly in marine, oceanic and coastal waters. Thecamoebae occur in freshwater plankton.

8.1.1   Spatial distribution and seasonal cycle

The composition and structure of micro-, meso- and macro-zooplanktonic communities vary greatly through space and time. In general, mesozooplankton (or metazooplankton) includes several dominant species, such as cladocerans and rotifers, which vary spatially and seasonally, depending on factors such as light intensity, dissolved oxygen levels, food availability, competition, predation, parasitism and the hydrodynamics of aquatic systems.

Spatial variation and distribution of zooplankton depend on various physical, chemical and biological factors. It is important to understand the spatial variation in aquatic ecosystems in order to assess and identify the distribution of organisms, prepare sampling programmes, and apply statistical methods (Legendre *et al.* 1989, 1990). According to Armengol *et al.* (1999), **spatial heterogeneity in reservoirs**, for example, is a structural and functional feature of ecosystems and not the result of a random process (Legende, 1993).

The distribution of plankton communities, zooplankton in particular was studied by Patalas & Salki (1992), Betsil & Van den Avely (1994). More recently in Barra Bonita reservoir Matsumura-Tundisi & Tundisi (2005) studied the zooplankton distribution in this artificial ecosystem. A detailed study of spatial distribution in the UHE Carlos Botelho reservoir (Lobo/Broa), by Bini *et al.* (1997), showed a high degree of spatial heterogeneity in the ecosystem. The number of copepods and nauplii of copepods increased in the zone of influence of the Lobo and Itaqueri Rivers, and the number of cladocerans increased toward the reservoir's limnetic zone. Abiotic factors such as levels of nutrients and suspended matter influence and affect spatial distribution.

Marzolf (1990) presented a theoretical model describing the abundance of zooplankton along a reservoir's longitudinal axis, as determined on two main factors: current velocity and the transfer of material (clay, nutrients, dissolved organic matter), in addition to available food (phytoplankton, specifically, nanophytoplankton). If current velocity is a significant factor in zooplankton distribution, there is an increase in the direction toward the dam. If the transfer of material is predominant, zooplankton density is greater in the region of the reservoir under the influence of the rivers. In cases where the two factors are equally important, distribution in the reservoir is similar to a 'frequency distribution' with positive asymmetry. The interacting current velocity and zooplankton distribution, and even transfer of material, appear to be the main factors influencing spatial distribution of zooplankton in UHE Carlos Botelho reservoir (Lobo/Broa).

Patalas and Salki (1992) determined that the morphology of lakes, the geology of the drainage basin, and the location of the main tributaries are key factors in the patterns of zooplankton distribution.

In the case of spatial distribution of zooplankton in the Barra Bonita reservoir, Matsumura Tundisi and Tundisi (2005) concluded that certain key factors – such as **current velocity, suspended organic particulate matter, the presence or absence of pollutants and contaminants** – were responsible for the relative abundance and composition of species in three study sites: the Tiete River, the Piracicaba River, and the confluence of the Tiete and Piracicaba Rivers. The communities of zooplankton and phytoplankton in these three areas present a mosaic of microhabitats, judged according to Margalef (1997) and Reynolds (1997). The study by Matsumura Tundisi and Tundisi (2005) also presents the hypothesis and theory that countless tributaries of the Barra Bonita reservoir contribute to a large amount of spatial heterogeneity, in which the discharge point of each tributary produces a frontier of water mass with varying density and nutrient levels, thus adding to the spatial heterogeneity and expansion capacity of feeding niches and favourable abiotic conditions.

8.1.2   Seasonal cycle

Fluctuations and the seasonal cycle of zooplankton in lakes and inland aquatic eco-systems depend on a range of factors. Burgis (1964), for example, considered variable precipitation (rainfall) to be the predominant factor affecting the biomass and suc-cession of species in Lake George (Africa). Matsumura Tundisi and Tundisi (1976) determined that precipitation is a decisive factor in the seasonal cycle of zooplankton in the UHE Carlos Botelho reservoir (Lobo/Broa).

Fluctuations in the level of the Amazon River affect the seasonal cycle of zoo-plankton: high densities of zooplankton are associated with lower water levels in the river. The same occurs in lakes in the Pantanal Mato-grossense. In these cases, an abundance of food occurs during periods of isolation of the lakes due to decomposi-tion of macrophytes and other organisms.

Seasonal variations in zooplankton can therefore be related to climatic factors (mainly precipitation and wind), hydrographic and hydrological factors (periods of flooding and large volumes of rivers and lakes, interspersed with periods of reduced volumes). In a study of Dom Helvécio Lake, Parque Florestal do Rio Doce (MG), Matsumura Tundisi and Okano (1983) found that several species of copepods had different seasonal cycles (see Figure 8.4). According to the authors, factors such as patterns of thermal stratification and circulation, which occur in summer (December-March) and winter (June-September) respectively, can cause seasonal fluctuations in species in addition to predation by *Chaoborus* larvae (mainly in the **limnetic** zone) on the nauplii of cyclopoids (and not on the nauplii of calanoids, which are more abun-dant near the coastal zone where *Chaoborus* larvae are less abundant). *Thermocyclops minutus* and *Tropocyclops prasinus* peak in abundance during different periods in the lake, mainly because they have different reproductive periods.

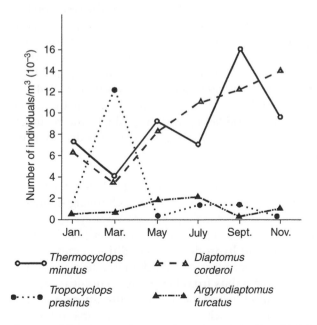

Figure 8.4   Pattern of seasonal fluctuation of copepod species in Lake Dom Helvécio in 1978.

The seasonal cycle of zooplankton, therefore, involves a number of biological factors, including reproductive period, co-existence with other species and impact of predation (intra-zooplanktonic or on other organisms, such as fish), as well as abiotic factors such as precipitation, stratification and vertical circulation. Water temperature clearly plays an important role in the reproduction and physiology of zooplankton organisms, and consequently, in the seasonal cycle. Ritzier (1995) verified that *Thermocyclops decipiens* and *Mesocyclops kieferi* are the dominant species in the Barra Bonita reservoir (SP.), coexisting throughout the year.

During periods of thermal stratification in summer, *M. kieferi* is abundant from 0 to 10 meters deep, and *T. decipiens* occurs from 5 to 10 meters deep; the two species are therefore in part spatially segregated during the season. In winter, during periods of thermal instability, organisms of the two species are more uniformly distributed. *Thermocyclops decipiens* and *Mesocyclops kieferi* are omnivorous organisms with high levels of detritus in their diet. Observations and laboratory studies show that the development phases, mortality rates and reproductive rates of both species are attuned to the environment, giving them competitive advantages and explaining their predominance in the Barra Bonita reservoir (Rietzler, 1995).

According to Espindola (1994), the presence of different *Notodiaptomus* species in the Barra Bonita reservoir throughout the seasonal cycle *(Notodiaptomus iheringi, Notodiaptomus cearensis,* and *Notodiaptomus conifer)* is due to qualitative variation in available food associated with climatic factors (such as precipitation), hydrographic factors (dissolved oxygen and temperature), and hydraulic factors (water flow and retention time). These are the main factors that determine the seasonal cycle of zooplankton and the population dynamics of these calanoid species. The pressure from predation by vertebrates (fishes) and invertebrates *(Mesocyclops* sp. and *Asplanchna* sp.) also contributes to fluctuations in the population density of *Notodiaptomus* spp. (including a new species not yet described found in the reservoir).

N. cearensis and *N. conifer* presented higher levels of production of eggs and greater **longevity** at 23°C (give or take 1°C). The temperature of 18°C (give or take 1°C) limited the development and growth of populations of the species. For these organisms, **diatoms** (*Aulacoseira distans*) and Chlorophyta (*Chlamydomonas* sp. and *Monoraphidium* sp.) were the preferred food. Espindola (1994) determined that the temporal segregation and the effects of temperature on the development of different species, and available food (along with hydrologic and hydrographic conditions), were all factors influencing the spatial and temporal distribution of species in the Barra Bonita reservoir.

Padovesi Fonseca (1996) presented similar conclusions on the population dynamics of zooplankton in a small reservoir (the **Jacaré-Pepira reservoir**, Brotas – SP). The reservoir is shallow, turbulent and subject to the changing influences of precipitation, wind and thermal variations throughout the seasonal cycle.

In conclusion, according to Lampert and Sommer (1997), abiotic factors (including precipitation, wind and **hydrodynamic conditions**) and biotic factors (including available food resources, reproductive rate, changing levels of competition and predation, and shifts in feeding habits from herbivore to detritivore) are key factors in the seasonal cycle of zooplankton and the dynamic alterations and succession.

Reproductive strategies such as rapid growth (maximum growth rate of the population – r_{max} – r strategists) with rapid dispersal and colonization capacity, as compared to slower growth and reproduction (k strategists), are factors that determine the seasonal cycle and appropriate and efficient use of energy for reproduction (r strategists) or defence against predators and minimization of mortality with reduction of the specific metabolic rate and reproductive rate (k strategists). The ability of populations to survive throughout the seasonal cycle thus depends on a balance between reproduction and mortality. **Phenotypic** and **genotypic variations** occur in these populations.

In an in-depth study on a tropical lake (Lake Lanao) in the Philippines, Lewis (1979) examined seasonal variations and abundance of phytoplankton, herbivores and carnivores in weekly samples from August 1970 to October 1971 (see Figure 8.5). The period of vertical circulation in the lake did not favour phytoplankton growth and development because the availability of underwater solar radiation was limited. This is a common process in many tropical lakes: limited underwater solar radiation due to circulation and excess suspended material resulting from precipitation and drainage, drastically reducing light penetration.

Calijuri and Tundisi (1990) recorded reductions up to 80% in underwater light penetration at the Barra Bonita reservoir (SP). In the lake region of Cananéia (SP), Tundisi and Matsumura Tundisi (2001) observed drastic reductions in the euphotic zone with marked reduction in the depth of primary production.

Still, according to the study by Lewis (1979), herbivores probably consume approximately 10% of the primary production annually. In the lake, the carnivores are almost entirely limited to *Chaoborus* larvae, which prey extensively on herbivorous zooplankton. The studies by Lewis show that the development cycles of different stages of copepods are closely interrelated (a calanoid species and a cyclopoid species), which generally does not occur in community structure and dynamics. Seasonal variations in the herbivore communities of Lake Lanao are attributed to variations in quantity and not quality of food (phytoplankton). Variations in the development of herbivores, as a result of the variability and fluctuations in phytoplankton, occurred

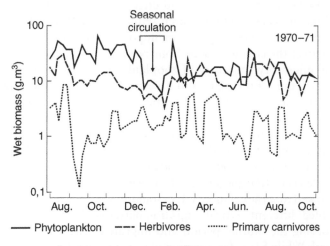

Figure 8.5   Quantities of phytoplankton, total primary herbivores and carnivores at weekly intervals in Lake Lanão. The curve of primary carnivores was slightly standardized (14-day period). *Source:* Modified from Lewis (1979).

as a result of the availability of various components, the size of available phytoplankton and their digestibility and capacity to support various herbivore communities.

According to Rocha *et al.* (1995), abundance and seasonal cycles in tropical and subtropical lakes and reservoirs in Brazil are related to fluctuations in hydrometric level, thermal structure, circulation, retention time (in reservoirs) and availability of food (phytoplankton and detritus).

8.1.3 Distribution and vertical migration

Zooplankton in both inland waters and ocean waters present a vertical migration pattern over a 24-hour period that varies from species to species and with the development stages of different species. Zooplankton generally spends the day towards the bottom of lakes and reservoirs, migrating at night toward the surface. Cases of 'reverse' vertical migration exist in which the zooplankton migrates to the bottom of an aquatic ecosystem during the night and remains on the surface during the day. The diurnal movements of freshwater zooplankton were first recorded by Neissman (1877b) in Lake Constance, followed by Pavesi (1882) in Italian lakes, Francé (1894) in Lake Balaton, Stever (1901) in small lakes along the Danube River, and Lorenzon (1902) in Lake Zurich.

Later studies to identify the causes of vertical migration led to experiments examining the role of light intensity in the phenomenon, variations in horizontal and vertical light rays (Bauer, 1909) and the positive phototactic response in relation to gradients of light intensity. Many researchers argue that zooplanktonic organisms tend toward a **negative geotaxis** in darkness and a **positive geotaxis** during periods of light. Tundisi and Matsumura Tundisi (1968) observed positive geotaxis in *Pseudodiaptomus austus* under illumination in experiments conducted in the lake region of Cananéia in the state of Sao Paulo.

Effects of polarized light (Baylor and Smith, 1953), increases in the specific density of organisms due to feeding (and subsequent sinking because of the inability to sustain themselves on the surface), and effects of hydrostatic pressure have been the subjects of much experimental research (Hutchinson, 1967). The diurnal cycle of light and darkness undoubtedly plays an important role in the vertical migration of zooplankton over the course of a 24-hour period.

The typical **migratory** trajectory of zooplankton is a vertical upward movement toward the surface during the night and a vertical downward movement toward the bottom during the day. The process of vertical migration appears to involve a relatively small expenditure of energy (Hutchinson, 1967).

The theory of an **endogenous rhythm** that regulates vertical migration and zooplankton movement was presented by Harris (1963), Hurt and Allanson (1976), and Zaret and Suffern (1967). In a detailed study of the vertical migration of zooplankton in Lake Dom Helvécio (Parque Florestal do Rio Doce – MG), Matsumura Tundisi *et al.* (1997) showed that the calanoids *Argyrodiaptomus furcatus* and *Scolodiaptomus corderoi* presented different vertical migration patterns. *A. furcatus* remained near the metalimnion during the diurnal period and migrated toward the surface during the nocturnal period. While *S. corderoi* remained near the metalimnion during the diurnal period, since part of the population also inhabits the metalimnion. Few individuals migrate at night toward the surface. During periods of vertical homogeneity in the

Collection and measurement of zooplankton

The collection of plankton and, more specifically, zooplankton, began over 150 years ago. Darwin was one of the first to use a plankton net on board the Beagle on his transoceanic voyage. Conical collection nets with mesh openings ranging from a few millimetres to centimetres were used to collect plankton for long-term oceanographic and limnological studies. These nets were used vertically, horizontally and obliquely to sample a taxonomically varied and diverse community, biologically and ecologically important.

The development of more sophisticated collecting methods began with the need to better characterize the plankton community and quantify it in terms of number of individuals, volume, or wet or dry weight. The evolution of collection technology and quantification of zooplankton included: conical-shaped nets adapted to flow meters to measure the volume of water sampled; closing nets to control the volume and the sampled depth for studies on vertical migration; suction pumps for sampling determined volumes with flow meters and hoses at certain depths.

Measures to calculate the biomass of zooplankton include: calculation of dry weight, total volume of sample (in cm^3 or mm^3); wet weight of plankton collected; total carbon sample collected. In all these techniques, microscopic observation of the sample is vital for classification and species composition. Plankton-collecting nets generally have 50-micron mesh for meta-zooplankton, $>100\,\mu m$ for macro-zooplankton, and $<30\,\mu m$ for micro-zooplankton. A variety of filtering systems and nets with different meshes (100, 50, 25, 10, 5 and $2\,\mu m$) are used for classifying the size of organisms and their quantitative importance. Figure 8.6 shows the water-flow patterns in some types of plankton-sampling nets.

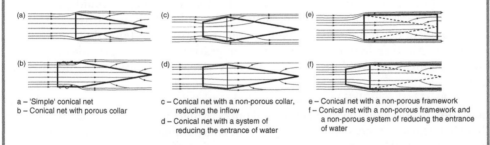

a – 'Simple' conical net
b – Conical net with porous collar

c – Conical net with a non-porous collar, reducing the inflow
d – Conical net with a system of reducing the entrance of water

e – Conical net with a non-porous framework
f – Conical net with a non-porous framework and a non-porous system of reducing the entrance of water

Figure 8.6   Patterns of flow associated with some basic types of plankton nets. Each type has a different filtering efficiency.

water column, the population is distributed homogenously, with a small accumulation on the bottom. *A. furcatus* presents a reverse migration toward the bottom of the lake during the period of vertical homogeneity in the water column.

Thermocyclops minutus presents vertical migration to the epilimnion-metalimnion interface during the lake's stratification period, remaining in the deepest part of the metalimnion during the day. *Tropocyclops prasimus cenidionalis* does not migrate vertically during the stratification period, but instead remains in the deepest part of the metalimnion during this period. During periods of vertical homogeneity in the lake, this species migrates to the depth of 5 metres during the day. Such examples of vertical migration in a lake that stratifies in summer and is thermally and chemically homogenous in winter (see Figure 8.7) clearly show that there are several different types of behaviour:

▶ species that inhabit the epiliminion and migrate vertically in that layer (*S. corderoi* and *A. furcatus*);
▶ species that inhabit the epilimnion-metalimnion interface with vertical migration in intermediate layers;
▶ species that remain in the metalimnion while presenting vertical migration, such as *T. prasinus*.

The vertical distribution of zooplankton varies with the intensity of thermal stratification and circulation. Figure 8.8 illustrates the vertical movements of different zooplankton species in Lake Dom Helvécio. While two species of calanoid copepods remain in the epilimnion during the stratification period, a cyclopoid species explores the epilimnion-metalimnion interface and another explores the metalimnion. This is a clear and interesting example of the exploitation of resources by different zooplankton populations through the presence or absence of vertical migration activity.

According to Lampert and Sommer (1997), the stimulus for the beginning and ending of vertical migration is the relative (and not absolute) change in light intensity. Phototaxis and geotaxis are essential in the regulation of vertical migratory behaviour in zooplankton.

Vertical migration of zooplankton must undoubtedly provide some advantages, such as more efficient use of energy during migration and more rapid population growth. The presence of zooplankton on the surface at night would provide the advantage of obtaining richer food, such as phytoplankton, and at the same time, would prevent deleterious effects on the population from mortality due to ultraviolet radiation. Another hypothesis on vertical migration is that it provides an important mechanism to avoid predation, especially visual predation by plankton-eating fishes.

Mechanical stimulation resulting from the presence of predators promoted rapid migration of zooplankton in experimental tanks (Ringelberg, 1991). One theory states that chemical substances released by predators may stimulate zooplankton to migrate (Loose *et al.*, 1993). Changes in the reverse migration of copepods also occurred when *Chaoborus* larvae, common in Lake Gwendolyne (British Columbia, Canada), were eliminated with the introduction of trout. During the day, individuals of *Diaptomus* were found on the lake's surface, and at night they migrated to deeper water to avoid predation by *Chaoborus*. With the disappearance of the larvae, the reverse migration stopped. When water containing *Chaoborus* larvae was introduced into experimental tanks, the organisms immediately began to migrate in reverse, indicating chemical stimulation (4-hour response period).

Other hypotheses on vertical migration involve genetic exchange during the migration period and an optimization of the grazing pressure on phytoplankton by zooplankton. During the day, the absence of grazing by zooplankton optimized the growth and reproduction of the phytoplankton biomass (Lampert and Sommer, 1997).

Many zooplanktonic species avoid the littoral zones of lakes and reservoirs, involving **horizontal migrations** that keep them away from the edges. A combination

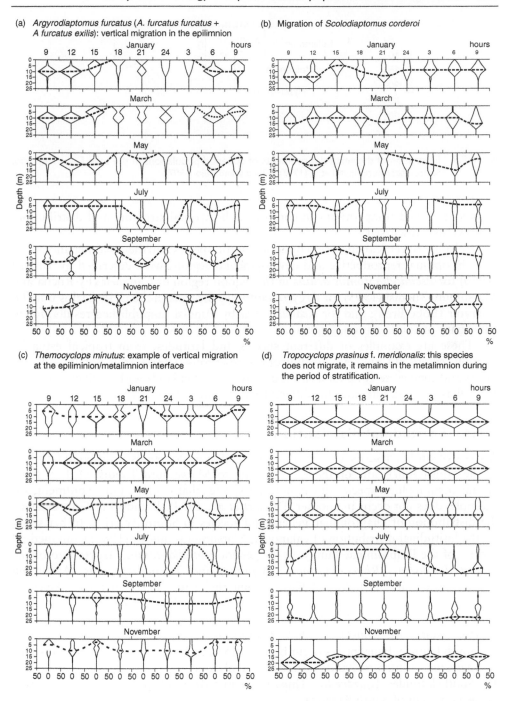

Figure 8.7 Patterns of vertical migration of copepod species in Lake Dom Helvécio during1979.
Source: Matsumura Tundisi *et al.* (1997) in Tundisi and Saijo (1997).

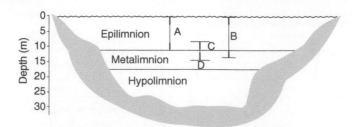

Figure 8.8   Illustration of the vertical movements of copepod populations in the Lake Dom Helvécio. *Source:* Matsumura Tundisi *et al.* (1997) in Tundisi and Saijo (1997).

of predation and competition can control the horizontal structure and distribution, as previously discussed.

Episodes of vertical migration by zooplankton species and their maintenance strategies in optimal reproductive and developmental regions were described by Tundisi (1970) for *Pseudodiaptomus acutus* in the lagoon region of Cananeia; by Rocha and Matsumura Tundisi (1995) for *Argyrodiaptomus furcatus* in the UHE Carlos Botelho reservoir (Lobo/Broa); and for *Acartia tonsa* by Tundisi (unpublished results) for in an estuary in Southampton, England.

These are examples of different species and latitudes, from tropical estuaries (lagoon regions of Cananeia), to subtropical reservoirs (UHE Carlos Botelho – Lobo/Broa) and temperate zone estuaries (Southampton, England) with similar reproductive behaviour and tolerance for different environmental factors.

8.1.4   Latitudinal distribution

A series of studies describe the distribution of calanoid copepods at different latitudes in the South American continent. According to Matsumura Tundisi (1986), there are no cosmopolitan species of calanoids; a species that occurs in one continent, for example, does not occur in another (Dussart *et al.*, 1984). Brandorf (1976) reviewed the distribution of Diaptominae genera and species in South America. In her study on the latitudinal distribution of calanoids, Matsumura Tundisi (1986) examined 20 species of the Diaptominae family, in the genera *Notodiaptomus*, *Argyrodiaptomus*, *Odontodiaptomus*, *Rhacodiaptomus*, *Aspinus*, and *Trichodiaptomus*. According to this author and Brandorf (1976), among the main genera in South America, *Argyrodiaptomus* species are the most common in the continent. Seven species belonging to this genus (*A. aculeatus*, *A. argentinus*, *A. bergi*, *A. furcatus*, *A. denticulatus*, *A. azevedo*, and *A. granulosus*) can be found between 25° and 40° South latitude. Four species occur between 15° and 25° South latitude: *A. aculeatus*, *A. neglectus*, *A. furcatus*, and *A. azevedoi*. Only one species occurs between 0° and 10° South latitude: *A. azevedoi*. Thus, the genus *Argyrodiaptomus* is characteristic of the continent of South America.

Odontodiaptomus paulistanus is extremely common in reservoirs in the state of São Paulo, between 20°S and 23°S latitude. The genus *Notodiaptomus* (with 22 species) is widely distributed throughout South America. *N. iheringi* is common in reservoirs in the state of São Paulo, especially in eutrophic systems, and *N. cearensis* is common

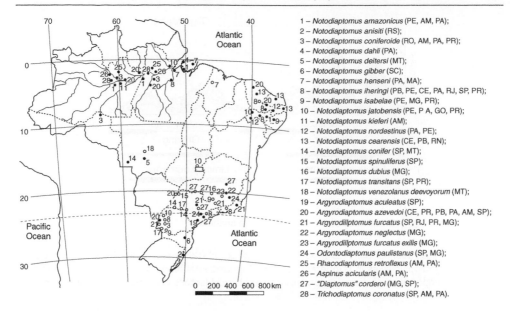

1 – *Notodiaptomus amazonicus* (PE, AM, PA);
2 – *Notodiaptomus anisiti* (RS);
3 – *Notodiaptomus coniferoide* (RO, AM, PA, PR);
4 – *Notodiaptomus dahli* (PA);
5 – *Notodiaptomus deitersi* (MT);
6 – *Notodiaptomus gibber* (SC);
7 – *Notodiaptomus henseni* (PA, MA);
8 – *Notodiaptomus iheringi* (PB, PE, CE, PA, RJ, SP, PR);
9 – *Notodiaptomus isabelae* (PE, MG, PR);
10 – *Notodiaptomus jatobensis* (PE, P A, GO, PR);
11 – *Notodiaptomus kieferi* (AM);
12 – *Notodiaptomus nordestinus* (PA, PE);
13 – *Notodiaptomus cearensis* (CE, PB, RN);
14 – *Notodiaptomus conifer* (SP, MT);
15 – *Notodiaptomus spinuliferus* (SP);
16 – *Notodiaptomus dubius* (MG);
17 – *Notodiaptomus transitans* (SP, PR);
18 – *Notodiaptomus venezolanus deevoyorum* (MT);
19 – *Argyrodiaptomus aculeatus* (SP);
20 – *Argyrodiaptomus azevedoi* (CE, PR, PB, PA, AM, SP);
21 – *Argyrodillptomus furcatus* (SP, RJ, PR, MG);
22 – *Argyrodiaptomus neglectus* (MG);
23 – *Argyrodillptomus furcatus exilis* (MG);
24 – *Odontodiaptomus paulistanus* (SP, MG);
25 – *Rhacodiaptomus retroflexus* (AM, PA);
26 – *Aspinus acicularis* (AM, PA);
27 – *"Diaptomus" corderoi* (MG, SP);
28 – *Trichodiaptomus coronatus* (SP, AM, PA).

Figure 8.9   Latitudinal distribution of calanoid copepods in Brazil's inland aquatic ecosystems.
Source: Tundisi Matsumura (1990).

in reservoirs in the Northeast. *N. conifer* occurs between 10°N and 36°S latitude, commonly in lakes and reservoirs, especially reservoirs along the Paranapanema River. Figure 8.9 shows the distribution of the most common calanoid species in Brazilian aquatic systems.

What factors determine the latitudinal distribution of these species? Physical and chemical conditions are important, and in particular the relationship with temperature and conductivity.

The association of temperature, salinity, and conductivity most likely determines the osmotic conditions needed for the establishment and development and colonization of calanoid species. According to Hutchinson (1967), the occurrence of endemic species is common among calanoids, given this group's tendency to create well-established latitudinal populations and exploit microhabitats. Calanoid copepods of inland waters present greater regional endemnicity than any other group of planktonic organisms. A small difference in tolerance to temperature, pH, and conductivity is probably enough to isolate these species. Even in lakes located close together, such as those in Parque Florestal do Rio Doce, there is an absence of some species, also observed by Lewis (1979) in Lake Lanao in the Philippines and adjacent systems.

In experiments on tolerance to conductivity/salinity and temperature of calanoid species in the state of São Paulo, Tundisi (unpublished results) obtained the following tolerance gradient for these factors: *Notodiaptomus iheringi* > *Argyrodiaptomus furcatus* > *Argyrodiaptomus azevedoi*, which may explain the dominance and succession of these genera and species in different ecosystems, as verified by Rietzler (1995) in the succession of *Argyrodiaptomus furcatus* and *Notodiaptomus iheringi* in the UHE Carlos Botelho (Lobo/Broa) and Barra Bonita reservoirs.

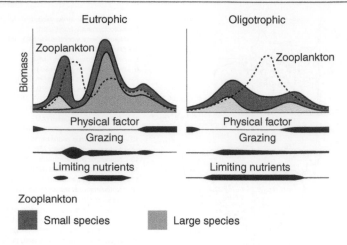

Figure 8.10   Illustration of the PEG model of seasonal zooplankton succession in eutrophic and oligotrophic lakes in temperate regions. The horizontal black symbols indicate the relative
importance of the selection factors, i.e., physical factors, predation or availability of food.
Source: Adapted from Lampert (1997).

8.1.5   Interrelationships phyto-zooplankton

The feeding habits of herbivorous zooplankton involve a selective process with a variety of structures and behaviours. Seasonal succession of phytoplankton and the corresponding succession of herbivorous zooplankton are the subject of many studies
on inland and marine ecosystems (Raymont, 1963; Reynolds, 1984, Sommer, 1989;
Lewis, 1979; Rocha and Matsumura Tundisi, 1997). Succession in phytoplankton and
zooplankton in temperate-region lakes was expressed using the PEG model (*Plankton
Ecology Group* – International Association of Limnology). Figure 8.10 shows the
succession model for temperate-zone eutrophic and oligotrophic lakes. To a certain
extent, however, patterns are much more complex and difficult to predict in tropical
lakes and reservoirs, where the physical controls of the process (such as hydrology
and hydrodynamics) can be much more effective and the cycles are faster, with interactions and overlapping of herbivores, detritivores and carnivores.

The concepts of **selective feeding** and **competition for resources** – strategies to
prevent the overlapping of feeding niches – apply equally to planktonic communities
in both temperate and tropical lakes. However, cycles in tropical lakes are faster and
the alternatives are much more important. Tropical systems – affected by hydrological
cycles, high temperatures and accelerated biogeochemical processes – are much more
dynamic and complex than temperate-region systems. Studies on several tropical
lakes, summarized by Talling and Lemoalle (1998), corroborate this hypothesis.

The authors outlined the processes and interactions of phyto-zooplankton and
predation by *Oreochromis niloticus* (tilapia) and *Haplochromis nigripinnis* in Lake
George (Africa) over 24-hour cycles (see Figure 8.11). Alternative predator-prey feeding cycles by zooplankton on phytoplankton and by fishes on zooplankton occur in
short periods, and these processes control the metabolism of lakes, since in tropical
regions they overlap the seasonal cycles and appear to be more significant (Barbosa
and Tundisi, 1980).

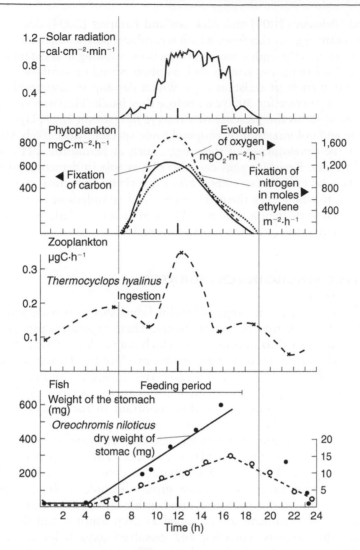

Figure 8.11   Diurnal cycles of solar radiation, fixation of carbon and nitrogen and food intake by zooplankton in Lake George.
Source: Modified from Talling and Lemoalle (1998).

8.1.6   Diapause

Resistant eggs and succession in zooplankton

Many freshwater calanoid species produce resistant eggs, especially in temporary lakes or waters. In South America, resistant eggs found in dry lake sediment that subsequently hatched to become *Argyrodiaptomus funcentus* adults were identified by Sars (1901), who also cultivated *Tropodiaptomus australis* in Australia with the same method, i.e., using dry sediment. The colonization of temporary waters is an adaptive strategy for a large number of calanoid species.

Laps and Hlekseev (2004) and Alekseev and Lampert (2004) described the production of resistant eggs in *Daphnia*, which reproduces parthenogenetically, but occasionally produces eggs through sexual reproduction, resulting in resistant eggs, carried by the female. Environmental stimuli such as photoperiod or availability of food are essential for production of resistant eggs, which develop in favourable conditions. Calanoid eggs can survive for 300 years before they hatch (Harrison *et al.*, 1995).

In many ecosystems the presence of a reserve of resistant eggs plays a key role in the succession and colonization of zooplanktonic species. Most likely chemical factors stimulate the development of these eggs, such as favourable pH conditions or salinity/conductivity. Resistant eggs can also play a role in detecting contamination and pollutants in lakes and reservoirs. Favourable chemical factors may promote the development of resistant eggs through biochemical stimulation; or physical factors may be associated with chemical factors. Physical factors include water temperature and sediment temperature, and length of photoperiod.

8.2 BENTHIC MACROINVERTEBRATES

As discussed in Chapter 6, the range of benthic fauna species is wide and varied, and includes herbivores, detritivores and predators. These organisms process the energy from internal and external sources in rivers, which is a product of the activities of periphyton, leaves, plant residues or organic matter produced by humans or animals.

In lakes, reservoirs or coastal and oceanic areas, these organisms largely depend on the production of organic matter that settles to the bottom of the ecosystem. A benthic invertebrate community is also important in the processing of organic material in rivers (and in the recuperation of rivers).

Aquatic insects are common in benthic macroinvertebrate communities in inland waters, with great diversity in rivers and streams. Species in the orders Ephemeroptera, Plecoptera, Trichoptera, Diptera and Odonata account for a majority of the biomass. Other important macroinvertebrate groups include molluscs, annelids and crustaceans. There are no macroinvertebrate benthic insects in marine environments.

In lakes, benthic organisms are classified as deep benthos and **littoral benthos**. Temperature, light intensity, currents, and dissolved oxygen levels all vary more widely in a 24-hour period in the littoral zone. The deeper zones in lakes are chemically and physically more uniform, except in eutrophic lakes during the stratification period, when anoxia and non-hypolimnic deoxygenation occur. Amphipods, larvae of chironomids and oligochaetes, molluscs and *Chaoborus* larvae are the most common organisms in the benthic animal community.

In the littoral zone, the presence of macrophytes provides rich and varied spatial heterogeneity, with niches that depend on the type of substratum and available food. Organic detritus is an important food source. Jónasson (1978) showed high species diversity in the littoral zone of Lake Esrom, Denmark, as compared with the benthic fauna in the lake's deep regions.

The structures of macro-vertebrate benthic communities can be altered by the production of fish, which greatly affects the community's biomass and species diversity. *Chaoborus* larvae are sensitive to chemical substances released by fish and may flee from predation, hiding in the sediment.

Table 8.1 Density (in individuals · m^{-2}) and biomass (g · dry weight · m^{-2}) of zoobenthos and chaoborids and chironomids in D. Lake Helvécio (Parque Florestal do Rio Doce – MG).

Animals	Station I	II	III	IV	V	VI	VII	VIII
Mollusca*		#			#	15		15
Hyridae				(61.93)		(130.81)		
Corbiculidae					(0.04)			
Planorbidae						15		
Oligochaeta						(0.09)		
Tubificidae					370	415		15
Odonata					(0.21)	(0.20)		(+)
Libellulidae						30		
						(7.34)		
Trichoptera					89	44	60	
					(0.07)	(0.03)	(0.04)	
Diptera								
Chaoboridae	44	30	44	15	74	15	60	44
	(0.03)	(+)	(0.01)	(0.01)	(0.11)	(+)	(0.04)	(+)
Chironomidae								
Tanypodinae		30	15	44	637	696	60	44
		(0.14)	(0.10)	(0.02)	(0.17)	(0.23)	(+)	(0.02)
Chironominae					578	830	119	119
					(0.43)	(0.28)	(0.09)	(0.07)
Total	44	60	59	59	1,748	2,075	299	237
	(0.03)	(0.14)	(0.11)	(0.03)	0.99	70.14	0.14	130.9
Planktonic Chaoboridae	1,380	354	113	0	n	n	n	n

+ <0.01 gm², * with shell; # only the shell, n – not sampled
Source: Fukuhara et al. (1997).

The life cycles of benthic macroinvertebrates include three or four stages in the case of insects: egg, nymph and adult, or egg, larva, pupa and adult. Most benthic organisms reproduce once a year, but in temperate climates some species require more than one year to complete the life cycle (Usinger, 1956).

Substratum type, current velocity and transportation of sediment are all factors that affect the composition, structure and functioning of benthic macroinvertebrate communities. In the different regions of rivers, current velocity and substratum type affect the composition, diversity and succession of different species of benthic macroinvertebrates (Welch, 1980), whose life cycle depends on and is controlled by the availability of food (Horne and Goldman, 1994).

Table 8.1 shows the density (in individuals · m^{-2}) and biomass (grams of wet-weight · m^{-2}) of zoobenthos and chaoborids and chironomids in Lake Dom Helvecio (Parque Florestal do Rio Doce – MG). The deep benthic fauna of this lake specifically include chaoborids and chironomids. *Chaoborus* occurs especially in deep regions with low (even near-zero) levels of dissolved oxygen. Near the littoral zone of this lake and in another lake (Lake Jacare), Planorbidae, Tubificidae, Trichoptera and Hirudinea are common.

Table 8.2 Density of *Chaoborus* and *Chironomus* larvae in several lakes in the tropical region.

Tropical lakes	Chaoborids (ind · m²)	Chironomids (ind · m²)	Notes
L. D. Helvécio	27–996	36–249	Jun, 1.3–33.0 m, Fukuhara et al., 1997
L. Jacaré	9–320	36–720	Jun, 1.5–8.5 m, Fukuhara et al., 1997
L. D. Helvécio	178–1,288	44–733	Aug, 10–23 m, Fukuhara et al., unpublished
L. Carioca	155–400	22–89	Aug, 3.5–8.0 m, Fukuhara et al., unpublished
L. Tupé (Rio Negro)	0–400	0	Aug–Apr, deep waters, Reiss, 1977b
L. Tupé (Rio Negro)	−2,180	15–570	Dec–Mar, littoral, Reiss, 1911b
Magalhas Lagoon	0–45	178–2,581	Dec, 0.2–3.5 m, Reiss, 1973
Ubaraha Lagoon	0	179–223	Dec, 0.2–1.5 m, Reiss, 1973
Cueiras River	0–44	0–2,729	Oct, 1.5–4.5 m, Reiss, 1977b
UHE Carlos Botelho (Lobo/Broa) Reservoir	1,909, 1,747	1,215, 1,014	1971 and 1979, Strixino and Strixino, 1980
UHE Carlos Botelho (Lobo/Broa) Reservoir	1,742	1,253	Jun, 10.4 m, Fukuhara et al., unpublished
L. Victoria (Ekumu Bay)	2,000–2,500	1,000	MacDonald, 1956

Table 8.2 shows the density of *Chaoborus* and *Chironomus* larvae in some tropical lakes. Fukuhara *et al.* (1997) showed the emergence *en masse* of *Chaoborus* (*Edwardsops*) *magnificus* in Lake Dom Helvécio, influenced by the lunar cycle. Hare and Carter (1986) also showed the lunar effects on the emerging adults of tropical insect species.

The physical, chemical and biological factors that control and regulate benthic macroinvertebrates and their physiology and distribution, in addition to those already mentioned (substratum type, current speed and predation), include water temperature and dissolved oxygen levels. These two key factors help determine the survival rate and optimum reproductive conditions of benthic species.

The respiration rates of these organisms depend on water temperature and availability of dissolved oxygen. The number of species that tolerate varying ranges of water temperature is a key feature of the aquatic environment.

Because of these characteristics – response to environmental factors and location on or in a substratum – benthic macroinvertebrates are excellent indicators of environmental conditions and contamination or pollution in rivers, streams, lakes and reservoirs.

Roldán (2006) described the processes of determining the 'ecological status' implemented by the European Union's Com-97 directive, which proposes that the water basin be the unit of study and that the ecological status of each water basin be determined and compared with reference data (Pratt and Rume, 1999).

The European standard cites the **communities of organisms** as indicators of the ecological status in different aquatic ecosystems. According to Roldán (2006), an organism is a good indicator of water quality when it is found in an ecosystem with defined characteristics and the population of the species is greater than or similar to the populations of other species that share this ecosystem or habitat. For example, in mountain rivers with clear oligotrophic water with an average temperature <15°C and 90–100% saturation of dissolved oxygen, the most common benthic invertebrates are likely to be Ephemeroptera, Trichoptera, and Plecoptera, with considerable

Chart 8.1   Indices of species diversity.

Shannon-Weaver (1949):

$$H' = -\sum_{i=1}^{S}(n_i/n)\ln(n_i/n)$$

H' – diversity index
n_i – number of individuals per species
n – total number of individuals
ln – natural logarithm

Simpson (1949):

$$I = -\sum \frac{n_i(n_i-1)}{N(N-1)}$$

where:
n_i – number of individuals per species
N – number of individuals

Margalef (1951):

$$I = (S-1)/\log_n N$$

S – number of species
N – number of individuals
$\log n$ – natural logarithm

Source: Roldan (2006).

numbers of crustaceans, Hemiptera, Diptera and Hymenoptera. In highly turbid eutrophic rivers with high levels of organic material and low levels of dissolved oxygen, dominant populations of oligochaetes, chironomids and certain species of molluscs occur, which are found in smaller proportions in uncontaminated water.

The type of benthic macroinvertebrates present can therefore be a measure of pollution levels, and Gletti and Bonazzi (1981) consider them to be the best indicators of water quality. A community's response to pollution or optimal water quality can be determined by the index of diversity, as seen in indices by Shannon-Weaver (1949), Simpson (1949) and Margalef (1951) (see Chart 8.1). Roldán (2006) applied the Biological Monitoring Working Party (BMWP, established in England in 1970) to rivers in Colombia to utilize benthic macro-invertebrates and other **aquatic invertebrates** to determine water quality. Table 8.3 shows the orders, families, genera and species used as indicators of water quality.

In one study, Marchese M. and Ezcurra de Drago (2006) used benthic macroinvertebrates as indicators of water quality, especially **macroinvertebrates** present in a eutrophication in the middle of the Paraná River. According to the study, the composition and structure of the macroinvertebrates indicate changes in water quality and the introduction of energy in the aquatic systems. The authors applied the Shannon-Weaver index to each aquatic ecosystem, comparing total and relative density of benthic organisms in each aquatic environment, as well as the degree of eutrophication and contamination. For example, the oligochaete *Tubiflex tubiflex* (*blanchardi* variety) is

Table 8.3 Classes of water quality, BMWP/Col values, meaning and colours for cartographic representations.

Class	Quality	Bmwp/Col	Meaning	Color
I	Good	>150 101–120	Very clean to clean water	Blue
II	Acceptable	61–100	Slightly polluted water	Green
III	Uncertain	36–21	Moderately polluted water	Yellow
IV	Critical	16–35	Very polluted water	Orange
V	Very critical	<15	Highly polluted water	Red

BMWP/Col – Biological Monitoring Working Party/Colombia index.
Source: Roldan (2006).

associated with highly conductive waters (Marchese, 1988). Chart 8.2, from Marchese and Ezcurra de Drago (2006), presents the gradient of indicator species typical in a specific trophic environment, from oligotrophic environments to eutrophic environments in aquatic ecosystems in the middle of the Paraná River.

8.3 COMPOSITION AND RICHNESS OF PLANKTON SPECIES AND ABUNDANCE OF ORGANISMS IN PELAGIC AND LITTORAL REGIONS OF LAKES AND RESERVOIRS

As already discussed, the origin of a lake, its trophic state, colonization process and the presence or absence of toxic substances and pollutants are all interdependent factors that control and limit planktonic composition and the abundance and diversity of species. Lakes and reservoirs can have fundamental differences, for example, from retention time, vertical circulation periods and contributions from tributaries. Patalas (1975) noted significant differences between the abundance of plankton species in small ($<24\,km^2$) and big ($>50\,km^2$) lakes. Plankton in the limnetic or pelagic region differs significantly from plankton in the coastal region. To a certain extent, the diversity of plankton species may also be the result of only collecting samples in certain regions. In reservoirs with marked spatial heterogeneities (Straškrába et al., 1993; Armengol et al., 1999), large sampling errors can occur because of the differences and the accumulation of plankton in the areas of tributary estuaries, for example.

A comparison between the Barra Bonita reservoir (SP.) and Lake Dom Helvécio (Parque Florestal do Rio Doce – MG) (Tundisi and Matsumura Tundisi, 1994) showed 37 species of zooplankton in the limnetic regions of the reservoir (20 rotifers, 8 cladocerans and 9 copepods), and only 16 species in the lake (6 rotifers, 5 cladocerans and 5 copepods). This difference is probably the result of the reservoir's large spatial heterogeneity, the multiple mixing with many annual circulations and the effect of 114 tributaries, each discharging from 2 to $15\,m^3 \cdot s^{-1}$, thus contributing to increased diversity. Temporal changes (due to eutrophication) are also observed in zooplankton succession in Barra Bonita reservoir (Matsumura Tundisi and Tundisi, 2003).

The **pelagic environment**, according to Margalef (1962) and Reynolds (1997), is made up of a mosaic of microhabitats with overlapping and constantly changing components, responding to forcing functions such as wind and retention time (in the case of reservoirs) (Matsumura Tundisi and Tundisi, 2003).

Chart 8.2 List of association of indicator species of a trophic gradient of environments of the middle Paraná River.

Oligotrophic Environments

Oligochaetes
 Narapa bonettoi

 Haplolaxis aedeochaeta
Turbellarians
 Myoratronectes paranaensis
Nematodes
 Tobrilus sp.
Diptera chironomids
 Tanytarsus sp.
 Parachironomus sp.
 Glyptotendipes sp.
Ephemeroptera
 Campsurus cf. notatus
Oligochaetes
 Paranadriluys descolei
 Bothrioneurum americanun
 Aulodrilus pigueti
 Pristina americana
 Paranais frici
Diptera chironomids
 Polypedilum spp.
 Cryptochironomus sp.
 Coelotanypus sp.
 Ablabesmyia sp.
Lamellibranch molluscs
 Pisidium sp.
 Corbicula fluminea
Oligochaetes
 Branchiura sowerbyi
 Limnodrilus udekermianus
 Nais variabilis
 Nais communis
 Dero multibranchiata
 Dero sawayai
Diptera chironomids
 Axarus sp.
 Goeldochironomus sp.
Gastropod molluscs
 Heleobia parchappei
Diptera chironomids
 Chironomus xanthus
 Chironomus gr. decorus
 Chironomus gr. riparius
Oligochaetes
 Tubifex tubifex
 (blanchardi form)
 Limnodrilus hoffmeisteri

Eutrophic environments

Source: Marchese and Ezcurra de Drago (2006).

According to Margalef (1991), the permanent re-organization of a system, as the result of **external energy**, interrupts the irregularities in the system and breaks down the horizontal axis in heterogeneous units in the space where phytoplankton and zooplankton communities gather. Horizontal mixture and vertical instability are factors that can contribute to increased planktonic diversity. In systems with continual instability, the **intermediate disturbance** hypothesis (IDH) (Padisák *et al.*, 1999) can be applied. The function of tributaries and their discharging waters is to add horizontal discontinuities in the aquatic system and thus lead to new groupings of communities along the horizontal axis, in short periods of time and in **mesoscale** structures (1 to 20 km).

In the case of zooplankton, the presence of **resistant eggs** of many species can affect species diversity in a certain time frame. Resistant eggs develop during favourable periods when water quality is good. They can influence succession and the abundance of zooplankton species. Likewise, resistant forms of phytoplankton (such as colonies or quiescent cells of *Aulacoseira* spp. in the sediment) can lead to a rapid increase of species in the water column, promoted by the action of forcing functions such as the wind. The interrelationships between phytoplankton and zooplankton are equally important in determining changes in the succession process of pelagic communities. Food availability, size, shape and nutritional status alter the succession of zooplankton species, in cases where predation (grazing) can significantly modify the composition of species and phytoplankton succession (Raymont, 1963; Reynolds, 1984, 1997).

In the shallowest littoral regions, the seasonal cycle and succession of phytoplankton and zooplankton are altered by the presence of species that periodically resuspend from the sediment, such as several *Aulacoseira* spp. – or the emergence of resistant copepod eggs, as demonstrated by Rietzler *et al.* (2004) and Matsumura Tundisi and Tundisi (2003).

The planktonic species inhabiting the pelagic and littoral regions in lakes and reservoirs have different physiological features, different morphological adaptations and different reproductive and growth requirements. An assemblage of species thus responds in different ways to environmental variability and the frequency of disturbances. Each planktonic community is the result of a combination of biological, chemical, and physical factors producing different associations that change through time and space.

Research studies are needed to identify the frequency, magnitude and direction of those changes. Knowledge of the biology and its responses with the limitations of different species is therefore essential for predicting the response.

8.4   FISH

Chapter 6 presented data on the composition of fish communities in inland waters. Fish play a significant role in the functioning of the ecological dynamics of aquatic communities, since their function is important, both quantitatively and qualitatively, in the food web and in planktonic, benthic and nectonic communities. Spatial movements of fish and migration complicate the quantitative calculation of their impact on food networks and structures of aquatic communities. Fish excrete detritus and ammonia and remove sediments, and therefore play an important role in the biogeochemical cycles of lakes, reservoirs, rivers and associated areas. Migratory fish such as salmon (**anadromous fish**), which live in oceans and reproduce in rivers, or catadromous fish, which live in inland waters and migrate to the ocean to reproduce, play an extremely important role in many aquatic ecosystems. For example, large balizador catfish (Barthem and Goulding, 1997) in Amazonian rivers (as described in the previous chapter), play an important role in the food chain and the structure of aquatic communities there.

Fish comprise nearly 40% of the vertebrate species on the planet. Communities of tropical and subtropical fish in the South American region have been studied extensively, especially in the Amazonian region (Goulding, 1979, 1981; Goulding *et al.*, 1988) and the São Francisco River (Sato and Godinho, 1999; Godinho and Godinho, 2003). Similarly, a series of extensive scientific works were conducted on water basins in the upper Paraná (Augustine and Ferreira, 1999, Agostinho *et al.*, 1987, 1991, 1993, 1994, 1999; Bonetto, 1986a; Menezes and Gery, 1983; Vazzoler and Menezes, 1992). Chart 8.3 outlines the main ecological features of tropical fish communities.

The seasonality of habitats, according to Lowe-McConnell (1999), affects fish behaviour and physiology in tropical aquatic systems, ranging from specific spatial dynamics (for example, flooding and drought) that create different habitats each year to the more stable conditions of some natural lakes and equatorial reefs. Seasonality and environmental variations affect fish and their feeding habits, life cycles, reproduction, and migrations. Migrations result from seasonal hydrological cycles and are used for reproduction, feeding and escape from predators.

Chart 8.3 Ecological attributes of communities of tropical fish.

Seasonality of the environment	Very seasonal ──────────▶	Not seasonal
Examples:	Floodplain Pelagic zone	Lacustrine littoral Coral reefs
Response of fish population:	Fluctuates greatly due to: (1) migration (high mobility), (2) rapid multiplication	Remains constant through the year and year to year
Life cycle:	Short, precocious maturation, low longevity	Long: Retarded maturation (frequent change of sex); high longevity
Growth rates:	Rapid	Generally slower?
Egg laying:	Seasonal, rapid response to supply of nutrient	Multiple through the year
Feeding:	Facultative, or specialized, adapted to low trophic levels	
Production/biomass ratio:	High	Very high
Behaviour:	Simple; uniform; formation of shoals	Complex, with learning, territoriality; symbiosis
Predominant selection:	Type-*r*, biotic and abiotic agents	Type-*K*; mainly biotic agents
Diversity:	Not very diverse, dominant species	Highly diverse, lacking dominants
Community:	Rejuvenated	Very mature
Implications:	Resilient?	Fragile?

Source: Lowe-McConnell (1999).

The migration and population dynamics of some fish species (including reproduction, mortality, and fluctuations in population levels) are related to the patterns of hydrological fluctuation in rivers, lakes and natural flood channels; productivity; and biogeochemical cycles in the flooded areas. Longer or shorter life cycles, more efficient use of **food resources**, early maturation and high fecundity are features of those species living in areas with widely oscillating hydrological variations on the **flood plains** (Agostinho *et al.*, 1999). Modes of reproduction in tropical freshwater fish vary from a single "big bang" spawn (as in the case of *Anguilla* species) to total and partial spawns (Chart 8.4).

The numbers of mature oocytes in ovaries of tropical freshwater fish are presented in Table 8.4.

Certain features of tropical fish, such as growth and maturation, depend directly on temperature: tropical species have faster growth rates, earlier maturation, and shorter lifespans than temperate-region fish. Some cichlid species in Lake Tanganyika, for example, have a life cycle of 1.5 to 2 years. Many species of small characins in Lake Chad in Africa do not live more than two years. Growth rates obviously vary according to feeding habits, environmental conditions, food availability, water temperature and population density (Lowe-McConnell, 1999). Growth rates of tilapia have been studied extensively because of the species' commercial importance (Pullin and Lowe-McConnell, 1982). The growth and maturation rates of these species (*Oreochromis niloticus* and *O. mossambicus*) vary widely in response to environmental conditions.

In addition to the studies by Barthes and Goulding (1997) on fish in the Amazon region, further studies were done on the tambaqui (Araujo-Lima and Goulding, 1997). Two volumes (published by Val, Almeida Val and Randall, 1996; and Val and Almeida Val, 1999) presented an overview of the physiology and biochemistry of tropical fish, particularly Amazonian species. The authors described the complex interactions and physiological role of **Amazon fish** species in widely varying environmental conditions, including **hydrological fluctuations**, food availability, and different bio-geochemical cycles and oscillating patterns in two physiologically important factors: dissolved oxygen and water temperature.

Amazonian fish fauna (in which there are nearly 2,000 species), according to Val *et al.* (1999), evolved in a highly varying ecosystem and, therefore, populations may periodically adjust their biochemical and physiological patterns to the widely varying conditions. These physiological adjustments make it possible for Amazonian fish to survive conditions of anoxia, high sulphurous gas levels and low ionic levels. According to Walker and Henderson (1999), the physiology of Amazonian fish can be viewed as an intermediary process between fish ecology and evolution, including reproductive rate, life cycle, general metabolism and behaviour.

The ability to survive at low levels of dissolved oxygen is an extremely important physiological feature of Amazonian fishes: respiration of air, metabolic depression, morphological specializations, and adjustments in oxygen transportation are some of the most common physiological adaptive features of Amazonian fish species (Val, 1999). Variations in dissolved oxygen levels depend on processes such as photosynthesis by aquatic plants; the morphometry of canals, rivers and lakes; decomposition of organic material; respiration by organisms; and hydraulic and hydrodynamic movements and shifts. During periods of flooding, anoxia or hypoxia can occur in

Chart 8.4   Types of reproduction in representation of fresh water tropical fish.

Type of fecundity	Seasonality in reproduction	Examples	Movements and parental care
"Big bang" ++++	Once in a life-time	*Anguilla*	Very long migrations, catadromes without parental care
Total egg-layers +++	Very seasonal with floods; annual or biannual	Many charcoides *Prochilodus* *Salminus* *Hydrocynus* Many cyprinids Some siluroids *Lates* (Lake Chad)	Fish 'spawning', very long migrations No parental care Local movements: pelagic eggs
Partial egg-laying ++	Prolonged season During high-water season(s)	Some cyprinids Some caracoids: *Serrasalmus* *Hoplias* Some siluroids: *Mystus*	Principal local movement Put eggs in plants (m, m + f) Puts eggs on the bottom (m) Guard eggs and young (m)
Classified as: egg-layers, small clutches +	High-water season, may begin at the end of the dry season or be non-seasonal End of rains	*Arapaima* Some anabantoids *Hoplosternum* *Hypostolllus* *Loricaria parva* [a]*Loricaria* spp. *Aspredo* sp. *Osteoglossum Cichlids:* [a] Most South American species [ab] Most African species [b]*Sarotherodon galilaeus* *S. melanotheron* sting rays [b] Pecilids *Anableps* Annual cyprinodont species	Guard eggs and young; nests on bottom (m + f) Guarding eggs, bubble nest superficial (m) Guard eggs in nests surface (m) Guard eggs; marginal burrows (gender?) Guards eggs under rocks (m) Carries eggs on the lower lip (m) Carries eggs in the womb (f) Offspring to Mouth (m) Guarding eggs and young (m + f) Brood eggs and young in the mouth (f) Brood eggs and young in the mouth (m + f) Brood eggs and young in the mouth (m + f) Viviparous Viviparous Viviparous Leave eggs in the mud during the dry season

m – male; f – female; [a] – end of the dry season; [b] – not seasonal
Source: Lowe-McConnell (1999).

some lakes, subjecting fish to increased stress. Chart 8.5 describes Amazonian fish species that electively or mandatorily breathe air.

Understanding these Amazonian fishes' tolerance to hypoxia is one of the great challenges in tropical fish physiology, and studies continue to show there are big opportunities for greater understanding of the physiological and adaptive processes involved in these conditions (Almeida Val, Val and Hochahka, 1993).

Table 8.4   Number of mature oocytes in the ovaries (fecundity) from
representative examples of tropical freshwater fish.

Protopterus aethiopicus	1,700–2,300
Arapaima gigas	47,000
Osteoglossum bicirrhosum	180
Mormyrus kannume	1,393–17,369
Marcusenius victoriae	846–16,748
Gnathonemus longibarbis	502–14,642
Hippopotamyrus grahami	248–5,229
Pollimyrus nigricans	206–739
Petrocephalus catostoma	116–1,015
Alestes leuciscus	1,000–4,000
Alestes nurse	17,000
Alestes dentex	24,800–27,800
Alestes macrophthalmus	10,000
Hoplias malabaricus	2,500–3,000
Salminus maxillosus	1,152,900–2,619,000
Prochilodus scrofa	1,300,000
Prochilodus argenteus	657,385
Labeo victorianus	40,133
Catla catla	230,830–4,202,250
Lates niloticus	1,140,700–11,790,000
[a]*Mystus aor*	45,410–122,477
Hypostomus plecostomus	115–118
Arius sp.	118
Loricaria sp.	c.100
Oreochromis leucostictus	56–498
Oreochromis esculentus	324–1,672
Pseudotropheus zebra	17–<30
Cichla ocellaris	10,203–12,559
Astronotus ocellatus	961–3,452
Anableps anableps	6–13 embryos

[a]Can spawn 5 times in a season, the majority of species above this
in the table (except *Arapaima* and *Osteoglossum*) are total number of
eggs; those below are in multiple spawnings.
Source: Lowe-McConnell (1975).

8.4.1   Fish production and limnology

Potential fish production is dependent on the trophic state of the waters supporting the
fish community. Other factors such as growth and reproductive rates and environmental features also will affect the yield. Use of indices such as the MEI (Morphoedaphic
Index) can help assess the potential biomass of fish in lakes and reservoirs (MEI =
STS/z, with STS = suspended total solids and z = average depth of lake or reservoir)
(Oglesby, 1982). Such a measure is useful for predicting the total production of fish
in an aquatic ecosystem.

In inland ecosystems, the fish production depends to a certain extent on the **ecological functioning** and primary productivity of each ecosystem. In some tropical
regions, including large inland deltas and flood valleys of the Amazon and Parana
rivers, fish production plays a significant role in the regional economy (Petrere, 1992;
Goulding, 1999; Roosevelt, 1999; and Barthem, 1999).

Chart 8.5 Amazonian fish that are facultatively or obligatorily air-breathing. The structures associated with oxygen uptake are indicated. Families are arranged according to trend from generalized structures to specialized structures.

Families and species	Optional	Obligatory	Structures BL	SK	SI	FBB
Lepidosirenidae						
Lepidosiren paradoxa		X	X			
Arapaimidae						
Arapaima gigas		X	X			
Erythrinidae						
Erythrinus erythrinus	X		X			
Hoplerythrinus unitaeniatus	X		X	X	X	
Doradidae						
Doras	X				X	
Callichthyidae						
Callichthys	X				X	
Hoplosternum	X				X	
Loricariidae						
Plecostomus	X				X	
Ancistrus	X				X	
Rhamphichthyidae						
Hypopopus	X					X
Electrophoridae						
Electrophorus		X				X
Synbranchidae						
Synbranchus marmoratus	X					X

BL – Swimming bladder and lung; SK – skin; SI – Stomach and intestine;
FBB – Pharyngeal and branchial diverticula, and in the mouth.
Source: Val et al. (1999).

The **construction of reservoirs** affects fishing and fish production in inland waters (Agostinho et al., 1994, 1999; Tundisi et al., 2006). Reservoirs can also provide interesting possibilities for raising species under controlled conditions. The introduction of exotic species into reservoirs to increase fish production can be an interesting solution, but also risks potential problems, especially in the case of pelagic predators (Leal de Castro, 1994).

Raising some fish species in aquaculture (ponds or networks of ponds) can be an appropriate solution to increase fish stocks and fishery production, but of course this technology can affect the ecosystem's water (reservoir or lake) due to rapid eutrophication.

The factors affecting fishery production or characteristics of inland ecosystems are related to those that impact the environment or the biology of the commercial fish species or their physiology, reproduction and survival. These factors include:

▶ introduction of exotic species, which alter food web (Figure 8.12);
▶ eutrophication;

(a)

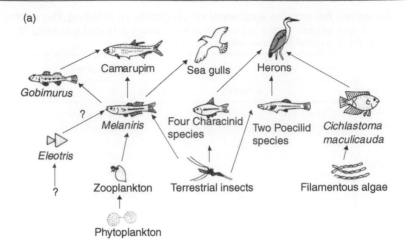

(b)

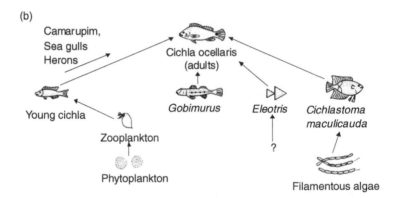

Figure 8.12 Impact of the introduction of exotic species in Lake Gatun and their effects of the food web.
Source: Horne and Goldman (1994).

▸ acute chronic water toxicity;
▸ effects of morphological changes on aquatic ecosystems (rivers, canals, reservoirs);
▸ heat pollution;
▸ invasive and haphazard use of water basins, affecting habitats and **riparian vegetation.**
▸ reforestation of riparian forests;
▸ changes in the fish species' reproductive areas.

8.5   FOOD CHAINS AND FOOD WEBS

In aquatic ecosystems, energy flows through a series of trophic levels. Primary photosynthetic production provides the quantity and quality of food available for herbivores and carnivores. Many aquatic organisms are omnivorous and vary their diet according to the time of year and availability of food. According to Horne and Goldman (1994), the concept of trophic levels is idealized and schematic. In greater complexity beyond the so-called **trophic chain** (food chain) is the **trophic food web**, which is **dynamic**. The most accurate terminology is **dynamic trophic food web** because it changes through time and space.

Food chain dynamics are complex, and determining structure and efficiency calls for special techniques and methods. The contents of zooplankton and fish digestive systems must be determined, as well as energy efficiency levels at each stage in the trophic food web.

Different aquatic organisms ingest different types of food, and their alimentary organ structures are morphologically varied, presenting different efficiencies. The different types of feeding groups of aquatic organisms include:

▸ **Herbivores:** feed on aquatic plants, phytoplankton, periphyton or macrophytes.
▸ **Carnivores:** feed on herbivorous aquatic organisms or other aquatic animals.
▸ **Detritivores:** feed on vegetal remains, sediment or animal remains.
▸ **Omnivores:** varied feeding habits, including eating plants, animals or detritus suspended in water or in sediment.

Aquatic organisms can capture or filter foods and be classified by feeding-system type:

▸ **Filter feeders** (or **suspension feeders**): the classic case of straining suspended particles (phytoplankton, bacteria or organic material) is planktonic copepods, especially calanoids (see Figure 8.13).

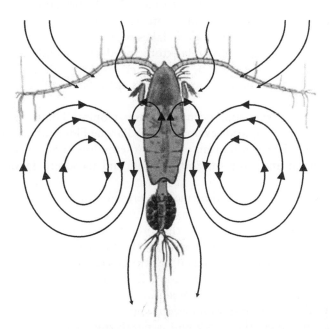

Figure 8.13   Filtering mechanisms of herbivorous copepods, represented by an original scheme for *Calanus finmarchicus* (Marshall, 1972) but valid for herbivorous copepods of inland waters in general (see also color plate section, plate 16).
Source: Modified from Marshall and Orr (1972).

- ▶ **Collectors:** organisms that **collect** different-sized suspended particles in water or on river and lake bottoms.
- ▶ **Scrapers:** organisms that **scrape** the surface and feed on microphytobenthos, bacteria, or aggregated organic matter or particles.
- ▶ **Sediment collectors:** organisms that collect, aggregate and consolidate sediment particles rich in organic material.

The feeding rates of these organisms and the energy content of each component are measured with a **calorimeter** pump. It is also important to identify the composition of each component of the trophic web: organisms are generally categorized by functional groups (such as herbivores, omnivores or carnivores).

Biological pyramids with the nutritional value of each trophic level can be identified, from phytoplankton and periphyton (at the base of the food network) to carnivorous predators (large fish or other vertebrates) at the top. Figure 8.14 illustrates a classic food network in a lake with a multiple-component trophic network.

The importance of vertebrate predators in the trophic network dynamics and the theory of trophic cascade (Carpenter *et al.*, 1985) have been described by many authors (including Kerfoot and Sih, 1987). The theory emphasizes the role of predatory fishes in trophic network structures, particularly in the composition and dynamics of the zooplankton community, whose primary prey is the photosynthetic phytoplankton.

Further evidence (presented by Dumont, Tundisi and Roche, 1990) shows that predation on zooplankton plays a significant role in the structure and dynamics of zooplanktonic communities. According to Lair (1990), predation by other invertebrates on zooplankters effectively controls micro-planktonic rotifers and small crustaceans in particular. According to Lair, in lakes where intra-zooplanktonic predation is intense, it can directly affect phytoplanktonic productivity and succession.

Blaustein and Dumont (1990) in Dumont *et al.* (1990) described predation by *Mesostoma* spp. (Platyhelminthes) on *Chydorus sphaericus* and *Moina micrura* (see Figure 8.15).

Planktonic *Mesostoma* spp. occur in many tropical lakes, including African lakes (Dumont *et al.*, 1973) and several lakes in Parque Florestal do Rio Doce – MG (Rocha *et al.*, 1990). These organisms establish themselves at various different depths in the water column and prey on zooplankton and mosquito larvae (*Aedes* spp.) (McDaniel, 1977).

In **rice cultivation fields** in California, the presence of *Mesostoma lingra* controlled mosquito larvae (*Culex tarsalis*) (Case and Washimo, 1979).

Dumont and Schoreels (1990) described the predation of *Mesostoma lingra* on *Daphnia magna*. Rocha *et al.* (1990) showed the impact of *Mesostoma* preying on zooplankton in lakes in Parque Florestal do Rio Doce, and in particular, the species' vertical migration in Lake Dom Helvécio in a 24-hour period. The nightly migration to the surface probably coincides with predation on zooplankton. In turn, *Mesostoma* is preyed on by *Chaoborus* and *Mesocyclops* spp.

Mesostoma spp. inject their prey with a paralyzing neotoxin and apprehend the prey with mucous. Predation on ostracods, according to Rocha *et al.* (1990), consists of introducing the pharynx between the two valves and a sucking motion of the entire contents of the organism. Matsumura Tundisi *et al.* (1990) compared the predation

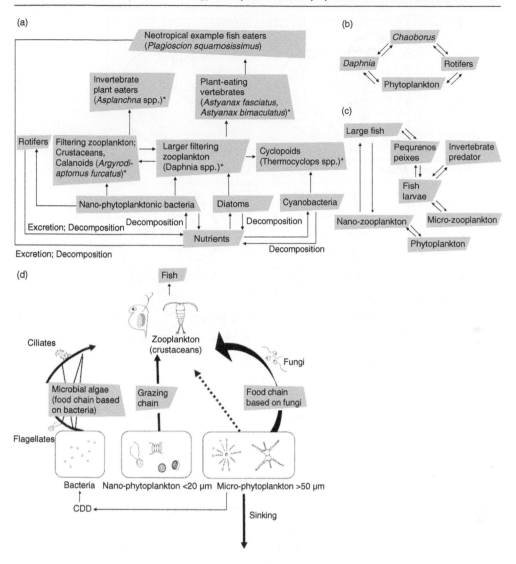

Figure 8.14   Classic food web in lentic aquatic ecosystems. (a) Interactions of components, with neo-tropical examples; (b) *Chaoborus* predating on *Daphnia*. The direction of the arrow indicates control of predators over prey; (c) Invertebrate predator and different stages of fish development; (d) Fungi-based food chain.

by *Mesocyclops* spp. on *Ceriodaphnia cornuta* and *Brachionus calyciflorus* in the Barra Bonita reservoir (see Figure 8.16).

The results clearly indicate intra-zooplankton predation is a **regulating factor** in food chains and food webs in lakes and reservoirs. The argument by Fernando *et al.* (1990) that the diversity of invertebrate predators is reduced in tropical lakes is no longer well accepted because of the wealth of information that now exists on intra-zooplanktonic predation.

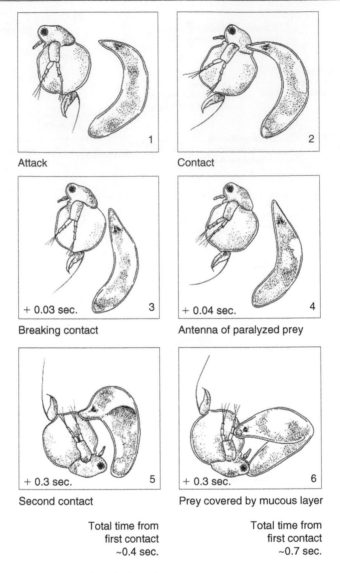

Figure 8.15   *Mesostoma* spp. attacking *Moina micrura*.
Source: Modified from Dumont, Tundisi and Roche (1990).

In addition to intra-zooplanktonic predation, zooplanktonic predation by invertebrates is probably a significant factor regulating zooplanktonic biomass, as shown by Perticarrari *et al.* (2004) in Lake Monte Alegre in Ribeirao Preto.

Arcifa *et al.* (1998) described the composition, fluctuations and interactions of a planktonic community in a shallow tropical reservoir (Lake Monte Alegre) that has been extensively studied over time. Through multivariate analysis, four periods over the course of a year were identified. Phytoplanktonic biomass was greater in periods III and IV (September and March) and zooplanktonic biomass was more abundant

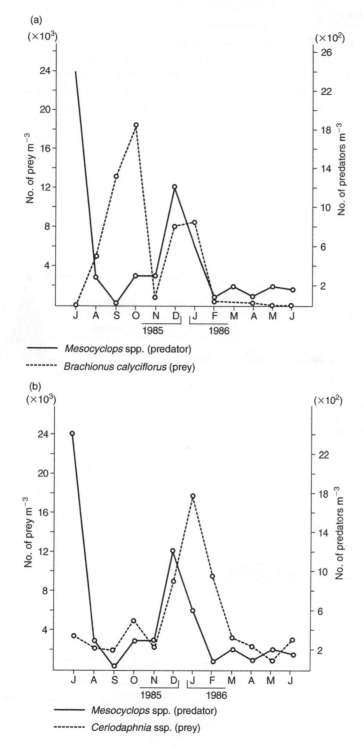

Figure 8.16   Seasonal fluctuation of predators *Mesocyclops longisetus* + *Mesocyclops kieferi* in the Barra Bonita reservoir (SP); (a) seasonal fluctuation of the prey *Brachionus calyciflorus*, and (b) of the prey *Ceridaphnia cornuta*.

in periods I, II and IV (April–August and January–March). In period III (September–December), zooplanktonic levels dropped as a result of factors such as predation by fish, excess suspended solids and predation by *Chaoborus*. Another possible cause for the drop was the abundance of *Aulacoseira granulata,* little used as food by zooplankton (see Figure 8.17). The authors also attributed the initial decline of zooplankton

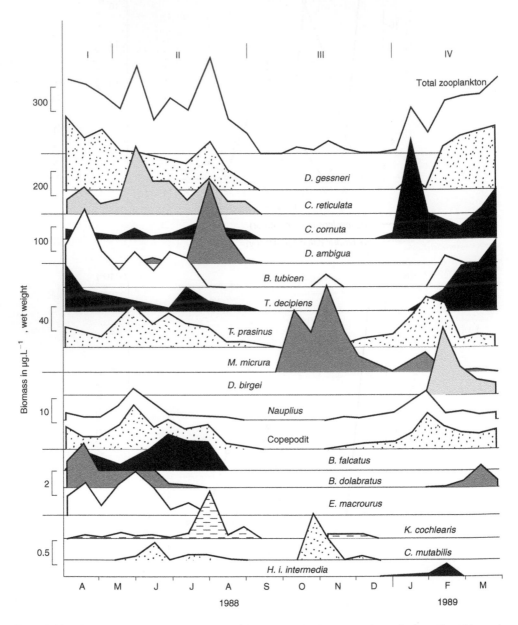

Figure 8.17   Fluctuations of biomass in μg · L⁻¹ (wet weight) of species of zooplankton (total in periods from I to IV).
Source: Arcifa *et al.* (1998).

during the period of depression to the effect of cold fronts and the cooling of the water column.

According to Arcifa (2000), who studied the eating habits of *Chaoborus* spp. in Lake Monte Alegre, the principal components of their diet in periods I and II were *Aeridium* and zoo-flagellates, while microcrustaceans of the *Bosmina* genus were the principal components in periods III and IV. Predation by chaoborids, according to Arcifa, could be an important quantitative factor in the control of the *Bosmina tubicen* population in the lake (Arcifa *et al.*, 1992). On the other hand, the reverse migration of zooplankton could be a possible attempt to escape *Chaoborus* larvae.

The quantitative determination of intra-zooplanktonic pressure still presents a basic methodological challenge. Mesocosm experiments that allow the manipulation of environmental factors, as demonstrated by Lazzaro *et al.* (unpublished), can help resolve and quantify this problem.

Hrbáček *et al.* (1961), Hrbáček (1962), Brooks and Dodson (1965), and Straškrába (1965) studied the predation of planktivorous fish on zooplankton and the impact on food webs. These pioneer works showed that with planktivorous fish, zooplankton communities were composed of small organisms in comparison with those in lakes with no planktivorous fish.

Lazzaro (1987) studied plankton-eating fish and their features, evolution, and feeding and selection mechanisms. These fish can be optionally or exclusively planktivorous. Fish that feed on particulate matter can feed by random selection of prey or by filter feeding. Fish can also use vision to select prey, or chemical-reception. In experiments Werner (1977) showed feeding by selection. **Filter feeders** pump water and filter material during swimming motion. In conclusion, fish have a range of behavioural mechanisms for feeding and predation.

An important problem to consider in **feeding chains** and webs is the concept of control of the chain or the trophic cascade from the top of the food chain, as theorized by Carpenter *et al.* (1985). The concepts of control or flow of energy from the top of the food web and control of this flow from the base of the food web (*top-down and bottom-up*, respectively) have long been a topic of discussion. The traditional concept of control from the base of the food web, which holds that each prey "can feed several predators", relies on the view that all trophic levels are positively correlated and control is exercised by nutrients (limiting factor).

Higher levels of nutrients result in greater primary productivity and in effect, a chain: greater biomass of phytoplankton, greater biomass of zooplankton, greater biomass of planktivorous fish and greater biomass of piscivores. At the same time, the concept of control of the food web from the top holds that a greater number of piscivorous fish leads to less planktivorous fish and, as result of a chain reaction, greater biomass of zooplankton, reduced biomass of phytoplankton and higher levels of available nutrients. Mesocosm experiments sought to prove the two hypotheses. In one, planktivorous fish were added, which led to reduced zooplanktonic biomass and increased phytoplanktonic biomass. When mesocosms were fertilized (Tundisi and Saijo, 1997), higher levels of phytoplankton were obtained, producing a controlling effect from the base of the food chain (Lampert and Sommer, 1997). The population dynamics of phytoplankton, zooplankton and fish therefore influence these two types of food web organization (Vanni *et al.*, 1997). The structure of the ecosystem is organized based on nutrient levels (involving external or internal load), and the predators present at the top of the food web. This hypothesis of control of the food web structure based on predators at the top is used in **biomanipulation** of lakes and reservoirs to control eutrophication.

Biomanipulation in management of lakes and reservoirs has been the subject of many studies over the last 20 years (De Bernardi and Giussiani, 2001). In Brazil, an important study was conducted on Lake Paranoá in Brasilia (Starling and Lazzaro, 2007), where an **eco-technology** was developed based on biomanipulation and the introduction of sterile silver carp for the biological control of cyanobacteria, converting a significant portion of the primary productivity into commercially valuable fish biomass. The authors found it problematical that in this conversion, **cyanotoxins** were incorporated into commercial fish and that the process was unsustainable over long periods of time. Other effects of the fish-zooplankton interactions on water quality were shown by Arcifa *et al.* (1986) and Attayde (2000), who conducted studies to demonstrate the direct and indirect effects of fish predation and excretion on food webs.

The two controls are interdependent and not exclusive, and both hypotheses, as well as more recent findings, point to a preponderance of one or another control during certain periods (Horne and Goldman, 1994; Lampert and Sommer, 1997). Bechara *et al.* (1992) showed the important effects of control from the top of the food web in streams, but the mechanisms of control from the base of the food webs are equally efficient in these ecosystems (see Figure 8.18).

Fish play an important role in trophic webs, as has been shown. In lakes and rivers in the Amazon region, two basic types of trophic networks can be found. Walker (1995) showed that protozoa such as amoebas, ciliates and fungi play an important role in preserving streams that receive litter from the forest. Goulding (1980) and Araujo-Lima and Goulding (1997) showed the importance of herbivorous fish that feed on fruits and seeds in Amazonia. According to the authors, species of fish such as tambaqui (*Colossoma macroponum*) and other herbivores play a key role in dispersing fruit and seeds in the region.

An important role played by bacteria in food webs is the so-called **microbial loop** initially described by Azam for marine ecosystems, and which is also applied to lakes, reservoirs, rivers, streams and associated areas of temperate and tropical regions. According to this theory and as confirmed in application of **experimental methods**, bacteria play a fundamental role in processing dissolved organic matter (DOM), as well as detritus from excretion and decomposition of planktonic organisms, fishes and benthic macroinvertebrates and microinvertebrates. The processing of all this material, part of which is excreted by photosynthetic phytoplankton (Tundisi, 1965; Vieira *et al.*, 1998), occurs through the action of bacteria that quickly recycle dissolved organic and particulate matter. Figure 8.19 shows the food web with the microbial loop included. Figure 8.20 illustrates a food web based on the **micro-lithosphere** and the relevant role of photosynthetic pico-phytoplankton in the flow of energy.

Various authors describe the predation by carnivorous aquatic plants on protozoa, rotifers and culicine (mosquito) larvae (Brumpt, 1925; Hegener, 1926). Such predation may have a significant quantitative effect of predation on zooplankton as shown by Rocha and Matsumura Tundisi (unpublished) for masses of *Utricularia* sp. feeding on zooplankton in Lake Verde (Parque Florestal do Rio Doce). Somenson and Jackson (1968) demonstrated significant quantitative effects of predation by *Utricularia gibba* on *Parenecia multimicronucleatun*. Predation by aquatic plants on

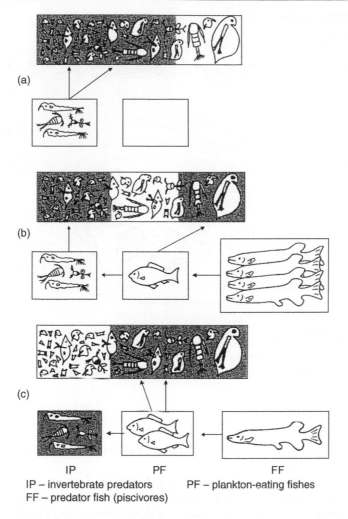

Figure 8.18   Effects of different vertebrate, invertebrate and zooplankton predators on the composition of zooplankton a) Without plankton-eating fish → larger zooplankton; b) Large populations of predatory fish, few planktophages → medium-sized zooplankton; c) Few predator fish, many planktophages → microzooplankton.
Source: Straškrába and Tundisi (2000).

zooplankton is considered a nutritional alternative for sources of nitrogen in environments with low inorganic nitrogen levels.

Detritivores play an important role in inland aquatic ecosystems. In small rivers in tropical forests, detritivores assume quantitatively important dimensions. Walker (1985), studying small streams in the Amazon, observed that the detritus includes remains of vegetation, organisms, fish and fruit. Such detritus is immediately attacked by decomposing fungi, which constitute the first food item. The detritus and litter along with the decomposers, bacteria and fungi, together with algae (desmids and

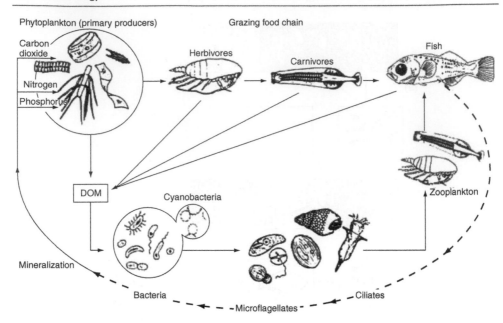

Figure 8.19 The "microbial loop". Classic figure produced initially to emphasize the role of bacteria in the mineralization of organic material. The reserve of dissolved organic material is used almost exclusively by heterotrophic bacteria and supports significant secondary production by bacteria. This figure shows see that the microbial loop is related to the grazing-feeding chain.
Source: Azam *et al.* (1983).

diatoms) are used as food by primary consumers such as flagellates, rotifers, cyclopoid copepods and ostracods.

The system of streams in the Amazonian forest receives an enormous quantity of vegetation remains (6–10 ton · ha^{-1} · year^{-1}; Klinge, 1977), which maintains a significant food chain with many alternatives. In these ecosystems, the diversity of food resources constitutes the base of stability, preventing a super- specialization of consumers. In addition to exploitation of the various alternatives, there are also various feeding techniques that allow improved and optimal exploration. In systems where the food chain essentially begins with detritus and decomposing fungi, algae and bacteria, the food of higher invertebrates and fish depend primarily on this source. The biomass thus produced in small streams feeds the great rivers of the system.

The detritus-based food chain is also important in flooded areas and floodplain lakes, where extensive growth and decomposition of aquatic **macrophytes** occur. In Amazonian reservoirs, where forest areas are flooded, the flooded areas present considerable accumulation of vegetation detritus (vegetation and decomposition), and, therefore, a detritus-based food chain quickly develops, resulting in rapid growth of a shrimp (*Macrobrachium* sp.).

Commercially important **detritivorous fish** are very common in South American rivers. Catella and Petrere (1996) discussed the importance of detritus in the diet of fish species in floodplain lakes. Through this **detritus chain**, the food web can contract,

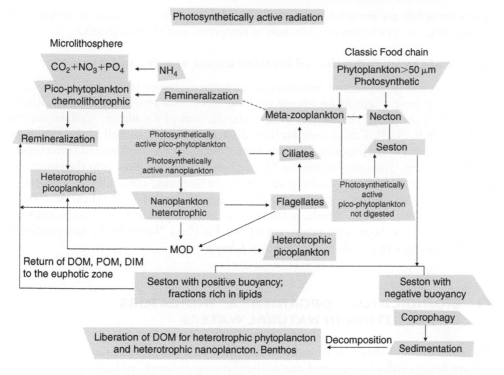

Figure 8.20 Major routes of transfer of organic matter under chemolithotrophic and photosynthetic conditions and *Chemolithotrophy* by different size fractions of phytoplankton: fractions >50 μm, nano-phytoplankton (<20 μm), pico-phytoplankton (0.2–3.0 μm). There is a difference between photosynthetically active pico-phytoplankton, chemolithotrophic pico-phytoplankton, **photosynthetically active nano-phytoplankton** and heterotrophic nano-phytoplankton. DOM – dissolved organic matter; POM – particulate organic matter; DIM – dissolved inorganic matter.
Source: Modified from Stockner and Antia (1986).

increasing the efficiency of communities in Neotropical floodplain lakes and rivers. The stomach contents of detritivorous fish present a large variety of different components, such as algae, bacteria, fungi, vegetative remains and unidentified portions of food items. Araújo-Lima *et al.* (1986) conducted studies that identified the sources of energy for detritivorous fish, and Forsberg *et al.* (1993) presented information on **autotrophic sources** of carbon for fish in the Central Amazon. Through the use of techniques to analyze the stomach contents of fish and ^{13}C carbon isotopes, Vaz *et al.* (1999) identified organic particulate matter as the main carbon source for characiform fish and catfish in the Jacaré-Pepira River and the Ibitinga Reservoir (state of São Paulo). In regions with fast-flowing rivers, the main source is allochthonous material, and in Ibitinga Reservoir, the main source is phytoplankton. In regions of the river with marginal lagoons, the principal food source is **particulate detritus material**, where bacteria can also be a significant source of carbon.

Araújo-Lima *et al.* (1995) conducted a broad review of the trophic relationships in fish communities in neotropical rivers and reservoirs. The authors concluded that omnivorous fish are abundant in streams, floodplain communities are dominated by detritivores, and piscivores are common in reservoirs and river channels.

Impact of the introduction of invasive exotic species in fresh water

The intentional or accidental introduction of aquatic species (fish, molluscs or crustaceans) can have a significant impact on food webs. In a recent publication, Rocha *et al.* (2005) described the impact of the introduction of *Cichla* cf. *ocellaris* on the food web in the UHE Carlos Botelho (Lobo/Broa) Reservoir, as well as the impact of the introduction of molluscs and exotic filter feeders into the food web in reservoirs in the Medio Tiete. The introduction of *Plagioscion squamosissimus* into the reservoirs caused substantial changes in the reservoirs' trophic webs, especially in the pelagic region where this species, a piscivorous fish, is an efficient predator (Leal de Castro, 1994; Stefani *et al.* 2005). Similarly, the introduction of golden mussels (*Limnoperna fortune*) in rivers, lakes and reservoirs of the La Plata Water basin has produced countless changes in the food web (see also Chapter 18).

8.6 BIOINDICATORS: ORGANISMS AS INDICATORS OF POLLUTION IN NATURAL WATERS

The structure and function of many freshwater and marine ecosystems have been and are being extensively altered due to the growing pressures of human populations and economic development in many regions, which can lead to significant changes in soil use and air pollution, with impacts on surface and ground water resources (see Chapter 18). These impacts are global, regional and local and range from climatic changes to deforestation of riparian forests, changes in the flow of rivers, and the introduction of exotic species. The spatial and temporal changes in these processes affect the structure and function of aquatic ecosystems and make it difficult to assess and predict the consequences under the effects of multiple **stress factors**. Researchers, decision-makers, planners and environmental managers now recognize the need to use scientifically based ecological approaches to monitor and predict the effects of these changes on the structure and function of ecosystems.

The sensitivity of an aquatic community (or populations of different species) can be a key indicator of environmental conditions (Loeb, 1994). Organisms and communities may respond to different changes in resources or alterations in environmental variables such as salinity/conductivity, water temperature, or organic and inorganic pollutants.

Hutchinson (1958) defined **hyper-volume** as the set of responses of an organism to all the factors affecting its ability to survive and reproduce. When changes occur in these factors, there is a corresponding shift in hyper-volume to a new spatial and temporal organization of favourable factors and the very survival of the organism. Stress factors affecting an organism or community can be physical, chemical or biological.

In order to define a set of stress factors in an environment, it is necessary to consider their attributes, that is, their variables and the hierarchy of factors affecting organisms, populations and communities.

Biological monitoring, and the evaluation of stress factors affecting organisms, populations and communities, are essential components for evaluating and predicting the responses of these organisms to physical, chemical and biological changes. The concepts of a niche and hyper-volume for each organism (based on environmental and biological variables) provide a theoretical basis for biological monitoring and the use of organisms, populations and communities as indicators to evaluate impacts. To a certain degree, biological monitoring enables the prediction of impacts and the evaluation of ecological risk and consequences of impacts. Any kind of stress occurring in an aquatic ecosystem directly affects organisms, populations and communities, which are the ecosystem's fundamental components.

The first researchers to study bacteria in an aquatic ecosystem to assess their response to organic pollution were Kolkwitz and Marsson (1909). Kolkwitz later expanded this concept – the *Saprobien system* – in 1950 (Hynes, 1994), followed by many studies that attempted to encode biological systems responding to the impact of mining and heavy metals, for example, or the effects of organic pollution.

Regional and local studies have produced many advances, and a mere list of species is not satisfactory, as Patrick (1951) already affirmed in a study on diatomaceous **organisms** as indicators. Macrophytes, benthic macroinvertebrates, planktonic organisms, crustaceans, and periphyton were all used to categorize local and regional indices, such as those proposed by Cairns *et al.* (1968) for North America and by Roldán (2006) for Colombia. Currently indices have been determined for the acidification of water, various **industrial effluents** and other stress factors.

What, then, are the basic methodological principles for efficient use of bioindicators? First, it is important to understand basic ecosystems, community structure and their interrelationships. Diversity indices applied to planktonic or nektonic communities are fundamental. It is also important to permanently maintain a local reference site that remains unaffected, to enable ongoing comparisons with the impacted ecosystem. Another essential aspect that should be considered is the **continuity** of evaluation of the impacted system, to enable ongoing comparisons.

The presence of certain pollution-indicating species is another fundamental requirement. These species often function as an early warning. Their disappearance may indicate alterations underway or significant stress factors at work in communities or populations (Matsumura Tundisi *et al.*, 2006) (see Figure 8.21).

Rocha *et al.* (2006) studied the biodiversity of reservoirs on the Tiete River under the effects of eutrophication and concluded that different **biological indicators** illustrated the eutrophic conditions in these reservoirs: greater abundance of oligochaetes, fish-eating birds and emergent macrophytes. A drop in the biomass of submerged macrophytes – as the amount of eutrophication increases and transparency of the reservoirs decreases – is another indicator.

Parasites on fish can also be indicators of eutrophication and environmental stress (Silva-Sousa *et al.*, 2006). The absence of ecto-parasites on fish is another indicator of stress factors related to increased **pesticides** in water.

The requirements for effective biomonitoring are therefore multiple, and are related to collecting and determining the biodiversity of organisms and species. In some cases, however, collection systems – for periphyton and macroinvertebrates, for example – can be used to monitor the growth, structure and impact of stress factors. Artificial substrata have been extremely useful in studies on the response of periphyton communities, auto-ecology of diatomaceous species (Patrick, 1990) or benthic macroinvertebrates (Pareschi, 2006).

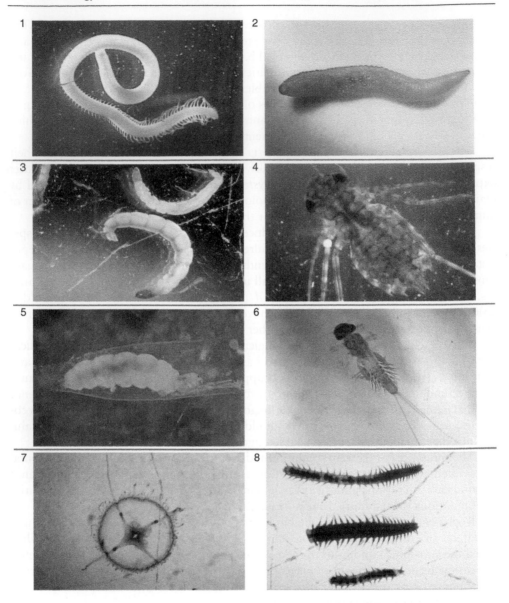

Figure 8.21   Organisms resistant to pollution: (1) *Branchiura sowerbyi* (Oligochaeta, Tubificidae, collected in the Ibitinga reservoir, middle Tietê, SP), (2) Hirudinea, Glossiphonidae (collected in the Xingu River, AM), (3) *Coelotanypus* sp (Chironomidae larvae, Tanypodinae); organisms tolerant to average pollution – (4) Libellulidae (Odonata larva, collected in the Ibitinga reservoir, middle Tietê, SP); organisms sensitive to pollution – (5) Trichoptera (larva inside the house, collected in the Xingu River, AM), (6) Ephemeroptera, Leptophlebiidae (larva, collected on the Xingu River, AM), (7) *Craspedacusta sowerbyi* (Cnidaria, rare, collected in the Tocantins river, TO), (8) Freshwater Polychaeta (often collected in the Xingu River, MA) (see also color plate section, plate 17).
Photos: Daniela Pareschi Cambeses.

Ghetti and Ravere (1990) used the following categories to describe biological monitoring of inland waters in Europe:

▶ analysis of natural communities (especially in rivers);
▶ toxicity tests to assess impact of discharges;
▶ bioassay for rapid assessment of effluent control;
▶ **bioaccumulation** tests;
▶ use of biological indicators in environmental impact studies.

Organisms used from both the structural/functional and taxonomic points of view are in the categories **plankton**, periphyton, microbenthos, macrobenthos and nekton. De Pauw *et al.* (1991) outlined seven saprophytic indices, 45 biotic indices, 24 diversity indices and 19 comparative indices.

Periodic toxicity tests using organisms include tests with *Daphnia, Phosphoreum* (photobacteria – inhibition of the bioluminescence of bacteria); tests with various fish species and toxicity tests with algae.

According to Cairns and Smith (1994), the main objectives of biological monitoring include:

▶ early assessment of the deterioration in the quality of the ecosystem in order to avoid deleterious effects;
▶ determining the impact of episodic events, such as accident spills of toxic substances, or the illegal dumping of residues and effluents;
▶ detecting tendencies or cycles;
▶ determining environmental effects from the introduction of genetically modified organisms.

More recently, systems such as microcosms and mesocosms have been introduced to evaluate impacts, along with the use of some species as early **information systems** to detect possible alterations (Cairns and Smith, 1994).

Fish as bioindicators

Fish can be effectively used as bioindicators. Because they are sensitive to many variables in water quality, they are used in bioassays to determine the toxicity of industrial chemicals or municipal effluents or other products resulting from human activity, such as mining. Initially these tests involved only acute toxicity and the immediate effect of pollutants. Currently, several species are used to determine chronic toxicity and near-lethal effects that include changes in behaviour and metabolism. The integrated approach is now better understood, including water chemistry, toxicology and its impacts on fish physiology, through the use of on-site tests with organisms submitted to various toxicological conditions and measuring the effects on behaviour, metabolism (renal excretion, for example, or accumulation of toxins in gills) and enzymatic response, such as determining the levels of the P450 enzyme in the cytochrome of the liver of fish.

Fishes as end-product of energy
flux in aquatic ecosystems.
Credits: José G. Tundisi

9 The flow of energy in aquatic ecosystems

SUMMARY

In this chapter, we describe the main mechanisms and processes involved in the flow of energy in freshwater ecosystems. The chapter presents the main methods and approaches used to measure the primary productivity of photosynthetic autotrophs, photosynthetic bacteria and the factors involved in primary production. Information and data on heterotrophic bacterioplankton and their productivity are also presented, as well as the principal techniques and processes for measuring secondary production in aquatic systems, including data on secondary production in several different ecosystems.

Comparative data are presented on primary production in oceans, lakes and reservoirs, including information on geography, volume, nutrient recycling capacity, organization of trophic networks, and levels of energy flow.

Also included is a brief summary on the importance of the relationship between primary production and fishery production, and the dimension of food webs.

9.1   DEFINITIONS AND CHARACTERISTICS

The flow of energy through inland aquatic ecosystems and oceanic ecosystems depends to a large degree on **photo-autotrophic primary producers** and **chemosynthetic bacteria**. Even these bacteria depend to a certain degree on **photo-autotrophic plants** because the reduced inorganic material – which they oxidize to obtain energy to synthesize organic matter – is originally produced by the activity of photoautotrophic plants. All organisms living on the Earth thus depend on the organic matter produced by plants through the process of photosynthesis. Primary production is not only photosynthesis, however; it also includes chemo-autotrophic processes.

Two factors are important in the study of energy flow: the efficiency of the process at each trophic level, and the structure/composition of the food web. In each ecosystem the food web may vary greatly, and so it is more accurate to refer to the dynamic food web, which includes the various processes of selective feeding and quantification.

Organisms and ecosystems find themselves in a thermodynamic **equilibrium**: the energy received by ecosystems and organisms is used for growth and maintenance, or is stored. Degraded energy refers to that energy which is dissipated in heat and excretion products.

The quantitative aspects (efficiency of processes, growth rates, saturation levels of photosynthesis) and qualitative aspects (selective feeding, main trajectory of the food network) can be studied in laboratories under controlled conditions. Transferring laboratory studies to actual field conditions is one of the major problems for greater understanding of food networks and efficiencies in ecosystems. In recent years, the use of huge experimental tanks (mesocosms) has to some extent facilitated the understanding of qualitative and quantitative synecological processes. However, for certain organisms, the very limitations imposed by the nature of these mesocosms can cause problems and alter results.

A large part of the quantitative methods used to study energy flow in aquatic systems was developed and standardized for comparative use during the International Biological Programme, and the series of IBP manuals on various techniques are useful as a basic reference (Worthington, 1975; Golterman *et al.*, 1978; Vollenweider, 1969, 1974).

Gross primary production and net primary production are expressed in mg $C \cdot m^{-2} \cdot day^{-1}$, or mg $C \cdot m^{-2} \cdot year^{-1}$, or g $C \cdot m^{-2} \cdot year^{-1}$, or ton $C \cdot km^{-2} \cdot year^{-1}$ (that is, the data is expressed in units of time and units of area or volume).

The chemical energy produced from photosynthesis and **chemosynthesis** flows through the various compartments formed by organisms, and is the energy used for the organisms' growth, reproduction and metabolism. Photo-autotrophic primary

---

Gross primary production – is the production of organic matter by photo-autotrophic or chemo-autotrophic organisms, without considering the organic matter used in respiration or other metabolic processes.

Net primary production – is the production of organic matter by photo-autotrophic or chemo-autotrophic organisms, minus the organic material consumed by respiration or other metabolic processes.

Chart 9.1   Comparison between primary phytoplankton producers and primary terrestrial producers.

Comparative properties	Phytoplankton	Terrestrial plants
Size of primary producers	Small	Large
Recycling of primary producers	Fast	Slow
Maximum quantity of chlorophyll-*a*	200–350 mg/m^2 (euphotic zone)	4×350 mg/m^2
Principal factor of selection that operates on plants	Passive sinking, grazing	Grazing, competition for light
Dependence on external energy	Total, except in very large agglomeration of biomass	Tendency to control microclimate
Ratio of biomass of animals/biomass of plants	High	Low
Control of transportation	Physical environment or animals	Plants
Detritus chain	Important	Very important

Source: Margalef (1978).

producers are fundamentally different in terrestrial ecosystems and aquatic ecosystems, according to Margalef (1978) (see Chart 9.1).

A substantial proportion of the primary production in aquatic ecosystems is located in the euphotic zone, defined as the region where light intensity is at least 1% of that on the surface. This is one of the basic differences between terrestrial and aquatic systems in terms of the photosynthetic production of organic matter.

Thus, in the case of photo-autotrophic phytoplankton, light intensity is best utilized at the site that is the closest possible to the surface for maximum potential use of the light energy available.

Lindeman (1942) conducted a classical series of studies on energy flow in a small lake (Cedar Creek Bog) in Minnesota (USA). He described the trophic features, the interrelationships of communities in the ecosystem and the annual productivity of each component of the food web. Annual productivity was presented in cal/cm^2 and the efficiency levels in each of the major groups of primary, secondary and tertiary producers were calculated. These studies were summarized by Lindeman in *The Trophic- Dynamic Aspect of Ecology*. This work laid the foundation and theoretical framework for studies on energy flow in inland aquatic ecosystems.

Figure 9.1 shows the original illustration from the Lindeman study, describing the principal relationships between the biotic components in the lake and their relationships with abiotic factors. Figure 9.2 illustrates the energy flow of primary producers as well as primary, secondary and tertiary consumers.

The production of organic material either by photosynthetic organisms or **chemosynthesizers**, is one of the defining events that has occurred on Planet Earth. The release of oxygen molecules during the photosynthetic process allowed the atmosphere to go from reduced conditions to oxidized conditions. It took millions of years before the current level of molecular oxygen in the atmosphere (approximately 21%) was reached.

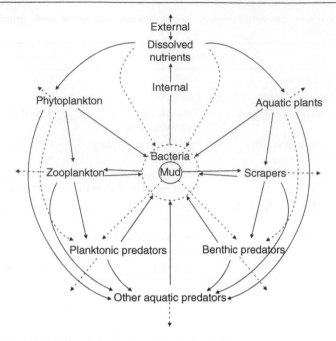

Figure 9.1   Original illustration by Lindeman (1941) describing the principal components of biota in Cedar Creek Bog (Minnesota, USA).

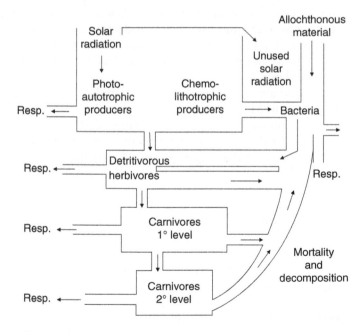

Figure 9.2   Flow of energy in an ecosystem.

The presence of oxygen in the atmosphere and ozone in the ionosphere reduced the amount of ultraviolet light reaching the earth's surface, enabling the evolution of life. At high ultraviolet radiation intensities (the short wavelengths in particular), life is not possible.

Life originated in the oceans, and photosynthetic organisms evolved later, also occupying inland waters, natural lakes, small ponds, pools and other inland aquatic systems at all latitudes and longitudes. All taxonomic groups of algae evolved before, during or after the Cambrian period, 1 million to 500 million years ago.

As already described in Chapters 6 and 7, organisms that perform photosynthesis belong to diverse plant groups in marine and inland waters. During photosynthesis, solar energy is converted into chemical energy and chemical compounds containing potential chemical energy accumulate in cells and are used to build structures in plants or to produce the energy needed to sustain numerous vital processes. In the early 20th century, it was discovered that photosynthesis includes two types of processes: photochemical and enzymatic. The photochemical processes in photosynthesis are proportional to the solar radiation and, in that phase, the photosynthetic rate is limited by the **photochemical** process. The photochemical part of photosynthesis is temperature-independent.

Photosynthesis is limited by the rate of **enzymatic reactions** at high radiation levels. In this case, temperature-dependent chemical reactions limit photosynthesis.

Therefore, the chlorophyll pigments that fix solar energy produce chemical energy from water and carbon (as dioxide) gas, which is stored in complex molecules. In addition to CO_2, as a source for carbon and water, photosynthesizing organisms need certain elements and substances for the construction of tissue and to utilize energy. The basic equation of photosynthesis is:

$$12H_2O + 6CO_2 + \text{solar energy} \xrightarrow[\text{enzymes}]{\text{chlorophyll}} C_6H_2O_6 + 6CO_2 + 6H_2O$$

In addition to other photosynthetic organisms – such as phytoplankton, aquatic macrophytes (fixed floating or submerged higher plants), **benthic microphytes** (periphyton) associated with substrata on the bottom of rivers, lakes and the surface of higher plants, which utilize CO_2 and H_2O – photosynthetic bacteria also produce organic material under certain conditions using H_2S (and not water) as a source of electrons.

Primary producers in aquatic ecosystems can be photosynthetic autotrophs or chemosynthetic autotrophs (which use energy released by chemical reactions).

Primary photosynthetic producers thus include:

▶ phytoplankton;
▶ aquatic macrophytes;
▶ micro-phytobenthos (periphyton);
▶ macro-phytobenthos;
▶ epiphytes (microscopic and macroscopic); and
▶ photosynthetic bacteria.

Chemosynthetic bacteria are also called chemo-autotrophic.

The relative importance of each of these components in the primary production of organic matter depends on multiple factors such as turbulence, circulation and vertical organization in the water column, nutritional conditions, transparency and depth of the eutrophic zone, amount of solar energy reaching the substratum and enabling photosynthesis, and oxido-reduction conditions for chemo-lithotrophs or photosynthetic bacteria that use H_2S. Studies on primary production measure the capacity of ecosystems to produce (from external light energy, CO_2 and H_2O) organic matter and organic compounds with high chemical potential, which are transported and flow to higher levels in the system (Vollenweider, 1974).

9.2   THE PHOTOSYNTHETIC ACTIVITY OF AQUATIC PLANTS

Techniques used to measure the photosynthetic activity of different photo-autotrophic organisms vary greatly due to differences in size, physiology and habitat of these organisms, such as aquatic macrophytes, phytoplankton and micro-phytobenthos. These techniques will be described in the following sections of this chapter.

9.2.1   Methods for estimating photosynthetic activity and primary productivity of phytoplankton

The structure and the spatial and temporal successions of phytoplankton communities were discussed in Chapter 7. The structure and dynamics are determined by several factors such as type of algae, size, buoyancy, temperature, water, adaptation to light, growth rate, demand for nutrients, effects of grazing by consumer organisms, water toxicity and relations with other organisms such as bacteria and fungi.

Phytoplanktonic primary production occurs in the **eutrophic zone** of lakes, reservoirs and oceans, compensating for the respiration rates that occur over the periods of light and dark. As already described in Chapter 4, the lower limit of the euphotic zone is the depth where 1% of the light intensity reaching the surface penetrates, and this depth approximates the **compensation depth**, or level of compensation, where there is an equivalence between production and consumption (Steemann-Nielsen, 1975).

Photo-autotrophic planktonic organisms depend on light intensity as a source of energy. The vertical profile of the light intensity, its spectral composition and the amount of underwater solar energy are basic ecological factors that determine the **primary production rate** of phytoplankton per cubic meter or square meter of water.

Photo-autotrophic organisms use solar energy for the production of organic material.

   Heterotrophic organisms use organic compounds such as glucose or pyruvate or others considered simpler.

   Chemo-autotrophic organisms use chemical sources for the production of energy and can be classified as chemo-organotrophic (those that use organic materials as donors of electrons), and chemo-lithotrophic (those that use inorganic materials as electron donors).

Source: Modified by Goldman and Horne (1994) and Cole (1994).

Measuring the primary production of phytoplankton in oceans and inland aquatic systems has involved a long history of quantitative experimental measurements underwater starting with seedlings in the 19th century (Regnard, 1891).

This history of measuring primary phytoplankton production, analysed by Talling (1984), presents several tendencies and procedures deriving from the following revelations and new concepts:

▶ The conclusion that periodic population censuses were not enough to determine primary production.

▶ The concept that measurement of the primary production of photosynthetic phytoplankton could provide the basis for determining an ecosystem's metabolism.

▶ The concept of such quantification as a descriptive characteristic of communities, which resulted in the mapping of the primary productivity in oceans as compared with productivity in lakes and reservoirs (Steemann-Nielsen, 1975; Sorokin, 1999).

▶ The confirmation of little conceptual interaction between work on primary aquatic production (marine and inland waters), physiology of vegetation, and terrestrial primary production, and between limnology and oceanography.

▶ A significant increase in the number of experiments, especially after introduction of the ^{14}C technique to measure primary productivity of phytoplankton in oceans, lakes and reservoirs.

▶ A growing trend to conduct experiments *in situ* and apply predictive **mathematical models** (Han and Straškrába, 1998).

A summary of the bases used to measure primary phytoplankton production is presented in Chart 9.2.

The chart shows that the fundamental basis for measuring primary phytoplankton production includes experimental methods, observation, on-site measurements, and methods based on measuring biomass and increase in biomass (cell division) to calculate primary production. Figure 9.3 illustrates the various different steps for measuring the primary production of phytoplankton and the progress over 100 years

Chart 9.2   General principles guiding the measuring of primary production in aquatic ecosystems.

		In cells	In the medium
Rate	Systems Experimental In small and large scale Observations and measurements in free water	^{14}C ^{15}N Synthesis of RNA Increased biomass Division in phase	O_2 Carbon Variation of O_2 (diurnal) Variations of CO_2 (diurnal) Silica (seasonal)
Quantities and compositions (correlations)	C:N:P ratios Fluorescence		Nutrients/trophy (Ex: chlorophyll/P_{total})

Source: Talling (1984).

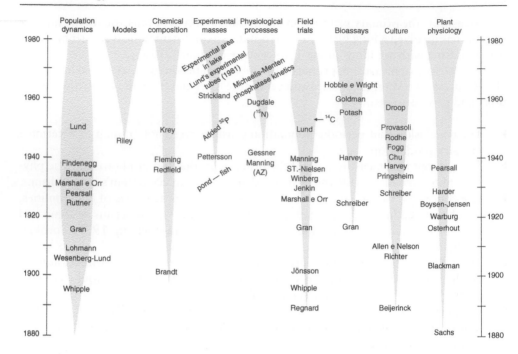

Figure 9.3 Principal stages of study of primary phytoplankton productivity during a 100-year period.
Source: Talling (1984).

(1880–1980), showing that a complex and comprehensive range of experiments, measurements and observations have been conducted by different researchers.

According to Talling (1984), for phytoplankton, the basic process involved in the term 'production' is **'replicative growth'**, basically exponential and quantifiable by a specific growth constant with dimensions of time. Since 1900, specific growth rates have been widely used by many researchers to measure productivity (the dynamics of planktonic populations).

Experiments on growth and on-site evaluation of the dynamics of partly phytoplanktonic populations were conducted by Ruttner (1924), Loose *et al.* (1934), Talling (1955) and Cannon *et al.* (1961).

Comparisons between population growth and photosynthesis to measure productivity are complicated because of the fact that phytoplanktonic populations are heterogeneous, making it difficult to evaluate the specific growth of each species present in the community. Assessing monospecific populations can also be difficult because of the potentially non-linear relationship between photosynthetic rates and growth.

Two experimental techniques have been widely used to measure the primary production of phytoplankton: assimilation of **radioactive carbon** (^{14}C), introduced by Steemann-Nielsen (1951); and production of dissolved oxygen during experiments to measure photosynthesis (Gaarder and Gran, 1927).

Both methods use the experimental technique (employed often in oceans, lakes and reservoirs in many countries) of paired transparent and dark bottles suspended in the euphotic zone during a specific period of time, which can vary from 2, 4, 6 to 24 hours (this last period used by some researchers in experiments with the

O_2 technique). The transparent-dark-container method was introduced by Gaarder and Gran (1927) and solely used until 1951 (when the ^{14}C method began to be applied). In Brazil, the ^{14}C method was introduced by researcher and professor Clovis Teixeira, who studied with Prof. Steemann-Nielsen and started applying the technique in 1962. A series of experiments based on this technology in Brazil resulted in a set of basic scientific publications (Teixeira and Tundisi, 1967; Tundisi, Texeira and Kutner, 1973; Tundisi, 1977; Tundisi et al., 1977; Tundisi et al., 1977; Tundisi, 1983; Barbosa and Tundisi, 1980; Henry et al., 1985; Tundisi and Saijo, 1977; Tundisi et al., 1997), which broadened the scientific understanding of primary production of phytoplankton in coastal waters, estuaries, lakes and reservoirs. A thorough review of primary productivity in phytoplankton was presented by Teixeira (1973).

^{14}C Technique

In this method, the incorporation of ^{14}C in organic material produced by phytoplankton is used to measure primary production. The ^{14}C is added to water samples in the form of $NaH^{14}CO_3$ (sodium bicarbonate). When attempting to measure the total CO_2 content in water and ^{14}C content in phytoplankton, the total quantity of assimilated carbon can be calculated as follows:

$$\frac{^{14}C \text{ available}}{^{14}C \text{ assimilated}} = \frac{^{12}C \text{ available}}{^{12}C \text{ assimilated}}$$

Sources: (Steemann-Nielsen, 1951, 1952; Wetzel and Likens, 1991).

A specific amount of radioactive bicarbonate is added to the samples with a known concentration of dissolved inorganic carbon (DIC). After a period of incubation in situ or under conditions of simulated illumination, the water samples are poured through 25-mm-diameter filters (e.g., Millipore) with 0.45-μm pores. Measuring the radioactivity in the filters (made by liquid scintillation equipment) enables the calculation and assimilation of inorganic radioactive carbon during the incubation period. It is therefore necessary to multiply the measured ^{14}C level by a factor corresponding to the ratio of total CO_2 to the amount of $^{14}CO_2$ at the beginning of the experiment, as shown in the previous formula. The following problems must be considered when preparing the samples from the filters to calculate the ^{14}C activity:

▸ Samples must be treated with HCl vapour to eliminate the ^{14}C fixed in structures (removal of radioactive carbonate in the cells).
▸ Conditions must be such that the $^{14}CO_2$ assimilation rate must be the same as that of $^{12}CO_2$. No $^{14}CO_2$ can be lost by respiration, and no organic material produced can be lost by excretion.

In practice, none of these conditions occur and certain corrections need to be applied in order to calculate the photosynthetic activity and primary productivity.

The **fixation in the dark** of phytoplanktonic algae is approximately 1–3% of the fixation at optimal underwater energies. However, the percentage can be much higher. Tundisi et al. (1997) calculated values of up to 20% in fixation in the dark at

Lake Dom Helvécio (Parque Florestal do Rio Doce – MG) as a result of the high bacterial level. Steemann-Nielsen (1975) calculated up to 40% of fixation in the dark in eutrophic lakes due to the presence of bacteria.

The ^{14}C assimilation rate is nearly 5% lower than the $^{12}CO_2$ assimilation rate. Data obtained by the ^{14}C technique are expressed in mg C $\cdot$ m^{-3} $\cdot$ h^{-1} or mg C $\cdot$ m^{-3} $\cdot$ day^{-1}.

The experimental procedure:

▸   Measure the depth of the euphotic zone (using Secchi disc ×2.7, hydrophotometer or underwater radiometer.)
▸   Determine the sampling depths from the surface; often roughly 100%, 50%, 25%, 10%, 1% penetration of solar energy are used as the sampling depths.
▸   Collect samples with non-toxic plastic container at the different depths.
▸   Place samples in 130-ml bottles with Pyrex tops. Generally, three transparent bottles and one dark bottle are used at each depth. Then, using a syringe, add 1-ml of the radioactive solution of NaH$^{14}CO_2$ to the bottom of each bottle.
▸   The amount of ^{14}C added varies according to various conditions, including phytoplanktonic levels, underwater radiation and water temperature. Generally, add 1–3 μCi, or in some cases, up to 5 μCi (μCi: unit of radioactivity).
▸   The samples are then suspended at different depths in the euphotic zone and incubated over a period of 1–4 hours (short in order to eliminate or reduce errors resulting from excretion of fixed ^{14}C or loss of $^{14}CO_2$ through respiration) (Vollenweider, 1965).
▸   After incubation, the samples are filtered and their activity is measured with liquid scintillation equipment.
▸   The final formula (Gargas, 1975) to calculate primary production of phytoplankton with the ^{14}C technique is:

$$^{12}Cassimilated = \frac{^{14}Cassimilated^{(a)} \cdot {}^{12}Cavailable^{(c)} \cdot 1.05^{(d)} \cdot 1.06^{(e)} \cdot K_1 \cdot K_2 \cdot K_3}{^{14}Cadded^{(b)}}$$

where:
[a] – liquid IPM = (transparent bottle IPM – background radiation – black bottle IPM – background radiation) (CPM = counts per minute)
[b] – specific activity of the NaH$^{14}CO_3$ added
[c] – mg C.L^{-1} available (calculated from total CO_2 estimation or measured directly) 1.05[d] – correction for losses of $^{14}CO_2$ from respiration during the experiment 1.06[e] – correction for ^{12}C/^{14}C isotopic discrimination
K_1 – correction for volume of the filtered aliquot
K_2 – correction for the factor of time exposed to the atmosphere
K_3 – factor for converting mg $\cdot$ L^{-1} to mg $\cdot$ m^{-3}

The ^{14}C technique is extremely sensitive and has been widely used in oceanography and limnology over the last 50 years. The incubation of samples, in addition to *in situ*, can be conducted under fully simulated conditions or simulated *in situ* (when

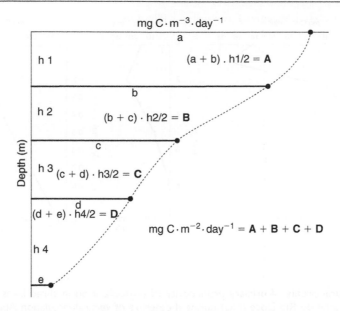

Figure 9.4 Procedure for measuring the primary production of phytoplankton in $mgC \cdot m^{-2} \cdot h^{-1}$ or $mgC \cdot m^{-2} \cdot day^{-1}$ based on the trapezoid technique: $h \times (a + b)/2$ = area of trapezium.

using incubation with natural solar energy but also various different types of light filters to simulate different depths).

Calculations for primary productivity of phytoplankton (in mg $C \cdot m^{-2} \cdot h^{-1}$ or mg $C \cdot m^{-2} \cdot day^{-1}$) are made using the trapezoid technique (Tundisi, Teixeira and Kutner, 1975; Gargas, 1975; Vollenweider, 1974). The technique consists of a proce-dure described below, based on experiments in which primary productivity is calcu-lated in mg $C \cdot m^{-3} \cdot day^{-1}$ (see Figure 9.4).

According to Steemann-Nielsen (1975), the ^{14}C technique provides an intermedi-ate measurement between **net photosynthesis** and gross photosynthesis. Ryther (1954) considers the ^{14}C technique to be a measure of net photosynthesis (remember: gross photosynthesis = net photosynthesis + respiration). Meanwhile, Steemann-Nielsen and Hansen's view (1959, 1961), which is currently widely accepted, is that the ^{14}C technique measures values intermediate between gross and net photosynthesis (Wetzel and Likens, 1991).

Figures 9.5, 9.6a and 9.6b and Tables 9.1 and 9.2 present results obtained from extensive research on primary productivity of phytoplankton using the ^{14}C technique in lakes in the Parque Florestal do Rio Doce (MG) (Tundisi *et al.*, 1997).

Technical problems with use of ^{14}C to measure primary productivity of phytoplankton

Practical problems that occur in the use of the ^{14}C technique include: obtaining a sample using **non-toxic containers**, excessive exposure of the sample to high solar energy; effect of the $NaH^{14}CO_3$ inoculum on photosynthetic rates, presence of toxic substances (such as Cu) in distilled water used in solutions of $NaH^{14}CO_3$ stock; long

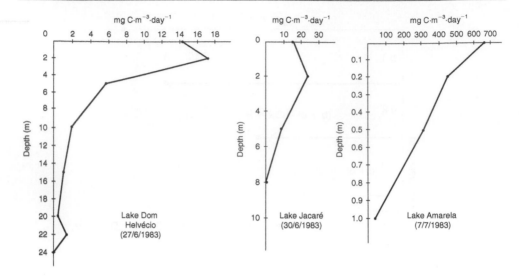

Figure 9.5   Vertical profile of primary productivity of phytoplankton in three lakes of the Parque Florestal do Rio Doce (MG) during the period of vertical circulation. Application of the ^{14}C technique.

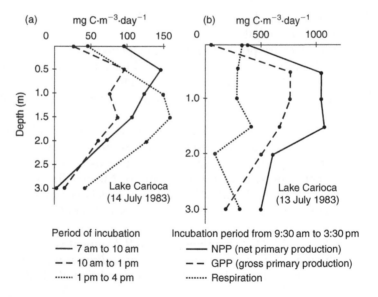

Period of incubation
— 7 am to 10 am
– – 10 am to 1 pm
······ 1 pm to 4 pm

Incubation period from 9:30 am to 3:30 pm
— NPP (net primary production)
– – GPP (gross primary production)
······ Respiration

Figure 9.6   a) Vertical profiles of primary productivity of phytoplankton in Carioca Lake (Parque Florestal do Rio Doce – MG) in three periods of the day during the vertical circulation period, using the ^{14}C technique. b) Vertical profiles of primary productivity of phytoplankton in Lake Carioca (Parque Florestal do Rio Doce – MG) during the period of vertical circulation, using the dissolved O_2 technique.

Table 9.1   Vertical profiles of the primary production of phytoplankton in lakes of the Parque Florestal do Rio Doce – MG.

	Depth (m)	Primary productivity $(mg\,C \cdot m^{-3} \cdot h^{-1})$	Chlorophyll-a $(mg \cdot m^{-3})$	Rate of assimilation $(mg\,C \cdot mgChl.a^{-1} \cdot h^{-1})$
Lake Dom Helvécio (17 JUNE '83)	0.0	1.44	1.3	1.11
	2.0	1.73	1.3	1.33
	5.0	0.57	1.1	0.52
	10.0	0.21	1.1	0.19
	15.0	0.12	1.6	0.07
	20.0	0.06	0.8	0.07
	22.0	0.15	1.3	0.12
	23.0		−0.04	0.5
Lake Jacaré (30 june '83)	0.0	1.52	6.4	0.24
	2.0	2.31	6.6	0.35
	5.0	0.92	8.5	0.11
	8.0	0.11	7.1	0.01

Source: Tundisi et al. (1997).

Table 9.2   Diurnal variation of the vertical profiles of primary phytoplankton production of Lake Carioca, Parque Florestal do Rio Doce – MG (14/7/1983)

	Depth (m)	Primary productivity $(mg\,C \cdot m^{-3} \cdot h^{-1})$	Chlorophyll a $(mg \cdot m^{-3})$	Rate of assimilation $(mg\,C \cdot mgChl \cdot al \cdot h1)$
Chlorophyll-a at the surface 26.0 mg · m⁻³ Productivity per unit area: 267.7 mgC · m⁻² · day⁻¹. From 7am to 10am: surface water incubated in: (collected at 6am)	0.0	9.98	26.0	0.36
	0.5	14.59	26.0	0.56
	1.0	12.48	26.0	0.48
	1.5	10.73	26.0	0.41
	2.0	7.06	26.0	0.27
	3.0	0.42	2.6	0.02
Chlorophyll-a at the surface 27.4 mg · m⁻³. Productivity per unit area: 326.3 mgC · m⁻² · day⁻¹. From 10am to 1pm: surface water incubated in: (collected at 9am)	0.0	2.51	27.4	0.09
	0.5	9.68	27.4	0.35
	1.0	14.91	27.4	0.54
	1.5	15.97	27.4	0.58
	2.0	12.83	27.4	0.47
	3.0	4.23	27.4	0.15
Chlorophyll-a at the surface 22.7 mg · m⁻³ Productivity per unit area: 326.3 mgC · m⁻² · day⁻¹. From 10am to 1pm: surface water incubated in: (collected at 9am)	0.0	4.62	22.7	0.21
	0.5	9.70	22.7	0.43
	1.0	7.66	22.7	0.43
	1.5	8.81	22.7	0.39
	2.0	6.16	22.7	0.27
	3.0	1.49	22.7	0.07

Source: Tundisi et al. (1997).

periods during filtration and preparing samples for calculating ^{14}C activity; and the quality of the filters used to filter phytoplankton samples. The pressure (vacuum) used to filter samples can damage cells and cause loss of material. Another problem is measuring the activity of filters in the liquid scintillation solution, since the activity should be conducted with the same efficiency throughout. Dissolved particulate matter containing ^{14}C can contaminate ^{14}C-CO$_2$ stock solutions; the levels of inorganic C in water samples must be measured as precisely as possible to prevent errors in identifying the ^{14}C fixed by phytoplankton (Peterson, 1980).

Estimates of phytoplankton respiration and the subsequent loss of ^{14}CO$_2$ during the incubation period have been widely studied. The results presented by Steemann-Nielsen and Hansen (1957) for oceanic waters were approximately 15% of the maximum rate of photosynthesis P$_{max}$. Respiration by the phytoplankton community can be inhibited by high-intensity underwater radiation, and excretion of dissolved organic carbon can occur at different intensities of underwater radiation (Tundisi, 1965; Viera et al., 1986, 1998).

Rodhe (1958) presented a classic study on the ^{14}C technique, which included the first results on application of the method to lacustrine phytoplankton communities. Rodhe looked at seasonal variations, regional differences in primary productivity, and effects of various incubation periods, including comparisons of 4-hour incubation periods with day-long incubations. Short incubation periods (2 to 4 hours) were recommended for experimental work.

In a series of experiments applying the ^{14}C technique, Tundisi (1983) showed that incubation periods of 2, 4, or maximum 6 hours can give results that vary as function of the chlorophyll-a levels. For oligotrophic lakes and reservoirs with chlorophyll-a levels under $10 \, \mu g \cdot L^{-1}$, 4–6-hour incubation periods can be used. In eutrophic systems with levels of chlorophyll-a ranging from $50–200 \, \mu g \cdot L^{-1}$, 2-hour incubation periods were recommended. Figure 9.7 shows the principal sources and reserves of carbon and the significant physiological processes in experiments measuring the photosynthetic activity of phytoplankton with the ^{14}C-CO$_2$ technique.

Based on all the problems and limitations involved in the experiments, we can conclude that the ^{14}C technique presents intermediate results between gross and net photosynthesis, resulting in estimates perhaps halfway between gross productivity and net productivity.

Dissolved oxygen method

Introduced originally by Gaarder and Gran (1927), the dissolved oxygen method for calculating phytoplankton productivity was used for many years, until and after 1951, when the ^{14}C technique was introduced by Steemann-Nielsen (1951, 1952).

Experimental procedure

Samples collected at different depths (generally at about 100%, 50%, 25%, 10% and 1% of the solar radiation reaching the surface of a lake or reservoir) are re-suspended in transparent and dark containers and completely sealed to keep bubbles from forming (using transparent 100 or 250-ml glass containers, preferably of quartz or Pyrex).

Generally, several different types of transparent and dark glass containers are used. Since the oxygen method is less sensitive than the ^{14}C technique, the incubation

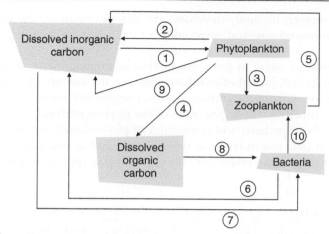

1. Photosynthetic carbon fixation
2. Respiration by phytoplankton
3. Food for herbivorous zooplankton
4. Excretion of dissolved organic carbon by phytoplankton
5. Zooplankton respiration
6. Respiration by bacteria
7. Fixation of CO_2 by bacteria
8. Fixation of dissolved organic carbon by bacteria
9. Photo-oxidation and photorespiration of phytoplankton
10. Food for zooplankton

Figure 9.7   Principal processes in calculating primary productivity with the $^{14}CO_2$ method. *Source:* Modified from Peterson (1980).

time needs to be long enough so that changes in the dissolved oxygen levels can be detected by the traditional Winkler method or by electrodes that measure dissolved O_2 in the water. In waters with low concentrations of chlorophyll-*a* ($<10\,\mu g \cdot L^{-1}$), changes in the dissolved oxygen levels may be insufficient to provide reliable results, even after 6 or 8 hours of incubation.

In general, 4–6 hours of incubation are needed in mesotrophic waters, and 2–4 hours in eutrophic waters (50–$100\,\mu g \cdot L^{-1}$ of chlorophyll-*a*). In hyper-eutrophic waters, 1 hour of incubation is generally sufficient.

Calculating primary productivity

In a traditional experiment, the initial dissolved oxygen level obtained in the profile (Ci) should decrease in the dark bottle (Ce). In the transparent bottle, the dissolved oxygen levels should increase (Ct), as a result of the difference between the photosynthetic production of oxygen and respiration.

Therefore:

▶ Ci − Ce = respiratory activity of the group of organisms, (phytoplankton, zooplankton, Bacterioplankton) per unit of volume per unit of time.
▶ Ct − Ci = net photosynthesis of phytoplankton per unit volume per unit of time.
▶ (Ct − Ci) + (Ci − Ce) = gross photosynthetic activity.

Gross photosynthesis can also be estimated directly by the difference Ct − Ce.

The dissolved oxygen method, therefore, enables estimates of the **gross photosynthesis, net photosynthesis** and **respiration** of each sample of the planktonic community.

Gross photosynthesis refers to **gross production** of organic matter arising from exposure to solar radiation. **Net photosynthesis** refers to the amount of organic matter after computing losses from respiration, death of organisms, excretion of organic matter, and possibly other metabolic processes of the phytoplankton.

To express the photosynthesis and respiration of phytoplankton in terms of carbon **fixed** or **respired** (and not in terms of oxygen produced or consumed), the photosynthetic quotient (PQ) or the respiratory quotient (RQ) is applied according to the following formula:

$$PQ = \frac{\text{molecules of oxygen produced during photosynthesis}}{\text{molecules of } CO_2 \text{ assimilated during photosynthesis}}$$

$$RQ = \frac{\text{molecules of } CO_2 \text{ produced during respiration}}{\text{molecules of } O_2 \text{ consumed during respiration}}$$

A PQ of 1.2 and RQ of 1.0 are commonly found in conditions of moderate solar radiation and chlorophyll-*a* levels between 10–$50\,\mu g \cdot L^{-1}$.

To convert the O_2 produced by photosynthesis into carbon, the following formula is used:

$$\text{Gross photosynthesis (mg C·m}^{-3} \cdot h^{-1}) = \frac{(Ct - Ce) \cdot 1{,}000 \cdot 0.375}{(PQ) \cdot t}$$

where:
PQ – photosynthetic quotient, normally 1.2
t – time of incubation
Ct – concentration of O_2 in transparent bottle
Ce – concentration of O_2 in dark bottle
0.375 – C/O_2 conversion factor
C – 12 (atomic weight)
O_2 – 32 (molecular weight)

Therefore, 12 mg C/32 mg O_2 = 0.375

$$\text{Net photosynthesis (mg C·m}^{-3} \cdot h^{-1}) = \frac{(Ct - Ci) \cdot 1{,}000 \cdot 0.375}{(PQ) \cdot t}$$

where:
PQ – photosynthetic quotient, normally 1.2
t – incubation time
Ct – O_2 in transparent bottle
Ct – O_2 in initial bottle
0.375 – C/O_2 conversion factor: 12 mg C/32 mg O_2

$$\text{Respiration } (\text{mg}\,C\cdot m^{-3}\cdot h^{-1}) = \frac{(Ci - Ce)\cdot RQ\,(1{,}000)(0.375)}{t}$$

where:

RQ – respiratory quotient: normally 1.0
t – time of incubation
Ci – O_2 in initial bottle
Ce – O_2 in dark bottle
0.375 – conversion factor:
C/O_2 – 12 mg C/32 mg O_2 = 0.375

The dissolved O_2 technique measures the metabolism of the community, represented by photosynthesis and respiration.

Various authors point out the fact that solar radiation can affect respiration (e.g., Steemann-Nielsen, 1975) resulting in overestimates of net photosynthesis. Because of diurnal fluctuations in photosynthesis and respiration, the technique of conducting multiple experiments over a day-long period can be used to offset these fluctuations, and in this case an average or daily sum of results can be used.

High sensitivity of the Winkler method in measuring dissolved O_2 is essential in the experiment. Measurements must have a precision of plus or minus $0.02\,\text{mg}\cdot L^{-1}$ dissolved oxygen (Wetzel and Likens, 1991). The dissolved O_2 method can be effective for values $>10\,\text{mg}\,C\cdot m^{-1}\cdot h^{-1}$ (Strickland and Parsons, 1972).

9.3 FACTORS LIMITING AND CONTROLLING PHYTOPLANKTONIC PRODUCTIVITY

Over the last 50 years, many experiments have been conducted on the productivity of phytoplankton in lakes, reservoirs and oceans.

Physical and chemical properties were measured simultaneously, adding significantly to the understanding of the factors limiting and controlling primary productivity of phytoplankton and the vertical matrix of the system.

9.3.1 Solar energy

Certain basic features of the relationship between underwater solar energy and photosynthesis are common to all phytoplanktonic species. At low levels of solar energy, the relationship between underwater solar energy and photosynthesis is linear. This linear response of the phytoplankton is a fraction of the photochemical component of photosynthesis. At higher levels of underwater solar energy, the photosynthetic rate 'saturates', and the saturation represents the maximum rate of the enzymatic processes at the temperature at which these processes occur. The intersection of the initial response of the photosynthetic process with the photosynthetic saturation rate, according Steemann-Nielsen (1975), describes the ratio of the two processes (photochemical and enzymatic). This ratio, introduced by Talling (1957) as I_K, is an important factor in describing the physiological adjustment of phytoplankton. There are significant differences among the different groups and species of phytoplankton in response to the underwater solar energy that saturates photosynthesis.

Figure 9.8 shows the classic photosynthesis-light curve with the location of I_K at the point of intersection of the rising extrapolation (which represents the photochemical process) and the horizontal line (which represents the enzymatic process) (Calijuri, Tundisi and Saggio, 1989).

In experiments with natural phytoplankton, it is difficult to obtain units related to the photosynthetic rate. Chlorophyll-*a* is one of the measures that can be used (Ichimura *et al.*, 1962). In laboratory experiments, it is not difficult to calculate the photosynthetic rate by different units: number of cells, dry weight, or chlorophyll-*a* level.

The phytoplanktonic community has the capacity to adapt to high or low levels of underwater solar energy. For example, if there is turbulence, then the various components of the phytoplanktonic community can be exposed to periods of low or high levels of underwater solar energy, depending on the location. This physiological behaviour is similar to the characteristics of 'shade' and 'sun' plants in terrestrial systems.

When underwater solar energy is low, values below I_K occur, suggesting a photochemical response lower than the phytoplankton (and still linear). When the underwater solar energy is higher and the phytoplankton develops under these conditions, the I_K values tend to be higher (see Figure 9.9).

Growth at low levels of underwater solar energy results in an increased level of chlorophyll-*a* per unit (e.g., cell). The change in chlorophyll-*a* levels is the phytoplanktonic community's response to growth at low solar energy levels.

Among the main mechanisms of photo-adaptation, increased chlorophyll-*a* and accumulation of accessory carotenoid pigments need to be considered as factors that facilitate adaptation to the underwater solar spectrum.

Figure 9.10 shows light-photosynthesis curves for natural phytoplankton in several lakes in Parque Florestal do Rio Doce (MG). Similar curves are also presented for Lake Cristalino (Amazonia) at different times of the day.

The response of phytoplankton to underwater solar energy depends on the climate of underwater radiation, on the period of time in which the phytoplankton populations remain in regions of high or low solar energies, and so on the 'photic history' of the phytoplankton.

At high solar energy flux, photosynthesis is interrupted when the enzymatic processes reach their maximum rates. High levels of underwater solar energy inhibit photosynthesis, and photo-autotrophic organisms suffer the effect of these high intensity energies. Species of higher plants do not normally suffer these effects; phytoplankton, however, is affected. Such results were observed in innumerable experiments in the higher-latitude tropics in summer.

The inhibition of photosynthesis is much more pronounced at high levels of solar radiation. For example, Tundisi (1965) observed inhibition and highly deleterious effects on cultures of *Chlorella vulgaris* subjected to solar energies $>1 \, cal \cdot cm^{-2} \cdot sec^{-1}$. The excretion rate of the *Chlorella vulgaris* culture also increases considerably at higher levels of solar radiation ($>0.8 \, cal \cdot cm^{-2} \cdot sec^{-1}$).

When the level of underwater solar radiation changes from low to high, the phytoplanktonic species' physiological behaviour changes. In *Chlorella*, for example, a substantial part of the photochemical mechanism is inactive and the photosynthetic

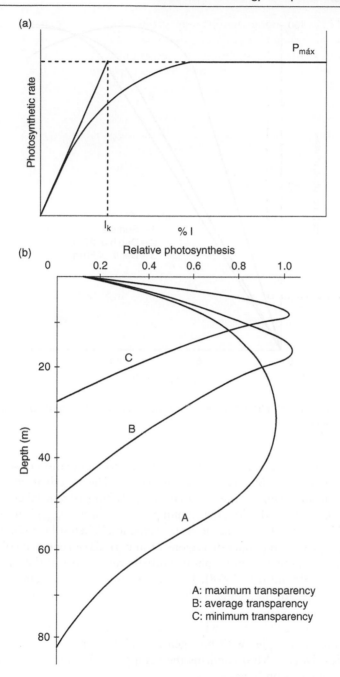

Figure 9.8   a) Interrelationship of the intensity of radiation and the productivity and maximum rate of photosynthesis; b) Vertical profiles of phytoplankton in waters with various transparencies.
Source: Morris (1974).

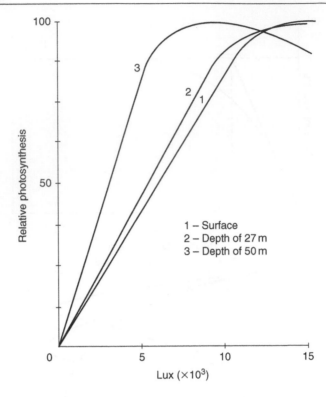

Figure 9.9   Relative rate of photosynthesis as a function of light intensity in phytoplanktonic populations.
Source: Morris (1974).

rate decreases, probably due to the photo-oxidative decomposition of several enzymes active in photosynthesis (Steemann-Nielsen, 1962). The inactivation of the photo-chemical mechanism results in the fact that in highly transparent lakes or oceans and on days with high solar radiation, maximum photosynthesis (P_{max}) is found at a depth where 30–50% of the radiation on the surface reaches. When solar radiation decreases (for example, in winter in temperate regions or in periods of repeated cold fronts with low solar radiation), the maximum photosynthetic rate occurs at the surface.

According to Straškrába (1978), I_K is a function of water temperature, phyto-planktonic biomass, and nutrients. Talling (1975) determined that the value of P_{max} (that is, the **photosynthetic capacity** in underwater radiation at saturation) increases with temperature; it can also vary with nutrient levels. The increase in I_K with experi-mental enriching of samples with **nitrogen** and **phosphorus** in Lake Jacaré (Parque Florestal do Rio Doce – MG) confirms the results of Ichimura (1958, 1968) on the effect of nutrient levels on I_K values.

9.3.2   Diurnal variations in photosynthetic activity

Diurnal variations in the photosynthetic activity of phytoplankton were observed by Doty and Oguri (1957) and Yentsch and Ryther (1957). These variations in photo-synthetic activity and in the production of chlorophyll have been observed at low

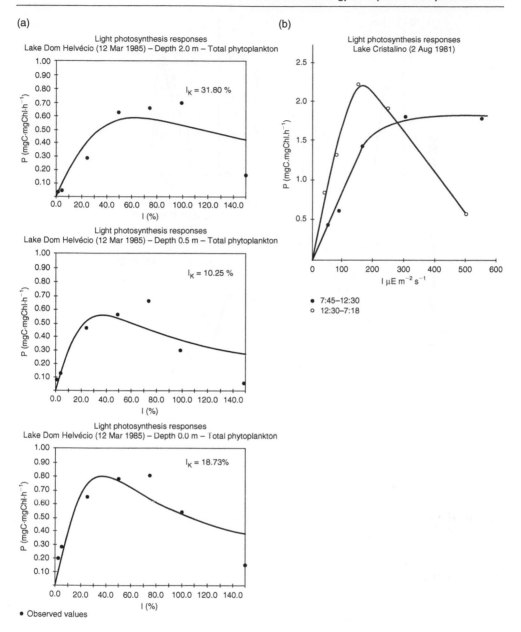

Figure 9.10 a) Light photosynthesis responses obtained for natural phytoplankton in nutrient-enriched ponds, in experiments on Lake Dom Helvécio (Parque Florestal do Rio Doce – MG); b) In Lake Cristalino (AM), using the technique of simulated *in-situ* [14]C incubation.

latitudes but also in lakes and oceans at higher latitudes. Several factors may contribute to these fluctuations, including an endogenous rhythm and synchronization between cell division, and responses to light-dark cycles. In general, a depression occurs during periods close to the highest radiation levels (midday). Saturation rates vary at different periods of the day (see Figure 9.11) (Sorokin, 1999).

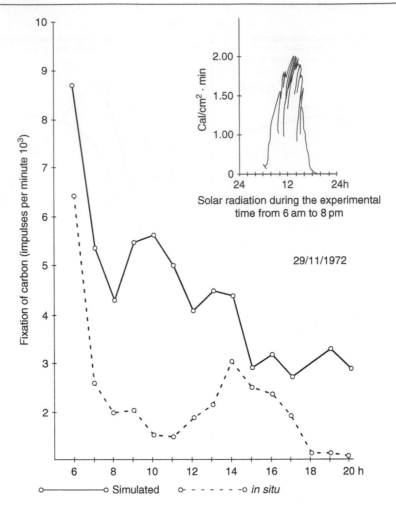

Figure 9.11   Diurnal rhythm of fixation of radioactive carbon (^{14}C) by phytoplankton in UHE Carlos Botelho reservoir (Lobo/Broa).

9.3.3   Effects of temperature

The hypothesis that prevailed for many years was that the prevailing temperature did not directly influence the primary productivity of natural populations of phytoplankton; and that the effects of temperature were relatively insignificant (Morris, 1974; Steemann-Nielsen, 1975). However, studies by Morris *et al.* (1971), Berman and Eppley (1974) and Peterson (1980) showed that temperature can affect primary productivity since the metabolism of any organism depends on temperature, according to the Q_{10} coefficient (approximately 2.2–2.5) (Sorokin, 1999).

Temperature can determine the upper limit of phytoplanktonic activity and regulate photosynthetic rate and primary production by regulating and controlling metabolism. Jewson (1976) and Jones (1977a, 1977b) observed the effect of temperature on respiration by phytoplankton. They found that a Q_{10} factor of 2.5 occurs. Adaptation to different temperatures can occur, and the time for this adaptation must be considered. It is also necessary to take into account that temperature affects not only photosynthesis but respiration as well.

Jørgensen and Steemann-Nielsen (1965) showed that cellular enzyme levels per cell can increase at low temperatures. Thus, at lower temperatures, the growth rate drops but the photosynthetic and respiration rates per cell do not (the result of maintenance of enzymatic processes).

9.3.4   Experimental Comparisons

Sakamoto *et al.* (1984) conducted experimental comparisons *in situ* of several methods of measuring the primary productivity of phytoplankton. The experiments were conducted in **Lake Constance** (Germany) under conditions of complete vertical mixing of the water column. The experimental methods, applied simultaneously, included: transparent and dark containers suspended *in situ*, with volumes varying from 120 ml to 570 ml; comparison of the primary productivity obtained by adding $NaH^{14}CO_3$ and $NaH^{13}CO_3$; dissolved oxygen technique with precision of 2–5 µg $O_2 \cdot L^{-1}$; experiment simulated *in situ* with the addition of ^{13}C; use of a 2-m-long and 19-cm-wide plexiglass tube as vessel, suspended *in situ*. The comparison gives an idea of the variety of techniques used to measure the primary productivity of phytoplankton. The results show considerable variations in productivity in the euphotic zone (P_z), most likely, according to the authors, due to the type of glass used in the containers, inconsistency in correcting for fixation in the dark and isotope discrimination, and variations in the dissolved oxygen method used, due to the fact that the conversion of O_2 produced during photosynthesis and the carbon fixed is indirect. Variations in the apparent photosynthetic quotient in the water column were observed.

I_K variations obtained experimentally and graphically were: $65\,\mu E m^{-2} s^{-2}$ (a measure of light-quantum flux) with the dissolved O_2 method; $102\,\mu E \cdot m^{-2} \cdot s^{-1}$, $\mu E \cdot m^{-2} \cdot s^{-1}$, $87\,\mu E \cdot m^{-2} \cdot s^{-1}$ and $147\,\mu E \cdot m^{-2} \cdot s^{-1}$ with the ^{14}C technique; and $237\,\mu E \cdot m^{-2} \cdot s^{-1}$ with the ^{13}C technique. All the methods used showed similar results for the predominant portion of phytoplankton in the experiment (<10-µm nano-phytoplankton). According to the authors, one important conclusion of this work is that the dissolved oxygen method, with high-sensitivity techniques to measure dissolved O_2, may be widely used in the future.

One of the methods used in turbulent waters, with an effective mixing zone and reduced euphotic zone (2–3 metres), is the use of vertical 2–3 metre-long transparent plastic pipes.

9.3.5   Influences of day-length and ultraviolet radiation

On long days with periods of prolonged solar radiation, phytoplanktonic growth can be inhibited (Sorokin and Kraus, 1959). Ultraviolet radiation can penetrate more deeply, depending on the water's optical features. Chemical substances can hinder the penetration of ultraviolet radiation into deeper layers. This occurs in coastal waters, estuaries, lakes and reservoirs. Steemann-Nielsen (1946a) showed that ultraviolet radiation has deleterious effects on phytoplankton; it may damage enzymes and photochemical mechanisms.

9.4   COEFFICIENTS AND RATES

Calculations of the primary productivity of phytoplankton in natural conditions can be optimized by applying several coefficients and rates that describe the efficiency of photosynthesis by phytoplankton. The most important are the coefficient of the specific production of phytoplankton per day (μ), the 'Assimilation Number' of chlorophyll-a per hour (AZ) and the photosynthetic quotient (PQ).

The coefficient of specific growth rate (μ) measures the activity of the photosynthetic production per unit quantity (e.g., $mg \ C \cdot m^{-3}$) of phytoplankton under certain environmental conditions. The coefficient is calculated as the primary production rate ($mg \ C \cdot m^{-3} \cdot day^{-1}$) in relation to the biomass (B) of a specific community of phytoplankton. These coefficients are high in the initial phase of the blooming of phytoplankton or in phytoplanktonic populations subject to intense grazing pressure by zooplankton, provided that pressure allows phytoplankton populations to keep growing. In aquatic systems in temperate regions in summer, these values can present a range of 0.8–1.5 per day, and in growing populations can reach 5–7 per day.

The Assimilation Number of chlorophyll-a (AZ) is another practical and useful parameter to identify the potential production of a phytoplanktonic population. The AZ value is calculated as photosynthetic (primary production) rate expressed in $mg \ C \cdot m^{-3} \cdot h^{-1}$, and the concentration of chlorophyll (in the same sample) in $mg \cdot m^{-3}$. Therefore, $AZ = mg \ C \cdot m^{-3} \cdot h^{-1}/mg \ Chl a \cdot m^{-3}$.

Some calculations express $AZ = mg \ C m^{-2} \cdot h^{-1}/mg \ Chl a \cdot m^{-3}$, that is, the activity in the euphotic zone. The AZ values vary from 3–10 mg $C \cdot$ mg $Chl a \cdot h^{-1}$. The highest AZ rates indicate phytoplanktonic populations in full growth, with more efficient photo-pigment systems (Sorokin, 1999).

Photosynthetic quotients (PQ) correspond to the molar ratios between oxygen produced during photosynthesis (O_2 mol L^{-1}) and CO_2 assimilated (CO_2 mol L^{-1}); that is $PQ = O_2/CO_2$. In general this molar ratio varies between 1.2–1.3.

This quotient can be used to compare the ^{14}C technique with the oxygen method to measure the photosynthesis of phytoplankton. The PQ is also used for the conversion of values of dissolved O_2, measured by the O_2 method for carbon fixed per m^2 or m^3 per hour or per day, as already described in the description of the O_2 method.

Another important and useful ratio is the relationship between the primary production in $mg \ C \cdot m^{-2} \cdot day^{-1}$ (P_t) and the primary production measured on the surface $mg \ C \cdot m^{-3} \cdot day^{-1}$ (P_s). This ratio may be relatively constant for a given body of water during certain periods of the year.

9.5   PHOTOSYNTHETIC EFFICIENCY

The photosynthetic efficiency of photo-autotrophic phytoplankton is low. It depends on several factors: the intensity of solar radiation, the physiological condition of the phytoplankton, and other aspects of the phytoplanktonic eco-physiology. In the lakes of Parque Florestal do Rio Doce (MG), Tundisi *et al.* (1997) measured photosynthetic efficiencies of 0.008% for Lake Dom Helvécio, 0.28% for Lake Carioca, 0.31% for **Lake Amarela**, and 0.007% for Lake Jacaré. Photosynthetic efficiency values calculated by Brylinsky (1980) ranged from 0.02–1%. Talling *et al.* (1973) calculated values of 0.15–1.6% for **Lake Kilotes** and 1.2–1.3% for Lake Aranguadi, both high-pH lakes in Africa. Tilzer *et al.* (1975) calculated values of 0.035% for **Lake Tahoe** in the United States and 1.76% for Loch Leven in Scotland.

Table 9.3 presents data on primary productivity of phytoplankton for systems with different trophic grades and varying conditions in temperate regions. Chart 9.3 presents an overview of different scientific studies on primary production of phytoplankton conducted over the last 50 years in the Neotropics, Africa and Australia.

To understand the qualitative and quantitative features of the different components of photo-autotrophic phytoplankton in the production of organic matter, experiments that study separately the various components the phytoplankton community are needed to identify each component's relative role.

Teixeira and Tundisi (1967) determined that for nutrient-poor waters of the equatorial Atlantic, the <20-μm component accounted for the majority of the primary productivity of phytoplankton. In studies on lakes in Parque Florestal do Rio Doce (MG), Tundisi *et al.* (1997) determined that the <10-μm component accounted for 80% of the primary production of phytoplankton.

These results were confirmed in more recent studies that showed that the photo-autotrophic pico-phytoplankton (<2 μm) accounts for the largest portion of primary production in some tropical lakes. Tundisi *et al.* (1977) also calculated the importance of the <20-μm component of phytoplankton in the phytoplanktonic primary production in the UHE Carlos Botelho reservoir (Lobo/Broa).

9.6   MODEL OF PRIMARY PRODUCTION OF PHYTOPLANKTON

The primary production of phytoplankton can be estimated using models based on the measurement of several variables, such as underwater radiation, chlorophyll-*a* and the coefficient of extinction of the water. The equation calculated by Talling (1970) is:

$$\Sigma a \ = \ M \frac{P_{max}}{Ke} f\left[\frac{I_o}{I_k}\right]$$

where:
Σa – photosynthesis or carbon fixed per unit of area
M – the population density
P_{max} – photosynthetic capacity per unit population in the underwater radiation at saturation determined experimentally in the lake
K_e – average extinction coefficient (see Chapter 4)
I_o – incidence of solar radiation
I_k – intensity of underwater radiation as measure of light-saturation
F – function of I_o/I_k ratio

The application of these models is useful only in the case where there is no possibility of ongoing direct experimental measurements of primary productivity. The model can be used to express the primary production of phytoplankton at a single point in a lake or reservoir in a single season, due to the variability of systems in time and space.

Table 9.3 Average values of density and production of phytoplankton in systems with different grades of trophy.

Trophic state and variations of annual primary production (pp)	Period of year	Camada	Population density — Numeric density of fractions: Pico, $10^6 \cdot l^{-1}$	Nano, $10^6 \cdot l^{-1}$	Micro, $10^6 \cdot l^{-1}$	Biomass of fractions: Pico, $mg \cdot m^{-3}$	Nano, $mg \cdot m^{-3}$	Micro, $mg \cdot m^{-3}$	Total, $mg \cdot m^{-3}$	Chla a, $mg \cdot m^{-3}$	Production — Primary production/day: $mgC \cdot m^{-3}$	$gC \cdot m^{-3}$	μ	Number/hour assimilation
Oligotrophic	AY	TLM	5	0.3	2	0.01	0.03	0.02	0.06	0.01	5	0.2	1.2	2
		DPM	10	1.8	5	0.02	0.18	0.05	0.25	0.5	2	–	–	3
Mesotrophic	PBS	TLM	50	9	1,400	0.1	0.03	14	15	5	300	3	0.6	
	MPS	TLM	50	3	10	0.1	0.18	0.1	0.5	0.5	30	0.5	1.0	
		MPD	100	13	100	0.2	0.9	1.0	3.0	3	–	–	–	
Eutrophic	MPA	TLM	150	170	400	0.3	0.3	4.0	6.0	1.5	100	1.2	1.2	
	PBSp	TLM	1,000	150	300	2.0	15	3.0	20	15	1,000	5	0.5	
		TLM	150	150	70	0.3	1.5	0.7	3.0	3	200	1.5	0.8	
	MPS	MPD	250	175	600	0.5	3.5	6.0	10	15	–	–	–	
	MPA	TLM	250	175	800	0.5	3.0	8.0	12	5	300	2.5	1.0	–
Hypertrophic	PBSu	TLM	1,000	1,300	1,000	2.0	13	10	25	30	1,500	4	0.8	4

Abbreviations: AY – all year; PBS – phytoplankton blooms in spring; MPS – minimum of phytoplankton in summer; MPA – maximum of phytoplankton in the autumn; PBSu – phytoplankton blooms in summer; MPD – maximum phytoplankton in the deep zones; Chl-a – chlorophyll; μ – coefficient of specific production per day; DPM – deep layer vertical mixing; TLM – top layer of the vertical mixing; NA – number of assimilation of chlorophyll per hour; average volumes of cells for each fraction of phytoplankton: pico – 2 μm^3; nano – 100 μm^3; micro – 10,000 μm^3.
Values for inland aquatic ecosystems, in temperate regions.
Source: Sorokin (1999).

Chart 9.3   Different scientific works on primary production by tropical phytoplankton.

Period	Notropics	Africa	Australasia
Pre-1960	Central America Deevey (1955)	Nile System Talling (1957a[1]); Prowse and Talling (1958)	
1960–80	**Amazonia** Hammer (1965) Schmidt (1973, 1973[2], 1976[2], 1982) Fisher (1979[1]) Melack and Fisher (1983) **Venezuela** Gessner and Hammer (1967) Lewis and Weibezahn (1976) **Central America** Gliwicz (1976b[2]) Pérez-Eiriz et al. (1976, 1980) Romanenko et al. (1979) **Titicaca** Richerson et al. (1977[2], 1986, 1992[2]) Lazaro (1981[2]) **Lakes and reservoirs of Brazil** Tundisi et al. (1978) Barbosa and Tundisi (1980) Hartman et al. (1981)	**South Africa, Sudan, Ethiopia** Talling (1965a[2]) Ganf (1972, 1975[1,2]) Talling et al. (1973) Melack and Kilham (1974[1]) Ganf and Horne 1975[1]) Melack (1979[2], 1980, 1981, 1982) Vareschi (1982[1,2]) Harbott (1982) Belay e Wood (1984) **West Africa, Zaire, Chad** Lemoalle (1969, 1973a, 1975, 1979a[1], 1981a, 1983[1,2]) Freson (1972[2]) Thomas and Radcliffe (1973) Karlman (1973[2], 1982[2]) Pages et al. (1981) Dufour (1982) Dufour and Durand (1982)	**India** Sreenivasan (1965) Hussainy (1967) Ganapati and Sreenivasan (1970) Michael and Anselm (1979) Kanna and Job (1980c[1]) **Malaysia** Prowse (1964, 1972) Richardson and Jin (1975) **Philippines** Lewis (1974[1,2])
1980+	**Central America** Erikson et al. (1991a[1], b) Lind et al. (1992[2]) **Venezuela** Gonzales et al. (1991) **Ecuador** Miller et al. (1984) **Titicaca** Vincent et al. (1984, 1986[2]) Richerson (1992) **Brazil** Reynolds et al. (1983) Tundisi (1983) Gianesella-Galvão (1985) Barbosa et al. (1989[1]) Forsberg et al. (1991) Tundisi et al. (1997[1,2]) Tundisi and Matsumura Tundisi (1990)	**Central and South America** Robarts (1979[2]) Hecky and Fee (1981) Degnbol and Mapila (1985) Cronberg (1997) **East Africa** Mugidde (1993[2]) Mukankomeje et al. (1993) Patterson and Wilson (1995[1]) **Ethiopia** Belay and Wood (1984) Kifle and Belay (1990[2]) Gebre-Mariam and Taylor (1989a) Lemma (1994) **West Africa** Nwadiaro and Oji (1986) **Malawi** Degnbol and Mapila (1985) Bootsma (1993a)	**India** Saha and Pandit (1987[2]) Durve and Rao (1987) Kundu and Jana (1994) **Sri Lanka** Dokulil et al. (1983[1]) Silva and Davies (1986[1], 1987[2]) **Bangladesh** Khondker and Parveen (1993[2]) Khondker and Kabir (1995) **Papua, New Guinea** Osborne (1991)

Note: [1]diurnal series, [2]annual series.
Source: Talling and Lemoalle (1998).

According to Margalef (1978, 1991), the external mechanical or kinetic energy that moves aquatic systems – that energy which does not depend on photosynthetic energy or on the energy flowing through organisms – is the result of interactions between the atmosphere and hydrosphere, and it accelerates the vertical and horizontal mixing processes. This external energy is qualitatively and quantitatively important in the distribution of planktonic communities, aquatic macrophytes and suspended material, adding physical and chemical factors that control and limit the productivity of photosynthetic organisms. Vertical dislocations generated by the intensity of winds can segregate or integrate different species of phytoplankton, bacteria, protozoa and flagellates, which not only fix organic matter through photosynthesis but also play an important in the recycling and decomposition of material (Margalef, 1978).

The two most important mechanisms that depend on external energy and support primary productivity in aquatic systems are **turbulence** (generated by the wind and which transports material to the surface or to deeper regions) and **resurgence** (which promotes the release of nutrients in the euphotic zone, accelerating primary production).

The principal difference between lakes and oceans does not have to do with the concentration of ions (NaCl) but rather with the different conditions of turbulence and mixing. It is thus very difficult to estimate the quantity of energy that supplies nutrients to the euphotic zone and promotes the primary production of phytoplankton. The relationship between primary production and available external energy cannot be linear or logarithmic because excessive energy promotes the dispersal of phytoplankton to the aphotic zone, making the production of organic matter unviable due to the transport of autotrophic photosynthetic productive material beyond the compensation point. Uniform distribution of external kinetic energy does not occur in aquatic ecosystems; it occurs in **vertical or horizontal cells** that promote the aggregation or dispersion of photosynthetically productive particles.

A drop in primary production of phytoplankton may be a natural consequence of segregation in the distribution of the productive factors, i.e., nutrients and light. The primary production of phytoplankton is basically controlled by the physical environment, i.e., advection and turbulence, which promote the aggregation of phytoplankton at the surface or in the euphotic zone and re-establish the supply of available nutrients for the phytoplanktonic community. Turbulence can also reduce the availability of solar radiation, either by carrying phytoplankton to deeper poorly lit layers or by transporting suspended matter, which blocks solar radiation and reduces the depth of the euphotic zone.

In tropical lakes, after intense rainfall, suspended material transported by drainage of surface water can rapidly reduce the euphotic zone and significantly decrease primary production of planktonic microbenthos and submerged aquatic macrophytes (Tundisi *et al.*, 2006).

9.7   METHODS FOR MEASURING THE PRIMARY PRODUCTION OF PERIPHYTON

Measuring the primary productivity of periphytonic communities is complex because they include a range of productive organisms that grow in sediments, on rocks, on different types of substrata and detritus, and even on living organisms. Distribution of these organisms is extremely heterogeneous, which needs to be taken into account in any measurement method used. Phototactic migration occurs, which can influence distribution patterns. The biomass is expressed per unit of water surface and unit of substratum area (Piecynska, 1968).

Counting the organisms can be done on the substratum itself, through examination with a microscope. Artificial substrata for collecting micro-algae and subsequent counting of cells or measuring chlorophyll-*a* are used to survey productivity (Panitz, 1978). Colonization of a substratum depends on its type and its roughness (Panitz, 1978).

Measuring the chlorophyll-*a* and number of organisms per m^2, in addition to total carbon, are quantitative techniques used (Wetzel, 1974).

Direct experimental methods to calculate the primary production include application of the dissolved oxygen method, based on the use of transparent and dark tubes in undisturbed sediment, with several hours of incubation, and the addition of $NaH^{14}CO_3$ in samples of sediment or other substratum containing periphyton, incubated under simulated conditions (Pomeroy, 1959; Grontved, 1960; Wetzel, 1963, 1964).

Periphytonic production after growth on glass plates of a specific size, using transparent and dark bottles, incubated *in situ* for four hours, using the dissolved oxygen technique, was extensively studied by Chamixaes (1994) in tributaries feeding into the UHE Carlos Botelho reservoir (Lobo/Broa).

9.8 MEASURING THE PRIMARY PRODUCTIVITY OF AQUATIC MACROPHYTES AND COMPARISON WITH OTHER PHOTO-AUTOTROPHIC COMPONENTS

The annual production of aquatic macrophytes can be estimated measuring maximum seasonal biomass. Samples taken over two seasons combined with samples of physical and chemical variables in the water help to calculate the biomass values of gross annual production. In general, the technique is to remove a certain amount of biomass from a specific number of areas in two different periods of sampling at various sites, and weigh the vegetation removed.

A preliminary selection of the number of areas and sizes (m^2) and formats should be made. Random sampling should be performed after the selection of areas, and this sampling is repeated on two or more occasions during the seasonal cycle. Both the aerial and rooted parts of the plants need to be sampled, and the collection in each sampling should include living material of macrophytes, roots and already dead material (Westlake, 1965).

For many aquatic plants, according to Westlake (1974), the organic carbon content is 44–48% of the organic weight. The energy content of many aquatic macrophytes is 4.3–4.8 kcal/gram of organic material (Straškrába, 1967).

Measuring the **total chlorophyll of macrophytes** can be useful in calculating the biomass during time intervals. The total biomass of macrophytes in one area can be calculated from aerial photographs or satellite imagery calibrated with field samples, and the total weight $(kg/m^2$ or $kg/ha)$ can be calculated from the field samples.

For some species of macrophytes, the **leaf area index** (= leaf area/base area) can be used to estimate the total biomass of chlorophyllous matter following the relation plants/unit area × leaves/plant × average leaf area, using a planimetric technique or by calculation for the leaf area.

Experimental techniques using plastic cylinders are also employed to calculate the production and consumption of dissolved oxygen in areas with submerged macrophytes, by estimating the volume of O_2 produced (or consumed in dark cylinders) per area (cm^2 or m^2) for a specific incubation period (Vollenweider, 1974). Figure 9.12 presents results obtained with this technique.

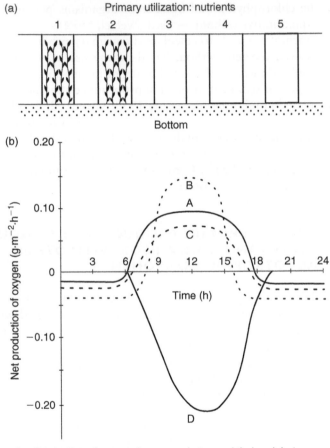

A – Production of oxygen by macrophytes and their epiphytes
B – Production of algae on the surface of sediment
C – Production by phytoplankton
D – Loss by diffusion of water into the atmosphere

Figure 9.12　Measurements of oxygen exchange in a bank of submerged plants dominated by *Mayaca sellowiana* in the UHE Carlos Botelho reservoir (Lobo/Broa): a) Distribution of five transparent plastic drums, open and closed for the sediment to the atmosphere and the presence or absence of plants; b) Patterns derived from diurnal variation in net oxygen exchange rates.
Source: Ikushima *et al.* (1983).

The primary production of aquatic macrophytes and micro-phytobenthos (periphyton) has been measured in different aquatic ecosystems. Primary production rates for micro-phytobenthos, using the ^{14}C technique and the production of O_2, were calculated for three African lakes: Turkana, Tanganyika and Malawi (Takamura, 1988). The rates obtained were $0.2\text{--}0.7\,\mathrm{g\,C\cdot m^{-2}\cdot day^{-1}}$ in shallow water with high transparency. The primary production of periphyton can even exceed $1\,\mathrm{g\,C\cdot m^{-2}\cdot day^{-1}}$. Calculations of the primary production of micro-phytobenthos in the **Limon River** in Venezuela (Weibezahm and Lewis, 1976) yielded more than $1\,\mathrm{g\,C\cdot m^{-2}\cdot day^{-1}}$.

For aquatic macrophytes, the measurements (based mainly on changes in biomass and dry weight) report values $>10\,\mathrm{g}$ dry weight $\cdot\,\mathrm{m^{-2}\cdot day^{-1}}$ for *Cyperus papyrus* (Westlake, 1975) and $5\,\mathrm{g}$ dry weight $\cdot\,\mathrm{m^{-2}\cdot day^{-1}}$ for *Lepironia articulata* (Furtado and Mori, 1982). Mercy *et al.* (1994) calculated $9.9\,\mathrm{kg}$ dry weight $\cdot\,\mathrm{m^{-2}\cdot year^{-1}}$ in the Amazon. Junk and Piedade (1993) calculated $7\,\mathrm{kg}$ dry weight $\cdot\,\mathrm{m^{-2}\cdot year^{-1}}$ for *Paspalum fasciculatum*.

Measuring the production of O_2 during 24-hour cycles, with periodic measurements of photosynthetic activity and respiration (using the dissolved O_2 technique), showed the production cycle of dissolved oxygen, loss of oxygen to the atmosphere, production of dissolved oxygen by macrophytes (and their epiphytes), and production by micro-phytobenthos on the surface of the sediment.

The relative importance of each of the various different primary producers largely depends on the physical organization of the ecosystem. One basic aspect in the study of the relative importance of primary producers is quantifying the various types of primary producers in the ecosystem, which is important for understanding the structure of the trophic web and how organic matter produced in early stages is transferred to the higher trophic levels.

Komárková and Markan (1978) compared the relative percentage of each of the primary producers in a eutrophic lake and in fish cultivation tanks (in two locations). The results are presented in Table 9.4.

In these ecosystems, the production of macrophytes outpaced the production of phytoplankton and periphyton. This type of situation is relatively common in the littoral zone of eutrophic lakes with a permanent mass of macrophytes covering a large portion of the lake. In some macrophytic communities, filamentous algae can be found in great masses, near the surface, utilizing the support provided by these vascular plants, which help sustain the algae at depths with high solar radiation.

Table 9.5 presents calculations of the primary production for various species of aquatic macrophytes in Brazil.

Table 9.4   Primary production by different photo-autotrophic organisms.

	Eutrophic lake	Ponds to raise fish (1)	Ponds to raise fish (2)
Macrophytes	57%	70%	53–83.5%
Phytoplankon	20%	7%	9–36%
Periphyton	23%	21%	5.5–11%

Source: Komárková and Markan (1978).

Table 9.5 Values of productivity ($t \cdot ha^{-2} \cdot year^{-1}$) of different species of aquatic macrophytes in various Brazilian aquatic systems.

Species	Ecological type	site	Productivity $t \cdot ha^{-2} \cdot year^{-1}$	Authors
Panicum fasciculatum	EM	Solimões River floodplain	70.0	Junk and Piedade (1993)
Paspalum repens	EM	Baixio Coast	31.0	Junk and Piedade (1993)
Luziola spruceana	EM	**Lake Camaleão**	7.6	Junk and Piedade (1993)
Oriza perennis	EM	Lake Camaleão	27.0	Junk and Piedade (1993)
Nymphoides indica	EM	UHE Carlos Botelho reservoir (Lobo/Broa)	7.6	Meneses (1984)
Pontederia cordata	EM	UHE Carlos Botelho reservoir(Lobo/Broa)	3.8	Meneses (1984)
Eichhornia azurea	EM	Lake Dom Helvécio	17.5	Ikusima and Gentil (1987)
Eichhornia azurea	EM	Lake Jacaré	6.6	Ikusima and Gentil (1987)
Eichhornia azurea	EM	Lake Carioca	8.4	Ikusima and Gentil (1987)
Eichhornia azurea	EM	**Infernão Lagoon**	3.5	Coutinho (1989)
Pontederia lanceolata	EM	Mato Grosso swamp	9.7	Penha (1994)

EM – Emergent
Source: Camargo and Esteves (1995).

9.9   INDIRECT MEASUREMENTS OF *IN SITU* PRIMARY PRODUCTION

The primary production of underwater macrophytes, phytoplankton and periphyton can be calculated, under certain conditions, by accurately measuring the pH and dissolved oxygen over a 24-hour period. Primary production rates can be calculated from the differences of pH, the levels of dissolved CO_2, and thus of the CO_2 'fixed' by the photosynthetic activity, or of the oxygen produced during the diurnal period and the oxygen consumed by respiration. In lakes or reservoirs with considerable biomass of these aquatic plants (periphyton and phytoplankton), the measurements can provide a reasonable approximation to calculate a community's photosynthetic, primary production and respiration rates.

The major limitation is the dependency on accurate methods for measuring pH and dissolved oxygen with sensitive electrodes or precise analytical methods.

9.10   MEASURING PRIMARY PRODUCTION IN DIFFERENT ECOSYSTEMS

The set of tables below compares data on primary production in aquatic and terrestrial ecosystems. The data were obtained through experimental work, estimates of the **maximum theoretical production** and global evaluations from experiments conducted on oceans, lakes, rivers and terrestrial systems. These comparisons are fundamental for evaluating the primary production capacity of aquatic ecosystems and their potential in the global production of organic matter, which in the final analysis may be used for exploitation. They show, for example, that coastal systems, shallow lakes and reservoirs – more than deeper areas – are more productive, since the nutritional conditions of the water masses enhance the production capacity and growth of living organic matter.

Since concentrations of **phosphorus, nitrogen** and **carbon** are critical for the development and growth of the phytoplanktonic community, **new production** (that is, production resulting from the contribution of nitrogen and phosphorus from **outside sources**) can be essential in this process. By contrast, **regenerated production** implies the internal recycling of nutrients distributed in the euphotic zone by vertical or horizontal circulation and processes of decomposition in the water and sediment (see Tables 9.6 to 9.9).

Table 9.6   Estimates of the theoretical maximum organic production in oceans.

1. Solar energy reaching Earth's upper atmosphere	$1-25 \cdot 10^{24}$ cal
2. Energy reaching Earth's surface after a loss of 60% by absorption and scattering	$5 \cdot 10^{23}$ cal
3. Energy reaching the ocean surface after the loss of 50% of the infrared radiation and 28% that reaches the terrestrial systems	$1.6 \cdot 10^{23}$ cal
4. Energy available for absorption by phytoplankton after 10% of reflection	$1.2 \cdot 10^{23}$ cal
5. Estimated efficiency of photosynthesis of 2%, considering maximum energy disposable for primary production	$2.5 \cdot 10^{21}$ cal
6. Energy required to assimilate 1 g of carbon in material	
7. Upper limit of organic material that can be produced in the ocean, that is: $\dfrac{2.5 \cdot 10^{21}}{1.3 \cdot 10^{4}}$	$1.9 \cdot 10^{17}$ g C $\cdot$ year^{-1} or $1.9 \cdot 10^{11}$ metric tons of carbon

Source: Morris (1974).

Table 9.7   Primary production in oceans by region.

Region	% of ocean	Area (km²)	Average productivity ($g\,C \cdot m^{-2} \cdot year^{-1}$)	Total productivity (10^9 ton C $\cdot$ year^{-1})
Open ocean	90	$326 \cdot 10^6$	50	16.3
Coastal zone	9.9	$36 \cdot 10^6$	100	3.6
Areas of upwelling	0.1	$3.6 \cdot 10^5$	300	0.1
Total				20.0

Source: Morris (1974).

Table 9.8   Comparison of primary production in oceans with production in inland waters and in terrestrial systems.

Community	Daily average productivity $g\,C . m^{-2} . day^{-1}$
Oceanic waters	1.0
Coastal zone	0.5–3.0
Forests, agricultures, inland lakes	3–10
Intensive agriculture	10–25
Grasslands	0.5–3.0
Deserts	0.5

Source: Morris (1974).

Table 9.9 Comparison of primary production in oceans and inland ecosystems with terrestrial systems.

Community	primary average annual production $g \ C \cdot m^{-2} \cdot year^{-1}$
Oceans	50
Coastal zone	100
Upwelling areas	300
Phytoplankton in inland waters	860 (maximum)
Marine phytoplankton	240
Macro-algae in littoral zone	1,600–2,100
Natural areas in temperate regions (terrestrial)	2,400
Tropical rainforest	5,900
Agricultural crops (cereals) in temperate regions	2,400

Sources: Westlake (1993), Ryther (1969), Odum (1959), Phillipson (1966).

Table 9.10 Net primary production by phytoplankton in rivers and floodplain lakes.

System	Production	Researcher
Amazon River		
Castanho Lake (white water)	0.35–1.5 (gC $\cdot$ m^{-2} $\cdot$ d^{-1})	Fittkau et al. (1975)
Cristalino Lake (black water)	0.05–1.04 (gC $\cdot$ m^{-2} $\cdot$ d^{-1})	Raí (1984)
Redondo Lake	0.29 (gC $\cdot$ m^{-2} $\cdot$ d^{-1})	Marlier (1967)
Negro River	0.06 (gC $\cdot$ m^{-2} $\cdot$ d^{-1})	Schmidt (1976)
Tapajós River	0.44–2.41 (gC $\cdot$ m^{-2} $\cdot$ d^{-1})	Schmidt (1976)
Paraguay River (E)		
Ferradura	0.08–1.25 (gC $\cdot$ m^{-2} $\cdot$ d^{-1})	Bonetto, C. (1982)
Vermelho Port	0.004–0.06 (gC $\cdot$ m^{-2} $\cdot$ d^{-1})	Bonetto, C. (1982)
Lower Paraguay	0.06–0.75 (gC $\cdot$ m^{-2} $\cdot$ d^{-1})	Bonetto, C. et al. (1981)
Lake Ferradura	0.01–0.45 (gC $\cdot$ m^{-2} $\cdot$ d^{-1})	Zalocar et al. (1982)
Paraná River		
Upper Paraná (Upper Ibate City) (E)	0.002–0.99 (gC $\cdot$ m^{-2} $\cdot$ d^{-1})	Bonetto, C. (1983)
Lower Paraná (Paraná City) (L)	0.001–0.8 (gC $\cdot$ m^{-2} $\cdot$ d^{-1})	Perotti de Jorda (1984)
Lower Paraná (city of Correntes) km.1,208 (E)		
right side	0.010–0.580 (gC $\cdot$ m^{-2} $\cdot$ d^{-1})	Bonetto, C. et al. (1983)
left side	0.003–0.285 (gC $\cdot$ m^{-2} $\cdot$ d^{-1})	Bonetto, C. et al. (1983)
Lower Paraná (city of Correntes) km. 1,208 (E)		
right side	0.000–0.120 (gC $\cdot$ m^{-2} $\cdot$ d^{-1})	Bonetto, C. et al. (1979)
left side	0.003–0.285 (gC $\cdot$ m^{-2} $\cdot$ d^{-1})	Bonetto, C. et al. (1979)
Lower Paraná (Esquina City) (E)	0.030–0.850 (gC $\cdot$ m^{-2} $\cdot$ d^{-1})	Bonetto, C. (1983)
Lower Paraná (city of Currentes) (L)	0.040 (mgC $\cdot$ m^{-2} $\cdot$ h^{-1})	Perotti de Jorda (1980b)
Lower Paraná (Bela Vista) (km 1,060) (L)	0.041 (mgC $\cdot$ m^{-2} $\cdot$ h^{-1})	Perotti de Jorda (1980b)
Lower Paraná (km 876) (L)	0.045 (mgC $\cdot$ m^{-2} $\cdot$ h^{-1})	Perotti de Jorda (1980b)
Lower Paraná (city of Diamante) (L)	0.195 (mgC $\cdot$ m^{-2} $\cdot$ h^{-1})	Perotti de Jorda (1980b)

Note: (E) – Field experiment; (L) – Laboratory experiment
Source: Various sources and Neiff (personal communication).

Table 9.10 presents data on the net primary production of phytoplankton from various aquatic ecosystems and rivers and lowland lakes of the Amazon and the Parana (Neiff, 1986).

Table 9.11   Net primary production (NPP) of aquatic macrophytes in floodplain lakes of the lower Paraguay-Paraná and in other tropical systems.

Species	Group Locale	Year	Author	NPP $(tn \cdot ha^{-1} \cdot year^{-1})$
Eichhornia crassipes	PNA–Lake Barranqueras	1977	Neiff and Poi de Neiff (1984)	12.46
Eichhornia crassipes	PNA– Santa Fe		Lallana (1980)	13.80
Eichhornia crassipes	India	1971	Gopal (1973)	6.75
Eichhornia crassipes	India		Sahai and Sinha (1979)	2.71
Azolla pinnata	India		Gopal (1973)	2.80
Nymphoides indica	PNA–Santa Fe-Correntes	1971/82	Neiff and Poi de Neiff	0.8–2.2
Nymphaea amazonica	PNA – Correntes (Ibera)	1977/78	Neiff and Poi de Neiff	1.1–2.6
Victoria cruziana	PNA – Barranquetas	1977/78	Neiff and Poi de Neiff	1.6–2.3
Typha latifolia	PGUY – Chaco – Formosa	1984/86	Neiff (1986)	14–23
Typha latifolia	PNA – Correntes	1977/78	Neiff (1986)	15–19.5
Typha dominguensis	Africa		Thompson (1976)	22.88
Typha dominguensis	Africa		Howard-Williams and Lenton (1975)	15.00
Cyperus giganteus	PNA – Chaco	1980	Neiff (1986)	12–20
Cyperus papyrus	Nile White		Pearsall (1959)	46–70
Cyperus papyrus	Africa		Thompson et al. (1979)	34–94
Hymenachne amplexicaulis	PNA – Southeast Chaco	1979/80	Neiff (1980, 1982)	16–21
Echinochloa polystachya	PNA – Southeast Chaco	1979/80	Neiff (1980)	4–6
Cynodon dactylon	South Africa		Heeg and Breen (1982)	8.39
Paspalum repens	Amazonia		Junk (1970)	3–5
Oryza sativa	India		Gopal (1973)	12.5

Note: PNA – Floodplain of lower Paraná; PGUY –Valley of Lower Paraguay
Source: Neiff and Pol de Neiff (1984).

Table 9.11 shows the net primary production of aquatic macrophytes in lowland lakes of the lower Paraguay-Parana, compared with aquatic systems in various other regions.

9.11   PRIMARY PRODUCTION IN TROPICAL REGIONS AND TEMPERATE REGIONS

A variety of ways to measure primary production of photo-autotrophic phytoplankton has developed over the last 30 years. When comparing photosynthetic capacities per unit biomass (P_{max}), tropical lakes present higher values than temperate-region lakes.

Lemoalle (1981) compared average rates of gross photosynthetic production in the euphotic zone of tropical and temperate-region lakes, and concluded that the higher rates in tropical lakes result from greater photosynthetic capacity. This is probably due to the warmer waters in these ecosystems. As in all aquatic ecosystems, factors that affect the photosynthetic capacity, P_{max}/B (B is biomass), in tropical lakes include: size of cells, suppression of nutrients and CO_2, and water temperature.

The alkaline lakes of Africa (*soda lakes*) typically have ample biomass and very high values of photosynthetic rate. This was attributed by Talling *et al.* (1973), Melack (1974) and Lemoalle (1981, 1983) to the reserve of CO_2 in the euphotic zones of these lakes.

Inhibition on the surface of tropical lakes and reservoirs and respiration rates (mg O_2) are considered as 'losses' in the photosynthetic process, according to Talling and Lemoalle (1998). High intensity solar radiation in the air (and also underwater), with a significant amount of ultraviolet radiation, may be the cause of this inhibition.

Respiration rates accompany photosynthetic rates and vary widely in tropical lakes and reservoirs. In some cases, such as the Amazon, a significant contribution of heterotrophic bacteria, resulting from accumulated masses of organic material, increases the respiration rate (Melack and Fisher, 1983).

Talling (1965) measured a mean photo-autotrophic respiration rate in the tropics of 1 mg $O_2 \cdot$ mg Chl $a^{-1} \cdot h^{-1}$. This rate depends on temperature (Ganf, 1972). The intensity of underwater radiation (e.g., by photorespiration) also affects respiration rates, but there is still a lack of information about these mechanisms in the tropics. Figure 9.13 shows the results by Rai (1982) for the primary production in several lakes in temperate and tropical regions.

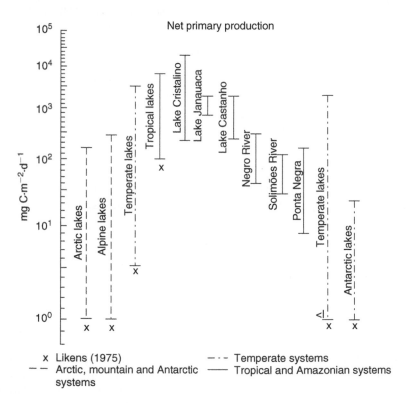

Figure 9.13   Results of net primary production for tropical lakes and temperate regions.
Source: Hill and Rai (1982).

Comparisons of the primary production in tropical lakes vs. temperate lakes must take into account seasonal and hydrological cycles, availability of solar radiation, water temperature and the biogeochemical cycles that are the result of factors such as water temperature, **recycling rate** and succession of organisms.

These comparison also take into consideration the annual primary production and climatologic factors that limit and control productive activity (see Figure 9.14).

Many methods to measure the primary production of photo-autotrophic phytoplankton have evolved over the last 30 years. Tables 9.12 and 9.13 present calculations

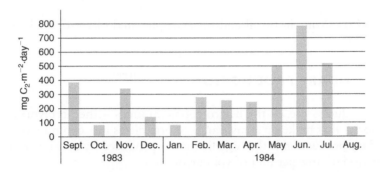

Figure 9.14   Primary production measured with the dissolved oxygen technique and converted to mg C · m^{-2} · day^{-1} in the Barra Bonita reservoir over a one-year period. The maximum primary production occurred in winter, during the more turbulent periods, which promotes regenerated production. In summer, solar radiation is greater, but there is heavy rainfall and primary production is limited by the decrease in depth of the euphotic zone due to the large amount of suspended material. It should be noted that 1983 was a year of intense El Niño activity.

Table 9.12   Data on chlorophyll, nutrients and primary production for a series of lakes and reservoirs in Brazil. Standard deviation in parentheses.

Lake or reservoir	TIN $\mu g \cdot L^{-1}$	TDP $\mu g \cdot L^{-1}$	Chlorophyll $\mu g \cdot L^{-1}$	Primary Production mg C · m^{-2} · h^{-1}	Researchers
R. Lagoa Dourada, SP	25.25 (9.48)	2.89 (2.09)	4.22 (0.82)	2.55 (1.5)	Talamoni (1995), Pompeo (1991)
R. Broa, SP	42.15 (10.25)	7.94 (2.91)	10.05 (2.41)	13.98 (4.84)	Matheus and Tundisi (1988)
R. Pedreira	788.51 (57.24)	5.67 (4.55)	43.70 (3.85)	156.40 (19.36)	Talamoni (1995)
R. Barra Bonita, SP	656.2 (811.50)	13.31 (3.98)	28.8 (45.92)	214.61 (150.91)	Tundisi et al. (1990)
R. Paranoá, DF	662.8 (366.01)	32.41 (7.18)	56.9 (21.10)	120.22 (77.18)	Branco (1991); Toledo and Hay (1988)
L. Dom Helvécio, MG	17.80 (18.12)	3.62 (2.97)	2.47 (1.16)	18.85 (8.67)	Okano (1980), Pontes (1980)
R. Caconde, RS	654.86 (206.64)	25.52 (4.88)	9.41 (3.91)	268.96 (222.04)	Guntzel (1998)
R. Jacaré-Pepira, SP	12.4 (4.99)	1.13 (1.07)	1.71 (0.62)	4.05 (1.39)	Claro (1978), Franco (1982)

TIN – Total Inorganic Nitrogen; TDP – Total Dissolved Phosphorus.
Source: Rocha and Matsumura Tundisi (1997).

Table 9.13 Net primary phytoplankton production in some oligotrophic and eutrophic lakes.

Lake	Trophic state	$g\,C \cdot m^{-2} \cdot year^{-1}$
Chan (Canada)	Ultraoligotrophic	1.3
Washington (USA)	Mesotrophic	96
Plüssee (Germany)	Eutrophic	186
Valencia (Venezuela)	Eutrophic	821

Source: Modified from Dodson (2005).

for lakes and reservoirs in Brazil and for eutrophic and oligotrophic lakes in several other countries.

9.12 SECONDARY PRODUCTION

Secondary producers are all the heterotrophic animals – the herbivores, carnivores and predators – that make up a vast range of organisms in a community.

Secondary production is the organic matter produced by consumer organisms in a certain period of time per unit of volume or area.

The organisms **ingest** food, **assimilate** a certain proportion of it and use the assimilated organic matter in processes such as respiration, excretion and reproduction. The organic matter that passes to the next trophic level is the net result of these processes.

An organism's body size and temperature are fundamental factors that influence the metabolic rate.

These secondary producers are interrelated in the food chain (see Chapter 8) and energy is transferred from one level to another through this complex food chain, more appropriately called *food web*. Measuring secondary production is much more difficult than measuring the primary production of phototrophic organisms and is based on metabolic processes and on the CO_2/O_2 exchange. In the case of primary production, the production/biomass relationship is much more direct, and hourly, monthly or annual rates can be determined.

The life cycles of secondary producers vary widely and it is also difficult to determine the biomass (for example, comparing fish, molluscs, and crustaceans).

Despite these problems, measurements of secondary production were made based on differences in biomass or numbers of individuals, and differences of these measurements in different periods. According to Cole (1975), for example, production in a specific time period can be determined by:

$$P = (N_1 - N_2) \cdot \frac{P_1 + P_2}{2}$$

where N_1 and N_2 represent the number of animals per unit of area or volume at the beginning and end of a year, for example; and P_1 and P_2 represent the organisms' corresponding wet weight or dry weight. Included in this calculation are the numbers of individuals eliminated through consumption by other trophic levels. This technique can be used for organisms with very long life cycles, but it is difficult to apply to organisms with extremely short life cycles that are constantly producing new biomass, such as planktonic organisms.

Edmondson (1960) and Edmondson and Winberg (1971) used reproductive rates to devise methods to measure the secondary production of aquatic organisms. In laboratory studies on female rotifers, Edmondson looked at their production rate of eggs at different temperatures. The numbers of eggs produced per female per day were calculated at various different temperatures, making it possible to determine a coefficient of increase of the population, counting the organisms. The coefficient can be obtained from the formula:

$$\Gamma = \frac{\ln N_0 - \ln Nt}{t}$$

where N_0 is the initial number of organisms and Nt is the number of organisms in time t and r is the coefficient of increase of the population.

From these calculations, a mortality rate can be established:

$$d' = b' - r'$$

where:
d' – mortality rate
b' – instantaneous births rate
r' – coefficient of increase in the population
b' can be estimated as follows:

$$b' = \ln (B + 1)$$

where B is the number of eggs produced per female per day, given by $B = E/D$, where E is the number of females per sample of plankton and D is the length of time of the embryonic stage.

This technique can be applied to other planktonic animals with ovarian sacs. Elster (1954) and Eichhorn (1957) applied it to calanoid copepods.

Other methods for estimating and measuring secondary production have been employed: **caloric equivalent** of dry weight, respiration rates of organisms at different stages of development, efficiencies of **growth of organisms**. Those methods can first be applied in the laboratory and then in natural conditions in lakes, reservoirs and rivers. In this way the flow of energy was calculated by Comita (1972) for the planktonic community. When an ecologically important species is studied (such as a secondary producer that is the main consumer in a lake), the annual flow of energy for the entire ecosystem can be calculated.

The individual requirements of organisms must be considered, such as: in calories, the sum of energy used in growth, metabolic losses through excretion and respiration, and unused energy eliminated by faeces.

Efficiencies of growth of organisms and ratio of production to biomass (P/B) can be applied, as well as the ratio between production/total food consumed (P/C) or production divided by food assimilated (P/A).

Table 9.14 presents data on the secondary productivity of zooplanktonic species, calculated by various authors, using different measures, such as dry weight of organisms and total carbon in the organisms, calculating daily or annual rates for each zooplanktonic species or group of organisms.

In a recent study, Wisniewski and Rocha (2006) presented a range of data on groups of calanoid and cyclopoid zooplankton. The production of copepods was greater during the rainy season (23.6 µg dry weight $\cdot$ m^{-3} $\cdot$ day^{-1}) and less during the winter and the dry season (14 µg dry weight $\cdot$ m^{-3} $\cdot$ day^{-1}). The production of cyclopoids was greater than that of calanoids, as commonly observed in tropical lakes, according to these authors.

Santos Wisniewski and Matsumura Tundisi (2001) presented a zooplanktonic production of 36.6–28.9 µg dry weight $\cdot$ m^{-3} $\cdot$ day^{-1} for the Barra Bonita reservoir (SP), which is eutrophic. These values are a higher order of magnitude than the results presented by Melao and Rocha (2001) for Dourada reservoir (Brotas-SP). They are lower, however, than the calculations presented by Hanazato and Yasuno (1985) for Lake Kasumigaura, a hyper-eutrophic lake in Japan. Rocha *et al.* (1995) calculated 173.85 µg dry weight $\cdot$ m^{-3}, 26.57 µg dry weight $\cdot$ m^{-3} and 10.88 µg dry weight $\cdot$ m^{-3}, respectively, for Lake Amarela, Lake Dom Helvécio and Lake Carioca in Parque Florestal do Rio Doce (MG). Lake Amarela is a shallow eutrophic body of water, whose main zooplanktonic component is rotifers.

Pelaez and Matsumura Tundisi (2002) calculated the production of rotifers in the UHE Carlos Botelho reservoir (Lobo/Broa) using several different techniques, including the recruiting method, presented by Edmondson and Winberg (1971), based on birth rates and dry weight of organisms, where $P = Pn \cdot W$, in which: P = production or dry weight of organic material; Pn = recruitment of new individuals; and W = average

Table 9.14 Secondary production of zooplankton organisms in various aquatic ecosystems

Species	Production	Lake/reservoir	Researcher(S)
Thermocyclops hyalinus	44 µgDW $\cdot$ m^{-2} $\cdot$ day^{-1}	George (Africa)	Burgis (1974)
Rotifer + Copepod	190 µgDW $\cdot$ m^{-3} $\cdot$ day^{-1}	Nakuru (Africa)	Vareschi and Jacobs
Cladocera	49.9 µgDW $\cdot$ m^{-3} $\cdot$ day^{-1}[1]	Chad (Africa)	Lévêque and Saint-Jean (1983)
Copepoda+Cladocera	6-10 gDW $\cdot$ m^{-2} $\cdot$ year^{-1}	Lê Roux (Africa)	Hart (1987)
Zooplankton	8 a 15 gC $\cdot$ m^{-3} $\cdot$ year^{-1}	Reservoir (Africa)	Robarts et al. (1992)
Copepod	26.92 µgDW $\cdot$ L^{-1} $\cdot$ day^{-1}	Lanao (Philippines)	Lewis (1979)
Cladocera	6.32 µgDW $\cdot$ L^{-1} $\cdot$ day^{-1}	Lanao (Philippines)	Lewis (1979)
Rotifer	1.16 µgDW $\cdot$ L^{-1} $\cdot$ day^{-1}	Lanao (Philippines)	Lewis (1979)
Chaoborus	6.33 µgDW $\cdot$ L^{-1} $\cdot$ day^{-1}	Lanao (Philippines)	Lewis (1979)
Herbifore	24,40 µgDW $\cdot$ L^{-1} $\cdot$ day^{-1}	Lanao (Philippines)	Lewis (1979)
Carnivore	6.33 µgDW $\cdot$ L^{-1} $\cdot$ day^{-1}	Lanao (Philippines)	Lewis (1979)
Thermocylops oblongatus	11.0 µgDW $\cdot$ m^{-3} $\cdot$ day^{-1}	Naivasha (Kenya)	Mavuti (1994)
Diaphanosoma excisum	6.0 µgDW $\cdot$ m^{-3}	Naivasha (Kenya)	Mavuti (1994)
Argyrodiaptomus furcatus	6.74 µgC $\cdot$ m^{-3} $\cdot$ day^{-1}	Carlos Botelho (Lobo/Broa), Brazil	Rocha and Matsumura Tundisi (1984)
Rotifer	0.022 gDW $\cdot$ m^{-3} $\cdot$ day^{-1}	Monjolinho (São Carlos, Brazil)	Okana (1994)

DW – Dry weight; C – Carbon.
Source: Modified from several sources.

individual dry weight. The authors' conclusions show that, for the rotifer species *F. pegleri* and *K. americana*, at temperatures of 20.9°C, development of the eggs took approximately 19 hours, compared with the calculations of Okano (1994) for *Brachionus falcutus*, *K. longiseta* and *K. cochlearis* of 20 hours at 20.4°C. Results of studies on rotifers by Edmondson (1960) were 42–43 hours for eggs to develop at temperatures below 20°C (between 10–15°C).

Rocha and Matsumura Tundisi (1984) measured the biomass and production of *Argyrodiaptomus furcatus* in the UHE Carlos Botelho reservoir (Lobo/Broa). The authors calculated the relation between the length of individual bodies and organic carbon in µg C (see Figure 9.15). The incubation period of *A. furcatus* is shown in Table 9.15.

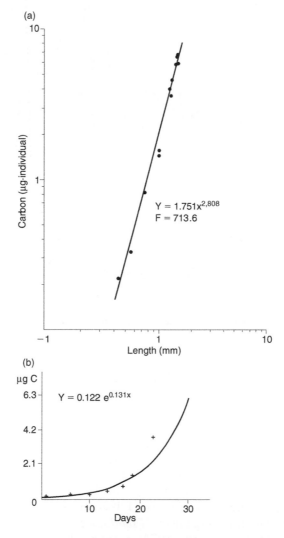

Figure 9.15   Relation between organic carbon and body length in *A. furcatus*. a) Length of body and carbon; b) Carbon in the body and development time.
Source: Rocha and Matsumura Tundisi (1984).

Table 9.15   Duration of various stages of A. furcatus.

Stage	Duration (Days)	Average duration
Egg	2	2
Nauplii (I–VI)	7–9	8
Copepodites (I–VI)	19–23	21
Egg to egg	27–35	31

Source: Rietzler et al. (2002).

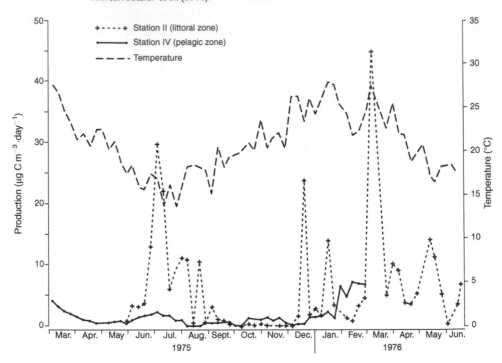

Figure 9.16   Annual cycle of secondary production of A. furcatus in UHE Carlos Botelho reservoir (Lobo/Broa).
Source: Rocha and Matsumura Tundisi (1984).

Based on calculating the relation between carbon and the length of individuals, sampling in the field and counting the various stages of organisms, the annual cycle of secondary production of the organism can be established (see Figure 9.16).

The maximum production for this species was $15–45 \, \mu g \, C \cdot m^{-3} \cdot day^{-1}$, in March, in a sampling site located near flooded areas (maximum depth – 2 m) in the ecosystem, which were extremely rich in organic matter and phytoplankton.

The maximum for the deepest sampling site, in a predominantly pelagic environment (maximum depth – 12 m) was $6.47 \, \mu g \, C \cdot m^{-3} \cdot day^{-1}$.

The daily P/B coefficient was 0.10. The P/B coefficients of 0.11 and 0.078 were obtained for other planktonic organisms (*Pseudodiaptomus lessei* (Hat et al., 1975) and *Thermocyclops hyalinus* (Burgis, 1974), respectively.

Melao and Rocha (2004) calculated that the production of copepods consisting exclusively of populations of cyclopoids ranged from 0.043 to 0.364 mg dry weight $\cdot$ m^{-3} $\cdot$ day^{-1} in the Dourada lagoon, a small reservoir located in Brotus (SP) near the UHE Carlos Botelho dam (Lobo/Broa).

Matsumura Tundisi and Rietzler (2004) calculated values of 42.2 mg dry weight $\cdot$ m^{-3} $\cdot$ day^{-1} and 53.55 mg dry weight $\cdot$ m^{-3} $\cdot$ day^{-1} in winter in the **Salto Grande reservoir,** a hyper-eutrophic ecosystem located in the state of São Paulo.

Eutrophic and hyper-eutrophic systems have higher levels of secondary production.

9.13 BACTERIA AND ENERGY FLOW

The **productivity of bacterioplankton** and bacteria that grow in sediments can be characterized by the production rate of bacteria (P $\cdot$ b) expressed in mg $\cdot$ m^{-3} $\cdot$ day^{-1} and by specific coefficients of daily production in bacterial populations (μ). Bacterial production of bacteria is, therefore, **P $\cdot$ b = B $\cdot$ μ,** where B is the **biomass of the bacteria** and μ is the coefficient of daily production. Estimating the productivity of bacteria helps to determine the rates of other processes in an aquatic ecosystem. Systems containing high organic loads have high bacterial productivity rates.

Bacterioplankton is an extremely important component of aquatic ecosystems, through which large quantities of energy pass. In regions with reduced zones and inorganic substances (such as methane or hydrogen sulphide gas), a large proportion of bacterioplankton may be chemosynthetic or photosynthetic organisms. Respiration by bacterioplankton is often significantly greater in plankton than that of other components such as protozoa or meso-zooplankton. The greater the amount of foreign material reaching lakes, rivers, and reservoirs, the greater the bacterioplanktonic activity. Organic matter resulting from decomposing planktonic or benthic organisms or fish (in addition to excretion from these organisms) contains a store of energy in detritus and dissolved organic matter that can be used only by bacterioplankton or bacteria that live on surfaces of sediments or substrata. Bacteria decompose and oxidize complex organic matter using about 15–35% for their own synthesis of organic matter. The stock of bacterial biomass is an important source of particulate carbon, with proteins and amino acids. Natural waters contain between 0.5 to 5×10^6 bacteria per ml (Sorokin, 1999). Some of these bacteria form aggregates around particles, such as decomposing phytoplanktonic cells, zooplanktonic faeces and the chitinous exoskeletons of zooplanktonic organisms. Sandes (1998) identified high levels of bacteria living in the mucus and cells of *Microcystis aeruginosa* after this cyanobacterium began decomposing. Fallowfield and Duft (1988) determined that bacteria play a role in utilizing the organic matter released by cyanobacteria. The biomass of bacterio-plankton in the pelagic zone plays an important role in the **tropho-dynamics** of pelagic communities and is used immediately by consumers such as copepods, filter feeders, molluscs and other organisms that filter particles 0.5–2 mm in diameter.

Sources of energy available for the growth of plankton or benthic bacteria in aquatic ecosystems include: organic material for heterotrophic bacteria; reduced inorganic compounds and gases (methane, hydrogen, ammonia and H$_2$S) for chemoautotrophic bacteria; and solar energy for anaerobic photosynthesizing bacteria. All groups of bacteria depend on phosphorus and nitrogen for their growth, and the availability of nitrogen and phosphorus is represented by the rate of flow of these elements in water and their concentrations. When there are high levels of reduced gas (CH_4, H_2S, H_2), the bacterioplankton and bacteria living in the sediment include chemosynthetic populations that use the energy produced by the anaerobic decomposition of organic matter, which can be as much as 10–20% of the total amount of energy entering the ecosystem. This efficiency, according to Sorokin (1999), can reach 15% in bacteria that use H_2S and 40–50% in bacteria that use CH_4 and H_2.

Sediment-dwelling bacteria play an important role in heterotrophic activity. They decompose organic matter, regenerate nutrients, and contribute to a stock of particulate material for benthic organisms and fish. Decomposition of benthic organisms, faecal material and particulate material results in the production of H_2S and rapidly consumes dissolved oxygen, inducing anoxia. A zone of anoxia is thus created in the sediment.

The composition of bacterial communities is very different in anoxia, oxic and suboxic conditions. In oxidized sediment, the following types can be found: *Pseudomonas*, *Corynebacterium*, *Arthrobacter* and Actinomycetes.

A contact layer on the sediment/water interface is formed by detritus, bacteria, filaments, cellulose-attacking bacteria and bacteria from the H_2S cycle, such as *Beggiatoa* and *Thiothrix*. Particles of detritus and sand are sites where Chlamydobacteria and Caulobacteria can develop.

The suboxic layer is composed of chemosynthetic bacteria that oxidize sulphur, methane, molecular hydrogen, manganese and iron. The density of these bacteria in the layers of sediment is extremely high, and the most common bacterial populations inhabit the first 2–3 cm of the sediment. The bacterial levels in this sediment can reach 5–$7 \cdot 10^9$ cells per cm^3 of sediment.

In this region there is also an extensive biomass of large numbers of ciliates. Populations of bacteria and ciliates are a common food source and play a significant role in the flow of energy and in supporting organisms that feed on detritus. In lakes and estuaries, the total number of bacterial cells in the oxidized sediment above the anoxic layer is 2–$3 \cdot 10^9 \cdot cm^{-3}$ (Sorokin, 1999).

In layers of sediment reached by underwater solar radiation, extensive layers of algae and bacteria form. Part of the bacterial population helps to oxidise compounds of sulphur, iron, manganese and methane.

Bacterioplanktonic and microbial communities in the sediment thus play a major role in the production of organic matter and in the trophodynamics of aquatic systems. Bacteria obtain the energy needed for reproduction and maintenance of metabolic processes through chemo-autotrophic processes, aerobic heterotrophic bacteria and photosynthesis under anaerobic conditions. The biochemical activity of bacteria is qualitatively and quantitatively important in the recycling of carbon, nitrogen, phosphorus, sulphur, manganese, iron and cobalt (Freire Nordi and Vieira, 1996).

The mechanisms of this biochemical activity and the way energy is obtained are determined by the specific characteristics of bacterial species, which are divided into physiological groups depending on the way they acquire energy for their growth and metabolism.

These physiological groups are identified depending on the substrate they use as source of energy, on the electron acceptors they use or other aspects, such as fixation of atmospheric nitrogen (see Charts 9.4 and 9.5).

Chart 9.4   Physiological groups of bacteria and their sources of energy (anoxic environment).

Energy sources	Electron acceptors	Physiological groups	Types of nutrition
Organic material	Various organic molecules and CO_2	Anaerobic heterotrophic bacteria	Chemo-organotrophic
Acetate and H_2 – gas	CO_2	Methane-producing bacteria	Chemo-lithotrophic
Organic material and H_2 – gas	SO_4^{2-}	Sulphate-producing bacteria	Chemo-organotrophic
	NO_3^{3-}	**Denitrifying bacteria**	
	CrO_4^{2-}	Chromate-reducing bacteria	
	Cl_4^-	Perchlorate-reducing bacteria	

Chart 9.5   Physiological groups of bacteria and their energy sources (oxic environment).

Energy sources	Specific type	Physiological groups	Types of nutrition
Organic material	Total stock of organic material in water	Heterotrophic bacteria	Chemo-organotrophic
	Proteins	Bacteria that fix nitrogen	
	Chitin	Chitinoclastic bacteria	
	Cellulose	Cellulolytic bacteria	
	Hydrocarbons	Bacteria that oxidize oils	
	Phenols	Bacteria that oxidize phenols	
Reduced inorganic substances	Hydrogen gas	Bacteria that oxidize hydrogen	Chemolithotrophic Chemosynthetic
	Methane	Bacteria that oxidize methane	
	$S_2O_3^-$ and H_2S^- (to SO_4)	Thiobacilli	
	H_2S (to SO_4)	Bacteria of the sulphur cycle	
	NH_4 (to NO_2)	Nitrifying bacteria	

Source: Sorokin (1999).

These physiological groups develop in the sediment and in the water column under anoxic conditions. One major group is the sulphate reducers, anaerobic heterotrophic bacteria, which use sulphur compounds as electron acceptors, reducing the compounds to H_2S. Table 9.16 details the population density, productivity and respiration rates of heterotrophic bacterioplankton.

9.13.1   Photosynthetic bacteria

In some environments photosynthetic bacteria play an important role in the production of organic matter. These bacteria can be subdivided into three groups, depending on the donors of electrons and hydrogen, using CO_2 during photo-assimilation.

▸ Photosynthetic bacteria that use water as the electron donor (cyanobacteria).
▸ Photosynthetic bacteria that use reduced sulphur compounds as electrons donors (green and purple bacteria).
▸ Anaerobic photosynthetic bacteria that use organic compounds as electron donors.

Table 9.16 Density of population, productivity and rates of respiration of bacterioplankton in aquatic systems with different trophic conditions.

Trophic state	Ecosystem	Density N_{Total}	Biomass (B)	Productivity P	μ(P/B) Gradient	μ(P/B) Average	Respiration RBP	Respiration TRM
Hypereutrophic	Polluted coastal lagoons	10–40	2,000–10,000	500–3,000	0.2–0.5	0.3	600–5,000	2,000–8,000
Eutrophic	Coastal lakes	5–10	2,000–3,000	500–1,500	0.2–0.5	0.3	600–1,700	1,500–3,000
	Ocean regions of temperate climate	2–5	400–2,000	200–500	0.4–0.8	0.5	250–600	300–1,000
	Lakes	3–8	600–2,000	300–800	0.3–0.7	0.5	380–900	500–1,500
	Organic layers with photosynthetic		2,000–8,000	20–150	0.01–0.03	0.02	–	–
Mesotrophic	Lakes, coastal lagoons	1.5–3.0	200–400	100–300	0.5–1.5	0.8	150–400	200–600
	Temperate-region oceanic areas	1.0–2.0	100–300	50–200	0.5–0.8	0.6	100–300	150–400
	Antarctic waters	1.0				0.1		
	Layers of photosynthetic bacteria	1.0–2.5	200–500	40–100	0.05–0.2		–	–
Oligotrophic	Lakes	0.5–0.8	40–70	15–50	0.4–1.5	0.8	20–60	30–80
	Tropical oceanic waters	0.1–0.4	10–30	10–40	0.8–1.5	1.2	15–50	20–60
	Antarctic waters in the Pacific	0.1–0.2	10–20	3–5	0.15–0.30	0.2	4–6	5–7
	Deep Antarctic waters	0.01–0.02	1–2	0.02–0.06	0.02–0.03	0.02	0.03–0.08	0.04–0.15

N_{Total} – number of bacteria (10^6 cells · mg^{-1}); B – biomass of bacterioplankton (mg · m^{-3} of total wet biomass); P – production of bacterioplankton (μg O_2 L^{-1} · day^{-1}); μ – coefficient of specific production P/B; TRM – total respiration of microplankton (μg O_2 L^{-1} · day^{-1}); RBP – respiration of bacterioplankton (μg O_2 · L^{-1} · day^{-1}).
Source: Sorokin (1999).

Anaerobic photosynthetic bacteria belong to the taxonomic group of Rhodospirillales, with two suborders: Rhodospirillinae and Chlorobiinae.

The first group includes the well-known families Chromatiaceae and Rhodospirillaceae. Purple bacteria oxidize H_2S and thiosulphate to molecules of sulphur, which are used as a source of protons for photosynthesis.

Green bacteria in the sulphur cycle, in the suborder, have bacterio-chlorophylls c, d and e in their cells. The most well-known and common genus is *Chlorobium*. Communities of photo-autotrophic bacteria in the sulphur cycle were studied extensively by Guerrero *et al.* (1986, 1987) and Takahashi and Ikushima (1970).

The level of H_2S is a major factor in the abundance of photo-autotrophic bacteria of the sulphur cycle in lakes. In meromictic lakes, a layer of H_2S is commonly found in the metalimnion, where light still penetrates. In these lakes there is often a layer of photosynthetic bacteria, usually of the genus *Chromatium*. In Lake Dom Helvécio (Parque Florestal do Rio Doce – MG), such a layer was observed in summer during a period of extensive thermal and chemical stratification (Tundisi and Saijo, 1997).

Photosynthetic bacteria support a very active food web in the levels where these bacteria are located. The main parameters controlling the presence of these bacteria, as well as their growth and vertical distribution in meromictic lakes or their intense stratification over long periods, are underwater solar radiation, the quality of radiation, and the level of H_2S (Pedros-Alio *et al.*, 1984).

The two main families of photo-autotrophic bacteria found in anoxic environments present two very different physiological strategies, according to Guerrero *et al.* (1987). The Chlorobiaceae family tolerates a wide range of solar radiation and H_2S levels. The Chromatiaceae family has more restricted tolerance limits to solar radiation and H_2S levels. The species in this bacterial family have several mechanisms and strategies to sustain themselves in rigid vertical positions: mucilage, gas vesicles, and flagella for mobility.

Primary productivity of green and purple bacteria is measured using the $^{14}CO_2$ technique. Productivity of these bacteria can reach $30–50 \, mg \, C \cdot m^{-3} \cdot day^{-1}$ in the stratified layers of meromictic lakes.

Chart 9.6 shows the most common Chromatiaceae species in different lakes, arranged according to their carotenoids.

9.14 EFFICIENCY OF FOOD WEBS AND TOTAL ORGANIC PRODUCTION

Energy flows through an ecosystem and its biological components and is used only once, whereas minerals and nutrients continually recycle between the ecosystem's biological and non-biological (abiotic) components. However, there is an ongoing interrelationship between the flow of energy and **nutrient cycling,** as discussed by Lindeman (1942). The biomass of organisms, meanwhile, results from the flow of energy, but it is not production. This latter involves concepts of **time and rates,** important components in production.

The second law of thermodynamics is important in the discussion on energy flow. According to this law, disorder or entropy increases as energy flows to higher trophic levels (e.g., to first- or second-level carnivores). According to the second law of

Chart 9.6 Dominant species of Chromatiaceae in different lakes, arranged according to their carotenoids.

	Species	Lake	Depth (m)	Source
	Chromatium minus	Vilar	4.2–6	Guerrero et al. (1980)
		Estanya	12–14	Guerrero et al. (1987)
		Cisó	1–2	Guerrero et al. (1980)
		Suigetsu	7	Jimbo (1938a)
		Nou	4	Guerrero et al. (1987)
		Banyoles III	15–16	Guerrero et al. (1987)
Okenona	Chromatium okenii	Fango	–	Bavendamm (1924a)
		Estanya	13–14	Guerrero et al. (1987)
		Cadagno	11–13	K. Hanselmann (pers. com.)
		Vechten	–	Parma (1978)
		Lunzer Mittelsee	–	Ruttner (1962b)
		Ritomsee	12.6	Duggeli (1924b)
		Belovod	–	Kusnetsov (1970b)
	Chromatium sp.	Vechten	6–8	Seenbergen and Korthals (1982)
	Lamprocystis M3	Cisó	1–2	Guerrero et al. (1987)
		Vilar	4.2–6	Guerrero et al. (1987)
	Thiopedia rosea	**Muliczne**	9–13	Czeczuga (1968a)
		Krummensee	8	Utermohl (1925)
		Plussee	5	Anagnostidis and Overbeck (1996a)
		Lago di Sangue	–	Forti (1932b)
		Wintergreen	2.6–3.5	Caldwell and Tiedje (1975)
		Negre I	2	This work
	Thiopedia sp.	Vechten	8-Jun	Steenbergen and Korthals (1982)
		Kaiike	4.8	Matsuyama (1987)
Spirilloxanthina	Thiocapsa sp.	Prévost Lagoon	0.5	Caumette (1986)
	Amoebobacter roseus	Konon'er	10.75	Gorlenko et al. (1983)
	Chromatium	Shigetsu	7	Jimbo (1938a)
	minutissimum	Repnoe	5.5	Gorlenko et al. (1983)
	Chromatium vinosum	Prévost Laggon	0.5	Caumette (1986)
	Thiocapsa sp.	**Repnoe**	5.5	Gorgolenko et al. (1983)
		Konon'er	10.75	Gorgolenko et al. (1983)
		Mara-Gel	17–19	Gorgolenko et al. (1983)
		'Fellmongery' Lagoon	–	Cooper et al. (1975b)
		Deadmoose	9–9.2	Parker et al. (1983)
Lycopenal, Lycopenol	Lamprocystis roseopersicina	Mirror	10–11	Parkin and Brock (1980a)
		Solar	2	Cohen et al. (1977)
		Medicine	3.2–3.7	Hayden (1972)
		Transjoen	–	Faafeng (1976)
		Gullerudtjern	–	Faafeng (1976)
		Plussee	5	Anagnostidis and Overbeck (1966a)
Rhodopinal	Chromatium violascens	Solar	2	Cohen et al. (1977)
		Faro	12.5–13	Truper and Genovese (1968a)
	Thiocystis violacea	Negre I	1–2	Guerrero et al. (1987)

Source: Guerrero et al. (1987).

thermodynamics, as the transfer of energy occurs, some of the energy is dissipated, and thus losses occur as energy moves up through the different levels.

The conversion of energy from one trophic level to another is very inefficient, and at each level organisms can lose much energy in the form of heat and through their metabolic processes. The concept of efficiency presented by Lindeman implies that the degree of utilization of energy at a particular level (λn for example) depends on the resources available in the previous level (λn – 1). A comparison of production and respiration rates at different trophic levels is essential for determining the efficiency of the process. Losses at each trophic level can be as high as 10%, and thus, at the end of the food chain, at the highest levels, barely a fraction remains of the energy fixed by the primary producers. In systems with imbalances, such as regions of upwelling or during periods of new production in lakes, reservoirs or rivers, the production of biomass increases at the higher trophic levels, but this is also accompanied by an increase in the loss of energy at these trophic levels.

Ryther (1969) compared the number of trophic levels in various oceanic communities along the continental shelf and in regions of upwelling. Even considering the scale involved in comparing the trophic levels of lakes with different depths, this work clearly illustrates the relationship between efficiency and organization of food webs. Food webs dominated by the phytoplankton/herbivore relationship have an efficiency of about 10% (Ryther, 1969). More complex food webs have an efficiency of about 15%, and food webs in areas of upwelling or eutrophic lakes with 'new production'

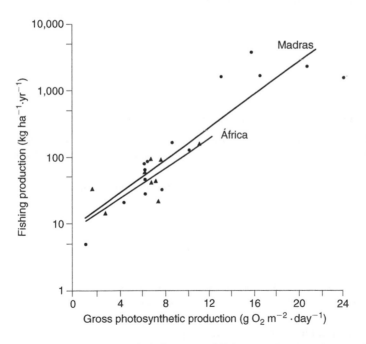

Figure 9.17   Relationship between annual production of fish (dry weight, logarithmic scale) and gross photosynthetic production by phytoplankton in lakes and reservoirs of Africa (insert symbol) and of India (insert symbol), close to Madras.
Source: Melack (1976).

Table 9.17 Average annual primary production and efficiency of trophic levels compared with the production of fisheries in different marine ecosystems.

Ecosystem	Average annual primary production $(g\ C \cdot m^{-2} \cdot year^{-1})$	Trophic levels	Efficiency (%)	Fish production $(mg\ C \cdot m^{-2} \cdot year^{-1})$
Oceanic	50	5	10	0.5
Continental shelf	100	3	15	340
Upwelling	300	1.5	20	36,000

Source: Ryther (1969).

Units used in studies on energy flow

Energy = Joule (J); cal (cal = 4.18 J)

Flow of energy = $J\,m^{-2}s^{-1} = W\,m^2 = kerg\,cm^{-2}s^{-1}$

$J\,m^{-2}s^{-1} = (60/4.18)\ 10^4\ cal \cdot cm^{-2} \cdot min^{-1}$

Flow of photons $1\ J\,m^{-2} \cdot s^{-1} \cong 5\,\mu mol \cdot m^{-2} \cdot s^{-1}$ (photosynthetically active radiation)

Biomass index = g dry weight of organic material

$1\,g \approx 5\,g$ wet weight

$1\,g \approx 0.45\,g\ C$

$1\,g \approx 20\,mg$ chlorophyll-*a* (phytoplankton)

$1\,g \approx 5\,cm^3\ (= 5 \cdot 10^{12}\,\mu m^3$ of the cellular volume of phytoplankton)

$1\,g \approx 4.5\,Kcal$ or $19\,kJ$

Units of primary and secondary production

$mg\ C \cdot m^{-3} \cdot h^{-1}$ = rate of photosynthesis per hour and unit of volume of water

$mg\ C \cdot m^{-3} \cdot h^{-1}$ = maximum rate of photosynthesis per hour and unit volume of water = P_{max}

$g\ C \cdot m^{-3} \cdot h^{-1}$ = rate of photosynthesis per day and unit of area

$g\ C \cdot m^{-3} \cdot h^{-1}$ or hourly rate of photosynthesis per unit of area

mg = biomass or biomass index

ΔB = increase in the biomass (mg)

ΣB = biomass per unit of surface area in the euphotic zone = $mg \cdot m^{-2}$

K = vertical coefficient of attenuation of solar incident radiation in water = m^{-1}

I_O = density of the flow of incident radiation on the water surface = $J\,m^{-2} \cdot s^{-1}$

I_K = density of the flow of radiation near the saturation of photosynthesis = $J \cdot m^{-2} \cdot s^{-1}$

P = production of biomass

PS = dry weight: mg dry weight $\cdot m^{-2}$

K_S = coefficient of specific attenuation per unit of concentration of biomass based on the increase of the spectral coefficient $k_{min} = m^{-1}(mg \cdot m^{-3})^{-1} = mg^{-1} \cdot m^2$

have an efficiency of 20%. Therefore, systems with more trophic levels yield a much smaller final production of organic matter usable by man.

9.15   FISHERY PRODUCTION AND ITS CORRELATION WITH PRIMARY PRODUCTION

According to Talling and Lemoalle (1998), several attempts were made to correlate the daily photosynthetic primary production per unit area with the commercial production of fish. Melack (1976) compared data from lakes in Africa and reservoirs in India (see Figure 9.17). The results showed a positive correlation, indicating that the logarithm of fish production increases linearly with gross photosynthetic production. Prowse (1964) showed that the production of tilapia in fish tanks reached 1–1.8% of net photosynthetic production. Other sources of photosynthetic primary production, such as aquatic macrophytes and periphyton, contributed directly to fishing production, or indirectly through the production of detritus that sustains invertebrates or detritivores. Table 9.17 shows the average annual primary production and efficiency of trophic levels compared with fish production in different marine ecosystems.

Biogeochemical cycles are involved in the interactions of terrestrial and aquatic organisms. Biogeochemical of terrestrial and aquatic organisms. interactions macrophytes in the Corrego do Geraldo. Aquatic Carlos Botelho reservoir (Lobo/Broa). UHE Carlos Botelho reservoir (Lobo/Broa).
Photo: J. G. Tundisi.

10 | Biogeochemical cycles

SUMMARY

The cycles of chemical elements and substances are interrelated with biological, geochemical and physical processes. The distribution and concentration of elements and substances in water depend on 'fixation' and the active concentration of macronutrients (carbon, hydrogen, nitrogen, phosphorus and sulphur) and micronutrients (magnesium, iron, copper and zinc). Macronutrients and micronutrients alike are found in living organic matter, particulate matter, decomposing matter, or dissolved in water. The nutrient recycling rate depends on the interactions between vertical and horizontal mixing and the activity and biomass of aquatic organisms.

The vertical distribution of nutrients is affected by the vertical circulation in lakes or reservoirs and depends on the type of circulation and its frequency. Bacteria with varying physiological and biochemical characteristics play a key role in biogeochemical cycles. Bottom sediment in rivers, lakes, reservoirs, estuaries and interstitial waters provide important quantitative and qualitative nutrient reserves. The availability of nutrients in sediment and interstitial waters depends on oxido-reductive processes and the anoxic or oxic layers of the sediment.

10.1   THE DYNAMICS OF BIOGEOCHEMICAL CYCLES

As shown in Chapter 5, the composition of natural waters largely depends on the geochemistry of the water basin and the main characteristics and processes in the basin: **soil types,** uses and agricultural practices. In inland waters, the distribution of nutrients is also influenced by **regenerative processes** in the lake's deepest layers and the sediment-water interface.

The principal nutrients are those that are important for all plants; these include carbon, nitrogen, and phosphorus. In combination with hydrogen and oxygen in various different configurations, these elements form the basis of metabolic processes and cell structure. Sulphur and silicon can also be added to this list, as silicon is found in the frustules of diatoms, and sulphur is an essential component in many proteins. These elements are called macronutrients since relatively high levels are needed for growth. Micro-nutrients, which are needed only in relatively low concentrations, include manganese, zinc, iron and copper. Their absence can restrict growth, but at higher concentrations these metals can be toxic.

The chemical composition of inland waters, when comparing data from many lakes, shows significant correlations between total dissolved solids (TDS) and HCO_3^-, and between Ca^{2+} and TDS (see Chapter 5).

Aquatic plants actively concentrate carbon, hydrogen, nitrogen, phosphorus and sulphur along with other micronutrients. The fixation of these elements occurs physio-logically. In general, average C:N:P levels by atoms occur stoichiometrically as 106C:16N:1P. This ratio is called the **Redfield Ratio** (Redfield, 1958). The cycles of these elements in inland waters are thus related to the biological processes in aquatic systems, and stoichiometric ratios in part reflect the way in which nutrients are found in water.

In general, **macronutrients** and trace elements are found in living organic matter, in dead and decomposing particulate matter, or dissolved in water. The amounts found in each of these compartments in the aquatic system are important and essential for understanding biogeochemical cycles. Therefore, the distribution and concentration of a specific nutritional element in a mass of water is a function of biological, geochemical and physical processes.

The nutrient recycling rate depends on the interactions between horizontal and vertical mixings (which determine temporal and spatial distributions), and also on the activity and biomass of the organisms present. Significant variables in these processes include the retention time of the water mass, the rates of transfer of elements between masses of water, and the recycling rates of the elements among the various compartments.A common distinction used in nutrient cycles is between a conservative vs. non-conservative substance. A **conservative substance** has a much longer retention time than the total mixing time of the lake and thus a more homogeneous distribution. The retention time of a non-conservative substance is much shorter than the total mixing time of the water mass and thus presents heterogeneous distributionsthrough space and time.

The fixation of nutrients by aquatic plants always occurs through soluble and diffuse sources in such a way that nutrients must pass through a semi-permeable membrane into the cell (Reynolds, 1984). The nutrient levels, however, are always much lower in the liquid medium than the levels occurring in cells. Passive diffusion is rare and a transport system (**ion pump**) is needed, creating by enzymes located close to the cell surface (cell wall).

10.2   CARBON CYCLE

Some of the factors involved in the availability of carbon in natural waters were already discussed in Chapter 5. The balance of carbonate (CO_3^{2-}), bicarbonate (HCO_3^-) and CO_2 determines the **acidity** or **alkalinity** in natural water. Carbon is an element consumed in large amounts by photosynthetic organisms and is thus one of the essential elements in the biogeochemical cycle in natural waters. The photosynthetic activity of aquatic plants removes carbon from surface waters. Rapid sedimentation occurs after the death of organisms, such that sedimentation of organic matter plays an important role in the **carbon cycle**. Due to the equilibrium system between dissolved CO_2 (and its diverse chemical forms) in the water and atmospheric CO_2, carbon is always available to **photosynthetic organisms**, so the main theories concerning the limitations of photosynthetic growth processes do not include carbon as a limiting factor (Goldman *et al.*, 1972). However, in conditions of high pH due to photosynthesis, carbon may limit the process in surface waters, with a shift in equilibrium toward **bicarbonate** and **carbonate**. In waters with high **nitrogen** and **phosphorus** levels, the ability to fix bicarbonate or carbonate gives an additional competitive advantage to some phytoplanktonic or aquatic plant species (Harris, 1978). Some phytoplanktonic species can maintain a high intracellular CO_2 level because of a '**bicarbonate pump**' in the cellular membrane, which, along with the activity of a carbonic unit, physiologically operates this system. The fixation mechanisms of carbon and its transport in aquatic plants continually interact with the chemical compounds of inorganic carbon dissolved in water, which has enormous implications for methods (experiments to measure photosynthesis), physiology (respiration, consumption of energy, photosynthesis, excretion) and ecology (control of chemical conditions by aquatic plants, interactions with the seasonal cycle, and the processes of decomposition and recycling).

The biochemical differences among different species greatly affect the interrelations between the seasonal cycle of aquatic plants and their succession in time and space, as was demonstrated for several phytoplanktonic species (Harris, 1980). In general, the flow of carbon into cells occurs photo-autotrophically, although in some cases, 'fixing' carbon in the dark can become important. Figure 10.1 illustrates the carbon cycle in natural waters.

10.3   THE PHOSPHORUS CYCLE

Phosphorus, as a component of nucleic acids and adenosine triphosphate, is an essential element in the internal processes and growth of aquatic plants. The **flow of phosphorus** into inland waters depends on the geochemical processes in a water basin. The most common forms of organic phosphorus are generally of biological origin. Dissolved phosphates are derived from the **leaching** of minerals, such as the apatite present in rocks. Phosphorus can also be found in different-sized particles, even in colloidal form. Sedimentation of particles and excretion from planktonic or benthic organisms contribute to the accumulation of sediment, forming an important reservoir of phosphorus. It accumulates also in the **interstitial water**, and depends, in great part, on the processes of circulation and oxido-reduction at the **sediment-water interface**.

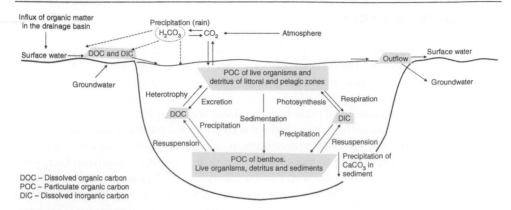

Figure 10.1 Generalized and simplified carbon cycle for a lake. In reservoirs with many compartments and shallow areas, the cycle can be more complex.

Hutchinson (1957) distinguishes the differences between **dissolved phosphate, sestonic phosphate** dissolved in **acid** (mainly ferric phosphate or calcium phosphate), **dissolved organic phosphorus** (colloidal) and **sestonic organic phosphorus**.

Of these various forms, **dissolved orthophosphate** is clearly the main source of phosphorus for aquatic plants, especially for phytoplankton. The use by phytoplankton of dissolved organic phosphate through the production of alkaline phosphatase has also been reported by Nalewajko and Lean (1980) and Reynolds (1984).

Undisturbed terrestrial systems retain **phosphorus**, while deforested **water basins** generally lose phosphorus. Moss (1980) recorded the following typical levels of phosphorus in inland waters:

▶ 1 µg P/litre – natural pristine lakes in high-altitude regions;
▶ 10 µg P/litre – natural lakes in forested lowlands;
▶ 20 µg P/litre –lakes in early stages of eutrophication in agricultural or deforested regions;
▶ 100 µg P/litre – eutrophic lakes in highly populated urban areas with sewer discharge;
▶ 1000 µg P/litre to 10,000 µg P/litre – stabilization ponds, highly fertilized fish-production ponds; endorheic lakes.

An important component of the **phosphorus cycle** in inland aquatic systems is in the sediments. A portion of the phosphorus undergoes a complexation process during periods of intense oxygenation in the sediment, thus periodically becoming unavailable. The cycles of phosphorus, iron, and the **oxido-reduction potential** in water and sediment are thus closely correlated.

Since phosphorus does not have a common gaseous phase, its availability depends on phosphated rocks and on the internal cycle of lakes, in which decomposition and excretion play an important role. Phosphorus plays a key role in regulatory and recycling processes in lakes. In stratified lakes, an important part of the phosphorus cycle can also occur in the metalimnion, where the regenerative process occurs through the

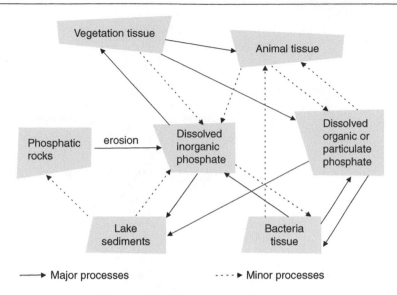

Figure 10.2 The phosphate cycle in aquatic ecosystems.
Source: Modified from Welch (1980).

reduction that occurs in metalimnetic layers with low oxygen levels (Gliwicz, 1979). Figure 10.2 illustrates the phosphorus cycle in aquatic ecosystems.

10.4 THE NITROGEN CYCLE

Aquatic plants use nitrogen mainly to synthesize **proteins** and **amino acids**. The primary sources of nitrogen are nitrate, nitrite, ammonium, and dissolved nitrogen compounds, such as urea, and free amino acids and peptides. **Atmospheric nitrogen** dissolved in the water can be 'fixed' by some species of cyanobacteria.

Inorganic nitrate is highly soluble and abundant in water with high nitrogen levels, often due to the discharge of household sewage or agricultural activity. High nitrate levels are found, for example, in the Bonita Barra reservoir in the state of São Paulo (1–2 mg $N-NO_3^- \cdot L^{-1}$), due to the discharge of domestic sewage and the drainage of fertilized agricultural soils (Tundisi and Matsumura Tundisi, 1990).

Tropical lakes and reservoirs generally present low nitrate levels, the result of the drainage from forests or savannas with nitrogen-poor soils. Stratified lakes may present low nitrate levels in the epilimnion (Tundisi, 1983).

In natural conditions, the ammonium level is also relatively low in epilimnion waters ($<100\,\mu g\ N-NH_4 \cdot L^{-1}$). In stratified lakes, ammonium levels can be higher, principally in anoxic conditions, where nitrate is reduced to ammonium (1–2 mg $N-NH_4 \cdot L^{-1}$). In eutrophic lakes, in the metalimnion and hypolimnion, ammonium levels can oscillate due to excretion and decomposition by organisms. In the epilimnion of tropical lakes, nitrogen may be regenerated through the excretion of ammonium by zooplankton or decomposition of organic matter by bacteria (Tundisi, 1983; McCarthy, 1980).

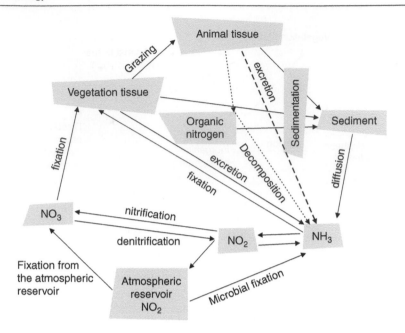

Figure 10.3   Simplified nitrogen cycle in inland aquatic ecosystems.
Source: Modified from Welch (1980).

The nitrite level is consistently low ($<60\,\mu g$ N-N$_3 \cdot$L^{-1}) since this ion can be reduced chemically or by bacterial activity that reduces nitrate or oxidizes ammonium. In tropical waters, nitrate levels are generally extremely low, frequently below the limit of detection. Nitrite occasionally accumulates in bubbles where the oxygen tension is below $1\,mg$ O$_2 \cdot$L^{-1} and in conditions of deep stratification.

Figure 10.3 shows the nitrogen cycle in lowland aquatic ecosystems.

The nitrogen cycle is also fairly complex because of the existence of a large reserve of nitrogen in the atmosphere (~80%). The transfer processes between the different compartments are extremely important for aquatic productivity. The transfer of atmospheric N$_2$ by microbiological fixation (e.g., cyanobacteria) and its return to the atmosphere via N$_2$O and denitrification are biologically and chemically important features unique to the nitrogen cycle (they do not occur in the phosphorus cycle). Micro-organisms accelerate the reaction and also store available energy from reduced compounds due to a series of chain reactions triggered and catalysed by enzymes. Since the energy sources are inorganic, the organisms are called **chemo-lithotrophic** (Welch, 1980).

The main processes in the nitrogen cycle are nitrification, denitrification, and biological fixation. Nitrification is the process through which NH$_3$ is transformed into NO$_2^-$ and NO$_3^-$. The process occurs under aerobic conditions, as a result of the activity of organisms such as *Nitrosomonas* and *Nitrobacter* bacteria.

$$2NH_3 + O_2 \rightarrow 2HNO_2 + 2H_2O + \text{energy}$$

$$2HNO_2 + O_2 \rightarrow 2HNO_3 + \text{energy}$$

Denitrification occurs mainly in the absence of oxygen or in conditions approaching anaerobiosis. *Thiobacillus denitrificans* is a denitrifying organism.

$$5S + 6NO_3 + 2H_2O \rightarrow 5SO_4 + 3N_2 + 4H + \text{energy}$$

Heterotrophic bacteria, such as *Pseudomonas* spp., are **facultative anaerobe** and can be found in sewage and waste-waters.

The **denitrification** process is the reverse of the nitrification process, i.e., bacteria reduce NO_2 and NO_2 to nitrogen gas N_2, which returns to the atmosphere, thus providing a mechanism to reduce nitrogen in waste-water or in excessively eutrophic waters. It is a basic process that commonly occurs in flooded and marshy waters.

Since nitrification needs an aerobic system and denitrification occurs in an anaerobic system, the aerobic/anaerobic shift is an effective process for the transfer of nitrogen to the atmosphere, and is a mechanism for treating waste-water in which an aerobic period precedes an anaerobic period. As a system becomes eutrophized, oxygen levels drop, making **denitrification** possible.

Another important process, the biological fixation of nitrogen, occurs in aquatic systems through bacterial activity (*Azotobacter* and *Clostridium*) and by the cyanophytes *Nostoc, Anabaena, Anabaenopsis, Aphanizomenon* and *Gloeotrichia*. Fixation of atmospheric N_2 can be high, as is the case in Clear Lake (California), in which 43% of the total entrance of nitrogen in the lake occurs through biological fixation (Horne and Goldman, 1994). The cyanophytes occur in cells that fix N_2 (heterocysts or heterocytes), whose number increase with the decrease of NO_3^- Reynolds (1972) showed that in N levels below $300\,\mu g \cdot L^{-1}$, the ratio of heterocysts to vegetative cells increases in an *Anabaena sp.*

Nitrogen-fixing cyanophytes are common in shallow inland waters and oceans, and practically absent from estuaries (McCarthy, 1980). They are important components of the nitrogen cycle in inland waters.

Thus the main processes involved in the nitrogen cycle in water are:

▶ fixation of nitrogen (N) – N_2 (gas) and chemical energy, transformed into ammonium (NH_4^-);
▶ nitrification – reduced forms, such as ammonium, are transformed into nitrite or nitrate$^-$;
▶ denitrification – nitrate, through reduction, is transformed into N_2 (gas);
▶ assimilation – dissolved inorganic nitrogen (ammonium, nitrate or nitrite) is incorporated into organic compounds;
▶ excretion – animals excrete nitrogen in the form of ammonium, urea or uric acid.

10.5 THE SILICA CYCLE

Silica is found in natural waters in the form of silicic acid and colloidal polymers of silicate, originating from soil or organisms such as diatoms, whose frustules do

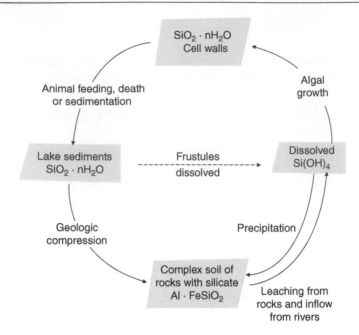

Figure 10.4   Silica cycle.

not present amorphous silica polymers. From this perspective, silica plays a significant ecological role, and its seasonal cycle is related to the growth of diatoms and
their subsequent dissolution (Reynolds, 1973a, 1984). Thus, the availability of silica
is related to the cycle of growth, decomposition and dissolution of diatomic frustules found in the sediment at the bottom of lakes and reservoirs. Lund (1950, 1964)
showed the relationship between the cycle of an *Asterionella* sp. and the silica cycle in
the Lake District in England.

Many studies have described this cycle of soluble reactive silicon and the seasonal
cycle of diatoms in inland water systems. The de-polymerization of silica forms $SiOH_4$
(silicic acid) that as 'soluble reactive silicon' is measured spectrophotographically by
the reaction described by Mullin and Riley (1955).

Levels of soluble reactive silica in natural waters range from a maximum of 200–
300 mg $SiO_2 \cdot L^{-1}$ to less than 1.2–10 mg $SiO_2 \cdot L^{-1}$. Silica levels can be important in the
development of *Aulacoseira* sp. For example, Kilham and Kilham (1971) showed that
Aulacoseira italica develops when silica levels are below 5 mg·L^{-1}, while *Aulacoseira
granulata* also grows at levels greater than 5 mg·L^{-1}. Figure 10.4 illustrates the
silica cycle.

10.6   OTHER NUTRIENTS

Various other elements are important for the growth, productivity and physiology of
aquatic plants. Among these are calcium, which occupies an important position chemically in the complex system of $pH - CO_2 - CO_3^{2-}$ in inland waters, and cyanobacteria

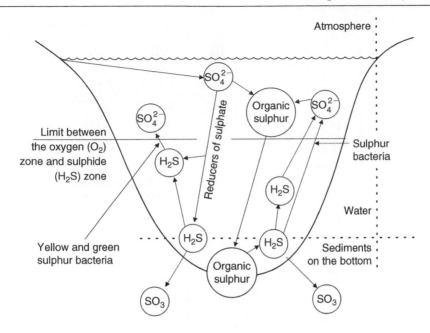

Figure 10.5   Sulphur cycle.
Source: Modified from Schwoerbel (1987).

in inland waters, which appear to have a great affinity with hard (higher alkalinity) water (Reynolds, 1984). Magnesium, found at the core of the chlorophyll molecule, is also important, but there is little evidence that it is a limiting factor in the growth of aquatic plants. Levels of sodium and potassium are higher than those normally needed by aquatic plants. The ratio of monovalent cations to bivalent cations appears to play a role in the phytoplanktonic cycle and succession of species. Major anions (such as chloride and sulphate) rarely are limiting factors due to their high levels in natural waters. The sulphide ion (as S^{2-} and HS^-) is important in an **anoxic hypolimnion** because it can be used by cyanobacteria as an electron donor or as a source of assimilable sulphur (Oréon and Pandan, 1978). Hino *et al.* (1986) found high levels of cyanobacteria in Lake Dom Helvecio (Parque Florestal do Rio Doce – MG) along with an anoxic hypolimnion and high H_2S levels (between 5–7 mg·L^{-1}).

In general, the patches of photosynthetic bacteria found in lakes are associated with the presence of H_2S in the metalimnion. These red bacteria are important in lakes, where they grow in the metalimnion with increasing levels of H_2S and light intensity, which although low, is utilized (between 0.5%–1% of the amount of light on the water's surface). Figure 10.5 illustrates the sulphur cycle.

Other elements that are important for the growth, productivity and physiology of aquatic plants are iron, manganese, molybdenum, copper and zinc, which at high levels can be toxic, and at low levels may be limiting factors in the processes of growth, as already discussed in Chapter 5.

The cycles of most of these metals are related to dissolved organic compounds in the water known as '**chelates**'; two well-known groups are humic and fulvic acids.

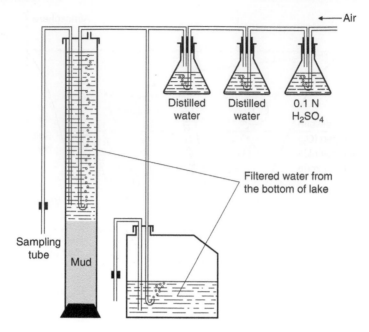

Figure 10.6   Experimental system for measuring the release of phosphorus or nitrogen from the sediment.
Source: Fukuhara *et al.* (1997).

These natural chelates release small quantities of chelated metal, which thus become available, and in the case of excess, function as a plug, removing quantities that could be toxic. Excretion of organic compounds by phytoplankton and aquatic plants also contributes to the reserve of chelates in natural waters.

Iron exists in particulate form and dissolved form and can be reduced (Fe^{2+}) or oxidized (Fe^{3+}). In aquatic systems, the phosphorus and iron cycles directly interact, due to the precipitation of ferrous phosphate during periods of oxygenation in the water column and re-dissolution of ferrous phosphate during periods of reduction. This cycle occurs in stratified lakes during summer, accumulating high levels of Fe^{2+} and PO_4-P in the hypolimnion (soluble), while during periods of intense circulation and re-oxidation, precipitation of $FePO_4$ occurs in sediment (insoluble). The iron and phosphorus cycles and the oxido-reduction (redox) potential in water are thus also interrelated with the circulation process and vertical distribution of oxygen. There is evidence that the dissolution of Fe^{3+} by reduction begins in the metalimnion of stratified lakes.

Experiments showing that the release of inorganic phosphorus or nitrate in anoxic or aerobic conditions can be performed with the equipment shown in Figure 10.6. Figure 10.7 shows the release of nitrate and ammonium (Fukuhara *et al.*, 1985). An experimental anoxia can be induced and inorganic phosphate or ammonium released from the sediment.

Chart 10.1 shows the interrelations between levels of dissolved O_2, redox potential and concentrations of iron, phosphate and sulphide gas.

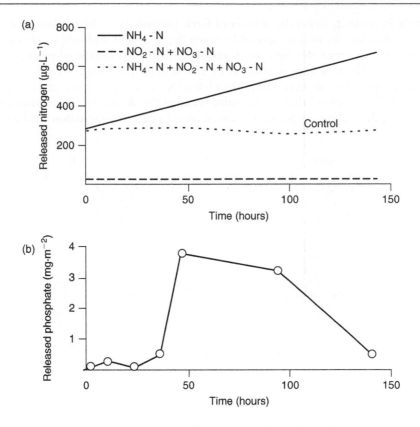

Figure 10.7   Release of inorganic nitrogen (NO_3), ammonia (NH_4), and phosphate in experiments conducted with the equipment shown in Figure 10.6.

Chart 10.1   Interrelations of O_2 levels, redox potential E_H and levels of iron, phosphate and hydrogen sulphide gas in stratified oligotrophic, mesotrophic and eutrophic lakes.

Trophic status	O_2 level	E_H	Fe^{2+} level	H_2S	PO_4-P
Oligotrophic	High	400–500 mV	Absent	Absent	Low
Eutrophic	Reduced	250 mV	High	Absent	High
Hypereutrophic	Reduced or absent	100 mV	Decreasing	High	Very high

Source: Modified from Wetzel (1975).

10.7   THE SEDIMENT-WATER INTERFACE AND INTERSTITIAL WATER

Sediment and the water-sediment interface play an important role in biogeochemical cycles. Depending on the oxido-reduction conditions of the water-sediment interface, precipitation and re-dissolution occur. For example, the oxido-reduction potential in the sediment-water interface determines the exchange rate of phosphate between the hypolimnion and the sediment. The influence of the oxidized or reduced layer is extremely important. Generally, a layer of ferric phosphate (oxidized layer) can form, which acts as a barrier to interactions between the sediment and subjacent water. The transport of phosphate through this layer depends on the degree of perturbation in the sediment, which is also developed by the activity of organisms (**bioturbation**). This **bioturbation** in part determines the characteristics of flux through the interface. Whitaker (1988) showed that rapid shifts occur in the degree of **oxidation** or **reduction** of iron, depending on the condition of anoxia or re-oxygenation in the system.

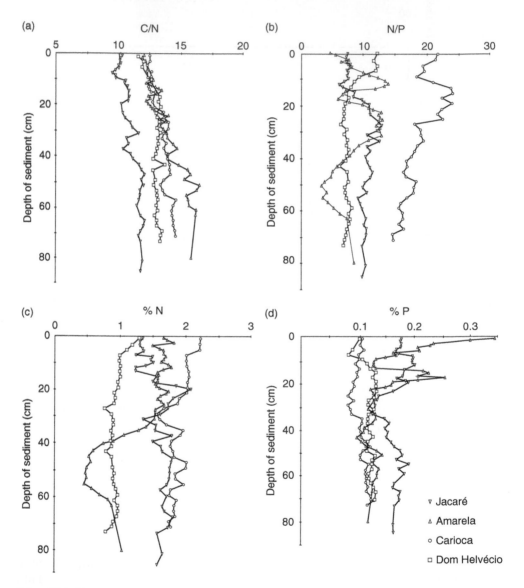

Figure 10.8   Vertical distribution of (a) the carbon/nitrogen ratio, (b) nitrogen/phosphorus ratio, (c) nitrogen, and (d) phosphorus in bottom sediments of four lakes of the Parque Florestal do Rio Doce (MG).
Source: Saijo *et al.* (1997) in Tundisi and Saijo (1997).

These shifts can occur in a few hours, and also partially depend on the diurnal stratification and destratification cycle and associated oxygenation.

Börstrom *et al.* (1982) described the following aspects of phosphorus transport mechanisms in the sediment-water interface:

▸ At the molecular level, the mobilization of phosphorus can be related to physical-chemical processes such as **dissolution**, and to biochemical processes resulting from the enzymatic decomposition of organic substances;
▸ At the compartment level, the transfer of phosphorus from the sediment to the hypolimnion is characterized by hydrodynamic mechanisms, mainly diffusion, ebullition, bioturbation and wind-induced turbulence.

Interstitial water can be an important reserve of nutrients. This interstitial water, which can be separated from the sediment by centrifugation, presents high levels of ammonium, phosphate and nitrate. Figure 10.8 shows the vertical distribution of nutrients in the interstitial water in the sediment at the bottom of Carioca lagoon (Parque Florestal do Rio Doce – MG).

Sediment and hypolimnion are interrelated with the oxidation-reduction system at the interface layer, the formation of a ferric phosphate barrier and the perturbation of this barrier by various processes, including biological processes.

The sediment thus provides a place for nutrients to concentrate; in particular, the phosphorus cycle is closely related to **sediment-water interactions**, circulation processes, stratification and destratification, and changes in the redox potential. Phosphorus can be released from decomposing particles in sedimentation. According to Golterman (1972), 80% of phosphorus can be regenerated during the sedimentation process. The availability of phosphorus in the **trophogenic zone** also depends on the regeneration and vertical distribution of oxygen. In the study of nutrient cycles in aquatic systems, it is important to determine a mass-balance that characterizes the concentration of each element in the various compartments, including inputs, outputs (losses), sedimentation and regeneration, based on the operating physical and chemical processes.

10.8 VERTICAL DISTRIBUTION OF NUTRIENTS

In aquatic systems, there is a process of **sedimentation of organic matter**, starting from the surface, causing consumption of oxygen mainly within the metalimnetic and hypolimnetic levels in stratified lakes. The vertical distribution of carbon, nitrogen and phosphorus is thus related to the processes of vertical stratification and circulation. In stratified lakes, nutrients accumulate in the hypolimnion, as shown in Figures 10.9 and 10.10.

The availability of nutrients for aquatic plants is fundamentally related to this spectrum of nutritional resources (along the vertical and horizontal axes of the system) and to the regenerative processes determined by chemical conditions, principally the **oxido-reduction** (redox) potential. This availability is also related to the absorption mechanisms of aquatic plants, at both the physiological and morphological levels. Undoubtedly, in the interfaces of discontinuity in the distribution of nutrients, the metalimnion is fundamentally important in the processes of regeneration and/or concentration of nutrients. The anoxic hypolimnion is a great reservoir of resources, which become available for plants in the epilimnion through processes of excretion and decomposition of organisms or through external sources (tributaries and rainwater).

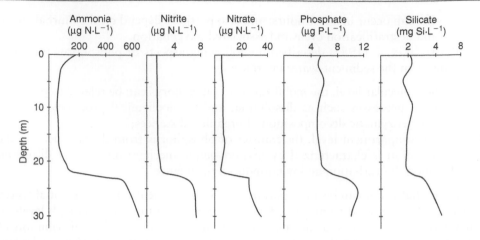

Figure 10.9 Vertical distribution of nutrients in the hypolimnion of Lake Dom Helvécio (Parque Forestal do Rio Doce – MG) during the period of stratification.
Source: Tundisi and Saijo (1997).

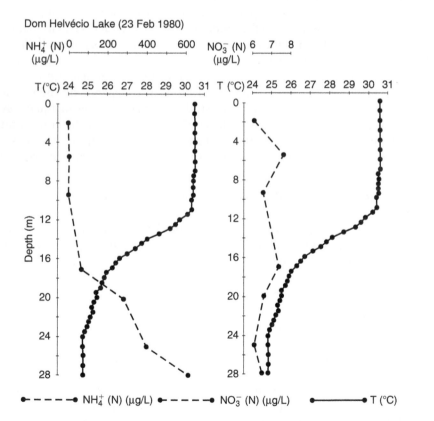

Figure 10.10 Vertical distribution of ammonia and nitrate in Lake Dom Helvécio (Parque Florestal do Rio Doce – MG) during the period of thermal stratification.

10.9 TRANSPORT OF SEDIMENTS FROM TERRESTRIAL SYSTEMS AND BIOGEOCHEMICAL CYCLES

The rapid transport rate of sediment from a terrestrial system to an aquatic system during periods of intense precipitation can be affected by deforestation or agricultural practices in the areas around reservoirs, lakes and rivers. Such transport produces a continuous influx of sedimentation with absorption of phosphate in the particles and immobilization of nutrients in the bottom sediment. This pulse of sediment also causes multiple changes in the aquatic system, affecting several ecological processes.

Desorption of phosphate from the sedimentation process can also occur during the movement of sediments through the water mass or during their settling.

In wetland areas or flooded fields, this transport of sediment by rivers plays an important role in the ecological cycle and metabolism of permanent or temporary lakes.

10.10 ORGANISMS AND BIOCHEMICAL CYCLES

Aquatic organisms are important in biogeochemical cycles for the following reasons:

▸ They excrete nitrogen, phosphate and organic compounds;
▸ They decompose after death and contribute nitrogen and phosphorus;
▸ They contribute to the active transport of nutrients along the vertical and horizontal axes of the system;
▸ Aquatic plants biologically fix elements.

Aquatic organisms can play a key role in recycling and transporting nutrients. Tundisi (1983), for example, considered that the contribution of **zooplanktonic excretion** in the epilimnion of tropical lakes is extremely important. In general the epilimnion has low levels of inorganic phosphorus and nitrogen (10–$20\,\mu g \cdot L^{-1}$ P-PO_4^{3-} and 20–$50\,\mu g \cdot L^{-1} N$-NO_3^{-}). One possibility for maintaining a biomass of phytoplankton in the epilimnion of stratified lakes, even if extremely low, is precisely the excretion of zooplankton, which, through the process of diurnal vertical migration, successively fertilizes several levels of water (see Figure 10.11).

In addition to the planktonic organisms, fish and benthos can also recycle considerable quantities of inorganic nutrients through excretion and active transport. Large vertebrates that live in the interface between terrestrial and aquatic systems, such as the hippopotamus in Africa and the capybara in South America, can supply aquatic systems with nitrogen and phosphate. For example, Viner (1975) estimated that close to 2–3% of the nitrogen and phosphate lost in the outflow of Lake George is recycled through hippopotamus faeces. A typical statistic shows, for example, that 1.6–2.2 kg C/animal/day is excreted, which represents 2,930–4,000 tons/year/total population. A large part of the contribution is dissolved or suspended (30%), which is more readily available.

In the case of lakes with many birds present, in wetland areas or permanent lakes, rapid recycling also occurs. For example, Tundisi (unpublished results) reported that in a hyper-eutrophic lake in the Mato-grosso swamp, a large portion of the phosphorus and nitrogen came from the excretion of birds nesting in vegetation on the lake.

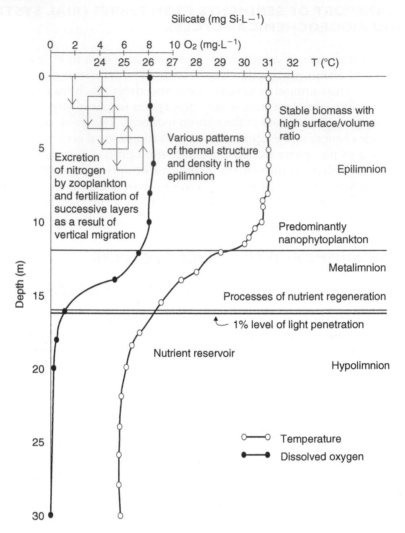

Figure 10.11 Possible mechanism of fertilization and nutrient recycling by zooplankton in the epilimnion of a tropical stratified lake (Lake Dom Helvécio – Parque Florestal do Rio Doce – MG).
Source: Tundisi (1983).

10.11 THE CONCEPT OF LIMITING NUTRIENTS

The supply of nutrients to aquatic plants can be below or above the levels required. Measuring a nutrient level, however, is not sufficient to identify whether or not it is a limiting factor. It is first desirable to determine the growth rate of the plants, the size of the reserve of nutrients, and the **recycling rate** among the various compartments. The total nitrogen and phosphorus dissolved in water control the organic-inorganic cycle. The compartmentalization of nitrogen and phosphorus at various levels of the food chain regulates the recycling rates. For example, nitrogen and phosphorus are retained longer in fish than in phytoplankton, because of the different **recycling rates** of these organisms (Allen and Starr, 1984). Phosphorus represents 12% of the dry weight of fish skeletons, which shows that this element can be retained outside the cycle.

Dugdale (1967) demonstrated that a hyperbolic relationship exists between the rate of fixation of a limiting nutrient by phytoplankton and the nutrient's levels in water. In general, the Monod equation can be applied to parallel the Michaelis-Menten enzymatic kinetics:

$$V_s = V_{s\ max}(V_s / V_s + V_{s'})$$

where:
Vs is the 'fixation' rate;
Vs_{max} is the maximum fixation rate;
S is the nutrient level;
Ks (coefficient of saturation) represents the nutrient level at which the 'fixation' rate, Vs, is half of the maximum rate ($Vs_{max}/2$).

Consider that Ks is specific for each type of nutrient. When nutrient levels are low, species with low Ks probably have a competitive advantage over those with high Ks. The concept of limiting nutrient is related to Liebig's Law of the Minimum, i.e., the growth of new biomass of aquatic plants is limited by one critical nutrient. The question related to the possible limiting nutrients in aquatic systems has been much discussed. In some studies, nitrogen is considered to be the main limiting factor, in others, phosphorus. One important conclusion is that it is difficult to generalize; nitrogen and phosphorus or nitrogen or phosphorus may constitute the limiting factor (or another nutrient), depending, of course, on the lake system under consideration and on its interrelationships. The individualism of lakes, in this aspect, is also very characteristic.

Chart 10.2 outlines several studies on limiting nutrients, and Chart 10.3 shows the set of experiments on enrichment conducted on many tropical lakes to identify the limiting nutrients for primary productivity and the growth of phytoplankton.

Figure 10.12 shows experimental systems for determining the response to enrichments. Figure 10.13 presents the results of an experiment conducted on artificial enrichments on the surface of the UHE Carlos Botelho reservoir (Lobo/Broa), obtained in response to consumption of dissolved oxygen in water.

Redfield (1934) and Fleming (1940) examined the content of organic matter in ocean water to determine the cellular content of carbon, nitrogen and phosphorus in phytoplankton and zooplankton. The average atomic ratio in the plankton samples was 106 to 16 to 1, that is, 106C:16N:1P. This ratio is generally seen as a reference standard to evaluate limiting factors in any mass of freshwater or ocean water.

Carbon can tend to be a limiting factor for phytoplanktonic growth when growth is saturated with nitrogen and phosphorus or when there is intense sunlight and high temperatures (even when the transport of CO_2 from the atmosphere to water is very slow). This limitation can also occur at low pH values with little dissolved bicarbonate in the water. Carbon as a limiting factor can also occur in super-fertilized tanks and wastewater in stabilization ponds (Schindler; 1977; Rast & Lee, 1978).

Chart 10.2 Several techniques to study limiting nutrients.

Parameters	Types of bioassay	C¹⁴ Technique to measure limiting nutrients	'Batch' bioassay	Continuous bioassay of culture	Enrichment in large tubes	Enrichment of the environments
	Locale of possible incubation	in situ, in vitro	in situ, in vitro	in vitro	in situ	in situ
	Biological inoculation used	Natural populations or test organism	Natural populations or test organism	Natural populations or test organism	Natural populations	Natural populations
	Number of possible treatments	Numerous	Numerous	One	Few (equal to the number of tubes installed in the locale)	One per lake
	Time of incubation	2 hours (2 to 5, could be more but never more than 24 hours)	Variable (from hours to days)	Variable (from hours to annual cycle)	Long (weeks up to a year)	Long (indefinite)

Source: Henry et al. (1983).

Chart 10.3 Addition of nutrients and responses of phytoplankton in tropical aquatic ecosystems.

Lake or reservoir	Locale	Limiting factors Primary	Secondary	Responses	References
Victoria	0–2°S 32–34°E	P	N	Cell count	Evans (1961)
Chilwa		N + P + S		Cell count	Moss (1969)
Mala		N + P + S		Cell count	Moss (1969)
Malombe		N + P		Cell count	Moss (1969)
Domabi		N		Chlorophyll	Moss (1969)
Makoka	15–17°S	N + P + S		Chlorophyll	Moss (1969)
Mpyupya	34–35°E	N		Chlorophyll	Moss (1969)
Mlungusi		N + P + S		Chlorophyll	Moss (1969)
Coronation		N + P + S		Chlorophyll	Moss (1969)
Shire		N + P + S		Chlorophyll	Moss (1969)
Malawi		N + P + S		Chlorophyll	Moss (1969)
Malombe		N + P + S		Chlorophyll	Moss (1969)
Domasi	15–17°S 34–35°E	N + P		Chlorophyll	Moss (1969)
Makoka		N		Chlorophyll	Moss (1969)
Mpympym		N		Cell count	Moss (1969)
Coronation		N + P + S		Cell count	Moss (1969)
George	0° 30–20°E	N + P		Cell count	Viner (1983)
Rietvlei	25°52.5'S 28°15.75'E	N – P		Potential for algal growth	Steyn et al. (1975a)
		N – P		Potential for algal growth	Steyn et al. (1975a)
		Microelement-n		Potential for algal growth	Steyn et al. (1975a)

Chart 10.3 Addition of nutrients and responses of phytoplankton in tropical aquatic ecosystems (continued).

Lake or reservoir	Locale	Limiting factors Primary	Secondary	Responses	References
Hartbeespoort	25°43'S 27°51'E	N – Fe		Potential for algal growth	Steyn *et al.* (1975a)
		N – P		Potential for algal growth	Steyn *et al.* (1975a)
		N – P		Potential of algal growth	Steyn *et al.* (1975b)
		N – P		Potential of algal growth	Steyn *et al.* (1975b)
		N – P		Potential of algal growth	Steyn *et al.* (1975b)
Roodeplast	25°37'S 28°23'E	P – N		Potential of algal growth	Steyn *et al.* (1975b)
		P – N		Potential for algal growth	Steyn *et al.* (1975a)
		N – P		Potential for algal growth	Steyn *et al.* (1975a)
		P – N		Potential for algal growth	Steyn *et al.* (1975a)
Vall	26°53'S 28°07'E	N – P		Potential for algal growth	Steyn *et al.* (1975b)
		P – N		Potential for algal growth	Steyn *et al.* (1975b)
		N – P		Potential for algal growth	Steyn *et al.* (1975b)
Ubatuba	23°45'S 45°01'W	P – N		Potential for algal growth	Teixeira and Tundisi (1981)
Kariba		N		C^{14} and chlorophyll	Robarts and Southhall (1977)
Henry Gallam		P – N		Potential for algal growth	Robarts and Southhall (1977)
Prince Edward		P – N		Potential for algal growth	Robarts and Southhall (1977)
Mazoe	27–32°S 16–22°E	P – N		Potential for algal growth	Robarts and Southhall (1977)
Little		P – Fe		Potential for algal growth	Robarts and Southhall (1977)
Connemara		N		Potential for algal growth	Robarts and Southhall (1977)
Umgasa		N		Potential for algal growth	Robarts and Southhall (1977)
UHE Carlos Botelho (Lobo/ Broa) dam	22°15'S 47°45'W	N + micronutrients		Respiration of phytoplankton community	Tundisi (1977)
Ebrié	0°10'N 4°E	N P C		Chlorophyll-*a*	Dufour *et al.* (1981)
Jacaretinga	3°15'S 59°48'W	N-P		Chlorophyll-*a*	Zaret *et al.* (1981)
Sonachi	0°47'S 36°16'E	P		Chlorophyll-*a*	Melack *et al.* (1982)
UHE Carlos Botelho (Lobo/Broa) reservoir	22°15'S 37°49'W	N-P		Cell counts and chlorophyll-*a*	Henry and Tundisi (1982a)
		N + P		Cell counts and chlorophyll-*a*	Henry and Tundisi (1983)
		N + Mo		Cell counts and chlorophyll-*a*	Henry and Tundisi (1982b)
		N		Cell counts and chlorophyll-*a*	Henry *et al.* (1984)
Dom Helvécio	19°10'S 42°01'W	N + P		Cell counts and chlorophyll-*a*	Henry and Tundisi (1986)
	19°10'S 42°01'W	N + P		Cell counts and chlorophyll-*a*	Henry and Tundisi (1983)
Barra Bonita	19°10'S 48°34'W	P		Cell counts and chlorophyll-*a*	Henry *et al.* (1985)
Jacaretinga	3°15'S 59°48'W	N		Cell counts and production of O_2	Henry *et al.* (1985)

Source: Henry *et al.* (1985a, b).

Figure 10.12   Experimental systems for study and response of limiting nutrients (see also color plate section, plate 18).

The concept of a limiting nutrient generally applies, therefore, to phosphorus and nitrogen levels in water and conditions of equilibrium (Odum, 1971). In this case, the limiting nutrient for phytoplanktonic growth is near the 'critical minimum' (according to Odum, 1971). Frequently, in the case of intermittent pulses of phosphorus and nitrogen, the concept of limiting nutrient is less useful.

Different phytoplanktonic species have different nutritional requirements, since they assimilate nutrients at different rates. Nutrient assimilation rates and fixation rates differ from species to species, conferring varying competitive advantages (Hutchinson's 'competition for resources' theory, 1961).

In conditions of nutrient deficiencies, small algal cells can be more efficient in assimilating nutrients than larger cells. The competitive ability of small cells ($<20\,\mu m$) to assimilate nutrients was examined theoretically more than 50 years ago by Munk and Riley (1952).

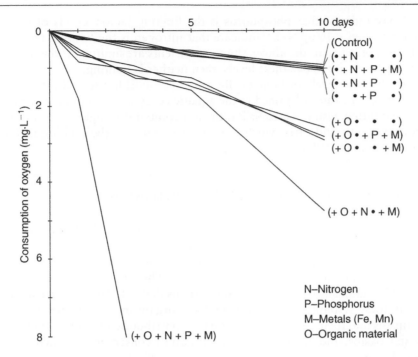

Figure 10.13   Experiment system to measure the response and enrichment in UHE Carlos Botelho
reservoir (Lobo/Broa).
*Source:*Tundisi *et al.* (1977).

The relative proportions of nitrogen and phosphorus levels in lakes and reservoirs
vary seasonally or annually, and therefore the limiting nutrients can vary through
time and space (Nakamoto *et al.*, 1976). A single nutrient required by algae could
be the limiting factor, according to Droop (1973) and Rhee (1978), who showed
experimentally that algal growth is stimulated by a single nutrient present at a lower
level relative to the alga's requirements.

Growth of phytoplankton in a specific aquatic ecosystem (reservoir, lake or river)
thus can be proportional to nutrient levels and these levels are dependent on the internal
and external loads, that is, the nutrients in the sediment. Phytoplankton can assimilate
nutrients in the water column in the mean atomic ratio of 106C:16N:1P. By identifying
the C:N:P levels and ratios in a water column and comparing them with the concept
of the Redfield Index (of 16N:1P), it is possible to deduce which nutrient is in excess and
which is a limiting factor. Nutrient levels much in excess of the 16N:1P ratio (necessary
for phytoplankton growth) should not be the limiting factor for the growth of algal
biomass. Therefore, it is important to determine not only total nitrogen and phosphorus levels (dissolved and particulate) in the water and tributaries, but also the forms
of biologically available nitrogen and phosphorus, in addition to total nitrogen and
phosphorus.

For example, it is desirable to determine the level of reactive soluble phosphorus and the levels of ammonium, nitrate and nitrite (if present). Practical experience
(Ryding and Rast, 1990) suggests that levels of biologically available phosphorus

<5 µg P may indicate that phosphorus is the limiting factor. Levels of biologically available nitrogen <10 µg N · L⁻¹ indicate that nitrogen is the limiting factor. If both present lower levels than the above, the two nutrients are limiting. If the nitrogen and phosphorus levels are above these levels, then neither is limiting.

If the absolute levels of biologically available phosphorus and nitrogen do not markedly drop, the rates identified in the aquatic ecosystem can indicate which nutrient will be limiting. Using the 16N:IP ratio as a comparative figure, any strong deviation from this ratio indicates which nutrient is potentially the limiting factor in the water mass (Stumm, 1985).

10.12 'NEW' AND 'REGENERATED' PRODUCTION

The concepts of 'new production' and 'regenerated production' refer to production of organic matter that depends on nutrients recycled in the aquatic system, based on organisms and sediments (regenerated production), or that production which depends on nutrients from external sources (new production). The 'new production'/'regenerated production' relationship in inland aquatic systems depends on the **internal load** accumulated in these systems, on the excretion from organisms, on the metabolic rates of the organisms and their decomposition, on the contribution of external nutrient loads from water basins, and on the use and condition of conservation or deterioration of the water basin.

The internal characteristics of circulation and redox potential also influence the recycling of nutrients for 'regenerated production'. Polymictic lakes, for example, tend to have patterns of homogenous vertical distribution of oxygen and phosphorus, both dissolved and precipitated in the sediment. Stratified lakes have a high internal load due to the release from sediment accumulated in the hypolimnion. The **internal load** of lakes and reservoirs can be determined through experiments on the varying oxido-reductive conditions, in addition to the estimates of nitrogen, phosphorus and other elements in the sediment. The relationship of 'new production' to 'regenerated

---

Phosphorus-chlorophyll ratio in lakes and reservoirs

In many lakes and reservoirs, there is a direct relationship between the concentration of phosphorus in water and the concentration of chlorophyll-*a*. In many lakes in temperate regions, this correlation can reach coefficients of regression of 0.9 (R^2). This means that the majority of total phosphorus in lakes is in the particulate state, and phytoplankton constitutes the greatest portion of this particulate phosphorus. When the ratio of total phosphorus/chlorophyll-*a* is extremely low, it means that other factors may control the abundance of phytoplankton. This correlation has been used to estimate the levels of phytoplankton based on the total phosphorus. Data for lakes and reservoirs in temperate regions function very well, as shown in Figure 10.14. However, for tropical lakes there is still some uncertainty for two reasons. First, there is still a lack of data compared with temperate regions. And second, the recycling of phosphorus is more rapid in tropical regions and these rates still need to be determined (see Chapter 7).

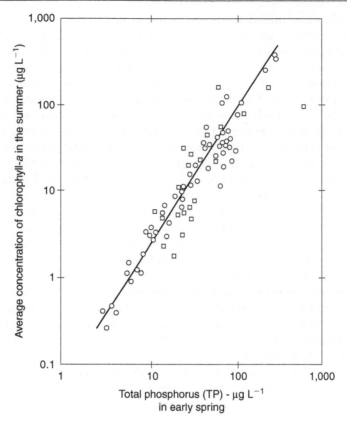

Figure 10.14   Ratio between total phosphorus (TP) and chlorophyll-*a* in some lakes of the temperate region.
Source: Horne and Goldman (1994).

production' can vary seasonally in the same lake, reservoir or other aquatic system. For example, in the UHE Carlos Botelho reservoir (Lobo/Broa), Tundisi *et al.* (1977) showed that in summer, the external load is higher due to precipitation, and 'new production' is thus efficient during this period. On the other hand, the production of organic matter in winter depends more on the internal load and processes of excretion and decomposition in the water column. Picocyanobacteria (<2 μm) play an important role, using both 'new production' and 'regenerative production' efficiently and in recycling organic matter.

10.13 GREENHOUSE GAS AND BIOGEOCHEMICAL CYCLES

Many gases – such ammonium, hydrogen, methane and **volatile gases** in the sulphur cycle – are found in anoxic oxygen-deficient environments. Some of these gases are the final products of microbial decomposition of organic matter. Evidence of a link between the accumulation of gases and climate changes and the chemistry of the atmosphere has increased (Adams, 1996). Gases are involved in oxidation/reduction and are produced in inland aquatic ecosystems, in anoxic conditions in water and sediments. Greenhouse gases and their emission into the atmosphere that have been studied in lakes, reservoirs, wetlands and sediment of surrounding areas are: hydrogen (H_2), ammonia (NH_3), volatile sulphide (H_2S, organic sulphur) and methane (CH_4). Decomposition of organic matter by micro-organisms provides the energy source for growth through fermentation and anaerobic respiration, promoting reactions of oxidation and reduction in cycles of carbon, nitrogen and sulphur. There are numerous groups of bacteria that use hydrogen (H_2) as a source of energy, and also a group of organisms that produces H_2 during the metabolism of carbohydrates, fatty acids and amino acids. Important sources of ammonia (NH_3) are wetlands. Anaerobic environments emit H_2S, carbon disulphide (CS_2) and **methyl mercaptan** (MSH, CH_3SH). H_2S gas is the most important emitted by anaerobic systems. Most of the carbon that is recycled in aquatic ecosystems and land is mineralized under anaerobic conditions, with products such as methane (CH_4). The decomposition of organic matter yielding methane in anaerobic sediments is the result of fermentation reactions involving organic compound > fatty acids > amino acids > carbohydrates > humic substances.

Emissions of greenhouse gases from the hydrosphere have been the object of numerous recent works in wetlands and anoxic sediments lakes and dams. Interactions between these cycles of gases and the effects of human activities, as well as other microbial processes, climatological and physical-chemical effects, have been intensively studied (Adams, 1996).

These gas emissions are fundamental in the biogeochemical cycles of carbon, nitrogen and sulphur and are extremely important for the chemistry of the atmosphere and its relationship with the hydrosphere.

Redox potential

Redox potential, or oxidation-reduction potential – Eh – represents alterations in the state of oxidation of many ions or nutrients. In pH 7.0 at 25°C, water saturated with oxygen presents a redox potential of +500 mV. Redox potential is measured in millivolts as an electrical voltage between two electrodes, one of hydrogen and another of the material whose status is being measured (iron, manganese or another metal). Gradients of redox potential are often found in nature; for example, in interfaces of anoxia and oxygenation, in systems of inland or marine waters, in sediment-water interfaces, or in soils. The transport of iron and manganese in sediment or in water receives much attention in limnology and oceanography, given the important role that these elements play in the biogeochemical cycles of other elements. In states of oxidation at high redox potential (400 or 500 mV), iron and manganese are insoluble (Fe^{3+} and Mn^{3+}). In reduced states (Fe^{2+} and Mn^{2+}), they are soluble and free of complexation. Thus, vertical profiles of iron and manganese related to the redox potential indicate their state of complexation or solubility. Vertical profiles in the sediment and interstitial waters of sediment illustrate these states of complexation and insolubility or solubility, and the transport processes at these interfaces occur through molecular diffusion. In sediment-water interfaces, the processes are rapid and depend on the turbulence or stratification and the grade of oxygenation or anoxia of the water.

Trace elements and their effects on the primary productivity and growth of phytoplankton

The effect of trace elements on the primary productivity and growth of phytoplankton can be illustrated by the study conducted by Henry and Tundisi (1982) at the UHE Carlos Botelho reservoir (Lobo/Broa). In this system, the surface water of the reservoir was enriched with different concentrations of potassium phosphate (KH_2PO_4), potassium nitrate (KNO_3) and sodium molybdate ($Na_2MoO_4 \cdot 2H_2O$). Various concentrations were utilized for each sample suspended at the surface of water for a period of 14 days in 2-liter Erlenmeyer flasks. After this period, the primary productivity of phytoplankton and the concentration of chlorophyll-*a* were determined. Molybdenum is present in the enzyme nitrate reductase and plays a key role in the assimilation of NO_2^- by phytoplankton, because NO_3^- is reduced to NO_2^- by the action of the enzyme nitrate reductase at the level of the cell wall. With the simultaneous addition of nitrate and molybdenum, a significant increase occurs in the growth and response of the phytoplanktonic community. These experiments show that, even with adequate concentrations of nitrogen for the stimulation or growth of phytoplankton, the presence of molybdenum is essential to promote the utilization of NO_3^- for the phytoplanktonic community.

Lakes as ecosystems

II

SUMMARY

Lakes, reservoirs, wetlands and rivers are complex ecosystems in continual dynamic interaction with the drainage basin in which they are found. The responses of these aquatic ecosystems to the forcing functions that act on them (including changes in winds, rainfall, solar radiation and air temperature) are diverse and depend on morphology, geographical location, latitude, longitude and altitude.

In this chapter we examine these responses, such as ongoing physical, chemical and biological processes, and look at the capacity of resilience and response based on the magnitude of the external forcing functions. Palaeolimnology is one of the main approaches that can demonstrate the changes of these forcing functions, such as climatology and erosion over geologic time. Vertical transport of particulate organic matter and dissolved organic matter depend on factors such as sedimentation, active transport by organisms, and vertical and horizontal currents.

The littoral zone of lakes provides an important filter for elements and substances originating from the drainage basin, and the relationship of the lake's littoral zone with the offshore limnetic zone is key in the exchange of organic matter between these two regions of the lake.

II.I   THE LACUSTRINE SYSTEM AS A UNIT

As already seen in Chapter 2, the lacustrine system as a unit acts effectively with the drainage basin and receives the impact of all human activities occurring in the basin.

The classification of lakes – taken as ecosystems and as important elements of the landscape – is directly related to their origin, which determines some of their general properties, such as morphometry and the water's basic chemical composition. Depending on these characteristics and the processes of circulation and stratification, lakes can also be classified according to vertical thermal patterns and their variations during the year.

Another lake classification system is based on processes of biological production and classifies lakes according to trophic status: oligotrophic, mesotrophic, eutrophic, and dystrophic (with high levels of dissolved humic material).

Criteria for classifying lakes are, of course, arbitrary and based on basic regional characteristics. Although each lake has its marked individuality, certain regional similarities can be found. In part, this is due to each lake's origin and morphology, which establish some clear operating patterns. Depending on the vertical depth and geographic location (altitude, latitude), for example, there is a significant interaction

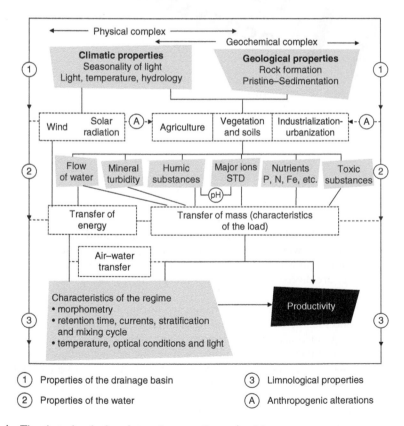

Figure 11.1   The three levels that determine aquatic productivity.
Source: Modified from Vollenweider (1987).

between external energy and the chemical, physical and biological processes of the lake's vertical structure. The influx of external energy (thermal balance or effects of wind) effectively promotes the renewing mechanisms of vertical mixing in the lake, which influence the production of organic matter and biological diversity.

Lakes closely reflect the range of interactions in drainage basins, including human activities. Along with the natural load of inorganic nutrients (nitrogen and phosphorus) and other essential elements, a lake may progressively suffer the impact of diverse human activities, which determine a certain oligotrophic-eutrophic trajectory. A lake's capacity to recycle introduced material clearly depends on circulation processes, compartmentalisation in the lake, and the structure of the biological community, which functions as an acceleration or brake system in the transport of organic material, as with the vegetation and the associated plant communities in the littoral zone.

Figures 11.1 and 11.2 show concepts of Vollenweider's (1987) on the lakes' features in a drainage basin, the anthropogenic effects, and key internal recycling processes, which, to a great extent, determine the level of biomass and trophic status. Figure 11.3 presents a picture of the controls on exchange of energy and water, and biogenic and geological determinants in shallow lakes in Africa.

As already seen in Chapter 1, Forbes' work (1887) on the 'lake as a microcosm' significantly advanced the science of limnology and the understanding of the interrelationships between the various compartments in a lake. Different approaches to the study of lakes – such as the processes of circulation and water chemistry

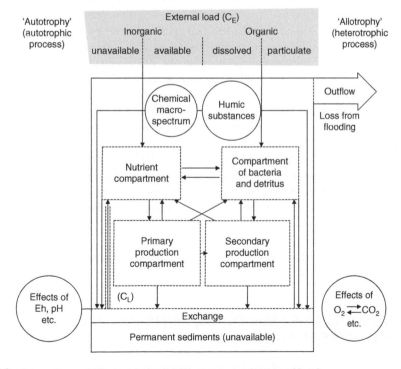

Figure 11.2   Internal recycling processes in lakes in terms of external load.
Source: Modified from Vollenweider (1987).

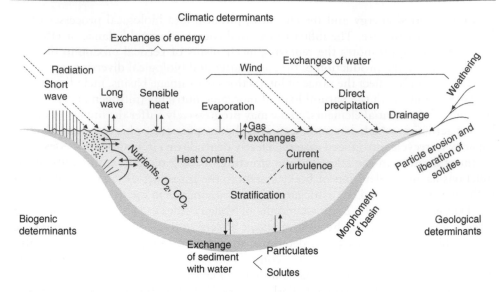

Figure 11.3 Talling's conception (1992) of the exchange of energy and water and biogenic and geological determinants in shallow lakes in Africa. This concept can be considered valid for a large number of shallow lakes in both the tropics and the temperate region.

(Birge and Juday, 1911), studies on benthic organisms (Thienemann, 1922), and subsequent classification of lakes (Naumann, 1924) – enabled great conceptual advances that involved a deepening of knowledge and establishment of a theoretical basis for ecology and limnology. However, consideration of the drainage basin as a unit and basic qualitative and quantitative component in the functioning of lakes has only come under examination since the 1960s, with work by Vollenweider (1968) and the need to quantify processes related to eutrophication of lakes and their causes. Hynes (1975) also described interactions between the ecology of rivers and their impact on lakes; and Likens (1983) identified, in a series of experiments, important relationships in the 'Hubbard Brook Ecosystem Study'.

The limnological studies by Margalef *et al.* (1976) on reservoirs in Spain represented an important step in the spatial study of drainage basins and reservoirs as sources of information. In Brazil, the studies of drainage basins and reservoirs advanced through studies on the UHE Carlos Botelho reservoir (Lobo/Broa) in the last 35 years (Tundisi, 1986), and, more recently, in the Tiete region (State of São Paulo) and the Paranapanema River (Henry, 1990; Nogueira *et al.*, 2005). Other recent studies deepened the understanding of the relationships between drainage basins, lakes and reservoirs (see Chapters 18 and 19).

11.2 ECOLOGICAL STRUCTURES, PRINCIPAL PROCESSES AND INTERACTIONS

Vollenweider (1987) summarized the main organizational levels regulating the operating mechanisms in drainage basins and processes related to aquatic productivity.

These levels include:

▶ properties of the drainage basin (geological and climatic);
▶ properties of the water;
▶ limnological properties;
▶ anthropogenic changes.

One basic problem is the size of the drainage basin, its compartmentalization (upstream/downstream), its interaction with other drainage basins and their geographical location, which implies climatic properties based on latitude, longitude, altitude, seasonality and cycles.

One of the key features in the approach to drainage basins and lakes is the morphometry of the system, which determines drainage patterns and spatial organization of the various compartments, such as rivers, lakes, reservoirs and wetlands. This spatial organization involves area/volume relationships between the drainage basin and rivers, lakes and reservoirs, as well as the distribution of these various sub-systems upstream and downstream. It also includes such ecotones as riparian forests and wetlands.

In identifying morphometric features, the following must be considered:

▶ geomorphological processes that give rise to the system and determine spatial patterns and the drainage system;
▶ soil types and vegetation cover that make up the mosaic of the landscape (extension, area and composition), which are basic elements in the drainage basin.

One of the key properties of the drainage-basin/lake relationship is the flow of water. The flow of water and its spatial-temporal distribution determine the characteristics of the transport of materials, elements and substances in the drainage basin.

It is useful to calculate the water cycle by studying precipitation-evaporation and the reserves of surface and ground water. Measuring the inflow and outflow of material based on the flow of water is also important. It is equally important to determine variations in the water level. In many regions, fluctuating levels cause extensive flooded valleys in the drainage basins, creating areas of temporary or permanent lakes subject to fertilization from the overflow of rivers, and also systems with increased genetic flow caused by the horizontal expansion of water masses and the transport of organisms. Extensive flooded forests in Amazonia, for example, constitute a source of food for fish and particulate matter for rivers (Goulding *et al.*, 1989).

Morphometry, flow and water cycle, soils and vegetable cover are important characteristics and structures of a drainage basin, and determine the basis for the functioning of the total basin-lake system.

A region's hydrogeochemistry is highly dependent on these structures. The basic chemical composition and electrical conductivity of water correlate with types of soil, vegetation cover and surface flow.

Vegetation cover affects the mechanisms of **transport of water**, reduces erosion and increases the potential of infiltration, fundamental for recharging aquifers. The composition of dissolved particulate matter in rivers depends on the vegetation in the drainage basin. Particulate matter is important in the recycling of nutrients and functions as a substratum and food. Dissolved matter affects the water's optical properties and can also be used as food.

11.2.1 The importance of rivers in drainage basins and ecological interactions

Rivers, distributed throughout a water basin according to drainage patterns and the geomorphologic slope of the land, play a fundamental role in transporting materials from diverse points. Rivers also function as detectors or accumulators of chemical, ecological and biological information in the various compartments.

Rivers can accumulate nutrients, such as phosphorus, and discharge them rapidly through the pulse of flow. The transport of material though rivers clearly varies with slope, flow and the varying situations in the continuum of the river. For example, in systems with upstream elevation and floodplains downstream in the basin, differences exist in the current's velocity, levels of dissolved oxygen and accumulation of organic matter, with greater retention capacity (Likens, 1983). Rivers are thus important intermediaries between the components of the terrestrial system and other aquatic systems in the drainage basins. They establish a basic structure in the spatial heterogeneity of the system and elements that connect the compartments of the drainage basin (see Chapter 3).

11.2.2 Ecotones in drainage basins

The main ecotones (component transition habitats) in drainage basins are the wetland areas and riparian vegetation along the length of the rivers. Another important region in the ecological relationship between the drainage basin and lakes is the shoreline of the lake, which is also an ecotone.

Wetland areas can be associated with rivers and lakes or they can be relatively isolated areas in the landscape. They may or may not undergo fluctuations in level, which implies particular hydrological conditions, retention system, nutrient transport system, rapid recycling of organic matter (Mitsch and Gosselink, 1986) and, in some cases, increased levels of primary production (Patten et al., 1992a, b) and biomass. Wetland areas can have a wide diversity of species, providing a reserve of key species for repopulating and colonizing other regions in the drainage basin and lentic aquatic systems. The preservation of wetland areas and their possible use in denitrification and accumulation of heavy metals are other aspects of the problem that must be considered. Hydrological conditions are essential for maintaining the structure and function of wetland areas, and the **hydroperiod** (the result of the water flow) influences biogeochemical cycles, accumulation of organic material, and species diversity. Because of this, maintenance of the pulses into wetland areas is important, and maintenance of level of fluctuation is essential for their protection (see Chapter 15).

Another important structure is the riparian vegetation along rivers. Made up of vegetation adapted to fluctuating levels with the capacity to tolerate high water levels and extensive periods of flooding, these interface zones have quantitative and qualitative interactions vital for the terrestrial and aquatic systems (Odum, 1981).

Gallery forests or riparian forests are important components of the landscape in any drainage basin. This riparian vegetation is located in sediment areas that support this highly differentiated vegetation (Ab'Saber, p. 21, 2001). These marginal channels and ridges generated in the upper basin of the rivers provide essential geo-ecological support for the development of riparian forests. The dynamics of **hydro-geomorphological processes** giving rise to different marginal structures in rivers is highly differentiated and varied, given the enormous range of seasonal processes, flow of water, sediment transport and the hydrodynamics of the water courses. According to Ab'Saber (2001), floodplains have an alluvial system based on sorting and type of sediment according to weight and size, and the sedimentation of these particles takes place according to their characteristics and the velocity and dynamics of the water flow in rivers and channels.

Coarser sediment is transported by tumbling or being dragged, while finer sediment is transported in solution and suspension. On these sediments deposited by different mechanisms roots the riparian vegetation, which includes all types of vegetation along the riverbanks. These riparian forests are a key element in the spatial heterogeneity and in the wide biodiversity of animals and plants.

Besides riparian forests, there are areas – such as the savannas of central Brazil – where clumps of biodiversity-rich forest are found in humid lowland depressions in hydromorphic soils (Wilhelmy, 1958; Cristoforetti, 1977). According to Jacomine (2007), soils in riparian forests are commonly organic (histosols), consisting of organic debris from deposits of the remains of the vegetation in 'various degrees of decomposition' (Jacomine, 2001). Other soils present in these areas are the **gley soils** (hydromorphic mineral soils), with an **organic layer** over a **gray layer**, mineral with a very loamy texture; **neosoils** (hydromorphic quartz sand); **plinthic soils** (**hydromorphic laterites**) in semi-hydromorphic soils, with varying textures; **fluvic neosoils** (alluvial soils), associated with semi-desert riparian vegetation (seasonal). These are very heterogeneous soils in terms of granulometry, structure, consistency and chemical properties.

Another type of soil are the **cambisols**, which are found in well-drained and moderately drained floodplains. According to Klinger (2001), cambisols develop from older alluvial sediments with good but imperfect drainage. The riparian forests growing on such varied types of soils have diversified and fundamental functions in the interactions between drainage basins, rivers and lakes. These functions include:

▶   Reserve of water – increase in storage capacity of water;
▶   Maintenance of water quality due to absorption of particulate matter, nutrients and **pesticides** (Lima and Zabia, 2001);
▶   Riparian vegetation contributes to an ongoing functional interaction between the **geomorphic** and **hydraulic** processes of the channels of rivers and aquatic biota. In addition to the stratification of the margins due to roots, riparian forests produce an endless supply of organic matter for rivers, used by organisms (aquatic invertebrates, fish and other aquatic vertebrates).

The geochemical recycling of nutrients depends on these functions of the drainage basin.

Riparian forests also attenuate the solar radiation reaching the canopy, contributing to a drop in temperature in the water, and reducing, due to shading, the primary production of phytoplankton and periphyton in areas with little circulation.

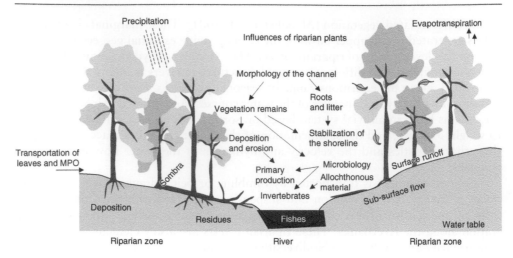

Figure 11.4   Conceptual illustration of a riparian area.
Sources: Modified from Likens (1992), Paula Lima and Zakia (2001).

Figure 11.4 presents the conceptual scheme of riparian forests. More studies of these interrelations with terrestrial systems are needed, according to Lima (1989).

The relationship between riparian vegetation, rivers and fish are presented and examined in the work of Barrella *et al.* (2002). From the perspective of the biology of fish, according to these authors, riparian vegetation has the following ecological functions, important for the ecophysiology, distribution and reproduction of fish species:

▶   limitation of the flow and flooding of water;
▶   cover and shade;
▶   maintenance of water quality;
▶   **filtering** toxic substances and suspended material reaching the river;
▶   supply of organic matter and substratum for fixation by algae and periphyton;
▶   supply of organic matter for fishes.

Decreased flow – due to the formation of backwaters, small dams and marginal lakes – leads to a heterogeneous fauna important in rivers, increasing regional biodiversity. Stumps, roots and branches in rivers can also be used as shelter for fish species. In addition, protozoa, algae and invertebrates grow in this substratum, providing food for **alevines** (Barrella *et al.*, 2002).

All these trophic relationships and eco-physiological responses of fish species are due to the existence of riparian forests. The destruction of such habitat and deforestation greatly increase the silting of rivers, reduce spatial heterogeneity in the drainage basin and interfere in basic ways with the mechanisms and processes operating in drainage basins and terrestrial and aquatic systems.

Gallery forests are thus key components in the landscape of drainage basins and greatly contribute to dynamic support of biodiversity and maintenance of evolutionary

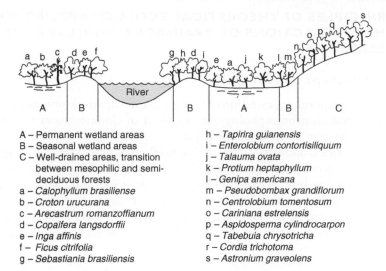

A – Permanent wetland areas
B – Seasonal wetland areas
C – Well-drained areas, transition
   between mesophilic and semi-
   deciduous forests
a – *Calophyllum brasiliense*
b – *Croton urucurana*
c – *Arecastrum romanzoffianum*
d – *Copaifera langsdorffii*
e – *Inga affinis*
f – *Ficus citrifolia*
g – *Sebastiania brasiliensis*

h – *Tapirira guianensis*
i – *Enterolobium contortisiliquum*
j – *Talauma ovata*
k – *Protium heptaphyllum*
l – *Genipa americana*
m – *Pseudobombax grandiflorum*
n – *Centrolobium tomentosum*
o – *Cariniana estrelensis*
p – *Aspidosperma cylindrocarpon*
q – *Tabebuia chrysotricha*
r – *Cordia trichotoma*
s – *Astronium graveolens*

Figure 11.5   Diagram illustrating the gradient of saturation (by water) of the soil and the occurrence of arboreal species in the basin of the Jacaré-Pepira River (Brotas, SP).
Source: Modified from Lobo and Joly (2001).

processes. They are essential components for the functioning of lakes, reservoirs and wetland areas, into which rivers feed from drainage basins. Riparian forests, provided they have the necessary size and characteristics, are also crucial in providing habitat for terrestrial and aquatic species.

Since gallery forests are periodically flooded with varying frequency and duration, there is a strong selective aspect to the flooding, especially in regard to saturation of oxygen. There is a control of oxygen level promoting a hypoxic or anoxic environment, affecting vegetation, physical-chemical and aquatic characteristics, and shifts in pH and redox potential of the soil (Lobo and Joly, 2002).

Figure 11.5 illustrates the gradient of water saturation in the soil and the presence of tree species in the drainage basin of the Jacare-Pepira River (SP). A thorough review of the characteristics of vegetation in riparian forests, especially from the physiological point of view, is presented by Lobo and Joly (2002).

Rodrigues and Leitao Filho (2002) presented a survey on the biological, ecological and physiological processes associated with gallery forests. Matheus and Tundisi (1988) showed that up to 50% of the nitrogen and phosphorus that could potentially be added to rivers is removed by the riparian vegetation. There is a well-defined association between riparian vegetation and topography in the areas next to rivers. The decomposition of organic matter in riparian forests is related to the intensity and duration of the flooding cycle.

These transition zones can absorb pollutants from **non-point sources**, which is why it is essential to protect and conserve them, for the maintenance of an adequate integrated operating mechanism between drainage basins and lakes. These areas also act as hydrological regulators because they reduce the velocity of water flow in rivers entering the lakes.

11.3 PRINCIPLES OF THEORETICAL ECOLOGY APPLIED TO THE INTERACTIONS OF DRAINAGE BASIN, LAKES, AND RESERVOIRS

11.3.10 Concept of succession

Succession and temporal organization of terrestrial and aquatic communities in drainage basins depend on geomorphology; interactions of climate, water and hydrogeochemistry; and on uses of the drainage basin.

Human activities, **industrial and agricultural uses,** and population growth all affect the natural processes of succession. Recuperation of drainage basins and lakes depends on understanding these processes and the scientific and technological investment needed to undertake the recuperation. The re-organization of riparian forests, for example, along rivers and lake-shores, depends on a sound understanding of the ecological interactions, dominant species and the role of each species in the overall subsystem.

11.3.2 The concept of pulses

In the ecological relationships of drainage basins and lakes, it is important to note that this joint functioning depends on pulses that, in many cases, are related to the forcing functions and **physical disturbances. Hydrological pulses** cause more rapid flow in rivers and variations in the flooding levels in marginal areas, riparian forests and marginal lakes. These pulses can accelerate biogeochemical cycles and affect the life cycles of aquatic organisms such as fish and zooplankton, and also terrestrial organisms such as insects. Human activity in drainage basins can cause pulses, such as the influx of suspended materials (from agricultural activity) and the discharge of pollutants. Pulses of biological activities must be considered, mainly the endogenous rhythms of organisms (migrations, photosynthesis, respiration, and reproductive cycles). The spectrum of pulses significantly affect the functioning of drainage basins and lakes.

11.3.4 Ecotones

The concept of pulses refers to **temporal variations** in the system. The concept of ecotone refers to spatial variations in the system. The several ecotones already mentioned play a key role in the functioning of drainage basins and their interactions with lakes.

In many wetland areas and littoral zones of lakes, vertebrates can be found, such as fish, amphibians, reptiles and birds, which play an important role in the active transport of organic material in the lake (fishes) and terrestrial system (amphibians, reptiles and birds). This mobile interface is important quantitatively, because the accumulated biomass can be as much as several tonnes per hectare (see Chapter 6).

The feeding, excretion and mechanical activities of these organisms influence terrestrial systems, aquatic ecosystems and their interface. Transportation of eggs and larvae by birds can function as an important colonization mechanism in drainage basins.

11.3.5   The theory of island biogeography and colonization

Lakes and reservoirs in drainage basins can be considered the 'aquatic equivalent of an island' (Baxter, 1977). Colonization of new reservoirs depends, in many cases, on the biota present in drainage basins, marginal lakes and flooded areas of rivers. Transport mechanisms of organisms, eggs or larvae in the drainage basin play an essential role in the colonization of new areas.

In arid or semi-arid regions, it is important to identify the fauna and flora in areas of drought and temporary waters. Such fauna and flora can re-populate rivers, lakes and reservoirs.

11.3.6   Spatial heterogeneity and diversity

The relationships between drainage basin and lakes depend on the compartments present in the two subsystems, on habitat diversity and distribution.

Spatial heterogeneity and the mosaic of subsystems in the drainage basin are related to geomorphology. Vegetation cover, extension of the littoral zone, and differences in altitude are basic elements in determining this heterogeneity and the grade of diversity and the spectrum of diversity in ecosystems.

11.3.7   Connectivity

Another important concept is that of connectivity, which enables the understanding of the degree of interdependence as well as **covariance** between the forcing functions, biological processes, and spatial and temporal distributions in the drainage basin and lakes. Applying the concepts of theoretical ecology to understand drainage basins and their interactions with lakes is useful, but they must be considered in light of the integrated management of the system, its recuperation and its rational exploitation.

11.4   FORCING FUNCTIONS AS EXTERNAL FACTORS IN AQUATIC ECOSYSTEMS

External forcing functions that provide an influx of kinetic energy, promoted by wind, inflow of rivers, or atmospheric warming by solar radiation, play a key role in the functioning of aquatic ecosystems, and in their responses, which vary in type and magnitude.

Chart 11.1 presents the forcing functions acting on lakes and the lakes' responses. Chart 11.2 outlines the relations of the principal forcing functions acting on lakes, wetland areas, floodplains and reservoirs in Brazil.

Figure 11.6 presents the basic features of a stratified lake in summer and the principal processes occurring in the vertical structure of the system. In this case, the lake is the D. Helvecio (Parque Florestal do Rio Doce – MG), a warm monomictic lake. The biological and chemical processes and the vertical distribution of organisms are spatially distributed and there is a separation in structures, which can be strong or weak, with spatial compartments, depending on the region of the lake under consideration. Figure 11.7 shows the same lake in the period of circulation caused by

Chart 11.1   Response of lakes to various physical forcing functions

Wind	Inflow of rivers	Atmospheric heating	Atmospheric pressure on surface	Gravity
Controlling factors: Area Volume of lake Depth Coriolis Effect Duration: Development of water basin	Controlling factors: Outflow Temperature Configuration of drainage basin	Controlling factors: Latitude Altitude Depth	Controlling factors: Area, volume of lake	Controlling factors: Volume and morphometry of lake
Resulting mechanisms: Waves Force of wind direction	Resulting mechanisms: Density currents	Resulting mechanisms: Alterations in density	Resulting mechanisms: Differences in pressure	Responses: Tides
Responses: Waves Circulation Upswelling	Responses: Advective currents	Responses: Turbulence and vertical mixing Stratification Internal waves	Responses: Internal waves	

Effective on small lakes and lakes with average area and volume	Effective on large lakes (for example: The Caspian Sea, North American Great Lakes)

Chart 11.2   Principal forcing functions in lakes, wetland areas and reservoirs in Brazil and their ecological effects.

System	Principal forcing function	Effects (examples of effects)
Amazon River (floodplains and lakes associated with the tributaries)	Wide fluctuations in level (10–15 meters) Winds (seasonal or diurnal)	Alterations in the succession of species, biodiversity and organization of communities Effect on biomass Dislocation from the littoral zone. Physiological effects on animals and plants
Wetland areas Swamp, floodplains of the Paraná River and tributaries	Fluctuations in level (2,5,7 meters) Winds (seasonal or diurnal)	Alterations in nutrient cycles and biomass Turbulence in lakes Diurnal alterations in the thermal structure of the system Alterations in the succession, organization of communities Physiological alterations
Lakes of the Rio Doce	Solar radiation – thermal heating and cooling	Thermal stratification and destratification (seasonal) Vertical distribution of organisms Diurnal patterns and vertical circulation
Reservoirs	Retention time (uses of water) Winds (seasonal or diurnal) Precipitation Cultural eutrophication and rapid influx of pollutants	Alterations in the vertical and horizontal **circulation patterns**. Modifications in the biogeochemical cycles Alterations in the vertical and horizontal distribution of organisms Gradients of density and temperature Effects of a catastrophe on phyto-, zoo- and bacteria plankton (opening of floodgates)

Source: Modified from Tundisi (1983).

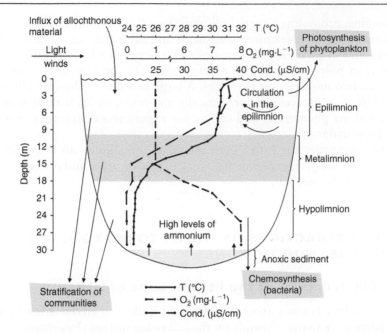

Figure 11.6 Basic characteristics of a stratified lake in the tropics in summer (Dom Helvécio Lake – Parque Florestal do Rio Doce – MG) showing the vertical compartmentalization of the ecosystem and stratification of the communities.

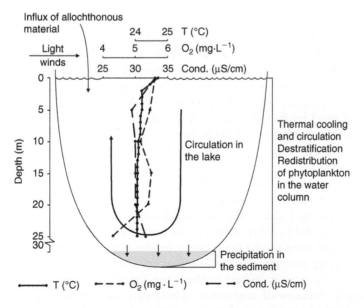

Figure 11.7 A non-stratified tropical lake in summer, showing a reorganization of the vertical structures (Lake Dom Helvécio – Parque Florestal do Rio Doce – MG).

thermal cooling, promoting a spatial re-organization of the system with redistribution of communities and complexation of elements and substances in sediment – especially phosphorus and iron. These two situations that occur in a hot monomictic lake can rapidly shift in polymictic lakes or remain relatively constant in meromictic lakes, that is, in the two most evident extremes. A large number of physical, chemical, biological and physical processes occur vertically and horizontally, making it challenging to understand the phenomena and calling for experimental strategies that can help elucidate these dynamics.

Effective strategies for better understanding the lake as an ecosystem include studying satellite imagery, experimental work in laboratories, and periods of intensive monitoring (see Chapter 20).

11.5 THE INTERACTIONS OF THE LITTORAL ZONE IN LAKES AND THE LIMNETIC ZONE

11.5.1 The functions of the littoral zone of lakes

The shore of a lake is other another ecotone with filtering characteristics, where the influx of particulate matter depends on the following factors (Jorgensen, 1990):

▶ slope;
▶ soil characteristics;
▶ climatic conditions;
▶ terrestrial vegetation;
▶ use of soil;
▶ use and management of water.

The shore of the lake, of which the littoral zone is part, is characterized by an active interaction between the sediment and the water, high levels of biological activity in the region of the dense growth of submerged macrophytes and by growth of micro-organisms (bacteria, fungi, algae, micro-zooplankton). As a result, there is a lot of biological productivity, with an increased biomass of benthic organisms. This region is also used by fish alevines and aquatic birds. Nutrients are recycled rapidly and denitrification activity in this area is extremely high. Therefore, this area's function as a filter is very important. In some regions, **waterborne disease vectors** develop in such areas (Löffler, 1990).

The transition zone between the terrestrial system and a lake is an ecologically important area because the metabolism, productivity, respiration and biological processes can be accelerated, due to the accumulation of biomass, biodiversity, decomposing organic matter and bacterial activity. The dissolved particulate organic matter that reaches the littoral zone can be made up of products drained by erosion of the water basin, atmospheric precipitation, falling leaves, underground streams, sewage and waste. Much of the organic matter in the littoral zone of lakes is produced by leaves and the organic remains of vegetation. There can also be autochthonous material (emerged and submerged macrophytes, periphyton, invertebrates) and where a dense diverse biomass of bacteria develops, with varying metabolism and physiology.

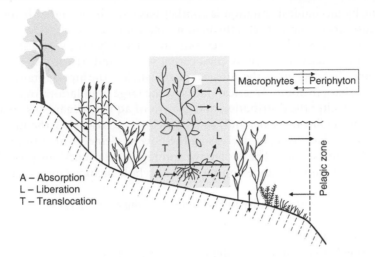

Figure 11.8   Simplified illustration of the role of macrophytes in the littoral zone of lakes and reservoirs in recycling of nutrients.
Source: Pieczynska (1990).

The terminology for the zonation of lake-shores varies in the literature (Hutchinson, 1967; Pieczynska, 1972; Wetzel, 2001). The extension of the littoral zone depends on the morphometry and the configuration of the lake or reservoir and on the terraces of the margins. The extension can vary depending on the periods of drought and flooding (if the water level of the lake or reservoir varies). Climate, soil, and physical and chemical properties of the water and sediment are factors that affect the littoral zone (Gunatilaka, 1988). The organic material processed or that enters the littoral zone can be consumed by animals, reduced, decomposed or deposited as sediment. Immersed, floating or submerged aquatic macrophytes play an extremely important role in the littoral zone, since, along with the processes of photosynthesis and respiration, which alter the chemical composition of the water, the macrophytes can provide a substratum for many aquatic organisms, periphyton and bacteria. These macrophytes are important in the biogeochemical cycles, in the formation of substrata, and in establishing permanent or temporary stands.

Aquatic macrophytes and the organisms associated with them play a significant role in the exchange of materials in the pelagic zone of the lake, by means of export of dissolved organic matter or the fixation of nitrogen and dissolved phosphorus from the limnetic zone (or the pelagic zone). The littoral zone also functions as an important filter controlling the lake's metabolism (Figure 11.8). The close proximity of macrophytes and periphyton implies a continuous exchange of nutrients and organic compounds between these components. The complex of micro-flora sometimes associated with macrophytes and aquatic higher plants may function as a highly integrated system in the littoral zone, according to Wetzel (1983b).

The horizontal and vertical movement of water in the littoral zone plays an important role in the transport of dissolved substances and elements and exchange of matter between the littoral and pelagic zones. Removal of sediment from the bottom of the littoral zone by mechanical agitation is another basic mechanism that helps distribute elements, substances and particles throughout the water column.

The littoral zones of lakes and reservoirs are extremely fragile and sensitive to the impact of human activities such as engineering works and irrigation. The impact of the vertebrates that depend on the aquatic system (such as the hippopotamus in African lakes, the capybara in South American lakes or the large populations of aquatic birds) can significantly alter the distribution and growth of aquatic organisms, as well as the metabolism in the littoral zone, and, as a consequence, also alter the metabolism in the limnetic zone, increasing or decreasing the flow of elements or substances for this zone from the littoral zone. The protective and regulatory functions provided by the littoral zone to the limnetic zone and the lake as whole include (Jorgensen 1995):

▶ maintenance of water quality in the transition zone of the lake;
▶ **reduction of erosion;**
▶ **protection against floods;**
▶ buffer zone between human settlements and the lakes;
▶ maintenance of the biodiversity and the genetic stock of plants and animals;
▶ population control of insects;
▶ establishment of habitats for the spawning and procreation of fish and reproduction of birds. Nursery areas for many species of fish and aquatic organisms;
▶ production of renewable resources used by man;
▶ aesthetic support for human beings.

The littoral zone needs integrated management.

11.6 LAKES, RESERVOIRS AND RIVERS AS DYNAMIC SYSTEMS: RESPONSES TO EXTERNAL FORCING FUNCTIONS AND THEIR IMPACTS

As already mentioned in previous chapters, in all aquatic ecosystems, the functioning of biogeochemical cycles, organization and dynamics of communities involve complex interrelations between physical, chemical and biological processes.

These processes are unique to each aquatic system and depend on the set of circumstances resulting from geography, latitude, altitude and origin of the system. For the purposes of the forcing functions, there are several functional responses whose magnitude is the result of the set of interactions between the processes.

Carpenter (2003) discusses changes in regimes, thresholds of disruption, and the resilience of ecosystems. Changes in regime and disturbances, according to the author, promote rapid shifts in the organization and dynamics of an ecosystem, with long-term effects on its structure and operation. These changes in regime involve multiple factors, such as alterations in the internal control mechanisms or in the external forcing functions (for example, increase in nutrients due to discharge from drainage basins).

Internal control mechanisms, when they undergo alterations, determine the threshold by which ecosystems shift from one regime to another (for example, changes from one trophic state to another, or the relations between species). The dynamics of the

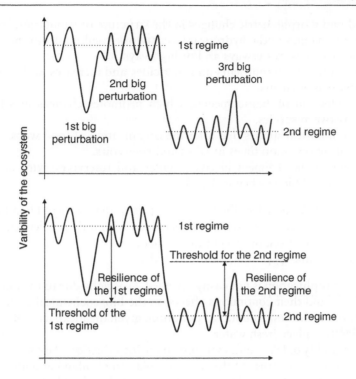

Figure 11.9 Fluctuations of an ecosystem over time based on two regimes (points).The thresholds, resilience and responses present alterations in function of the magnitudes of perturbations. *Source:* Modified from Carpenter (2003).

ecosystem are being changed, and according to Carpenter (2003), changes occur following a distribution probability based on a specific regime, which can present cycles in variations. A disturbance may cause a regime change if it exceeds the threshold, and thus promote alterations in the ecosystem's dynamics and organization.

Carpenter *et al.* (2002a) defined resilience as the magnitude of the disturbance needed to cause a regime shift to surpass the threshold. According to some authors, resilience is the return of ecosystem to a particular regime after the disturbance, and this process is also called stability (May, 1973). Disturbances in ecosystems can be caused by extremely slow changes in the threshold or by rapid external perturbations that drive the system to other regimes (Scheffer, 1998; Scheffer *et al.*, 2001a).

Figure 11.9 presents hypothetical alterations that occur in the ecosystem over time, with disturbances that can alter the functioning regime and the ecosystem's organization and structure.

What are the alterations that can provoke regime changes in lakes, rivers and reservoirs? They are many and can be related to the following forcing functions and internal auto-regulatory mechanisms:

▶ Climatic changes that alter cycles of precipitation, wind or solar radiation, modifying circulation patterns and provoking increased drainage.
▶ Alterations in soil use and deforestation of drainage basins.

▶   Physical and morphometric changes in the structure of tributaries, lakes and reservoirs, or changes in the hydrographic network resulting from the construction of drainage canals, **waterways** and/or highways.

▶   Changes in the use of nutrients and pesticides and increases in the external load into aquatic ecosystems.

▶   The introduction of exotic species, which promotes changes in self-regulation regimes between species.

▶   Removal of critical species that alter the organization of food webs.

▶   Alterations in retention times in lakes and reservoirs.

▶   Alterations in the sediment in rivers, lakes and reservoirs, with impact on the biodiversity and biogeochemical cycles.

Carpenter (2003) also described some 'surprises' that occurred in aquatic ecosystems in the twentieth century, events that occurred with little prediction, or that were not a motive of prediction, or inaccurate predictions.

Some of these unexpected behaviors include:

▶   **cultural eutrophication** is not easily reversed, especially due to the internal load of the systems and their continuous effects on the biogeochemical cycles.

▶   dilution is not the best solution to pollution, given the incapacity of some systems to efficiently replace fresh water.

▶   the **vulnerability of lakes and reservoirs** to the introduction of exotic species is quite serious, probably because of the relative isolation of inland aquatic ecosystems.

▶   the construction of dams increases the geographic distribution of waterborne diseases in the tropics.

▶   uses in drainage basins have significant indirect effects on the functioning of lakes and reservoirs and the integrated functioning of water resources.

11.7 PALAEOLIMNOLOGY

All processes occurring in lakes during different geological periods are recorded in the sediment. Palaeolimnology addresses this area of research.

The following components and materials can be identified in lake sediment: the remains of organisms, pollen, diatom frustules, chitin debris from zooplankton, chironomids, scales and vertebrae from fish, sponge spicules, vegetation debris and carbon, organic substances (such as pigments and their degraded products), as well as organic carbon, phosphorus and nitrogen. The dating of the different strata collected is generally done with ^{14}C. Radioactive caesium is used in some lakes to date events after the atomic explosions of the 1950s and 1960s.

One of the most extensive palaeolimnological studies was conducted on Lake Biwa in Japan (Horie, 1984), where work on more than 1,000 meters of sediment provided much relevant scientific information for the development of limnology. Through palaeolimnology and the study of accumulated sediment, the age of Lake Biwa was calculated (approximately 5 million years old), as well as the significance of geological alterations in its natural conditions and functioning.

Palaeolimnological studies can also help to identify the climatic changes that occurred in the region under study and subsequent alterations in the drainage basin, such as changes in vegetative cover, soil use and human activities next to the lake. For example, studies by Cowgill and Hutchinson (1970) showed alterations in sediment accumulated in Lake Monterossi (Italy). These authors showed that construction of the Cassia road by Romans 2000 years ago altered the trophic state in the lake, due to deforestation of the drainage basin in the era when the construction on the road took place.

Cowgill (1977a, 1977b, 1977c) discussed the importance of the sediment in the **chemical record** of alterations in drainage basins. According to the author, there is clear evidence that calcium, strontium, potassium and sodium in lakes are indicators of agricultural activities in the surrounding drainage basins.

Absy (1979) conducted palaeolimnological studies on Holocene sediments in the Amazon River valley and several lakes in the Amazon region. Pollen types and pollen diagrams were studied in the sediment, as well as data obtained on current vegetation and pollen deposited in the region under study. The research also included data on pollen from **recent sediment**. According to the author's conclusions, the changes in the vegetation recorded in the diagrams resulted from the local sedimentation processes, and changes in the water level, which promoted changes in the terrestrial and aquatic vegetation – in this case, especially floating grasses.

Information on pollen deposited in sediments, analysis of the age of sediments by ^{14}C and the succession of vegetation permits the elaboration of curves of fluctuation of climate for the Amazon River, stressing 'dry' periods above 4,000 years before the present. These palaeolimnological data permit detecting periods of substitution of pluvial forests for savannahs of Gramineae in the Amazonian region.

Another important study was conducted by Rodriguez Filho and Muller (1999) on Lake Silvana (Parque Florestal do Rio Doce – MG). In the study, the authors showed evidence of palaeoclimatic changes through identifying the minerals, labile minerals and pollen in the sediment, deposited over 10,000 years. Results showed that mineral-sediment-geochemical studies are valid to identify periods of a general climatic change in grasses (prairies) to savannas and semi-deciduous tropical forests (current), as result of climatic change in that period, a shift from dry to humid climate. The results also enable the verification of the interrelations between data on lake sediments and the history of **erosion in drainage basins**.

Salgado-Luboriau *et al.* (1997) studied changes in the vegetation of savannas and wetlands with palm trees in Central Brazil, from the Upper Quaternary.

Other studies showing the importance of palaeolimnology in detecting climatic changes affecting drainage basins and lakes were conducted by Dumont and Tundisi (1997) on four lakes in the Parque Florestal do Rio Doce. These studies identified periods of increased salinity in the lakes, indicated by the presence of halophilic diatoms. Higher salinity corresponds with drier periods. The episodes of higher salinity (dry periods) could be demonstrated not only by the presence of halophilic diatoms but also by the presence or decrease in the number of cladocerans (*Bosmina* and *Alona* species).

In periods of more increased salinity, the number of these species in the sediment declined but they did not totally disappear, indicating that the salinity, although high, did not reach extremely high levels, since these species can tolerate moderate salinity or brackish waters. Sponges are also present during periods of higher salinity (see Figure 11.10 and Figure 11.11).

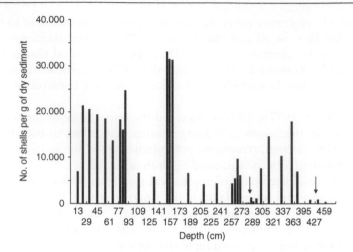

Figure 11.10   Quantitative distribution of *Bosmina* spp. in sediment in Lake Carioca (Parque Florestal do Rio Doce – MG). Saline intervals are marked by the arrows.
Source: Dumont and Tundisi (1997).

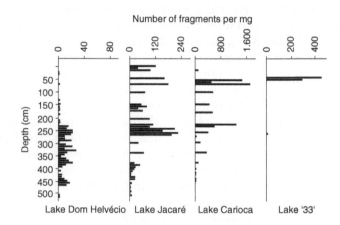

Figure 11.11   A comparison of presence of sponges in four lakes in Parque Florestal do Rio Doce (MG).
Source: Dumont and Tundisi (1997).

All of these studies on lakes and wetland areas show that palaeolimnology can be an important indicator of climatic events and human-produced alterations in drainage basins, including the rate of erosion and the transport of sediments, elements and substances.

11.8   TRANSPORT OF DISSOLVED PARTICULATE ORGANIC MATTER AND VERTICAL AND HORIZONTAL CIRCULATION IN AQUATIC ECOSYSTEMS

Figure 11.12 presents the principal processes by which particulate matter originating from organisms in the water column is moved and transported. In addition to

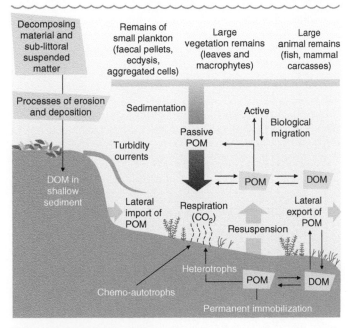

DOM – Dissolved organic material
POM – Particulate organic matter
Remains of old skin or cuticle after moulting in organisms

Figure 11.12   Principal processes of dislocation and transport of particulate and dissolved materials in a lake, in function of biogeochemical processes, and the participation of organisms in this transportation.
Source: Modified from Barnes and Mann (1991).

sedimentation, there is active transport, caused by vertical migration of organisms and the dislocation promoted by horizontal currents, diffusion and by re-suspension caused by currents and hydrodynamic dislocation at the level of sediment. There is lateral transport by density currents that occur in periods of intense drainage. All these dynamic processes contribute to the active circulation of dissolved particulate organic matter, especially in lakes, reservoirs and estuaries.

12 Man-made reservoirs

SUMMARY

Man-made reservoirs are strategically important aquatic ecosystems, since in addition to the theoretical limnological and ecological basis they provide, they are used for many and varied purposes that affect the water quality, operating mechanisms and succession of aquatic communities in rivers and drainage basins. Of key importance in the functioning of reservoirs and their physical, chemical and biological properties are type of construction, retention time, the filling period, and impact of the multiple uses on the water quality in these ecosystems.

The fish species found in a reservoir depend on colonization from the surrounding drainage basin and the introduction of exotic species into the system. Management of fisheries and fish species in reservoirs is thus a complex process, requiring a sound scientific base and long-term comparative studies.

Management of reservoirs should be based on ongoing monitoring and evaluation of the operating mechanisms, solid understanding of the limnology of these ecosystems, and application of innovative techniques based on low-cost eco-technologies and eco-hydrology integrated into the functioning of the system.

Man-made reservoirs have ecological, economic, and social interactions with their drainage basins. A reservoir is a complex system with many components and subsystems that interact and vary through space and time. To understand all the problems and operating mechanisms of these ecosystems requires an integrated approach including observation, experimentation and measurement. Networks, interactions, and direct and indirect effects all need to be qualitatively and quantitatively studied.

As complex systems, reservoirs present a hierarchy of functions, regulatory mechanisms, controls and feedback. Thus, the importance of developing models for research and **management of reservoirs** is discussed in this chapter, as well as use of eco-technology for reservoir management, with examples and models. Community structure and composition in reservoirs are described, as well as factors affecting diversity, productivity and biomass of aquatic organisms.

12.1   GENERAL FEATURES AND POSITIVE AND NEGATIVE IMPACT

The human experience of constructing reservoirs goes back thousands of years. The earliest reservoirs were initially built to store several cubic meters of water for drinking supply or irrigation. With the advances in construction technology, these aquatic eco-systems have grown to be enormous and expensive high-tech projects, used for mul-tiple purposes. All continents now have reservoirs on major rivers (see Chapter 18), causing several negative impacts, but also providing opportunities for work, power generation, and new social and economic development. Chart 12.1 presents the major positive and negative impacts of the construction of dams and reservoirs. Figure 12.1 shows the location of the main reservoirs in Brazil.

Currently, on the entire planet, the total volume of water behind dams surpasses 10,000 km^3, and occupies an area of approximately 650,000 km^2.

In addition to the practical use of reservoirs, the study of these artificial eco-systems can contribute to a deeper understanding of basic problems in ecology and limnology: succession of communities in systems with rapid changes in limnological functioning, the effects of natural and artificial pulses on aquatic ecosystems and their biota, and the interaction of physical, chemical and biological systems upstream and downstream from the reservoir.

Since reservoirs have multiple uses, it is essential to determine the water quality and monitor and assess the future impact in order to understand the interactions that occur among the uses of the drainage basin, the reservoir's multiple uses, and the preservation or deterioration of water quality.

Management of reservoirs is another important activity in this context, since arti-ficial reservoirs, unlike natural lakes or rivers, are built for various uses, and manage-ment should incorporate and optimize these multiple uses and their direct and indirect costs and impacts.

12.2   TECHNICAL ASPECTS OF CONSTRUCTING RESERVOIRS

The technology used to construct reservoirs greatly influences the physical, chemi-cal and biological features of these projects. This technology has different dimen-sions and also different processes, since it arises from the multiple uses programmed, the hydrological regime of the rivers feeding into the reservoirs, and even from the existing engineering experience in each country or region. Such technology has gone through various stages of development, culminating in enormous dams and reservoirs with extremely diversified structures used for generating electrical power, navigation and irrigation.

Most reservoirs were built for a single purpose. The earliest reservoirs were built next to rivers. After preparation of the chosen site, a channel was dug from the river to flood the area. From the perspective of water quality management, these polders dif-fered significantly from most modern reservoirs, which are formed in dammed rivers. This book focuses on modern reservoirs and not the original type.

Historically, early reservoirs were built for irrigation, then for flood prevention, and later for other uses, including increased flows for crop irrigation downstream,

Chart 12.1   Construction of dams/reservoirs: positive and negative effects.

Positive effects	Negative effects
Production of energy – **hydroelectricity**	Dislocation of populations Excessive human emigration
Establishment of low-energy water purification	Deterioration of the conditions of the original population
Retention of water in the site	Health problems by spread of transmittable waterbourne diseases
Source of drinking water and supply systems	Loss of native fish species in rivers
Representative biological diversity	Loss of fertile lands and trees Loss of floodplains and terrestrial/aquatic ecotones – natural useful structures
More prosperity for sectors of the local population	Loss of flooded lands and alterations in habitats of animals
Creation of opportunities for **recreation** and **tourism**	**Loss of biodiversity** (unique species); dislocation of wild animals
Protection against flooding of downstream areas	Loss of agricultural land cultivated for generations, such as rice fields Excessive human immigration to the region of the reservoir, with the accompanying social, economic and health problems
Increase in fishing possibilities	Reducing the downstream flow of the reservoir and increasing their variances The need to compensate for the loss of agricultural lands, fishing sites and recreation and subsistence activities **Degradation of the local water quality** Reduced flow downstream and greater variations
Storage of water for periods of drought	Reduced temperature and suspended material in flows released downstream
Navigation	Reduction of oxygen at the bottom and in the flows released (zero in some cases) Increased H_2S and CO_2 on the bottom and in the flows released Barrier to fish migrations
Increase in irrigation potential Creation of jobs	Loss of valuable water and cultural resources. For example, the loss, in Oregon (USA), of many indigenous cemeteries and other sacred sites, compromising the cultural identity of some tribes
Promotion of new regional economic alternatives	Loss of aesthetic values
Control of flooding	Loss of terrestrial biodiversity, especially around dams/reservoirs in the Amazon region Increase in the emission of greenhouse gases, principally in reservoirs where the native flora is not deforested
Increase in production of fish for aquaculture	Introduction of exotic species in aquatic ecosystems **Impacts on aquatic biodiversity** Excessive extraction of water

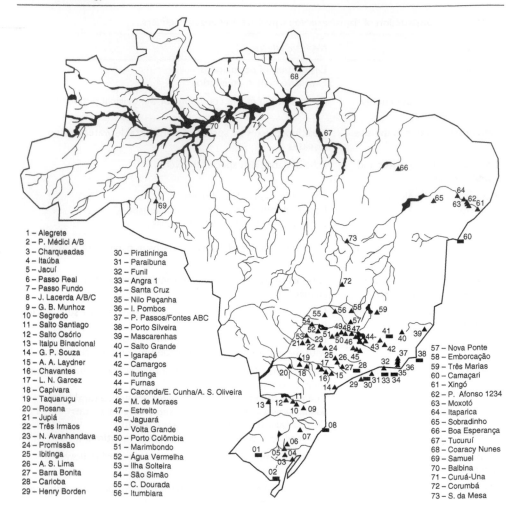

1 – Alegrete
2 – P. Médici A/B
3 – Charqueadas
4 – Itaúba
5 – Jacuí
6 – Passo Real
7 – Passo Fundo
8 – J. Lacerda A/B/C
9 – G. B. Munhoz
10 – Segredo
11 – Salto Santiago
12 – Salto Osório
13 – Itaipu Binacional
14 – G. P. Souza
15 – A. A. Laydner
16 – Chavantes
17 – L. N. Garcez
18 – Capivara
19 – Taquaruçu
20 – Rosana
21 – Jupiá
22 – Três Irmãos
23 – N. Avanhandava
24 – Promissão
25 – Ibitinga
26 – A. S. Lima
27 – Barra Bonita
28 – Carioba
29 – Henry Borden

30 – Piratininga
31 – Paraibuna
32 – Funil
33 – Angra 1
34 – Santa Cruz
35 – Nilo Peçanha
36 – I. Pombos
37 – P. Passos/Fontes ABC
38 – Porto Silveira
39 – Mascarenhas
40 – Salto Grande
41 – Igarapé
42 – Camargos
43 – Itutinga
44 – Furnas
45 – Caconde/E. Cunha/A. S. Oliveira
46 – M. de Moraes
47 – Estreito
48 – Jaguará
49 – Volta Grande
50 – Porto Colômbia
51 – Marimbondo
52 – Água Vermelha
53 – Ilha Solteira
54 – São Simão
55 – C. Dourada
56 – Itumbiara

57 – Nova Ponte
58 – Emborcação
59 – Três Marias
60 – Camaçari
61 – Xingó
62 – P. Afonso 1234
63 – Moxotó
64 – Itaparica
65 – Sobradinho
66 – Boa Esperança
67 – Tucuruí
68 – Coaracy Nunes
69 – Samuel
70 – Balbina
71 – Curuá-Una
72 – Corumbá
73 – S. da Mesa

Figure 12.1   Locations of main hydroelectric (electricity-generating) reservoirs in Brazil (height of dams greater than 15 m).

navigation, **drinking-water supply**, fishing, industrial water supply, and most recently, generation of electrical power and recreation.

Fishing resources are a byproduct introduced in temperate regions for recreational purposes and in the tropics for food production. In time, most reservoirs have come to serve secondary functions.

Storage of a certain quantity of water is usually the primary goal of reservoir management. With the increase in environmental degradation and multiple uses of water, there are growing issues on water quality in these systems. For supplies of drinking water, the most stringent restrictions on water quality apply. In addition, some technical procedures require water quality to comply with certain standards. Fish cannot develop and provide food for humans in heavily polluted waters. Recreational activities, another traditional use, also call for relatively clean water.

Chart 12.2 Characteristics of reservoirs constructed for various primary purposes.

Primary use	Size	Depth	Retention time	Depth of outlets
Protection against floods and control of flow	Small to medium	Shallow	Depends on the region	Surface
Water storage	Small to medium	_	Variable	Below the surface
Hydroelectricity	Medium to large	Deep	Variable	Near the bottom
Drinking water	Small	Better deep	Long	Medium to deep
Aquaculture	Small	Shallow	Short	Surface
Reservoir of water for pumping	Small to medium	Deep	Variable	Near the bottom
Irrigation	Small	Shallow	Long	Surface
Navigation	Large	Deep	Short	Totality
Recreation	Small	Shallow	Long	Superficial

Source: Straškrába and Tundisi (2000).

Both aspects – quantitative and qualitative – are interrelated. We cannot use more water than the volume available, and low levels cause deterioration in quality. This relationship is a typical problem in reservoirs, and is the source of numerous issues for management. Similarly, possible damage to the domestic or industrial supply can affect fishing, recreation and multiple uses downstream. Deterioration occurs in the flows released by the reservoir even when the water itself is not the direct cause: the cause may be a low flow, rich in nutrients. The multiple uses of many tropical reservoirs create conditions for the spread of waterborne diseases.

In addition to the main uses for which reservoirs are constructed, other uses include:

▸ They help purify water, since they eliminate impurities and retain sediment, organic matter, excess nutrients and other pollutants.
▸ They often serve as places of leisure, with lake activities (such as swimming, canoeing, motor-boating, sailing, waterskiing, fishing, rowing and ice-skating) and land activities (such as fishing, hiking, bird-watching, sun-bathing and camping).
▸ They represent a biological resource that can be local, providing the following agricultural activities: nursery for fish, aquaculture and production of aquatic plants, such as reeds or other species.
▸ Some parts of reservoirs can be used for aquatic plants, birds or other animals, or even as areas of aesthetic value.

Aspects of the reservoir's construction (for example, the overall volume in relation to stream flow or the position of the intakes and spillways) affect the water quality. The construction features are related to the primary purpose for which the reservoir was built. The purpose affects its size, the site chosen for construction of the dam, the height determined by the morphometry of the valley, the storage volume and the capacity in relation to the inflows, a factor which determines retention time (see Chart 12.2). However, these parameters only represent averages, and deviations in a specific reservoir can be quite significant.

Dussart (1984) distinguishes six basic types of reservoirs in terms of **water storage**: reservoirs that store water for various purposes, reservoirs for fluvial regulation and navigation, reservoirs for short-term storage of low volumes of water, reservoirs with several inflow sources from the drainage basin (several rivers), reservoirs for water quality control, and reservoirs with water pumped from downstream.

Some of the main purposes for constructing reservoirs have already been discussed. Other purposes include irrigation, navigation, recreation, and site for sewage disposal. Recently, however, most reservoirs have multiple uses. Nowadays it is common for all types of reservoirs to be used for recreation, to generate electricity, and for many other uses. These multiple uses can cause conflicts among the different users, and such conflicts need to be addressed by the managers.

From the perspective of water quality, the location and type of discharge mechanism (into the river downstream or outlets for different purposes) are the most important technical aspects in a reservoir project.

12.3   IMPORTANT VARIABLES IN THE HYDROLOGY AND FUNCTIONING OF RESERVOIRS

12.3.1   Effects of construction of a reservoir on the fluvial regime

In a river in its natural condition, there is a continual gradient of physical conditions from the river's sources (springs) to its mouth. The conditions of the construction of a reservoir, and of its biota, depend on the reservoir's position within the hydrographic network. Ward and Stanford (1983) developed a classification system that distinguishes 12 types of rivers. According to their method, the first type represents streams immediately after the spring. The second is the union of two creeks of the first type. The third is the joining of two rivers of the second kind. and so on. When a reservoir is constructed along a river, the river's physical, chemical and biological characteristics are affected to a greater or lesser degree. The effects on areas downstream from a reservoir are determined by the position of the dam in relation to the course of the river; consequently, by its classification. At some distance below the dam, the river returns to its natural characteristics, as if it had not been dammed. The distance for this recuperation is called the recovery distance, i.e., that distance at which a certain number of variables recover their previous status; it also expresses the degree of interference in the current conditions of the river (Ward *et al.*, 1984).

In terms of the reservoir's water quality, both the location of the dam in relation to the course of the river (its type) as well as the dam's height are determined by several important hydrological characteristics. These include the flow, form of the valley, temperature of the tributary waters, solar radiation, turbidity, and the chemistry of the nutrients that affect the biota. For example, Figure 12.2 illustrates some of the main differences between reservoirs located in rivers of different types:

▶   A reservoir built on the first type of river is located in mountainous areas undisturbed by development and is fed by a small stream with the following predictable characteristics: low flow rate and temperature, low levels of organic matter

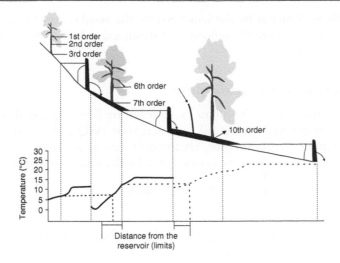

Figure 12.2 Effects of the type of river on the characteristic of the reservoir and distance of re-initiation downstream from the dam/reservoir.

and nutrient salts, scarce plankton and fish that typically feed on benthos. The reservoir will typically be located in a deep valley with steep slopes. Such a mountainous location usually has low temperatures and high levels of humidity, precipitation, and solar radiation. The reservoir is deep, stratified, with longitudinal flow, and oligotrophic. There are no horizontal gradients, or when there are, they are small. Any differences between two such reservoirs in the same geographic region will be a function of **geological characteristics** (calcareous or non-calcareous rocks) or environmental factors, such as degree of exposure to the sun and wind (which affect temperature and mixing).

▶ A reservoir built in the middle section of a river is fed by a course of water with the following characteristics: an average flow rate, moderate slope, medium temperatures, higher levels of organic matter and nutrient salts, occasional turbidity, a well-established phytoplanktonic community, and fish that can survive in standing water. The limnology of an unpolluted reservoir is largely determined by the morphology of the valley. **Shallow reservoirs** are usually unstratified, unlike deep ones. Another important factor is the **theoretical retention time**, determined by the specific flow rate and volume of the hydric body. The theoretical retention time can vary greatly. In small reservoirs, stratification is not very pronounced, and the planktonic biomass in not very developed. Larger reservoirs, with longer retention times, present horizontal and vertical gradients in the well-developed physical and chemical variables, a reasonable growth of plankton and species of fish normally found in lakes.

▶ Reservoirs built on the lower stretches of rivers typically have gentle slopes and are characterized by the flooding of large areas with enormous horizontal variability, well-developed wetland-area communities, with large shoals and natural vegetation. These reservoirs are generally eutrophic and carry a large organic load that contributes to the formation of an anoxic bottom. Shallow reservoirs are usually well mixed by the wind: hence, the conditions of stratification only develop in areas where the influence of depth exceeds that of the surface layers affected by the wind.

12.3.2 Flow and retention time

The quotient of the reservoir's volume, V, to tributary flows, Q (per day or per year), determines the reservoir's theoretical retention time (V/Q), also known as the residence time, hydraulic retention time, retention rate or wash rate.

The theoretical retention time is calculated by the following equation:

$$R = V/Q \text{ (days)}$$

where:
Q = average daily flow (m^3/s) multiplied by the number of seconds in a day (86,400)
V = volume of the reservoir (in m^3)

For greater accuracy, the time is calculated per year or shorter but sufficient time period. If the water level (and thus the volume of the reservoir) varies significantly, R is calculated separately for each sub-period (week, month) and then the average value is calculated.

The theoretical retention time is calculated during the 'filling' of the reservoir, and is measured by the number of days it takes to reach full capacity (through the flow and rainfall occurring during this period, which may differ from the averages over long periods of observation). R does not provide information on the current average retention time or volume of water in the reservoir. There may be cases where certain volumes of water pass through the reservoir in a much shorter time than the theoretical value calculated (in a 'shortcut' or submerged current). Fluctuations in water level can not only cause changes in retention time, but also increase the erosion of the banks, leading to greater turbidity and other negative effects on water quality.

Retention time is one of the main differences affecting water quality in reservoirs. This axiom is more pronounced in deep and stratified reservoirs than in shallow non-stratified reservoirs. Add to this the fact that inflowing rivers cause greater mixing in the former group than in the latter.

Reservoirs in dammed rivers usually have longitudinal zones caused by the one-way flow of the water.

12.3.3 Depth, size and shape of the drainage basin

The **depth of the reservoir** greatly influences water quality. Depth is extremely relevant in relation to surface area and the intensity of the wind in the region. These elements are important because they affect the intensity of the mixing in the reservoir. A reservoir can be classified as **hydrologically shallow** when it is completely mixed by wind action, and **hydrologically deep** when the mixing is insufficient and cannot prevent stratification of the liquid mass (Straškraba *et al.*, 1993).

The conditions of vertical and horizontal mixing in a reservoir are also related to its volume and size. Table 12.1 categorizes reservoirs according to size.

Table 12.1 Categories of reservoirs based on size.

Category	Area (km²)	Volume (m³)
Large	10^4–10^6	10^{10}–10^{11}
Medium	10^2–10^4	10^8–10^{10}
Small	1–10^2	10^6–10^8
Very small	<1	$<10^6$

Source: Straškrába and Tundisi (2000).

The morphology of the reservoir basin is determined by the natural characteristics of the valley to be dammed. The normal form is triangular, with the shallow part by the influx of the river and the deepest near the dam. The location of each reservoir in the context of its surrounding drainage basin is unique, differing from natural lakes, which usually occupy the central part of the basin.

12.3.4 Location of discharge mechanisms in reservoirs

Reservoirs with primary and secondary functions, such as water storage for different purposes, are classified according to the following types of discharge mechanisms: those with a single outlet that releases water downstream from the reservoir, and those with discharge mechanisms designed to meet a specific goal. In both cases, the location of the reservoir, the design of the discharge structures and their operation are **hydrological factors** influencing water quality. The design of these mechanisms affects the conditions of stratification in the reservoir. Water quality can shift rapidly in reservoirs with marked stratification when large quantities of water are drained from specific levels, which is why these variations must be considered in selecting a specific water level based on previous observations on its quality.

Typically, water can flow out from a reservoir from one of three levels: from the surface (pouring over the crest of the dam), from the bottom (**bottom discharges**) and through outlets into turbines or the river downstream. Figure 12.3 illustrates the depth of the outlets chosen from a series of reservoirs. In some cases, the mechanisms are located at a certain depth, as in dams in which the primary purpose is generating electricity. Those outlets are typically quite large, in some cases sufficient to drain the entire reservoir. These features are important since they help to determine the stratification of water quality in the reservoir. Differences between the water quality in lake and reservoirs are mainly explained by the fact that lakes drain from the surface while reservoirs typically drain from deeper layers or from the bottom.

Multiple-outlet structures are sometimes built in a dam to provide the choice of withdrawal from the optimal layer to be treated for human consumption. These modifications enable higher-quality water to be extracted from different depths at different times. However, stratification of water quality within the reservoir depends, among other things, on the withdrawal of certain sections of water. Extensive withdrawals from a certain level can lead to major changes in stratification.

Thus, even if a certain layer with good-quality water is identified, its position can shift during withdrawals of large volumes.

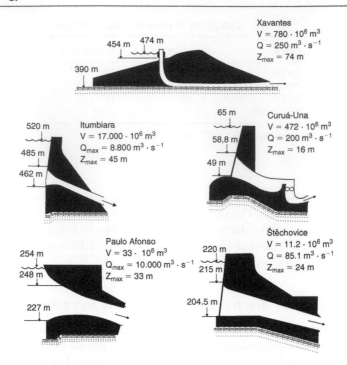

Figure 12.3   Profiles of several reservoirs in Brazil and the Czech Republic, with the depths and shapes of water outlets.
Source: Straškrába and Tundisi (2000).

12.4   INTERACTIONS OF RESERVOIRS AND DRAINAGE BASINS – MORPHOMETRY OF DAMS

An important feature in the interaction of a reservoir with its drainage basin is the modification of 'ecological filters' that act as selective factors for communities, populations and species. The mosaic terrestrial-aquatic system is modified into an aquatic system in which spatial, vertical and longitudinal **micro-heterogeneity** is a predominant forcing function in the distribution of organisms and the spatial organization of communities.

Sedimentation is an extremely important process, since it limits the lifespan of the reservoir, reducing the hypolimnion. This sedimentation is a result of the use of the drainage basin; for example, deforestation accelerates sedimentation either through the effects of rain or wind. The initial influx of material into the reservoir is a result of the drainage basin's geochemistry and activities occurring in the drainage basin before the closing of the reservoir, such as deforestation and the use of fertilizers and pesticides.

The changes produced by the reservoir in the regional ecosystem depend, however, on the reservoir's morphometry and the morphometric characteristics of the area to be flooded. There are two basic types of reservoirs in terms of morphometry:

▶   Reservoirs with a marked dendritic pattern and complex morphometry – presenting a high shoreline development index. In this case, the number of compartments,

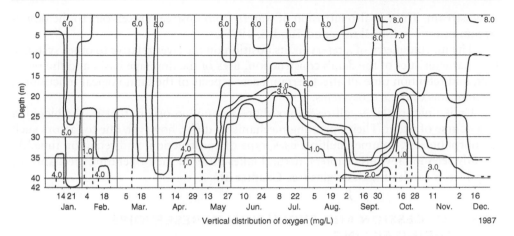

Figure 12.4   Seasonal cycle of the vertical distribution of dissolved oxygen in the Tucuruí reservoir, in the Tocantins River, in the first years of the flooding phase, during which massive decomposition of the flooded forest occurred.
Source: Tundisi *et al.* (1993).

as well as the processes of accumulation of material and **compartmentalized circulation,** are very important.

▶ Reservoirs with a simple morphometric pattern along the longitudinal axis – with few compartments and low shoreline development index. In this case, circulation processes may be less complex, the mechanisms of accumulation of material and transport of sediments occur along a more accentuated longitudinal axis, and the retention time of elements and substances is generally shorter (on the order of several days or maximum a few weeks).

The **morphometry of reservoirs** greatly affects the dynamics of the processes occurring in the water and sediment, including wind action and circulation mechanisms induced by wind, cooling and thermal heating. Reservoir morphology also has important consequences for the operating mechanisms affecting eutrophication. In general, eutrophication begins in the upper parts of the reservoir, where movement is reduced and retention time is longer. Processes of cultural eutrophication are most common along the many channels or compartments of the reservoir, resulting in a progressive eutrophication from the various compartments. In cases with several levels of eutrophication, the ratio of carbon:nitrogen:phosphorus (C:N:P) differs in the water and in the composition of phytoplankton.

Compartmentalization in reservoirs produces a large number of subsystems, which can significantly affect the water quality in the major axis of the reservoir: a process of anoxia can occur in the compartments with poor circulation, due to the reduced circulation and the accumulation of decomposing organic material (see Figure 12.4).

There are other fundamental characteristics of a reservoir in terms of its position on the river and in relation to other reservoirs. For example, in the state of São Paulo and in southern Brazil, there are many series of dams on rivers, which create important interactive mechanisms, including increase of phytoplankton toward the reservoirs downstream as well as dilution due to the greater volume of reservoirs downstream. There is also intrusion of waters from upstream reservoirs to the downstream reservoirs (Straškraba and Tundisi, 1999).

The morphometry of a reservoir, its position on the river relative to other systems, the characteristics of the dam's construction, uses of the drainage basin and the reservoir, and retention time all affect certain basic aspects of the limnological functioning mechanisms of these ecosystems.

Unlike natural lakes, all reservoirs share a common origin, that is, the blocking of a course of water. Their operating mechanisms differ, however, due to factors such as morphometry, volume, multiple uses, type of construction, and retention time. As already seen in previous chapters, natural lakes differ in their operating mechanisms, largely as a result of their origins (see Chapter 3).

12.5   SUCCESSION AND EVOLUTION IN RESERVOIRS DURING FILLING

Immediately after completion of a dam, a series of changes occur in the river, beginning with a significant decrease in the current and a progressive increase in lacustrine conditions. The decrease in dissolved oxygen can be rapid and drastic, especially in cases where flooding of large masses of vegetation occurs. The following sequence of conditions can be observed in the phase of filling the newly forming lake:

▶ **River conditions:** Dissolved oxygen levels close to those in the river, and turbulence also similar to that in the river.
▶ **Transition conditions:** In regions where the current has already slowed, dissolved oxygen levels and turbulence drop sharply.
▶ **Lacustrine conditions:** In deep areas, thermal stratification develops with a generally anoxic hypolimnion, with no turbulence.

Generally, the composition of plankton reflects these conditions, and the transition zone progressively decreases during filling.

12.6   RESERVOIR SYSTEMS

The term 'reservoir system' refers to the construction of a series of dams, hydrologically connected, whose operations are interrelated with a common goal, such as the storage of water or the generation of electricity (see Figure 12.5). **Cascade reservoirs** are chains of reservoirs located on the same river. **Multiple reservoir systems** are groups of reservoirs located on different stretches of one river or on several river systems with interconnected flows. **Pump reservoir systems** are characterized by pumped water that circulates between the reservoirs. **Hydric transfers** occur in one or more reservoirs when water is withdrawn and pumped to another fluvial system in order to increase the flow in the latter.

Cascade reservoirs: In terms of water quality, cascade reservoirs are characterized by the fact that the effects of a reservoir upstream are transferred to a reservoir located downstream. The water quality in the upstream unit is usually similar to other isolated reservoirs. The water quality in the second reservoir, or reservoirs, is, as a rule, altered. The capacity of a reservoir to affect another one downstream depends on

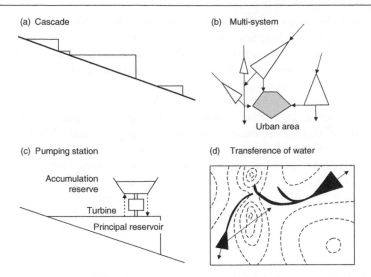

Figure 12.5 Types of reservoir sytems.
Source: Straškrába and Tundisi (2000).

its characteristics, whether it is a deep stratified reservoir (marked effects) or a shallow reservoir (lesser effects). The degree of influence also depends on the classification (type) of the stretch of river connecting the two hydric bodies, the trophic level of the upstream reservoir, and the distance between the two. A reservoir located on a river with a higher classification has a longer retention period and its effects are greater on the river downstream. The distance between the two reservoirs is also significant. Hundreds of kilometers downstream from the first reservoir, the river returns to its natural state and the effects of that system are no longer active. Effects are thus more significant when the two reservoirs are closer together.

Multiple reservoir systems: These are complex multi-purpose infrastructures used to store a water supply in places and periods where there is water shortage, especially in countries with water deficits. The water quality in these systems varies greatly, and is a function of the varying flow rate. Especially in cases where the participating reservoirs are located in different geological formations (and thus have different nutrient levels), coordinating the management of the quantitative and qualitative aspects of the water in each reservoir can be a challenging task.

Pump reservoir systems: These systems are built where the demand for electric power is uneven throughout the day and throughout the week on different days. There is an excess supply of electrical power during certain periods and a shortage in other times. During a period with excess supply, water is pumped to a reservoir (often smaller) located at a higher altitude. The difference in quotas will be used to increase energy production during periods of higher demand. The pumping and storage will not significantly affect the water quality. Therefore there are generally no substantial differences between the two hydric bodies, although in some cases differences can occur.

Hydric transfers: Long ago large and extensive aqueducts were built. The total volume of water transferred through these ancient systems to other basins was not very high. Today, however, many systems have a huge transfer capacity, and this can affect not only water quality, but also the entire hydrological equilibrium of the region. For example, the Aral Sea was transformed from a clear lake into a filthy puddle as a result of poor management that withdrew enormous amounts of water for an ambitious irrigation project on large cotton farms. Drastic changes occurred in the hydrological regime of the entire region.

Hydric transfers can cause many changes. They sometimes become the vehicle for spreading waterborne diseases in addition to causing water quality to deteriorate and complex chemical changes to occur, affecting local populations. When these transfers are linked to irrigation, **salinization** of crops can occur.

In the semi-arid regions of southeastern Australia during the 1920s, several systems were constructed to transfer water from the abundant rivers of the Australian Alps, which flow to the **Pacific Ocean,** to large arid areas in New South Wales and southern Australia. The subsequent salinity documented in the crops created many problems for agriculture, and many areas are now considered 'dead'.

12.7 PRINCIPAL OPERATING PROCESSES AND MECHANISMS IN RESERVOIRS

The **spatial organization of reservoirs,** in most cases, presents great heterogeneity, with a wide range of physical and chemical conditions of the water and modifications of these along the horizontal and vertical axes. Operating conditions that depend in part on the multiple uses and type of construction include the existence of *vertical and horizontal gradients, retention time* (an essential parameter in the control of physical and chemical conditions of water in reservoirs), *hydraulic stratification, transport of sediment, sediment-water* interactions, the *vertical and horizontal transport system* and the *composition, diversity* and *structure of biological communities* (Straškraba *et al.,* 1993; Straškraba and Tundisi, 1999; Tundisi and Straškraba, 2000).

12.7.1 Circulation in reservoirs

The vertical and horizontal circulation mechanisms in reservoirs depend on factors involved in the following processes:

▶ Inflow of tributaries into reservoirs, which produce distinct longitudinal gradients and spatial heterogeneity. For example, in the Barra Bonita reservoir in the state of São Paulo, Matsumura Tundisi and Tundisi (2003, 2005) showed the existence of 114 tributaries, which profoundly altered the gradients in the system (vertical and horizontal).
▶ Climatic factors such as wind or precipitation, which can cause turbulence and vertical gradients. Rainwater flowing down from the slopes of tributaries produces **vertical heterogeneity** in the system, affecting the distribution of phytoplankton and zooplankton. In many cases, this intrusion fertilizes the eutrophic layers in a reservoir, creating conditions for the blooming of certain phytoplanktonic species, and, in other cases, promotes cyanobacterial growth (Matsumura Tundisi and Tundisi, 2003).
▶ Effects of spillways and outlets into turbines.

Reservoirs differ from lakes in that they have a surface outlet or bottom outlet for the spillways and turbines, which creates horizontal and vertical gradients, including the aforementioned effect of hydraulic stratification.

The operational strategies in reservoirs affect the limnological processes, according to the physical and chemical properties of the masses of water, as well as the conditions and characteristics of the biological communities (Kennedy and Walker, 1990; Armengol *et al.*, 1999; Kennedy, 1999).

In a study by Kennedy (1984) on three reservoirs in the United States (the Degray, Red Rods and West Point reservoirs), there are distinct differences in the development of horizontal gradients. Agostinho and Gomes (1997) identified similar longitudinal gradients in the Segredo reservoir in the state of Paraná in Brazil.

Henry (1999) conducted an in-depth study on thermal balances, thermal structure and dissolved oxygen in Brazilian reservoirs. The results showed a strong **latitudinal dependence of temperature** of the surface water, based on the latitudinal variations in solar radiation. Stable thermal stratification was identified in the lacustrine zone of reservoirs with retention time >40 days. Climatic and morphological factors affected the thermal equilibrium in the reservoirs.

Table 12.2 presents the morphometric parameters and retention time in two reservoirs studied by Henry (1999). Table 12.3 shows the **thermal amplitude** in the water column for ten reservoirs at different latitudes in Brazil, where the **effects of altitude and seasons** were also fundamental for the thermal structure in the reservoirs.

In some reservoirs, thermal stratification occurs over long periods, such as in the Segredo reservoir (Thomaz *et al.*, 1997), or it can be sustained for many months, such as in the **Jurumirim reservoir** in the state of São Paulo (Henry, 1993a).

Dissolved oxygen deficits occur in reservoirs with long-term stable stratification, as shown in Table 12.4 (Henry, 1999a).

Table 12.2 Morphometric parameters and theoretical retention times of several reservoirs in Brazil.

Reservoir	Latitude	Longitude	Elevation M.H.S.L (m)	Area (km^2)	Zmin (m)	Zmax (m)	Theoretical retention time (days)
Tucurui	3°43'S	49°12'W	72	2,430	17.3	75	51
Boa Esperança	6°45'S	43°34'W	304	300		~35	196
Paranoá	15°48'S	47°45'W	1,000	40	14.3	38	300
Três Marias	18°15'S	44°18'W	585	1,120	6.8	~30	29
Pampulha	19°55'S	43°56'W		2.4	5.0	16	120
Volta Grande	20°10'S	48°25'W		222	10.2		25
Monjolinho	22°01'S	47°53'W	812	0.05	1.5	3.0	~10
Dourada	22°11'S	47°55'W	715	0.08	2.6	~6.3	
Jacaré	22°18'S	48°01'W	600	0.003	0.9	~2.2	11
Jacaré-Pepira	22°26'S	48°01'W	800	3.7	3.0	12	
Jurumirim	23°29'S	49°52'W	568	446	12.9	40	322
Das Garças	23°39'S	46°37'W	798	0.09	2.1	4.6	69
Itaipu	25°33'S	54°37'W	223	1,460	21.5	140	40

M.H.S.L. – Maximum height above sea level
Source: Henry (1999a).

Table 12.3   Thermal amplitude (ΔT, annual average) in the water column (Δz) for reservoirs in Brazil.

Reservoir	ΔZ (m)	ΔT (°C)	Year	Author(s)
Tucurui	72	1.27	January–December 1986	Henry (1999a)
Paranoá	11	2.09	March 1988 to March 1989	Branco (1991)
Três Marias	30	3.10	March 1982 to February 1983	Esteves et al. (1985)
Pampulha	12	3.64	November 1984 to November 1985	Giani et al. (1985)
Monjolinho	2.5	3.64	March 1986 to March 1987	Nogueira and Matsumura Tundisi (1994)
Jacaré	1.4	4.21	January 1990 to March 1991	Mercante and Bicudo (1996)
Jacaré-Pepira	6.5	0.98	August 1977 to November 1978	Franco (1982)
Jurumirim	30	2.38	March 1988 to March 1989	Henry (1992)
Das Garças	4.6	1.49	January to December 1997	Henry (1999a)
Itaipu	140	5.30	May 1985 to June 1986	Brunkow et al. (1988)

Source: Henry (1999a).

Table 12.4   Comparison of deficits of dissolved oxygen in Lake Kariba (Africa) and some reservoirs in Brazil.

Lake/Reservoir	Year	O.D. $(mg\ O_2 \cdot cm^{-2})$	Reference
Kariba-Basin III	1964–1965	4.47	Coche (1974)
Kariba-Basin II	1964–1965	10.14	Coche (1974)
Dom Helvécio	1978	1.73–2.37	Henry et al. (1989)
Jurumirim	1988–1989	0.03–0.72	Henry (1992)
Das Garças	1997	0.40–1.52	

O.D. – Oxygen deficit.
Source: Henry (1999a).

In summary, circulation in reservoirs depends on a hierarchy of climatic and hydrological factors and operating rules and mechanisms related to the multiple uses of these man-made ecosystems. Since many reservoirs are shallow (<30 m), they are subjected to wind action, which frequently causes circulation throughout the complete system. Thus the majority of these reservoirs are polymictic, especially in southeastern Brazil, where the effects of cold fronts play an extremely important role (see Chapter 20).

Reservoirs with *thermal stratification* resulting from thermal heating, or with *hydraulic stratification* resulting from operating systems or structures, can have periods of anoxia in the hypolimnion, causing serious problems in management of the river or the reservoirs downstream and affecting structures with turbines (Junk et al., 1981; Straškraba et al., 1993).

Tundisi et al. (2001) conducted studies related to the Wedderburn number, vertical distribution of phytoplankton, and circulation in shallow reservoirs in southeast Brazil.

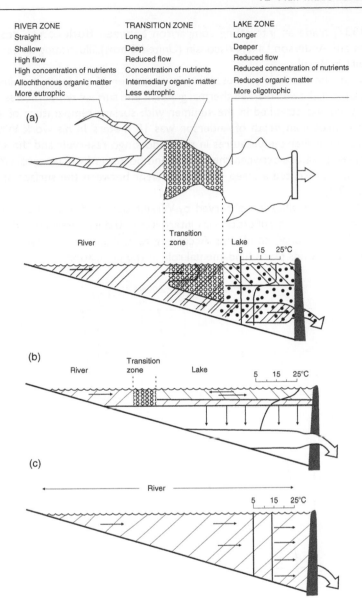

RIVER ZONE
Straight
Shallow
High flow
High concentration of nutrients
Allochthonous organic matter
More eutrophic

TRANSITION ZONE
Long
Deep
Reduced flow
Concentration of nutrients
Intermediary organic matter
Less eutrophic

LAKE ZONE
Longer
Deeper
Reduced flow
Reduced concentration of nutrients
Reduced organic matter
More oligotrophic

Figure 12.6   Longitudinal zones of a reservoir (Kimmel and Groeger, 1984) and alterations in the extension of the zones, flow and mixing pattern for different values of R (retention time). a) $10 < R < 100$ days; b) RR > 100 days; c) R 10 days.
Source: Straškrába and Tundisi (2000).

Essential features of reservoirs include the existence of horizontal and vertical gradients and a continuous flow toward the dam (Armengol *et al.*, 1999) (see Figure 12.6 a, b, and c). These gradients present temporal variations that depend on the flow of water toward the reservoir and on seasonal differences in level occurring during

Wright (1937) made an interesting comparison between **Bodocongó reservoir** in the Paraiba and Anderson Lake, Wisconsin (United States), illustrating an early interest in reservoirs in Brazil.

Although the Bodocongó reservoir is shallower (6 m depth maximum) than Anderson Lake, it showed marked thermal gradient and surface temperatures of 28.5°C. Anderson Lake was stratified in the summer, with surface temperature of 20.3°C. At the time, the maximum depth of Anderson was 18 meters. In his work, Wright drew attention to the higher temperatures in the Bodocongó reservoir and the smaller differences in temperature between the surface and bottom of the reservoir. Meanwhile, Anderson Lake presented a steep thermal gradient between the surface and bottom (see Figure 12.7).

The thermal stratification observed by Wright occurred in all the reservoirs he studied in Paraíba. This stratification was intermittent and its persistence depended on temporary (not seasonal) conditions. According to the author, in these reservoirs of Northeast Brazil, winds play a fundamental role in the circulation.

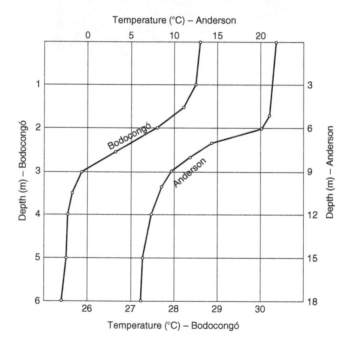

Figure 12.7   Comparison between the **Bodocongó reservoir** (Northeast Brazil) and **Lake Anderson** (Wisconsin, USA).
Source: Wright (1937).

the year. Vertical gradients are more pronounced if advection currents are distributed at the various different depths as a result of stratification produced by the inflow of denser, colder water through from tributaries feeding into the reservoir (Imberger, 1985).

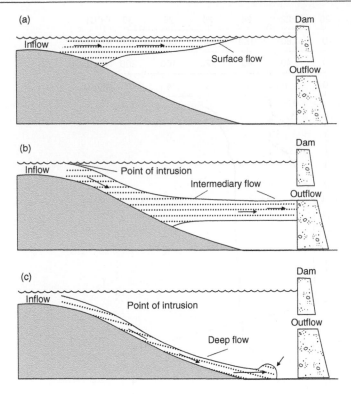

Figure 12.8   Three types of density currents in a stratified reservoir: a) surface, b) intermediate and
c) deep. The entrance point of the flows is indicated.
Source: Straškrába and Tundisi (2000).

There are three main types of inflows of water into reservoirs. These types of advection currents are illustrated in Figure 12.8. Horizontal zonation – characterized by physical, chemical and biological gradients – may be more pronounced in reservoirs than in lakes. For example, the Slapy reservoir in Czechoslovakia presents horizontal zonation caused by the different flows in the rivers and in the transition zone, a region of more stagnant water downstream and another zone with more rapid flow resulting from the effects of the turbines. In reservoirs, these horizontal gradients can affect the composition of the community and the reproductive phase of phytoplankton and zooplankton, imposing specific conditions with selective significance.

In addition to vertical and horizontal gradients, a reservoir, depending on its type of operation, can present different retention times during the seasonal cycle. Retention time is an important forcing function in light of the modifications that can occur in a reservoir's vertical and horizontal structures and vertical distribution of phytoplanktonic populations. In addition, fluctuations and alterations in retention time affect the spatial and temporal succession of phytoplankton, frequency of cyanobacterial blooms, and chemical composition of the sediment.

Another important operating mechanism is 'hydraulic stratification', where thermal and chemical vertical stratification occurs not as a result of the interaction of

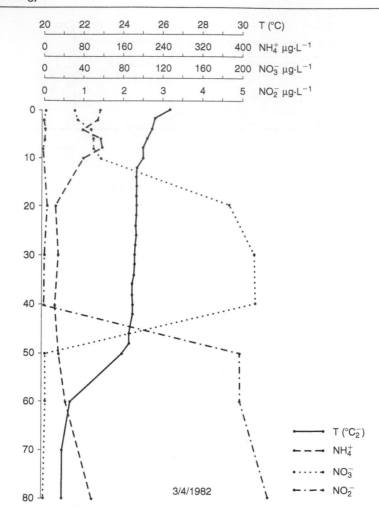

Figure 12.9 Consequences of hydraulic stratification in the Furnas reservoir illustrated by the deep accumulation of ammonia and distinct vertical thermal gradients. The water flows out through turbines at 50 m depth.
Source: Tundisi (1986a).

climatic/hydrographic processes but due to the height of the outlets for the turbines. The varying heights produce changes in the vertical axis, including density gradients. Increased anoxia and presence of H_2S are two important consequences of hydraulic stratification in the artificial hypolimnion (Tundisi, 1984) (see Figure 12.9). The height of the water outlet in the reservoir is therefore a major forcing function, and is also important in the vertical and horizontal circulation processes.

   In addition to hydraulic stratification, which can be pronounced in small reservoirs ($20–100 \cdot 10^6$ m^3 in volume), this 'selective withdrawal' of water can lead to additional turbulence and mixing, which in turn results in **micro-compartmentalization** in the horizontal and vertical gradients.

12.8 THE BIOCHEMICAL CYCLES AND CHEMICAL COMPOSITION OF RESERVOIR WATER

Nutrient cycles in reservoirs present very particular characteristics. In a **series of reservoirs** located on a single river, each dam/reservoir system eliminates a portion of the nutrient cycle, causing a progressive drop in dissolved phosphorus and nitrogen levels in the water. Retention time is important in this cycle, as well as the interactions of climatic and hydrological processes. The uses of the system must also be taken into account. For example, when the reservoir's level drops – whether due to regulation of the flow or water shortage – large portions of the littoral area end up exposed, which can cause rapid decomposition. After the first rains, the decomposed matter is carried into the reservoir, fertilizing it with a pulse of nutrients that interrupt the nutrient cycle with increased dissolved particulate phosphorus and nitrogen.

Major factors affecting the nutrient cycle in reservoirs include:

▸ contributions from feeder rivers with surface runoff;
▸ contributions from advective processes;
▸ retention time;
▸ height of the reservoir's water outlets;
▸ stratification and turbulence;
▸ control of flow, change in level;
▸ interactions of the biological communities, including decomposition, excretion, removal of sediments by actions of benthic or nektonic organisms.

The *transport of sediments* differs significantly from one reservoir to another, depending on the *flow*, the transversal section of the rivers that form the reservoir, and the type of sediment transported. Transport of sediment affects the *penetration of solar energy* into the reservoir and the nutrient cycle. Sediment particles generally provide substrates for bacteria, due to the accumulation of organic matter on the surface. In addition, depending on the granulometry of the sediment, these particles can affect the vertical distribution and penetration of solar energy and the primary production of phytoplankton (Kirk, 1985; Rodrigues, 2003).

These interactions influence the production potential and nutrient-recycling rate potential, since the relationship z_{eu}/z_{af} is fundamentally related to the vertical thermal structure. Slow circulation rates in the euphotic zone indicate slow nutrient-recycling rates and slow phyoplanktonic transport rates from the upper portion of the water column to the less-illuminated regions and aphotic zone. High chlorophyll levels (for example, during the eutrophication process) are also a factor in limiting the penetration of solar radiation.

In addition to the importance of the absolute instantaneous values of z_{eu}/z_{af} and z_{eu}/z_{max}, temporal (daily, monthly, annually) and spatial variations in these interactions should also be considered. Among the operating mechanisms in reservoirs, it is important to take into account *sediment-water* interactions and the *chemistry of the sediment*. These two systems are related by the following processes: *absorption, retention and release downstream*; *availability and fractionation*; and *denitrification*.

All these chemical processes involve the metabolism of PO_4^{3-} Fe^{3+}, NO_2^-, NO_3^- and NH_4^+, in addition to the oxygenation and deoxygenation processes resulting from turbulence and stratification. Advection also has to be taken into account, which significantly affects the distribution of chemical substances in the vertical axis of reservoirs and produces stratification due to increased density.

The chemistry of sediment in reservoirs, as in all aquatic systems, directly correlates with the effects of organisms on the biochemical cycle and the pH of the water. For example, Ca^{2+} and the pH of the water limit the solubility of phosphate in eutrophic waters (Golterman, 1984).

The **hydrogeochemistry of interstitial water** is also an important feature, given the potential for fertilization of this water and storage of inorganic nutrients and ions. These mechanisms are also related to the potential for oxido-reduction, water pH, granulometry of the sediment, and the biogeochemical processes produced by the organisms, mainly the bacteria found in or on the sediment. In a reservoir, a significant portion of the biogeochemical cycles of carbon, nitrogen, phosphorus, sulphur and iron occur in the sediment and interstitial water.

The chemical characteristics of sediment, interstitial water and their 'metabolism' are also fundamentally related to circulation and thermal stratification and destratification. In lakes, the sediment is clearly a reserve and important deposit of biological material from the upper layers that starts be deposited immediately after death. In addition, the sediment is affected by the drainage basin, considering the contribution of plant material and suspended particles. In reservoirs, although the mechanisms are theoretically the same, the outflow of material at the bottom needs to taken into consideration, due to the location of the water outlet leading to turbines and the degree of instability, which is much greater than in lakes, causing permanent oxygenation of the sediment in many reservoirs.

Among the main operating mechanisms in a reservoir, **macroscale**, *mesoscale* and microscale patterns can be identified. Macroscale patterns occur in drainage basins where heterogeneities exist in the soil, geology, land use and climatic patterns (Thornton *et al.*, 1990). These are patterns that in the macroscale of the drainage basin influence and direct, to some extent, the responses of the reservoirs in terms of the hydrology, biogeochemical cycles and the diversity and succession of the aquatic biota.

Cascade reservoirs also function as a continuous horizontal gradient with macroscale patterns, for example, those located on the Tiete River in São Paulo, or the São Francisco River in Northeastern Brazil (Barbosa *et al.*, 1999).

In addition to these macroscale patterns, there are patterns in reservoirs that can be considered to be mesoscale. **Longitudinal patterns** in reservoirs definitely present mesoscale variations, as shown by Kennedy *et al.* (1982, 1985) and Armengol *et al.* (1999).

The three zones in reservoirs, already described in this chapter, are typical of mesoscale patterns. Their functioning and distribution in the reservoir depend on the relations between the operation of the reservoir and the inflow of the rivers (Straškraba *et al.*, 1993).

Microscale patterns also occur (at the level of meters or centimeters): convective processes, reactions to reduction or oxidation in the sediment, primary productivity of phytoplankton, or predator-prey relationships.

Lakes and reservoirs present similar characteristics in terms of microscale patterns. In reservoirs, however, this horizontal and vertical variability is eventually much greater, given the particular aspects of reservoirs. For example, the differences in the ionic composition of the water level and retention times are much more marked in reservoirs than in lakes (Ryder, 1978).

From the standpoint of vertical and horizontal circulation, reservoirs (as well as lakes) present patterns that vary with latitude, longitude and altitude. However, along with these patterns, it is important to consider the reservoir's operating system and multiple uses, which in many cases, can be determinant factors. For example, Straškraba (1999) showed that the short retention time of shallow reservoirs promotes a process of continuous circulation, renewing the thermal stability and making the system much more vertically dynamic. Henry (1999) conducted an extensive study on the vertical distribution of temperature and thermal patterns in reservoirs in Brazil, showing these latitudinal variations and also the effects of morphometry.

12.9   PULSES IN RESERVOIRS

Pulses are defined as drastic changes, of natural or man-made origin, which can affect any variable – physical, chemical or biological – in reservoirs. Pulses in a reservoir can be the result of an influx (such as precipitation or wind) or an outflow (such as withdrawal of water through the spillway).

Natural pulses result from climatic changes such as precipitation and wind, and can have direct or indirect effects. They tend to be seasonal, and may be frequent and repetitive, or infrequent, such as an especially strong wind or heavy rainfall).

Pulses caused by human action can be the result of the manipulation of the spillways or a change in water level (Kennedy, 1999). These pulses can be frequent and repeated according to the reservoir's operating system. However, in some cases, the rapid opening of gates results in a highly intensive pulse, with extreme effects on the planktonic populations and biogeochemical cycles.

The effects of pulse-produced fluctuations in reservoirs can be qualitative and quantitative, in both the reservoir and river downstream. For example, Calijuri (1988) describes the effect on the Barra Bonita reservoir (SP) of the rapid influx of **seston** into the system produced by heavy rainfall, which reduced the photic zone to 20% or less of its original depth.

Inorganic particles introduced *en masse* (high concentrations) into the reservoir can drastically interfere with the water's chemical composition and biogeochemical cycles.

Frequent **seasonal pulses** caused by climatic conditions can affect the water's chemical composition due to the introduction of advective currents and enrichment of the euphotic zone.

Rapid and frequent changes in water temperatures downstream can occur when reservoir water is withdrawn from specific depths by operational means.

In the case of cascade reservoirs or pumped-storage plants, significant increases in water temperature (3–4°C) or chemical levels can occur.

Margalef (personal communication) describes the effect of pulses in pumped-storage reservoirs in the Pyrenees, functioning in coordination at different altitudes. In this case, pumping water to the upper reservoir produced temporal compartmentalization of water temperatures. In the upper reservoir (located at 1500 m), the water was colder, and temporary stratification occurred whenever masses of water were received from the lower reservoir (situated 800 m).

Occasional pulses can occur with the breakdown of the thermocline from the action of strong winds, producing changes in the vertical distribution of dissolved oxygen, nutrients and plankton. The composition of phytoplankton is eventually altered by the effects of the wind, which can redistribute filaments of *Aulacoseira* sp. in the water column (Lee *et al.*, 1978) and cause attenuation in the colonies of *Microcystis aeruginosa*. Downstream pulses can also occur, for example, in the case of rapid openings of the gates, which may lead to the supersaturation of oxygen downstream and the death of many fish.

The magnitude of natural pulses differs geographically, depending on local climate, altitude, rainfall and winds. The magnitude of artificial pulses, in turn, follows the operating rules of each system and the multiple uses of reservoirs.

Indirect effects of pulses can be significant. For example, the opening of floodgates to low-oxygenated water produces the downstream release of phosphorus from the sediment in reservoirs.

From the point of view of reservoir management, it is essential to understand the causes of natural pulses and their frequency, as well as the reservoir's operating system and the possible effects of pulses on downstream reservoirs (to determine the ecological flows and hydrograms).

An example, with qualitative and quantitative results, is the intermittent pumping of the Pinheiros River into the Billings reservoir. Pulses of nitrogen and phosphorus and other substances enter into the Billings system, with drastic consequences on the reservoir's water quality and levels of suspended matter, organic load, nitrogen and phosphorus and dissolved oxygen. The magnitude of these pulses produced various impacts on the Billings reservoir and the biological communities. Many mass fish kills occurred.

The frequency of pulses from pumping, coupled with the reservoir's circulation and climatic conditions, can be used as an emergency mechanism to maintain water volume, and is thus a possible management mechanism in the case of the Billings reservoir (Tundisi *et al.*, unpublished results).

12.10   COMMUNITIES IN RESERVOIRS: THE AQUATIC BIOTA, ITS ORGANIZATION AND FUNCTIONS IN RESERVOIRS

A reservoir presents a varied spatial structure with great variation in its vertical and horizontal circulation and hydrodynamics, which depend on the morphometry of the inflowing tributaries and the effects of climatic and hydrological conditions (Matsumura Tundisi and Tundisi, 2005). These physical conditions, which also have biogeochemical consequences, influence the distribution succession of organisms, and the productivity and biomass of communities.

Figure 12.10 presents the main interactions between components of the aquatic biota and the reservoir's physical and chemical properties. This figure illustrates the complexity of the reservoir's internal processes and the interactions of the different compartments. From the standpoint of the ecosystem, reservoirs have very particular conditions related to aquatic biota: planktonic organisms are significantly affected by the retention

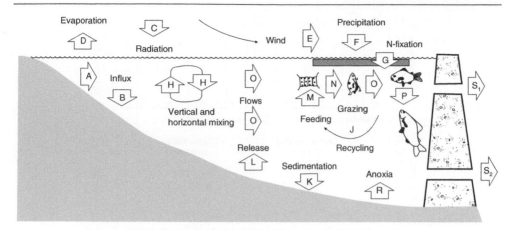

Figure 12.10   Internal processes in reservoirs. The processes A, B, D, E, H and S are part of the physical subsystem; processes F, G, K, L, C and R form part of the chemical subsystem; and J, M, N, O and P form part of the biological subsystem. The three subsystems are interconnected.
Source: Straškrába and Tundisi (2002).

time and succession of species; the biomass depends on this retention time. Since the retention time in reservoirs is usually shorter than in lakes, the processes of primary productivity are influenced by the availability of nutrients and the phytoplankton's capacity to reproduce, thus replacing the biomass lost downstream in the outflow.

The planktonic, fish and benthic communities are all significantly affected during the reservoir's filling phase. Changes occur in the available substratum in the old river flooded by the reservoir and in the system's hydrodynamics and underwater radiation regime. The changes in the conditions of underwater radiation cause periphyton to disappear or be drastically reduced, leading to the loss of a key component in the food web.

Increased pressure and the lack of currents near the bottom change the physical and chemical conditions, promoting new structures that enable benthic organisms with suitable physiological and reproductive conditions to colonize the bottom environments. In many reservoirs during the filling phase, anoxia occurs on the bottom, which leads to changes in the fauna and flora from the old river bottom.

Flooded reservoirs with submerged vegetation present new substrata for many organisms. Tundisi *et al.* (1993) described how the periphyton effectively provided food for the *Macrobrachium amazonicum* shrimp in the Tucuruí reservoir on the Tocantins River (Pará) immediately after filling, when large masses of periphyton established themselves and thrived in the trunks of submerged vegetation. Ecotones that develop in relation to reservoirs, usually in different compartments, provide another opportunity for enriched biodiversity of the populations of plankton, and benthos, fish and aquatic birds. In some cases, reservoirs provide the opportunity for colonization processes to occur and biodiversity to recuperate. The problem of **biocoenosis** in reservoirs and the relations of the forcing functions with the fauna and

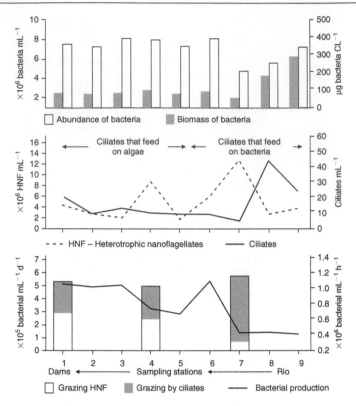

Figure 12.11   Distribution of bacteria on the longitudinal axis of the reservoirs of SAU, Spain, Catalonia, as well as the longitudinal distribution of heterotrophic flagellates and Culados (Portug.). *Source:* Armengol *et al.* (1999).

flora were addressed in two recent publications: Nogueira *et al.* (2005) and Rodrigues *et al.* (2005).

12.10.1 The role of bacteria in reservoirs

In reservoirs with long retention times and moderate nutrient loads, it is generally assumed that autochthonous primary production is the principal source of organic carbon, supporting the production of bacteria, which reaches nearly 20–30% of the photosynthetic primary production. In cases where there is an **allochthonous load** of organic material, there is a contribution from the rivers in terms of bacterial population, so mesotrophic and eutrophic reservoirs present different biomasses of bacteria that flow in. Protists, especially heterotrophic nanoflagellates and flagellates, are the main consumers of bacteria. However, depending on the food web structure in the reservoir, other components of the biota (such as rotifers and cladocerans and phytoflagellates) are important consumers of bacterioplankton (Sanders, 1989; Simek *et al.*, 1999). Figure 12.11 shows the distribution of bacteria along the longitudinal axis of the **SAU reservoir** in Spain and the distribution of heterotrophic and ciliated flagellates in the same reservoir.

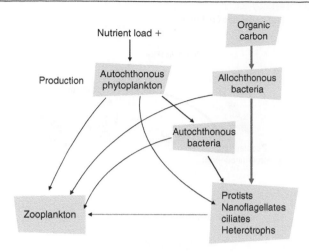

Figure 12.12   Simplified diagram of the cycle and main flows of carbon in the upstream portion above a reservoir showing the influx of the river with nutrient load and organic carbon.
Source: Simek *et al.* (1999).

Figure 12.12 and Figure 12.13 present the principal flows of carbon in reservoirs. Figure 12.12 presents the two main sources of carbon supporting the growth of bacteria: the allochthonous source and the production of organic matter by autochthonous photo-autotrophic phytoplankton.

12.10.2 Phytoplankton and primary productivity

The main primary producers in reservoirs are the same components of photo-autotrophic producers in rivers and lakes: phytoplankton, photo-autotrophic bacteria, periphytonic algae and floating macrophytes, immersed or submerged or fixed on roots.

The relative contribution of each of these photo-autotrophic components depends on the reservoir's average depth, shoreline development index, transparency, and fluctuations in level, in addition to the intrusion of suspended matter (resulting from erosion in the drainage basin), which can reduce the growth of periphytic algae and macrophytes. Macrophytic and epiphytic communities can contribute a big component to reservoirs with stable levels.

Tundisi *et al.* (1993) described extensive periphytonic communities growing on trunks of flooded vegetation in the **Tucuruí reservoir** on the Tocantins River. These autotrophs support a biomass of protozoa and are used as food by the shrimp *Macrobrachium amazonicum*, as previously described.

In addition to the phytoplankton that develops in the reservoir, the contribution of tributaries and reservoirs upstream must be taken into consideration, if there is a series of reservoirs.

The biomass and composition of phytoplanktonic species in reservoirs depend on the interactions of physical factors (such as temperature and circulation), chemical factors (such as nutrient levels and the relative distribution of different dissolved ions), and biological factors (such as interaction of species, and effects of predation

(a) Low stock of fishes

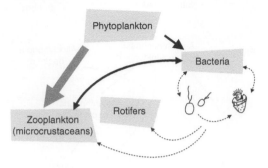

(b) High stock of fishes

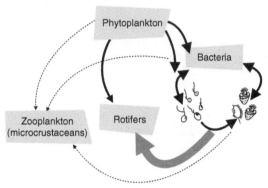

Figure 12.13   Summary of carbon cycle and the principal flows of carbon in reservoirs. The figure shows two contrasting aspects of the flow of carbon between microcrustaceans of zooplankton, rotifers, and phytoplankton and the basic components of microbes (bacteria, heterotrophic nanoflagellates and ciliates). The wider more solid arrows indicate the relative importance of the main flows, and the dotted arrows indicate the less-important flows.
Source: Simek *et al.* (1999).

and parasitism). Rodrigues *et al.* (2005) evaluated the phytoplanktonic composition in 30 reservoirs in the Paraná River basins and tributaries.

The composition and diversity of species and their biomass were observed during the dry season (July–August 2001) and rainy season (November–December 2001). There were 171 species of phytoplankton in the reservoirs, in nine taxonomic classes. Low species richness was found, and the phytoplanktonic biomass values were also low, less than $2\,\mathrm{mm}^3 \cdot \mathrm{L}^{-1}$ (measured by biovolume). According to Rodrigues *et al.* (2005), phytoplanktonic growth in these reservoirs was limited due to nutrient sedimentation during the dry and rainy and seasons.

The authors' findings showed that retention time, vertical mixing and pulses produced by wind and intrusions of water with high levels of suspended material were the determinant factors in the composition of phytoplanktonic species in the reservoirs.

Table 12.5 Primary productivity in reservoirs in the state of São Paulo.

Reservoirs	Lat. (S)	Long. (W)	Alt. (m)	Primary prod. $(mg\ C \cdot m^{-2} \cdot d^{-1})$	Chl-a	Rate of assimilation $(mg\ C \cdot mgChla \cdot h^{-1})$
Barra Bonita	22°29'	48°34'	430	398.27	15.9	2.56
Bariri	22°06'	48°45'	442	521.85	20.3	2.64
Ibitinga	21°45'	48°50'	460	483.94	29.8	2.16
Promisão	21°24'	49°47'	410	584.08	68.7	0.83
Salto de Avanhandava	21°13'	49°46'	360	268.74	14.9	1.60
Capivara	22°37'	50°22'	520	188.67	12.7	3.40
Rio Pari	22°51'	50°32'	420	105.19	13.3	1.43
Salto Grande	22°53'	49°59'	405	102.80	5.7	2.07
Xavantes	23°08'	49°43'	400	193.79	20.8	0.95
Piraju	23°11'	49°16'	571	100.94	12.9	0.91
Jurumirim	23°11'	49°16'	571	103.05	9.7	1.02
Rio Novo	23°06'	48°55'	755	60.87	12.1	0.79
Limoeiro	21°27'	47°01'	650	225.89	22.3	2.26
Euclides da Cunha	21°36'	46°54'	700	25.99	3.8	0.96
Graminha	21°32'	46°38'	800	582.98	34.4	0.94
Estreito	20°32'	47°24'	1000	126.71	25.1	0.61
Jaguará	20°11'	47°25'	536	154.08	22.3	0.70
Volta Grande	20°05'	48°02'	510	340.23	31.7	1.22
Porto Colômbia	20°10'	48°48'	500	318.86	40.2	1.00
Marimbondo	20°18'	49°11'	390	262.10	37.5	0.80
Água Vermelha	19°58'	51°18'	452	232.47	32.5	0.80
Ilha Solteira	20°24'	51°21'	356	248.35	20.2	1.73
Jupiá	20°58'	51°43'	260	301.61	15.5	2.15

Data from 1979
Primary production determined with the ^{14}C technique.
Source: Tundisi (1983).

These results corroborate data in the literature on phytoplankton in lakes and reservoirs (Reynolds, 1997, 1999) and their succession. Lima *et al.* (1978) showed that much phytoplanktonic succession in the UHE Carlos Botelho reservoir (Lobo/Broa) resulted from the effects of wind, which fostered the development of *Aulacoseira italica* in winter, and of precipitation, which promoted regeneration of nutrients during the summer.

The presence of different types of phytoplankton in reservoirs varies. Tundisi and Matsumura Tundisi (1990) and Tundisi *et al.* (1993) calculated the varying components in the primary productivity, predominantly in the $<5\ \mu m$ and $>20\ \mu m$ components. However, recent data (Simek *et al.*, unpublished) shows that in the UHE Carlos Botelho reservoir (Lobo/Broa), **photo-autotrophic picophytoplankton** $(<2\ \mu m)$ is an important primary producer (approximately 40–50% of the total primary production by photo-autotrophic phytoplankton). However the pelagic environment of reservoirs can be dominated by those components $<20\ \mu m$ and $>2\ \mu m$ in terms of the primary production (Henry, 1993b). Table 12.5, Table 12.6 and Table 12.7 show the primary productivity in reservoirs in the state of São Paulo, measured with the ^{14}C technique (Tundisi, 1981, 1983), and include data from Thornton *et al.* (1990) on reservoirs in other latitudes and comparative data from rivers, lakes and 64 man-made reservoirs.

Table 12.6   Comparison of reservoirs, lakes and rivers at various different latitudes.

	Year	Primary production $(mg\ C \cdot m^{-2} \cdot day^{-1})$	Method and trophic state
Turtle Greek, Kansas (USA)	1970–1971	67	^{14}C Oligo
De Gray, Arkansas (USA)	1979–1980	199	^{14}C Oligo
Lake Mead, Arizona, Nevada (USA)	1977–1978	810	^{14}C Meso
Norris, Tennessee (USA)	1967	360	^{14}C Meso
Gorky (Russia)	1956	456	^{14}C Meso
Slapy (Czech Republic)	1962–1967	501	O_2 Meso
Kainji (Nigeria)	1970–1971	2,434	O_2 Eu
Volta (Ghana)	1966	2,547	O_2 Eu
Stanley (India)	–	2,329	O_2 Eu

Table 12.7   Comparative data of primary production.

	$(mg\ C \cdot m^{-2} \cdot day^{-1})$
Tropical lakes	100–7600
Lakes in temperate regions	5–3600
Arctic lakes	1–170
Antarctic lakes	1–35
Alpine lakes	1–450
Rivers in temperate regions	1–3000
Rivers in tropical regions	1–150
102 natural lakes	3–5529
64 reservoirs	67–3975

Sources: Likens (1975); Wetzel (1983); Tundisi (1983); Kimmel *et al.* (1990).

The control and limitation of primary production in reservoirs have the same characteristics that occur in lakes and oceans in terms of available energy and nutrients. In the case of reservoirs, the basic factors determining the magnitude and seasonal variation of the primary productivity of phytoplankton (temperature, light intensity, and availability of macro- and micronutrients) (Steemann-Nielsen, 1975) depend on the interactions of these factors with the drainage basin (which provides, by intrusion, the regeneration of nutrients), as well as on the major forcing functions (wind, precipitation) and the interactions of vertical mixing with the depth of the euphotic zone (z_{eu}/z_{mix}). In the case of reservoirs, retention time is an important regulating factor from the standpoint of nutrient recovery as well as biomass levels and species succession. Tundisi *et al.* (2004) showed that the impact of cold fronts is a major factor in temporal succession, regeneration of nutrients, and primary productivity of phytoplankton in the UHE Carlos Botelho reservoir (Lobo/Broa) and other reservoirs in southeast Brazil. Cyanobacterial blooms occur during periods of thermal stability and high solar radiation (Reynolds, 1999).

Control of retention time in reservoirs can thus be a key factor in the control of the phytoplanktonic biomass, succession of species, and primary productivity, and can be used for effective management, especially in eutrophicated reservoirs. In addition, physical controls, such as increased turbulence and vertical mixing, can help reduce the intensity and volume of blooms and reduce the availability of solar radiation to phytoplankton. Chemical controls, such as the use of $CuSO_4$ to reduce cyanobacterial biomass, are not indicated, especially in drinking water supplies, since $CuSO_4$ accumulates in the sediment and possibly has subsequent toxic effects. These chemical controls can be applied in tributaries, at the site of origin of the blooms, and not in the main body of the reservoirs, especially near water intakes.

12.10.3   Periphytic algae

The biomass and diversity of species of periphytic algae depend on the available substratum, the contribution of species from the drainage basin in which the reservoir is located, and the effects of abiotic factors, such as light penetration (which can be controlled by aquatic macrophytes), nutrient levels and water temperature. Turbidity can significantly affect the succession and development of the periphytonic biomass, as well as species richness.

The predominance of Bacillariophyceae and Cyanophyceae periphyton in reservoirs (Rodrigues, 2005), was shown to be a competitive advantage based on the low levels of particulate phosphorus and the turbidity during the summer, also given the higher water temperatures during this period. The most common class in the three reservoirs studied by the author was Bacillariophyceae (59.6% in the **Mosão reservoir** and 87% in the **Rosana reservoir**).

In a study conducted on periphytic colonization of different substrata in the UHE Carlos Botelho reservoir (Lobo/Broa), Panitz (1980) concluded that qualities of the substratum, especially roughness, are basic factors in the colonization and future development of the perphytic community (see Chapter 7).

12.10.4   Aquatic macrophytes

Aquatic macrophytes can develop extensively in reservoirs, in many cases spreading over a large area and hindering navigation and power generation. On the other hand, the roots of these organisms can sustain a wide range of flora and fauna, with abundant biomass and great diversity (Takeda *et al.*, 2003). The most common macrophytic species that colonize reservoirs in Brazil are *Eichhornia crassipes, Eichhornia azurea, Salvinia molesta, Salvinia* spp., *Egeria densa* and *Egeria najas*, found floating freely in eutrophicated reservoirs. *Typha* sp. and *Eleocharis* sp. are emergent macrophytes commonly found in some reservoirs (Thomaz *et al.*, 2005).

Abiotic factors (such as availability of underwater radiation and alkalinity) are essential for colonization and occupation of space by aquatic macrophytes. In addition, hydraulic variables (retention time, morphometry) and biological variables, such as competition and predation, can influence the succession and spatial distribution of the various different species. For example, Tundisi (unpublished results) noted alterations in the dominance of *Pistia stratiotes* and *Eichhornia crassipes* in the Barra Bonita reservoir (SP), based on phosphorus and nitrogen levels in the water. *Pistia stratiotes* was common at higher nitrogen and phosphorus levels.

Thomaz *et al.* (2005) found no correlation between a reservoir's age and the wealth of macrophytic species. Analyses conducted on 30 reservoirs showed the existence of 37 taxa, *Eichhornia crassipes* being the most common free-floating species. In some reservoirs, immediately after stabilization of water level in the filling phase, there is rapid colonization by floating macrophytes, especially of the genus *Salvinia,* as well as *Pistia stratiotes* and *Eichhornia crassipes.* However, this period is relatively short and depends on nutrient levels. In many reservoirs, in wetland areas and at the entrance of tributaries, extensive banks of rooted or floating macrophytes (*Pontederia* sp., *Eichhornia crassipes* or *Eichhornia azurea*) develop, which are effective in removing phosphorus and nitrogen from the inflowing waters (Whitaker *et al.*, 1995). *Eichhornia crassipes, Salvinia molesta* and *Pistia stratiotes* develop rapidly in reservoirs after the filling phase (Bianchini, 2003).

12.10.5 Zooplankton

The richness and diversity of zooplanktonic species in reservoirs have been extensively studied over the last 20 years, especially in the tropics and subtropics (Matsumura Tundisi *et al.*, 1990; Rocha *et al.*, 1995, 1999; Lansac-Tôha, 1999, 2005; Nogueira, 2001; Sampaio *et al.*, 2002). The composition, structure, dynamics and succession of zooplanktonic species in reservoirs are influenced by physical conditions (temperature, electrical conductivity), chemical conditions (dissolved oxygen and ionic levels) and biological conditions (predation, parasitism, colonization after the filling phase). Matsumura Tundisi and Tundisi (2005) showed that the diversity of phytoplanktonic and zooplanktonic species in the Barra Bonita reservoir (SP) is determined by the reservoir's trophic state, horizontal gradients, and degree of vertical mixing and stratification in the water column. The relationship between the areas under the influence of the river, the transition area and the lacustrine area in reservoirs is a major factor in the spatial distribution of the different zooplanktonic groups.

The impact of tributaries on the diversity of zooplanktonic species in the reservoir must be taken into account, since each tributary produces an element (mosaic) of microhabitats (Margalef, 1967, 1991) that can significantly differentiate a specific region and alter hydrodynamic and chemical characteristics that promote the concentration and/or dispersion of certain groups or species in the reservoir.

Transport of suspended material through the intrusion of rivers can affect the primary production by phytoplankton and the species composition and succession of zooplankton. According to Thornton *et al.* (1990), the availability of resources such as photo-autotrophic phytoplankton, detritus and bacteria affect the succession of zooplankton (Nauwerck, 1963). Alternative sources, such as detritus and bacteria, can be important, as shown by Matsumura Tundisi *et al.* (1991) and Falótico (1993) in the filling phase of the Samuel reservoir (Rondônia). Dissolved organic material may be a resource for zooplankton, and the effect of predation on zooplankton is another important factor in the control of species development and distribution.

The result of studies (by Lansac-Tôha *et al.*, 2005) on 31 reservoirs located on tributaries of the Paraná River showed that the zooplanktonic community presented increased diversity of rotifers and small cladocerans and copepods during the rainy season. The authors also found significant correlations between the phytoplanktonic and zooplanktonic biomasses. The abundance of typical planktonic groups is associated with the productivity of the reservoirs and the trophic status (Arcifa *et al.*, 1981). Matsumura Tundisi *et al.* (1990) verified that the increased number of rotifers is an indicator of the trophic status, and Branco and Senna (1996) also concluded that the predominance of small cladocerans, rotifers and copepods is a result of the predominance of cyanobacteria (Bonecker, 2001). When there is a reduced phytoplanktonic density in small shallow reservoirs, higher densities of Testacea (Protozoa) can occur (Lansac-Tôha *et al.*, 2005).

In a comparative study conducted between the D. Helvécio Lake (a monomictic lake in Parque Florestal do Rio Doce) and the Barra Bonita reservoir (SP), Tundisi and Matsumura Tundisi (1994) showed that the reservoir presented a greater number of zooplanktonic species (20 rotifers, 8 cladocerans and 9 copepods, 37 in total) than the lake (6 rotifers, 5 cladocerans and 5 copepods = 16). The authors considered the hypothesis of 'intermediary disturbance' (Legendre and Demers, 1984) as the mechanism driving this difference in composition between the two systems, since the Barra Bonita reservoir is less stable physically (thermal structure and conductivity) and chemically (dissolved oxygen and nutrients).

The longitudinal distribution of zooplankton in reservoirs depends on the different current velocities in the river-influenced zone, the transition zone and the lacustrine zone, and depends on each reservoir and the period of the year (Matsumura Tundisi and Tundisi, 2005). Machado Velho *et al.* (2005) showed that the predominance of rotifers as a major zooplanktonic component in reservoirs can in fact vary, as microcrustaceans are more common in some conditions and events. The sediment in reservoirs can play an important role in zooplanktonic diversity and in seasonal or spatial succession of species. In the UHE Carlos Botelho reservoir (Lobo/Broa), Rietzler *et al.* (2002) found that the alterations in the components of *Argyrodiaptomus furcatus* and *Notodiaptomus iheringi* plankton occurred because there was a pool of resistant eggs of these species in the sediment. The dominance of one species or another is due to development of the egg after the start of favourable processes – in this case most likely involving electrical conductivity in the water associated with ion and nutrient levels. *Argyrodiaptomus furcatus* is a species that does well in more transparent water with lower conductivity.

These two species can be used as indicators of conditions of pollution, contamination or eutrophication. Alterations in the composition of zooplanktonic species over a 20-year period in the Barra Bonita reservoir was confirmed by Matsumura Tundisi and Tundisi (2003). Alterations in the cyclopoid/calanoid ratio and the relative abundance of several calanoid copepod species, according to the authors, were the result of eutrophication processes. Another species common in oligotrophic waters with low conductivity is *Argyrodiaptomus azevidoi* (Matsumura Tundisi, 2003).

The presence and development of aquatic macrophytes in reservoirs provide the opportunity for alternative energy sources (detritus and bacteria) for zooplankton, as well as increased availability of food. Rocha *et al.* (1982) observed that *Argyrodiaptomus furcatus* (calanoid copepods) occurred in densities ten times greater upstream from the UHE Carlos Botelho reservoir (Lobo/Broa), where macrophytes of the genera *Mayaca*, *Pistia*, and *Nymphaea* were established. Kano (1995) examined the zooplanktonic density and production in reservoirs and observed that the littoral zone was a reproductive region with dense populations, portions of which shifted to inhabit the limnetic zone.

12.10.6 Aquatic macroinvertebrates

Chironomid larvae occur in many reservoirs, partly due to their adaptive strategies of expanding to multiple habitats and their diversified feeding habits (Strixino and Trivinho Strixino, 1980, 1998). The distribution and abundance of chironomid larvae were recently studied in 13 reservoirs of the Iguaçu River basin in Paraná State, by Takeda *et al.* (2005), who found 2741 chironomid larvae from 31 *taxa* distributed in the subfamilies Chironominae, Orthlocladiinae and Tanypodinae. The most common genera found were *Tanytarsus*, *Polypedilum*, and *Dicrotendipes*.

Several factors affect the distribution and composition of benthic invertebrates: levels of organic matter in the sediment, dissolved oxygen in the water, fluctuations in the level (Brandinerte and Shrimizu, 1996) and the velocity of currents near the reservoir's bottom sediment. Retention time, reservoir age and position of the dam in the cascade of dams can affect the biomass and diversity of the benthic community (Barbosa *et al.*, 1999). The fluctuations in level and intermittent discharges can affect benthic fauna in reservoirs, reducing diversity and biomass. These fluctuations in levels affect populations downstream from the reservoir. Bottom discharges and high turbidity can also drastically reduce the biomass and diversity of the benthic invertebrates found downstream (Petts, 1984).

Bivalve molluscs were studied in 31 reservoirs of the upper Paraná by Takeda *et al.* (2005). The most common species found were *Corbicula fulminea* and *Limnoperna fortunei*, both exotic species, and *Pisidium* sp., the only native species. In the case of *L. fortunei*, extensive proliferations were found in these reservoirs (see Chapter 18) (Takeda *et al.*, 2003).

12.10.7 Fish species in reservoirs

Reservoirs significantly affect the fish fauna, due to the interruption of migrating routes for reproduction and the changes resulting from the river being diverted into a lentic system. Reservoirs have an important potential for the exploitation of extensive and intensive fishery production. In several regions around the world, commercial fishery in reservoirs has grown substantially, for example, in Russia, the United States, Southeast Asia and Africa (Fernando and Holcik, 1991, Shimanovskaza *et al.*, 1977; Gothschalk, 1967; Beadle, 1991). A review of fishery management in reservoirs on the Paraná River was recently published by Agostinho and Gomes (2005). In it, the authors considered that main problems affecting fishery management in reservoirs were the lack of information on the fishing system (environment, fish and fishermen), lack of ongoing monitoring, and the naturally high variability in the resources.

When the structural and fluvial organization and development of the fish fauna in a reservoir is analyzed along space/time axes, five groups of processes must be considered:

▸   productivity of the reservoir after it is filled;
▸   eutrophication and nutrient enrichment from the drainage basin;

▶ development of complex biotic interactions in the reservoir (feeding, competition, predation);
▶ hydrologic regime;
▶ reservoir management.

The two most dramatic aspects in construction of dams are the alteration that occurs in fish faunas as result of the river's change in regime for a system with slower flow and greater depth; and the shift from a **lotic environment** to a **lentic environment** produce new habitat types, to which many of the fishes that inhabit rivers are not adapted.

According to Kubecka (1993), the first two groups of processes listed above affect fish stocks and biomass, while species composition is affected by the third group (biological interactions). In addition, another important factor, as confirmed by Petrere (1994), is the formation of floodplains and wetlands around the reservoir. This topic was recently reviewed by Henry *et al.* (2005).

In many reservoirs, a deep pelagic area forms, which in many cases is not exploited by any fish species in the region. In 1965, the introduction *of Limnothrissa miodon* from Lake Tanganyika into Lake Kariba (a large tropical reservoir on the Zambezi River), led to a significant increase in fisheries in the reservoir. In 1969, large shoals of this species were observed, and intensive fishing began that year (Beadle, 1991).

According to Fernando and Holcik (1991), the ichthyofauna in reservoirs mainly depends on species from the drainage basin. If the region has fish species adapted to lacustrine system, some habitats and feeding niches can be filled in the reservoir. This ichthyofauna can quickly colonize the reservoir and exploit the potential of the newly formed lentic environment.

Once a reservoir becomes operational, the river ichthyofauna decreases drastically. Two factors appear to cause this reduction: the extensive and deep pelagic zone and the drastic reduction in current velocity. Colonization of the pelagic zone occurs only if species from lacustrine habitats already inhabiting the drainage basin exploit this zone. The reservoir is gradually colonized by these species, and in many cases, the fish concentrate near the tributaries.

Of the 38 species of fish existing in the river before the construction of the Cabora Bassa dam on the Zambezi, many disappeared almost immediately, for example, *Bariliens zambeensis* and the small species of *Barbus,* common in rivers with fast currents. Cichlids such as *Tilapia rendalli* and *Sarotherodon mortimeri* remained in the reservoir but at a lower density. Some species of siluroids, characins and cyprinids presented explosive population growth immediately after the construction of the dam and filling of the reservoir (Jackson and Rogers, 1976).

In the case of Lake (reservoir) Volta (Ghana), a mass fish kill occurred immediately after construction of the reservoir because of accelerated deoxygenation. Of the many species of the *Alestes* genus, only two (*Alestes baremose* and *Alestes dentex*) disappeared (Pek, 1968b; Kebek, 1973). On the other hand, there was a significant increase the talapine species *Sarotherodon galileus* (which feeds on phytoplankton and periphyton), *Tilapia zillii* (which feeds on detritus), and *Sarotherodon niloticus* (which feeds on macrophytes and algae), precisely because of the diversification of food during and after completion of the filling of the reservoir.

The fish species found in reservoirs after dams are completed migrate to areas under the influence of the tributaries or to the very tributaries themselves. Current velocity, which varies in different zones of the reservoir, is a major factor in the distribution of different fish families.

In addition to the change in current velocity, the lack of dissolved oxygen in some reservoirs immediately after filling can lead to large-scale mortality and the disappearance of a fish fauna. Inhibition of **fish migration** upstream, because of the obstacle of the dam, is another extremely important effect. Fish can find tributaries through detecting changes in water quality (Jackson *et al.*, 1988). In the case of reservoirs, maintaining a flow of water in the system, in the old riverbed, can serve as a stimulus to migration.

Migration of fish on old riverbeds was detected in reservoirs off Lake Victoria, where *Barbus altianalis* enters and reproduces kilometers upstream by following the old bed of the Kajera River into the lake (Greenwood, 1977). The surviving population of migratory fishes can persist in the reservoir, using the tributaries, as was shown in the Cabora Bassa reservoir and Kariba reservoir in Africa (Jackson and Rogers, 1976).

Construction of a reservoir thus affects the ecology of the ichthyofauna in the following ways:

▸ Serves as a barrier to longitudinal movement of fish, whether they are migrating to feeding areas or breeding areas.

▸ Causes changes in the river's hydrological regime, modifying the magnitude of floods and interfering in horizontal and transversal movements of fish; can lead to flooding or drought in marginal lagoons important for development of fingerlings.

▸ The strictly rheophilic (running water) species do not have sufficient survival conditions in lentic waters and quickly disappear or their populations significantly decrease in the reservoir, leaving only a few species that can still colonize reservoirs but that need tributaries for reproduction.

The filling of the reservoir produces a spatial re-organization of the system and can result in new wetlands and swamps, in addition to producing an extensive and deep pelagic zone. In some cases, it was shown that a few years after filling, there were more fish species in the reservoir than in the river, given the variety of niches resulting from the spatial reorganization (Jackson *et al.*, 1988). This also depends on the evolution of the reservoir and the succession process on the space-time axis.

Agostinho *et al.* (1999) summarized the patterns of colonization and aging in a series of neotropical reservoirs located in the upper Paraná River basin. After the filling phase, anoxia occurred, which produced displacements of fish into tributaries or transition regions between the reservoir in formation and the river upstream. Species that remained in the reservoir concentrated in littoral areas and near the mouths of tributaries. After stabilization of the level and with gradual colonization by submerged macrophytes, a change in the fish species occurred: species diversity was much greater in the littoral zone than in the pelagic zone. Upon reaching stability, after a period of increase in biomass, the diet of the fish shifted to utilize **autochthonous resources**. In general, detritivorous and iliophagous species decreased, while herbivores and zooplanktivores increased their density. The main changes occurring in the fish fauna after construction of the reservoir, as compared with the rivers, were the drop in the number of predators, in average species size and in species diversity.

12.10.8 The aquatic vegetation in reservoirs and fish fauna

As already shown in this chapter, ecological succession in reservoirs depends on several factors, mainly on the status of the drainage basin before the reservoir was filled. In many reservoirs, massive bloomings of *Anabaena* and *Microcystis* occur, in addition to an increase of epiphytic and epilithic algae, which form dense substrata in trunks of the vegetation. This dense vegetation is colonized by an extraordinary and varied invertebrate biomass, forming the 'aufwuchs' – a mass of periphytic algae, bacteria, fungi, particulate organic matter, animals such as protozoa and turbellarians, found on substrata (stones, remains of higher plants) – which is used extensively by fish for food (Petr, 1970).

The removal of higher plants from the flooded area can result in loss of productivity. Extensive banks of macrophytes and high levels of periphytonic algae are essential for the development of a wide source of food for fish, as seen in Lake Kariba in Africa (McLachlan, 1969, 1970) and the Tucuruí reservoir in Brazil (Tundisi *et al.*, 1993). Figure 12.14 shows the organization of the food web in the Tucuruí reservoir on the Tocantins River, Pará, resulting in increased biomass of *Cichla ocellaris* (peacock bass). The increased biomass and diversity of macrophytes and periphyton are partly responsible for the 'trophic upsurge', i.e., the increase in biomass during and immediately after the filling of the reservoir. The development of extensive banks of macrophytes is certainly beneficial for the fish fauna in these ecosystems. In addition, dense areas of aquatic macrophytes can retain large amounts of nutrients in reservoirs. When the littoral zone and marginal areas of reservoirs are colonized by aquatic macrophytes, the survival rates of juveniles and young of many species increase. The vegetation provides food, protected areas for reproduction, and a haven for prey to avoid large-scale predation.

The reproduction of fish in reservoirs is thus related to the possibility of travelling upstream in wetland areas, which are recognized as areas with greater diversity,

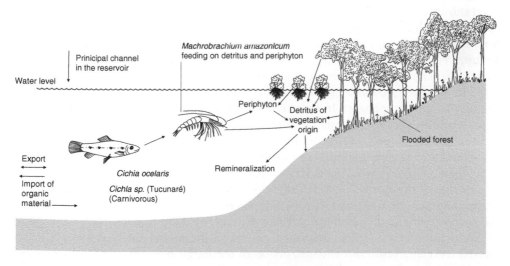

Figure 12.14 Organization of the food web in the Tucuruí reservoir (Tocantins River); the substratum includes periphyton and detritus of vegetation origin.
Source: Tundisi *et al.* (1993).

Figure 12.15   Terrestrial vegetation flooded and in decomposition in the UHE **Luis Eduardo Magalhães** Lageado reservoir, Tocantins River (see also color plate section, plate 19).

accumulation of detritus and possibly more varied food for the fish (see Figure 12.15). In reservoirs with short retention time, there is extensive localization of ichthyofauna upstream, near the entrance of the tributaries into the reservoir.

12.10.9   Pelagic zones and ichthyofauna in reservoirs

Many reservoirs form an extensive pelagic zone that provides a great opportunity for colonization by planktophagous fish and their predators. In some reservoirs, it has been shown that immediately after the dam is constructed, countless species of plank-tophagous fish could colonize it, greatly increasing their biomass in comparison to the lotic environment that was replaced. This increase in biomass (for example, in **Lake Kainji**, *Pellonulla afzeliusi* and *Sicrrathrissa leonensis* present an average biomass of 3140 tons) can form a basis for extraction fishing in reservoirs, which is significant.

The predators colonizing the pelagic zone feed on these planktophagous fish (mainly **zooplanktophages**). In Lake Kariba, for example, the nile perch (*Lates niloticus*) and the tiger fish (genus *Hydrocynus*) developed appreciable biomass due to the availability of food. In the case of *Hydrocynus* sp, 70% of its food is provided by the freshwater sardine *Limnothrissa miodon*. In this case, it is a new feeding habitat developed by *Hydrocynus* sp., described in Lake Kainji and Volta Lake.

Examples of species that were generalist feeders in the old river becoming special-ists in the reservoir are common, for example, in Lake Volta (a reservoir), the Cabora Bassa reservoir and the reservoir of Tucuruí in the Amazon basin.

Agostinho *et al.* (1994) described extensive alterations suffered by fish in the Paraná river as a result of the construction of the Itaipu reservoir. Of the 110 species of fish found in the region, 83 continued present in the reservoir. Some commercially important species, such as the pacu (*Piaractus mesopotamicus*) and the piracanjuba (*Brycon orbignyanus*), disappeared completely from the reservoir area. As these are species that fed on the remains of vegetation from riparian forests, their survival and reproduction was highly affected.

Migratory species such as *Leporinus elongatins*, *Leporinus obtusidens*, *Prochilodus scrofa* and *Pseudoplatystoma corruscans* remained in the Itaipu reservoir (Paraná River) and used the floodplains of the river upstream during part of their life cycle. *Salmino maxilosus* (golden dorado) is rarely caught in this reservoir. A species introduced upstream, the South American silver croaker, or pescada (*Plagioscion squamosissimus*), thrived in the Itaipu River. The same occurred with a planktophagous species, the Highwaterman catfish (*Hypophtalanus edentatus*), which took advantage of the extensive pelagic zone in the reservoir. Another successful species is the insectivore *Auchenipterus muchalis*, which matures rapidly and has a short life span.

The main detritivorous species were affected by the Itaipu reservoir, according to Agostinho *et al.* (1994), suffering a drop in density.

One of the most important conclusions of the studies conducted on the Itaipu reservoir has to do with the interaction of the reservoir with the wetland areas upstream. This interaction makes it possible to maintain a population of fish that extensively use this region to feed and reproduce, and, according to Agostinho *et al.* (1994, 1999), it has a direct effect on the substitution of the stocks of fish in the reservoir.

Several species of tilapia have thrived in reservoirs in Brazil, in many cases producing large biomasses. The introduction of various exotic species in reservoirs in Brazil has produced extensive alterations in the organization of the food chain of reservoirs (Rodrigues *et al.*, 2005).

12.11   BIOMASS AND FISHERY PRODUCTION IN RESERVOIRS

As previously mentioned, biomass increases after the construction of a reservoir is completed, although diversity often decreases significantly. The ecosystem, in the case of a reservoir, is extremely dynamic and in constant reorganization, at least during the first five years after completion of the dam. There is continual colonization by different species and, during the early years, the reservoir undergoes morphometric changes affecting the water chemistry and the biogeochemical cycles.

Evaluation of the production of fishery stock in reservoirs is a difficult task that calls for the use of various methods. One method commonly used for lakes and reservoirs is the MEI (Morphoedaphic Index), which, according to Ryder (1965) can be expressed:

$$MEI = \frac{total\ dissolved\ solids\ (mg \cdot L)^{-1}}{average\ depth}$$

Henderson and Welcomme (1974) substituted total dissolved solids for conductivity. For the African continent, Bayly (1979) showed that the biomass of fish is greater in reservoirs than in natural lakes. The higher productivity can be masked by the growing number of fishermen who are attracted to the new reservoir and intensively fish during the 'trophic upsurge'. This occurred in the Tucuruí reservoir on the Tocantins River and the Samuel reservoir in Rondônia, where *Cichla ocellaris* (tucunaré) developed large stocks.

Henderson and Welcomme (1974) conducted an intensive study showing that the biomass of fish captured is related to limnological variations and involves a variety of fishing methods. Schelesinger and Regier (1982) examined the hypothesis that differences observed between fishery production in higher and lower latitudes (more fish in tropical regions) is due to the higher temperatures in lower-latitude ecosystems. These authors found a positive correlation between the biomass captured and the average air temperature, with other **limnological variations** being equal or almost equal. The results were:

$$\log \varphi = 0.0236 - 0.280 \log \text{MEI} - 0.050\,\text{T}$$

where:

φ = estimate of the maximum sustained production per unit area in the lake or reservoir.

MEI = morphoedaphic index (already cited)

T = average air temperature (in °C)
(logarithms are base 10)

Other methods used in fishery production, applied to reservoirs in Nigeria by Ita (1978), are fishing in canoes, net fishing and biological/experimental sampling in nets (results in kg/1,000 m^2).

Other results are presented in Table 12.8, compiled from data gathered by Petrere (1994), Petrere and Agostinho (1993) and Paiva *et al.* (1994), which shows the capture and production of fish in many reservoirs in Brazil, compared with other reservoirs in several continents.

For more in-depth information on fishery production in reservoirs, many years of comparative and sequential data are needed, including more in-depth studies over longer periods, even from before completion of the dam. Once the reservoir is operational, it is crucial to continue monitoring the fish fauna upstream and downstream from the reservoir.

The development of stable fish stocks in reservoirs is part of the biological component of limnological succession, according to Holcik *et al.* (1989) and Kubecka (1993). During the 'evolution' or aging process of the reservoir, Vostradrosvshy *et al.* (1989) described possible succession in reservoirs in central Europe: the early stages

Table 12.8   Fish capture and production in reservoirs in South America and Africa.

	Capture (t · year^{-1})	Production (kg · ha^{-1} · year^{-1})	Author
7 reservoirs in the Parana River basin	4.51	–	Petrere and Agostinho (1993)
17 reservoirs in Northeast Brazil	151.8	–	Paiva *et al.* (1994)
Reservoirs in Africa	99.5	–	Marshall (1994)
Lakes in Africa	58.4	–	Bayley (1988)
Sobradinho	24,000	57.1	Petrere (1986)
Itaipu	–	11.6	Petrere (1994)
Guri (Argentina)	300	10	Alvarez *et al.* (1986)

Source: Petrere (1994).

after the dam is completed are dominated by **salmonids,** the second stage is dominated by **perch,** and the third by **cyprinids.**

The **management of fish fauna** and maintenance of fish stock in reservoirs are complex tasks because they involve not only sound ecological and limnological knowledge of the system and biology of the species, but they also depend on the reservoir's operating parameters and its multiple uses. Thus, ecological knowledge is inseparable from the system and the management of stocks, with the goal of maintaining sustainable fishing. The introduction of exotic species into the reservoir can be an extraordinarily complex element of management. Often, lack of scientific knowledge on the food web structure and the interactions of populations can lead to extremely complex and irreversible situations with the dominance of non-commercial species and little chance of commercial production.

Fishery management thus begins by identifying the species in the reservoir, the diversity, the role of the various species in the feeding network and the regulatory functions, such as **predator-prey relations.** In the management of stocks, it is necessary to take into account estimated fishing levels, the fishing effort and the biomass captured based on the number of fishermen. Evaluation of the existing stocks can be done not only by estimating the fish caught and the fishing effort, but also by echo-sounding.

Another important factor is biological understanding of the species in order to identify the breeding season, the interaction of the biological characteristics of the populations with climatic and hydrological forcing functions, and the capacity for replenishing the stock. Unexploited populations (single or multiple species) maintain their equilibrium. The first consequence of commercial production of a monospecific population is loss of equilibrium, with the following consequences: reduction in biomass, increased mortality rate (due to natural mortality and the result of fishing) and decrease in the average age of the population. In multispecies populations, a more complex process occurs, since optimal fishing effort is different for each species.

Therefore, the rational exploitation of fish stocks implies a sound initial understanding of the biological and ecological diversity of the species present, and also the gathering of statistical data on production, fishing efforts, catch by fishermen, number of fishermen, and market data. It is also essential to monitor the **aging process** or 'evolution' of the reservoir and changes in the structure of the fish fauna and biomass, either from the effects produced by exploitation of fishery resources or other factors, such as contamination and pollution.

Another extremely important problem is determining the distribution of the various species and fish stocks, according to the different regions of the reservoir. For example, in Sobradinho reservoir in the state of Bahia (Protan/CEPED, 1987 in Petrere, 1994), three regions were distinguished in terms of the reservoir's limnology:

▶ **Lentic region:** which comprises 60% of the area of the reservoir, with lower production and larger-sized specimens captured in this region.
▶ Transitional region: where 75% of the fish are captured. This region comprises 30% of the area of the reservoir, with an extensive region of macrophytes associated with submerged vegetation. In this reservoir, as in others, this is an area of growth and feeding.
▶ **Lotic region:** a typical region of a river with high levels of suspended solids, where migration and reproduction of fish occur in the marginal lagoons.

The transition region, in many reservoirs, appears to be the area with the highest levels of fish production, which could be associated with greater primary productivity, since the euphotic zone is deeper and external sources of nutrients are carried in from the lotic region, in addition to the autochthonous regeneration that occurs in this area. In Itaipu, Okada *et al.* (1994) showed that 50% of the catch comes from the transition region. In the case of reservoirs in northeast Brazil, there was an increase in the production of a *Tilapia* sp. after its introduction (*Tilapia rendalli* and *Tilapia niloticus*) (Fernando, 1992). *Tilapia* species contribute 30% of the total fish production from these reservoirs.

The introduction of pescada (*Plagioscion squamosissimus*) in reservoirs is one of the great examples of species adapted to lentic conditions that can be successfully cultivated in many reservoirs located at different latitudes (Petrere and Agostinho, 1993). Along with *Cichla* spp., *Plagioscion* spp. are the most common species in Tucuruí reservoir (Tocantins River) and are also a major part of the fish fauna in the Barra Bonita reservoir (Tietê River) and Itaipu reservoir (Paraná River) (Petrere and Agostinho, 1993). *Plagioscion* spp. are piscivores that thrive in the pelagic zone, feeding mainly on small fish that inhabit this area.

12.12   'EVOLUTION' AND AGING OF THE RESERVOIR

The aging process or 'evolution' of a reservoir basically depends on the characteristics of the filling phase of this man-made ecosystem and on the interrelations established over time between the reservoir and drainage basin. If during the filling phase, vegetation decomposes and is not removed from the future basin, then from the start the reservoir has an accumulation of organic particulate dissolved matter, causing alterations in its functioning, in the biomass of species, in the colonization by fish fauna, and in the structure of the food network (Matsumura Tundisi *et al.*, 1991). In the more advanced phases of interactions between the reservoir and the drainage basin, there may be inflows of organic matter, nutrients from domestic sources (untreated sewage), or agriculture nutrients – fertilizers that accelerate the eutrophication process, as shown by Tundisi *et al.* (2005) in the Billings reservoir in the metropolitan region of São Paulo.

The spatial organization of the reservoir over time presents changes that in some cases can be extensive. For example, the development and consolidation of ecotones change the colonization and succession of species in the pelagic and littoral zones (Henry, 2003).

The 'aging' process of reservoirs has some consequences defined by Straškraba *et al.* (1993). There is a profound change in the system's thermodynamics, once the mechanisms of primary production are modified. For example, in the Barra Bonita reservoir, Tietê River, primary production of phytoplankton increased 15 times in 25 years (Matsumura Tundisi and Tundisi, 2004). The decomposition of organic matter can become a major factor, increasing oxygen consumption. Organic material in the sediment increases significantly during the 'aging' process. Entropy thus may increase along with indirect effects, affecting the food web and the structure of the system (Margalef, 1983).

In reservoirs that reach a certain degree of **hyper-eutrophication**, for example, the phytoplanktonic biomass increases considerably. Decomposition of biomass produces a quantitatively important indirect effect (Sandes, 1998), since it provides a substratum for a variety of bacterial flora that develop associated with *Microcystis aeruginosa*. In some reservoirs, after several years the stabilization reached is a positive factor for the mainte-nance of diversity and biomass, especially if the retention time remains constant and the conditions of pollution and contamination are controlled in the drainage basin.

Large fluctuations in the reservoir's level can also alter succession and organiza-tion during this 'aging' phase. Extensive exposed areas where grasses can grow (up to 50–60 tons per hectare) contribute large amounts of organic matter in the phase following the seasonal filling of the reservoir.

Large influxes of suspended matter from the drainage basin can occur. Tundisi (1994, unpublished results) showed the impact of large masses of suspended inorganic material in Barra Bonita reservoir, resulting in extensive fish kills due to the complete depletion of dissolved oxygen in the water column during the unloading period of this suspended material.

Therefore, during the reservoir's 'aging' process, there is a set of pulses that inter-feres with the reorganization process and hampers its acceleration. Rapid discharges from bottom outlets in reservoirs can significantly alter the chemical composition of the sediment and reduce the system's internal load, altering biogeochemical cycles in the reservoir and downstream.

Figure 12.16 shows the hypothetical 'evolution' of a reservoir with examples of several reservoirs in different drainage basins in Brazil related to their stage of

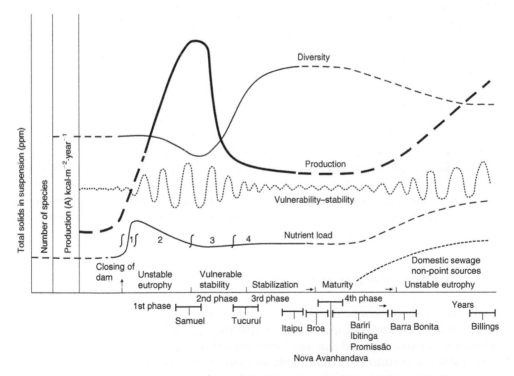

Figure 12.16 Processes of 'evolution' or 'aging' in several reservoirs from different drainage basins in Brazil and at different stages.
Source: Modified from Balon and Coche (1974).

development. In the reservoir's 'aging' process or 'evolution', interactions with the drainage basin play a key role.

12.13   MULTIPLE USES AND MANAGEMENT OF RESERVOIRS

Initially constructed for one or more specific purpose (for example, hydroelectric power or drinking water supply), large reservoirs currently have multiple uses, including promoting **regional development**. Therefore, management of the reservoirs is a complex task, calling for interdisciplinary expert teams that can minimize impacts, promote optimization of multiple use and effectively manage the man-made ecosystem and its evolution within the basin. Since reservoirs are more sensitive than lakes (because of the often larger size of the drainage basin as compared to natural lakes), it is important to analyze the impact on the entire drainage basin in order to determine responses and trends.

The various stages in management strategy of reservoirs are presented in Chapter 19.

12.14   URBAN RESERVOIRS

Many metropolitan regions in countries in both the Northern and Southern Hemispheres have reservoirs located in densely populated areas. These reservoirs are submitted to countless pressures from multiple uses, and are also ecosystems providing multiple environmental and social 'services'. For example, the metropolitan region of São Paulo is served by 23 water-storage reservoirs, which, along with other functions, are used for *recreation*, *fishing*, *hydroelectric power* (in some cases), and *tourism*. These systems are under constant pressure from the following influences: point and non-point sources of phosphorus, nitrogen, degradation of shores and the littoral zone, deforestation, discharge of residual solids, sedimentation, influx of toxic substances, air pollution, and extensive urban housing (see Figure 12.17).

In addition, **production of toxins** from blooms of *Microcystis aeruginosa* increases toxicity in the system. Management of these reservoirs is a complex task that is difficult to achieve, involving structural and non-structural factors. The non-structural measures require ongoing conflict resolution and coordination with the committees in charge of the basin, increasing the governance of the system (see Chapter 19).

12.15   RESEARCH ON RESERVOIRS

The study of these man-made systems, mainly in developing countries, is important to establish operational processes and enable the planning of new reservoirs with a minimum of environmental damage. Limnological research and adequate monitoring also enable the optimization of multiple uses and control of the chemical composition of water in the desired condition. For example, it is important that the study provide opportunities for control of water quality downstream from the reservoir. Control of the water quality and the physical and chemical conditions, associated with flow, can

Figure 12.17 Man-made urban reservoir – Paranoá Lake, Brasilia (see Annex 6) (see also color plate section, plate 20).

be achieved with a good understanding of the reservoir and its compartments on a temporal and spatial scale. This control should be extensive and involve information from before the dam was constructed in order to provide a basis for the thorough study of the alterations and developments in the reservoir under the influence of the regional conditions (Tundisi, 1993a, b, c).

It is essential to understand the **interrelations of the drainage basin** with the reservoir. A limnological study is thus complex, especially considering the countless interactions of a system covering thousands of square kilometers, and even more considering the flooded area. A solid understanding of these systems is based on limnology, botany, zoology and biology, and helps advance the field of tropical limnology (Henry, 1999b).

Gathering a large amount of data on reservoirs in a specific geographical area can facilitate, through comparison and classification, regional planning and management of drainage basins. For such a classification, it is necessary to understand a reservoir's principal operating mechanisms (Magalef *et al.*, 1976; Tundisi, 1981).

The biological community – in its composition, diversity and biomass – can be considered to be an expression of these interactions. Comparative studies can thus enable the introduction of appropriate techniques for the management and control of biomass.

Tributary of the São Francisco River, Três Marias region.
Photo: J. G. Tundisi

13 | Rivers

SUMMARY

In this chapter rivers are presented as aquatic ecosystems with permanent water flow and also permanent (and intense) interactions with the basins in which they operate. Benthic invertebrates and fish dominate river fauna. The aquatic biota in these lotic systems are adapted to the unidirectional flow of the water and to the structure, type and chemical composition of the bottom sediment.

Rivers provide transport systems for organic and inorganic matter. The contribution of allochthonous matter means that the energy flow is largely dependent on the steady contribution of organic and inorganic remains of vegetation, other organisms, fine suspended matter and sand. Autochthonous primary production is largely maintained by periphyton, aquatic macrophytes, and phytoplankton, located in areas of deposition and low circulation.

Organic material in a river – transformed by fish, bacteria, and the larvae of aquatic insects – moves downstream in a 'nutrient-spiral'.

Characteristics of rivers include drift – on which the survival of many organisms, especially insects, depends – and zonation. We will discuss various theories on zonation and the concept of *continuum* in the river.

Rivers are subject to the impacts of ongoing human activities, which have several levels of magnitude, ranging from construction of channels and deforestation of gallery forests to the discharge of heavy metals, herbicides, pesticides, and countless organic substances soluble in water.

The regeneration and restoration of rivers needs a scientific **database** in which temporal and spatial series (hydrological, physical, chemical and biological) enable promotion of scenarios, analysis of trends, and restoration of water quality and drainage basins.

13.1   RIVERS AS ECOSYSTEMS

A river differ from a lake, wetland area, reservoir or pond (lentic systems) in two fundamental ways: the first is the permanent horizontal movement of a river's currents, and the second is the interaction with its drainage basin, from which a perpetual contribution of allochthonous material arises – mainly terrestrial organic matter: leaves, fruits, remains of vegetation and aquatic insects. The steady contribution occurs in streams where there are well-structured and well-preserved gallery forests whose canopy shades the small streams. In cases with greater light penetration, periphyton and macrophytes are common and the production of organic matter is autochthonous. Plankton can occur in rivers, especially in areas with slow current.

The dominant invertebrates in rivers are benthic, and the dominant vertebrates are fish. The perpetual **unidirectional movement** of the water is the main feature in rivers and it controls the structure of the bottom and the materials found in the sediment. The aquatic biota – the lotic fauna and flora – are adapted to this unidirectional flow and to the structure of the sediment on the bottom of the river – chemical type and composition.

The velocity of the water in the river channel (usually expressed in $m \cdot s^{-1}$) varies significantly across its cross section, and the friction between the current and sediment can vary greatly. In the vertical section of a river, the velocity of the current is faster on the surface (in shallow rivers). In deeper rivers, the surface water can also flow faster than the waters at greater depths. There are exceptions, however. The relationship between a current's velocity, depth, and the physical structure and distribution of sediment determine important physical characteristics in terms of the horizontal structure of the system and the transport of particulate and dissolved materials. The total volume of water that passes a given point in the river in unit time can be calculated by $Q = LPU$, where L is the width of the river, P is the average depth at the point under consideration, and U is the average velocity of the current. The resulting volume flow is given in $m^3 \cdot s^{-1}$.

The volume of a river's discharge (in $m^3 \cdot s^{-1}$) is an important characteristic; based on this information, it is possible to determine the material load transported by the river (in $kg \cdot s^{-1}$, for example, or $tonne \cdot year^{-1}$). Rivers flow can be *laminar*, in which there is a parallel displacement of water layers on the vertical axis; or *turbulent*, with complete mixture (see Chapter 4).

13.2   TRANSPORT PROCESSES

Fluvial geomorphology determines the main characteristics of the water network, in which varying drainage patterns are established depending on latitude, longitude, altitude, slope, and soil type. Studies began in the second half of the 18th century on the evidence that this water network is determined by the physical relationships and agents that interact in the drainage basins.

Physical factors affecting the transport of matter and load include: width and depth of the river channel, the velocity of the flow, the degree of roughness of the sediment, and the degree of sinuosity of the river and its major tributaries (Allan, 1995). In certain places the slope of the river is also important because it establishes different velocities of the current and different transport mechanisms for particulate and dissolved matter. The spatial/temporal characteristics of rivers depend on their interactions with drainage basins and the fluctuations in the **regional hydrology,** which determine differential patterns of flow that can vary seasonally and even over short periods (hours or days). These characteristics are also important in the spatial/temporal distribution of biota. In periods of heavy rainfall or drought, for example, the spatial distribution of micro-habitats in rivers is affected by variations in current velocity. Such micro-habitats can range in size from a few centimetres to several metres long.

13.2.1 Transport of organic and inorganic matter

The inorganic and organic sediments transported by rivers come from the erosion of riverbanks and the erosion of soils in drainage basins. Deposition of such sediments occurs in different parts of the river, especially in floodplain areas, backwaters and zones with low current velocity. Particles of suspended matter are deposited according to their size and density. The size of particles that a river carries depends on the velocity of the current and the river's morphological characteristics. A critical velocity of erosion for sand particles, for example, is $20\,cm \cdot s^{-1}$. Extreme natural events (such as heavy rains or droughts with low current velocity) play an important role in transport and deposition of sediment. The impact of human activities can alter the flow of the transport of suspended materials.

As a rule, rivers deposit larger matter upstream and fine particulate matter ($<20\,\mu m$) downstream in regions with lower velocity, such as backwaters and stretches with meanders.

Figure 13.1 shows the velocity of the current and size of the particles carried by rivers. In addition to organic and inorganic suspended particles, there is also a continuous transport of mainly vegetative remains (leaves and detritus), which is used as food by a range of organisms that process this material. The clearest impact of the processing of this material is the transformation of the organic matter along the river, such that downstream there are higher levels of fine particulate organic material and dissolved organic matter. The *transport* or displacement of such organic matter includes detritus of various origins, algae, and invertebrates.

13.3 LONGITUDINAL PROFILE AND CLASSIFICATION OF THE DRAINAGE NETWORK

Streams, small rivers and brooks can be distinguished from rivers by their size. Large rivers may be thousands of miles long and dozens of miles wide. The longitudinal profile of rivers and stream starts with a steeper slope and sinuosity downstream (see Figure 13.2). The velocity of the current and the deposition of matter vary according to the stretch of the river under consideration.

Rivers and streams in drainage basins are classified according to their order. Small streams and headwater springs are first order (see Figure 13.3). When two small first-order streams come together, they become a second-order stream and so on. The main

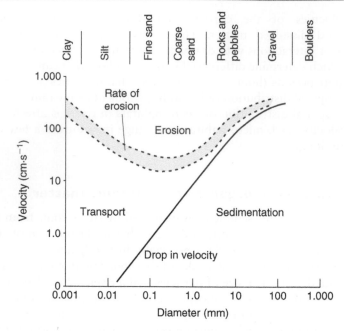

Figure 13.1   Relationship between average velocity of current in 1.0 m deep water and size of mineral particles that can be eroded from a bed of similar-sized material.
Source: Modified from Allan (1995).

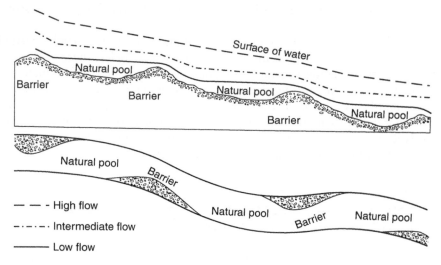

Figure 13.2   Longitudinal profile of a river, showing the various components.

channels of large rivers can be 10th- or 12th-order before reaching the ocean. Table 13.1 shows the classification of rivers based on various characteristics of size and discharge.

The drainage basin through which the main river and its tributaries run varies greatly in shape and slope with a certain regularity typical to some regions. For example, the small streams in the *cerrado* of Brazil display dendritic patterns that are

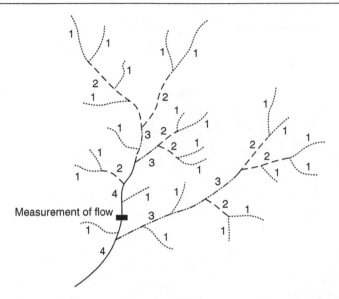

Figure 13.3   Illustration of the classifications of tributaries in a drainage network.

Table 13.1   Classification of rivers based on discharge, drainage area and length.

Size of river	Average discharge $(m^3 \cdot s^{-1})$	Drainage area (km^3)	Length of river (km)	Order of river
Very large rivers	>10,000	>10^6	>1,500	>10
Large rivers	1,000–10,000	100,000–10^6	800–1,500	7–11
Rivers	100–1,000	10,000–100,000	200–800	6–9
Small rivers	10–100	1,000–10,000	40–200	4–7
Streams	1–10	100–1,000	8–40	3–6
Small streams	0.1–1.0	10–100	1–8	2–5
Small creeks from springs	<0.1	<10	<1	1–3

Source: Chapman (1992).

related to topography and slope (see Figure 13.4). These small rivers and their gallery forests include a wide variety of wetland areas that maintain the biodiversity of the *cerrado*, not only aquatic fauna, but also the terrestrial fauna associated with it. For example, gallery forests provide areas for nesting and refuge for aquatic birds.

13.4   FLUCTUATIONS IN LEVELS AND DISCHARGE CYCLES

The discharge patterns in rivers determine the main properties of these lotic systems. Seasonal and diurnal variations in rivers that depend on the climatic and water cycles control the physical, chemical and biological processes. Discharge depends on the rainfall, geology and geomorphology of the drainage basin, the slope of the rivers, the presence of vegetation or dams and river characteristics of the bottom sediment.

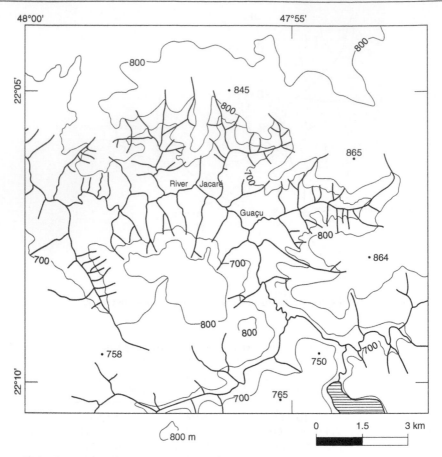

Figure 13.4   Dendritic pattern in spatial organization of small rivers of the *cerrado* in the state of São Paulo (Brazil).
Source: Tundisi (1994).

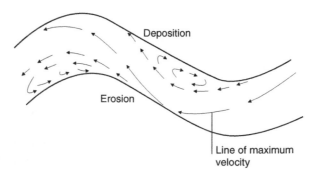

Figure 13.5   Areas of maximum velocity of deposition and erosion in meanders of the river.
Source: Modified from Allan (1995).

Extremely rapid discharges after heavy rains increase the transport of matter and organisms downstream, which accumulate mainly in wetland areas and meanders (see Figure 13.5).

13.5   CHEMICAL COMPOSITION OF THE WATER AND THE BIOGEOCHEMICAL CYCLES

Rivers receive a large quantity of organic and inorganic material from the drainage basins and drainage networks through which they run, affecting the chemistry of the water and the biogeochemical cycles. Thus, in addition to the water, a river carries a range of materials that, according to Berner and Berner (1987) and Horne and Goldman (1994), consists of:

▶ Suspended inorganic matter: compounds of aluminium, iron, silicon, calcium, potassium, magnesium, sodium, phosphorus;
▶ Dissolved ions: principally Ca^{2+}, Na^+, Mg^{2+}, K^+, HCO_3^-, SO_4^{2-}, Cl^-.
▶ Dissolved nutrients: nitrogen, phosphorus, silicon.
▶ Dissolved and particulate organic matter.
▶ Gases (N_2, CO_2, O_2).
▶ Trace metals (particulate and dissolved).

Other elements should also be added (resulting from human activities in drainage basins), such as aluminum, mercury, lead, cadmium, zinc, cobalt, copper and chromium, which are present in dissolved or particulate form and are incorporated into the food chain, damaging both the fauna and flora. Other components to be considered, depending on whether the rivers and drainage basins are located in agricultural or industrial areas, are pesticides and herbicides, oils and greases.

The dominance of weathering, in which many carbonate components are present, is very common. More than 50% of total dissolved solids (TDS) are composed of bicarbonates, chlorides and sulphates.

Table 13.2 shows chemical composition of river waters. As a comparison, Table 13.3 shows chemical composition of rainwater. It should be noted that rainwater near coastal areas present a greater contribution of certain ions, such as sodium, magnesium, potassium and chlorine; in rivers, chlorine and sodium are mostly derived from the weathering of rocks; sulphate derives from volcanic activity or acid rain in industrial areas.

The combination of the discharge (in $m^3 \cdot s^{-1}$) and concentration of organic or inorganic constituents (in $g \cdot m^{-3}$) makes it possible to estimate the load of these substances or elements, which is expressed in $t \cdot year^{-1}$ or in $kg \cdot day^{-1}$. This load varies in space and time and depends on the periods of greater flooding, according to seasonal changes of the water cycle. Thus, it is essential to conduct a detailed seasonal study to estimate the load in different periods of the year, according to climatic changes.

Relations between *evaporation*, *precipitation*, and *predominance* of *rock type* have already been described in Chapter 5, in terms of the basic chemical composition of inland waters.

Table 13.2   Chemical composition of river water (mg · L^{-1}).[a]

	Total dissolved solids	Ca^{2+}	Mg^{2+}	Na^+	K^+	Cl^-	SO_4^{2-}	HCO_3	SiO_2	Discharge $(km^3 \cdot year^{-1})$	Rate of surface runoff[b]
Global average											
Present	110.1	14.7	3.7	7.4	1.4	8.3	11.5	53.0	10.4	37.4	0.46
Natural	99.6	13.4	3.4	5.2	1.3	5.8	8.3	52.0	10.4		
North America											
Present	142.6	21.2	4.9	8.4	1.5	9.2	18.0	72.3	7.2	5.5	0.38
Natural	133.5	20.1	4.9	6.5	1.5	7.0	14.9	71.4	7.2		
South America											
Present	54.6	6.3	1.4	3.3	1.0	4.1	3.8	24.4	10.3	11.0	0.41
Natural	54.3	6.3	1.4	3.3	1.0	4.1	3.5	24.4	10.3		
Europe											
Present	212.8	31.7	6.7	16.5	1.8	20.0	35.5	86.0	6.8	2.6	0.42
Natural	140.3	24.2	5.2	3.2	1.1	4.7	15.1	80.1	6.8		
Africa											
Present	60.5	5.7	2.2	4.4	1.4	4.1	4.2	26.9	12.0	3.4	0.28
Natural	27.8	5.3	2.2	3.8	1.4	3.4	3.2	26.7	12.0		
Asia											
Present	134.6	17.8	4.6	8.7	1.7	10.0	13.3	67.1	11.0	12.5	0.54
Natural	123.5	16.6	4.3	6.6	1.6	7.6	9.7	66.2	11.0		
Oceania											
Present	125.3	15.2	3.8	7.6	1.1	6.8	7.7	65.6	16.3	2.4	–

[a]The present levels include data on human activities. The natural values were corrected to exclude pollution.
[b]Runoff rate – average runoff per unit area/average rainfall.
Source: Berner and Berner (1987).

Table 13.3   Typical levels of the principal ions found in rain water (mg.L^{-1}).

Ion	Inland rain	Rain over Ocean and Coast
Na^+	0.2–1	1–5
Mg^{2+}	0.05–0.5	0.4–1.5
K^+	0.1–0.5	0.2–0.6
Ca^{2+}	0.2–4	0.2–1.5
NH_4^+	0.1–0.5	0.01–0.05
H^+	pH 4–6	pH 5–6
Cl^-	0.2–2	1–10
SO_4^{2-}	1–3	1–3
NO_3	0.4–1.3	0.1–0.5

Source: Berner and Berner (1987).

Rivers transport nitrogen in the form of nitrate, nitrite or ammonia, and silicate in the soluble form. Phosphate is also associated with particulate matter, as shown by Likens (1977, 1997) especially for streams in forested areas (see Table 13.4). The proportion of each component of the cycle varies according to the climate, season of the year, and the geology of the drainage basin.

Table 13.4 Particulate fraction and dissolved fraction for various elements transported in rivers.

Element	Particulate fraction (%)	Dissolved fraction (%)
P	63	37
N	3	97
Si	26	74
Fe	100	0
S	0.2	99.8
C	32	68
Na	3	97
K	22	78
Ca	2	98
Mg	6	94
Cl	0	100
Al	41	59

Source: Modified from Likens et al. (1997).

Table 13.5 Average monthly electrical conductivity ($\mu S \cdot cm^{-1}$), dissolved oxygen (mg $\cdot$ L^{-1}) and pH in the Riberão do Lobo (Itirapina, SP).

	Conductivity	Dissolved oxygen	pH
April (1985)	26.0	8.5	6.8
May	18.0	8.7	6.8
June	14.0	7.8	7.0
July	26.0	9.0	6.7
August	12.0	9.0	6.9
September	15.0	7.7	6.9
October	15.0	7.0	6.1
November	17.0	6.9	6.2
December	14.0	7.5	6.2
January (1986)	14.0	7.1	6.5
February	15.0	7.4	6.4
March	18.0	7.6	6.5
Average	17.0	7.8	6.6

Source: Matheus and Tundisi (1988).

Table 13.5 presents the level of dissolved oxygen, pH and conductivity in Ribeirão do Lobo (*cerrado*), in the central region of the state of São Paulo (Matheus and Tundisi, 1988). Rivers in the *cerrado* tend to have low conductivity, high levels of dissolved oxygen, areas with rapid currents, and slightly acidic pH.

Annual variations in the chemical components – such as phosphorus, nitrogen, silica and other ions – are dependent on and controlled by the drainage basins, the **discharges during the hydrological cycle** and other factors, such as fixation of nitrogen by aquatic plants, erosion, decomposition of vegetation and retention by the humic layer in the sediment. Due to these factors, the seasonal variations in the biogeochemical cycles of the rivers are much more pronounced than in lakes and the concept of limiting nutrients, as applied to lakes, does not apply very closely to rivers.

In rivers, there are regions of accumulation of nutrients, especially phosphorus, nitrogen or silicon, which can be released by biochemical or physical processes, such as the effect of currents or high discharges in periods of intense rainfall. The presence or absence of carbonates defines rivers with hard water or acidic water with low levels of carbonate ions. This chemical composition determines different types of **lotic fauna and flora**. For example, in the *cerrado* of Brazil, streams generally have acidic water with low pH and low levels of carbonates or bicarbonates. Throughout much of the Amazon region, streams are acidic and have low pH.

The distribution of molluscs, for example, is to some extent closely related to the level of calcium and alkalinity in the river. Not only are some species limited, but the thickness of the shell is related to the level of calcium in the water.

13.5.1 Biogeochemical cycles and particulate and dissolved organic components

From both the qualitative and quantitative viewpoint, dissolved and particulate organic matter is much more important in rivers than in lakes. This organic matter includes: allochthonous material with particles larger than 1 mm in diameter called coarse particulate organic matter (CPOM), as compared with fine particulate organic matter (FPOM) which consists of particles smaller than 1 mm in diameter.

The dissolved organic matter (DOM) in rivers derives from the decomposition of particulate matter, the excretion of organisms such as fish and invertebrates, and the steady recycling of particulate organic matter downstream. Such organic matter, either allochthonous or autochthonous in origin, is immediately attacked by bacteria and fungi that add nutritional value to the detritus, which in turn is eaten by other invertebrates or detritus-eating fish. This dynamic system is very important in small streams, where the decomposition of particulate matter from leaves and other remains of vegetation and organisms result from the processing of such material, which is finally transported as dissolved organic matter to larger rivers downstream in the basin (Walker, 1978). The author observed small streams in the Amazon basin and how a community of amoebae, ciliated protozoa, rotifers, and nematodes process organic matter, thus establishing a relatively complex food network, based on predator-prey relationships and an organization of up to five trophic levels, in terms of protozoa, from the processing of particulate organic matter of allochthonous origin.

Figure 13.6 shows the process of 'nutrient spiralling' in a river and the relationships between aquatic biota and the transport of materials. It is important to consider the rate of transfer of dissolved substances between the components of the system: biota, sediments, and water. Figure 13.7 describes the nitrogen cycle in a river.

As far as the heavy metals that occur in rivers due to natural phenomena (soil composition, for example) or anthropogenic processes, it is important to point out that their physical and chemical form (speciation) is essential in cycling and in the 'spiral of elements' found in rivers. The complexation of metals by dissolved organic substances (humic acids, for example) is not completely defined and further experiments under controlled conditions are needed to better define the interactions of copper, zinc, mercury, cadmium and lead with dissolved organic substances in the various physical and chemical situations – redox potential, electrical conductivity, rate of oxygenation – to which rivers and streams are submitted (Hart and McKelvie, 1986).

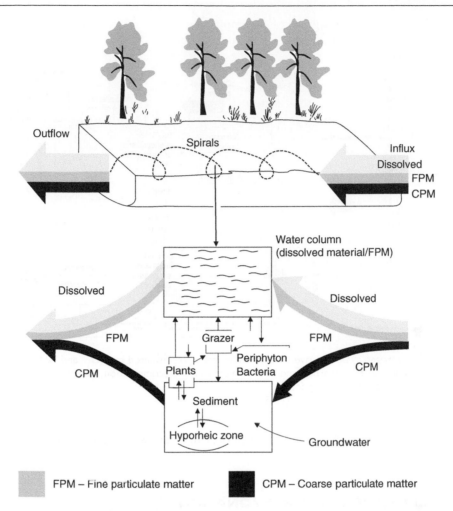

Figure 13.6   Nutrient-spiralling in a river.
Source: Hart and McKelvie (1986).

In Brazil, where there are regional and local differences in the chemical composition of the water and the 'background' of metals and humic substances in different regions, there is certainly a complex differentiation in various regions, with speciation and chemical reactions that differ depending on water temperature, degree of oxygenation and other physical and chemical processes.

The effect of events (such as flooding, for example) on the **transport of phosphorus** and nitrogen is well known and documented. The transport of phosphorus occurs mainly in the form of particulate matter (between 70–90%) (Cullen *et al.*, 1978a; Horne and Goldman, 1994).

Natural humic substances are the result of the decomposition of vegetation, and many rivers and streams with abundant riparian vegetation present high levels of dissolved organic matter (DOC – dissolved organic carbon, in many cases ranging from 2.0 to 30 or $50 \, \text{mg} \cdot \text{L}^{-1}$). This dissolved organic matter plays a key role in the

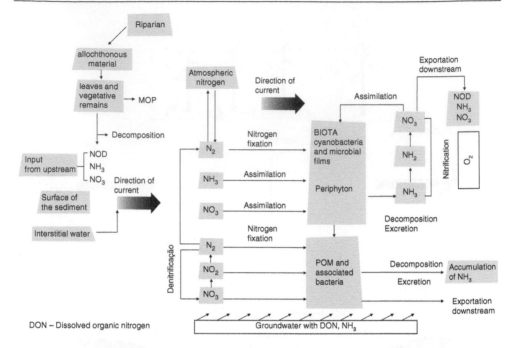

Figure 13.7   Nitrogen cycle in a river. Available nitrogen is represented by NO_3^- and NH_3, which are immediately absorbed and directly assimilated. Decomposition, excretion and exudates are ways of recycling food.
Source: Modified from Allan (1995).

recycling of metals, nitrogen, and phosphorus in rivers with high levels of dissolved organic carbon.

13.6   CLASSIFICATION AND ZONATION

Attempts to identify and measure zones in rivers have taken place since the middle of the 19th century (Borne, 1877).

Rivers and streams are complex ecosystems, especially because they present great spatial alterations from their springs to the large floodplain areas in the fluvial plains. According to Horne and Goldman (1994), some early attempts to classify rivers based on ichthyofauna were made, as by Schindler (1957), who considered the various features of rivers in relation to current and dissolved oxygen as the basis for the classification of different species of fish that inhabit the different parts of the river.

The zonation of rivers was part of an effort by limnologists to characterize lotic systems as opposed to the classification system for lakes based on trophic levels proposed by Thienemann (1925) and Naumann (1926).

Nowicki (1889) studied the zonation of the **Vistula River** in Poland. Thienemann (1912, 1925) also sought to present the zones of rivers based on physical characteristics, fish and invertebrate fauna for the rivers of Germany, describing successive zones in small streams.

Principal physical factors of rivers that affect aquatic biota

The aquatic biota in rivers are adapted to a number of factors that influence the biota's structure and function. One key factor defining the physical and chemical environment of rivers and streams is the current (Margalef, 1983). The main factors acting on the lotic fauna and flora include:

Velocity of the current and associated physical forces – The current transports food and displaces organisms (drift), and the velocity of the current affects the deposition of particles. Morphological adaptations to currents occur in the flora and fauna of rivers. The hydrodynamic forces affect organisms in different ways. The regime of the current is extremely variable.

Flow in the water and next to the sediments – Laminar, turbulent or transitional.

Substrata – Type and quality of substratum: sand, pebbles, fine clay material, rocks, rocky sediment, organic substrata (tree trunks, leaves) and inorganic substrata. The substratum affects the abundance and diversity of organisms (Allan, 1995).

Water temperature – The water temperature in lotic systems varies both daily and seasonally due to factors such as climate, altitude, type and extent of gallery forest, and contribution of groundwater. Temperature establishes limits to the geographical distribution and physiology of organisms, influencing their reproduction, survival and life cycle.

Dissolved oxygen – The level of dissolved oxygen plays a key role in the distribution, survival and physiology of lotic fauna and flora. The decomposition of masses of vegetation or the discharge of residual organic matter (i.e., sewage) substantially alters the diversity and biomass. Fish located downstream from large waterfalls in rivers are adapted to survive in water with higher concentrations of dissolved oxygen (up to 120% saturation) (Tundisi, 1992, unpublished data).

In Europe, the work of Carpenter (1928) sought to classify mountain streams in the region of Wales, based on physical characteristics and the fauna, especially fish species, but also crustaceans, insects, annelids, planarians and coelenterates. Other classifications and zonation proposals were made by Huet (1949, 1954) for rivers in certain areas of Belgium; Müller (1951) for rivers in Germany, relating zonation to benthic communities; and Illies (1958) for streams in Northern Europe.

In North America, the pioneering work of Burton and Odum (1945) aimed to present zonation in streams in Virginia based on fish species, and studies by Funk and Campbell (1953) were conducted on tributaries of the Mississippi River. Other studies on the zonation of rivers have been conducted in Africa (Harrison and Elsworth, 1958), New Zealand (Allen, 1956) and Brazil (Kleerekoper, 1955). Illies (1964) studied the distribution of invertebrate fauna in one of the tributaries of the Amazon River in Peru.

One of the discussions on zonation in rivers refers to the use of fish species to represent biocoenoses, which is why a series of studies were undertaken on the longitudinal distribution of benthic fauna in rivers in Europe, North America and Africa between 1925 and 1970 (Hawkes, 1975). The aim was to establish associations between fish species indicative of different conditions (physical, chemical and biological) of rivers and benthic macro-invertebrates (Hallam, 1959). Illies (1953) conducted studies specifically targeting aquatic insects, comparing the longitudinal distribution of Ephemeroptera, Plecoptera, Trichoptera and Coleoptera. The causes for this longitudinal distribution, described in many studies, were listed by Hawkes (1975) as follows:

▸ current and type of substratum;
▸ velocity of current and laminar or turbulent flow;
▸ temperature;
▸ level of dissolved oxygen;
▸ dissolved nutrients and level of carbonates;
▸ interactions with other organisms (predator-prey relationship, parasitism).

Another type of zonation studied was the **classification of biotypes** in different stretches of the river, as described in Berg (1948). The classification of biotypes takes into account the physical characteristics of the river bottom (rocks or sand, pebbles, presence of algae or macrophytes, organic sediment). The author listed the characteristic species in each site and described the velocity of current typical for each part of the river:

▸ Very strong: $>0.1\,\mathrm{m\cdot s^{-1}}$
▸ Strong: $0.05–0.1\,\mathrm{m\cdot s^{-1}}$
▸ Moderate: $0.025–0.05\,\mathrm{m\cdot s^{-1}}$
▸ Low: $0.01–0.025\,\mathrm{m\cdot s^{-1}}$
▸ Very low: $<0.01\,\mathrm{m\cdot s^{-1}}$

Marlier (1951) defined 'synecological units' based on associations of benthic animals. Another type of classification considers the pH of rivers: acidic (pH 5.0–5.9) not buffered; weakly acidic (pH 6.0–6.9), alkaline (pH 7.0–8.5) (Hawkes, 1975).

Hierarchical classifications and zonations of stretches of rivers have been proposed based on the density of benthic fauna and the presence of indicator organisms. The presence of species or genera of macrophytes and benthic algae have also used as the basis for classification (Butcher, 1933; Tansley, 1939; Lagler, 1949; Macan, 1961). Several authors (for example, Hawkes, 1975) considered that the zonation and classification of rivers based on fish species have an important applied value in terms of the conservation and restoration of these ecosystems.

One widely applied classification and zonation of rivers is that of Illies (1961a), who based his zonation proposal on data obtained through works done on rivers in various continents: South America, Europe, and Africa (especially South Africa) (Illies, 1961b; Harrison and Elsworth, 1958).

The two main categories proposed by Illies are:

i) **Rhithron** – defined as a zone of high current velocity; volume of a few cubic meters; regions where the average annual water temperature does not exceed 20°C; substratum with rocks, stones, pebbles and fine sand.

ii) **Potamon** – defined as a zone of low current velocity, predominantly laminar; average annual temperature above 20°C, or in tropical latitudes maximum temperature above 25°C; substratum with organic sediment, small pools and natural ponds with low levels of dissolved oxygen.

According to Illies and Botosaneanu (1963), Rhithron organisms are cold water stenothermic, associated with abundant oxygen (> 80% saturation) and aerated. Potamon organisms are eurythermic or stenothermic in warm waters, with development of plankton in the various branches or ponds associated with this region. The epi-, meta, and hypo-rhithron zones were also considered by Illies (1961a) to be extensions of the original proposal of zonation and **crenon-eucrenon** (sources) and **hypocrenon** (headwaters) as regions above the rithron.

The classification and zonation of rivers are useful for conservation strategies and ecological studies. However, the adoption of one or more classification method depends on the region (latitude, longitude and altitude) and comparative studies. According to Marlier (1951), the synecological approach is necessary and certainly will be very useful.

A classification method based on characteristics of the associations and assemblages of organisms combined with physical characteristics should be considered to be the most appropriate. Chart 13.1 presents the association of insect families with the Rhithron and Potamon zones in rivers. Different species occur in different

Chart 13.1   Insect families associated with the rhithron zone (steep upper course) and potamon zone (slower-flowing lower course) in rivers.

Order	Family Rhithron	Potamon
Ephemeroptera	Ecdyonuridae Ephemerellidae Leptophlebiidae	Siphlonuridae Potamanthidae Polymitarcidae Caenidae
Plecoptera	Capniidae Leuctridae Nuemouridae Gripopterygidae	Perlodidae Perlidae
Diptera	Blepharoceridae Simuliidae Podonomidae Psychodidae	Chironomidae Culicidae Tabanidae Stratiomyidae
Coleoptera	Elmidae Psephenidae Holodidae Hydraenidae	Dysticidae Haliplidae
Heteroptera		Corixidae Notonectidae
Trichoptera	Rhyacophilidae Odentoceridae Glossosomatidae Philopotamidae (except *Chimarrha*)	Leptoceridae Hydroptilidae
Hydrachnellae	Hygrobatidae Protziidae	

Source: Hawkes (1975).

biogeographic regions, but as a rule, families of insects tend to be represented in rivers in different continents.

13.6.1   Influence of human activities in zonation

The use of drainage basins by humans produces alterations in the zonation of rivers, especially when it comes to deforestation, land use and erosion. Every alteration in a river affects the zonation (Hynes, 1961).

The introduction of exotic species, either accidentally or for the purpose of aquaculture or increase in biomass, alters the food network and biocoenosis. The construction of dams affects the zonation of rivers and affects the composition of a biocoenosis, as discussed in Chapter 11. The channelling of rivers affects the zonation and enables migration of species that move between different drainage basins, especially when there are connections between the basins. Such is the case that could happen in Brazil if a channel is constructed connecting the Tocantins and São Francisco rivers comes to pass.

13.6.2   The concept of *continuum* in the river

A more substantial understanding of the dynamics of lotic systems was presented by Vannote *et al.* (1980). This approach is based on the order of the rivers, the type of particulate organic matter and the type of benthic invertebrates present. The basis of this approach is the alteration that occurs from the headwaters of the river to its outlet in another river or estuary. Together with the modifications and physical gradients, according to this concept, a series of associated biotic adjustments occurs.

This concept, according to Callow and Petts (1996), states that the structure and function of benthic communities, from the headwaters of the river to its mouth, is provided by the gradient of allochthonous and autochthonous organic matter. The relative importance of each group of invertebrates, shredders, collectors (gatherers and filter feeders), herbivores, predators and scrapers shifts in relation to food supply (see Figure 13.8a). In small streams of order 1 to 3, coarse particulate organic material (CPOM) is predominant, which is the basic food for shredders, such as freshwater crabs and invertebrate larvae. Fine particulate organic matter (FPOM), resulting from this activity, dominates rivers of order 4 to 7. Sediment collectors or filter-feeders dominate this region. Several types of aquatic insect larvae feed on this FPOM. In rivers of order 8 to 12, the autochthonous primary production begins to dominate (due to the algae of the microphytobenthos and aquatic macrophytes). In these rivers, there are components of FPOM and DOM (dissolved organic matter) that are used by herbivores and scrapers, such as molluscs and aquatic insect larvae (see Figure13.8b and 13.9).

The concept of *continuum* in a river has been studied by many researchers, such as Minshall *et al.* (1983), and applies to many rivers in temperate and tropical regions.

In floodplain regions located in the fluvial plains, the ecological integrity of the system depends on the connectivity between the natural channels of the rivers and the floodplains. This connectivity is represented by the concept of flood pulse, described by Junk *et al.* (1989), and stresses the importance of the flood pulse in the dynamic ecology of terrestrial and aquatic communities and in the spirals of carbon, phosphorus and nitrogen.

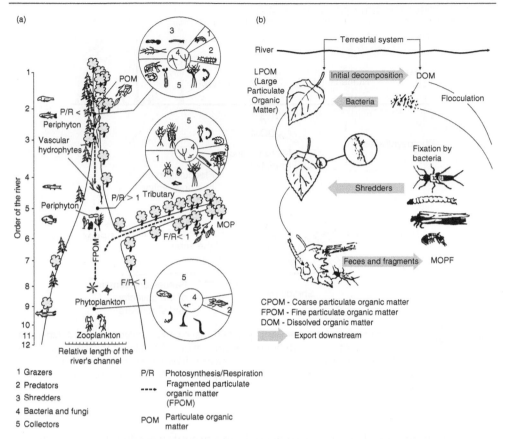

Figure 13.8   a) The interactions of shredders, particulate organic carbon, fungi and bacteria, in small streams in temperate regions; b) Food web in rivers and components of the processors of organic material.
Sources: Modified from Vannote *et al.* (1980) and Allan (1995).

The flow of water in rivers provides a variety of habitats, and the varying patterns of the velocity of the current affect the benthic communities of invertebrates and microphytobenthos. Therefore, within this physical context, rivers are structured by food chains and their configurations and structures in different regions. Disturbances such as floods or erosion play an important role in the organization and reorganization of these communities (Townsend, 1989), which can present mosaic structures in different regions.

In a study conducted in central Minas Gerais (19°20'S and 43°04'W) in a river in the Serra do Cipó region (Córrego Indaiá – 1st to 4th order; and Córrego do Peixe– 5th and 6th orders), which belongs to the Doce River basin, Callisto *et al.* (2004) assessed the structure, diversity and functional trophic groups of the benthic community of macro-invertebrates of these systems, and categorized the rivers from a physical, chemical and biological perspective (benthic micro-invertebrates, faecal coliforms, heterotrophic bacteria and yeast). Some 60 taxa of benthic macro-invertebrates were identified, of which the dominant group was aquatic insects, with 50 families in eight orders. These results of the study show that the structure, diversity and composition

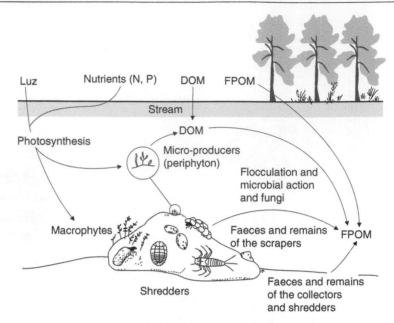

Figure 13.9 Trophic relationships between periphytic and macrophytic shredders and scrapers in a stream with fragmented allochthonous organic matter. The film of microbial periphytic-bacterial-organic matter on substratum surfaces is fragmented or partitioned. The figure presents a common amoeba that plays an important role in the processing of matter in Amazon streams.
Source: Modified from Allan (1995).

of the communities of benthic macroinvertebrates are influenced by the availability of food resources, by seasonality, and by the heterogeneity of the sediment.

In 1st-order regions, the river bottom consists of rocks. During the rainy season the river presents a steady flow of water, while in the dry season there are isolated ponds with large quantities of FPOM. The stretches of 2nd to 4th order have bottoms made up of c. 70% rock, with sequences of riffles and well-defined pools; rapids in steep areas; and deep pools with pebbles and coarse sand near the banks of the river. The 3rd-order stretches have bottoms with rocks, pebbles and stones, and stretches with rapids and shallow pools.

In these rivers, 5th-order stretches have a rocky bottom formed by pebbles, rocks, and sand. Deposits of erosion occur along the banks and in the channels of the river. In turn, the 6th-order stretches have a rocky bottom covered with pebbles, sand and silt deposits, in the series of riffles and pools. These components of the substratum are replaced by zones of deposition of erosion located in the sand and meanders of the river, with lower velocity of current.

This example clearly illustrates the longitudinal differences that occur in the structure and function of streams in which the benthic communities, algae of the micro-phytobenthos, bacteria and yeasts are quantitatively important in the processes and functioning of the system, in addition to the physical conditions. Callisto *et al.* (1998, 2001) present suggestions on the structure of the habitat, its diversity and the diversity of benthic functional trophic groups, indicating the importance of using benthic macro-invertebrates as a tool to assess the health of a stream (Callisto *et al.*, 2001).

In Brazil, recent works on the distribution and zonation of communities of invertebrates and fish include studies by Huamantico and Nessimian (2000), Camargo and Florentino (2000), Resende (2000), Callisto *et al.* (2000a, b), Oliveira *et al.* (2000), Schulz *et al.* (2001), Higuti and Takeda (2002), Mazzoni *et al.* (2002), Araujo and Garutti (2003), Garavello and Garavello (2004), Cusatti (2004), Callisto *et al.* (2005), Cetra and Petrere (2006), Pedro, Maltchik and Bianchini (2006).

These studies conclude that the distribution pattern of the lotic fauna depends on the interaction between the geomorphology of the river or stream, the type of substratum, the hydraulic conditions, the water temperature, and the biological interactions such as predation and parasitism. Silveira *et al.* (2006) determined the spatial and temporal distributions of the benthic invertebrate fauna of the Sana river (in the drainage basin of the Macaé River in Southeast Brazil). In this study, the highest total species richness occurred in the litter substratum of the flow, while the litter substratum on the bottom presented the highest number of unique niches. This appears to be a characteristic pattern of the Atlantic forest rivers (Kikuchi and Uieda, 1998).

Bishop *et al.* (2006) studied the influence of environmental factors on the distribution of the larvae of Ephemeroptera, Plecoptera, and Trichoptera and concluded that altitude, order of river, and vegetative cover were the most important factors in the distribution of the larvae of these organisms. Third- and fourth-order streams were the most susceptible to rainfall-related variations, which influenced the abundance of the organisms. First-order streams are less susceptible to the effects of precipitation.

In summary, a wide range of factors interact to order and consolidate the characteristics of biological communities in rivers, especially those that depend on the substrata and their composition: heterogeneity of substratum; the 'hydraulic habitat' of the lotic flora and fauna; levels of proteins in sediment biofilms; the processing of leaves and detritus by invertebrates; levels of heterotrophic bacteria in sediments, levels of suspended material (organic and inorganic) in the water; relations between periods of flooding and droughts; the sources of available energy (allochthonous or autochthonous) for the consuming lotic organisms; and the general characteristics of the 'physical habitat' available for the fauna (Bretschko and Helesic, 1998).

13.7   INTERMITTENT RIVERS AND STREAMS

On many continents, in arid and semiarid regions, there are intermittent rivers in which, during periods of precipitation, there is a considerable flow of water that otherwise does not exist in the dry season. Organisms in these intermittent and temporary systems display great capacity for resilience. Sudden floods disperse organisms that present diverse mechanisms for resistance and survival to endure desiccation. Prolonged droughts and rapid floods are common in the rivers of Northeast Brazil, causing alterations in the dynamics of aquatic macrophytes. Temporary rivers can have a dry cycle from 200 to 300 days, while ephemeral rivers have a dry cycle of c. 350 days. According to Pedro *et al.* (2006), the duration of the dry period is an important factor in the survival of different groups of species of aquatic macrophytes.

The intensity and flow of flooding after periods of drought have an impact and affect the colonization, the maximum biomass of the aquatic macrophytes, and their productivity.

More advanced and long-term studies are needed on the aquatic flora and fauna of ephemeral and temporary rivers and streams in Brazil (Maltchik and Pedro, 2001). Fischer *et al.* (1982) studied temporal succession in a temporary river in the desert.

13.8   PRIMARY PRODUCTION

Irregularities in the fluctuations of the physical and chemical variables of rivers, as well as the great heterogeneity and spatial variability of aquatic biota (especially micro-phytobenthos, periphyton, phytoplankton and aquatic higher plants), make it very difficult and complex to measure primary productivity in rivers. According to Wetzel (1975), almost all abiotic and biotic variables influence the primary production with daily, seasonal or irregular variations.

Variations in the velocity of the current, underwater solar radiation and nutrient levels occur in each section of the river, which makes it difficult to compare equivalents of different stretches. Variations in the **micro-distribution** of the components of primary producers make it technically difficult to calculate *in situ* the primary productivity, as is done in lakes, reservoirs or coastal waters. Despite these difficulties, analysis of the primary production (gross primary production) of communities has been conducted as well as studies on experimental lotic systems under controlled conditions (Warren and Davis, 1971).

In general, autotrophic organisms in lotic systems include a diverse community of algae, angiosperms and bryophytes. It is known that fast-flowing waters in the headwaters of rivers are dominated by **heterotrophic metabolism** with large quantities of organic matter present, especially in rivers with overhanging vegetation, as in tropical forests (Gessner, 1955).

Macrophytes, bryophytes and macro-algae are important contributors to the primary productivity in rivers and streams. In stretches of rivers with fast-flowing current ($>1\,\mathrm{m\cdot s^{-1}}$), macrophytes with special attachment mechanisms are common (Gessner, 1959). In calmer stretches of rivers with slower currents and high levels of nutrients, macrophytes and micro-phytobenthos can be quantitatively very important components of primary production (Neiff, 1997) (see Table 13.6).

Periphyton, macrophytes and phytoplankton are the principal producers found in rivers. Periphytic algae grow on rocks (epilithon), on soft sediments (epipelon), or on other plants (epiphyton) as discussed in Chapter 6. The composition of periphytic species varies seasonally and with the water cycle, especially in areas where the water flow varies greatly, in periods of precipitation and drought. Periphyton in the sediment of rivers, or located in/on soft substrata and on other aquatic plants, is distributed in relatively well-defined spatial groups (Margalef, 1983). The author also identified groups upstream and downstream that depend on the velocity of the currents. This scale of clustering depends on micro-habitats in the rivers. Light intensity, nutrients, and predation by herbivores are factors that influence the periphyton in rivers. The supply of nutrients can be an important limiting factor for periphyton, as shown by several authors (Allan, 1995).

Table 13.6 General distribution of the flora of rivers and streams in relation to the velocity of the current.

Velocity (m · s⁻¹)	Type of community	Dominant forms
<0.2–1	Algae fixed on substratum	Epipelic and epiphytic algae: *Navicula, Oscillatoria, Oedogonium*
>1	Fixed algae	Epiphytic algae: diatoms, *Ceratoneis*
0.2–1	Macrophytes	Angiosperms: Elodea, *Potamogeton* Macro-algae: *Chara*
0.5–2	Macrophytes	Some angiosperms: *Ranunculus, Fontinalis*
<0.5–1	Phytoplankton velocity	Small unicellular diatoms: cyanobacteria in nutrientenriched river water
>1	Phytoplankton	Volvocales, *Cryptomonas*

Source: Hawkes (1975).

One of the most common methods used to study periphyton in rivers and determine its growth rate and response to chemical contaminants (heavy metals, herbicides and pesticides) is by experimenting with plastic, glass or ceramic plates. Chamixaes (1997) measured the primary productivity of periphyton by exposing plates at the bottom of rivers in a *cerrado* region (Lobo/Broa reservoir) and later measuring the dissolved O_2 in the water using the method of dark and transparent bottles. Substratum type, temperature, and light intensity are all limiting factors in the productivity of periphyton in rivers.

Quantitative methods to measure primary production in rivers include measuring the changes in biomass; many samples are required. Artificial substrata are used to reduce the degree of heterogeneity and measure the primary productivity under experimental conditions (in transparent and dark bottles, using the ^{14}C technique or dissolved O_2 as described in Chapter 9). Alterations in biomass are determined based on sequential analysis and calculations of the increment in biomass as dry weight or number of organisms or concentration of chlorophyll-*a* per cm² or m².

Measurements *in situ* of primary productivity in rivers are made based on adaptations of the original techniques to measure photosynthetic productivity with the dissolved O_2 method or by ^{14}C fixation.

Because of the difficulty of conducting measurements *in situ* of the primary production of the various components of the photosynthetic autotrophic community, a series of measurements of the metabolism of communities were made using the alterations in the levels of dissolved oxygen or CO_2 or pH of the water.

Odum (1956, 1957) developed this method to be applied to streams, which was subsequently applied in a large number of rivers (Odum and Hoskin, 1957; Hoskin, 1959; Hall, 1972; Wetzel, 1975). The relations between the levels of dissolved oxygen, loss or addition of oxygen to and from the atmosphere, and increase of oxygen from drainage resulted in the following generalized formula, which represents the shift in dissolved oxygen per unit area and unit time:

$$\Delta C = P - R \pm D + A$$

where:

ΔC – Rate of change in levels of dissolved oxygen in water during flow

P – primary production

R – Respiration of the community

D – Loss or addition of oxygen to the atmosphere

A – Advection of oxygen linked to the current and turbulence

The estimates of the primary productivity of the various components of the system of autotrophic producers present a wide range of results. For periphyton, the data obtained by Mann (1975) show values from 920 to $8,176 \, \text{kcal} \cdot \text{m}^{-2} \cdot \text{year}^{-1}$; for aquatic macrophytes, net primary production varied from 0.1 to $8,833 \, \text{kcal} \cdot \text{m}^{-2} \cdot \text{year}^{-1}$; and for phytoplankton in lotic waters, from 2,810 to $4,388 \, \text{kcal} \cdot \text{m}^{-2} \cdot \text{year}^{-1}$. Also according to Mann (1975), the patterns for secondary production varied from 70 to $614 \, \text{kcal} \cdot \text{m}^{-2} \cdot \text{year}^{-1}$ for herbivores and detritivores, and from 3 to $60 \, \text{kcal} \cdot \text{m}^{-2} \cdot \text{year}^{-1}$ for carnivores (data from several rivers and streams in emperate regions).

The data obtained are based on estimates of changes in biomass, variations in dissolved oxygen over a 24-hour period, and changes in pH with estimated alterations in CO_2 content.

13.9   ENERGY FLOW

Few studies exist on the subject cited above, the biological flow of energy in rivers. The classic work in this subject was conducted by Odum (1957) in Silver Springs. A long-term review of the flow of energy in rivers and streams between all biological components was published by Mann (1975).

13.10   THE FOOD CHAIN

Food chain in rivers are dominated by benthic invertebrates and fish. As already mentioned, sources of organic matter in rivers are either allochthonous or autochthonous. The food network thus depends on the ratios and contributions of allochthonous and autochthonous material in different parts of the rivers. The organization of trophic networks is complex and considerable overlap exists between the various different diets of invertebrates and fish. Many species show little difference in their dietary items, and even the classic hierarchy (herbivore, carnivore, detritivore) can cause confusion if based exclusively on analysis of stomach contents. Thus, the division of invertebrates into functional groups (Cummins, 1973) can be useful, and the method of collection is more important than the available resource.

A complete mapping of the food networks in rivers needs to include the microbial loop, which is extremely important, especially in terms of particulate and dissolved organic matter. Bacteria present in microfilms on substrata with layers of organic matter make up – along with fungi, protozoa, algae, enzymes and polysaccharides – an active and highly nutritious food source for detritivores, herbivores, and carnivores. Figure 13.10 presents some of the components of organic matter in rivers and

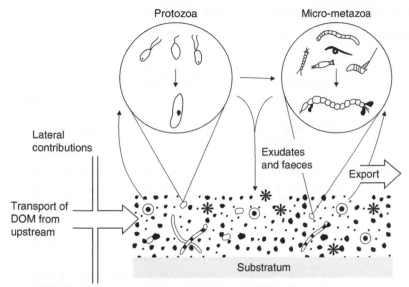

Figure 13.10 Microbial food-chain on a substratum of small streams or large rivers.
Source: Modified from Allan (1995).

their interactions. Collection of 0.5 μm bacteria is usually done by flagellates (with ⩾5 μm mesh) and ciliates (with approximately 25 μm mesh), which enables the flow of energy to larger consumers. The number of trophic levels can be very large or just one or two levels, depending on the velocity of the current, accumulation of organic matter, and the compaction of sediment with a very fine layer of organic matter on the sediment or between the particles.

Chart 13.2 shows the main functions of the invertebrate consumers in rivers, the resources used, and the feeding mechanism, with examples.

Due to the process of **convergent evolution**, organisms that make up the feeding network in rivers are very similar across all continents. Species diversification occurs, but Trichoptera, Plecoptera, Ephemeroptera and Odonata, oligochaete worms and molluscs are dominant. Simuliidae larvae (gnats) are dominant in many rivers with stronger currents.

13.11 LARGE RIVERS

Over the last 30 years, many studies on small rivers and streams have been published (Hynes, 1970). Data on the ecology and functioning mechanisms of large rivers is more recent. The exceptions are books published on the Nile (Rzóska, 1976; Dumont, 2009), the Volga River (Morduchai-Boltovskoi, 1979), the Amazon River (Sioli, 1984; Whitton, 1984) and a more recent work that includes tropical rivers (Payne, 1986). The literature on small streams is extensive and varied (Zaret, 1983; Caramaschi *et al.*, 1999).

Chart 13.2   Principal functions and types of food for invertebrates in rivers.

Role in the food chain	Food resource	Feeding mechanism	Examples
Shredders	CPOM; leaves and associated microflora: bacteria and fungi	Cut, grinding of material, chewing	Many families of Trichoptera, Plecoptera and crustaceans; some molluscs
Shredders	CPOM; fungi and surface layers of leaves and detritus	Cut, grinding of material	Diptera; Coleoptera; Trichoptera
Collectors/filterers	FPOM; bacteria and organisms in suspension in the water	Collect particles using nets or secretion for aggregation	Simulidae, Diptera, some species of Trichoptera, some Ephemeroptera
Collectors	FPOM; Bacteria and organic microfilm	Scrape material on the surface of sediments; bury in soft sediment	Many Ephemeroptera and chironomids
Herbivores/ scrapers	Periphyton, especially diatoms and organic microfilmt	Scrape material on the surface	Many families of Ephemeroptera, Trichoptera, some families of Diptera, Lepidoptera and Coleoptera
Predators	Macrophyte, prey animal	Apprehension and partition	Odonata; Megaloptera, Trichoptera, Diptera and Coleoptera

Source: Allan (1995).

Davies and Walker (1986) published an important work on large rivers, comparing scientific data and information on the Nile, Niger, Orange Vaal, Volta, Zaïre and Zambezi (Africa); Colorado and Mackenzie (North America); Amazon, Paraná and Uruguay (South America), Murray-Darling (Australia) and Mekong (Southeast Asia).

The biogeography and zonation in these ecosystems have been studied on the continental and regional level. Dumont (1986a, 1986b, 2009) explored the subject of the Nile as 'a very ancient river' and the similarities of its zooplankton with that in other rivers in North Africa. The fish fauna of the Nile, for example, consists of relatively few species, because of the series of palaeo-climatic shifts that affected the climatic and morphological conditions of the river. The authors concluded that the greater a river's **hydrological stability** over a long period of time, the greater the endemism. This is how Lowe-McConnell (1986) explained the approximately 1,300 species of fish in the Amazon River.

According to Davies and Walker (1986), the concept of *continuum* in the river (Vannote *et al.*, 1980) does not directly apply to streams and small rivers, due to the permanent alterations in space (large internal deltas, temporal re-organizations) and anthropogenic influences.

Table 13.7 General characteristics of the rivers of South America.

River	Discharge $(m^3 \cdot s^{-1})$	Area $(\times 10^6\ km^2)$	Length (km)	Surface runoff $(L \cdot s^{-1} \cdot km^{-2})$	TDS/RMT $(\times 10^6 t \cdot year^{-1})$	TSS/RMT $(\times 10^6 t \cdot year^{-1})$
Amazon	175.000	6.3	6.577	28.0	290	900
Paraná	15.000	2.8	4.000	5.3	38.3	80
Orinoco	36.000	1.0	2.150	32.7	30.5	150
São Francisco	3.760	0.63	2.900	6.0	–	6
Magdalena	6.800	0.26	1.316	26.5	20	220
Uruguay	4.600	0.24	–	16.0	6(?)	11(?)

TDS – Total dissolved solids;
TSS – Total suspended solids;
RMT – Rate of mass transport.
Sources: Depetris (1976); Ducharne (1975); Furch (1984); Milliman and Meade (1983); Meybeck (1976); Paolini *et al.* (1983); Paredes *et al.* (1983).

Winterbourn *et al.* (1981) suggested that the river continuum concept is not even applicable in the case of small streams, especially in systems with highly variable climate and low retention capacity for allochthonous matter.

In large rivers, the influences of **flood pulses** on the floodplains and the river itself are extensive and are quantitatively important in nutrient cycles, reproduction of fish, and migration (Welcomme, 1986); this has also been confirmed by Junk (2006) for the Amazon (see Chapter 16).

The big river systems are evolutionarily important areas (Margalef, 1983), since these systems are active centres of evolution, promoting biodiversity with the dynamism of their physical, chemical, hydrological and geomorphological characteristics.

Allanson *et al.* (1990) published an extensive review of inland aquatic ecosystems in the south part of the African continent and examined in detail the ecology of rivers, looking at data from the 1930s on, concluding with studies on biodiversity (Harrison and Elsworth, 1958), biogeography (Oliff, 1960), erosion and effects of sequential deposition of sediments (Chutter, 1967) and invertebrate communities (King, 1983).

Other studies on large rivers in South America have been published by Bonetto (1986a, 1986b), Neiff (1986, 1996), Di Pérsia (1986) for the Paraná River (see Chapter 16 for the Upper Paraná River and Paraná River) and Di Pérsia and Neiff (1986) for the Uruguay River. According to these authors, the Uruguay River basically differs from the Paraná River in the relative absence of wetlands and the extreme dependence on allochthonous matter in the Uruguay River.

Di Pérsia and Olazarri (1986) presented studies of the zoobenthos in the Uruguay River. The biogeochemistry of the large rivers of South America was studied by Richie *et al.* (1980) and DePetris (2007). Table 13.7 presents the main characteristics of these rivers. Studies on the Orinoco River (Weibezahn *et al.*, 1990) provide basic data on hydrology and transport of suspended matter and carbon particles. Table 13.8 shows the main characteristics of the largest rivers around the world, and Table 13.9 compares the levels of phosphate and nitrate in major rivers.

Table 13.8   Principal characteristics in discharge and drainage of the world's major rivers.

River	Discharge (D) $km^3 \cdot year^{-1}$	Drainage area (A) $km^2 \cdot 10^6$	Ratio $(D/A) \times 10^{-3}$
Tropical rainforest			
Amazon	5,500	7	0.79
Zaire	1,800	4	0.45
Mekong	4,800	0.787	6.1
Temperate or subtropical rainforest			
Reno	70	0.22	0.32
Paraná	730	3.2	0.23
Uruguay	124	0.37	0.34
Moderately dry, all climates			
Mississippi	560	4.8	0.12
Mackenzie	333	1.8	0.19
Niger	220	1.1	0.19
Volga	238	1.3	0.18
Desert rivers			
Colorado	18	0.6	0.03
Nile	90	3.0	0.03
Murray-Darling	22	1.1	0.02
Orange-Vaal	12	0.65	0.02

Source: Horne and Goldman (1994).

Table 13.9   Levels of nitrate and phosphorus in large rivers ($\mu g \cdot L^{-1}$).

River	$NO_3 - N$	$PO_4 - P$	References
Niger	1.100–6.300	500–3.100	Welcomme (1986)
Orange-Vaal	300–1.400	30–100	Cambray et al. (1986)
Colorado			Day and Davies (1986)
Mackenzie	600	16	Brunskill (1986)
Parana	>500	>100	Bonetto (1986)
Volta	0–5.000	20–160	Petr (1986)
Volga	50–4.000	1–250	Payne (1986)
Nile	10–1.000	1–40	Rzóska (1976)
Mississippi	700–3.000	40–440	Fremling et al. (1989)
Amazon			
White water	4–15	15	Payne (1986); Forsberg et al. (1988)
Clear water	>1	>1	Payne (1986); Forsberg et al. (1988)
Black water	36	6	Payne (1986); Forsberg et al. (1988)
General average			
Africa	170	nd	Payne (1986); Forsberg et al. (1988)
Europe	840	nd	Payne (1986); Forsberg et al. (1988)
North America	230	nd	Payne (1986); Forsberg et al. (1988)
South America	160	nd	Payne (1986); Forsberg et al. (1988)

nd – No data
Source: Horne and Goldman (1994).

13.11.1 The economic importance of major rivers

Major rivers have an enormous economic, ecological and social importance. They are ecosystems with great biodiversity and sources of food for millions of people. They also provide a means of transport and stimulate local and regional economies. For example, Pantulu (1986) estimated that in the Mekong River 500,000 tons of fish are fished annually, for an annual contribution of US$225 million to the economy. Fish comprise 40–60% of the population's consumption of animal protein (Pantulu, 1986). Fishing in the Amazon River accounts for US$90 million in revenues each year and feeds about 200,000 people (Petrere, 1978).

The great rivers of South America are also economically important. In addition to helping meet the protein needs of the local and regional populations, the rivers are also used for navigation, irrigation, recreation, and sport fishing. The impact of the construction of dams/reservoirs on large rivers was examined in Chapter 12.

13.12 FISH COMMUNITIES IN LOTIC SYSTEMS

In Chapter 6 the composition of the fish fauna in the São Francisco and Amazon rivers is described (Araújo-Lima and Goulding, 1998; Barthen and Goulding, 1997; Sato and Godinho, 1989). The fauna of Brazil's major rivers has been extensively studied, including biological diversity, reproduction, distribution, and feeding relationships (Araújo-Lima et al., 1984, 1986, 1990, 1994; Goulding et al., 1988; Bayley and Petrere, 1989; Menezes and Vazzoler, 1992).

Santos and Ferreira (1999) presented a review of the fish species in the Amazon basin and Agostinho and Ferreira Julio Jr. (1999) of the fish in the Paraná River basin.

A special volume on fish in streams in Brazil was edited by Charamaschi Mazzoni and Peres Neto (1999), with studies on the diversity, composition, behaviour and biology of species inhabiting small streams, providing information for conservation, management and restoration of these ecosystems.

Buckup (1999) presented a systematic biography for fish in streams. Chart 13.3 shows the systematic classification of teleost fish families in these streams. Information on taxonomic composition of different families, their presence and geographical distribution are based on existing literature (Menezes, 1988), in particular the relationship between endemism and events causing geographic isolation in different periods.

Menezes' analysis (1988) was based on Oligosarcus species, but according to Buckup (1999), there are several groups of fish that fit the biogeographical evolution model presented. Chart 13.4 shows regions of Oligosarcus endemism and the events of geographic isolation associated with their origin.

Contributions from Britski (1997a, 1997b), Britski and Garavello (1980) and Buckup (1993, 1998) complete the descriptions of species, phytogenetic inter-relationships and geographical distribution of the ichthyofauna of lotic systems in Brazil. The works of Menezes (1972, 1987a, 1987b, 1988, 1992) and Menezes et al. (1983, 1990a, 1990b) complete the set of information. Evolutionary processes of the ichthyofauna in streams of South America were examined by Castro (1999), who defined two types of coastal streams and two types of inland streams as the basis for an analysis that could provide the foundation for an extrapolation to other regions of

Chart 13.3   Systematic classification of the families of teleost fish in Brazilian streams.

Clupeocephala
  Clupeomorpha
    Clupeiformes
      Clupeidae
      Pristigasteridae
      Engraulidadidae
Eutelostei
  Ostariophysi
    Characiformes
      Paradontidae
      Chilodontidae
      Anostomidae
      Curimatidae
      Crenuchidae
      Hemiodidae
      Gasteropelecidae
      Characidae
      Erthyrinidae
      Lebiasinidae
      Ctenoluciidae
  Siluriformes
    Siluroidei
      Aucenipteridae
      Pimelodidae
      Cetopsidae
      Aspredinidae
      Trichomycteridae
      Callichthydae
      Loricariidae

Gymnotoidei
  Gymnotidae
  Electrophoridae
  Hypopomidae
  Rhamphycthyiidae
  Apteronotidae
  Sternopygidae
Neoteleostei
  Synbranchiformes
    Synbranchichidae
  Cyprinodontiformes
    Rivulidae
    Anablepsidae
    Poeciliidae
  Perciformes
    Cichlidae
    Gobiidae
    Nandidae

Source: Buckup (1999).

Chart 13.4   Endemic regions of *Oligosarcus* and events of geographical isolation associated with the origin of new species.

Region	Geological age	Event
Andean	Tertiary	Uplift of the Andes
Upper Paraná	Lower Tertiary	Isolation of São Francisco
Upper Uruguay	Miocene	Isolation of the Upper Paraná
Jequitinhonha	?	?
Doce River	Pre-Quaternary?	Capture/isolation of the São Francisco
Lakes of the Doce River	Quaternary	Isolation of the Doce River
Southern Coast	Upper Tertiary	Formation of the lower Paraná/Uruguay
Central Coast	Flandrian inter-glacial	Increase in sea level
Northern Coast	Flandrian inter-glacial	Increase in sea level

See Menezes (1988) for further information.
Source: Buckup (1999).

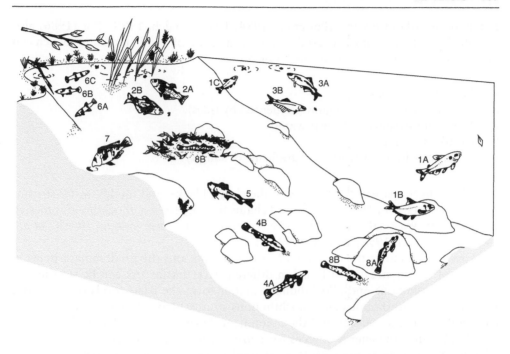

Figure 13.11   Schematic illustration with typical features of a stream in the Atlantic forest, including the spatial distribution and principal feeding tactics of the fish species found there. The left side of the illustration corresponds to a stretch of backwater and the right side to a stretch of current. The sizes of the fish are not drawn to scale. *Deuterodon pedri*: 1A – grabs dragged items; 1B – prunes, grazes or grabs small prey; 1C – grabs on water surface. *Hollandichthys multifasciatus*: 2A – grabs on water surface; 2B – grabs dragged items. *Mimagoniates microlepis*: 3A – grabs on water surface; 3B – grabs dragged items. *Characidium japuhybensis*: 4A – lurks; 4B – scraping of substratum. *Rahmdioglanis* sp: 5 – scraping of substratum. *Phalloceros caudimaculatus*: 6A – prunes; 6B – grabs small prey or dragged items; 6C – grabs on water surface. *Geophagus brasiliensis*: 7 – collects substratum and separates prey. *Awaous tajasica*: 8A – grazes; 8B – collects substratum and separates prey.
Source: Modified from Sabino and Castro (1990).

Brazil. In this analysis, the author notes that in the evolutionary patterns examined, the predominance of small species proved to be a common pattern in all four types of coastal streams studied (see Figure 13.11).

Several authors cited in Esteves and Aranha (1999) undertook the task of describing the characteristics of and classifying streams. According to these authors (Knopell, 1970; Soares, 1979; Vieda, 1993; Garutti, 1988; Sabino and Castro, 1990), streams are classified as small rivers, with non-permanent flood areas, velocity of the current ranging from 0.1–1.7 m.s^{-1}, high levels of dissolved oxygen, transparent, with pH and conductivity related to the hydrogeochemistry of the drainage basin (Araújo-Lima et al., 1995; Salati, 1998).

Streams in Brazil present wide diversity, distinct characteristics and different degrees of complexity. This complexity varies from streams in the coastal plain to mountainous island streams (Por *et al.*, 1984, 1986, Covich, 1988). Por (1986) also classified the streams of the coastal region in the state of São Paulo State (which can be a good example for other regions in Brazil) as:

▸ Black-water streams.
▸ Brackish black-water streams (influenced by tides).
▸ Mountain streams with clear water.
▸ Clear-water rivers.
▸ Estuaries with clear water and **gradients of salinity**.

These diverse streams – along with the clear-, black-, and muddy-water streams of the Amazon (Sioli, 1984) – have food chains that depend on allochthonous matter, which suffers seasonal variations in the contribution to these systems (Rocha *et al.*, 1991, Henry *et al.* 1994; Walker, 1992).

Chart 13.5 shows the diversity of feeding habits and the predominant items in the diet of fish in streams in Brazil in different river basins. As can be seen in this table, autochthonous and allochthonous matter form the basic foods of these fish. Feeding on autochthonous and allochthonous sources depends, clearly, on the contribution of organic matter and the amount of detritus matter. Araújo-Lima *et al.* (1986) identified the energy sources for detritivorous fish in the Amazon River (see Chapter 8 for a more in-depth look at food chains and the role fish play in the chains).

13.13 DRIFT

Adrift in rivers, a wide range of living organisms and detritus move downstream along with the current, providing a potential source of food for many other organisms. The detritus includes algae, bacteria, invertebrates, and fragments of roots and leaves. A large amount of drifting material consists of insect larvae, which dislocate during the night to avoid predators. In some cases, drift also consists of masses of microphytobenthic algae carried downstream by the currents (observations by researchers from the International Ecology Institute, on the São Francisco River).

Many insect larvae present spatial attachment structures to avoid drifting, while at the same time they filter food carried by the drift. Many other insect larvae move along with the drift and later emerge and fly back upstream to lay their eggs in the headwaters (Horne and Goldman, 1994).

For some insect species, the numbers of larvae displaced by the drift (per day per unit area of the stream) are much larger than the biomass found in a determined area (Waters, 1966). This author found very different drift values in winter and summer for a *Baetis vagans* (Ephemeroptera, Baetidae) population. It is possible that variations in the number of organisms that occur during the drift in rivers in the tropics and subtropics are influenced not only by water temperature but also by periods of drought and rainfall, which present great variations in volume and velocity of the current (Henry *et al.*, 1994) (see Figure 13.12) (Petts and Amoros, 1996).

Chart 13.5   Predominant feeding habits and items in the diet of fish in Brazilian streams.

Author	Basin	Number Of Species	Predominant Category Of Food	Predominant Items	Origin
Knoppel (1970)[1]	Amazon (AM)	49	Omnivore	Insect larvae, vegetative remains	Allochthonous
Soares (1979)[1]	Amazon (AM)	20	Carnivore	Terrestrial insects, molluscs, crustaceans	Allochthonous
Uieda (1983)[1]	Parana (SP)	18	Insectivore-herbivore-planktophage	Insects, crust-aceans, higher plants	Autochthonous and allochthonous
Costa (1987)[1]	Leste (RJ)	17	Insectivore	Aquatic insects	Autochthonous and allochthonous
Teixeira (1989)[1]	Leste (RJ)	25	Insectivore	Insects, micro-crustaceans	Autochthonous and allochthonous
Sabino and Castro (1990)[1]	Leste (SP)	8	Omnivore-insectivore	Aquatic and terrestrial insects, algae	Autochthonous and allochthonous
Uieda (1995)[1]	Leste (SP)	24	Omnivore-insectivore	Aquatic and terrestrial insects, vegetation, algae	Autochthonous and allochthonous
Melo (1995)[1]	Amazon (MT)	82	Omnivore-insectivore	Aquatic and terrestrial insects, fruit and seeds	Autochthonous and allochthonous
Aranha (1991)[2]	Leste (RJ)	4	Algivore	Algae	Autochthonous
Gomes (1994)[2]	Leste (RJ)	*Deuterodon sp Astyanax Janeiroensis*	Omnivore	Insects	Autochthonous and allochthonous
Buck and Sazima (1995)[2]	Leste (SP)	4	Algivore	Algae	Autochthonous
Trajano (1989)[3]	Leste (SP)	*Pimelodella kronei P. transitoria*	Carnivore	Invertebrates	Autochthonous and allochthonous
Lobon-Cervia et al. (1993)[3]	Parana (SP)	*Crenichichla lepidota*	Carnivore	Insects and micro-crustaceans	Autochthonous
Aranha et al. (1993)[3]	Parana (SP)	*Corydoras aeneus C. gr. carlae*		Invertebrates and algae	Autochthonous
Porto (1994)[3]	Leste (RJ)	*Pimelodella lateristriga*	Omnivore	Insects	Autochthonous

[1]Studies on communities; [2]Studies on taxocenoses; [3]Auto-ecological studies.
Source: Esteves and Aranha (1999).

(a)

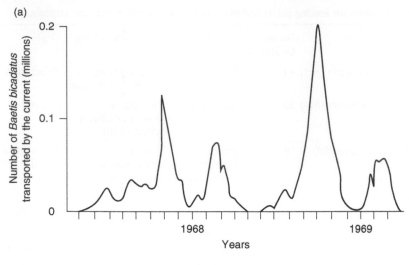

(b)

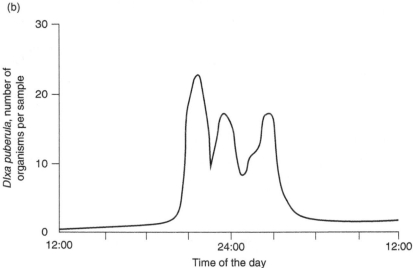

Figure 13.12   Drift of invertebrates in rivers. Number of organisms in two seasons of the year and drift during the night. a) Logan River, Utah (USA); b) Wilfin Back (England).
Source: Modified from Horne and Goldman (1994).

13.14   IMPACT OF HUMAN ACTIVITIES

Similar to other aquatic and terrestrial ecosystems, rivers are affected by innumerable human activities, many of which have been occurring for a long time. For example, irrigation channels were built in Egypt in 3200 BC, and there is evidence that dams were built as early as 2759 BC (Petts, 1989). Channel engineering and '**water engineering**' developed by the Romans resulted in major modifications in water courses, specially altered to supply drinking water and water for irrigation.

In the 19th century, in both the United States and Europe, modifications in the flow of water, morphometry of rivers, wetland areas and floodplains around rivers were of major consequence. At that time, this procedure was actually the end of a process of alteration and construction of channels for navigation, **flood control**, and use of floodplains, which began in the second half of the 18th century (around 1750).

Starting in 1900, first in the United States and later in Europe and Asia, large dams began to be built to produce hydroelectric power; and by the 1980s, such construction had spread throughout all continents (see Chapter 18).

In general, lotic systems are affected by the following modifications:

▶ pumping water for irrigation or for public or private supply (farms), altering the flow and structure of rivers;
▶ organic and inorganic pollution from industrial and agricultural sources (both point and non-point sources). Pesticides, herbicides, heavy metals and discharge of untreated sewage are some of the sources of pollution threatening the integrity of rivers;
▶ **intensive soil use**, leading to an increased suspended material and run-off of large quantities of substances and elements into lotic systems;
▶ introduction of exotic species, which can alter the food network and the natural process of interactions between communities;
▶ removal of riparian vegetation, which is extremely important for the river in maintaining the buffer conditions. Removal of the vegetation, aside from reducing the organic matter available to fish and invertebrates, leaves the riverbanks and embankments unprotected, thus altering the morphometry;
▶ construction of dams for generating hydroelectricity and public water supply. The effects of such construction were analyzed in Chapters 12 and 18;
▶ modification of floodplains and wetland areas associated with dams for agriculture, construction of water channels or for urbanization;
▶ construction of channels, bridges and passageways that interfere with the functioning of rivers, altering the substratum (physical and chemical composition) and removing and affecting organisms;
▶ construction of large areas for irrigation, with withdrawal of substantial amounts of water to perform this activity. Classic cases include the **Colorado River** (California, USA), 400 km in length, whose waters are used for irrigation; and the Aral Sea (see Chapter 18), whose tributary waters have also been diverted for irrigation;
▶ drainage of agricultural regions and **urban drainage** – contaminated by household waste (sewage) and industrial waste – are the two greatest threats to lotic systems. In the United States, 80% of the 49 billion metric tonnes of soil added to the rivers come from agricultural soils. Agriculture accounts for 46% of the sediment, 47% of the total phosphorus and 52% of the total nitrogen discharged into the country's rivers and streams (Gianessi *et al.*, 1986). In Brazil, in agricultural regions in the state of São Paulo alone, removal of topsoil reaches 20 tonnes per hectare per year.

All these actions produce a series of major alterations in rivers, some already described:

▶ physical alterations in the **morphometry of rivers**;
▶ **modifications in the habitats** of aquatic macro-invertebrates and fish.

▶ modifications in the flow of energy in rivers: in many cases, a heterotrophic to autotrophic shift occurs;

▶ modifications in water temperature (heating) due to the removal of vegetation;

▶ alterations in the hydrology and flow of water, affecting biodiversity;

▶ increased levels of nitrogen and phosphorus, resulting in eutrophication. In many rivers around the world, nitrate levels have doubled over the last 30 years (Pringle *et al.*, 1983). This has resulted not only in increased eutrophication, the growth of periphyton and macrophytes has also increased, through the greater flow of energy based on autotrophs;

▶ removal of riparian vegetation, which reduces the flow of decomposing vegetation and, thus, the organic matter to be broken down by the various organisms that depend on this resource as food;

▶ **loss of aquatic biodiversity** can be characterized primarily by the **loss of habitat** for fish and aquatic invertebrates, caused by various types of intervention in the structure and dynamics of rivers. Following the loss of biodiversity, the biomass of some species usually increases;

▶ **alteration in the sediments** of rivers is another fairly common consequence of human activities in drainage basins and can affect lotic systems in several ways: increase of fine sediment; interference in sediment-water interactions, and release of gases and nutrients from the sediment;

▶ Modifications in the substratum available to benthic macro-invertebrates and periphyton.

13.15 RESTORATION OF RIVERS

The innumerable human activities that cause damage to lotic ecosystems affect not only water quality and operating mechanisms of rivers, they also physically alter the structures, floodplains and capacity for restoration of these systems. Thus the restoration of rivers is currently one of the most important management goals for drainage basins and lotic systems. It is important to keep in mind that this process presents wide variations in terms of various components of a lotic system (see Figure 13.13) (Petts and Amoros, 1996).

The **restoration of rivers** includes the following actions:

▶ **rehabilitation of riverbanks** and gallery forest to control the biogeochemical cycles, restoring the natural functioning, to retain and particulate matter and absorb organic and inorganic pollutants. The rehabilitation of gallery forest preserves and promotes the rehabilitation of the terrestrial and aquatic fauna and flora, which depend on the corridors of vegetation along the river (Large and Petts, 1996). This rehabilitation restores the natural characteristics of the water cycle and promotes the recovery of buffer zones;

▶ **restoration of the corridors** of vegetation along the river;

▶ rehabilitation of habitats and restoration of biodiversity;

▶ recovery and rehabilitation of the substratum of the river for diversification of habitats and restoration of biodiversity;

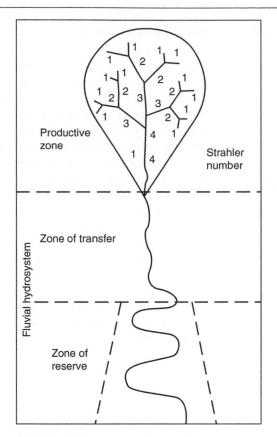

Figure 13.13 The different spatial organizations of a fluvial system and the various zones of components, implying that different processes of re-vitalization can occur along the length of the river.
Source: Modified from Petts and Amoros (1996).

▶ re-oxygenation of rivers, in the case of oxygen depletion;
▶ restoration of the floodplains, marginal lagoons and ecological structures along the rivers.

The corridors of gallery forest along rivers are particularly important because they:

▶ are biologically diverse;
▶ are biologically productive;
▶ provide a refuge area;
▶ are sources of dispersion of species;
▶ include refuges from the pre-industrial era (Petts and Amoros, 1996).

The recovery process of rivers should include the **restoration of functions,** control of the *principal influxes* (nutrients, primary production) and control of the *disturbances* in the system – alterations in flow, buffer zones and the riparian forest.

Collecting material and conducting experiments in rivers

As seen in this chapter, river ecosystems are highly complex and diverse. Collecting biological material is very complex and calls for the use of different types of special equipment, including equipment to collect the sediment in rivers with soft bottoms (Eckman Birge or Petersen instruments – see photos in Chapter 20), special nets to collect insects (adult or larvae), and different types and shapes of nets for collecting fish.

The scraping of stones or pebbles, with subsequent collection and fixation of biological material, can be useful in the study of invertebrate fauna and periphyton. Artificial substrata (glass or ceramic plates) are another resource commonly used to study the attachment and colonization of bacteria, periphyton, and invertebrates, in the evaluation and characterization of benthic communities. Water is usually collected in non-toxic Van Dorn bottles. To calculate physical and chemical variables (including pH, O_2, water temperature, conductivity, redox potential, turbidity, and total suspended solids), multi-parametric probes are used, which, if properly calibrated, allow these variables to be rapidly calculated. Another problem to consider is the network of sampling sites, which should encompass the various zones of the river, defined according to type of substrate, current, and other physical characteristics.

Calculating the velocity and direction of currents in rivers can provide essential data to describing the functioning of these ecosystems. Finally, the collection and processing of leaf litter and decomposing vegetation from the bottom of rivers can be essential in the study of bacteria, fungi, algae and invertebrate larvae.

The study of daily variations in rivers can greatly help in understanding the dynamics of these ecosystems. The construction of **artificial channels** and small stretches of rivers can also be very useful for conducting experiments: different velocities of currents can be simulated, as well as nutrient levels and substratum type. Artificial channels simulating rivers are also used to study distribution and the impacts of air pollution and particulate pollutants.

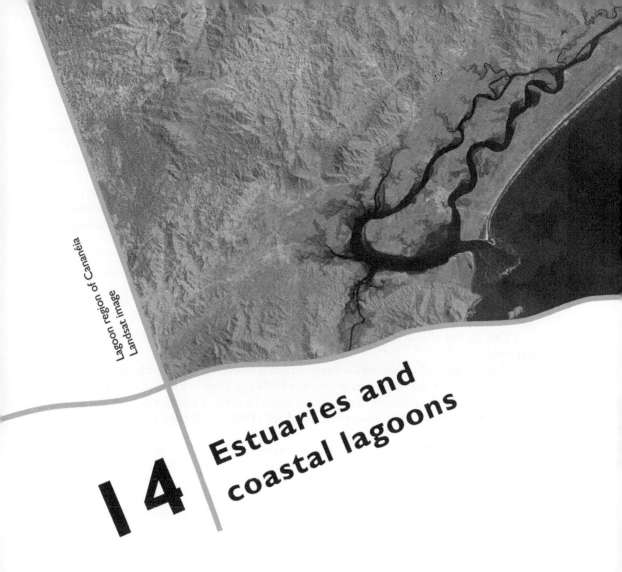

Lagoon region of Cananéia
Landsat image

14 | Estuaries and coastal lagoons

SUMMARY

In this chapter we present the main features of estuaries and coastal lagoons, the factors that determine their structure and function, as well as the operating mechanisms of these intermediary ecosystems (between inland aquatic systems and marine systems). The **differences between estuaries and coastal lagoons** are discussed along with the role of salinity, horizontal gradients, morphometry and seasonal and spatial fluctuations that occur in these systems.

Case studies are presented featuring four different types of estuarine and coastal ecosystems in Brazil and South America: the Cananéia lagoon region in the state of São Paulo; the coastal lakes in the state Rio de Janeiro; the **Patos Lagoon** in Rio Grande do Sul, and the estuary of the Plata River (Argentina/Uruguay).

Estuaries and coastal lagoons are essential for sustaining aquatic biodiversity. These are transition regions with elevated biological productivity and alternative food chains. These systems are also subjected to numerous impacts, especially human actions.

The chapter also discusses the impacts and the measures to mitigate and recover and protect these ecosystems.

14.1　GENERAL FEATURES

An estuary can be defined as an aquatic ecosystem in which the waters of a river mix with marine waters, producing measurable gradients of salinity (Ketchum, 1951b). The word 'estuary' comes from the Latin *aestuarium, aestus* (tide) and *aestuo* (foam that floats).

A coastal lagoon is defined as a shallow lake or body of water connected to a river or ocean (Latin: *lacuna; lacus* – lake). These definitions, however, are not exclusive; a coastal lagoon connected to the ocean can also be affected by the tide, like an estuary. Kjferve (1994 p. 3) defined coastal lagoons as:

> 'A body of shallow coastal water, separated from the ocean by a barrier, connected at least intermittently with the ocean by one or more restricted connections and usually parallel to the coast'.

This author also defines estuaries, coastal lagoons, fjords, basins, tidal rivers, and narrows.

Physiographically, estuaries are semi-isolated bodies of water with varying salinity and tidal influence that produces gradients of salinity. They are **transitional ecosystems** with highly varied conditions and **transient states** of vertical and horizontal circulation. The classic definition of an estuary by Pritchard (1955) in Tundisi (1970, p. 1) is:

> 'An estuary is a semi-enclosed body of water with a free connection to the open ocean in which tidal water is measurably diluted by freshwater originating from the inland drainage'.

The **discharge of fluvial waters,** either from rivers (fluvial discharge) or precipitation, must be greater than the volume of water transferred evaporation by the process of precipitation (Miranda *et al.*, 1998).

Basic differences between estuaries and coastal lagoons have been identified by Emery *et al.* (1957) and are presented in Chart 14.1.

Tides and salinity are factors that make the structure of estuaries very complex, more than those of rivers or stratified lakes. The **physical and physiographical conditions of estuaries** (such as canals, stretches of beaches, coastal water, organic sediments at the headwaters of the rivers which constitute estuaries, and barriers of sediment) make the environment rich in varied **ecological niches** that present organisms with different combinations of salinity, water temperature, dissolved oxygen levels and circulation, all of which vary extremely.

Salinity, one of the most important features of estuaries, alters daily and periodically throughout the seasonal cycle, in which the interactions of precipitation/evaporation/tides also shift. For example, in the Cananéia lagoon region (an estuary with **mangrove swamp**), during periods of heavy rain and low tide salinity levels can reach 5–10‰ (parts per thousand), while in winter, when pressure from cold fronts forces coastal waters into the estuary, salinity can reach 20–25‰.

Estuaries are formed by movements of **submergence** or *emergence* along coastal areas, resulting from plate-tectonic movements and local effects, such as current direction and force, wave action, deposition of sediments transported by rivers, glaciation, and tidal effects. Alterations caused by tectonics, glaciation or climate produce the

Chart 14.1   Difference between estuaries and coastal lagoons.

Estuary	Coastal lagoons
Initial stage: deeply closed	Initial stage: straight line of coast, without coastal lagoons
'Young' estuary: headwaters with well-defined shoreline	'Young' coastal lagoons: barriers separating shallow lagoons from open ocean
Estuary in more advanced stage: formation of beaches, barriers of sediment, early coastal vegetation (*Spartina* spp. or early formation of mangrove forest)	Coastal lagoon in more advanced stage: predominance of coastal vegetation (*Spartina* spp.)
Estuary in mature stage: large number of barriers, near the coast, sediment along the shoreline	Coastal lagoon in mature stage: deposition of sand, predominance of aquatic vegetation
'Mature' estuary with great movement of sediments, barriers in large numbers, abundance of vegetation in shallower areas	'Mature' coastal lagoon: expanded area, continuous migration and alteration in the barriers with the ocean, decreased vegetation

Source: Modified from Emery *et al.* (1957).

initial form of an estuary. **Coastal patterns** follow, resulting from the mechanical actions of the sea on land masses (see the sequence of forms in Figure 14.1).

Some estuaries do not present a direct connection from the river into the ocean, but they form **depositional embankments**, producing bays with salinity gradients that function as distinct heterogeneous spatial systems. In 'pure' estuaries, rivers flow directly into the coast. An extreme example of glaciation affecting estuaries can be found in the fjords in Norway and other temperate regions with deep V-shaped valleys and pronounced vertical salinity gradients.

The continual mixing of fresh water with saltier waters presents physiological problems for estuarine plants and animals. Suspended materials brought by rivers accumulate in banks, producing areas rich in food for many organisms. However, low oxygen levels or even anoxia can occur.

Estuaries are highly productive transitional regions or ecotones, diverse in fauna and flora with many feeding niches for herbivores, carnivores, and scavengers. Because of these features, these regions have great potential for exploitation by man, mainly for fish and shellfish. However, physiologically, these faunas and floras are quite specialized.

The patterns of dilution of the higher salinity coastal water (33‰) with fresh waters (0.5‰) present a gradient that can vary from 35–33‰ to 0.5‰. This brackish water, varying with each tidal cycle, presents different characteristics in each estuary, depending on the salinity of the coastal water and the volume of fresh water discharged daily into the estuary.

Circulation patterns in estuaries vary: estuaries with **positive circulation** or *positive estuaries* are those where a vertical gradient produced by the entrance of fresh water moves over the denser saltier water, forming a counter current. In this case, there is a gradual blending of fresh water with saltier water. On the surface there is a continuous up flow of the less salty water. In positive estuaries, the volume of fresh water entering the estuary is greater than evaporation.

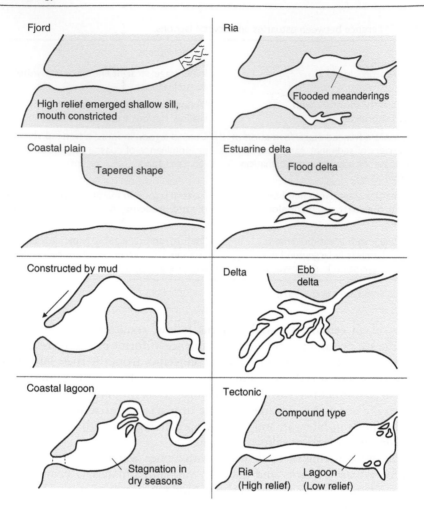

Figure 14.1   Physiographical types of estuaries.
Source: Adapted from Fairbridge (1980).

In *negative estuaries*, evaporation is greater than the entering fresh water. In this case, salinity on the surface increases and, the surface water sinks, forming a salty current leaving the estuary. There are cases where evaporation equals the input of fresh water in the estuary and then salinity does not vary greatly. In this case, the estuary is called *neutral*.

Figure 14.2 summarizes the three types of estuaries with their general patterns of circulation. These circulation patterns depend on the flow of fresh water, the estuary's average depth, its length and the direction of prevailing winds, which, to a degree, can affect vertical and horizontal mixing patterns.

The **morphometry of estuaries** varies greatly, depending on the transport of suspended material, the general direction of deposits and the organization of the

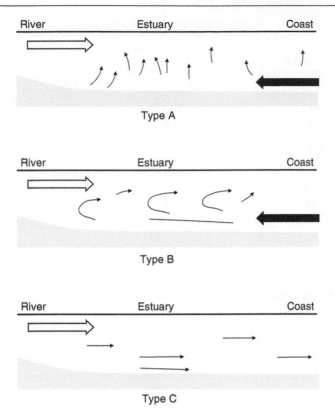

Figure 14.2   Different patterns of circulation in estuaries.
Source: Adapted from Pritchard (1955).

higher-plant community, which can gradually change the morphology of the estuary and the patterns of horizontal circulation.

The degree of vertical mixing in positive estuaries can be *highly stratified*, *partially homogeneous* or **completely homogeneous**. In estuaries with irregular bathymetric shapes, pockets with saltier waters can form, with longer retention times. In such cases, anoxic conditions can occur in relatively deep valleys with little circulation, as is the case in fjords.

Especially large estuaries are affected by the Coriolis effect, which can cause differentiated horizontal distribution, with a more horizontal and less vertical flow.

The variations in high and low tides in estuaries depend on the height of the tide at the entrance. Estuaries where the height of the tide is low at its entrance present smaller intertidal zones. Greater differences in the tide at the entrance produce large intertidal zones within the estuary.

Circulation in estuaries is a key factor in the distribution of plankton, benthos and nektonic organisms, periphyton and macroscopic algae. Circulation determines the transport of sediment and patterns of distribution and recycling of nutrients, which are important in the biogeochemical cycles and ionic composition of water.

Table 14.1 Velocity of sedimentation of particulates.

Material	Average diameter (μm)	Sinking velocity ($m \cdot day^{-1}$)
Fine sand	250–125	1,040
Very fine sand	125–62	301
Silt	31.2	75.2
Silt	15.6	18.8
Silt	7.8	4.7
Silt	3.9	1.2
Clay	1.95	0.3
Clay	0.98	0.074
Clay	0.49	0.018
Clay	0.25	0.004
Clay	0.12	0.001

14.2   SEDIMENT IN ESTUARIES

The sediment in estuaries is an example of the complex and dynamic nature of this ecosystem. The sediment deposited comes from the flow of rivers, the action of the coastal water and the pattern of currents in the estuary. Near the coast, the sediment is predominantly sandy, while further in the estuary, it is finer, often clayey, with high levels of organic matter. The deposition of sediment naturally depends on the drainage basin of the river(s) flowing into the estuary, as well as the land uses in these basins and erosion rates.

There is, thus, a formation in the sediments in the estuary. Sedimentation rates vary according to particle size and current velocity. Also, the mixture of seawater with fresh water produces **flocculation of fine particles** that adhere together to form aggregates (McHusky, 1981). Thus, many estuaries have high levels of suspended matter and low light penetration. The gradient of suspended material depends on circulation, particle sedimentation rate (see Table 14.1), discharge of material by rivers, and the mechanical differences between the tidal force and river velocity.

In estuaries with mangroves, there are high levels of POM and DOM, with humic substances that form complex molecules. The humic substances and decomposing matter in estuaries produce a characteristic colour that occurs mainly in rivers and tributaries and is important as a filter for blocking certain wavelengths of solar radiation that reach the water's surface (Tundisi, 1970).

14.3   CHEMICAL COMPOSITION AND PROCESSES IN BRACKISH WATERS

The dilution produced by fresh water coming in contact with seawater does not follow a predictable theoretical linear pattern. For example, bicarbonate levels barely drop in the dilution of seawater with fresh water. The levels of chloride present approximately linear dilutions.

Table 14.2   Concentration (mg · L^{-1}) of organic carbon in estuaries, rivers and coastal areas.

	River	Estuary	Coastal area	Sewage
DOC	10–20	1–5	1–5	100
POC	5–10	0.5–5	0.1–1.0	200
Total	15–30	1–10	1–6	300

DOC – Dissolved organic carbon
POC – Particulate organic carbon
Source: Modified from Head (1976).

The distribution of trace metals (such as iron, manganese, cobalt, zinc, copper and cadmium) depends on the amounts and vertical distribution of suspended matter.

In addition to inorganic sediments in estuaries, there is a wide range of suspended organic matter, from excretion and decomposition of organisms. Also, in addition to POM, there are high levels of DOM, which can be converted into POM by the effects of salinity and flocculation. Many areas of deposition in estuaries show high levels of organic matter. Table 14.2 shows the levels of dissolved and particulate organic carbon (DOC and POC) in estuaries as compared to other ecosystems.

The levels of POM and DOM in estuaries are therefore critical in triggering processes related to the production of organic matter, sedimentation and level of dissolved oxygen.

The saltwater/freshwater interface stimulates the flocculation of countless organic and inorganic particles, and dissolved organic matter is condensed, adsorbed with phosphates. This mass of coagulated matter provides the base for the growth of a prolific and diverse **microbial flora**, which in turn provides an important food source for plankton and benthic organisms. The zone with the densest presence of benthic organisms is the intersection zone between the fresh water and saltwater and brackish waters in estuaries (Horne and Goldman, 1994).

Waves, in many cases, are smaller in estuaries, since these areas are sheltered from the wind. However, in some estuaries in open regions close to coastal waters, the waves can be an additional factor in the vertical and horizontal mixing of fresh water and brackish waters. Currents are generated by tides and the discharge of fresh water from tributaries. Currents can vary greatly, depending on the tidal action and volume of water discharged into the estuary.

The circulation and mixing in estuaries do not depend on tidal and freshwater forces only, but also on the size of the estuary and the ratio between length and width of the system. **Tidal propagation** in an estuary involves intense processes of lateral transfer or advection, as discussed by Miranda *et al*. (1998).

Based on saline levels, estuaries were classified (Emery *et al*., 1957) as marine, brackish or **hypersaline**. The chemical composition of estuarine waters thus varies depending on the prevalence of each of these types. In hypersaline water, due to evaporation, there is an excessive level of salts deposited and a differential deposition of salt. In general, in many estuaries, there is a higher ratio than in sea water of carbonate and sulphate to chlorine and calcium to sodium, and these ratios depend on the volumes and proportions of fresh water and coastal waters in the estuary. Thus, **biogeochemical cycles in estuaries** depend on circulation processes, saline gradients and oxido-reduction patterns for the different types of horizontal and vertical circulation.

14.4   COMMUNITIES IN ESTUARIES

In estuaries, physical compartmentalization and vertical and horizontal spatial heterogeneities are much more pronounced than in large lakes or oceans. Planktonic, nektonic and benthic communities have to adapt to the widely fluctuating environmental conditions in these ecosystems.

Estuarine fauna is predominantly marine in the regions closest to the ocean. The number of freshwater species naturally increases in the headwaters of the estuary, where low-saline conditions prevail (up to 5‰). Optimal development of species and colonization of estuaries depend on the interaction of many factors, according to Day (1951). A number of interacting factors can determine the colonization of certain areas. Figure 14.3 shows the factors that limit the distribution of organisms in an estuary.

Despite the wide variations in environmental factors in estuaries, such fluctuations are characteristic of estuarine ecosystems. They enable the maintenance of a stock of biodiversity highly adapted to these fluctuations (Simpson, 1944); **estuarine species** are descendants of conservative forms with a long evolutionary history. Some species of oysters common in estuaries are originally from the Cretaceous Period, and many gastropods (originally from the Triassic Period) also colonize these environments. **Euryhalinity** (tolerance of a wide range of salinity) is an extremely conservative feature physiologically, and is more typical of families than of genera or species.

The broad tolerance of **estuarine organisms** may explain their ability to colonize estuaries or even fresh water ecosystems in many continents. For example, the tolerance of a *Dreissena* sp. in the Caspian Sea may explain its flexibility in colonizing (as an invasive species) the Great Lakes in North America.

Many physiological adaptations occur in organisms in estuaries, including structural changes, which enable them to tolerate low salinities (Yonge, 1947).

According to Panikkar (1951), the greatest wealth of species in estuaries in tropical and subtropical regions (as compared with estuaries in higher latitudes) is

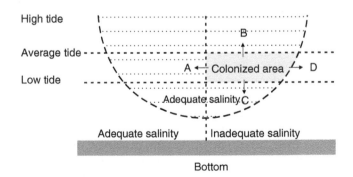

Figure 14.3   Limiting factors of colonization in estuaries.
Sources: Day (1951); Emery *et al.* (1957).

probably due to the increased **capacity for osmoregulation** at higher temperatures. Tundisi and Tundisi (1968) discussed the distribution of *Acartia lillejborghi* in the lagoon area of Cananéia (São Paulo) and concluded that its penetration in this estuarine region was due to the higher water temperatures. The species *A. tonsa*, typical of temperate-region estuaries, has less penetration in estuaries due to lower temperatures and reduced capacity for osmoregulation.

The most common organisms in the fresh water of estuaries are several larvae of insects and some adult forms of aquatic insects, oligochaetes and some lunged molluscs. These organisms are very common in freshwater estuaries that are more stable. Penetration of **marine fauna** in estuaries depends on the size of tides and the saline gradients. Seasonal migrators are common in estuaries. These migrations are part of the reproductive process or feeding patterns of juveniles.

Estuaries are considered areas for reproduction and maintenance of a high biomass of species (*nursery grounds*), especially in tropical and subtropical latitudes. Figure 14.4 shows the comparative distribution of *marine species*, *brackish-water species*, and *freshwater species* in estuaries (Remane, 1934).

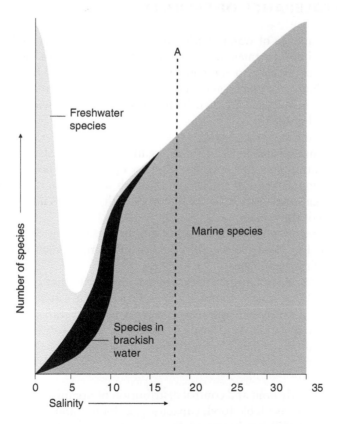

Figure 14.4   Comparative distribution of marine species, freshwater species and species in the brackish waters in estuaries.
Source: Modified from the classic illustration by Remane (1934) in Emery *et al.* (1957).

A similar reduction in the number of aquatic plant species in estuaries also occurs. There is a rather broad distribution of *Spartina* spp., *Salicornia* spp. and *Scirpus* spp., which are typical of brackish wetlands. Filamentous algae are common in estuaries (*Enteromorpha*, *Chaetomorpha*, and *Cladophora*) and present dense growth and development in these areas. The distribution and biomass of these algae depend on the substratum and control due to the height of the tide. Phytoplankton in estuaries presents lower species density and greater biomass. Nutrient enrichment is highly dependent on the rivers that feed the estuary.

The flora of estuaries – whether higher plants, macroalgae or phytoplankton – depends on factors such as salinity, turbidity, and characteristics of the substratum for colonization by benthic algae and higher plants. In some estuarine regions, especially in the tropics, phytoplankton can reach high levels of biomass and productivity (Tundisi, 1970).

The **phytoplankton in estuaries** is composed mostly of marine species, with an occasional freshwater species (Cowles, 1930; Tundisi, 1970; and Teixeira and Kutner, 1963).

14.5   DISTRIBUTION OF ORGANISMS IN ESTUARIES AND TOLERANCE OF SALINITY

In estuaries, a sequence of waters with varying levels of salinity and sediments can lead to horizontal compartmentalization. As a result, there can be a gradient from the entrance to the estuary to the source of the rivers, and subsequent diversified horizontal distribution of planktonic, nektonic and benthic organisms. The following categories of organisms generally can be found in an estuary with a relatively well-defined horizontal gradient.

▸ **Oligohaline organisms:** most of these organisms do not tolerate salinities much above 5‰, although a few tolerate salinities up to 5‰. These organisms live in the upper compartments of the estuary, in fresh water from rivers or lakes.
▸ Estuarine organisms: usually found in waters with 5–18‰ salinity, in the intermediary region of the estuary.
▸ **Euryhaline marine organisms:** these are marine species that can survive in salinities down to 18‰ and are found in the central compartment of estuaries. Some survive in salinities of 5‰.
▸ **Stenohaline marine organisms:** these are marine species that can survive in salinities down to 25‰; usually found at the entrance of the estuaries.
▸ Migratory organisms: these organisms, mainly fish and crabs, inhabit the estuary only during a specific period in their life cycle. Typical examples in temperate regions are salmon and eel. In the tropics, some species of shrimp characteristically migrate to the estuary in adulthood.

Factors that clearly limit and control distribution of organisms in estuaries include salinity, temperature, available food, capacity to colonize, and inter-specific competition (Emery *et al.*, 1957) (see Figure 14.5).

According to Jeffries (1969), **estuarine communities** are controlled by physical conditions, due to effects of intense fluctuations. The physiological responses of

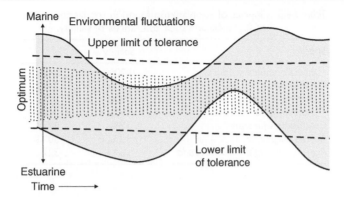

Figure 14.5 Degrees of tolerance, upper and lower limits and the optimum for estuarine, marine and freshwater species in estuaries.

estuarine organisms vary not only during a single phase of the life cycle, but responses also vary from one phase to another. For example, larval stages of benthic organisms may be more tolerant than the adult stages, as was discussed. In addition to temperature, which can alter physiological patterns, the water's density and viscosity depend on salinity and temperature. Many stenohaline marine organisms have osmoregulatory capacity, involving a change in internal osmotic levels when the organism enters the estuary. However, for stenohalines, depending on the species, there is a lower limit.

Another mechanism consists of an organism maintaining its internal osmotic level independent of that of the medium, which involves maintenance of vital functions with higher internal levels than those in the diluted waters of the estuary.

Estuarine animals, therefore, have the capacity to tolerate shifts in the medium's salinity, based on several mechanisms that in some cases include active osmoregulation (hyperosmotic or isosmotic).

Planktonic species' spatial distribution and tolerance to various conditions of salinity and temperature in estuaries are well illustrated by the studies on the zooplankton's tolerance to salinity in the Cananéia lagoon area (Tundisi and Tundisi, 1968). The studies were conducted in two steps: the first was a series of laboratory experiments in which species of planktonic copepods common in the Cananéia lagoon region were subjected to saline gradients over a six-hour period. The survival rates of those organisms in different salinities showed lethal salinity for 50% of the organisms. The salinity gradient used was 0.0 to 35‰.

In addition to these experiments, water samples were collected in the estuary, from the regions with the lowest salinity (0.00‰) up to 35‰, from the coastal waters and at the entrance into the estuary. The results show the distribution of planktonic species in estuaries with pronounced salinity gradients. Those species showing greater tolerance for high salinity are presented in Table 14.3.

These experiments and the simultaneous sampling showed the diverse characteristics typical of estuarine plankton, also common to other benthic or nektonic estuarine organisms:

▶ a reciprocal (opposite) **biological gradient** was identified between *Pseudodiaptomus acutus* and *Acartia lillejborghi*.

Table 14.3   Species of zooplankton found in the lagoon region
of Cananéia and their tolerances of salinity.

Species	Salinity	
Pseudodiaptomus acutus	0.00‰	24‰
Oithona ovalis	7‰	24‰
Euterpina acutifrons	13‰	30‰
Temora stylifera	17‰	30‰
Centropages furcatus	18‰	32‰
Acartia lillejborghi	13‰	30‰
Acartia lillejborghi – aclimatizada	6‰	32‰

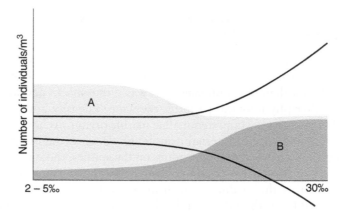

Figure 14.6   Reciprocal biological gradient. A and B – species with optimum tolerance in fresh waters
and salty/brackish waters, respectively.

▸ a range of typical freshwater species such as *Pseudodiaptomus acutus* or Cirripedia larvae;

▸ a range of species typical of coastal and marine waters such as *Euterpina acutifrons, Temora stylifera* and *Centropages furcatus*;

▸ a range of species that tolerate moderately saline water such as *Oithona ovalis* and *Acartia lillejborghi*.

The acclimatization experiments conducted on *A. lillejborghi* (in which the animals were periodically subjected to decreasing saline gradients) showed that some species have a **capacity for progressive adaption** to lower (or higher) salinities when subject to slow shifts in salinity, which is what occurs in many estuaries during each tidal cycle. Figure 14.6, taken from Tundisi (1970), illustrates the concept of 'reciprocal biological gradient' in an estuary.

According to Jeffries (1962a, 1962b, 1967), Chart 14.2 presents an ecological classification of the main species of estuarine holoplankton, based on the characteristics of distribution in relation to environmental factors and the species' centre of dispersion.

Chart 14.2   Category of distribution of organisms in estuaries.

Category	Characteristics
Estuarine	Propagation only in brackish waters. Tolerance for reproduction between 5–30‰ Found in oceans, casually
Estuarine and marine	Propagation in large parts of the estuary Limited by salinities less that 10‰ Reproduction in estuaries and non-coastal waters
Marine euryhaline	Found in the estuary; reproduction in coastal waters Maintenance of the population depending on continual supply from the ocean
Marine estenohaline	Occasionally propagated in the entrance of the estuary. Characteristic of neritic waters

Chart 14.3   Classification of estuarine waters and ecological classification of organisms.

Estuarine regions	Venice system		Ecological classification Types of organisms and approximate gradient of distribution in the estuary, relative to divisions and salinity
	Gradients of salinity (‰)	Zones	
River	<0.5	Limnetic	limnetic
Head of the estuary	0.5–5	Oligohaline	oligohaline
Upper estuary	5–18	Mesohaline	mixohaline typical estuarine
Middle estuary	18–25	Polihaline	
Lower estuary	25–30	Polihaline	estenohaline euryhaline migrating
Area of outflow into ocean	30–40	Euhaline	marine marine

Chart 14.3 presents the classification of estuarine waters according to the **Venice** system and the ecological classification of organisms according to salinity.

14.6   MAINTENANCE OF STOCKS OF PLANKTONIC AND BENTHIC POPULATIONS IN ESTUARIES

Bousfield (1955) showed that in the Miramichee River estuary in the United States, the distribution of Cirripedia (barnacles, *Balanus* spp.) is related to the movements of fresh water to the surface and the counter-current in the depths associated with those movements, and to higher-saline waters moving upriver. Through processes of **vertical migration** at different stages of the tide, the population can remain in certain areas of the estuary (Lance, 1962).

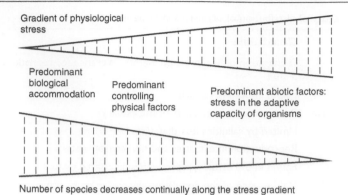

Number of species decreases continually along the stress gradient

Figure 14.7   Representation of the hypothesis of stability in time by Sanders (1969), showing the numeric relationship of species/stress and the predominance of abiotic factors in regions upstream from the estuary.

A similar mechanism was proposed for the Cananéia lagoon region by Tundisi and Tundisi (1968). It was verified that *Pseudodiaptomus acutus* was most common in the upper portions of the estuary (near the headwaters of the rivers) and that the range of tolerance to salinity for adult females was broad (from 1–27‰). In laboratory experiments, it was found that adult females of this copepod moved away from light when subjected to light coming from the surface of the water. Samples collected at high and low tides showed that at high tide, as the light intensity increases on the surface, the animals migrate toward the bottom, and are then dragged again by the counter-current toward the upper portions of the estuary. A maintenance mechanism of this species in certain regions of the estuary is thus achieved by means of a negative response to light and tolerance to wide variations in salinity (Tundisi and Tundisi, 1968).

This type of mechanism to retain the population in certain areas of the estuary is relatively common. However, the particular case of *Pseudodiaptomus acutus* concerns the coupling of two factors and two responses in terms of light and salinity simultaneously. Margalef (1974) looked at this process as an endogenous rhythm to maintain estuarine species in regions with optimal temperatures and salinity, enabling reproduction as well as survival of the population.

For benthic populations, larvae and young organisms exhibit similar behaviours, making it possible for adults to remain in optimal regions for reproduction, and consequently, dispersal of eggs and larvae with a capacity to tolerate wide saline gradients.

The extent of the regulatory capacity of estuarine species varies greatly. Figure 14.7 shows variations in gradients of physiological stress according to Sanders (1969), in which, as physiological stress increases, the number of species decreases.

In estuaries, the benthic fauna plays an extremely important role, much more so than zooplankton, in terms of the food chain and transfer of energy. Shallow estuaries with extremely rapid currents near the bottom and variable bottoms are similar to rivers in which the zoobenthic community is extremely important. The estuarine *infauna* consists of truly estuarine benthic fauna and *Polychaetes*, *oysters*, *crabs* and *shrimp* living on or depending on the sediment. These organisms are extremely important in the food web dynamics in these ecosystems.

Many bird species also inhabit estuaries and feed on these benthic forms, which can reach high biomasses (>100 g dry weight per 0.1 m^2), particularly in areas with intermediate salinities (Horne and Goldman, 1994). In many estuaries, **macroinfauna** (unable to pass through nets with 0.5-mm mesh holes) are predominant.

14.7 PRIMARY PRODUCTIVITY IN ESTUARIES

The **primary producers in estuaries** include a wide variety of organisms, from micro-phytoplankton, nano-phytoplankton and micro-phytobenthos to higher plants and benthic macro-algae. This variety of primary producers and the enrichment from rivers are important factors in primary productivity, which in many aquatic ecosystems is greater than in deep lakes and oceans. Exposure of sediment to desiccation significantly increases the temperature and decomposition rate of organic matter, which recycles nutrients, especially phosphorus and nitrogen. They are immediately absorbed by benthic macro-algae, micro-phytobenthos and phytoplankton.

A large portion of the nutrients in estuaries comes from allochthonous sources, such as coastal vegetation and vegetation from mangrove swamps in tropical estuaries. These allochthonous sources can sustain extremely high primary production. However, nutrients from shallow sediments can also be an important reserve, as shown by Tundisi (1969) for the Cananéia lagoon area. In this region, the brackish or freshwater rivers that discharge particulate and dissolved organic matter into the main estuary and many channels in the mangrove region ('marigots') play an important role in the recycling of nutrients back into the estuary. In addition, these channels act as collectors of particulate matter (leaves and vegetable debris from mangrove trees). The matter then undergoes a rapid process of decomposition, providing a source of inorganic nutrients in the channels, stimulating the primary productivity of phytoplankton and phytobenthos growing in the mangrove roots.

Some tropical estuaries have extensive areas of mangrove vegetation (*Rhizophora* sp., *Laguncularia* sp. or *Avicenia* sp.), which play an important role in primary production of phytoplankton and microbenthos in the estuaries. According to Prakash and Rashid (1969), humic substances play an important role in phytoplanktonic growth and in the primary productivity of phytoplankton and phytobenthos in estuaries with high levels of dissolved humic substances. Droop (1966) and Aidar-Aragão (1980) showed the effect of humic substances on the growth of *Skeletonema costatum*.

In the Cananéia lagoon area, Tundisi, Teixeira and Kutner (1965) hypothesized that even small quantities of dissolved humic substances play a key role in the growth of phytoplankton in the summer. This is particularly true for *Skeletonema costatum*, which in summer plays an important role in the primary productivity of micro-phytoplankton (Tundisi, 1969). Humic substances affect the growth of phytoplankton, shortening the doubling time of phytoplankton populations, most likely due to physiological stimuli resulting from chelation processes, making ions available or not for cells or cell chains. Photo-reductive and photo-oxidative processes are also extremely important in making these humic substances available for phytoplankton and microbenthos in estuary regions with mangrove vegetation (Droop, personal communication with Tundisi, 1965).

Table 14.4   Primary production compared for various ecosystems in relation to the productivity
of estuaries.

Ecosystem	Gross primary production $kcal \cdot m^{-2} \cdot year^{-1}$	World total $10^{11} \cdot kcal \cdot year^{-1}$ (all systems)
Estuaries and coral reefs	20,000	4
Tropical forests	20,000	29
Fertilized agricultural areas	12,000	4.8
Eutrophic lakes	10,000	–
Eutrophic wetland areas	10,000	–
Unfertilized agricultural area	8,000	3.9
Coastal region of upwelling	6,000	0.2
Grassland	2,500	10.5
Oligotrophic lakes	1,000	–
Oceans (oceanic waters)	1,000	32.6
Deserts and tundras	200	0.8

Source: Horne and Goldman (1994).

Vannucci (1969) summarized the three main factors affecting productivity in estuaries: i) characteristics of water: nutrient turbidity, concentration of dissolved matter; ii) the characteristics and depths of sediment: phosphorus levels, particle size, the iron-to-phosphorus ratio in the sediment; and iii) availability of nutrients in the water and sediment and availability of organic matter in general, the decomposition of which accelerates primary and secondary productivity.

Of the major factors limiting the production of primary phytoplankton in estuaries, one important factor can be low levels of underwater solar radiation resulting from large amounts of suspended matter. The degree of turbulence and the critical depth, which produce rapid changes in the availability of solar energy, are also factors affecting the vertical and horizontal circulations that contribute to these changes.

The seasonal succession of phytoplankton depends on stability or instability of the environmental conditions. Margalef (1967) considered several factors important, including the differential influx of nutrients, the feeding by herbivorous zooplankton, and processes of circulation and vertical mixing. In estuaries with wide fluctuations in salinity and nutrient levels, there is great variation in the patterns of succession and periodic interruptions in succession, with changes in the predominance of each association of species. An important factor in phytoplanktonic succession in estuaries is the fluctuating salinity, which in less-tolerant species can cause mass kills and shifts in the sequence of succession.

Table 14.4 presents the primary productivity in estuaries as compared with other terrestrial or aquatic ecosystems (Horne and Goldman, 1994).

Estuaries are clearly among the most productive systems on the planet, because of the following factors: available nutrients, high biomass levels, rapid cycles and rapid decomposition in the water column and sediment, and food chains with several alternatives that stimulate production of organic matter and nutrients.   It is important to stress the adaptation of flora and fauna, which enables **optimal estuarine conditions** in those regions with mixing of nutrient-poor coastal water with nutrient-rich fresh water, and there provides optimal conditions for biomass production.

The range of interactions between estuaries and waters of the nearby marine ecosystem is presented in Figure 14.8.

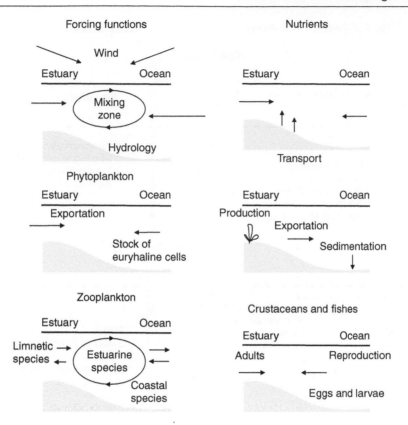

Figure 14.8   Interaction of physical, chemical and biological factors with the estuary and coastal marine waters.
Source: Modified from Abreu and Castello (1997).

14.8   THE FOOD WEB IN ESTUARIES

The food web in estuaries depends on the energy fixed photosynthetically by primary producers (which, as shown, have diverse sources in estuaries), as well as the transport and transformation of this organic material by advective processes, interactions of fresh water and brackish water, and **water-sediment interactions**. Organic matter is decomposed rather quickly in estuaries, and the experiments on and measurements of biomass in the seasonal cycles have proven this rapid conversion.

Figure 14.9 presents a typical estuarine food chain, which shows that birds, especially species that feed on sediment or fish, play a quantitatively significant role in the estuarine food chain. Also important are detritus-feeding organisms and those that feed on higher aquatic plants, for example, the roots and leaves of *Spartina* sp. or mangrove vegetation (decomposing leaves and roots).

Detritivores are very important in estuaries. Detritus is available throughout the year, even in temperate-zone estuaries. Zooplankton and benthos are primary consumers, equally important in estuaries. Beyond detritus, live particulate matter, such

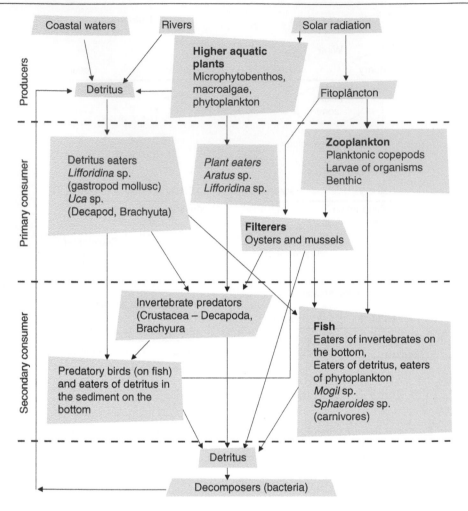

Figure 14.9   Food web in an estuary. At several trophic levels, species typical of estuarine regions and
mangrove regions of Brazil were located.
Source: Modified from Por (1944).

as phytoplankton and bacteria, can be an important source of food. Benthic animals
can filter suspended material or feed on matter deposited in the sediment.

14.9   DETRITUS IN ESTUARIES

In many estuaries, the contribution of higher plants to the formation of debris is very
large. All the primary producers previously mentioned provide different-sized organic
debris, which immediately begins to be altered by microorganisms. Micro-organisms
increase the nutritional value of the detritus and accelerate its decomposition and
transformation into humic substances. Organic debris accumulates in the sediment

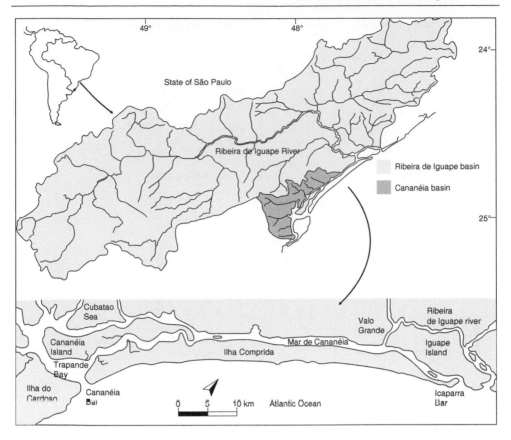

Figure 14.10   The drainage basin of the Ribeira de Iguape River and the Cananéia lagoon region with estuarine system.

and is removed by animals in a process known as *bioturbation* (biological disturbance by mechanical action).

Animals and their excretion also contribute to organic detritus. In addition to particulate matter in the form of waste, estuaries have high levels of dissolved organic substances that can be metabolized by bacteria, thus increasing live particulate matter. In many estuaries, these dissolved organic substances come from the decomposition of mangrove vegetation or aquatic plants (Tundisi, 1970).

14.10   THE CANANÉIA LAGOON REGION

The Cananéia lagoon region is an estuarine lagoon complex with channels, islands and rivers characteristically surrounded by mangrove vegetation. This 110-km² region is located at 25°S 48°W and is connected to the Atlantic Ocean through channels in its north and south areas (see Figure 14.10). The climate of this region shows a predominance of masses of polar or tropical air. The polar air is prevalent in autumn and

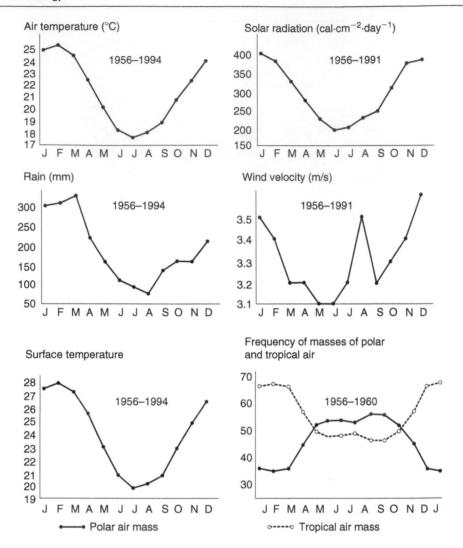

Figure 14.11   Climatic characteristics of the Cananéia lagoon region.
Source: Modified from Garcia Occchipinti (1963).

winter (March to September) and the tropical air in spring (September) until the end of summer (February).

Cold fronts during winter months are common, causing alterations in the water's salinity and accumulation of coastal water in the lagoon region (Garcia Occhipinti, 1963). In the summer, there are heavier rains and increased runoff of humic substances, from the decomposition of mangrove vegetation. These humic substances play a significant role in the functioning of the lagoon system, stimulating high primary productivity in summer, as seen in the predominance of *Skeletonema costatum*, a diatom typical of this region. Figure 14.11 shows the distinctive climatic characteristics distributed during this period of the year in the Cananéia lagoon region.

The geomorphology was formed in part by large-scale alterations that occurred during the Quaternary, and changes in the hydrodynamics of they system that continually caused spatial re-organization of the coast of the channels and the distribution of sediment in the lagoon region (Petri and Sugio, 1971).

Figure 14.12 shows the distribution of salinity in the main channels of the lagoon region during high tide and low tide.

14.10.1 Communities

Mangrove vegetation is distributed around the islands, along rivers and 'marigots' (the French word for channels in the mangroves). Mangrove vegetation is composed of *Rhizophora mangle*, *Avicennia shaweriana*, *Laguncularia racemosa* and *Conocarpus erecta* (the least-common species). *Spartina* sp. grows in areas with high levels of finer organic matter. A rich diversity of fauna invertebrates lives in its roots, especially nematodes and other invertebrates with tolerance of brackish waters between 10–25‰.

A transitional vegetation occurs between the mangrove vegetation and the Tropical Atlantic Forest (Cintron and Schaeffer, in Novelli, 1983). The spatial distribution and structural variability of the edge vegetation depend on complex interactions between fluctuating water levels, processes of decomposition, the salinity of the soil, and contributions of nutrients from the forest (Adaime, 1985).

The *planktonic community* presents more pronounced **biomass cycles** in interior and estuarine waters than in adjacent coastal waters (Teixeira and Kutner, 1963; Teixeira, Tundisi and Kutner, 1965; Tundisi, 1969, 1970). The primary productivity of phytoplankton is also higher, mainly in summer, reaching $1\,g\,C \cdot m^{-2} \cdot day^{-1}$ (as compared with coastal waters, where production is $\sim 0\,g\,C \cdot m^{-2} \cdot day^{-1}$). Inhibition of approximately 20% of photosynthetic activity is common in the lagoon region (Teixeira *et al.*, 1969). The growth of the diatom *Skeletonema costatum* in summer and the increase in biomass and primary productivity were attributed to the stimulus provided by the humic substances coming from the decomposition of mangrove vegetation (Aidar-Aragão, 1980).

The vertical distribution of phytoplankton is related to tidal cycles. During periods of high tide, populations of *Chaetoceros* and *Skeletonema costatum* are common in the deepest water in the lagoon area (Brandini, 1982).

The zooplanktonic population of this region is dominated by copepods, including *Oithona hebes*, *Acartia lillejborghi*, *Pseudodiaptomus acutus*, *Euterpina acutifrons*, *Paracalanus crassirostris*, *Oithona oswaldocruzi*, *Acartia tonsa* and *Temora turbinata*. These organisms are the principal components of the zooplanktonic community, and their spatial distribution (horizontal and vertical) is influenced by salinity. The daily production of planktonic copepods measured in the lagoon area ranges from 2.08 to $44.76\,mg\,C \cdot m^{-3} \cdot day^{-1}$ (Ara, 1998).

Table 14.5, Table 14.6 and Table 14.7 show the chemical content of the main species of planktonic copepods, the P/B rate (daily production/biomass) and the average annual accumulation of biomass of these species, respectively (Kara, 1998). The abundance of the copepod populations is related to temperature, salinity and chlorophyll levels, according to Matsumura Tundisi (1972) and Kara (1998).

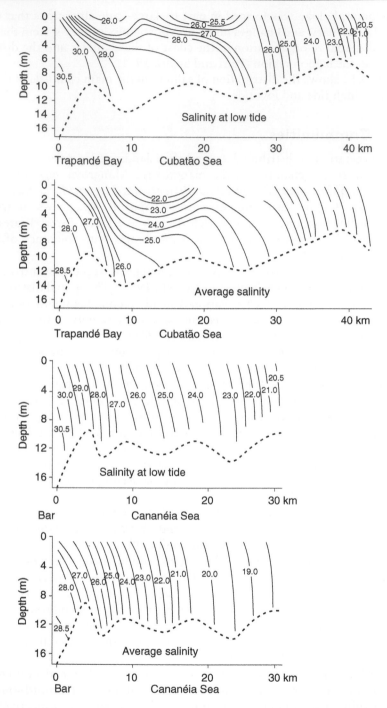

Figure 14.12   Distribution of salinity in the main channels of the Cananéia lagoon region.
Source: Miranda *et al.* (2002).

Table 14.5   Chemical content (carbon, nitrogen and hydrogen) expressed as percentage of dry weight of the 10 main species of copepods in the Cananéia estuary-lagoon complex (SP).

Species	n	Carbon (%)	Nitrogen (%)	Hydrogen (%)
Acartia lillejborghi	2	45.33±0.09	11.71 ± 0.03	6.72 ± 0.05
Acartia tonsa	2	44.21±0.08	11.35 ± 0.01	6.78 ± 0.01
Pseudodiaptomus acutus	2	46.11±0.04	11.64 ± 0.00	7.05 ± 0.02
Paracalanus crassirostris	2	46.26±0.01	10.90 ± 0.01	7.03 ± 0.04
Paracalanus quasimodo	2	45.56±0.21	11.26 ± 0.03	6.90 ± 0.05
Temora turbinata	2	44.57±0.02	11.60 ± 0.06	6.79 ± 0.03
Labidocera fluviatilis	2	45.21±0.04	12.11 ± 0.06	6.94 ± 0.01
Oithona hebes	2	46.11±0.05	11.69 ± 0.04	7.02 ± 0.02
Oithona oswaldocruzi	2	46.37±0.06	10.96 ± 0.05	7.16 ± 0.03
Euterpina acutifrons	2	46.04±0.14	11.30 ± 0.02	7.01 ± 0.00
Copepoda total (average)	20	45.58±0.73	11.45 ± 0.36	6.94 ± 0.14

n – number of observations
Source: Ara (1998).

Table 14.6   Daily P/B rate of copepods in estuarine-lagoon complex of Cananéia (SP), from February 1995 to January 1996.

Species	Daily P/B rate (day^{-1})		
	Min.	Max.	Average
Acartia lillejborghi	0.20	1.44	0.44
Acartia tonsa	0.29	1.61	0.73
Pseudodiaptomus acutus	0.18	0.94	0.40
Paracalanus crassirostris	0.33	1.30	0.65
Paracalanus quasimodo	0.28	1.54	0.52
Temora turbinata	0.20	1.02	0.44
Labidocera fluviatilis	0.13	1.05	0.29
Oithona hebes	0.38	1.72	0.81
Oithona oswaldocruzi	0.38	1.60	0.74
Oithona oculata	0.38	2.10	0.99
Oncaea media	0.40	1.87	0.89
Euterpina media	0.29	1.23	0.58
Less frequent species	0.15	1.65	0.53
Total copepods	0.21	1.44	0.55

Source: Ara (1998).

The *benthic community* of the Cananéia lagoon area is represented by 73 taxa, most commonly annelids, crustaceans, molluscs, and especially polychaetes.

The distribution of macrobenthic fauna depends on the substratum (type and physical-chemical variables) (Jorcin, 1997). High density and diversity of benthic fauna can occur, according to Jorcin (1997), when there are high levels of nutrients, which support diversity and biomass. Insects and oligochaetes (Naididae) are common in the freshwater benthic fauna found in the upper portions of the estuary. Seasonal changes in abundance and diversity of macrobenthic organisms result from changes in salinity, redox potential, the granulometry of the sediments and the levels of organic matter.

Table 14.7 Average annual accumulation of biomass, productivity rate and P/B rate of copepods in the lagoon-estuary complex of Cananéia (SP), from February 1995 to January 1996.

Species	Accumulated biomass (annual)				Annual productivity rate		
	$(mgDW \cdot m^{-3})$	%	$(mgC \cdot m^{-3})$	%	$(mgC \cdot m^{-3} \cdot year^{-1})$	%	P/B $(year^{-1})$
Acartia lillejborghi	3,889.44	24.7	1,762.95	24.4	721.72	18.1	149.4
Acartia tonsa	498.23	3.2	220.10	3.1	199.91	5.0	331.5
Pseudodiaptomus acutus	3,783.23	24.0	1,744.34	24.2	767.18	19.2	160.5
Paracalanus crassirostris	1,092.81	6.9	505.53	7.0	259.75	6.5	187.5
Paracalanus quasimodo	141.99	0.9	64.61	0.9	24.81	0.6	140.2
Temora turbinata	1,325.68	8.4	590.94	8.2	210.26	5.3	130.0
Labidocera fluviatilis	513.19	3.3	232.14	3.2	51.47	1.3	80.9
Oithona hebes	2,557.46	16.2	1,179.32	16.4	1,067.14	26.8	330.3
Oithona oswaldocruzi	805.92	5.1	373.76	5.2	347.08	8.7	338.9
Oithona oculata	32.49	0.2	14.97	0.2	13.54	0.3	330.1
Oncaea media	20.08	0.1	9.13	0.1	6.12	0.2	244.7
Euterpina acutifrons	1,043.54	6.6	480.34	6.7	306.00	7.7	232.5
Less frequent species	71.54	0.5	32.49	0.5	11.93	0.3	134.0
Copepod total	15,776.76	100.0	7,210.94	100.0	3,987.12	100.0	201.8

Source: Ara (1998).

The fish community in the Cananéia lagoon region includes 68 species belonging to 52 genera and 23 families. Most species are of marine origin, but can tolerate lower-salinity brackish waters (Zani Teixeira, 1983).

The bacterial communities can thrive due to the content of the organic matter. According to Mesquita (1994), the presence of bacteria (46% in detritus) is related to the senescent biomass of the phytoplankton.

14.10.2 Microbial metabolism and nutrient cycles

Nearly 10% of the gross primary production is consumed by microbial metabolism. Due to the large amounts of dissolved and particulate organic matter from mangrove vegetation, a detritus food chain is probably common in the estuary's interior regions of brackish or fresh water while in the areas closer to the coastal waters, phytoplankton and zooplankton are more common in the food chain. In winter (June/July), microbial metabolism is less active, according to Mesquita (1983).

Nutrient cycles in this region are dominated by the processes of decomposition and recycling of organic matter from the high levels of vegetation litter of the mangroves – $9.02 \, ton \cdot ha^{-1} \cdot year^{-1}$ (Schaeffer-Novelli et al., 1990). The high temperatures of the inland waters (up to 36°C on the surface of 'marigots') are influential in the **biogeochemical cycles of decomposition**. Levels of particulate carbon (POC) vary from $40–324 \, \mu g \cdot L^{-1}$.

Sediment can be an important source of nutrients where anoxic conditions prevail, especially in inland channels with 'anoxic pockets' associated with poor circulation (particularly during low tide), high temperatures and decomposing organic matter (Menezes, 1994).

The **decomposition of litter** in the mangrove region plays a major role in nutrient recycling (Gerlach, 1958; Kato, 1966).

14.10.3 Impacts

In addition to the destruction of mangrove forests due to the urban sprawl of condos and marinas, the Cananéia lagoon region suffers effects from deforestation and mining in neighbouring water basins, especially the destruction of the **Tropical Atlantic Forest**. These effects have caused changes in the operating mechanisms of the region, particular in nutrient cycles, increased toxins in water and sediment, and shifts in biodiversity. For example, the removal of mangrove vegetation, as in other regions of Brazil, reduces the substratum available for crustaceans, snails and periphyton.

Another important impact in reducing human intervention was creation of the Valo Grande channel (see Figure 14.10) between 1828 and 1830, which facilitated shipping and transport of goods. The opening of the channel and its subsequent widening turned the **Ribeira de Iguape River** into the principal freshwater contributor into the Cananéia lagoon area ($435 \, m^3 \cdot s^{-1}$), drastically reducing salinity, introducing sediment, and altering the ecological functions and biological structure of the region. Later the channel was closed in the mid-1970s, causing other alterations in the region: increased salinity, changes in the velocity of currents, and other physiographical and ecological alterations.

14.10.4 Comparisons with other estuarine tropical regions with mangrove vegetation

Rodriguez (1975) compared ten major tropical river regions and their estuarine regions. They all have significant discharges, and the freshwater and estuarine regions are extensive, penetrating into the different coastal regions for many miles. The rivers, all in the tropics, include the Amazon (6275 km), east coast of Brazil, the Atlantic Ocean; the Congo (4600 km), west coast of the African continent, the Atlantic Ocean; the Niger (4160 km), **Gulf of Guinea**, west coast of Africa, the Atlantic Ocean; the San Francisco (3160 km), east coast of Brazil, Atlantic Ocean; the Orinoco (2900 km), northern South America (Venezuela), the Atlantic Ocean; Brahmaputra (2700 km), Bay of Bengal, Bangladesh; Zambezi (2750 km), Mozambique, the Indian Ocean; the Ganges (2480 km), India, Bengal Bay; the Irrawaddy (200 km), **Burma** (Bengal Bay); the **Senegal** (1689 km), northern part of the African continent, the Atlantic Ocean.

Low salinities are found up to 80 kilometres east of the Orinoco in the Atlantic Ocean, and Teixeira and Tundisi (1967) found low salinities (12‰) from the Amazon estuary extending out 80 km into the Atlantic Ocean. The authors also noted fertilization on the surface waters of the ocean from nutrients carried by the Amazon River.

Vertical salinity gradients occur in tropical estuarine regions affected by large rivers. The plankton in these regions is subjected to estuarine conditions, but the benthic communities and big fish live under the influence of saltier waters and conditions, close to those of ocean waters (35‰). Ocean currents can extend the influence of estuarine waters, enriching the regions into which they flow and also fertilizing other more distant oceanic regions. For example, the less salty waters of the Orinoco River enrich some Caribbean islands, and the waters of the Amazon River discharge

Other studies in tropical mangrove forests and estuaries of Brazil

Neuman Leitão (1994) described a series of studies on coastal and estuarine zooplankton conducted by Oak (1939, 1940), Oliveira (1945), Lopez (1966), Bjornberg (1972), Paranaguá and Nascimento (1973), Almeida Prado (1972, 1973, 1974) Paranaguá *et al.*, (1979), Pecala (1982), Paranaguá and Nascimento Vieira (1984), Por *et al.* (1984), Almeida Prado and Lansac-Toha (1984), Paranaguá *et al.* (1986), Nogueira *et al.* (1988), Silva and Bonecker (1988), Lopes (1989), Neuman Leitão *et al.* (1992a), Nogueira Paranaguá and Paranhos (1992), Neuman Leitão *et al.* (1993).

These studies showed the importance of the populations of copepods in the tropical estuarine plankton, their distribution, based on salinity and temperature, the seasonal cycles and the role of this planktonic community in the biogeochemical cycles and the food webs in estuaries. They were also important for the consolidation of taxonomic information and systematic knowledge of estuarine zooplankton.

sediment north of **French Guyana** in the Atlantic Ocean. In some estuaries, the seasonal cycle causes discharge of large volumes of fresh water during the rainy season in summer, as is the case of the small rivers on the coast of Ecuador that drain into the Pacific Ocean on the west coast of South America.

Productivity in tropical estuaries is due to the discharge of nutrients (which also stimulates the productivity in adjacent coastal waters). The tidal movements of masses of fresh water and coastal waters determine the displacement and survival of planktonic populations, the larvae of benthic organisms (essential for re-colonisation) and fingerlings.

Mangrove forests – in addition to stimulating the primary productivity of phytoplankton and influencing the seasonal cycle due to discharge of humic substances (Smayda, 1970; Tundisi and Matsumura Tundisi, 1972) – also play an essential role by providing many niches in their structures, in their support roots and trunks as well as providing substrata for periphyton and bacteria in the colonization of the leaves and vegetative remains that are continually added to the channels in lagoon regions. An example of diversification in this niche is the distribution of genera and species of crabs and shellfish in the roots and trunks of *Rhizophora mangle* (Por, 1994).

14.11 COASTAL LAGOONS

Restinga ecosystems (according to Lacerda *et al.*, 1993, in Araújo *et al.*, 1998), occupy about 79% of the Brazilian coast, covering extensive areas of up to 3000 km² (the northern littoral zone in the state of Rio de Janeiro) or in narrow littoral belts. In these coastal ecosystems, lake depressions occur, the result of the closing of the river mouth, fed by rainwater or small streams. *Restingas* are important systems for the metabolism of coastal lagoons and maintenance of specialized vegetation, which is important in the supply of nutrients, humic substances and detritus that are significant in the ecological functioning of coastal lagoons and maintenance of biogeochemical and biological processes (Esteves, 1988).

Araujo *et al.* (1998) studied the Jurubatiba Restinga located between 22°23'S and 41°15'–41°45'W, and described seven vegetation types: *praial grass* (halophile and psammophile); grass with shrubs (herbaceous swamps); post-beach; *Clusia* (open shrubs of *Clusia*), *Ericaceae* (open shrubs of Ericaceae); **swamp forest** (permanently flooded forest), and *restinga* **forest** (periodically flooded forest). In addition, another type is the open scrub of **Palmae**, sandy forests, and aquatic vegetation. Araújo's 1992 definitions were adopted.

The publication by Ayala Castañares and Phleger (1969) is an important international landmark for its elucidation of the structure and function of coastal lagoons. Kjerfve (1994) edited another important publication on processes in coastal lagoons. In Brazil, there are two volumes that present extensive studies on coastal lagoons: Esteves (1998) edited a fairly comprehensive volume on the functioning of coastal lagoons in the state of Rio de Janeiro (coastal lagoons in Restinga de Jurubatiba National Park and in the municipality of Macaé), and Seeliger *et al.* (1997) edited a volume on the lagoon of Patos in the **subtropical convergence**.

The coastal lagoons of the state of Rio de Janeiro, according to Soffiati (1948), can be classified in three categories: i) **board lagoons**, which are the result of water courses barred by spillovers of the collector rivers and also by stretches of sandbars; ii) floodplain lagoons, partly formed by *restinga*; and iii) **sandbar-plain lagoons**. These lagoons were extensively studied by Esteves (1998) and collaborators, and the body of work on these ecosystems has made an essential contribution to scientific knowledge and management of these tropical ecosystems.

Soffiati (1998) highlights the relationship between these ecosystems and the various different human societies that have used them since early development in this region in the South American continent, especially after the human invasions in the 16th, 17th, and 18th centuries. The author also describes human interventions, which resulted in the drainage of many lakes, reducing their area or drying them completely.

The origin of coastal systems grouped into *tabuleiro* lagoons, floodplain lagoons and *restinga* lagoons was discussed by Soffiati (1998).

Esteves (1998) divided coastal lagoons into two groups according to their origin: i) those formed by processes of sedimentation and erosion of geomorphological origin, isolating old marine bays – which gives rise to brackish lagoons with clear waters (**Maricá, Saquarema and Araruama** are the best known examples in the state of Rio de Janeiro); and ii) lagoons formed from sedimentation at the mouth of rivers draining into the ocean, creating freshwater or slightly brackish coastal lakes. The author also describes another type of mixed-origin lagoon, i.e., one which results from the isolation of bays and blocking of river mouths.

The average annual productivity of lagoons is high (as much as $300\,\text{g}\,\text{C}\cdot\text{m}^{-2}\cdot\text{year}^{-1}$) in contrast to shelf and oceanic waters (maximum $100\text{--}150\,\text{g}\,\text{C}\cdot\text{m}^{-2}\cdot\text{year}^{-1}$), which make them productive systems with high levels of fishery production. The *dark-water coastal lagoons*, according to Esteves (1998), drain rivers that run through sandy stretches of the *restinga* or waters that come from the water table in sandy areas. Fulvic and humic compounds resulting from the partial decomposition of plant accumulate in these soils and are characteristic of these lagoons. These compounds are chemically and biologically important, with some physical effects on light penetration in these environments.

In these lagoons, as in all waters with high levels of dissolved humic and ful-vic material (for example, in the 'marigots' waters of mangrove regions), the species present are highly selective. For example, in the **Comprida lagoon**, of the coastal lagoon region in the state of Rio de Janeiro, there were 7 freshwater species and 1 marine species (Aguiano and Carameschi, 1995 in Esteves, 1998). Coastal lagoons also have low biodiversity of zooplankton (Branco, 1998) and aquatic macrophytes, such as *Typha dominguensis* and *Nymphoides humboldiana*, which occur in isolated individual lagoons, in contrast with the richer flora in lagoons with lower levels of humic substances.

Salinity, according to Esteves *et al.* (1984) and Esteves (1998), is one of the most important environmental factors affecting colonization and biodiversity in coastal lagoons. Salinity also varies greatly during the seasonal cycle due to the influx of rain-water during the summer. Esteves *et al.* (1984) defined four types of coastal lagoons in northeast of Rio de Janeiro:

▸ From fresh water to typical euryhaline (30‰).
▸ From fresh water to oligohaline (0.5‰–5‰).
▸ From oligohaline (0.5–5‰) to mesohaline (5–18‰).
▸ Euryhaline throughout the year (>30‰).

In addition to rainfall, which affects the salinity of coastal lagoons, the salin-ity also varies as a function of the influx of ocean water (during high tide) entering through natural or man-made open channels; or blown by the wind (sea spray).

Man-made openings in the sand bars, as an attempt to manage the system, can result in interactions that produce economic and ecological losses, since they alter the system's salinity, the young communities of marine fish, and the phytoplankton and zooplankton of the coastal lagoons, and cause increased siltation in the lagoon and a drop in tourist potential, according to Esteves (1998).   Such human interven-tion alters the ecological functioning of coastal lagoons, with economic and social consequences.

These coastal lagoons have different volumes, shapes, bottom topography, and prevailing winds, which determine varying patterns of horizontal and vertical circu-lation (Panosso *et al.*, 1998) (see Figure 14.13). For example, the prevailing winds

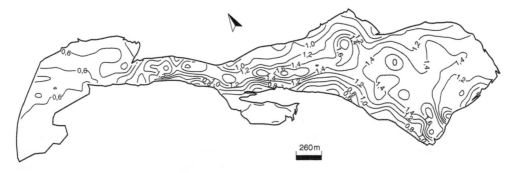

Figure 14.13   Bathymetric map of Imboassica lagoon (Restinga de Jurubatiba National Park, Macaé – RJ).
   Source: Panosso *et al.* (1998).

around the Imboassica lagoon, according to these authors, tend to evolve, transport, disperse and accumulate sediments. Littoral region and wind action are two important forcing functions in these lagoons. The dynamics of the substratum and the accumulation of sediment and its deposition are essential, according to these authors, in the spatial organization of benthic communities, and eutrophication in lagoons depends on their morphometry, volume and available nutrients from the tributaries.

Melo and Suziki (1998) studied the phytoplanktonic community in the Imboassica, Cabiúnas and Comprida lagoons (a lagoon complex in the state of Rio de Janeiro). They concluded that spatial differences arise from gradients of salinity and nutrients, so spatial distribution differed in each lake studied. The Imboassica and Comprida lagoons had the greatest number of taxa of Bacillariophyceae, while Zygnemaphyceae was common in the **Cabiúnas lagoon**. According to the authors, the seasonal dynamics of phytoplankton and the differences between the various ecosystems are caused by the opening of the sandbar, the discharge of **domestic wastewater,** and the biomass of aquatic macrophytes. Spatial and temporal variations in nutrient levels, due in part to human action, contribute to the structuring of phytoplankton in these environments.

Temporal variation in salinity is another factor influencing the seasonal succession of phytoplankton, for example, in the Imboassica lagoon. As salinity decreases (<5‰), there was a predominance of cyanobacteria, chlorophytes and dinoflagellates. With salinity of approximately 20‰, a predominance of diatoms occurred.

Roland (1998) studied phytoplanktonic production using the ^{14}C technique and onsite incubation. The study focused on two lagoons, Imboassica and Cabiúnas. In the first, rates of $4.83\,mg\,C\cdot m^{-3}.day^{-1}$ (fraction $<1\,\mu m$) to $143.0\,mg\,C\cdot m^{-3}\cdot day^{-1}$ (fraction $100\text{–}35\,\mu m$) were found. For Cabiúnas lagoon, the rate varied from $0.93\,mg\,C\cdot m^{-3}\cdot day^{-1}$(fraction$>100\,\mu m$)to$11.23\,mg\,C\cdot m^{-3}\cdot day^{-1}$(fraction$3\text{–}1\,\mu m$ average value and SD ± 1.64) and $11.0\,mg\,C\cdot m^3\cdot day^{-1}$ (average value and SD ± 0.97; fraction $<1\,\mu m$).

Carbon fixation in the dark was high in the Imboassica lagoon and low in the Cabiúnas lagoon, except for the fraction $<1\,\mu m$, suggesting heterotrophic bacterial activity.

According to Roland (1998), the levels of physiological activity in these lagoons are determined by different nutrient levels and the salinity/eutrophication ratios are essential to the primary productivity process of phytoplankton, bacterial activity, and the chemical characteristics of the food available for primary consumers. For comparison, Table 14.8 presents data from Margalef (1969) on coastal lagoons on several continents. Table 14.9 (from Knoppers, 1994) presents data on dissolved inorganic nitrogen and the percentage supplied in the primary production in a series of coastal lagoons.

Silva (1998) studied primary producers, such as aquatic macrophytes, focusing in particular on the growth and production of *Typha dominguensis* in the Imboassica lagoon. The average results of live and dead biomass in three 1-meter-square sites, according to the author, were 1663 and $938\,g\,dry\text{-weight}\cdot m^{-2}$. Lower water levels affect this species, causing mortality of the ramets of *Typha dominguensis*, which presents a maximum net primary production rate of $5.92\,g\,dry\text{-weight}\cdot m^{-2}\cdot day^{-1}$.

Table 14.8   Primary production of coastal lagoons (representative values).

Country	Lagoon	Author	Method	production
Italy	Venice lagoon	Vatova (1961, 1963b)	^{14}C	$79–87\,gC \cdot m^{-2} \cdot year^{-1}$ (1960) $147\,gC \cdot m^{-2} \cdot year^{-1}$ (1959)
Italy	Marano lagoon	Vatova (1965)	^{14}C	$19–28\,gC \cdot m^{-2} \cdot year^{-1}$
Egypt	Hydrodrome	Vollenweider (in Elster, 1960)	^{14}C	$21\,gC \cdot m^{-2} \cdot year^{-1}$
Egypt	Lake Mariut	id.	^{14}C	$340–2,140\,gC \cdot m^{-2} \cdot year^{-1}$
Egypt	Lake Edku	id.	^{14}C	$68\,gC \cdot m^{-2} \cdot year^{-1}$
Mass., USA.	Eel Pond, Woods Hole	Teal (1967)	^{14}C	$80–400\,mgC \cdot m^{-2} \cdot year^{-1}$
Georgia, USA	Sapelo Island	Ragotzkie and Pomeroy (1957)	^{14}C	$180–270\,mgC \cdot m^{-2} \cdot year^{-1}$ $2.18–13.7\,gC \cdot m^{-3} \cdot day^{-1}$ (nanoplankton)
Venezuela	Manglar lagoon in Margarita	Ballester (personal communication to Ramon Margalef)	^{14}C	$306–1,200\,mgC \cdot m^{-2} \cdot day^{-1}$
Mexico	Laguna de Alvarado	Margalef (unpublished)	^{14}C	$5–34\,mgC \cdot m^{-3} \cdot hour^{-1}$ (Dec 1967)
Mexico	Madre lagoon	Copeland and Jones (1965)	O_2	$1.11–2.14\,gC \cdot m^{-2} \cdot day^{-1}$
Texas, USA	Madre lagoon	Odum *et al.* (1963)	O_2	$1.0–15.8\,gC \cdot m^{-2} \cdot day^{-1}$
Texas, USA	Galveston Bay	id.	O_2	$6.4\,gC \cdot m^{-2} \cdot day^{-1}$

Source: Margalef (1969).

The opening of the basin of the Imboassica lagoon causes detritus of *Typha dominguensis* to be carried out of the system and variations in the lagoon's level causes mass deaths of the aerial part of this macrophytic species.

Another community of primary producers studied in this region was periphyton, found by examining submerged leaves of *Typha dominguensis* in the Imboassica lagoon. In this community, variations resulting from the artificial opening of the sand bar were also important in reducing the number of taxa. Silva (1998) pointed to shifts in salinity and nutrient levels as causing the response of the periphytic community to these fluctuations, in these environmental factors.

The most common types of periphytic algae were Bacillariophyta, Chlorophyta and cyanobacteria. Spatial variations that caused changes in the periphytic composition, according Fernandes (1998), were due to the discharge of domestic sewage into the lagoon, the salinity gradient, and the presence of substrata in the lagoon, such as aquatic macrophytes.

Table 14.10 presents comparative data on the net primary production in different aquatic ecosystems, compared with primary production in estuaries.

According to Knoppers (1994), an equilibrium occurs between autotrophic and heterotrophic metabolism in many coastal lagoons. However, in some cases, heterotrophism is predominant. The author classifies coastal lagoons according to the base of primary production there – phytoplankton, aquatic macrophytes, macro-algae, or algae of the microphytobenthos.

Table 14.9 Principal loads of nitrogen, demands of primary production and demands met by a series of coastal lagoons.

Coastal ponds	Principal sources	Load of dissolved inorganic nitrogen (nmol N·m⁻²·year⁻¹)	Demand by primary production (nmol N·m⁻²·year⁻¹)	Demand supplied (%)	References
Harrington Sound (Bermuda)	GW, S, P	136	3.86	4	Bodungen et al. (1982)
Charlestown Pond (USA)	GW, S, P	561	3.12	18	Nowicki and Nixon (1985); Lee and Olsen 91985)
Ninigret Pond (USA)	GW, S, P	340	2.98	11	Nixon and Pilson (1985);Thorne-Miller et al. (1983)
Potter Pond (USA)	GW, S, P	710	3.18	22	Lee and Olsen (1985);Thorne-Miller et al. (1983)
Pamlico Pond (USA)	R, M, P	860	4.41	20	Davies et al.; Nixon and Pilson (1983)
Long Island Pond (USA)	R, S, P	400	2.58	15	Nixon and Pilson (1983); Riley (1959)
Apalichola Bay (USA)	S	560	4.53	12	Nixon and Pilson (1983)
Barataria Bay (USA)	S	570	4.53	12	Day et al. (1978); Nixon and Pilson (1983)
Laguna de Terminos (Mexico)	R, M, P	20	2.87	1	Day et al. (1988); Stevenson et al. (1988)
Guarapina Pond (Brazil)	R, P	313	5.18	6	Moreira and Knoppers et al. (1990)
Urussanga Pond (Brazil)	R <S	26	5.89	<1	Costa-Moreira (1989); Carmouze et al.(1991)
Fora Pond (Brazil)	R, S	156	5.73	3	Carmouze et al. (1991); Knoppers et al. (1991)
Mauguio Lagoon (France)	R, S	291	2.57	11	Vaulot and Frisoni (1986)
Thau Lagoon (France)	R, S	582	2.84	20	Vaulot and Frisoni (1986)
Sem-Dollard (Holland)	R, S, M	414	3.77	11	Barreta and Ruuardij (1988); Cadeé (1980)
Ebrié Lagoon (Ivory Coast)	R, S	410	2.97	14	Dufour and Slephoukha (1981); Dufour (1984)

R – River; S – Sewage; P – Precipitation; M – Marine; GW – Groundwater
Source: Knoppers (1994).

Table 14.10   Net primary production for different aquatic ecosystems and comparisons with the primary production of estuaries.

System	Area ($10^6\,Km^2$)	Net production ($gC \cdot m^{-2} \cdot year^{-1}$)	Total production $10^{12} \cdot kgC \cdot year^{-1}$
Ocean	332	125	41.5
Upwelling	0.4	500	0.2
Continental platform	33	183	4.1
Estuaries	1.4	300	0.4
Coastal lagoons	0.3	300	0.1

Source: Knoppers (1994).

Branco (1998) studied zooplankton in the coastal lagoons of Rio de Janeiro (Imboassica, Cabiúnas and Comprida), and determined the community's composition and structure. The author reached the following conclusions:

▸ Some taxa (such as species of rotifers in the genus *Hexarthra*, *Lecane bulla* and nauplii of planktonic copepods) were found in all three environments, despite the differences in salinity, morphology, and presence or absence of aquatic macrophytes.

▸ Considering these similarities in composition, the author also classified a different set of species for each lagoon. For example, in the Imboassica lagoon, the presence of polychaete larvae, molluscs and gastropods was fairly common. In the Comprida lagoon, *Bosminospsis deitersi* are common, as well as juvenile and adult forms of *D. azureus*, *Lecane leontina hilunaris* and chaoborid and chironomid larvae. While in the Cabiúnas lagoon, along with *B. deitersi* and *D. azureus*, *Brachionus falcatus*, *Keratella lengi*, *Polyarthra dolichoptera*, *Diaphanosoma birgei*, and *Moina minuta* are commonly found.

Some taxa, according to Branco (1998), were common to the three lagoons. The presence of hydromedusae and several coastal marine species is due to the connection between the Imboassica lagoon waters with coastal areas. The openings and closings of the sand bar in the lagoon show how the conditions of salinity and circulation can influence the composition of zooplankton. The connection to coastal waters or the isolation of the lagoon is a key factor in the zooplanktonic species composition in coastal lagoons. Zooplankton thus is another indicator of the conditions of salinity and circulation, and in these coastal lagoons, according to White (1998), it is also an important link in the food chain, consumed by invertebrates and fish.

In studies on the benthic macroinvertebrates in Imboassica, Cabiúnas and Comprida lagoons, Callisto *et al.* (1998) found that polychaetes, bivalves and the gastropod *Heleobia australis* were common in Imboassica lagoon, while In Cabiúnas and Comprida lagoons, aquatic insect larvae were most common, especially chironomids, chaoborids and the tricopteran *Oxythira hyalina*. The density of organisms dropped in Imboassica compared to Cabiúnas and Comprida. In Imboassica, the authors identified the discharge of sewage and the openings of the sandbar as key factors in the distribution of benthic invertebrate communities in the lagoon.

Gonçalves Jr. *et al.* (1998) studied the granulometric composition of the sediment and the communities of benthic macro-nvertebrates in the three lagoons and

Chart 14.4 Constancy of the captured species in Cabiúnas lagoon and Comprimida lagoon in two sample periods.

	Species	1st period	2nd period
Cabiúnas lagoon	Cyphocharax gilbert	Constant	Constant
	Astyanax bimaculatus	Constant	Constant
	Hoplias malabaricus	Constant	Constant
	Geophagus brasiliensis	Constant	Accessory
	Oligosarcus hepsetus	Constant	Constant
	Centropomus parallelus	Constant	Accessory
	Lycengraulis grossidens	Constant	Accessory
	Parauchenipterus striatulus	Constant	Accessory
	Rhamdia sp.	Consultancy	Accidental
	Cichlasoma facetum	Consultancy	Accessory
	Eucinostomus argenteus	Accidental	Accessory
	Anchovia clupeoides	Accidental	Accessory
	Hoplerytrinus unitaeniatus	Accidental	X
	Strongylura timucu	Accidental	X
	Genidens genidens	Accidental	X
	Citharichthys spilopterus	Accidental	X
	Micropogonlas furnieri	X	Accessory
Comprida lagoon	Haplias malabaricus	Constant	Constant
	Geophagus brasiliensis	Constant	Constant
	Hopleythrinus unitaeniatus	Consultancy	Constant
	Centropomus parallelus	Accidental	X
	Cichlasoma facetum	Accidental	Accessory

Source: Reis et al. (1998).

concluded that sediment type significantly affects the structural patterns of the benthic macroinvertebrate communities, especially in the Imboassica lagoon, which has a more heterogeneous granulometric distribution.

Reis *et al.* (1998) studied the ichthyofauna in the coastal lagoons of the state of Rio de Janeiro. Their study showed the presence of some incidental species resulting from sporadic contact with the sea. Chart 14.4 shows the species in the Comprida and Cabiúnas Lagoons as an example of composition and **structure of fish species**.

The man-made opening of the sandbar in the Imboassica lagoon, according to Frota and Caramaschi (1998), is the main reason for the presence of **freshwater species** and marine species (most common). In terms of both number and biomass, marine species are predominant in this lagoon. Examples include *Mugil liza* – Mugilidae – mullet; *Diapterus lineatus – Gerreidae*–caratinga; *Panalichthys brasiliensis – Bothidae*– sole; and *Lycengranlis grossidens* and *Auchvia clupeoides* – Engraulidae – large anchovies. The authors classified estuarine-dependent species from marine origin; **occasional estuarine species** of marine origin, and freshwater species.

Albertoni (1998) studied the occurrence of penaeid and palaeomonid shrimp in the Imboassica, Cabiúnas, Comprida and Carapebus lagoons. Eight species of penaeid and palaeomonid shrimp were identified, including species from both marine and inland waters, distributed according to each lagoon's characteristic salinity and contact with ocean water from the opening of bars. For example, marine shrimp (*Penaeus paulensis* and *P. brasiliensis*) were found in Imboassica lagoon, and *Macrobrachium potiuna* and *M. iheringii* in Cabiúnas lagoon. These are river species common in isolated lagoons

Chart 14.5   Shrimp species in four coastal lagoons of the state of Rio de Janeiro.

	Species	References
Imboassica lagoon	Penaeus (Farfantepenaeus) paulensis	Perez–Faefante (1967)
	Penaeus (Farfantepenaeus) brasiliensis	Latreille (1817)
	Penaeus (Litopenaeus) schimitti	Burkenroad (1936)
	Macrobrachium acanthurus	Wiegmann (1836)
	Macrobrachium olfersii	Wiegmann (1836)
	Palaemon (Palaemon) pandaliformis	Stimpson (1871)
Cabiunas lagoon	Macrobrachium potiuna	Muller (1880)
	Macrobrachium acanthurus	Wiegmann (1836)
	Macrobrachium iheringii	Ortmann (1897)
	Palaemon (Palaemon) pandaliformis	Stimpson (1871)
Comprida lagoon	Macrobrachium potiuna	Muller (1880)
	Palaemon (Palaemon) pandaliformis	Stimpson (1871)
Carapebus Lagoon	Palaemon (Palaemon) pandaliformis	Stimpson (1871)
	Macrobrachium acanthurus	Wiegmann (1836)
	Penaeus (Farfantepenaeus) brasiliensis	Latreille (1817)
	Penaeus (Litopenaeus) chimittii	Burkenroad (1936)

Source: Albertoni (1998).

not connected with coastal waters. *Macrobrachium acanthurus* is a species that inhabits fresh and brackish waters. Marine shrimps and their distribution and biology were also studied in coastal lagoons by Albertoni (1998) (see Chart 14.5).

The studies on these coastal ecosystems, summarized by Esteves (1998), show basic features of coastal lagoons, which include: fluctuations in level and volume of water from rainfall and coastal marine waters when there is contact and interaction with water from the sea; highly selective and dynamic ecosystems with morphometric, physical, and chemical features that affect the biodiversity, community structure, seasonal and spatial succession, and the primary and secondary productivity. Many fluctuations and alterations are the result of human actions.

In Esteves' overview (1998) of the coastal ecosystems in the state of Rio de Janeiro (which can serve as a comparative example for other similar systems in Brazil), the author cites the importance of these ecosystems for conservation and maintaining a unique and rich biodiversity of species; their importance as a freshwater reserve; and the services provided by these ecosystems: high-quality leisure areas, **flood control**, receptor of treated industrial waste, value of real estate in surrounding area, scenic beauty, and tourism development in the region.

14.11.1   Man-made impacts

These coastal lagoons have undergone a series of eutrophication processes that can be considered as examples for other ecosystems and coastal lagoons in Brazil: discharge of domestic and industrial wastewater, landfill on the shores, sedimentation from the drainage basin, removal of sediment and calcareous deposits, deforestation in the area around the lagoons, introduction of exotic species of fish, and construction in the surrounding area (Esteves, 1998).

The author discussed the range of consequences of these impacts, especially for the Imboassica lagoon, proposals to mitigate these effects, including the process of **artificial eutrophication** resulting from the dumping of untreated domestic sewage, and the use of indicator organisms to identify changes in levels of nutrients and dissolved oxygen, as well as the possible presence of cyanobacteria. Deterioration of sanitary conditions in Imboassica lagoon resulting from eutrophication was presented as an example of a potential effect in other lakes in the region.

14.11.2   Mitigating measures in Imboassica lagoon

The mitigating measures recommended for Imboassica lagoon by Esteves (1998) and Lages Ferreira (1998a, 1998b) include:

▸ Control of the artificial openings into the lagoon; such contact with coastal waters significantly alters ecological conditions and spatial and temporal succession, as shown by several authors. Changes were observed at all trophic levels, from primary producers to fish.
▸ Control the release of domestic wastewater and treated sewage into lagoons. In this case, the use of aquatic macrophytes to help remove nutrients can be effective, as proposed by Lages Ferreira (1998) for these ecosystems.
▸ Control of the spill channel of the lagoon, especially during the rainy season.
▸ Restoration of ecological stability in the lagoon, with regulation of the openings of the basin to contact with coastal waters.

14.12   PATOS LAGOON

Ecological studies in the Patos Lagoon (Rio Grande do Sul) were presented in a volume edited by Seeliger, Odebrecht and Castello (1997). Included are works on climate, geology, geomorphology, productivity, biodiversity and nutrient cycles in the Patos Lagoon, as well as detailed information on fish, fishing and ecosystem management. This section provides an overview of these studies.

According to Seeliger and Odebrecht (1997), the South Atlantic (between South America and Africa) is subject to high-pressure centres from the Atlantic anticyclone, which controls the climate and large-scale hydrodynamics of ocean circulation. The subtropical convergence on the coast of South America extends from 32 to 40°S.

The basin of the Patos Lagoon receives the influx from a 201,626 km² drainage basin. The subtropical convergence affects the coastal waters and inland aquatic systems of southeast Brazil, and the influence of cold fronts and its relation to the **Atlantic anticyclone** establishes the principal climatic and hydrodynamic forces, which have repercussions on the biochemical cycles and the structure of the biological processes (Tundisi *et al.*, 2004). According to Seeliger and Odebrecht (1997), coastal areas and coastal and inland waters are interdependent. The regional climate in the area of the

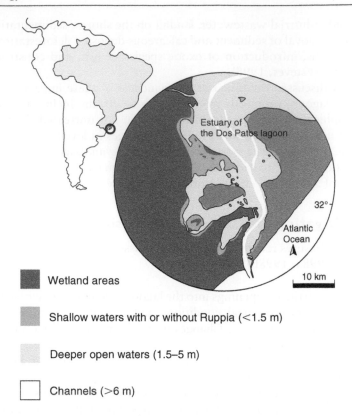

Figure 14.14   Geographic location and principal habitats of the estuary of the Dos Patos lagoon.
Source: Modified from Seeliger (2001).

Patos Lagoon depends on the number and intensity of cold fronts, and the annual precipitation is the result of the frequency of cold fronts (1200–1500 mm). Precipitation and evaporation result in a hydric surplus of 200–300 mm annually.

With a surface area of 10,227 km^2, Patos Lagoon can be divided into five biological units (Asmus, 1997) (see Figure 14.14), of which the **Guaíba River** stands as the largest freshwater tributary. The coastal areas of the Patos Lagoon are dominated by wetland areas with freshwater vegetation and sandy beaches. Ten percent of the Patos Lagoon is estuarine area, which empties into the Atlantic Ocean through a channel. Villwock (1978) and Paim *et al.* (1987) described the geomorphology and geology of the region, whose main characteristics are Tertiary and Quaternary deposits in the estuary and a complex of multiple barriers with aeolian deposits, stable and active fields of dunes and lagoon terraces (Calliari, 1997).

14.12.1   Estuarine hydrography and circulation

The discharge of fresh water is a feature of the estuary and lagoon area of the Patos Lagoon (85% of the Guaíba and Camagua Rivers and the São Gonçalo channel), with wide seasonal variations (41–25,000 m$^3 \cdot$ s^{-1}, for example, in the Guaíba River).

Circulation in the estuary and the Patos Lagoon has as a main force the wind regime, which, during winter, presents average speeds of $5.7-8.2\,\mathrm{m\cdot s^{-1}}$ with prevailing southwesterly winds. Winds control circulation, distribution of salinity and the water level (Garcia, 1997).

According to Niencheski and Baumgarten (1997), variations in the physical and chemical parameters are due to the wind regime, the characteristics of the sediments and human activity. According to Niencheski et al. (1986), the estuary of the Patos Lagoon is chemically unstable due to the variability of the sediment-water interfaces, with specific regions of physical-chemical interactions of water and sediment.

The main sources of suspended material are the rivers from the northern and central regions flowing into the Patos Lagoon and they depend on the rainfall patterns in the region around Patos Lagoon and **Mirim Lagoon**. Suspended material in the lagoon ranges from $30-50\,\mathrm{mg\cdot L^{-1}}$. The environment is permanently saturated with oxygen.

14.12.2 Biochemical nutrients and cycles

Nutrient cycles are dependent on sediment-water interactions, the pressure of forces (wind, for example), the contributions from the drainage basin during the rainy season, salinity, and the re-suspension of elements and substances from the sediment. The effects from various different human activities, including mining, indicate concentrations of zinc, lead, lithium, chromium, manganese, copper, cadmium, arsenic and silver (Niencheski et al., 1994).

14.12.3 Primary producers

According to Costa (1997), vertical and horizontal salinity gradients and the formation of intertidal areas (islands and banks) promote the establishment of plants that are typical of estuarine wetlands and areas with brackish waters. This flora is transitional between tropical and subtropical and temperate vegetation (for example, *Paspalum vaginatum*: tropical, and *Limonium brasiliensis*: cold temperate). *Spartina alterniflora*, *Spartina densiflora* and *Scirpus americanus* occur in some floodplains.

Other secondary species are *Typha dominguensis* and *Acrostium aerum*. *Salicornia grandichaudiana* also occurs in areas with development of *Spartina densiflora*, but the former is dominant in permanently flooded wetland areas and areas with wide fluctuations in salinity. Submerged higher vegetation, such as *Ruppia maritima* (Seeliger, 1997), is common. *Ceratophyllum demersum* occurs in periods of low salinity (Moreno, 1994).

Ruppia maritima depends on light intensity, temperature and salinity for growth and development, which peaks in summer. The underwater light regime drops and limits this species' growth. Average production is $25\,\mathrm{g}$ dry weight $\cdot\,\mathrm{m^{-2}}$ but during the summer peak of development, it can surpass $120\,\mathrm{g}$ dry-weight $\cdot\,\mathrm{m^2}$. Changes in hydrodynamic factors and penetration of underwater solar radiation affect the development and annual production of this plant, which according to Moreno (1994) can reach 5200 metric tons.

There are 94 species of micro-phytobenthos, including 40 species of cyanobacteria, 26 species of chlorophytes, 3 phaeophytes, 1 xanthophyte and 24 red algae

(Coutinho, 1982). Species distribution patterns depend on salinity. The dominance of cyanobacteria in the micro-phytobenthos is possibly due to environmental fluctuations in the estuary and in the Patos Lagoon itself. The substratum is also affected by the differential distribution of microphytobenthic flora.

The total production of benthic flora is affected by the saline conditions and abiotic environment in general. Seasonal variations in the growth pattern of microflora and **macroflora** occur, and seasonal succession depends on variations in environmental conditions during the year. The annual productivity cycle varies in relation to climate, salinity, circulation and available nutrients.

Odebrecht and Abreu (1997) found the most common phytoplankton were <20 μm, predominantly cyanobacteria and dinoflagellates, between periods of low and high salinity. Diatoms, cyanobacteria and dinoflagellates exhibit distinct seasonal patterns due to variations in nutrients, salinity, and the underwater radiation regime. Primary production of phytoplankton varies from $25 \, mg \, C \cdot m^{-3} \cdot h^{-1}$ (minimum) to $160–350 \, mg \, C \cdot m^{-3} \cdot h^{-1}$ during winter and summer, respectively. Approximately 70% of the primary production occurs in the <20-μm fraction.

Epiphytic micro-algae are common, using as substratum the submerged or emergent aquatic higher plants. Figure 14.15 shows the production of each group of primary producers in Patos Lagoon. According to Seeliger, Costa and Abreu (1997), the primary producers in this ecosystem include emergent and submerged macrophytes;

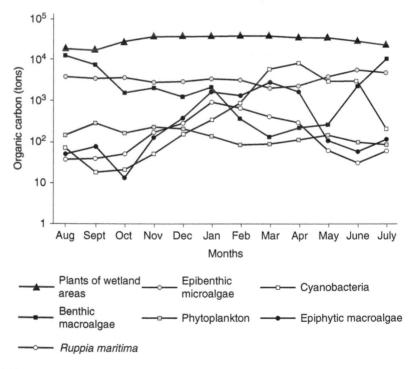

Figure 14.15 The total amounts of carbon that contribute monthly to the different primary productions in the estuary of the Dos Patos lagoon.
Source: Modified from Seeliger (2001).

benthic and floating macro-algae; cyanobacteria, and planktonic, epibenthic and epiphytic micro-algae.

According to Odebrecht and Abreu (1997), bacteria affect the production and flow of carbon in the food chain, and Abreu (1992) showed that the biomass of phytoplankton, heterotrophic and bacterial flagellates and the biomass of ciliates are closely related.

14.12.4 Zooplankton, zoobenthos, vertebrates

Proto-zooplanktonic organisms are represented in the Patos Lagoon by several groups of heterotrophic flagellates (2–3 μm), dinoflagellates and ciliates, such as loricates (tintinnids) and *loricated oligotrichs* (Strombiidae). Proto-zooplankton plays an important role in the food web, as predator and food supply.

Estuarine zooplankton includes marine species of freshwater planktonic and pleustonic species and their spatial and seasonal distribution are strongly influenced by the distribution and **variation in salinity** and hydrodynamics of the water masses. With the entrance of seawater, an influx of marine species into the estuary occurs (such as *Acartia tonsa*, *Oncaea conifera*, Cirripedia larvae, and echinoderms).

During periods of heavy rainfall and discharge of fresh water, freshwater species such as *Notodiaptomus incompositus* and *Mesocyclops annulatus* are common, as well as pleustonic cladocerans. Periods of mixing of fresh water and marine water present marine species (*Paracalanus parvus* and *Euterpina acutifrons*) and freshwater species (*Moina micrura*). Water temperature and variations in salinity affect the seasonal distribution pattern of zooplankton (Monte *et al.*, 1997).

Invertebrates typical of estuarine conditions dominate the benthic fauna (Bemvenuti, 1997). There are 15 species of **estuarine invertebrates** and only three limnic species. **Epifaunal organisms** occur in the marginal areas of brackish waters filled with vegetation. For example, the gastropod *Heleobia australis* and euryhaline decapods such as *Callinectes sapidus* inhabit marginal bays during summer, their breeding season. *Penaeus paulensis* is the most important commercial decapod in the estuary.

Wetland areas with vegetation are full of insects, and amphipods and isopods also occur. The distribution of benthic organisms is related to the type of substratum, the presence or absence of vegetation, the benthic biomass (12,927 individuals $\cdot$ m^{-2} for *Heleobia* sp., for example, or $281\,\mathrm{g}\cdot\mathrm{m}^{-2}$ for *Enodona* sp.) occurs mainly in marginal areas associated with the development of submerged or emergent aquatic higher plants.

Processes of **recolonization of benthic fauna** in the estuary are influenced by the nature of the substratum and spatial and temporal variations in abiotic factors and the production that affects the epifauna of isopods, amphipods and tanaids (Nelson, 1979). Fish are important predators of the epifauna and the infauna of the estuary.

The ichthyofauna of the Patos Lagoon includes at least 110 species of fish (Vieira and Castello, 1997). However, few are abundant or occur frequently. **Resident species** in the estuary are represented by the genera *Bleniidae*, *Gobiidae* and *Poecilidae*. Marine species are represented by *Mugil plantanus* and corvina *(Micropogonias furnieri)*. Larvae and postlarvae of these species use the estuary. Some marine species penetrate the estuary occasionally under favourable conditions (*Umbriva canosai*, *Pepritus paru*). Anadromous species, such as *Netuna barba* and *Netuna* *planifrons*, spend most of their life cycle in the ocean but migrate to the limnic areas of the Patos Lagoon for breeding. The immature stages of these species use the estuary for feeding and development. Occasional freshwater visitors (cichlids and characins) and some tropical marine species occur in the estuary.

Most species that make up the fish community of the Patos Lagoon are of marine origin. Vieira and Musick (1994) divided estuarine fauna into pelagic associations from the bottom and from shallow waters. **Epibenthic** fish are found near the bottom, and are important as commercial species in the region.

According to Vieira and Musick (1994), the spatial and temporal distribution of fish in the Patos Lagoon is controlled by environmental factors and competition for food. According to the authors, production is not a major factor in this region.

Bird fauna is abundant and diversified (Vooren, 1997), with six species of piscivorous birds (for example, *Phalacocorax olivaceus* - cormorant). Uncovered wetlands in the innermost regions of the Patos Lagoon are inhabited by birds that feed on benthic organisms.

Several species of marine mammals also visit the Patos Lagoon. Examples include *Tursiopsis truncatus* (dolphin) (Pinedo *et al.*, 1992) and *Otainha flauencens* (sea lion).

The food web in the Patos Lagoon is supported by a diverse range of primary producers (see Figure 14.16) that provide various feeding alternatives for herbivores (grazing), carnivores and detritivores. Organic detritus plays an important role in these areas, particularly given the diversity of primary producers and the range of decomposing

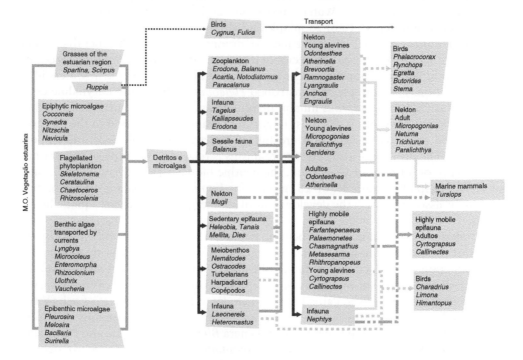

Figure 14.16   Conceptual diagram of the trophic components of biotic components in the Dos Patos lagoon.
Source: Modified from Seeliger (2001).

matter from phytoplankton, microphytobenthos, epiphytic algae and submerged and emerging macrophytes. In this estuary, these food webs can also be diversified in the function of different zones of production and decomposition of organic matter.

The abundant vegetation of aquatic macrophytes provides numerous alternatives for feeding, based on different-sized organic particles at different stages of decomposition. Polychaetes, molluscs, amphipods, crabs, shrimp and young fish live in regions rich in organic matter. For example, the gastropod *Heleobia australis* feeds on the dense populations of bacteria that develop in the leaves of *Ruppia maritima*.

14.12.5 Impacts

The Patos Lagoon is subjected to multiple impacts from multiple human activities in the regional drainage basins that contribute to the lagoon and the estuaries. A 13% reduction in discharge of fresh water (Costa and Seeliger, 1997) occurs during periods of drought, because of dams and use of water for irrigation. The reduction in the flow of fresh water can have an increasingly important quantitative impact, similar to eutrophication from untreated sewage waste and the addition of nutrients from agricultural activities.

Blooms of *Microcystis aeruginosa* develop in the limnetic region of the Patos Lagoon and are transported to the estuary (Odebrecht *et al.*, 1987; Yunes *et al.*, 1994). Metals, pesticides and hydrocarbons can potentially cause degradation of and impact on the water and biota. Other impacts include sedimentation, drainage, destruction of wetland areas, cutting of wetland vegetation, deforestation and erosion.

According to Seeliger and Costa (1997), the Patos Lagoon is subject to global changes including a possible rise in sea level and salinization of the upper portions of the estuary. Marine plankton and benthic communities may be subject to effects of increased ultraviolet radiation.

14.12.6 Management and prognosis

According to Asmus and Tagliani (1997), the region of Patos Lagoon has ecological, economic and social importance, because of the biodiversity and the **potential for rational development** (fishing, tourism, agriculture, industry and navigation – port facility).

Thematic maps of the coastal plains allow an integrated vision with natural and man-made components, as 33 environmental units have been identified (Asmus *et al.*, 1991). An matrix of environmental functions described by Asmus *et al.* (1989, 1991) enabled the setting up of different units of preservation, conservation and development. This process applied to the Patos Lagoon is an extremely important example of the use of scientific information and development of mathematical and ecological models that can be utilized in management.

14.13 THE PLATA ESTUARY (ARGENTINA/URUGUAY)

This estuary extends over a wide coastal area in the South Atlantic (35–36°S latitude). The Paraná and Uruguay Rivers flow into the estuary, which measures 280 km from its headwaters to its outflow in the South Atlantic Ocean, and 230 km across.

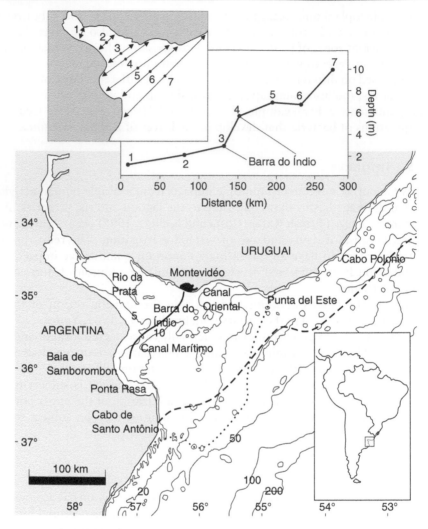

Figure 14.17 Estuary of the River Plata.
Source: Modified from Mianzan et al. (2001).

The basin that discharges into this estuary is about 14,000 km². The **mixoha-line region** is nearly 38,000 km², based on the average isohaline position of 30‰ (Mianzan et al., 2001).

The salty/brackish/freshwater boundaries play a significant role in the reproduction of fish species and the zooplanktonic biomass. The discharge of fresh water, with an annual average of 22,000 m⁻³·L⁻¹, is significant in the dynamic of the estuary, causing a saline-water intrusion with vertical stratification and a reverse halocline about 5 meters deep. According to Guerrero et al. (1997), the dynamics of the estuary are controlled by waves from the tide under forcing actions such as wind and the effect of continental drainage modified by topography and the Coriolis effect. These forces act across extremely large estuaries such as that of the Plata basin.

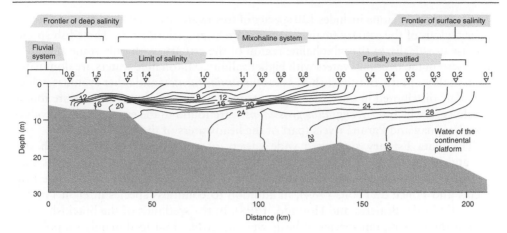

Figure 14.18   Salinity gradient along the main axis of the estuary of the River Plata under typically stratified conditions.
Source: Modified from Mianzan *et al.* (2001).

Figures 14.17 and 14.18 show the Plata Estuary as described by Mianzan *et al.* (2001) and a salinated section along a distance of approximately 200 km.

The primary producers in the estuary of the Plata basin present a diversity of components that depend on the salinity gradient and type of substratum. In regions with lower salinity (0.2–5.0‰), diatoms of the *Aulacoseira* genus are common in the phytoplanktonic community, as well as the cyanobacterium *Microcystis aeruginosa* (especially in more polluted rivers) (Gomes and Bauer, 1998). Chlorophyll-*a* values in oligotrophic environments are generally below $4 \, \text{mg} \cdot \text{m}^{-3}$ and in mesotrophic regions reach $4–10 \, \text{mg} \cdot \text{m}^{-3}$. Benthic algae such as *Ulva lactuca, Enteromorpha* sp. and *Chondria* sp. are common in the littoral zone. Coastal wetlands are dominated by *Spartina alterniflora, Salicornia ambigua, Juncus acutus* and *Scirpus maritimus* (Scarabino *et al.*, 1975).

The benthos in the low-salinity brackish-water region is dominated by *Helobia piscium, Corbicula fluminea, Limnoperna fortunei* and *Chilina fluminea*, which are characteristic of habitats with soft bottoms. The last three species, according to Darrigan (1993), were introduced. Oligochaetes, nematodes, leeches, harpacticoid copepods and chironomids are common and associated with sediments rich in organic matter (Rodrigues *et al.*, 1997).

The soft sediments and the organic substrata in the brackish areas of the estuary are characterized by abundant populations of lamellibranchs, gastropods (*Turbanilla uruguayensis*), detritus-eating and carnivorous crabs, which can reach high densities (*Misabelleana* – lamellibranchs – for example, can reach $1500–2700 \, \text{ind} \cdot \text{m}^{-2}$) (Mianzan *et al.*, 2002). The supralittoral and littoral regions are inhabited by lichens, cyanobacteria of micro-phytobenthos, crustaceans, molluscs, Cirripedia and polychaetes. Crabs, such as *Uca uruguayensis*, inhabit consolidated substrata in mixohaline areas and feed on detritus among the roots of *Spartina* sp. and on small invertebrates (Ringuelet, 1938; Botto and Irigoyen, 1979).

The nektonic fauna includes 120 species of freshwater fish that are distributed in the upper regions of the estuary (mainly Cypriniformes and siluriforms). These fish species decrease drastically in the mixohaline region of the estuary, with only some species of *Pimelodus* in the fronting waters with higher salinity. Mixohaline waters are dominated by euryhaline species of fish, such as *Micropogomias furnieri* and other cyanides, which are widely distributed in the estuary. In areas with higher salinity, anchovies are found, the adults of which – *Lycengraulis grossidens* – are found in the estuary and reproduce in the Uruguay and Paraná rivers (part of the headwaters of the Plata Estuary).

The Plata Estuary sustains a wide variety of marine and freshwater fish species, and is therefore an important natural resource for the region. Mammals such as *Pontoponia blainvilla* and the fur seal *Otani florecsent* are also found there (Vaz Pereira and Ponce de Leon, 1984), in addition to countless species in eight families of aquatic birds (Bonetto and Hurtado, 1998). In the wetlands of the brackish upper parts of the estuary, ten species of birds were described that feed mainly on polychaetes in the benthos and bivalve shellfish. The crab *Uca uruguayensis* is also an important food source for some species of seabirds (Iribarne and Martinez, 1999).

Planktonic species found in freshwater systems, coastal mixohaline regions and marine systems, including the high-salinity area at the mouth of the Plata Estuary, present representatives typical in each region with different salinities and circulation systems. According to Mianzan *et al.* (2002), one of the estuary's most important features is the presence of horizontal and vertical interfaces, areas where intensive ecological processes occur. The **density discontinuities** occur for 200 km in the estuary and some species can cross them, for example, the anchovy *Lycengraulis grossidens*.

The Plata Estuary is an enormous classical estuary with distinct ecological gradients, which enables the distribution of many species of benthic and nektonic plankton, horizontally and vertically distributed depending on the movements of the water masses, tolerance to salinity, and established density gradients. For example, many planktonic species, such as the copepod *Acartia tonsa* or the ctenophore *Mnemiopsis mccradyi*, gather in the saltwater/freshwater fronts, supporting food chains based on these planktonic species.

14.13.1 Human impacts in the Plata Estuary

The Plata Estuary sustains extensive commercial use for fishing, large-scale navigation, and recreation. The cities of Buenos Aires and Montevideo together have 13 million inhabitants, and their activities impact the estuary. Fishing is a $30-million-per-year business in the estuary. It is the drainage area of the *La Plata* basin (see Chapter 16) and thus receives water contaminated by the agricultural, domestic and industrial discharges of approximately 120 million people.

As an international estuary, its management can only succeed through the interaction of international interdisciplinary teams that minimize the impacts and consequences of human activities upstream from the estuary, promoting integrated management actions and encouraging authorities in several countries – Argentina, Bolivia, Brazil, Uruguay and Paraguay – to work to address impacts and protect estuary resources. For example, treatment of sewage throughout the La Plata basin and in the estuary is necessary and will certainly produce positive results. Protection and regulation of fishing and control of navigation are other key measures.

14.14   IMPORTANCE OF ESTUARIES AND COASTAL LAGOONS

The four examples presented here – the Cananéia lagoon area, the coastal lagoons in the state of Rio de Janeiro, the Patos Lagoon, and the Plata Estuary – illustrate several important mechanisms in the functioning of these ecosystems on the east coast of South America, in Brazil. These comparisons verify that:

▸ All systems have interfaces between inland systems and coastal marine areas, influenced by both fresh water and coastal waters.
▸ Spatial variability, seasonal cycle and distribution of organisms are influenced by the cycles of salinity, water temperature and the hydrodynamics of the ecosystem. Climatic forces such as wind and precipitation play a role in these regions, leading to alterations in the seasonal cycles of primary producers, biogeochemical cycles, and distribution of organisms.
▸ Organisms present special tolerance-to-salinity mechanisms and use various different strategies for reproduction and dispersal in the estuary.
▸ Detritus-based food webs are often common in these ecosystems.
▸ Coastal vegetation – mangroves, vegetation typical of brackish wetlands, vegetation of submerged macrophytes – plays a key role in the biogeochemical cycles in these ecosystems, in the feeding niches of benthic organisms, and in the regeneration of nutrients in detritus-based food chains.
▸ These estuarine regions of marginal lagoons and large estuarine complexes play an important role in maintaining aquatic biodiversity. They are highly productive biological ecosystems, nurseries for aquatic organisms. They are a significant regional economic factor, because they allow fishery exploitation and aquaculture (fish, molluscs, crustaceans) and are relatively easy to access. Their sustainability is crucial for the coastal regions of Brazil and for many countries in the tropics and subtropics.
▸ Estuaries, coastal lagoons and lagoon areas in Brazil and in many continents are subject to a range of human impacts: the impacts of navigation and fishing, the dumping of domestic sewage, aquaculture; extensive exploitation of fish stocks, introduction of exotic species, **pollution from heavy metals,** and eutrophication, along with the loss of mangrove vegetation and removal of coastal vegetation.
▸ The coastal waters of Brazil are affected by the estuaries and coastal lagoons, especially by the contribution of nutrients and low-salinity waters, and contamination. Much of the primary and secondary productivity in coastal waters depends on fertilization from the estuaries.
▸ The management of these ecosystems is complex and calls for some structural and non-structural actions: **management of inland drainage basins** whose tributaries empty into the estuaries; control of the use and occupation of the land in the inland basins and in the basins of the tributaries feeding into the estuary; control of the multiple uses of the estuary: fishing, navigation, recreation, aquaculture, occupation and operation of dredging stations, sewage treatment in the adjacent municipalities. It is crucial to have improved education and community participation by municipalities in the management of the estuary and water resources. A sound scientific base with comprehensive data on estuaries

and coastal lagoons is critical for the promotion of conservation measures, restoration and management, as described in works already cited: Seeliger, Odebrecht and Castello (1997); Esteves (1998); Seeliger (2001); and Tundisi and Matsumura Tundisi (2001).

14.15 EUTROPHICATION AND OTHER IMPACTS IN ESTUARIES

Estuaries receive the influx from many rivers and streams that contribute to the hydric network of the inland water basins that drain into the system. When domestic sewage or industrial wastes are allowed to flow into the rivers or streams, the pollution reaches the estuary. In some estuaries, installation of industrial factories or thermo-electric power plants to generate energy accelerates eutrophication. The effects of eutrophication can be minimized by dilution with coastal waters, the effect of the tides. But this depends on the mechanisms of circulation and on the physiography of the system.

Many estuaries are used as ports, and the constant presence of ships can cause eutrophication and contamination. Also, many different estuaries have aquaculture projects, which can lead to an increase in nitrogen and phosphorus (from rations used and excretion from organisms, especially from fish farming). Because of the **accessibility of estuaries**, various different types of industries have built installations nearby, due to the facility of transport.

In estuaries, due to the constant movement of water masses from tides, deeper waters tends to be eutrophic, and as a result, the entire estuary becomes eutrophic, and not only the less salty surface waters in positive estuaries. Thus, the deepest waters can fertilize coastal waters and expand the geographic distribution of eutrophication. In general, the impacts on the continents and the estuary itself of human activities in the estuary include:

▸ eutrophication by untreated domestic sewage and runoff and discharge of industrial and agricultural activities;
▸ pollution and contamination produced by industries on the coast and in estuaries;
▸ eutrophication due to aquaculture (fish, molluscs and shrimps);
▸ pollution from ships and large-scale shipping activities;
▸ thermal pollution (in some estuaries) resulting from power plants and nuclear plants;
▸ penetration and colonization by invasive species;
▸ radioactive pollution (in some estuaries);
▸ alterations on the coast and filling of the estuarine region to build marinas, gas stations or industries;
▸ destruction of mangrove vegetation in estuaries causing increased sedimentation;
▸ increase in suspended material transported by actions on the coast, in the estuary or in the inland drainage basins that empty into the estuary, from deforestation or construction of marinas, buildings or condominiums.

Anaerobic sediments in estuaries, producing anoxia or hypoxia in the bottom, occur when there is an influx of high levels of organic matter from domestic sewage, agricultural activities (fertilizers) or food processing industries. In such cases, a gradient of benthic fauna occurs with the complete absence of macrofauna in the **anoxic regions** of the estuarine sediment.

A classic example of a successful invasive species is the colonization of the Southampton (England) estuary by *Venus mercenaria*, a lamellibranch mollusc native to Florida (USA), which proliferated in Southampton after World War II (Raymont, 1963).

14.16 MANAGEMENT OF ESTUARIES AND COASTAL LAGOONS

Management plans and estuarine coastal waters are essential for preserving the mechanisms of primary production and effective commercialization and protection of native species, fundamental to productivity and estuarine food chains. Especially in tropical regions along the coast of Brazil, estuaries are also important in the fertilization of coastal waters and in maintaining an inventory of commercially important species for the production of seafood (molluscs, fish, crustaceans and algae).

Due to their importance as an ecosystem interface (ecotone) between coastal regions and inland water basins, some estuaries are economically crucial and their management and conservation are a strategic priority, especially in Brazil (Tundisi and Matsumura Tundisi, 2001).

The accessibility of estuaries – for recreation, production and commercialization of food, and the installation of ports and industries – makes them vulnerable to impacts from such activities, and integrated management is therefore essential. Management alone, however, is not enough. The control and management of inland drainage basins contributing to the estuary are extremely important, since there is a continuum between ecosystems and estuaries that needs to be preserved and managed (Vannucci, 1969).

Major threats to the biological and ecological integrity of estuaries

The main impacts produced by human activities on estuaries, according to Kemmish (2004), include:
- loss of habitat and alteration of structures;
- eutrophication: degradation of water quality, growth of toxic algal blooms, increased turbidity, increased mortality of benthic organisms, hypoxia and anoxia of estuarine waters;
- Over-fishing;
- **chemical pollution:**
  - Oils and grease;
  - metals;
  - synthetic organic compounds;
  - radioactive substances and elements:
- alterations in the water cycle and in inland drainage basins;
- introduction of exotic species;
- changes in sea level;
- modifications of the littoral zone: **loss of wetland areas**, changes in the discharge of fresh water and transport of sediments.
- Solid waste: degradation of habitat due to the accumulation of solid waste (plastic bottles and bags, cans) and other types of waste.

Exploitation of estuaries by man

The accessibility of estuaries allows for ample exploitation, particularly for the production and extraction of food. As a result of high levels of primary production and food chains with several alternatives, estuaries are used for intensive fishing in the estuarine region or in the adjacent coastal regions that are fertilized by the estuarine waters. Many species of fish such as mullet (*Mugil cephalus*) breed in the coastal waters, and the young migrate to the estuaries, where there is plenty of food and diverse feeding niches.

Many species of brachyuran decapods such as *Callinectes* spp. and *Ucides cordatus* feed on detritus in the mangrove areas and are used commercially. Lamellibranch molluscs such as *Anomalocardia brasiliensis* (cockles), *Crassostrea rhizophorae* (oysters) or *Mytella falcata* (mussels) are also exploited commercially. The exploitation of mullet (*Mugil brasiliensis*) is an important fishing industry in estuaries in Brazil. Recently, estuaries have begun to be used in the intensive aquaculture of fish, shrimp and other crustaceans.

Veredas in the drainage basin of the São
Francisco River (MG)
Photo: J.G. Tundisi.

15 Wetlands, temporary waters and saline lakes

SUMMARY

This chapter looks at the distribution and functioning mechanisms of wetlands, temporary lakes, and saline lakes.

Wetland areas are found on all continents and coastal regions. They are ecotones that have important quantitative effects on hydrological cycles and biodiversity. They function as important regulatory systems in biogeochemical cycles.

A classification system for wetland areas is presented and their role in the cycling of carbon, nitrogen and phosphorus is discussed. Organisms' adaptations to these areas are presented and methods to evaluate the impacts that occur in them are discussed.

Wetland areas are used by humans as sources of renewable natural resources (fishing, harvesting products, and aquaculture).

Temporary waters occur on all continents, in the form of lagoons, ponds, and intermittent rivers. The flora and fauna in these waters present special characteristics and adaptations for tolerance to drought and flash floods.

The production of resistant eggs that can tolerate long-term droughts is one of the important characteristics of organisms in temporary waters.

Saline lakes are found in arid and semi-arid regions, in endorheic areas. The chemical composition of these lakes and their fauna and flora and trophic relationships are presented.

15.1 WETLANDS

15.1.1 Definitions and classification

Wetlands are a very common type of aquatic system located along ecotones between aquatic and terrestrial systems in the interior of continents. These include numerous epicontinental types and in coastal regions they cover about 6% of the land surface. These wetland areas or swampy areas are found on all continents, in arid and semi-arid regions, in temperate and tropical latitudes, and even along altitudinal gradients. In many coastal regions in the tropics, wetland areas are surrounded by mangrove vegetation. These ecosystems have been intensively used for agriculture and fishing, exploitation of peatlands and extraction of timber and tannins (mangrove).

The definitions and classification of wetland areas or swampy areas are difficult and imprecise. There are many regional terms that characterize types and subtypes. In this book, the term wetlands or wetland areas (or floodplain) will refer to the same ecosystem that is either permanently flooded in shallow areas or undergoes floods (periodic or not) with fluctuations in level.

Wetland areas fill an intermediate position between terrestrial and aquatic ecosystems. These areas constitute a continuum of different types of communities, making it difficult to set defined boundaries. There are many definitions of wetlands. The International Union for the Conservation of Nature and Natural Resources (IUCN) defined wetlands are areas where the soil is saturated with water or submerged, natural or man-made, permanent or temporary, where water can be static or flowing, saline, brackish or fresh. Water-dominated areas include swamps, coastal marshes, bays, ponds, lagoons, lakes, rivers and reservoirs. Where marine and coastal waters are included, depths to 15 meters define the wetlands (Gopal *et al.*, 1992).

The International Biological Programme already defined wetlands (Westlake *et al.*, 1988) as areas dominated by herbaceous macrophytes where productivity occurs in the aerial environment above the water level, while the plants survive the excess water that would be detrimental to many higher plants with aerial roots (Gopal *et al.*, 1992).

Figure 15.1 shows the classification of wetland areas, according to the programme of the Scientific Committee on Problems of the Environment (SCOPE), of the International Council of Scientific Unions (ICSU) (Patten *et al.*, 1992).

Cowadin *et al.* (1979) stressed that there is no completely correct ecologically based definition, first because of the diversity of wetlands, and then because the dividing line between dry areas and flooded areas is difficult.

In addition to these permanently flooded areas, with or without fluctuations in level (Table 15.1), it is necessary to distinguish between other equally important aquatic systems in continents: temporary areas and saline lakes, which are also included in the study of limnology and have enormous theoretical and practical importance, as will be seen below.

Despite the difficulty of defining swampy areas, some common features stand out:

▸   the presence of water and special types of soils differs from those in the nearby higher and drier areas.

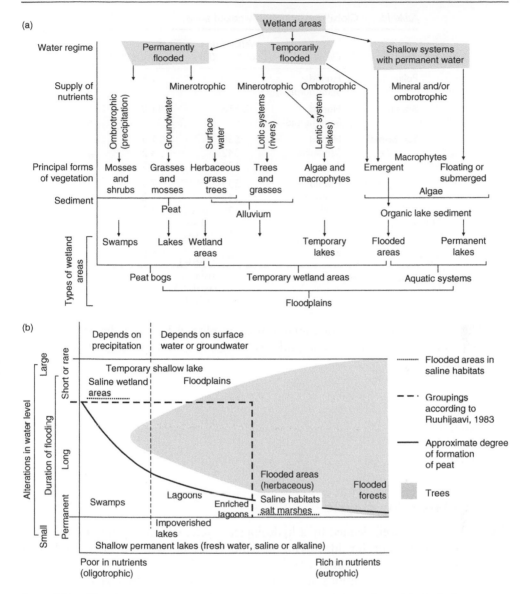

Figure 15.1   Classification of wetland area according to Scope, along a gradient of the hydrologic regime and nutrients: a) Types of wetland areas; b) Classification based on the water level.
Source: Patten *et al.* (1992)

▶   These are intermediate systems between terrestrial and aquatic ecosystems that support vegetation (**hydrophytes**) at least temporarily adapted to permanently flooded conditions or to periodic fluctuations in level. These areas are generally shallow.

▶   The variation in the fluctuation of the water level in wetland areas is quite large, which makes definitions difficult.

Table 15.1   Global distribution of wetland areas.

Zone	Climate	Wetland area (km² · 1,000)	Total percentage (%)
Polar	Humid	200	2.5
	Semi-humid		
Boreal	Humid	2,558	11.0
	Semi-humid		
Sub-boreal	Humid	539	7.3
	Semi-arid	342	4.2
	Arid	136	1.9
Sub-tropical	Humid	539	7.3
	Semi-arid	342	4.2
	Arid	136	4.5
Tropical	Humid	2,317	8.7
	Semi-arid	221	1.4
	Arid	100	0.8

Source: Mitsch and Gosselink (1989).

The classification system of wetland areas published by Cowadin *et al.* (1979) includes 'systems with similar biological, hydrological, geomorphological and chemical characteristics':

▶ Marine
▶ Estuary
▶ Riverine
▶ Lacustrine
▶ Marsh

15.1.2   Hydrologic cycle

Wetland areas are well defined by a hydrological cycle, which is probably the most important determinant for identifying specific types of wetland areas and their processes (Mitsch and Gosselink, 1986).

The hydrological conditions determine the changes in the physical and chemical conditions of the water, such as pH, nutrient availability, and presence or absence of dissolved oxygen. The nutrient balance caused by the influx and outflow of water, the intensity of the flow of matter, and the energy cycle are determined by the hydrological cycle. Alterations in the hydrology produce rapid changes in the species diversity and biomass (see Figure 15.2). This, in turn, produces a certain control over the conditions, accumulating sediments, altering the direction of flow, and by accumulation producing bogs. Transpiration by the plants in these regions can also be affected by the hydrologic cycle.

The hydrologic cycle defines the hydroperiod or **hydropulse**, which represents the seasonal pattern of the water level. The cycle and water level are particular for each wetland area and are influenced by the physiographic characteristics of the area.

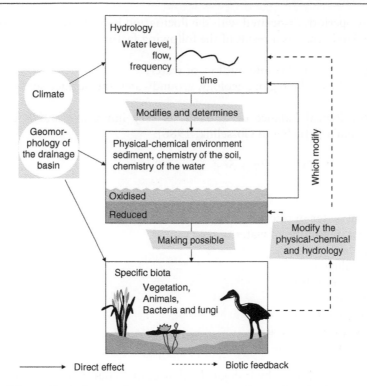

Figure 15.2   **Effects of hydrology on the physio-chemistry and biota of wetland areas.**
Source: Mitsch and Gosselink (1986).

The duration and frequency of flooding should be determined. Cowadin *et al.* (1979) defined the hydroperiod of wetland areas as follows:

Wetland areas with tide

▸ permanently flooded with tidal waters;
▸ irregularly exposed to tidal variations, during <1-day periods;
▸ regularly flooded and exposed (at least daily);
▸ irregularly flooded.

Wetlands without tide

▸ permanently flooded;
▸ intermittently exposed – with rare flooding during dry periods);
▸ seasonally flooded;
▸ semi-permanent flooding during some periods of the year;
▸ saturated – substratum saturated for long periods, but without surface water;
▸ temporarily flooded – flooded for short periods;
▸ intermittently flooded – varying and irregular periods of flooding without a characteristic seasonal pattern.

The hydroperiod, associated with the fluctuation in level, varies considerably for different wetland areas as a result of the following factors:

▶   balance between influx and outflow of water;
▶   Physiography of the region, geology, groundwater and surface soil.

The hydrological balance of wetlands is very important and is given by the following formula (Mitsh and Gosselink, 1986):

$$DV = Pn + Si + Gi - ET - So - Go \pm T$$

where:
V – Volume of water in reserve
DV – Changes in volume of water in reserve
Pn – Net rainfall
Si – Surface influx
Gi – Groundwater influx
ET – Evapotranspiration
So – Outflow by surface
Go – Outflow by groundwater
T – Influx (+) or outflow (–) by tide.

Calculating the annual hydrological balances in wetland areas is very important, in order to determine the frequency of events and, by association, the nutrient balances and export and import of materials.

As already stated, the hydrological cycle directly or indirectly affects the functioning mechanisms of the biotic components, so it is an extremely important forcing function in the system.

15.1.3   Biogeochemical cycles

Biogeochemical cycles include transformation and transport processes between the wetland areas and their surrounding ecosystems.

The extent and rates of biogeochemical cycles depend on the accumulation of biomass in the wetland areas and on the type of vegetation. With predominantly easy-to-decompose vegetation and accumulation of aquatic macrophytes, the cycle is accelerated.

Some flooded forests, such as temperate-zone cypress forests, contribute very little biological material, as the principal exchange is of gases. Flooded areas, such as forests in the Amazon, contribute large amounts of organic matter into the water. Soils can be mineral or organic; Chart 15.1 presents a comparison between the two types. **Organic soils** receive, of course, contributions from the vegetation.

In general, flooding causes reduced oxygenation in the soil, resulting in anaerobic conditions. The rate of oxygen loss from the soil and subsequent reduced levels depend on temperature, availability of organic substrates and the demand for oxygen molecules, which depends on the reductors present (Patrick *et al.*, 1972, 1972, 1976). The absence of dissolved oxygen affects nutrient availability and alters biogeochemical

Chart 15.1   Comparison of mineral soils and organic soils in wetland areas.

	Mineral soils	Organic soils
Organic content (%)	<20–35	>20–35
pH	Almost neutral	Acid
Density	High	Low
Porosity	Low (45–55%)	High (80%)
Hydraulic conductivity	High	Low and high
Capacity to retain water	Low	High
Availability of nutrients	Generally high	Generally low
Ion exchange capacity	Low, dominated by principal cations	High, dominated by hydrogen
Representative wetland area	Cilliary plants and swampy regions	Temperate bog areas

Source: Mitsch and Gosselink (1986).

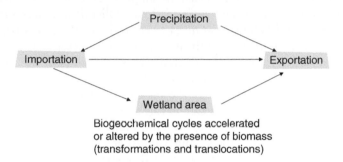

Figure 15.3   Import and export of material, effects of precipitation and regulatory role of wetland areas.

cycles, and as a consequence plants have a number of specific adaptations to this type of system. Generally, a thin layer of oxygen is present on the soil, resulting from photosynthesis by algae, the effects of the wind and the oxygen exchange though the air-water interface. Changes in redox potential in this vertical profile of only a few inches of soil are important to the cycle of certain elements (such as manganese, iron and sulphur).

The transport of nutrients in wetland areas is related to the hydrobiological cycles and includes *influxes from surface water, atmosphere, rainfall* and *tides* (in the case of tidal wetlands); *outflows* from the atmosphere, surface water and ground water and *losses* from fixation in the bottom sediment. Transformation and translocation are determined by the biomass present.

Wetland areas receive material from adjacent ecosystems and export it by means of various processes, as shown in Figure 15.3.

Recycling of nutrients

The retention of sediments, suspended material and the remains of dissolved organic material is characteristic of wetland areas and plays an important role in biogeochemical cycles. The mechanisms that help retain nitrogen in wetland areas include:

▶   sedimentation
▶   fixation by plants
▶   denitrification.

In wetland areas with long retention time, sedimentation is especially significant (Jansson *et al.*, 1994). In addition to the sedimentation of organic material with high levels of particulate nitrogen and phosphorus, macrophytes and epiphytes assimilate nitrogen. In addition to this re-mineralization mechanism, other processes can occur in wetlands, making more inorganic nitrogen available, which is reassimilated or carried downstream. The most important nitrogen-loss mechanism in wetlands is the bacterial process of denitrification in which nitrate (NO_3^-) and nitrite (NO_2^-) are converted via nitrous oxide (N_2O) into atmospheric nitrogen (N_2). Atmospheric nitrogen can be fixed by some plants and bacteria through the process of biological fixation. However, in terms of energy, N_2 fixation is expensive and occurs only when the supply of ammonia and nitrate is low.

Denitrification is an important nitrogen-loss mechanism in wetlands, as shown by Whitaker (1993) and Whitaker *et al.* (1995), in the UHE Carlos Beltho reservoir (Lobo/Broa). Approximately 30% of the nitrogen that arrives is lost into the air by the denitrification process. In a study conducted on six reservoirs of the Middle Tietê in the state of São Paulo, Abe *et al.* (2003) showed that denitrification is a quantitatively important process in these reservoirs. The bacterial community in free water or in wetland areas plays an instrumental role in this process (Abe and Kato, 2000).

Other studies on denitrification in wetland areas in Brazil include Enrich-Prost and Esteves (1998), Abe *et al.* (2002), and Tundisi *et al.* (2006).

In a study conducted on a floodplain of the **Ribeirão do Feijão** (state of São Paulo), Sidagis-Galli (2003) analysed the physical and chemical characteristics of water to determine the nitrification and denitrification rates in sediments. The nitrification rate varied between 0.145 to 0.068 μmol $N-NO_3 \cdot g^{-1} \cdot day^{-1}$ and the predominant metabolic route, according to this author, was heterotrophic, in which the bacteria used ammonium as a substrate. Denitrification rates in this wetland area were on average of 0.0082 μmol $N-NO_2 \cdot g^{-1} \cdot day^{-1}$.

There is a considerable reduction of nitrogenated compounds, mainly ammonia, which shows the important quantitative role floodplains provide as a filter and purification system for surface waters feeding into the river. These results confirm the studies by various authors who calculated nitrification and denitrification rates associated with bacterial composition (Feresin and Santos, 2000; Gianotti and Santos, 2000) in wetland areas of Brazil. Figure 15.4 shows the steps in the biogeochemical cycle of carbon in wetland areas and the processes involved in the cycle, i.e., sedimentation, respiration, leaching, deposition, absorption, **methanogenesis**, oxidation, flocculation, fixation and suspension (resuspension) of sediments.

Wetland areas as retention systems for nitrogen, phosphorus, heavy metals and organic material

The use of natural or man-made wetland areas for the retention of nitrogen, phosphorus, and heavy metals has been reported by many authors (Novitski, 1978; Mitsch

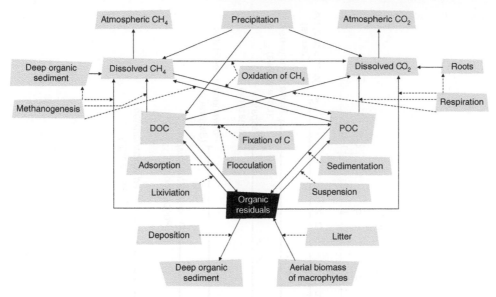

DOC – Dissolved organic carbon   POC – Particulate organic carbon

Figure 15.4   Biogeochemical cycle of carbon in the aquatic subsystem of the wetland area of the Okefenokee, Florida (USA).
Source: Modified by Patten (1988).

and Gosselink, 1986; Weisner *et al.*, 1994; Leonardson, 1994; Hendricks and White, 2000; Hill, 1996; and Whitaker and Matvienko, 1998). Abe *et al.* (2006) analyzed the retention potential for nitrogen in a wetland area of the Parelheiros region of São Paulo and confirmed the importance of the conservation of this wetland area for the initial treatment of water in the Billings reservoir of the Guarapiranga dam. Studies and projects on the use of man-made wetlands in Brazil as a purification technique for large quantities of water were introduced by Salati (personal communication), Manfrinato (1989) and Salati *et al.* (2006).

In several regions of Brazil, especially in the state of São Paulo and the metropolitan region of São Paulo, wetlands have been used extensively as initial treatment systems. Tundisi (1977) showed that all the wetlands in the tributaries of the UHE Carlos Botelho reservoir (Lobo/Broa) play an important role in maintaining the mesotrophic-oligotrophic conditions in the reservoir.

Retention of nutrients, elements and dissolved material is now used worldwide for the treatment of industrial waste and for the primary or initial treatment of domestic waste or for recycling nutrients from agricultural sources (fertilizers or sewage from sheep farms, for example). However, the wetland's capacity to recycle and retain nutrients is not infinite. Wetland soils retain toxic substances, including metals, and these can be absorbed by aquatic plants and mobilized through the food web by the consumption of birds, fish and invertebrates.

The conservation of natural wetlands is an important measure to control nutrients and recycle pollutants and heavy metals. This capacity to recycle and the extensive biodiversity of wetland areas led to the organization and study of a set of values for

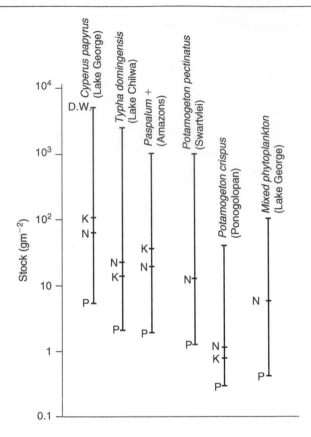

Figure 15.5   Stocks (per unit of area) of potassium, nitrogen, and phosphorus in vegetation in wetland areas in tropical and subtropical regions.
Source: Talling and Lemoalle (1998).

wetlands. These values relate to the 'services' provided by the ecosystems. In addition to these services, wetlands have another set of very important functions: regulation of the hydrological cycle and the ability to control flooding, given their retention capacity, with large surface area, reducing the downstream flow (Howard-Williams, 1983).

The **control of eutrophication** in Lake Biwa, Japan (Nakamura and Nakajima, 2002) and in Lake Balaton, Hungary (Istvanovics, 1999) was achieved with the intensive use of aquatic macrophytes in wetland areas.

Figure 15.5 shows the stocks (per unit area) of potassium, nitrogen and phosphorus in tropical and subtropical wetland areas with floating or submerged macrophytes as compared with phytoplankton in a shallow tropical lake (Lake George, in Africa).

Most of the organic matter produced in wetlands undergoes decomposition, and different stages of the decomposing material are consumed by animals. However, a considerable portion of organic matter (sometimes more than 30%) is transported from adjacent areas. It is difficult to distinguish between allochthonous and autochthonous material decomposing in wetlands, unless isotope tracers (^{13}C) are used, which can reveal the origin of the decomposing matter (Gopal, 1992) and its transport route.

Through the use of ^{13}C, Martinelli *et al.* (1994) studied the dynamics of carbon in the Amazon region, comparing the entire complex of vegetation that develops in channels, lakes and flooded forests; all these subsystems consist of **herbaceous vegetation,** large forests, submerged vegetation, periphyton, phytoplankton and emergent macrophytes. The authors' conclusions show that the main sources of carbon are atmospheric CO_2 and fluvial CO_2. Decomposition transfers carbon to the sediments. The amount of carbon transferred to the various reserves results from hydrological changes produced by the seasonal cycle and the flow and movement of water masses. The great spatial variability observed in the composition and dispersion of carbon reserves is, according to these authors, a complex factor in the collection and interpretation of results.

The frequency and intensity of the decomposition are directly related to the hydrocycle, its magnitude and the velocity of the current. Fungi and bacteria play an important role in these areas (Clymo, 1983).

In summary, in terms of the biogeochemical cycles and recycling of nutrients in wetlands (Richardson, 1992), the following aspects can be summarized:

▸ Wetland areas act as effective transformers of nitrogen, phosphorus and carbon.
▸ Wetland areas release significant levels of N_2 to the atmosphere through processes of denitrification.
▸ Phosphorus is absorbed by plant roots, precipitated in the form of iron or aluminium and fixed by bacteria, fungi and algae.
▸ Carbon is reduced and oxidised and its transport in the hydrological cycle is critical.
▸ Nutrient retention in wetland areas varies with the season and the duration and intensity of the hydrocycle.
▸ Wetland areas can act as a sink or source of elements, depending on the type of wetland, the season, and the duration of the hydrocycle.
▸ Wetlands are not efficient sinks for potassium or sodium and retain less phosphorus than riparian forests.

15.1.4   Major biological adaptations

The organisms that thrive in wetland areas present several special adaptations (see Chart 15.2), due to the many stressful effects of these ecosystems:

▸ Fluctuations in level involve periods of drought and loss of water.
▸ There are periods of intense anoxia, in many cases associated with high temperatures (surface temperatures of water in the tropics can reach 35°C).
▸ In wetland areas near the coast or in inland regions with intense evaporation, salinity levels vary greatly.

Therefore there are adaptations to anoxia, low O_2 levels, variations in salinity and drought. In addition to these adaptations, floating and submerged plants and trees in the wetlands area provide structures for periphyton, aquatic invertebrates and fish fingerlings as protection against predators. Also, the extensive network of roots and leaves and decaying material acts as a filter, retaining dissolved and particulate organic matter. Gopal (1992) published an extensive review on the main adaptations

Chart 15.2   Characteristics and adaptations of communities in wetland areas.

Adaptations to high salinity	
	1. High intracellular osmotic concentration, produced by accumulation of salts (NaCl) or by organic compounds (such as glycerol, for example).
	2. Accumulation of potassium and extrusion of sodium.
Adaptations to anoxia	
Plants	1. Structural mechanisms of roots with aerenchyma, allowing aeration of the root from the aerial portions of the plant.
	2. Anaerobic respiration and production of ethanol.
	3. Elevated enzymatic activity of catalyzing enzymes for reproduction of ethanol.
	4. Production of adventitious roots (in *Avicenna*).
Animals	5. Modified or altered regions for the specific function of exchange of gas (animals); gills in fish and crustaceans.
	6. Intense vascularization and efficient circulatory system (animals).
	7. Modifications of respiratory pigments and reduced locomotive activity.
	8. Physiological adaptations, including alterations in metabolism.

of organisms to the regime of wetland areas and extensively analysed the adaptations to these ecosystems.

Adaptations in reproduction and feeding have been verified in many organisms in wetland areas. For example, reproduction can be related to periods of flooding and drought. Benthic organisms produce a large number of offspring to facilitate distribution. In flooded areas with tidal cycles, some molluscs require a salinity shock to release gametes.

There is also a wide variety of feeding habits that translate morphologically into special appendices for feeding on microscopic particles (greater development of cilia and darts). Absorption of amino acids and other dissolved organic substances has been demonstrated by Vomberg (1987).

Wetland areas are still inhabited by many organisms such as reptiles, birds and mammals, which extensively use the aquatic environment and swampy areas for feeding. These organisms provide a transport system for nutrients and decomposing vegetation and can play a role in the dispersal of aquatic plants and animals (see Chapter 6).

Plants in wetland areas present another important metabolic alteration, which is the fact that some are C_4 plants, which is the result of incorporating CO_2 into the plant as oxaloacetate acid, instead of phosphoglycerate acid, common in C_3 plants, because they use low levels of atmospheric CO_2 and present low photorespiration rates (Gopal, 1992).

15.1.5   Primary production and species diversity

The data on primary production in wetland areas has been extensively studied by Mitsch and Gosselink (1986), who demonstrated the sequence of results presented in Chart 15.3.

Chart 15.3   Primary production in wetland areas.

Swampy areas with running waters	>	Swampy areas with little flow of water	>	Swampy areas with standing water, without flow

Decreasing primary production

These results are most likely to apply to forested swampy areas. In regions with small lakes and standing water, extensive cyanobacterial blooms and large amounts of macrophytes are often noted. Tundisi *et al.* (unpublished) found high chlorophyll levels (up to $200\,\mathrm{mg}\cdot\mathrm{m}^{-3}$) in lakes of the Mato Grosso Swamp.

Species diversity is also high in these regions, due in some cases to a mosaic of swamps, flooded forests, small streams, lakes and gallery forests. The diversity includes fish, amphibians, reptiles, birds and mammals, which are found in sheltered wetland areas with appropriate food and conditions for breeding.

The food webs in wetland areas are complex and diverse. The reasons for this high diversity are related to the great diversity of niches and high productivity of organic matter. Invertebrates in wetland areas include a large number of insects that in the adult phase breathe oxygen through air bubbles, although in their larval phases they have gills. Lunged molluscs are common in some tropical wetlands. Air-breathing fish with adaptations of highly vascularized swim bladders, such as Arapaima, are common in many wetlands of the Amazon region.

In all wetland areas, periphytic, epiphytic and epipelic communities are important contributors to primary productivity. The primary productivity of phytoplankton in wetlands can vary from 2 to $10\,\mathrm{g\,C}\cdot\mathrm{m}^{-2}\cdot\mathrm{day}^{-1}$ (Westlake, 1980).

Studies on periphyton in wetland areas show that this community can contribute up to 30% of the total production in beds of macrophytes (Wetzel, 1965).

The relative contributions of each component in communities of primary producers in wetland areas (phytoplankton, periphyton, epiphytic and epipelic algae, emergent and submerged macrophytes, and hydrophytes) vary in relation to the hydrocycle, fluctuating levels, and the availability of nutrients (see Table 15.2).

Simões Filho *et al.* (2000), in studies conducted on the Mogi Guaçu River (**Lake Inversão**, Jataí), confirmed that the duration of the flooding pulse appeared to be significantly more important than its intensity in terms of particulate matter, which implies the recycling of nutrients and the primary productivity of submerged macrophytes, phytoplankton and periphyton.

Few studies have been conducted on the total flow of energy in wetland areas, which is still an important challenge, particularly in tropical regions.

Direct herbivory by animals is generally considered to play a relatively insignificant role in the functioning of wetlands (Gopal, 1992). The vegetation is often made up of non-palatable plants. According to Odum (1957) and Teal (1957, 1962), the detritus food chain plays an important quantitative role in wetlands. However, in some wetland areas with floating macrophytes and abundant periphyton, herbivores constitute an important factor in the transformation and use of organic matter produced by primary producers.

Birds are an extremely important component in wetland areas and can consume macrophytes directly. Smith (1982), for example, listed 50 different bird species

Table 15.2 Annual productivity of several aquatic plants from wetland areas compared with phytoplankton. Values in free dry weight of ash (grams of organic material per m^2 per year).

	Average	Gradient	Maximum
Phytoplankton of inland waters	–	1–3,000	–
Submerged plants			
Temperate regions	650	–	1,300
Tropical regions	–	–	1,700
Floating plants			
Salvinia spp.	150		1,500
Aguapé		4,000–6,000	
Papiro		6,000–9,000	15,000
Plants with roots			
Typha (taboa)	2,700		3,700
Phragmites	2,100		3,000
Wetland areas with flooded vegetation			
Cypress		692–4,000	
Various plant species	1,600	695–4,000	
Humid tropical forest	2,250		
Boreal forest	900		
Savannah	790		
Herbaceous vegetation in temperate regions	560		
Oceanic phytoplankton	140		

Sources: Modified from Teal (1980), Westlake (1982) and Moss (1988).

(*Dendrocygna*, Anser, *ANA*, Branta and others) that feed on *Paspalum, Polygnum, Nymphaea, Typha, Scirpus* and *Naias*.

Molluscs and various other invertebrates feed on aquatic macrophytes; aquatic insects (collectors, predators, scrapers or partitioners) feed on a wide variety of algae, macrophytes, detritus, periphyton, and small animals (fishes, especially fingerlings and eggs). The dynamics of the food chain in wetland areas are complex and diverse, and many studies have shown the important role of detritus in these ecosystems (Gopal, 1992).

15.1.6 Assessment of impacts

A series of impacts from various sources and with differing intensities affect wetlands. Evaluation is complex, and includes:

▸ Short-term and long-term effects.
▸ Propagation of effects due to the connectivity of systems.
▸ Regional processes related to impacts: regulation of flow, regulation of biogeochemical cycles and loss of biodiversity.
▸ The impacts of excessive toxicity on wetland areas.
▸ Loss of the hydrocycle and fluctuations in level.

15.1.7   Ecological modelling

Ecological modelling of wetland areas is critical for managing these ecosystems and implementing conservation programmes. **Hydrological models** and biogeo-chemical-cycle and management models have been developed, which are described in books edited by Mitsch and Gosselink (1986), Mitsch *et al.* (1988) and Patten *et al.* (1992). Basically, these models take into account the diversity of different types of wetlands, the hydro-chemical complexity of these transitional ecosystems, the transient processes, the interfaces (for example, the sediment-water interface), and the processes of sedimentation, resuspension, denitrification and leaching. They also take into consideration the capacity for exchange of substances and elements with adjacent systems, such as terrestrial areas, rivers, other wetland areas, floodplains and riparian forests. Ecological modelling has been applied to coastal wetlands, estuaries, mangrove swamps, forested wetlands, shallow lakes and reservoirs.

Many processes were applied to organize an extremely useful set of forecast systems using ecological modelling in unmanaged wetland areas:

▸   hydroperiod;
▸   water quality;
▸   efficiency of drainage;
▸   efficiency of flood control;
▸   morphology;
▸   density of vegetation;
▸   density of invertebrates;
▸   surface area;
▸   type of substratum;
▸   trophic conditions;
▸   fluctuation in level;
▸   depth;
▸   drainage network feeding the wetland area.

Sound management of wetland areas is important to enable their recuperation and conservation, and to optimize their use. For example, Mitsch and Gosselink (1986) pointed to the following positive aspects of managing these areas for multiple uses:

▸   enables the maintenance of water quality;
▸   enables reduction in erosion;
▸   protects and regulates in the case of flooding;
▸   provides a natural buffering system for processing air pollution.
▸   provides a suitable buffering system between urban and industrial areas;
▸   provides breeding areas for fish and shrimp species;
▸   maintains a varied reserve of wetland plants, due to the great diversity of special-ized plants;
▸   controls insect populations;
▸   maintains examples of ecosystems with complete natural communities.

15.1.8   Valuation

The value of the flood plains of the high Paraná River was calculated by Rosa Carvalho (2004) using values such as 'cost of trip' for recreational uses, which had a value of US$234 million.

The value of the wetland areas should include the following functions:

- ecological function;
- flood control;
- water quality control;
- biodiversity;
- productivity;
- wildlife;
- cultural values;
- recharge of aquifers;
- dissipation of erosive forces;
- reproductive and feeding habitats and niches for invertebrates, fish and mammals;
- recreational opportunities;
- aesthetic value.

15.1.9   Other studies in Brazil

Two recent Brazilian books authoritatively address the problems of wetland areas, consolidating a set of interrelationships (Santos and Pires, 2000; Henry, 2003). An important contribution was also made by Wetzel et al. (1994).

The volume organized by Henry (2003) (with contributions from many authors) presents an important set of assessments and conclusions, descriptions of the functioning of various different types of wetland areas and their characteristics in south and southeastern Brazil. It makes an important methodological and conceptual contribution to the understanding of the problem, with original information on tropical and subtropical wetland areas.

In his book, 'Ecotones in the Interfaces of Aquatic Ecosystems' (2003), Henry emphasizes that these transitional areas have structural and functional attributes such as morphometry, position in relation to rivers, reservoirs, horizontal and vertical gradients. He discusses **littoral ecotones** that recycle, conserve and export nutrients and act as a buffering system; riparian ecotones (water adjacent to lotic environments), and ecotones in humid flooded areas and in the water-sediment transition. In characterizing wetlands as ecotones and transitional zones, Henry (2003) stresses the importance of these ecosystems in maintaining biodiversity and biological diversity, also shown by Bini et al. (2001) and Neiff et al. (2001) in the floodplains of the Paraná River. Neiff (2003) also highlights the spatial heterogeneity of wetland areas and the pulsatile nature of these environments, as a result of variations in the hydrological cycle. The flood pulse is characterized as an important driving force in the richness of the macro-invertebrate community, which increased after the flood pulse in a lagoon associated with the Sinos river (Rio Grande do Sul) (Stenert et al., 2003).

Zooplankton communities and limnological characteristics of the Paraná River floodplain were studied by Sendacz and Monteiro (2003). The authors concluded that

---

Wetlands areas in urban regions

In many metropolitan and urban areas of Brazil, there are wetlands with herbaceous vegetation or riparian forests. These wetlands are associated with large or small urban rivers (as is the case of the Tietê River in the São Paulo metropolitan region, where wetland areas are common) and play a key role in nutrient recycling, particularly in reducing diffuse inputs of loads of nitrogen, phosphorus and heavy metals. For example, Abe *et al.* (2006) showed the important role played by the floodplain of the Parelheiros region in metropolitan São Paulo in reducing point loads from the Taquacetuba branch of the Billings Reservoir due to the process of denitrification (nitrogen cycle) and fixation of phosphorus on the roots of herbaceous vegetation in this flooded area. The preservation of these regions in the urban areas is thus essential for the maintenance of cycles and conservation of water quality. Estimates of the economic value of these areas in urban regions, given the "services" they provide, should contribute to the decision to protect, maintain and expand them (Tundisi, 2005a).

---

zooplanktonic diversity is high, compared with other hydrological systems, especially in terms of calanoid copepods.

15.1.10   Use of wetland areas by humans

In tropical regions, many wetland areas have been used by local populations as sources of rationally exploited renewable resources. In pre-Columbia America, indigenous populations used wetland areas for fishing, limited agriculture of flood-tolerant plants, gathering products (wild rice, medicinal plants) and hunting.

15.2   TEMPORARY WATERS

Water can be used for temporary purposes, domestic use, seasonal agriculture, and watering livestock in semi-arid regions. For example, in Brazil's Northeast, after the flood of the São Francisco River in the rainy season, many temporary saline lakes form. These lakes are used as sources of water and salt for cattle.

In many regions, lagoons or pools of various different sizes occur, which present high fluctuations in level, with periods of total desiccation. These areas are generally shallow (1–2 m), occupy depressions in arid or semi-arid regions, and are greatly affected by the surrounding terrestrial system, especially in terms of chemical composition (salinity), turbidity and retention time of the water (Alonso, 1985). They undergo large fluctuations in water level, which in part determine the variations in salinity and turbidity.

Alonso (1985), who studied Spanish lagoons, described some basic characteristics that enable calculation of the *periodicity*, *mineral content*, and *turbidity* of small inland waters in Spain. The description of these factors and their interactions are a good example of the type of approach used to measure different variables in the study of limnology (see Figure 15.6). Table 15.3 shows the percentage of the principal ions,

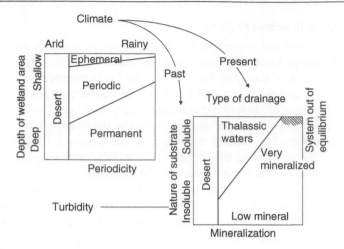

Figure 15.6   Relationship between climate, nature of substrate and depth in ephemeral periodic lakes and permanent lakes.

Table 15.3   Percentages of the main cations and anions in fresh waters, mineralized lakes and hypersaline lakes (sum of ions above $100\,\mu eq \cdot L^{-1}$).

	Ca^{2+}	Mg^{2+}	$Na^+ + K^+$	HCO_3^-	SO_4^{2-}	Cl^-
Freshwater lakes	38.9	18.2	42	53.5	23.8	22.3
Reservoirs in Spain	38.2	27	35	59.8	22.2	14.3
World average (rivers)	63.5	17.4	19	73.9	16	10.1
Central Europe	68.2	25.4	6.4	85.2	10.8	3.9
Mineralized lakes	36.5	20.5	42.9	18.3	43.2	38.6
Brackish/saline lakes	4.5	46	49	1.2	52	46

Source: Alonso (1985).

particularly in mineralized pools, saline lagoons and freshwater lagoons (Alonso, 1985), and can be useful as a comparative reference for other regions.

The greater the evaporation, the greater the precipitation of salts, and the proportion of different ions varies according to the initial geochemical composition of the hydrographic basin and the waters. Therefore, sequential precipitations occur in these lakes.

The divalent/monovalent cation ratio, which is around 2.5 (Margalef, 1975), can vary widely in these temporary waters. Turbidity can also vary enormously, depending on factors such as wind, suspended matter, inorganic solids, and clay particles.

The fauna and flora of these small bodies of water present very interesting features and important adaptations. These waters are of great interest in terms of the processes of geographical distribution, the **flora/fauna composition,** and succession. The principal adaptations of the fauna and flora in these lagoons have to do with the life cycle (coordinated with periods of drought and flood) and the production of resistance mechanisms that ensure germination and reproduction.

For the lagoons of Spain, Alonso (1985) reported the presence of Diaptomidae, Euphyllopods and cladocerans. These organisms produce extremely resistant eggs, capable of long-term survival through periods of drought, followed by a short life cycle. Most of these organisms can feed on detritus and have varying degrees of tolerance to salinity. Some produce latent forms that remain buried in sediment during periods of drought. These stages can, in the case of copepods, be related to copepodites III, IV, V.

Another important characteristic of communities in temporary lagoons is the tolerance to fluctuations in water level, salinity, and dissolved oxygen. The study of communities in temporary waters also has an important practical aspect: in many cases, organisms that live in these waters can be cultivated by hydrating the sediment. One of the most common organisms in wetland areas is *Streptocephalus* (Anostraca), which has extremely resistant eggs.

Temporary waters in Brazil's Northeast, for example, or in coastal regions such as Maranhão shallow lakes = Lençóis Maranhenses, are important from the point of view of ecology as well as the use of the flora and fauna.

From the evolutionary standpoint, these ecosystems present communities that have adapted extremely well in terms of physiology and reproduction. The adaptations are related to the fluctuating conditions of the temporary waters, from the water phase to drought. From an ecological standpoint, the mechanisms of dispersal, colonization and propagation of the fauna and flora in temporary waters are also innovative and diversified, given the fact that these ecosystems vary greatly in their physical and chemical conditions and morphometry. In terms of application, it is important to note that all aquatic fauna and flora in areas with intermittent rivers and temporary waters have well-established mechanisms to handle drought, and can be cultivated and thus result in large biomasses. This is true of several species of Anostraca, which can be hatched from drought-resistant eggs, generating a large biomass used as food for fish and crustaceans.

Food webs in temporary waters can be simple or complex, depending on the stage of flood, volume, salinity and drought conditions. Turbidity can limit phytoplankton growth. Anostraca take advantage of suspended clay particles, feeding on the fungi and bacteria adhered to the particles. Among carnivores, cyclopoids are common, as are insect larvae (Coleoptera and Odonata).

The persistence of the water is a major factor in the level of biodiversity and community composition in temporary waters. Continued persistence enables continual colonization, as well as expansion of ecological niches and food webs (Alonso, 1985).

The capacity to adapt to temporary waters occurs in three ways:

▶   abandonment of the environment under adverse conditions;
▶   production of resistant forms;
▶   production of drought-resistant eggs.

Odonata and Coleoptera abandon the environment under adverse conditions; some species remain in the sediment whenever a certain level of moisture occurs (decapods, amphipods, isopods).

Cladocera and Anostraca (*Streptocephalus, Temnocephalus, Dendrocephalus*) produce drought-resistant eggs. These eggs can survive for several years of drought, and then hatch immediately upon conditions of hydration. In some species, hatching also depends on water salinity.

The **biota of temporary waters** are adapted to unstable conditions. This explains, according to Bayly (1967), why Anostraca do not survive in marine waters. The biota develop rapidly after hydration and quickly colonize the rehydrated environments.

Cladocerans, copepods of the genus *Diaptomus*, and cyclopoids *(Eucyclops, Tropocyclops)* commonly inhabit these temporary waters. Occasionally some species of birds feed on fish and crustaceans during the **period of drought**.

15.3   SALINE LAKES (ATHALASSIC WATERS)

15.3.1   Features and definitions

Saline lakes occur in dry desert areas on all continents, in endorheic basins where evaporation exceeds precipitation. Hammer (1986) defines saline lakes as those with no connection to the ocean in recent geological times, or where evaporation occurred after flooding by marine waters and then subsequent flooding, with salinities greater than or equal to $3\,g \cdot L^{-1}$ and mainly terrestrial fauna and flora.

Saline lakes are found in endorheic (internal drainage) areas, and are closed lakes fed by surface drainage and rain. They are generally found in climates where evaporation exceeds precipitation. Excessive evaporation leads to a concentration of salts and results in saline lakes.

Figure 15.7 shows the distribution of arreics and endorheic sites. Geographical and climatic conditions determine the regions where drainage does not reach the ocean. Figure 15.8 shows the main arid regions in South and North America, and the latitudes where saline lakes occur in these continents.

The drainage basins of saline lakes vary in surface area. The largest is Lake Eyre, Australia (1.3 million km^2). Other large basins include the **Dead Sea** ($31,080\,km^2$), **Lake Niriz** ($26,440\,km^2$) and **Lake Chilwa** ($7,000\,km^2$) (Hammer, 1986).

Saline lakes can originate from tectonics, **volcanism,** or glaciation. In some cases, the origin is related to soluble rocks or fluvial mechanisms.

The morphometry of these lakes varies widely, and can undergo alterations due to fluctuations in level, which are more common than in exorheic systems. In shallow saline lakes, these fluctuations are high, as in the case of Australia's Lake Eyre (0–6 m).

Hammer (1986) listed 25 saline lakes in South America with areas ranging from $1\,km^2$ to $50\,km^2$. Most saline lakes in Bolivia and Chile are 1–2 m deep. In general, depths greater than 50 m are rare on all continents.

Saline lakes have been described in the Mato Grosso, in the interior of Brazil (Cunha, 1943), although little scientific information is available on them (Mourão, 1989). Saline lakes and temporary wetlands occur in Brazil's Northeast, in the middle and lower regions of São Francisco.

The classification types of inland saline lakes have been extensively debated. The main point is the distinction between saline and 'brackish water' such as those found in estuaries and coastal lagoons. This difference is the factor that begins to distinguish inland waters as 'saline' compared with 'fresh' water. The term 'athalassic' was proposed by Bayly (1964) to distinguish saline lakes of non-marine origin in the interior of continents. Although the term has been criticized (Hammer, 1986), it has been commonly used, and the term 'brackish' is no longer used in limnology. The lower

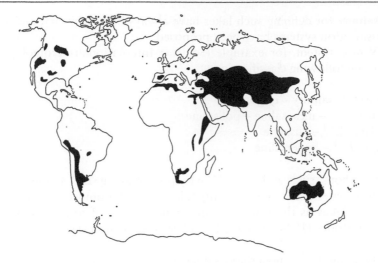

Figure 15.7   Geographic distribution of principal areas with saline lakes.

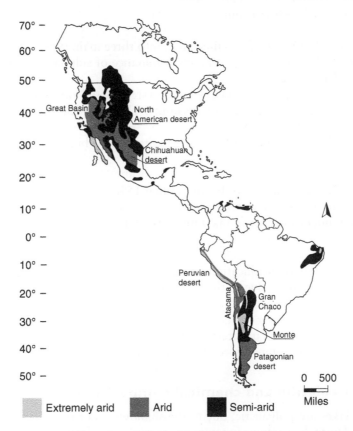

Figure 15.8   Distribution of arid, semi-arid and extremely arid climates in North America and South America, and regions where saline lakes occur.
Source: Modified from Williams (1996).

limits of salinity for defining such lakes have always been arbitrary, and various different classification systems have been proposed.

The Venice system, for example (Societas Internationalis Limnologiae, 1959), considered the following classification:

0–4 mg · L^{-1} salinity – oligosaline
4–30 mg · L^{-1} – mesosaline – polisaline
30–40 mg · L^{-1} – eusaline
>40 mg · L^{-1} – hypersaline

Other authors, such as Löffler (1961), proposed 1 g · L^{-1} salinity as the upper limit for fresh water based on the salinity tolerance of Entomostraca. Williams (1964) defined saline lakes as those with greater than 3 g · L^{-1} of total dissolved material. Ramson and Moore (1944) presented another classification system:

300–1000 mg · L^{-1} – moderately saline
1,000–10,000 mg · L^{-1} – saline
10,000–30,000 mg · L^{-1} – highly saline
>30,000 mg · L^{-1} – hypersaline

Hutchinson (1957) classified saline lakes into three main types: those with a predominance of carbonates, those with a predominance of sulphates, and those with a predominance of chlorides.

Beadle (1943) presented a classification system based on penetration of freshwater species into lakes with increasing salinity. The classification system was based on the species' tolerance to salinity.

Beadle (1959) presented the following limits in this 'biological classification' of saline lakes:

▶ Maximum upper limit for freshwater fauna: 15‰.
▶ Average limit with salty water preference: 15–50‰.
▶ Maximum upper limit for fauna with a preference for salty waters: >50‰ up to saturation.

The many different classification systems for saline lakes reveal the difficulty of establishing defined limits, due to the gradients and overlaps that occur between different types of saline lakes. Also, classification using physiological and ecological data based on the upper or lower limits of tolerance to salinity by freshwater species and saline-tolerant species must to a certain extent take into account the scarcity of data on tropical lakes, since temperature affects tolerance to salinity.

15.3.2   Circulation and chemical composition

Most saline lakes are polymictic, due to the shallowness of the lakes and the effects of the wind. Diurnal variations in temperature can occur, forming secondary thermoclines (Vareschi, 1982). Hutchinson (1973a, 1973b) described thermal stratification in deeper saline lakes. Meromictic saline lakes were described by Hutchinson (1973b) for Big Soda Lake (USA), and by Melack (1978) and MacIntyre and Melack (1982) for the African continent.

The inorganic chemical composition of the saline lakes and their salinity are determined by the following factors:

▶   the geochemistry of the drainage basin and weathering;
▶   the chemical nature of rainwater (composition of the rain);
▶   selective dissolution and precipitation of salt that depends on evaporation and the continuing process;
▶   the contribution of groundwater.

According to Hutchinson (1957, p. 553), the salinity of inland waters should measure the concentration of all the ionic constituents present. The term 'salinity' thus refers to the following ions: Na^+, K^+, Ca^{2+}, Mg^{2+}, Cl^-, SO_4^{2-}, HCO_3^- and CO_3^{2-}.

Salinity can be measured by total dissolved solids (TDS), electrical conductivity (μmho or μSiemens$\cdot cm^{-1}$) or salinity (mg$\cdot L^{-1}$).

The **chemical composition of saline lakes** varies widely, although composition within a single region is mostly uniform. In terms of ionic composition, saline lakes are divided into predominantly *carbonate*, *chloride* or *sulphate*. Predominantly carbonate lakes occur in Africa. Predominantly chloride lakes occur on all continents but mostly in Australia and South America. Predominantly sulphate lakes occur in North America and eastern Russia.

Predominantly sulphate saline lakes are unknown in the literature for Australia, Antarctica and South America, although some recent information shows high levels of sulphates in the lakes of the Mato Grosso swamp.

Some subtypes of these lakes have a mixed dominance of anions. For example, the chloride-carbonate subtype occurs on five continents and the sulphate-chloride type on three continents.

In terms of cations, there are three main types (Na^+, Mg^{2+}, Ca^{2+}), with a predominance of Na^+, and various intermediary types of different ions (for example, Na^+, Mg^{2+}, NaCa) (Hammer, 1986).

As evaporation occurs, it alters the levels of the various anions and cations, and differential precipitation proceeds. In the final stage, salts accumulate, depending on the sequence of the initial proportion of chemical elements present in the water mass. The sequence of precipitation of salts and evaporation includes carbonate to sulphates and chlorides, depending on the proportion of the existing ions, mainly of Ca^{2+}. The sequence of precipitation of salt from evaporation is very important and depends, as already stated, on the initial chemical composition of the water and the rate of evaporation.

Table 15.4 shows the current distribution of the major saline lakes larger than $500\,km^2$, and Chart 15.4 lists the main salts precipitated in saline lakes and some solubilities.

15.3.3   Fauna and flora

In many saline lakes with high levels of H_2S, photosynthetic bacteria can be significant primary producers. In eight saline lakes in Africa and Australia, the total number of bac-

Table 15.4 Current distribution of principal saline lakes with area under $500\,km^2$, in descending order of mass of dissolved salts.

Type of lake	A_O ($10^3\,km^2$)	V (km^3)	TDS ($kg \cdot m^{-3}$)	M_{sal} ($10^{15}g$)	Reference
Endorheic lakes					
Caspian Sea	374	78,200	13	1,016	Herdendorf (1984)
Dead Sea	1.02	188		56	Herdendorf (1984)[b]
Aral	64.1	1,020		10.7	Herdendorf (1984); Hammer (1986)
Urmia	5.8	45		10.35	Hammer (1986); Fairbridge (1968)
Issyk-Kul	6.24	1,730	5.8	10.0	Herdendorf (1984)
Kara Bogaz[b]	10.5	20	350	7.0	Hammer (1986); Fairbridge (1968)
Great Salt Lake[c]	4.36	19	285	5.4	Herdendorf (1984)
Van	3.74	206	22.4	4.6	Herdendorf (1984)
Eyre	7.7	23	100	2.3	Herdendorf (1984)
Σ Others[a]	101	915	29.5	27	Herdendorf (1984); Fairbridge (1968)
Total with the Caspian Sea	578	82,360	13.9	1,149	
Total without the Caspian Sea	204	4,160	31.9	133	
Exorheic Lakes					
Σ Coastal lagoons	40.0	128	5	0.64	Hammer, 1986

TDS – Total dissolved solids; M_{sal} – Total dissolved salts in mass; Saline lakes are those that with TDS $\geqslant 3g \cdot L^{-1}$ (Williams, 1964).
[a] 42 lakes larger than 500 km², including Turkana and Balkhash; [b] In the 1950s; [c] Variation in size with the hydrological balance.
Sources: Hammer (1986), Williams (1964).

Chart 15.4 Principal salts precipitated in saline lakes and some solubilities.

Aragonite	$CaCO_3$
Gypsum	$CaSO_4 \cdot 2H_2O - 1.93$
Dolomite	$CaMg\,(CO_3)_2$
Mirabilite	$Na_2SO_4 \cdot 10H_2O - 88.7$
Espomite	$MgSO_4 \cdot 7H_2O - 305$
Halite	$NaC \cdots\cdots 357$
Bischophyte	$MgCl_2 \cdot 6H_2O - 536$
Trona	$Na_2CO_3 \cdot NaHCO_3 \cdot CO_2 \cdot 2H_2O$
Calcite	$CaCO_3$
Borate	$Na_2B_4O_7 \cdot 10H_2O$
Carnalite	$KgMgCl \cdot 6H_2O$
Bloedite	$Na_2Mg(104)_2H_2O$
Sepiolite	$Mg_2SiO_3 \cdot nH_2O$
Tenardite	Na_2SO_4
Termonatrite	$NaCO_3 \cdot H_2O$
Glauberite	$Na_2SO_4 \cdot CaSO_4$

Source: Hammer (1986).

teria in each lake ranged from $0.02 \cdot 10^6 \cdot mL^{-1}$ to a maximum of $40–270 \cdot 10^6 \cdot mL^{-1}$ (Hammer, 1986). *Halobacterium halobium* is a bacterial species common in highly saline lakes (>200‰) and high levels of dissolved organic matter (Post, 1981).

Information on the productivity of bacterioplankton is scarce. Drabkova *et al.* (1986) calculated production of bacterioplankton in **Lake Shantropay**, which showed a salinity gradient of $1.1–11.9 g \cdot m^{-2}$. Data also shows the presence of *Chromatiaceae* and *Chlorobiaceae,* green and purple bacteria, bacteria that can photosynthesize under anoxic conditions (Pfennig and Tiuppa, 1981).

High biomass levels of **halophilic bacteria** were found in all saline lakes studied: the Dead Sea, the Great Salt Lake, and saline lakes in East Africa (**Natron** and **Nakuru**).

A large number of researchers (Hammer, 1986) have studied phytoplankton in saline lakes. Talling *et al.* (1973) observed the predominance of *Spirulina fusiformis* (ex. *platensis*), a cyanobacterium common in many African saline lakes, where species of *Fragilaria, Botryococcus, Anabaena,* and *Microcystis* were also rather common.

Species found in the saline lakes of South America include *Microcystis, Pediastrum, Coscinodiscus, Pleurosigma, Botryococcus braunii, Lyngbya* sp. and *Chlamydomonas* sp. (Olivier, 1953; Pollingher and Serruya, 1983). A certain degree of cosmopolitanism prevails in the distribution of phytoplankton in saline lakes, with about 29 species in hypersaline waters. A *Dunaliella* sp. is common in many saline lakes on all continents. Melack (1979) compared the primary productivity in saline lakes, studying unialgal populations in **Lake Simbi** (Kenya), identifying values from 0.62 to $5.22 g O_2 \cdot m^{-2} \cdot h^{-1}$. Chlorophyll-*a* measured $200–600 mg \cdot m^{-2}$. Melack *et al.* (1982) attributed the limited primary production of phytoplankton in one of these saline lakes to phosphorus deficiency.

Values obtained for primary productivity of phytoplankton vary (for 22 lakes on several continents) from a minimum of 233 to $58,160 mg C \cdot m^{-2} \cdot day^{-1}$, with peak chlorophyll-*a* levels of $2,170 mg \cdot m^{-3}$ and photosynthetic efficiency from 0.18 to 8.04% (Hammer, 1986).

Of the zooplankton studied in saline lakes, only five species of cladocerans are common in hypersaline waters, *Daphnia similis, Moina hutchinsoni, M. microcephala, M. mongolica* and *Daphniopsis pusilla.* There is extensive literature on the copepods of thalassic water produced by Löffler (1961), in which he discusses the distribution of 51 species of copepods. Bayly (1972) described the distribution of eight species in saline lakes, in many of which *Arctodiaptomus bacillifes* is a common calanoid species.

Arthropods, crustaceans and amphipods of saline lakes were described in the littoral community, including *Hyallela azteca* (amphipod), very common in the Northern hemisphere and *Asellus aquaticus*, a species of isopod, in the European continent.

Insects are very common in the littoral zone of saline lakes, with a predominance of beetles and some Diptera of the Culicidae family.

These lakes each present a diversified community of macrophytes of the genera *Salicornia, Juncus* and *Carex*, including some species of *Typha*, which predominate in hyposaline lakes ($<100 meq \cdot L^{-1}$ salts).

The flora and fauna of the littoral zones of saline lakes are extremely important and highly specialized, especially in conditions of hyper-salinity. Löffler (1956) and Bayly and Williams (1966) studied the benthic fauna of saline lakes.

Table 15.5   Species with osmotic regulation capacity in athalassic waters.

Species	ISO-osmotic point	Hypo-osmotic regulation	Hyper-osmotic regulation
Insects			
Aedes detritus	8‰	++	++
Ephydra riparia	8‰	+++	++
Ephydia cinerea	22–33	+++	++
Chironomus salinarius	15‰	+++	+++
Chironomus halophilus	10.5‰	++	+
Chironomus plumosus	500mOsm*	+	+
Tanypus nubifer	<600mOsm*	+	+
Enallagma clausum	15‰	+	+
Sigara stagnalis	13‰?	+	+
Trichocorixa v. interiores	12‰	+++	++
Hygrotus salinarius adults	23‰	++	++
Hygrotus salinarius larvae	–	–	+
Crustaceans			
Artemia salina	9‰	+++	+++
Parartemia zietziana	10‰	+++	++
Haloniscus searlei	19‰	+++	++
Fish			
Gasterosteus aculeatus	310mOsm*	+*	+*
Pungitius pungitius	310mOsm*	+*	+*
Salmo gairdneri	320mOsm*	+*	+*
Taeniomembras microstomus	15‰	+++	+*
Aphanius dispar	500mOsm*	+++	–

Osmotic pressure of internal liquids (fish blood and other liquids in invertebrates); +? Uncertain information; +++ Strongly regulated; + Weak regulation; + Little information.
Source: Hammer (1986).

The number of benthic species decreases with salinity and in some **hypersaline lakes**, chironomids are common (Moore, 1939), with a few dominant species (*Chironomus tentans, C. athalassiars*). Molluscs are common benthic organisms in saline lakes. Generally, the benthic biomass varies in lakes with moderate and lower salinity or in the lakes with very low salinity (**hyposaline**). Table 15.5 (Hammer, 1986), presents a list of species capable of osmoregulation in thalassic saline lakes.

Fish and bird communities of saline lakes have been extensively studied (Moore, 1939; Mendis, 1956b; Hammer, 1986). *Orestias agassizi* is a fish species common in some saline South America lakes.

Introduction of exotic species in saline lakes on all continents, but especially in African lakes, have been successful, particularly those lakes that are, in cations, predominantly sodium or magnesium. Fish do not tolerate predominantly sulphate lakes.

Amphibians also occur in saline lakes. *Bufo vulgaris, Rana temporaria* and *Rana pipiens* have been reported in some saline lakes of Northern Hemisphere.

Birds use saline lakes as a resource for feeding, breeding and nesting. Since they do not live directly in the water, they are not restricted by the salinity. However, since they eat organisms and of course some are salty, they may present physiological problems.

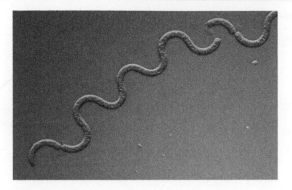

Figure 15.9   Spirulina platensis.

In all the saline lakes on all continents, species of birds are described in these in these ecosystems. **Flamingos** were described in the **Colorada Lagoon** (Bolivia) by Hulbert (1978, 1981) – *Phoenicopterus chilensis, Phoenicoparrus andinus*. The lakes studied presented salinities ranging from 5–300‰. The flamingo on the Chilean lakes feeds on invertebrates (*Artemia salina*, chironomid larvae, amphipods, and insects). All species feed on the sediment-water interface. Hulbert (1982) described the existence of 500,000 *P. chilensis* in some shallow 5-km² lakes more than 1 meter deep.

According to Hulbert (1982), *P. andinus* feeds on diatoms (>80 m) and other benthic microorganisms such as amoebae, ciliates and nematodes. High levels of *Phalaropus tricolor* were observed (>100,000) in lakes in the Bolivian highlands (Hulbert *et al.*, 1984).

Phoeniconaias minor is a flamingo species that is common in Lake Nakuru in Kenya (Vareschi, 1978) (1.5 million flamingos). It feeds on a *Spirulina* sp. (see Figure 15.9). Well-developed salt glands help the flamingo to regulate salt ingested in food. These birds have specialized plankton-filtering systems (10,000 plates, according to Jenkin, 1957) and can filter $31.8 \, L \cdot h^{-1}$ of water (Vareschi, 1978). Other bird species that inhabit Lake Nakuru are the larger flamingo (*Phoenicopterus roseus*) and pelicans (*Pelicanus ruficollis*).

Birds, therefore, play an important role in saline lakes, affecting their functioning and removing organisms at the same time contributing large amounts of fertilization due to excretion. Vareschi (1974) calculated that nearly 2,700 tons of wet-weight were consumed by birds around Lake Nakuru.

15.3.4   Food Webs

Food webs are simplified in saline lakes, and Figure 15.10 shows two examples of feeding webs, one with the presence of fish, and one with the absence of fish.

15.3.5   Impacts, uses and management

The uses of wetland areas, temporary waters and saline lakes are very diverse and present important regional characteristics. Wetland areas are used to supply food, and

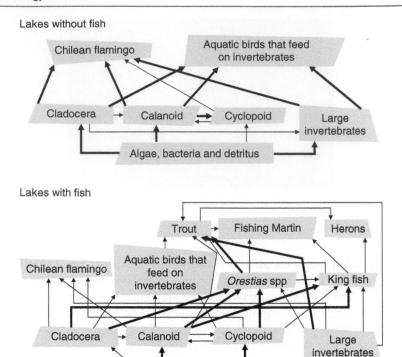

Figure 15.10   Simplified food chain in lakes in the Bolivian high plains in South America. The direction of the arrows indicates the relative rates of feeding.
Source: Modified from Hammer (1986).

the exploitation of fish, mammals, reptiles and birds. In some wetland areas, shallow lakes contain increased biomass of freshwater shrimp and agricultural activities occur in the hydroperiods.

Saline lakes can be a source of minerals (sodium chloride, sodium sulphate, magnesium sulphate and chloride), which are exploited commercially. Heliothermic saline lakes are used to provide energy (Hammer, 1986).

Food supply can come from algae, macrophytes, insects, fish and birds. *Artemia salina* is also commonly found in saline lakes and used as a source of food for fish.

Recreational uses of saline lakes and health applications are common in several continents.

Due to these many functions, it is very important to have a scientific understanding of the principal limnological and ecological mechanisms regulating the functioning of these special types of aquatic systems, as well as their structure. For example, Williams (1972) argues that saline lakes are needed for demonstrations and training purposes, due to their simplified food chain, reduced diversity and broad homogeneity. This would facilitate studies and demonstrations in the structure and function of these ecosystems.

The various human activities in wetland areas, temporary waters and saline lakes cause numerous impacts that result from the following activities:

▶ deforestation that converts forested wetland areas into pastures, reducing diversity of habitats and niches and the structure of mosaic structure in forested wetland areas;
▶ intensive use of pesticides during agricultural activities and contamination with these residues during accumulation;
▶ construction of reservoirs, which interfere with the hydrological cycle, with fluctuations in the levels and recycling of nutrients;
▶ dumping of industrial or domestic waste;
▶ intensive use for agriculture, aquaculture and fishing.

Lake Jacaretinga (AM)
Photo: Thomas Zaret

16 | Regional limnology in Central and South America

SUMMARY

This chapter compares and summarizes data from limnological studies conducted in major internal deltas, lakes, rivers, wetland areas and reservoirs at various different altitudes and latitudes in Central and South America. The principal features of aquatic ecosystems are presented as well as the methods to approach the study of limnology.

The chapter begins with a discussion of the functioning mechanisms of aquatic systems in the tropics and in the temperate zone, and the principal theories and hypotheses to address the similarities and differences in climatic, hydrological and limnological processes.

16.1 COMPARATIVE REGIONAL LIMNOLOGY AND ITS ROLE IN THEORETICAL AND APPLIED LIMNOLOGY

Studies in lake districts that attempt to describe and compare lakes, reservoirs, rivers or wetland areas are extremely important for the deepening of scientific knowledge in limnology. Such research – which involves studies on lakes, reservoirs, rivers and wetlands that differ in their morphometric characteristics, area, volume, depth and trophic condition but that are in regions with similar climate, soil and vegetation – help identify the basic forcing functions that control the ecological functioning mechanisms of inland waters.

The development of a comparative regional limnology, therefore, is basic to understanding these mechanisms. This approach has been applied at various latitudes, leading to important contributions in basic limnological knowledge.

One important aspect in the comparative study is addressing and resolving the concepts of time-space and scale. According to Vollenweider (1987), data should cover both axes (space-time) and the complete range of frequencies occurring in aquatic systems. Some of the weaknesses include the lack of availability of automatic operating systems, time scales and the delay of biological events due to physical and chemical events. An important question to address is the optimization of studies to reduce as much as possible the **noise** and **redundancy** in the results. The various **biological responses** and cycle periods and frequency of different trophic levels must be taken into account in the development of regional limnological programmes.

Regional limnology is an important study. It enables the identification of the principal forcing functions that act on the system and the comparison of operating mechanisms and processes between different regional systems, enriching and deepening scientific knowledge, allowing ample basis for practical applications in restoration, recovery and management of systems.

Such studies have been undertaken in many lacustrine systems. In this chapter and the next, basic information on inland aquatic systems in the Americas (the Great Lakes) and Central and South America will be summarized, especially the **Amazon systems**, the Pantanal and the Paraná, in Brazil; also African lakes, the lake districts in England and in Japan, and comparative studies on reservoirs in Spain. Included are some special lakes such as Baikal in the Soviet Union, Tanganyika in Africa, and Titicaca in South America. Limnological studies on reservoirs in Czechoslovakia are also summarized in this chapter, as well as some studies on lakes in China.

To begin a discussion on regional limnology, it is necessary to compare the different approaches and functioning mechanisms of inland aquatic ecosystems at different latitudes. Undoubtedly, such comparisons involve the discussion of temperate aquatic ecosystems/tropical aquatic ecosystems and their similarities and differences.

Margalef (1983) noted that every lake, reservoir or river is a unique ecosystem that has its own operating mechanisms and historical process resulting from the regional geology and geomorphology, with implications for limnological/ecological and biological/evolutionary functioning. However, according to Likens (1992), Margalef (1997), Reynolds (1997) and Tundisi (2003a), it is also necessary to seek **unifying principles** that make it possible to identify functioning mechanisms in these ecosystems and to verify that, although these principles are the same, the magnitude and rates of the processes vary because of the dominant forcing functions and their characteristics.

Classic comparisons between Lake Victoria in Africa and the lake Windermere in England's Lake District clearly show the cycles of solar radiation, water temperature (at 0 m and 60 m in both lakes), and the levels of chlorophyll-*a*.

According to Talling (1965), the photosynthetic capacity in Lake Victoria was very much higher than in **Lake Windermere**, most likely due to the difference in temperature between the two lakes, which was 10–12°C. However, other factors, such as nutrient availability, alkalinity and incident solar radiation can also influence the photosynthetic capacity, as shown by Tundisi and Saijo (1997), for lakes in the Doce River lake system (Brazil).

Lake Windermere presented a net production of approximately $20 \, g \, C \cdot m^{-2} \cdot year^{-1}$ (which, corrected for respiration, would be two to three times greater for gross production), while Lake Victoria presented a much higher gross production of $950 \, g \, C \cdot m^{-2} \cdot year^{-1}$. However, it is necessary to consider that, given the high temperatures of Lake Victoria (tropical), the respiration rate of planktonic and benthic organisms and fish might be extremely high, which in large part dissipates the primary production in the food chain.

Studies on Lake George (Beadle, 1981) show a shallow tropical environment with high rates of photosynthesis and respiration as a function of temperature, with strong influences of the bottom sediment on the ecological processes (i.e., for example, the nutrient cycle and increased turbidity, with limited solar energy available partly due to agitation and suspension of sediments in water).

These two examples show the features of relatively deep and shallow lakes in inland waters, whose operating functions vary based on morphology, climatology and hydrogeochemistry. Small lakes in major internal deltas subject to fluctuations in level, isolation, and effects of precipitation and winds present different mechanisms (Tundisi *et al.*, 1984).

A brief summary (see also Chapter 7 and Chapter 10) includes the following aspects with reference to tropical lakes:

▸ Biological diversity – in the large lakes in Africa (Tanganyika, Malawi and Victoria), there are more than 500 endemic species or groups of species, indicating how old the lakes are, with emphasis on the dimension of time in cumulative evolutionary changes. In no other tropical region is there such an accumulation of endemic species. Lowe-McConnell (1975) reported 25 endemic species (cyprinids) in Lake Lanao in the Philippines. Clearly, **environmental and evolutionary alterations** are interconnected, possibly influenced by latitude, as discussed by Talling and Lemoalle (1998). It is important to consider that alterations in the water level over time can influence speciation (Margalef, 1983).

▸ Response to the **variability of environmental factors** (solar radiation regime, water balance, and winds) can be one of the key factors in a systemic response of organisms, including cyclical frequencies and responses.

Talling and Lemoalle (1998) identified the following interactions in tropical systems:

▸ **Physical-physical interactions** – solar radiation and temperature cycles, for example, also wind cycles and changes in water density.

▶ **Physical-chemical interactions** – thermal and chemical stratifications, pulses of nutrient levels from circulation promoted by wind or thermal cooling.

▶ **Physical-biological interactions** – diurnal, lunar, and annual cycles controlling physical and chemical factors and speeding up or slowing down biological processes, such as reproductive cycles and physiological reactions and responses. These physiobiological interactions may involve a trigger mechanism; an **acceleration of cycles** based on several factors together; or a direct **regulating influence** (breathing or reproduction, for example).

Basic tropical characteristics, according to Talling and Lemoalle (1998), are:

▶ absolute magnitude of environmental factors;
▶ variability over time of these factors;
▶ responses of the biota.

Comparative regional limnology can expand the capacity of perception and synthesis of those phenomena, which are fundamental to the development of the processes of primary and secondary productivity, nutrient cycles, and life cycles of organisms.

Lakes and reservoirs in subtropical regions and lakes and reservoirs in high-altitude tropical regions present other seasonal processes and responses of the biota (Serruya and Pollingher, 1983; Lewis, 1987; Löffler, 1964).

Straškrába (1993) analyzed the latitudinal distribution of the seasonal and stochastic components of the principal physical variabilities that control processes in inland ecosystems. Based on the latitudinal distribution of components such as solar radiation, photoperiodicity, air and water temperature, precipitation and hydric balance, depth of mixing in the lakes and reservoirs, and turbidity, the author defined three limnogeographic regions: tropical region (0–15°); **dry region** (15–35°) and temperate region (35–60°). In the tropics, solar radiation presents little seasonality, the photoperiodicity is nearly constant, and the depth of mixing in lakes is more extensive. In the dry region, the hydric balance is negative, and the **mineral turbidity** and chemical composition of the water are extremely variable. In the temperate region, solar radiation presents high seasonal variability, air and water temperatures have high annual variability, and the surface temperature of lakes and reservoirs is close to 0° in winter.

16.2   REGIONAL LIMNOLOGY IN SOUTH AND CENTRAL AMERICA

There are many aquatic systems in Central America and South America, important in **basic limnology** and its application. These aquatic systems are also important for theoretical ecology, since the study of different biological problems contributes greatly to advance concepts and in the application of comparative data with lakes and rivers of temperate regions.

The lake systems of Central America have special features due to the presence of barriers formed by marine waters (II and III in Figure 16.1) in the isthmus of Panama and in the oceanic connection in southern Nicaragua (I, II and III in Figure 16.1).

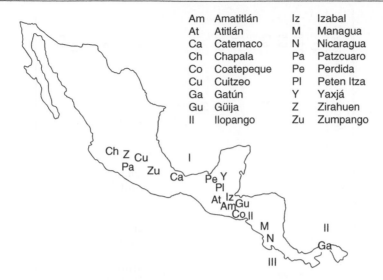

Am	Amatitlán	Iz	Izabal
At	Atitlán	M	Managua
Ca	Catemaco	N	Nicaragua
Ch	Chapala	Pa	Patzcuaro
Co	Coatepeque	Pe	Perdida
Cu	Cuitzeo	Pl	Peten Itza
Ga	Gatún	Y	Yaxjá
Gu	Güija	Z	Zirahuen
Il	Ilopango	Zu	Zumpango

Figure 16.1   Map of Central America showing the major lakes of this region.
Source: Zaret (1984).

According to Deevey (1957), Myers (1966) and Zaret (1984), the **Central American continent** of today at one point in the geological past was a large island, which explains why 55% of the fish fauna are saline-tolerant species (which implies potential dispersal by marine waters). In addition, Central American lakes, mainly tectonic in origin, had their characteristics altered secondarily by volcanic activity. Recent glacial activity, responsible for two-thirds of all lakes in the world, was not experienced in Central America (Hutchinson, 1957).

Secondary volcanism caused the formation of many lakes in the Mexican Plateau, located 2120 m above sea level. Also, the two main lakes in Guatemala (Atitlán, 1555 m in elevation; Amatitlán, 1189 m) are the result of secondary volcanic activity. Guatemala's **Lake Izabal** (717 km²) has special characteristics. It is connected to the Gulf of Honduras, in the Caribbean, and presents marine species (Brinsow, 1976). Table 16.1 (from Zaret, 1984) summarizes data from **Central American lakes.** Many small Central American lakes, especially in Mexico, have been historically important due to their use as a supply of water, production of fish and aquatic birds, transportation and recreation. Decomposing aquatic plants are used as fertilizer. In the last 40 years, many alterations have occurred in these lakes, mainly resulting from the introduction of exotic fish species that prey on endemic populations. For example, the well-known crab of **Lake Atitlán**, *Potamocarinos guatemalensis*, almost completely disappeared after introduction of *Minopterus salmoides* in 1950.

Another important lake system is Lake Gatún/Chagres (Panama). Table 16.2 shows data on the chemistry of the water in Lake Gatún. Chart 16.1 lists species of aquatic macrophytes and Chart 16.2 lists the aquatic insects.

The introduction of the South American cichlid *Cichla occelaris* (tucunaré) in Lake Gatún has caused several alterations in the food chain in the lake, especially in fish and zooplankton. The appearance of a cyclopoid species (*Cyclops sp.*) and changes in the behaviour of *Diaptomus gatunensis* (which modified its vertical migration

Table 16.1 Chemical and physical values for lakes in Central America.

	Chapala	Pátzcuaro	Catemaco	Amatitlán	Atitlán	Izabal	Güija	Coatepeque	Ilopango	Managua	Nicarágua
Maximum depth (m)	9.8	15	18	34	342	16	25	120	248		43
Surface temperature (°C)	20.2	19.0	23.4	24.8	23.1	30.0	26.7				27.8
Secchi (m)	0.5	1.5	1.0	2.4	6.1		1.2	12.5	10.5		0.5
pH						6.5				8.5	6.3
O_2 (mg·L^{-1})	7.8	7	8.6	6.3							
CO_2 (mg·L^{-1})	1.0		5.0	6.3	0						6.0
HCO_3^- (mg·L^{-1})	165	460	55.0	165	194		90.5			231	65.0
CO_3^{2-} (mg·L^{-1})										13	
Total N (mg·L^{-1})	240			96			237	36			
SO_4^{2-} (mg·L^{-1})											11
Na^+ (mg·L^{-1})											16
Ca^{2+} (mg·L^{-1})	2			17.3			14.6	22.2			2
K^+ (mg·L^{-1})											6
Mg^{2+} (mg·L^{-1})	1.1			7.8			trace				
Fe^{2+} (mg·L^{-1})	0.175										
Cl^- (mg·L^{-1})	17.0	21.5		25			4.4			8.8	19
Maximum primary production (gO_2·m^{-2}·day^{-1})	10.0	10.0		5.14		7.3					9.3

Source: Zaret (1984).

Table 16.2 Main features of the chemistry of the water in Lake Gatún. Data on chlorophyll-*a* are included.

	Maximum annual value[1]	Surface water (April 1972)[2]
pH	7.2	7.56
Dissolved oxygen	7.78	8.0
$NH_4^+ - N$ (mg·L^{-1})		0.01
$No_2^- - N$ (mg·L^{-1})	0.004	0.00
Total N (mg·L^{-1})	0.06	0.05
Total P (mg·L^{-1})		0.022
Ca^{2+} (mg·L^{-1})	10.2	4.1
Mg^{2+} (mg·L^{-1})	3.8	3.2
Cl^- (mg·L^{-1})	5.0	5.4
Fe (mg·L^{-1})		5.2
SO_4^{2+} (mg·L^{-1})	6.0	
SiO_2 (mg.L^{-1})	16.2	
Conductivity (20°) (µS·cm^{-1})	90	98
Chlorophyll-*a* (mg·m^{-3})	4.1	

[1]Data from Gliwicz (1976a); [2]Data from the Panama Canal Company.

Chart 16.1 Aquatic macrophytes in Lake Gatún.

Submerged plants	Free-floating plants
Cabomba aquatica *Ceratophyllum demersum* *Chara* spp. *Hydrilla verticillata*[1] *Najas guadalupensis* *N. marina* *Utricularia vulgaris*	*Azolla caroliniana* *A. filiculoides* *Eichhornia crassipes* *Lemna mínima* *L. minor* *Pistia stratiotes* *Salvinia rotundifolia* *Spirodela oligorhiza*
Emergent plants	Shoreline plants
Eichhornia azurea *Hydrocotyle umbellata* *Marsilea polycarpa* *Nymphaea ampla* *Polygonum hydropiperoides* *Pontederia rotundifolia* *Sagittaria* spp. *Scirpus* sp. *Typha angustifolia*	*Ceratopteris pteridoides* *Jussiaea subintegra* *Luziola subintegra* *Paspalum repens*

[1]Introduced species
Source: Pasco (1975).

patterns) result from predation by *Cichla occelaris* on *Melanurus*, a zooplanktonic predator (Zaret, 1984). Lake Atitlán was studied by Birge and Juday (1911).

16.2.1   Limnology in Nicaragua

Lake Cocibolca is the largest body of water in Central America and a natural reserve for the future's drinking water. It is the lake with the largest area and volume of water

Chart 16.2   Aquatic insects of Lake Gatún.

Coleoptera	Heteroptera
Dysticidae *Chaetarthria glabra* *Laccophilus g. gentilis* *L. ovatus zapotecus* *Thermonectus margineguttata*	Belostomatidae *Belostoma micontulum* *B. subspinusum cupreomicans* *Lethocerus colossicus*
Gyrinidae *Gyretes acutangulus* *G. centralis*	Gelastocoridae *Gelastocoris major* *Nerthra raptoria* *N. rudis*
Haliplidae *Haliplus panamanus*	Naucoridae *Ambrysus geayi* *A. horvathi* *A. oblongulus* *Pelocoris nitidus*
Diptera	
Chaoboridae *Corethrella ananacola* *C. blanda* *C. dyari* *Sayomyia brasiliensis*	Nepidae *Ranatra zeteki*
Chironomidae *Cantomyia cara* *Chironomus aversa* *C. fulvipilus* *Coelotanypus humeralis* *C. naelis* *C. neotropicus* *C. scapularis* *Corynoneura spreta* *Cricotopus oris* *C. tanis* *Polypedilum pterospilus*	Notonectidae *Buenoa pallipes* *B. platycnemis* *Martarega hondurensis* *M. williamsi*
	Ochteridae *Ochterus manii* *O. viridifrons*
	Pleidae *Plea puella*

Source: Hogue (1975).

between Lake Titicaca and the Great Lakes in North America. Studies on this lake and its hydrographic basin have been conducted by CIRA (Aquatic Resources Research Center of Nicaragua) involving eutrophication and management plans (CIRA, 1996, 2004; Montenegro, 2003; Vammen, 2006).

16.2.2   Limnology in Mexico

Munawar *et al.* (2000) published a large volume on limnology in Mexico with lots of data on functional processses.

16.3   INLAND ECOSYSTEMS IN SOUTH AMERICA

The continent of South America has many **biogeophysiographic differences,** occupying a vast latitudinal north-south gradient and an important east-west altitudinal gradient. The principal hydrographic basins in South America are shown in Figure 16.2. For

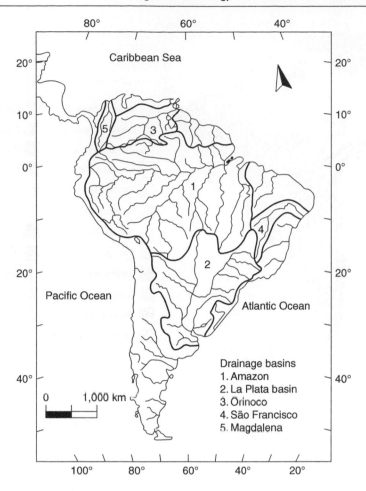

Figure 16.2   Major drainage basins in South America.
Source: Tundisi (1994).

comparison, and considering the importance of interactions between terrestrial and lacustrine systems, Figure 16.3 shows the biogeographical regions in South America. In these morphoclimatic dominions, the principal hydrographic basins are found: the Orinoco basin, the Amazon basin, and the La Plata basin. Large inland deltas, extensive flooded areas with floodplain lakes occur in these three basins. In addition to the large inland deltas of the Amazon basin, wetland areas with periodic fluctuations in level are significant in inland limnology. Altitudinal gradients are important in the Andean lakes, the **endorheic system** with saline lakes in Bolivia, and the Araucanian lakes in glacial valleys. A brief description is included with the work on Lake Titicaca.

Information on regional limnology in South America will be described here for the Amazon system, the Pantanal, the La Plata basin and the lake system of the medio Rio Doce, which has special characteristics and should thus be mentioned as an essential contribution to regional and tropical limnology. Comparative studies of man-made

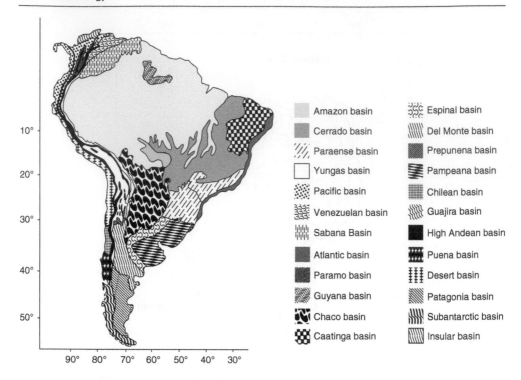

Amazon basin
Cerrado basin
Paraense basin
Yungas basin
Pacific basin
Venezuelan basin
Sabana Basin
Atlantic basin
Paramo basin
Guyana basin
Chaco basin
Caatinga basin
Espinal basin
Del Monte basin
Prepunena basin
Pampeana basin
Chilean basin
Guajira basin
High Andean basin
Puena basin
Desert basin
Patagonia basin
Subantarctic basin
Insular basin

Figure 16.3   Biogeographic provinces of South America.

reservoirs in the state of São Paulo (Tundisi, 1981, 1988) are included in this chapter, and a brief description of Lake Titicaca.

16.3.1   Lake Titicaca

Lake Titicaca is a large high-altitude navigable lake (3809 m in elevation). It is a deep lake, with large volume, located in the Altiplano of the Andes. Due to its geographical location, it is subjected to weather conditions typical of the tropics (14°S latitude), particularly the incidence of solar radiation.

As it is located at a high altitude, the lake is affected by the high-altitude climate (high light intensity, low temperature, low humidity). Lake Titicaca has a retention time of 63 years and high levels of evaporation. It is part of a set of lakes in the Altiplano, and the entire hydrological system for this region (200,000 km² in area, altitude of 3700–4600 m) is endorheic (see Figure 16.4). Today's lake system of the Altiplano is the result of the evolution of a much older system that began in the lower Pleistocene, with the transition from a relatively warm climate to a humid and cold climate at the end of the Pliocene.

A summary of limnological studies conducted on Lake Titicaca was published by Dejoux and Iltis (1992). Although localized, the impacts on this lake are the result of eutrophication and industrial waste that affect fauna and flora, stimulating the growth of macrophytes, zooplankton and phytoplankton in limited areas of the lake (Northcote, 1992).

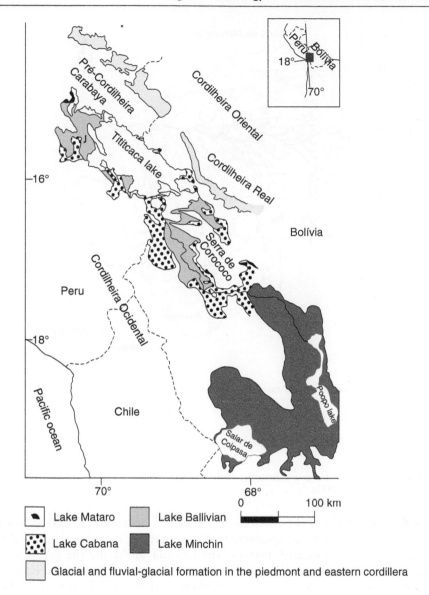

- ⬛ Lake Mataro   ▢ Lake Ballivian
- ▦ Lake Cabana   ■ Lake Minchin
- ▢ Glacial and fluvial-glacial formation in the piedmont and eastern cordillera

Figure 16.4 Water systems of the Andes' highland.
Source: Lavenu (1941).

16.3.2 The Amazon system

The development of regional limnology includes important studies on floodplain lakes of the Amazon and the interactions of these lakes with the rivers. Sioli (1975, 1984, 1986) summarized the major limnological studies conducted in the Amazon basin. A study on the Negro River (Goulding, Carvalho and Ferreira, 1988) detailed the main ecological features of this system.

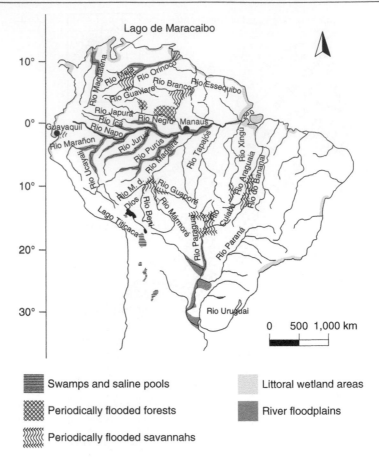

Figure 16.5   Principal tributaries and features of the Amazon basin.
Source: Junk (1993).

The drainage basin of the Amazon region covers an area of 7 million km². The lowland area, a large zone of **Quaternary sedimentation,** is where the fluvial valley of the Amazon River is located, between the Guyana shield and the Brazilian shield (see Figure 16.5). The chemical composition of the waters of these rivers is related to the regional hydrogeochemistry, geomorphology, and the pedological and lithological conditions of the headwaters (Sioli, 1968). Many tributaries of the Amazon river drain older formations of the Cretaceous, Tertiary and Quaternary in the fluvial valley. For example, the Negro River, a tributary that has its headwaters in the Guyana Shield, flows through Tertiary and Quaternary formations.

The location of the **origin of rivers** in the Amazon region includes two basic aspects: the composition of the water, and the second, transport of sediment.

The first observations on the origin of the colour of water in black-water rivers of the Amazon region date from the 18th century (Goulding *et al.*, 1988). Humboldt (1852) also reported observations on 'black water' and Russel Wallace (1853) was the first naturalist to categorize the waters of the Amazon region into three classical types: white water, black water, and transparent or crystalline water. White-water

Table 16.3   Concentrations of principal elements and nutrients in the main types of Amazonian waters.

	Principal elements ($mg \cdot L^{-1}$)				Nutrients ($\mu g \cdot L^{-1}$)	Nutrients ($mg \cdot L^{-1}$)		
	Na	Ca	Mg	Cl	PO_4–P	SO_4	NO_2–N	SiO_2
Andean rivers	2–3	1–2 (23)*	1–2	?	?	4–6	3–4	35
White waters	1.5–4.2	7.2–8.3	1.2–8.3	?	15	1–6.4	4–15	<9
Clear waters	1–2	<2	<1	0–3	<1	0	<7	3–9
			0.1–2.1				0–0.5	
Black waters	0.55	<0.46	?	?	5.8	?	0.036	2.4

Sources: Fittkau (1964); Greisler and Schneider (1976); Oltmann (1966); Schmidt (1970, 1972a, b, 1973, 1976, 1982); Ungemach (1972a); Turcotte and Harper (1982).

Chart 16.3   Processes of the influx of nutrients, losses and recycling in Amazonian wetlands.

Sources	Recycling	Losses
Influx of nitrogen and phosphorus by advection (rivers)	Decomposition by bacteria	Sedimentation
Desorption in lakes	Excretion of zooplankton	Losses from discharge of water
Influx of nitrogen and phosphorus by rain	Sediment-water interaction	Denitrification
Fixation of N_2 in lakes	Macrophytes: Absorption of nitrogen and phosphorus and decomposition	
	Absorption of nitrogen and phosphorus by phytoplankton	
Surface drainage into lake	Exchange between epilimnion, metalimnion, hypolimnion	

Sources: Tundisi *et al.* (1984); Fisher and Parsley (1979); Junk (1986); Melack and Fisher (1979); Fosberg (1981); Zaret *et al.* (1981).

rivers, Wallace noted, have high loads of suspended sediment, while black-water rivers and crystalline waters carry little sediment. Wallace considered that the black water most likely was the result of decomposing vegetation (leaves, stems and roots). Sioli (1951, 1956) hypothesized that periodic flooding of the forest by the river during the annual cycle produces humic and colloidal substances similar to those in decomposing vegetation. Black water can also develop in flooded forests or in soils with low calcium levels.

Therefore, it is important to know the origin of Amazonian rivers in order to understand the regional hydro-geochemistry and its interacting processes, mainly in terms of the biogeochemistry and productivity of the water. Table 16.3 shows the major ionic and nutrient levels in three types of Amazonian waters (Day and Davies, 1986). Chart 16.3 presents the inflow processes of nutrients, losses and recycling in floodplain lakes of the Amazon region.

In some regions, rain water has higher ionic levels than the rivers. In small streams located on firm ground in the forest, there is an enormous contribution of alloch-

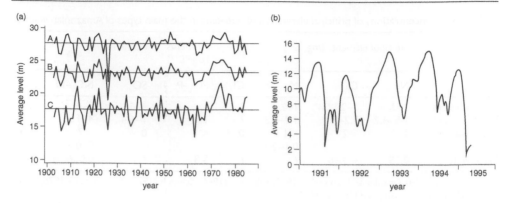

Figure 16.6   a) Fluctuations in average level of the Negro River, 18km. from Manaus. (A) Annual average for each year; (B) Average of all the values for each year; (C) Annual minimum for each year; (b) Variations in the water level for five consecutive years.
Sources: Modified from Tundisi (1994).

thonous material, which is essential for the maintenance of feeding networks with a predominance of detritivores. In addition, this material significantly affects the chemical composition of rivers. While the Amazon region is dominated by large rivers, extensive and varied **biological, biogeochemical,** and **ecological phenomena** occur in the small rivers of the Amazon system.

Fluctuations in level, flooded forests, and floodplains

Due to the enormous fluctuations in the outflow of rivers (see Figure 16.6), the Amazon and its tributaries produce large flooded areas, causing major changes in the ecological functioning of the system, resulting in increased levels in floodplain lakes and flooded forests. These floods have two main results: the first is the transport of nutrients from rivers to floodplain lakes with effects on the succession of communities, primary production and biogeochemical cycles. The second is the flooding of forests and the contact of water with the forest, which increases the capacity to feed fish with different types of food and feeding niches being exploited more efficiently (see Figure 16.7).

Primary productivity, biogeochemical cycles and aquatic communities

The primary productivity of the rivers and **lakes of Central Amazonia** was measured by Braun (1952), Hammer (1965), Mulier (1965, 1967), Sioli (1968), Schmidt (1973, 1976), Fitkau *et al.* (1975), Junk (1979, 1983), and Rai (1984).

Primary production in the small rivers in the forest is very low, because of the **shading from vegetation,** turbulence and low nutrient levels. In white-water rivers, primary productivity is also low because of the poor penetration of light. Primary productivity of phytoplankton is therefore relatively low in Amazonian rivers, and in black-water rivers (for example, the Negro), it is similarly low. Primary productivity of phytoplankton is higher in lakes and limited to the first few meters of the water column. In Castanho Lake, for example, Schmidt (1973) determined that the primary production of phytoplankton occurs only at depths of 0.5–6.0 meters.

During periods of low water levels, primary production increases. Primary production of periphyton is high, and bacterial production dominates the production

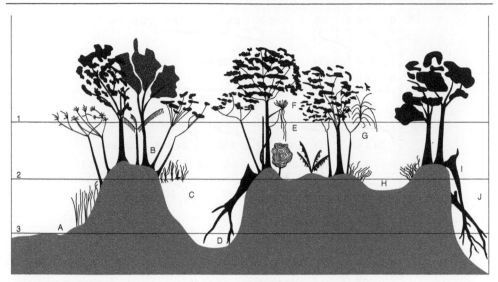

1. Extremely high water level
2. Intermediate water level
3. Water level in extreme drought

A. Exposed pools on beaches in the dry season
B. Forest underwater
C. Open water in a channel or pipe
   that never completely drains
D. Branches fallen from trees along the shorelines
   of lakes or channels.

E. Clearing in the forest of a beach
F. Little pools of water in plants or trunk of trees
G. Floating macrophyte roots
H. Floodplain lake that can drain completely during the dry season
I. Fallen trees in an area of erosion on the shorelines
   of a white-water river
J. Open water of the main channel
   of a white-water river

Figure 16.7   Side view of a floodplain showing variations in the water level and the main aquatic habitats.
Source: Queiroz and Crampton (1999).

of plankton in most lakes studied. Figure 16.8 shows the vertical profiles of primary productivity of phytoplankon in Lake Castanho (Schmidt, 1973).

The primary productivity of phytoplankton is controlled by differences in water level, light intensity, and nutrients. Both nitrogen and phosphorus are limiting nutrients in black-water lakes.

Tables 16.4a and 16.4b show the net primary production of phytoplankton for Amazonian lakes as compared with other tropical lakes. In addition to phytoplankton, periphyton and bacterioplankton, aquatic macrophytes are also important primary producers in lakes of the Central Amazon, as shown by Junk and Howard-Williams (1984). Communities of aquatic macrophytes depend on periods of drought and flooding. *Paspalum fasciculatum* and *Echinochloa polystachya* are, respectively, aquatic and semi-aquatic macrophytic species.

In the aquatic phase, during flooding, *Paspalum repens, Oryza* sp., *Scirpus cubensis* and species that float (such as *Eichhornia crassipes, Salvinia* spp., and *Pistia stratiotes*) can be found. The occurrence and distribution of different species are affected by the nutrient levels and nutrient cycles associated with the fluctuations in level. Generally, in black-waters with low nutrient levels and acidic pH, growth of these plants is limited, and there is little colonization of these systems. Some species that

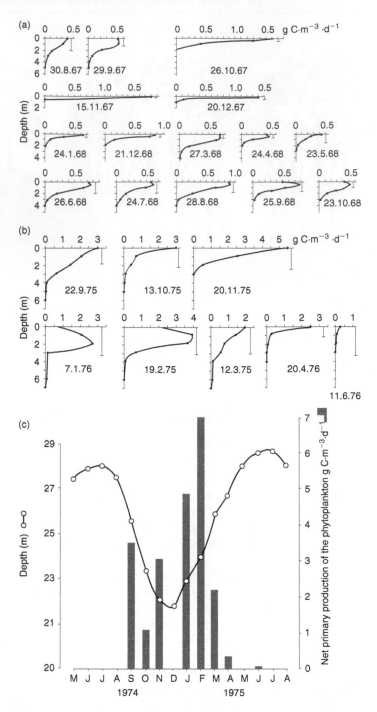

Figure 16.8 a) Seasonal variation of the net primary production in Castanho Lake and Secchi-disc readings; b) Seasonal variation in the net primary production and Secchi-disc readings in Cristalino Lake; c) Maximum primary production, which occurs during periods of low water levels.

Source: Schmidt (1984).

Table 16.4a Net primary production of phytoplankton in lakes in the Central Amazon. The production in Lake Redondo is annual. Comparisons with other tropical aquatic ecosystems are included.

Lake	$mg\ C \cdot m^{-2} \cdot day^{-1}$	Reference
Castanho	350–1,500	Fitkau et al. (1975)
Cristalino	53–10,451	Rai and Hill (1984)
Tupé	100	Rai (1979)
Tapacura	410–1,300	Hartman et al. (1981)
Redondo	52,000[a]	Marlier (1967)
Aranguadi	13,000–22,000	Baxter et al. (1965)[b]
		Talling et al. (1973b)
Bunyoni	1,800	Talling (1965b)[b]
Chad	700–2,700	Lemoalle (1969)[b]
George	5,400	Ganf (1970)[b]
Kivu	1,440	Degens et al. (1971)[b]
Mariut	10,800	Vollenweider (1960)[b]
Mulehe	960	Talling (1965)[b]
Victoria	1,080–4,200	Talling (1965b)[b]

[a] Annual production ($mg\ C \cdot m^{-2}$)
[b] *Source:* Beadle (1974).

Table 16.4b Gradient of primary production of phytoplankton in lakes of the Central Amazon compared with lakes of different trophic categories.

Trophic type	Primary productivity ($mgC \cdot m^{-2} \cdot day^{-1}$)	Chlorophyll-a ($\mu g \cdot L^{-1}$)	Total organic carbon ($mg \cdot L^{-1}$)
Oligotrophic[a]	50–300	0.3–3	<1–3
Mesotrophic[a]	250–1,000	2–15	<1–5
Eutrophic[a]	>1,000		5–30
Dystrophic[a]	<50–5,000		3–30
Lakes of the central amazon			
Clear waters	350–1,500	1.3–92	8–23
Mixed waters	820–3,500	0.7–47	7–23
Black waters	53–10,451	0.5–27	5–17

[a] Wetzel (1975), modified by Likens (1975).

occur in water with low nutrient levels have roots in the sediment, possibly as a mechanism to supply additional nutrients: the decomposition of aquatic macrophytes in Amazonian lakes and rivers is rapid, most likely due to high temperatures and the composition of the cell wall. Therefore, these plants contribute significantly to the addition of nitrogen, phosphorus and other elements to the water. Many vertebrates use aquatic macrophytes as food in the Amazon River, floodplain lakes, and tributaries. These plants also contribute significantly to increased detritus in the water column and sediment (Howard-Williams and Junk, 1977).

The microbiology of Amazonian waters was studied by Rai (1979) and Hill and Rai (1982), who published a summary (1984).

Interrelationships between the seasonal cycle of phytoplankton and the cycle of bacterial populations were demonstrated by counting the bacteria, using the plate

Table 16.5   Comparisons of lakes in the Central Amazon and other aquatic systems in terms of heterotrophic parameters.

System	$Kt + Sn\ (\mu g \cdot L^{-1})$	$V_{max}\ (\mu g \cdot L^{-1}/h)$	Total hours	References
Black-water lakes	226–2,030.6	0.039–79	49–9,821	Rai and Hill (1980)
Mixed white- and black-water lakes	876.9–1,282.1	1.16–32.9	53–416	Rai and Hill (1980)
White water lakes	222.8–3,485.3	0.22–27.75	12–1,655	Rai and Hill (1980)
Subarctic Pacific	1.6	0.11	–	Vaccaro and Jannasch (1966)
Crooked Lake (USA)	20–80	1–10	100–400	Wetzel (1967)
Char Lake	0.5–5	0.001–0.008	40–1,700	Morgan and Kalff (1972)
South Creek Estuary (USA)	2,124–50,04	0.149–24.12	0.2–22.4	Crawford, Hobbie and Webb (1974)
Plussee (Germany)	3.8–46.9	0.2–1.2	6–202	Overbeck (1975)
Tokyo Bay (Japan)	21.96–66.96	0.396–17.98	8.7–23	Yamaguchi and Ichimura (1975)

Source: Rai and Hill (1984).

method, total coliform, faecal streptococci, total bacterial count and heterotrophic activities.

The **production of heterotrophic bacteria** is significant in the lakes of the Central Amazon, and in most lakes studied by Rai (Rai and Hill, 1981), the peaks of heterotrophic activity coincide with low-water periods. The patterns of bacterial activity appear to correspond to nutrient cycles, the water level, primary production and the concentration of the natural substrate. Table 16.5 (from Rai and Hill, 1984) draws a comparison between the lakes of Central Amazon and other systems in relation to the measures of heterotrophy.

Robertson and Hardy (1984) described the **zooplankton of Amazonian rivers and lakes**, listing 250 species of rotifers, the most common component of zooplankton. There are about 20 species of cladocerans and 40 copepods, predominantly calanoids. Among the calanoid species, *Notodiaptomus amazonicus* and *N. coniferoides* are common; among cyclopoid species, *Mesocyclops longisetus*, *M. leuckartii*, *Thermocyclops minutus* and *Oithona amazonica* are common.

As in other communities of Amazonian lakes, the water level affects zooplanktonic biomass and species succession. Zooplankton in Amazonian lakes and rivers provides an important food source for many fish species (Carvalho, 1984; Zaret, 1989).

As for the benthic community, Roiss (1976) found predominantly chaoborids and ostracods, with an average annual biomass of $0.136\ m^{-2}$. In the littoral zone, chironomids were common.

The fish fauna of the Amazon has been described by Lowe-McConnell (1984), and Goulding (1980) presented a series of studies on fish distribution, migration and feeding. Goulding, Carvalho and Ferreira (1988) described the relations and trophic diversity of fish species in the Rio Negro. Goulding (1980) identified the following direct and indirect food items for Amazonian fishes: fruits and seeds, leaves, flowers, decaying vegetation, arthropods, faeces, terrestrial arboreal vertebrates (small birds and rodents), aquatic insect larvae, crustaceans, molluscs, zooplankton found in the stomach of tambaqui, mainly cladocerans and copepods, algae, detritus and some fish species. The author also found that fluctuation in the water level is a key factor in the feeding behaviour of Amazonian fish.

According to Lowe-McConnell (1987), there are approximately 1300 species of fish in the Amazon basin. The author classified fish species by their feeding habits:

▶ Fish species that in the adult stage feed exclusively on vegetation, such as *Colossoma, Mylossoma, Myleus* and *Brycon*. Fruits and seeds account for 89% of the total volume of food consumed by these fish. During the feeding period, these fish species develop extensive fat reserves.

▶ Fish species feed mostly on vegetation but also on animals. In these species, at least 20% of the total fish diet includes animals. *Serrasalmus serrulatus* (piranha) and *Pinelodus blodii* are two species with these characteristics.

▶ Fish that feed on fine detritus. The most important genera in this category are *Curimatus, Semaprochilodus* and *Prochilodus*, which make up a large part of the fish fauna in the nutrient-poor rivers of the Amazon region. These fish feed on fine detritus, which basically includes decaying vegetation.

The studies on the feeding and distribution of fish in the Amazon region showed the importance of fluctuating levels in migration and exploitation of feeding niches in the forest during **periods of flooding,** illuminating the complexity of interactions between terrestrial and aquatic systems. Goulding (1980) estimated that 75% of commercial fishing comes from the food chain that begins in flooded forests.

Figure 16.9 describes the principal impacts of flood pulses on the fish fauna in the Amazonian flood valleys. Figure 16.10a (from Junk, 1982) shows the principal interactions that occur in the system based on fluctuations in water levels and in the floodplains. It is clear that the pulses produced by fluctuations in level are critical for the inflow of nutrients into the lake, the succession of aquatic communities and the magnitude of biomass and species diversity. Figure 16.10b illustrates these exchanges and flows.

The vertical circulation in flood-plain lakes is essential for the recycling of nutrients, particularly taking into account the diel cycles of thermal stratification and destratification (Tundisi *et al.,* 1984). Daily vertical circulation affects the vertical distribution of nutrients, functioning as a pulse mechanism for the inflow of nutrients into the euphotic zone.

MacIntyre and Melack (1984) studied vertical circulation in Lake Calado, showing that frequent variations occur during the flood cycle in the vertical mixing down to the bottom of the lake. Periods of **diel circulation** are followed by periods of complete stratification for several days. If the force of the wind is sufficient to cause alterations in the density gradient, vertical mixing can occur daily (during the day). Therefore, responses of the lakes to incident radiation, the force of the wind and nocturnal cooling produce frequencies of stratification and destratification that influence the vertical distribution of nutrients and metabolism in the lake.

With respect to nutrient cycling in flood-plain lakes, Forsberg (1984) confirmed that in Lake Cristalino (located in the Rio Negro system), the likely limiting factor is phosphorus, while in Lake Jacaretinga (located on the Solimões River), nitrogen is the limiting factor. The probable cause of limiting nutrients (nitrogen or phosphorus), according to Forsberg, is the deficiency in nitrogen or phosphorus in river water entering the lake. The TN/TP ratios are very high in the Negro River/Cristalino Lake

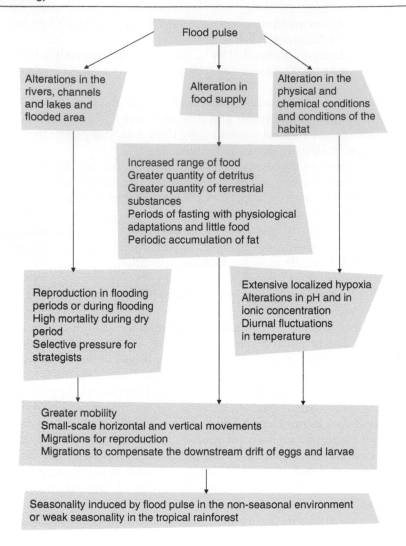

Figure 16.9   Impacts of flood pulse on fish in the flood valleys of the Amazon.
Source: Junk (2006).

system and low in the Solimões River/Jacaretinga Lake system. Forsberg's studies also show the role of macrophytes as primary producers and important components in the nutrient cycle. The high turbidity of the lakes (especially in the Solimões system) prevents high levels of primary production by phytoplankton. Aquatic macrophytes are important as primary producers and in the recycling of nitrogen and phosphorus (mainly the 'floating grass' *Paspalum repens*, which has an annual cycle related to fluctuation in level) (Junk, 1973).

According to Devol *et al.* (1984), the **sediments in floodplain lakes** provide important reserves of nutrients, accumulating organic carbon in large quantities. These floodplain regions are utilized as areas for intensive exploitation of biomass (fishing, limited aquaculture) and production of food in the terrestrial systems (crops in the floodplain during low-water periods).

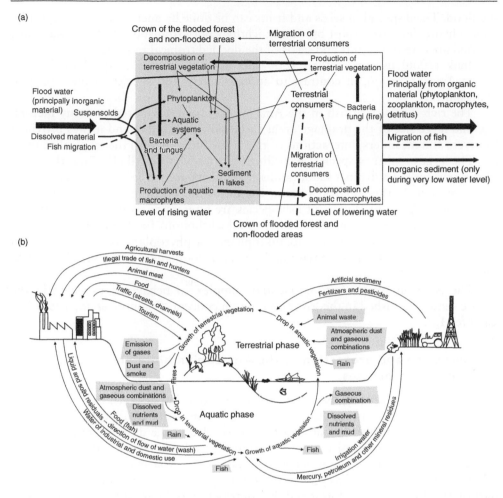

Figure 16.10   a) Principal interactions of nutrient cycle in floodplains and in terrestrial and aquatic systems of the Amazon; b) Flow of nutrients and energy in the Amazonian floodplains and exchanges between the atmosphere and the unflooded land.
Sources: a) Junk (1982) and b) Junk (2006).

Aquatic invertebrates in the Amazon

Aquatic invertebrates in the flood valleys, according to Junk (2006), present several adaptive strategies for drought: **reproductive strategies** (large numbers of eggs; life cycles, asexual propagation, parthenogenesis, partition); **resistance to drought** (dormancy, resistant eggs; larvae and adults in dormant periods); amphibious life (parenting).

Terrestrial vegetation and flood periods

Terrestrial vegetation, during flood periods, presents diverse adaptive strategies, such as pneumatophores and adventitious roots in the flooded areas, internal transport of oxygen; xeric leaves; and maintenance of leaves under water, or loss of leaves during the flood. The dispersal of seeds and fruits can be done by anemochory (transport by wind); hydrochory (transport by water); ichthyochory (transport by fish). Many fish in flood areas feed on fruits and seeds of the flooded tropical rainforest.

Junk (2006) lists more than 1000 species of vegetation adapted to flooding in the Amazon region, in contrast to approximately 100 species in the Northern Hemisphere.

The rich diversity of fish in the Amazon has been attributed to many factors: age and large size of the drainage area; succession of habitats and niches provided by rivers and meanders; interactions of rivers and streams with the tropical rainforest. The large number of species, according to Lowe-McConnell (1987), is the result of the existence of a distinct set of faunistic elements in each section or specific area of the system. In the Amazon system, the headwater streams are upstream from the forests, and the predominant fish species are filterers or scrapers. In the central Amazon basin, there is a predominance of allochthonous food from forests; detritivores play a bigger role than zooplanktophages or phytoplanktophages, although some species also feed on phytoplankton and zooplankton (Lowe-McConnell, 1986).

The Amazon system is too vast and complex to behave as expected in the classic sense established in the 'river continuum concept' (Vannote *et al.*, 1980) (see Chapter 13).

Conservation and protection of the Amazonian biota and the principal processes

As shown in this summary, there is a complex relationship between rivers, flood valleys, lakes and forests in the Amazon region. The majority of allochthonous material produced is transformed into waste and recycled in the different terrestrial and aquatic systems. The diversity of ecological niches and subsystems is therefore maintained by the flow of energy in terrestrial and aquatic systems, which channel and maximize the processes of recycling. The diversity of aquatic fauna depends in large part on the diversity of the forest and the contribution of biomass to rivers and lakes. The fluctuation in river level is essential as a forcing function and input of external energy to maintain the processes and mechanisms.

Any changes to the system, either by cutting vegetation, modifications in the floodplain, or interference in the fluctuations of level (for example, in reservoirs in the Amazon), will significantly alter the system, with loss of diversity, extinction of species, and disruption of interactions built over long periods of geological time. According to Sioli (1984), the main problems of land development in the Amazon include the following:

▶ loss of nutrients in the closed vegetation/soil cycle;
▶ reduced annual precipitation;
▶ increased suspended matter entering rivers;
▶ increased erosion;
▶ increased surface discharge;
▶ instability in the regime of rivers;
▶ low river levels during dry periods.

Some of the principal human activities affecting the main aquatic and terrestrial ecosystems include over-exploitation of forests, **impacts on aquatic systems** in the Amazon region (deforestation, increase from overfishing, development of intensive aquaculture in floodplains), expansion of agriculture and deforestation, and the effects of mining on rivers.

Multiple uses of aquatic biodiversity and aquatic systems in the Amazon

Because of the sheer magnitude of the Amazon region, its diversity of environments and great biodiversity in aquatic and terrestrial ecosystems, the ecosystems there are used for many purposes (especially the aquatic ecosystems). These uses include food supply (fishing and agriculture in the floodplains), transportation and navigation; exploitation of the forest and the terrestrial biodiversity. Aquaculture is another growing activity in the Amazon, especially of commercial fish species. Examples of sustainability in the exploitation of forested floodplains and regional biodiversity are the project of the Mamirauá **Sustainable Development Institute** and countless other development projects in different regions of the Amazon region in Brazil and other countries of the Amazon basin (Ayres and Ghosh, 1999).

The hemispheric and global importance of the Amazon

The hydrological cycle in tropical rainforests is interdependent with climate, the chemistry of the atmosphere, biogeochemical cycles and growth and degradation in the rainforest ecosystems. The Amazon region and its complex ecosystem of flooded forests, lakes and natural channels play an important role in regulating biogeochemical, aquatic and terrestrial and atmospheric cycles over the entire continent. Any alterations in this complex and dynamic ecosystem (such as deforestation, construction of dams, or disorderly occupation of floodplains) produces changes in the emission of greenhouse gases, regulation of regional and continental hydrological cycles, and the energy balance of the planet (Dikinson, 1987).

16.3.3   The Pantanal

The Pantanal is the largest wetland area on the planet (138,183 km²). It consists of a variety of floodplains, lakes, canals, rivers and flooded forests (see Figure 16.11). The region is the result of effects of flooding by the Paraguay River. Located in the central region of South America between the latitudes 16–22°S and longitudes 55–58°W, it includes parts of Brazil, Paraguay and Bolivia. The basin of the upper Paraguay drains an area of 496,000 km² shared by these three countries.

The Pantanal region is situated in an area of bas-relief, which during the rainy season is flooded by the Paraguay River and its tributaries. Precipitation varies between 1.25 m in the north and 1.10 m in the south. The region presents a water deficit 6–12 months of the year. The type of vegetation is savannah (closed).

Flooding levels in the Pantanal vary as a function of the vegetation and complex hydrological systems. Fluctuations occur in space and time, affecting the advance and retreat of water, the flora, fauna and also modulating, to some extent, the uses of the land (Da Silva 2000).

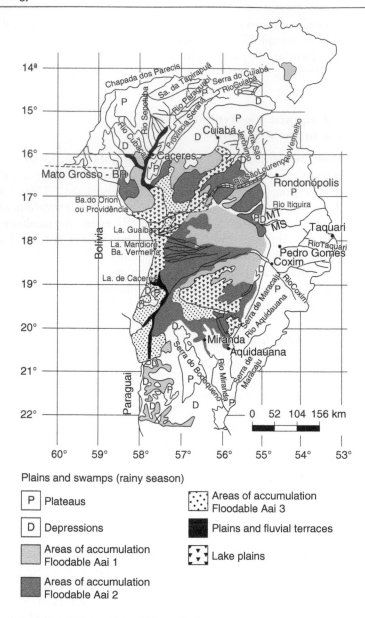

Figure 16.11 Pantanal wetlands and its drainage basin.
Source: Modified from Ab'Saber (1988).

In addition to the **annual flood pulse**, other interannual episodes of drought and flooding occur, resulting in functional and structural alterations in aquatic and terrestrial ecosystems, affecting the spatial organization in the mosaics of vegetation and lakes and the aquatic and terrestrial diversity. Located in an inland depression, it is defined by Ab'Saber (1988) as a complex plain of dentritic-alluvial headwaters that includes savannah-based ecosystems as well as biotic components of the dry Northeast and the peri-Amazonian region.

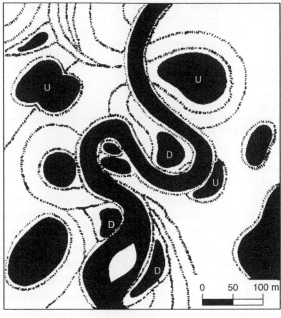

U – Umlaufsun   D – Darmufersen

Figure 16.12   The lakes of the Mato Grosso swamp and their origin from meanders, according to
Wilhelmy (1966).
Source: Tundisi (1994).

The Pantanal depression is located in an area that was a vast vaulted shell ('boutonnière' or crystalline domes – which designate areas of bulging or vaulted shields). Analyses of the **palaeogeographic behaviour** confirm the theory of emptying accompanied by eversion, pediplanation and detrital filling (Ab'Saber, 1988), or, that is, after the emptying of the vault, the detrital basin of the Mato Grosso Patanal was formed.

With the emergence of the Brazilian plateau, between the Cretaceous and the Pliocene, extremely large faults occurred in these vaulted shells, which can be observed in recent satellite imagery. The sedimentation basin has 400–500 m of accumulated sediment. Obtaining data on the **Pleistocene sedimentation** of the plain confirmed the formation of large alluvial fans.

Climatic changes and alterations in the regional hydrology shifted **semi-arid subtropical conditions** to humid tropical conditions in different geological periods. According to Ab'Saber (1988), the principal aquatic, subaquatic and terrestrial contours and ecosystems have developed over the past five or six millennia. The lakes of the Mato Groso Pantanal were classified by Wilhelmy (1958) (see Figure 16.12), who presented the hypothesis of lakes formed by dams and lakes formed from flooding of internal meandering lobes.

The Pantanal region is biologically diverse (see Table 16.6), as a result of paleoclimatic **fluctuations in the Quaternary** that caused the formation of areas of refuge shared by pockets of forests with greater biological diversity; and consequently these areas, with the return of tropical conditions, enabled the expansion of wild fauna and

Table 16.6   Species richness in the Mato Grosso swamp.

Communities	Number of species
Plants	1,863
Mammals	122
Reptiles	93
Fish	264
Birds	656

Source: Da Silva (2000).

flora from the refuges. In these areas, according to the refuge theory formulated by Braziläian researchers (Ab'Saber, 1977; Ab'Saber and Brown, 1979; Vanzolini, 1973, 1986), the increased competition enabled evolutionary patterns typical of Neotropical America. The expansion of dry climates to the edges of the depression of the Mato Grosso Pantanal is due to the presence of caatinga trees and shrubs. The dry areas began gradually to be colonized by different species of tropical vegetation during the shift from dry climates and return to humid tropical climates.

Diurnal variations in the Pantanal lakes are caused by thermal warming and cooling and the formation of layers with different densities during the day, with the presence of temporary thermoclines and the vertical transport of nutrients during the isothermal period. Periods of intense wind can also cause rapid circulation in shallow lakes; in some of these lakes, the primary production depends almost entirely on aquatic macrophytes (*Eichhornia azurea*, *Eichhornia crassipes*, and *Pistia stratiotes*). Vangil Pinto-Silva (1991) reported diel variations in water temperature (especially in shallow lakes) and in the processes of photosynthesis and respiration and cycles of dissolved oxygen and carbon dioxide. However, in some lakes with low macrophytic biomass, the levels of phytoplanktonic chlorophyll-*a* can reach $200 \, mg \cdot L^{-1}$. Banks of macrophytes in the lakes can serve as a source of food for benthic animals and insect larvae and as a breeding area for fish. Some fish species in the Pantanal, such as pacu (*Colossoma* sp.), feed on decomposing vegetation, fruits and seeds, and are important in the dispersal of vegetation. According to Por (1995), the Pantanal presents a wide diversity of mollusc species in the Ampullariidae family, especially South American bivalve molluscs. Freshwater crabs are also common in the Pantanal, especially from the Tricholactylidae family.

Aquatic birds, very common in the Pantanal, play an important role in the trophic network, transport of biological material, and colonization of permanent and temporary lakes. Large populations of aquatic birds in breeding sites produce increased fertilization around the lakes and ponds. Tundisi (unpublished data) found total phosphorus levels up to $800 \, mg \cdot L^{-1}$ in a pond located next to a nesting area by the **Cuiabá River**.

Da Silva (1990) summarized these effects and identified alterations in the ecological processes, primary productivity, decomposition, nutrient cycles, and succession of species and communities. Wide variations in electrical conductivity were identified in the lakes and rivers of the Pantanal, as the result of periods of drought and flooding, regional hydrogeochemistry, and periods of precipitation. Junk and Furch (1990) demonstrated variations from 5 to $350 \, \mu S \cdot cm^{-1}$ in the tributaries of the Pantanal's northern region. Mourão (1989) reported variations of 1,594 to $5,200 \, \mu S \cdot cm^{-1}$ for saline lakes in the southern region, and Da Silva and Oliveira (1998) measured

$850\,\mu\text{S}\cdot\text{cm}^{-1}$ in lakes located in areas with birds' nests. Por (1995) reported values of up to $4,000\,\mu\text{S}\cdot\text{cm}^{-1}$ for saline ponds in the Pantanal.

Changes in ionic levels and suspended matter are related to periods of precipitation and evaporation (Da Silva and Esteves, 1995).

The effects of drought and flood, with the resulting dilution/concentration of biomass, nutrients and dissolved ions, also are reflected in the cycles and biomass of phytoplankton, zooplankton, and communities of fish and birds (Espindola et al., 1999a). The decomposition of aquatic macrophytes and their growth and cycle depend on periods of drought and flooding, as shown by Da Silva and Esteves (1993).

Por (1995) estimated that more than 405 fish species live in the Pantanal, much less than the number of species in the Amazon basin.

Of the total number of bird species, 156 live or depend on the wetland areas, while 32 preferentially feed on fish (Contra and Artas, 1996).

In the wetland areas of the Pantanal, there is an abundance of semi-aquatic fauna, such as large mammals – capybara (*Hydrochoerus hydrochaeris*) – and also many reptiles (for example, *Caiman latirostris*), which play an important role in the food webs of lakes and wetland areas. Fluctuations in water level affect the behaviour and physiology of fish species, especially the breeding, feeding, migration and growth, as shown by Ferraz de Lima (1986) and Da Silva (1985).

The region has a large volume of water distributed in shallow waters and shallow lakes, mostly permanent and some temporary. Saline lakes described by Mourão (1989) have sodium carbonate (Na_2CO_3) as the main component.

The pH level can reach 10 and inorganic combined nitrogen is present mostly in the form of ammonia. The Patanal has several sub-regions, with patterns corresponding to different geomorphological conditions (Klammer, 1982; Adamoli, 1980); different phytosociological patterns, or even based on the occurrence and duration of floods (Adamoli, 1986).

The seasonal cycle of fluctuating levels controls the processes, cycles, succession and productivity of terrestrial and aquatic communities. The accumulation of biomass in shallow waters directly affects the physical and chemical processes in lakes, influenced by the respiration and photosynthesis of aquatic plants and by decomposition.

Junk and Da Silva (1995) published an extensive review comparing the Mato Grosso Patanal and the floodplains and flooding areas of the Amazon River and its tributaries. The authors concluded that in terms of regional hydrogeochemistry, major differences exist between the Amazon's floodplains and tributaries and the Pantanal and its tributaries. Most likely the reason is the diversity in the Pantanal's geological and mineral formations.

The Amazonian rivers directly affect the waters of the lakes and flood channels of the Amazon, whereas in the Pantanal, water from rainfall, evaporation, groundwater and tributaries greatly affect the hydrogeochemistry. In the Amazon region, the enormous discharges of the rivers connected to the floodplain and lakes, as well the presence of ample tropical rainforest, affect the hydrogeochemistry of the rivers in the region. In both the Pantanal and the Amazon floodplains, the influence of biomass is fundamental, especially in the biogeochemical cycles of nutrients such as nitrogen and phosphorus.

The Amazonian rivers have a major mechanical affect on the floodplains, because of their volume, and in the white-water rivers, sedimentation permanently modifies the structure of aquatic systems. The Pantanal depression functions as a giant sink of nutrients, and the physical performance of rivers is more apparent in or near the channels. The differences between the flooded forests of the Amazon and its floodplain and those of the Pantanal are striking. In the Pantanal, the predominant vegetation is herbaceous, with forested areas along the 70 riparian zones by the rivers.

Impacts on the Mato Grosso Pantanal

A series of impacts affect the functioning of the lakes and natural channels of the Pantanal. These impacts affect the ecosystem, fauna, flora and biodiversity, and include: deforestation; excessive agricultural activities that cause erosion; intensive and predatory fishing; hunting and predation by man of vertebrates, from fish to mammals; threats to birds' nesting areas; pollution from navigation; and industries and various industrial activities. Tourism and related activities also cause an impact. Mining, **waterway transport,** hydroelectric construction and urbanization are other impacts that occur.

These impacts have the following consequences: increased transport of sediment, increased toxicity, more exotic species, **erosion of river-banks, mercury contamination,** increased sedimentation, **alteration of flooded areas,** conversion of seasonal wetland areas into permanently flooded areas, pollution of water from untreated sewage, changes in the regime of flood pulses and consequences in the flood area (Da Silva, 2000).

16.3.4   The lakes of the Rio Doce

The lakes in the lake system of the middle Rio Doce are typical inland aquatic ecosystems, mainly because they do not connect with the Rio Doce and its tributaries. According to De Meis and Tundisi (1986), these lakes were formed in a period of 3000–10,000 years by the blocking of tributaries of the Rio Doce, after periods of intense precipitation and sedimentation, which gave rise to an unusual system of shallow lakes and wetland areas located in the tropical Atlantic forest regions (see Figure 16.13).

Of the 150 existing lakes, approximately 56 are in the protected area of the Parque Florestal do Rio Doce. Fifteen or more of these lakes have been the subject of ongoing studies, since the area has become a site of the long-term Ecological Research Program under the direction of Professor F. A. Barbosa.

A volume edited by Tundisi and Saijo (1997) summarized the studies conducted, and numerous scientific papers have also been published on primary productivity (Barbosa and Tundisi, 1980), circulation and deficit of dissolved oxygen (Henry *et al.*, 1997), and the structure and distribution of zooplanktonic organisms (Matsumura Tundisi *et al.*, 1997).

Average annual rainfall in the region is 1.5 meters, and average annual temperature is 22°C. The rainy season lasts from October to April, and there is a dry winter period with lower temperatures (<20°C). The main conclusions of the studies include: The lakes of the Rio Doce, in the midst of an Atlantic tropical forest, are subject to the permanent influx of allochthonous material. Decomposing organic matter settles in the sediment of the lakes, and its mobilization into the euphotic zone and epilimnion depends on the circulation process during winter. Since the lakes are protected by

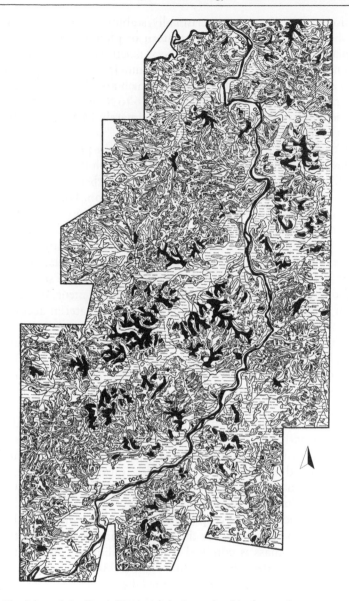

Figure 16.13   The lakes of the Doce River and the interplanaltic depression.
        Source: De Meis and Tundisi (1986).

'a sea of hills', the thermal stratification is well defined and stable, even in shallower lakes (for example, Carioca Lake, with a maximum depth of 12 m). The stratification, which reaches its maximum in January or in summer, starts gradually in August and, after the initial thermal heating, stabilizes with the input of rainwater, which increases the density difference of the deepest layers, completing the process (Tundisi, 1997).

Stratification results in a characteristically stable metalimnion in several lakes, which establishes a vertical distribution pattern of phytoplanktonic, zooplanktonic and bacterioplanktonic species. Photosynthetic bacteria occur in the lower layers of the metalimnion, where light penetration is about 1% of the solar radiation reaching the surface and where H_2S and CH_4 occur, which are abundant in the hypolimnia of the lakes during the long stratification period (6 to 8 months). The organic carbon content of the sediment is extremely high (approximately 14.5–22% of the dry weight of the sediment surface).

Organic nitrogen in the sediment is also high, but the phosphorus content is relatively low. The availability of nutrients from the hypolimnion into the epilimnion is controlled by the intense stratification period, when there is limited nitrogen and phosphorus for the primary production of phytoplankton. During winter, the availability increases because of diffusion and gradual thermal cooling.

The lakes located outside the Parque Florestal do Rio Doce (Lake Amarela and Jacaré Lake) were submitted to the influence of allochthonous matter from *Eucalyptus* sp. This allochthonous matter, less rich than that from the Atlantic tropical forest, results in lower levels of organic matter in the upper layers of the sediment in the lakes and increased organic matter in greater depths of the sediment, due to the previous contribution of the Atlantic forest to the allochthonous material accumulated in the sediment.

The diversity of aquatic macrophytic species is relatively low in four of the lakes studied (Dom Helvécio, Lake Carioca, Lake Amarela and Jacaré Lake) and includes: *Eichhornia azurea*, *Typha dominguensis*, *Salvinia auriculata*, several species of *Nymphaea* sp., as well as the submerged species of *Cabomba piauhyensis* and *Najas conferta*.

In Dom Helvécio and Jacaré Lakes, the larvae of *Chaoborus* and *Chironomus* dominated the zoobenthos. A typical **diel vertical migration** has been described for these lakes. The co-existence of four different *Chaoborus* species, identified by their larvae, can be explained by interspecific spatial separation, partition of resources and selection of prey. Fukuhara *et al.* (1997) observed synchronization between the emergence of *Chaoborus* adults and the lunar cycle.

The primary production of phytoplankton is limited by nutrients, because of the absence or low level of circulation in the period of intense stratification. Much of the phytoplanktonic biomass is confined to the metalimnion, where there are high levels of cyanobacteria, as described by Hino *et al.* (1986).

Primary production of organic matter in the lakes occurs from photo-autotrophic carbon fixation, emergent and submerged macrophytes in the shallow lakes, and photosynthetic bacterioplankton in the metalimnion of some lakes. The **heterotrophic fixation** of carbon by bacteria may also be important, as well as quantitatively significant **chemosynthetic processes** in the hypolimnion (Barbosa and Tundisi, 1980).

Studies on the fish fauna in four lakes showed the presence of 27 species of fish belonging to 11 families. The introduction of *Cichla ocellaris* (tucunaré) and *Pygocentrus* sp. (piranha) in Dom Helvécio and Jacaré Lakes drastically reduced the native fish fauna, especially the population of an important engraulideo, *Lycengraulis* sp. *Hoplias malabaricus* (traira) and *Geophagus brasiliensis* (acara) were also found in the lakes studied. Caracoids dominated the fish fauna in the lakes prior to the introduction of tucunaré and piranha.

Impacts on the lakes of the middle Rio Doce

The lakes of the middle Rio Doce suffer the impact of following human activities:

▶ deforestation of the tropical Atlantic forest;
▶ construction of roads and erosion;
▶ removal of wetland areas;
▶ introduction of exotic fish species;
▶ tourism activities that affect the lakes: excessive fishing, deforestation;
▶ plantations of *Eucalyptus* sp. that replace the natural vegetation.

According to Barbosa, Esteves and Tundisi (1982), the litter of the tropical Atlantic forest plays an important role in the metabolism of the lakes and in the biogeochemical cycles. Removal of this forest therefore causes problems in the biogeochemical cycles of lakes and the structure and chemical composition of the sediment.

16.3.5 The La Plata basin

The Paraná, Paraguay and Uruguay Rivers

Four rivers – the Paraguay (2550 km), Uruguay (1612 km), Paraná (2570 km) and Plata (250 km) – form the La Plata basin, which covers 3 million square kilometers. The basin, covering portions of southeastern Brazil, northeastern Argentina, southeastern Uruguay, southeastern Bolivia and all of Paraguay, is the most developed in South America, with a population of approximately 150 million inhabitants. It is the second largest basin in South America and the fifth largest in the world (Welcomme, 1985).

The basin presents four well-defined areas: the *upper* basin, the *high*, the *middle* and the *lower* (Bonetto, 1994). The main river of the La Plata basin is the Paraná, formed by the **Paranaíba** and Rio Grande Rivers, in Brazilian territory. Over the last 80 years, studies on the limnology, ecology and aquatic biology of the four geographic areas have been conducted by researchers from Brazil, Bolivia, Argentina, Paraguay and Uruguay. The upper Plata basin, formed by the upper Paraná River and its tributaries, has undergone major alterations due to construction.

A group of researchers from Nupélia, of the State University of Paraná, extensively studied the upper and middle Paraná River and its flood valleys. A recent volume edited by Thomaz, Agostinho and Hahn (2004) summarizes the principal physiographic, physical, ecological and conservation mechanisms of this region. There is an extensive and varied literature on the region's geomorphological and geological aspects, including stratigraphic, palynological, and sediment-transport studies and the effects of dams on the natural processes in these flood valleys (Rocha *et al.*, 1999, 2001; Fernández, 1990; Crispim, 2001; and Thomaz *et al.*, 2004). The upper Paraná river basin, according to these authors, covers an area of 802,150 km². The main tributaries of the Paraná are presented in Table 16.7.

The **fluvial channels** present a dynamic range of processes that control the transport and deposition of alluvial material, with differences in each tributary, due to the velocity of the current, type and origin of soil, and geological structure. The presence of large islands in these stretches of the Paraná River is a result of the transport and deposition of sediments. The surface of the interfluvial averages (Bigarella

Table 16.7   Principal tributaries of the upper Paraná River.

River	Length of river (km)	Area of the basin (km²)	Annual average discharge (m³·s⁻¹)
Paranaíba[r]	1,075	222,000	3,000
Grande[l]	1,227	143,000	2,100
Tietê[l]	1,150	74,100	602
Iguaçu[l]	1,320	69,000	1,542
Ivaí[l]	860	34,000	727
Ivinheima[r]	444	31,100	287

[r]Right bank; [l]Left bank

and Ab'Saber, 1964) forms the divisors of the waters in the basins of the small and large tributaries of the Paraná River. The deposits from these tributaries are the main sources of sediments in the basin. These deposits can contain sand, clay, and also quartz (see Figure 16.14).

The floodplain of the Paraná River covers an area of approximately 450 km in length, which is characterized by two areas: low terrace and fluvial plain (Sousa Filho and Stevaux, 2004). The fluvial plain varies along its extension, because of the presence of palaeochannels, with spatial patterns of large intertwined branches, natural channels, small lakes and basins. In depressions in the floodplains, there are lakes and ponds associated with the main channel of the Paraná river or natural active channels with many diversified flood areas. According to Filho and Stevaux (2004), the Paraná River is a multichannel river with multiple channels with interconnected branches.

All this dynamic and varied structure, which undergoes periodic flooding, presents an enormous and constant challenge for the survival, colonization and perpetuation of the aquatic biota.

This ecosystem can thus be considered to represent a set of dynamic processes of short-term (several days) and seasonal fluctuations and variations, depending on the level of fluctuation of the water, the velocity of the current and its characteristics.

It is important to note, in addition to this situation, major climatic and hydrological changes occurred during the Quaternary, featuring four climatic events, according to Santos and Stevaux (1998): The first period was arid, 40,000 years ago; the second period was humid, 7500–8000 years ago; the third period was arid, 3500–4500 ago; and the fourth period was humid, approximately 1500 years ago and continuing to the present. Climatic change that occurred during the Quaternary must have been accompanied by **neo-tectonic movements** that resulted in incisions, new channels, fluvial deposits, formation of small terraces, and resurgence of semi-arid conditions (Iriondo, 1988; Souza Filho, 1993).

Alterations in the cycles of precipitation, erosion, drought, growth of forests and development of savannas (closed) reflect the changes that occurred in the geomorphology, geology, hydrography, hydrodynamics, water temperature, pH, transport of sediment, and precipitation. These changes clearly had a significant impact on the aquatic biota and its diversity, on the spatial organization, and on the succession in the entire drainage basin.

The construction of dams in the region of the upper Paraná had a significant impact on these complex fluvial systems of the river and its dynamics. The impacts

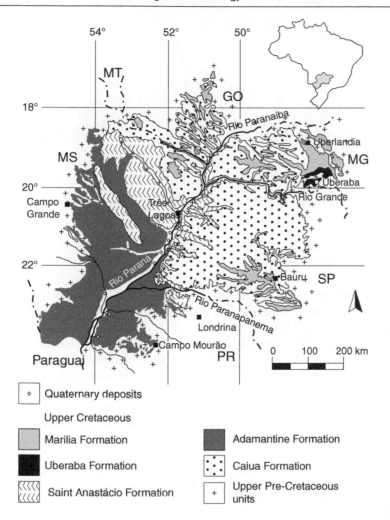

Figure 16.14   Geological map of the basin of the upper Paraná, with emphasis on Cretaceous rocks. *Source:* Stevaux (1997) in Vazzoler *et al.* (1997).

of the **Porto Primavera** dam, Ilha Solteira dam, Jupiá dam, Tres Irmãos dam, Itaipu dam, and Rosana dam were examined in the study by Souza Filho *et al.* (2004). One of the major impacts has been alterations in the hydrological regime, the outflow, and transport of sediment by the river. Transport of sediment involved its concentration change from $24.9\,\text{mg}\cdot\text{L}^{-1}$ to $14.74\,\text{mg}\cdot\text{L}^{-1}$ immediately upon construction.

Fernandez (1990) found changes in the margins and the pattern of erosion. The alterations produced by the construction of dams and their impact downstream occurred within a decade, with the modification of the hydrological regimes. This provoked new processes, spatial and temporal, of adjustments in the anastomosed structures of channels, with effects on the aquatic biota.

The hydrological regimes of the Paraná River (see Figure 16.15) present flood pulses of varying magnitudes, which play an important role in the functioning of the

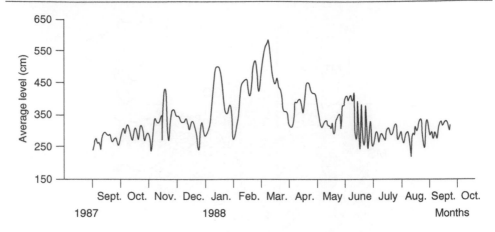

Figure 16.15   Fluctuations in the water level in the Paraná River throughout the year.
Source: Thomaz *et al.* (1997).

flood valleys and permanent lakes. The permanent ponds in the Paraná River flood valleys and other associated ecosystems (rivers, temporary ponds, streams) present the variables listed in Table 16.8.

The water levels play a key role in the dynamics of the permanent and temporary lakes and ponds in the flood valleys, which depend, according to Junk *et al.* (1989), on the level and volume of flooding and the degree of connectivity with the rivers and main channels. There is clearly a delay in the response time of the limnological variables, especially dissolved oxygen, total phosphorus, total nitrogen, transparency and chlorophyll-*a*. This last variable presents a bimodal pattern that depends on the hydrometric level, with lower values during periods of flooding (maximum 20 µg chlorophyll-*a* · L^{-1}). The flood pulses have a diluting effect on the phytoplankton (Huszar, 2000), which reduces the biomass. During periods of high water level, the processes of respiration and decomposition exceed the net primary production (Paes da Silva and Thomaz, 1997).

Despite the effects of construction of dams, Thomaz *et al.* (2004) determined that a certain degree of seasonality occurs, depending on the existing regulated hydrological cycle.

According to Neiff (1990, 1996), aquatic communities in the great flood valleys are regulated by periods known as potamophase (river phase) and limnophase (lake phase, dry phase). The hydro-sedimentological pulse, with dilution or concentration of phytoplanktonic and zooplanktonic organisms, occurs in these phases. The flood pulse is the main forcing function acting on the upper, middle, and lower Paraná River.

Differences in the hydrological cycle and the periodic disturbance of the **system of fluvial valleys** and lakes may explain the high diversity of phytoplankton, and the intensity, frequency and regularity of flooding depend not only on the degree of connectivity between the channels and the main river but also on the length of retention time, morphometry and topographical position (Garcia de Emiliani, 1993; Train, 1998). The shallow ponds of the flood valley of the Paraná present great complexity in the composition of phytoplankton, with marked seasonality due to the pulses of

Table 16.8 Values of some limnological parameters recorded in different habitats of the floodplain in the Upper Paraná River.

Environment	Temperature (°C)	Secchi D·(m)	pH	Electric cond. (µS·cm⁻¹)	Total alk. (meq·L⁻¹)	DissolvedO₂ (% Sat.)	N-kjeldahl (mg·L⁻¹)	Total-P (µg·L⁻¹)	Chlorophyll-a (µg·L⁻¹)
Lakes	23.7 (3.9)	0.90 (0.40)	6.6 (0.5)	30.8 (10.0)	0.26 (0.09)	61.9 (31.8)	0.70 (0.34)	65.4 (42.4)	8.6 (9.1)
	15.8–31.7	0.25–2.85	5.1–9.1	16–55	0.08–0.49	6.4–116.0	0.20–2.59	9.3–262.2	0.2–64.7
	n = 116	n = 57	n = 116	n = 107	n = 107	n = 115	n = 91	n = 107	n = 114
Semi-lotic environments		0.91 (0.26)	6.9 (0.4)	27.0 (9.7)	0.24 (0.09)	88.3 (20.4)	0.46 (0.17)	44.0 (11.3)	7.8 (8.2)
		0.45–1.80	5.8–7.6	16–58	0.11–0.52	40.0–126.4	0.18–1.07	17.4–66.5	0.4–35.7
		n = 57	n = 58	n = 57	n = 42	n = 57	n = 45	n = 40	n = 58
Ivinheim River	24.0 (3.8)	0.7 (0.4)	7.0 (0.3)	41.3 (4.4)	0.40 (0.07)	88.5 (16.7)	0.36 (0.14)	51.2 (19.9)	1.8 (1.3)
	16.8–30.5	0.15–2.95	6.3–7.6	32–55	0.22–0.62	43.7–116.7	0.10–0.68	27.8–132.3	0.1–4.9
	n = 46	n = 46	n = 46	n = 46	n = 42	n = 46	n = 33	n = 38	n = 35
Paraná River	24.2 (3.3)	1.1 (0.5)	7.4 (0.3)	58.4 (6.2)	0.44 (0.05)	104.4 (8.7)	0.32 (0.11)	23.5 (11.6)	2.5 (1.4)
	18.3–30.0	0.35–2.15	6.7–8.2	42–74	0.27–0.57	67.8–125.7	0.14–0.6	4.9–53.6	0.1–6.3
	n = 69	n = 68	n = 68	n = 68	n = 64	n = 51	n = 49	n = 53	n = 46
Temporary lakes	24.0 (2.7)	0.30	6.2 (0.5)	53.6 (24.7)	0.31 (0.20)	69.2 (30.4)	2.08 (1.06)	223.0 (113.3)	–
	18.2–27.7	0.05–1.55	4.9–6.8	24–131	0.06–0.87	4.0–139.0	0.36–5.38	28.0–348.5	–
	n = 24	n = 24	n = 24	n = 24	n = 23	n = 24	n = 24	n = 24	–
Streams	23.5 (3.2)	–	6.1 (0.4)	57.2 (7.1)	–	99.8 (26.2)	0.45 (0.24)	60.3 (46.7)	–
	18.2–29.3	–	5.1–6.8	41–74	–	88–172	0.14–1.14	16.0–202.0	–
	n = 24	–	n = 24	n = 24	–	n = 24	n = 24	n = 24	–

Average values are presented, with standard deviation (in parentheses), the range of variation and the number of observations (n).
Source: Thomaz et al. (1997).

energy and matter associated with flooding. The phytoplanktonic diversity studied in several lakes varied from 139 to 272 species (Train and Rodrigues, 2004).

The responses of phytoplankton to different patterns of flooding, vertical mixing and concentration of nutrients vary, according to these authors, in a function of seasonality, ability to maintain stocks of diatomaceous algae in the sediment (Reynolds, 1994) and their resuspension. The distribution pattern in size and biovolume of phytoplankton vary according to the limnophase and potamophase.

Colonization and succession of periphyton are also related to the hydrometric cycle and level, the existing resources and the regime of environmental disruption/ stress produced by the flooding or receding of water. Physical disturbances, according to Rodrigues and Bicudo (2004), play a controlling role in community dynamics, favouring the development of Chlorophyceae; the increase in periphytonic biomass was greater during the high-water period. The content of chlorophyll-*a* tends to increase in the shift from lentic to lotic environment. This may be related, according to Rodrigues and Bicudo (2004), to increased phosphorus. According to these authors, however, changes in habitat type are the principal regulating factors in the cycle of the periphyton community, in addition to alterations in the hydrological cycle. The **dynamics of nutrients**, especially phosphorus, also appear to be a decisive factor in periphyton succession.

A number of authors (Lansac-Tôha *et al.*, 1997) extensively studied the zooplankton of the Paraná River and its flood valleys, including testate amoebae of the genus *Difflugia* with 27 species and *Arcella* with 18 species; *Centropyxis* with 8 species (which are dominant in the zooplankton, comprising 75% of the species found); rotifers (25 families); and microcrustaceans. The increased species richness during high water levels was related to the increased number of habitats as a result of the flooding, increasing the availability of food and transport of zooplanktonic organisms during the flood, increasing homogenization of fauna and connecting habitats. The presence of aquatic macrophytes in permanent ponds can influence the diversity of zooplankton (Lansac-Tôha *et al.*, 2004).

The **communities of zoobenthos** in these floodplains were studied by Takeda *et al.* (1997). According to the authors, an important component in the ecosystems is the vegetation, which is composed mainly of Gramineae and Polygonaceae. The influence of riparian vegetation (which provides food of all kinds, particularly detritus, for invertebrates, mainly scavengers) varies with each phase of the hydrological pulse. The composition of zoobenthos also varies depending on the substratum (often fine organic matter or sandy matter from inorganic sources and decomposing vegetation, such as leaves and twigs). The benthic macro-invertebrates found in floodplains are mainly holometabolic insects, and the factors that influence the spatial distribution of these benthic invertebrates are: 1) substratum (consolidated or not), 2) yield of the main river, 3) available food, 4) alterations in the terrestrial system, 5) flood pulse. The temporal variations that affect the physical and chemical conditions of rivers, channels, lakes and floodplains also affect the temporal variations in benthic invertebrates, especially those in the littoral zone, according to Takeda *et al.* (1997).

The abundance and dominance of the taxa of benthic macroinvertebrates depend on the morphometry, the volume of lakes and channels, the presence or absence of macrophytes, and the phase of the flooding cycle. For example, Takeda *et al.* (1991a, 1991b) reported potential increases in chaoborids during the high-water phase. Rivers,

basins, channels and floodplain ponds present different compositions of benthic invertebrates, and the associations are related to various determining factors. Substratum is
one of the key components in which nematodes, oligochaetes, chironomids and harpacticoid copepods live (in interstitial spaces in the sandy sediment). In areas with mud
and fine substratum material *Campsaurus* sp. predominates, burrowed in the mud.

It is difficult to recognize a determinant pattern, due to differences in substrata,
the hydrological cycle, and the level and stage of flooding. As in other communities of
floodplains, the range of forcing functions is large and complex, determining spatial
and temporal patterns on a much larger scale than the spatial and temporal scale of
a single floodplain.

The ichthyofauna was studied by Agostinho *et al.* (1997), who identified 170 species of fish, six of which were introduced from other basins (corvina: *Plagioscion squamosissimus*; peacock bass: *Cichla monoculus*; tilapia: *Oreochromis niloticus*; wolf
fish: *Hoplias lacerdae*; tiger oscar: *Astronotus ocellatus*; and tambaqui: *Colossoma
macropomum*).

In the channel of the Paraná River, there are 100 species. Its tributaries Ivinhema
and Iguatemi present 91 and 77 species, respectively. In the floodplain of the Paraná
River, there are 103 species in the ponds, due to the diversity of habitats, shelter, and
food. Many species use the ponds for development, and adults are adapted to fluctuations in the dissolved oxygen, water temperature and other physical and chemical
conditions, as also occurs with fish in flooded areas of Africa (Beadle, 1981) or in the
Amazon (Junk, 2006).

The predominance of Characiformes and Siluriformes (85% of both orders) is
typical in these environments. Some species are located in certain regions or temporary ponds, and others, such as *Astyanax bimaculatus*, are widely distributed, probably due to greater tolerance to environmental conditions and physical and chemical
fluctuations.

Table 16.9 shows several indices of numbers of species, diversity and even distribution of the various components of the flood valleys. Figure 16.16 shows the
catch, per unit of effort, of the 15 main species in different periods. Agostinho *et al.*
(1997) listed 35 vulnerable species of fish in this stretch of the Paraná River basin.
Among these there are several commercially valuable species, such as the pintado
(*Pseudoplatystoma coruscans*), pacu (*Piaractus mesopotamicus*), dourado (*Salminus
maxillosus*) and jau (*Paulicea luetkeni*).

The reproductive success of these species is affected by barriers in the region,
especially for highly migratory species – potamodromous species that migrate to

Table 16.9 Simpson's index of diversity (H), equitability (E) and number of species (N) of fishes in
different environments and periods shown.

| Environments | 1986–87 | | | 1987–88 | | | 1992–93 | | | 1993–94 | | |
	N	H	E	N	H	E	N	H	E	N	H	E
Lakes	53	0.840	0.856	52	0.862	0.878	51	0.916	0.934	48	0.907	0.926
Channels	54	0.895	0.856	58	0.919	0.935	48	0.930	0.949	49	0.884	0.902
Rivers	63	0.954	0.912	62	0.936	0.951	72	0.932	0.945	62	0.900	0.915

Source: Agostinho *et al.* (1997).

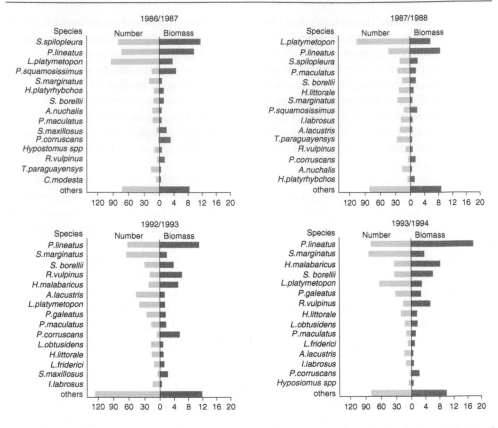

Figure 16.16   Capture per unit of effort, in number and biomass (number of individuals or kg/1000 m² net/24 h) of the 15 main species of fish in different years of sampling.
Source: Agostinho *et al.* (1997).

preferred habitats for spawning and early development. According to Agostinho *et al.* (1992), dams affect the floodplain with a set of impacts ranging from hydrodynamic alterations and transport of sediment to the interruption of species' migratory routes and reduction and/or alteration in feeding and reproductive niches.

The **diet** of the fish fauna in the Paraná River floodplains showed great trophic variety and adaptability. The diet of 57 species analyzed by Hahn *et al.* (1997) includes molluscs, insects (aquatic and terrestrial), detritus, zooplankton, algae, other fish species, sediment and vegetation (decomposing aquatic and terrestrial vegetation). The principal food sources are insects, fish and microcrustaceans. Detritus-eating fish (scavengers) are common, particularly loricariids, as detritus is the predominant food resource.

The insects most exploited by fish in these ecosystems are the chironomids, followed by the ephemeroptetids. Feeding activities of fish vary with the characteristics of the environment and species. The floodplain and fluvial valleys are widely utilised for feeding; various fish species utilise all the environments at the same time. Patterns of diurnal and nocturnal feeding activities occur, showing circadian rhythms, with light/dark cycles as synchronizing or stimulating factors. The trophic relationships between the components of the floodplain's aquatic biota are complex and varied, with extensive exploitation of feeding niches. Agostinho *et al.* (1997) described eight trophic categories for fish, according to their preferred or dominant food: herbivores, planktophages, insectivores; iliophages; detritivores, benthophages, piscivores, and omnivores. Aquatic and terrestrial food resources are utilised by these fish species.

Lotic and semi-lotic environments play an important role in reproduction, as they provide reproductive habitats for small and medium-sized species. Non-dammed tributaries of the Paraná River provide important spawning areas for floodplain species, where the young of these species are common.

Most commercial species use river channels (lotic environments) to reproduce and ponds and canals as areas for growth and recuperation. Hence the need exists, according to Agostinho *et al.* (1993, 1995), to keep the floodplain intact in its complexity. Vazzoler *et al.* (1997) reported environmental affects on the seasonality of reproduction, especially water temperature and light intensity, together with flooding cycles and day length.

The close relationship of **reproductive periods, reproductive intensity**, day length, temperature and flood cycles is such that when the eggs hatch, the larvae can make use of food and shelter provided by the environment of the plains.

Vazoller *et al.* (1997) considered **fluvometric levels** to be proximate synchronizing factors, flood peaks to be proximate finalizing factors, and expansion of floodplain environments to be terminal factors. The authors propose (p. 278 op. cit.) that: (1) temperature and daylength are predictive triggers initiating gonad maturation, (2) onset of flooding is a synchronizing factor for spawning, (3) flood peak is one of the final triggers finishing the reproductive process.

Nakatami *et al.* (1997), in studies on fish eggs and larvae in the Paraná River floodplain, showed that some species develop all phases of their life cycle in the flooded areas, while migratory species use these areas for only part of the cycle. The floodplain/pond regions are essential for the growth, feeding and early development of fish species. Larvae that occur together with aquatic macrophytes are more pigmented; larvae from more pelagic zones are less pigmented.

A large number of researchers studied the fluvial system of the middle and lower Paraná, including Bonetto (1970, 1975, 1976, 1978, 1994), Bonetto *et al.* (1968, 1969, 1972, 1981, 1983, 1994), Neiff (1975, 1986), Paggi (1978), Poi de Neiff (1981, 1983) and Zalocar *et al.* (1982).

The middle Paraná functions as a typical river with large flood valleys and a large and expanded fluvial valley. The chemical composition of the water in this stretch differs from that in the upper Paraná. Average conductivity is $90\,\mu S \cdot cm^{-1}$, twice that of the upper Paraná. Bicarbonate is the most common anion in the middle Paraná (average $>35\,mg \cdot L^{-1}$). The pH values tend to be around 7.5 in the middle Paraná, and sulphate levels are reduced. In the lower Paraná, average conductivity increases ($120\,\mu S \cdot cm^{-1}$), and the most common ions are $HCO_3 > SO_4^{2-} > Cl^-$ with $Na^+ > Ca^{2+}$ and increased levels of sodium and sulphate.

Studies along the middle and lower Paraná clearly show the influence of flows of the river in the functioning of the permanent and temporary ponds and natural channels.

According to Bonetto *et al.* (1969), the dynamics of the river are constantly shifting, during both high-water and low-water periods, altering the composition, biomass

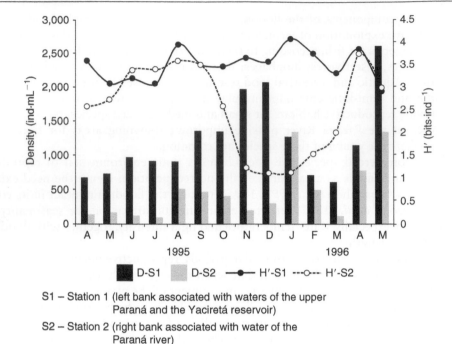

S1 – Station 1 (left bank associated with waters of the upper
      Paraná and the Yaciretá reservoir)

S2 – Station 2 (right bank associated with water of the
      Paraná river)

Figure 16.17   Fluctuations in phytoplankton in the Paraná River. Variations in density (D) and diversity
(H′ = Shannon-Weaver index) in two stations.
Source: Zalocar *et al.* (2007).

and productivity of the communities of phytoplankton, zooplankton and periphyton. Figure 16.17 shows fluctuations in the density of phytoplanktonic populations in the middle Paraná. During periods of flooding, a diluting effect was also observed in the upper Paraná. The Paraguay River receives high levels of suspended matter from the **Bermejo River.**

The fauna associated with floating or submerged aquatic plants is abundant and varied. Groups such as nematodes, oligochaetes, cladocerans, copepods, ostracods, amphipods, decapods and molluscs can have as many as 70,000 individuals.kg^{-1} dry weight. In the case of the fauna associated with *Eichhornia crassipes*, there were 100,000 individuals·kg^{-1} dry weight.

The fauna associated with these macrophytes varies according to the species of macrophytes and their permanence or temporality in the seasonal cycle. Neiff (1986) reviewed the aquatic plants of the Paraná River and classified them according to seven functional units, based on their dependence on the flow of the river and currents:

▸ Aquatic plants that live in the river and its tributaries. These plants are all **Podostemaceae** – according to Dugand (1944), *tachyreophyton*. The plants occupy a specialized, high current, niche.
▸ Aquatic plants associated with low to moderate flows, occurring along the banks in permanent wetland areas. These **rheophytic** plants are mainly dominated by

Paspalum repens, *Panicum elephantines*, and *Echinochloa polystachya*. They are plants with roots and their total biomass can reach $15\,t \cdot ha^{-1}$.

▶ Aquatic plants in seasonal wetland areas (*Polygonum stelligerum*, *Ludwigia peploides*). They grow in wetland areas, but not permanently flooded areas.

▶ Aquatic plants that occur in areas where water reaches only during exceptional flooding. *Typha latifolia*, *Typha dominguensis* and *Cyperus giganteus* are all common. Typically, these plants form islands.

▶ Aquatic plants in permanent lakes subjected to annual flooding: *Salvinia* spp., *Eichhornia* spp., and *Myriophyllum brasiliensis* are common.

▶ Aquatic plants in lakes that undergo only occasional exceptional floods (the floating species *Eichhornia crassipes*, *Pistia stratiotes*, *Salvinia* spp. and the submerged *Cabomba australis*).

▶ **Amphibious plants** that are subjected to three or four months of flooding each year. These include *Panicum prionitis* and *Andropogon lateralis*.

The spatial and temporal fluctuations of these aquatic plants depend on the velocity of the current, flood valleys, frequency of flooding and the physiographic, morphological and morphometric characteristics of rivers. The principal forcing factors or functions acting on the spatial and temporal distributions of aquatic macrophytes include the interactions of the components of flow and sedimentation and of the regime of the river.

Therefore, the dynamics of the major fluvial valleys of the Paraná River, which are quite pronounced in the middle and lower Paraná, are closely related, quantitatively and qualitatively, to flood pulses. These dynamics regulate biogeochemical processes, geo-morphological processes and community structure. The production of organic matter depends on flood pulses, and the organization of communities and its metabolism are related, in large part, to the concentration of dissolved organic carbon, particulate organic carbon, and the decomposition rates and activity in the bacterial communities (Drago, 1973; Depretis and Cascante, 1985, Power *et al.*, 1995).

The presence of detritivorous fish species that feed on particulate organic matter is characteristic of the Paraná River flood valleys. Aquatic macrophytes play a key role in the functioning of the ecosystem, particularly in the cycling of organic matter and nutrients. Floating macrophytes provide shelter for fish, alter the water quality and provide food and shelter for aquatic invertebrates (Poi de Neiff *et al.*, 1994).

The fish fauna of the Paraná River includes 540 to 550 species, although Bonetto (1986a) estimates the total number closer to 600. The upper Paraguay and the Pantanal have the richest community of fish species (about 300). The number of species declines toward the south, downstream. The fish fauna of the upper Paraná River has little in common with the middle and lower Paraná. In the upper Paraná, there are close to 130 species, but a significant number are endemic.

The **seasonal behaviour of fish** can be **sedentary** or migratory. The sedentary species remain in one region and explore a habitat for breeding and growth without undertaking migrations. These are small and medium-sized species. Since the migrating species are potamodromous, they travel upstream for hundreds of miles and after reproduction return to the main channel of the river. The fingerlings and larvae move downstream into floodplain lakes and ponds where they complete their development. The migratory movements of fish on the Paraná River are attributed to reproduction,

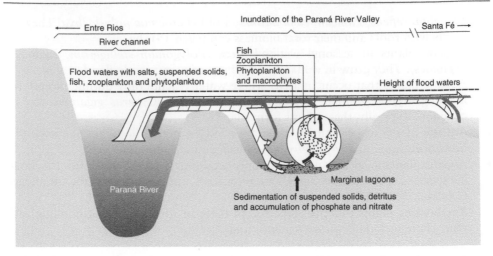

Figure 16.18   Exchanges between lotic and lentic waters in the middle Paraná River and the different components of the trophic network.
Source: Bonetto *et al.* (1969).

Authors' note

The number of species of fish in the various different sub-basins of the Paraná River varies from researcher to researcher. This is due to different collection methods, problems of identification, periods of collection and fishing efforts. Although inconsistent, the orders of magnitude of the number of fish species, when compared, are reasonable and within the margins of statistical sampling error.

food and also temperature. Bonetto *et al.* (1963, 1971, 1981a) followed fish migrations, particularly commercially valuable species (see Figure 16.18).

Migrations involve distances of more than 2000 km (upstream and downstream). Migration rates of $1016 \, km \cdot day^{-1}$ and $2122 \, km \cdot day^{-1}$ were observed for *Prodilodus platensis* and *Salminus maxilosus*, respectively (Godoy, 1975). Characiformes and Siluriformes are the main components of the fish fauna in the Paraná River; the Characiformes account for 40% of the total fauna.

Multiple uses of water resources in the Plata basin

The water resources of the Paraná, Paraguay, Plata and Uruguay Rivers and their tributaries are used intensively for a wide range of economic activities and the rivers play an essential role in the development this region. The 'pioneering frontiers' of urban development, deforestation, and agricultural production have accelerated since the second half of 19th century, first to the northwest of São Paulo, with the railroads, and then toward the south, in the state of Paraná, and west, in the state of Mato Grosso.

From 1920, the frontiers intensified even more, occupying previously undisturbed areas, barely explored by anyone but indigenous people. The final occupying force of the space and the large agricultural areas, which increasingly used the water resources and biota, was the construction of highways and dams, particularly in the upper Paraná and its tributaries. Dense human populations in the western part of the state of São Paulo and the states of Paraná, Santa Catarina, and Mato Grosso produced rapid and irreversible urbanization, which caused a growing demand for water resources in the second half of the 20th century, when rapid industrialization and agribusiness started demanding increasing investment in energy, irrigation and public supply.

Currently, the following multiple uses occur in the La Plata River basin and its main tributaries:

▶ urban and rural public water supply;
▶ irrigation for agribusiness;
▶ waterway transport – commercial navigation;
▶ energy production (hydroelectric);
▶ fishing in rivers and dams;
▶ intensive aquaculture in reservoirs;
▶ recreation – parks and marinas;
▶ tourism – tourist boats.

This set of multiple uses occurs to a greater or lesser degrees in all areas of the Plata River basin down to the estuary of the Plata River (Rosa, 1997).

Impacts and consequences of multiple uses

One of the major impacts that occurred in the water systems of the Plata basin was the development of an intense energy production programme, with the construction of dozens of hydroelectric dams (more than 60) in 80 years (Tundisi *et al.*, 2006). These dams caused major irreversible changes in the flood valleys of the whole system of the upper Paraná, with consequential effects on the middle and lower Paraná as well (Tundisi, 1993).

In addition to the impacts of hydroelectric development, other complex impacts affecting the spatial organization of the fauna and flora system are worth mentioning:

▶ Withdrawal of water for public supply and dumping of untreated sewage.
▶ Irrigation and discharge of waters with excess fertilizers and pesticides.
▶ Industrial waste. Uses of water in industries and dumping of waste.
▶ Navigation and its impacts.
▶ Recreation and tourism, which cause various different impacts throughout the entire basin.
▶ Excessive commercial fishing and introduction of exotic species, especially in reservoirs.

These impacts cause the following processes of degradation:

▶ **Alteration of the hydrological regime** by the dam, increased evaporation and impacts on aquifers.
▶ Eutrophication of reservoirs and rivers, with large-scale growth of aquatic macrophytes.

▶ Excessive growth of cyanobacteria and the impact on water quality, organisims and public health. Effects on the structure and dynamics of food webs.
▶ Removal of native fish species and introduction of exotic species that alter the structure and operation of the food web.
▶ Introduction of exotic species, for example, the mollusc *Limnoperna fortunei* (see Chapter 18).
▶ Loss of aquatic biodiversity.

All these impacts have serious consequences on public health, the structure of fish communities and aquatic plants, and economic repercussions due to increased costs of water treatment, loss of aquatic and terrestrial biodiversity, and the general increase in toxicity and costs for society.

One of the greatest impacts and consequences, with huge economic losses, is the loss of the ecological integrity of these great rivers, represented by the economic losses related to ecosystem services. The loss of evolutionary capacity must also be taken into account, as well as the deterioration in the dynamics of the gene flow between the system's components, with both quantitative and qualitative consequences that severely impact the aquatic biota and processes (Tundisi, 2007).

16.3.6   Limnology in Argentina

In addition to the extensive studies conducted on the basin of the Plata River by Bonetto (1986, 1993), Bazan and Amaja (1993), Bechara (1993) and Martinez (1993), Argentine limnologists produced a set of very consistent information about the Neotropical and subtropical aquatic biotas (Ezcurra de Drago, 1972, 1974, 1975). Studies on the geographical distribution of cladocerans and zooplankton in reservoirs and ponds were done by Paggi (1978, 1980, 1989, 1990). Benthic fauna as indicators of trophic conditions or contamination was studied by Marchese (1987) and Marchese and Ezcurra de Drago (1992, 2006). Quirós (1990, 2002) and Quirós *et al.* (2006) conducted further studies on reservoirs and ponds in the pampas, and eutrophication processes were studied by Cirelli *et al.* (2004, 2006). Research at S.C. de Bariloche has examined the chemistry and biology of natural highly acidic waters on slopes of the Andes.

The ecology and biology of the fish communities in the Paraná River and its tributaries were the subject of extensive research by Baigun *et al.* (2005) and Oldani (1990). Neiff (1978) and Poi de Neiff *et al.* (1994, 1997) studied the aquatic macrophytes and associated fauna in the Paraná River under the effects of periods of flooding and drought. The hydrodynamics of the Paraná River and its tributaries, sediment transport and biogeochemical cycles were studied by Drago (1973) and Drago *et al.* (1981, 1998), and the biogeochemical cycles of the floodplains by Bonetto *et al.* (1983, 1994). In addition, a series of works on the ecology and physiology of the protozooplankton and zooplankton of lakes in southern Argentina was developed by Balseiro (2002).

Regional limnology in Argentina has presented many advances, particularly in community dynamics, commercially valuable fish species, the impacts on rivers and reservoirs and modifications in the fluvial dynamic and biogeochemical cycles.

16.3.7 Limnology in Uruguay

Conde and Sommaruga (1999) studied the development of limnology in Uruguay, a country in the Plata basin, located in the temperate zone, with temperatures of 17°C in spring, 25°C in summer, 18°C in autumn, and 12°C winter. Average annual precipitation is 1.1 m in the southern region and 1.3 mm in the northern region. Rainfall is evenly distributed throughout the year. Figure 16.19 illustrates Uruguay's five major hydrographic basins.

One of the main features in Uruguay is an extremely diversified network of rivers, is 200,000 hectares of wetland areas and three large rivers (Plata, El Negro and Uruguay). Coastal ponds are important ecotones already intensively studied (Jorcin, 1993, 1996).

The limnological studies on urban reservoirs describe their trophic state and chemical and biological makeup (Pintos and Sommaruja, 1984; Fabian, 1995). Carbon flows between bacteria, flagellates, ciliates, rotifers and macrozooplankton were

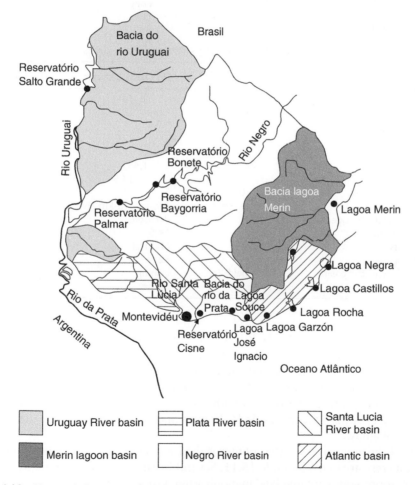

Figure 16.19 Uruguay's five major drainage basins including the most important freshwater systems. *Source:* Conde and Sommaruga (1999).

Chart 16.4   Regions in Uruguay with higher risk of eutrophication.

Region	Consumption of phosphate fertilizers	Level of eutrophication	Productive activities and human settlements	Related cases of eutrophication
Centre-South	1,700 (11%)	Moderate to high	Land used for agriculture, advanced industrial development and three large cities	Colonization of macrophytes in rivers during the summer
West	5,475 (45%)	Moderate to low	Land for agriculture and the largest human after Montevideo	Blooming of *Microcystis* in bays and reservoirs
Southeast	4,777 (13%)	Moderate to low	Extensive rice fields and high discharge of fresh water into the ocean; no large city	Periodic occurrences of yellow tide in coastal ocean areas

Source: Modified from Sommaruga *et al.* (1995).

Table 16.10   Comparative characteristics of Uruguay's three coastal lagoons.

Feature	Negra lagoon	De Castillos lagoon	De Rocha lagoon
Distance from the coast (km)	4	10	0.1
Area of lagoon (km²)	142	100	72
Average depth (m)	2.9	1	0.56
Maximum depth (m)	3.8	2	1.4
Volume (km³)	0.42	–	0.04
Basin area (km²)	720	1,453	1,312
Area of basin/area of lagoon	5.1	14.5	18.2
Important tributaries	0	1	4
Influence of the ocean	Marine wind	Along a 10-km river	Direct
Connection with the ocean	Low	Periodic	Periodic
Average flow (m³·s⁻¹)	Low	14.6	570
Hydrologic regime	Modified	Natural	Natural
Population	None	Rare	30,000
Human activity in the basin	Rice fields	Fishing	Fishing
Industries	None	None	Few

studied in another reservoir (Lake Rochó) by Sommaruga (1995). Conde *et al.* (1995, 1996a, 1996b) conducted research on polluted urban rivers and large hydroelectric reservoirs, such as the Salto Grande and Bonete reservoirs. Sommaruga and Conde (1990), Sommaruga and Pintos (1991), and Jorcin (1996) studied coastal ponds.

Chart 16.4 presents the regions with higher risk of eutrophication in Uruguay (Sommaruga *et al.*, 1995), and Table 16.10 presents comparative data on three coastal ponds of Uruguay.

One of the great rivers in the country is the Uruguay River, whose main limnological parameters are given in Table 16.11. Several institutions, municipal governments and public companies are involved in monitoring, limnological studies, sanitation and environmental policy in Uruguay. The Departments of Limnology and Oceanography

Table 16.11   Selected limnological features of the Uruguay River.

Parameter	Value
Annual flow (km^3)	145.0
Drainage (mm · year^{-1})	443.0
Conductivity (μS · cm^{-1})	47.0
pH	7.35
Alkalinity (mg · L^{-1})	23.1
Total sediment in suspension (mg · L^{-1})	76.0
Sodium (mg · L^{-1})	5.0
Potassium (mg · L^{-1})	2.0
Calcium (mg · L^{-1})	7.0
Magnesium (mg · L^{-1})	2.0
Chloride (mg · L^{-1})	3.0
Sulphate (mg · L^{-1})	5.0
Bicarbonate (mg · L^{-1})	36.2
Reactive silica (mg · L^{-1})	15.0
Dissolved organic carbon (mg · L^{-1})	6.6
Particulate organic carbon (mg · L^{-1})	1.15
Chlorophyll *a* (μg · L^{-1})	2.6

Source: Conde and Sommaruga (1999).

have contributed significantly to the scientific knowledge on the limnology and ecology of reservoirs, rivers, estuaries and basins in the inland and coastal areas of Uruguay.

Limnological studies in Uruguay have increasingly applied scientific information to the management and recovery of eutrophicated reservoirs, for the **recovery of hydrographic basins**.

16.3.8   Limnological studies in Venezuela

Lakes, reservoirs and rivers in Venezuela have been studied in Venezuela for a long time, including aquatic ecosystems, hydrographic basins, geological and geomorphological formations. An important contribution to regional and global limnology, based on studies of inland aquatic systems in Venezuela, was the research on the Orinoco River (Weibezahn *et al.*, 1990). In turn, a significant contribution to the scientific work on aquatic ecosystems in Venezuela was made by Infante (1988, 1997) and Infante and Riehl (1984). In the latter work, the authors highlighted the effects of cyanobacteria and their toxins on the zooplankton in **Lake Valencia**. On this lake there have been detailed studies by Lewis and his associates. Further comparative studies were done on reservoirs in Venezuela and Nicaragua (Infante *et al.*, 1992, 1995).

More recent studies on the phytoplankton in reservoirs, primary productivity and the impacts of enrichment on the functioning of the phytoplanktonic community were presented by Gonzalez *et al.* (2000, 2003, 2004) as well as studies of zooplankton (Gonzalez *et al.*, 2002) and comparisons of the composition of zooplankton and phytoplankton with the trophic state (Gonzalez *et al.*, 2002, 2003).

More dynamic studies with experimental microcosms were developed by Gonzalez *et al.* (2001), as well as research on the **Pao-Cachinche reservoir**, a tropical hyper-eutrophic reservoir used for public water supply and irrigation. Continuing with this line of investigation, Gonzalez *et al.* (2006) presented a project of recuperation and management of this reservoir, using modern aeration methods. Limnological studies in Venezuela show that the transfer of basic knowledge to management systems is consolidating and quickly becoming one of the channels for competent management of lacustrine systems, especially reservoirs.

Historically, important contributions to aquatic ecology and aquatic biology have been made by Rodriguez (1973), through studies on the Maracaibo system. Aquatic systems in Venezuela are subject to impacts such as discharge of pollutants, phosphorus, nitrogen, toxic substances and suspended matter, making it difficult to manage and treat the water.

16.3.9   Limnological studies in Colombia

A large number of researchers, including Mathias and Moreno (1983), Roldán *et al.* (1988, 1992, 1999, 2001a, 2001b, 2003), Zuniga Cardozo *et al.* (1997) and Gavilán-Diaz (1990), have conducted limnological research in Colombia. The studies gave emphasis to biological indicators of water quality, and the data collected clearly advanced experimental methods and scope in Colombia's rivers, lakes and reservoirs.

Roldán (1992) produced an important work on Neotropical limnology. The use of wastewater treatment systems was implemented with *Eichhornia crassipes* (Floresz, 1990; Eicheverri *et al.*, 2006) and new areas of study began recently in Colombia (Bolaños and Pelaez-Rodriguez, 2006).

Limnology in Colombia clearly shows how the development of a line of scientific investigation, applied to regional limnology, can be useful for the whole continent, using the methodology developed in these regional systems.

16.3.10   Limnology in Chile

Lakes and reservoirs in Chile are in temperate dry climates, at altitudes that cause them to receive only low concentrations of nitrogen. Primary production in these lakes is related to bottom temperature and nitrogen levels. The lakes and reservoirs in Chile are subject to the Humboldt Current, which influences the climatic processes and the functioning of inland aquatic systems (Pardo and Vila, 2006). Cabrera (1984) published a large volume of information on the primary productivity in Chilean lakes and physical and chemical characteristics of Araucanian lakes (Campos, 1984; Campos *et al.*, 1978, 1983, 1987, 1993, 1998). Other studies were conducted on primary productivity, biogeochemical cycles and the importance of sediment in these cycles (Pizarro *et al.*, 2003; Pizarro and Rubio, 2006).

Cabrera *et al.* (1995) studied UV radiation in Chile and the impact on phytoplankton and primary productivity. Detailed and important studies on the primary productivity of phytoplankton in Rapel reservoir were published by Cabrera *et al.* (1977), Caraf (1984), Bahamonde Cabrera (1982) and Montecino (1981). Reynolds *et al.* (1986) published an extensive study of *Melosira* (*Aulacoseira* spp.) in the same reservoir.

16.3.11   Glacial-valley lakes in the Southern Hemisphere

Araucanian lakes studied by Thomasson (1963) are an interesting example of lakes in glacial valleys formed during the last glacial period. These lakes are located between

Table 16.12   General features of 11 Araucanian lakes.

Lake	Year of discovery (A.D.)	Altitude (m)	Area (km²)
Cólico	–	–	57.2
Calbuco	–	–	53.0
Villarrica	1550	230	172.2
Calafquén	1576	240	121.3
Panguipulli	1576	140	114.6
Rimihue	1576	117	82.8
Ranco	1552	70	407.7
Puyehue	–	212	153.3
Rupanco	1553	172	224.1
Llanquihue	1552	51	851.1
Todos los Santos	–	184	180.7

Source: Margalef (1983).

the Andes and the Pacific, 39–41°S latitude. The region is known as Araucania, because of the old name of the inhabitants. Table 16.12 shows some characteristics of 11 of Araucanian lakes.

The hydrological equilibrium of these lakes depends on the water supply from the Andes, resulting from the thaw and heavy rainfall. In Lake Villarrica, extensively studied, the most common phytoplanktonic species were in the genus *Melosira* (*Melosira ambigua*, *Melosira granulata*, *Melosira* spp. – now all *Aulacoseira* spp.) and *Rhizosolenia eriensis*. In zooplankton, protozoa were identified (mainly of the genus *Stentor*), rotifers (*Hexarthra fennica*), cladocerans (*Bosmina chilensis* and *Ceriodaphnia dubia*). Few species of cyanophytes were found. Thomasson (1963) divides the lakes studied into two groups: lakes with a predominance of *Melosira granulata* (now *Aulacoseira granulata*) and better nutrient conditions, and lakes with predominantly *Dinobryon* sp. with poorer nutrient conditions and small drainage area.

The absence of predator zooplankton (cladocerans or cyclopoid) in these lakes (Margalef, 1983) is notable.

16.3.12 Typology of reservoirs in the state of São Paulo

In 1978–79, a FAPESP-sponsored limnological study was conducted on 52 reservoirs in São Paulo. The purpose of the project was to produce a comparative study of reservoirs, which could represent a gradient of limnological conditions and regions of preservation/impact in the state of São Paulo. Another goal of the project was to help standardize the methodology used in studying reservoirs and strengthen mechanisms to train support staff at the graduate, Ph.D. and post-doctorate studies.

The project enabled a broad limnological assessment of 52 reservoirs and contributed to the advancement of aquatic biology in reservoirs in Brazil and in subtropical regions. The main scientific findings, descriptions and summaries were published over the last 25 years (Tundisi, 1981, 1983, 1993, 1994; Arcifa *et al.*, 1981; Esteves, 1983). The most important conclusions of all this work were:

▶ Most reservoirs in the state of São Paulo are polymictic, with intermittent periods of stratification and circulation, and many circulations annually. In some cases, stratifications occur due to the specific conditions of construction, such as, for example, 'hydraulic stratification' (Tundisi, 1984) (see Chapter 12).

▶ Sediments contain high levels of organic matter, nitrogen and phosphorus, as a result of contamination processes in the water basins and the polymixis, which, especially in the case of phosphorus, forces the precipitation of ferric phosphate in the sediment.

▶ The distribution of planktonic organisms, mainly zooplankton, and the calanoid/cyclopoid ratio are tied to the electrical conductivity and ionic levels in water. Alterations in the composition of phytoplankton and zooplankton over the last 20 years are due to variations-promoted by differences in levels of anions and cations (Tundisi *et al.*, 2003).

▶ The primary production by phytoplankton was moderate to high (0.5–3.0 g $C \cdot m^{-2} \cdot day^{-1}$), reflecting processes of eutrophication and contribution of nitrogen and phosphorus from water basins.

▶ The project provided a base of limnological data that was essentially monitoring the alterations due to the degradation of water basins and the increased contributions of nitrogen and phosphorus from untreated domestic sewage and agricultural activities. Subsequent studies in reservoirs already sampled revealed a 15-fold increase in primary production by phytoplankton in 20 years (Barra Bonita reservoir) (Tundisi, 2006).

▶ The fish fauna has undergone major changes because of the introduction of exotic species, such as, for example, *Plagioscion squamosissimus* in the Tietê area, which greatly impacted the pelagic fauna of these reservoirs.

▶ The types of reservoirs generated an essential database for scientific information, since recent comparative data are based on this initial database.

The work done at the UHE Carlos Botelho reservoir (Lobo/Broa) since 1971 (see Figure 16.20) culminated in the development of the project 'Types of Reservoirs in the

Figure 16.20   UHE Carlos Botelho reservoir (Lobo/Broa), where limnological studies began with methods that led to the project 'Typology of reservoirs in the state of São Paulo' and the other projects in Brazil, especially in the Southeast region (see also color plate section, plate 21).

State of São Paulo' (Tundisi and Matsumura Tundisi, 1995). These works pioneered the approach of studying 'hydrographic basins' and the introduction of technologies and methods in studies on populations and communities (Panitz, 1979; Tundisi, 1983; Chamixaes, 1994).

16.3.13   The program Biota/FAPESP

In the Biota/FAPESP programme, 220 aquatic ecosystems were studied in the state of São Paulo, the majority reservoirs. Physical and chemical parameters were collected, as well as zooplankton from the littoral and pelagic zones, for comparison. The distribution of these zooplanktonic species was investigated in function of the trophic state of the system, the degree of contamination, and the species' tolerance to certain physical and chemical conditions of aquatic systems. The studies also examined the geographic distribution of zooplanktonic species in the state of São Paulo.

Results of this program include the data presented in Figure 16.21, which shows the distribution of Cyclopoida in State of São Paulo.

According to the study by Matsumura Tundisi and Tundisi (2003, 2005), the factors that influence the distribution of zooplankton and phytoplanktonic succession, especially in reservoirs, are: the nature of the trophic process, the electrical conductivity of water, presence or absence of predators, intra-zooplankton predation and the degree of water contamination (heavy metals, dissolved organic substances, substances and other components in the water and sediment).

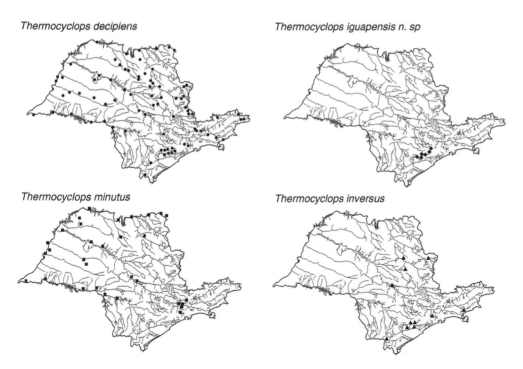

Thermocyclops decipiens

Thermocyclops iguapensis n. sp

Thermocyclops minutus

Thermocyclops inversus

Figure 16.21   Distribution of Cyclopoida (species of *Thermocyclops*) in the State of São Paulo.
    Source: Silva and Matsumura Tundisi (2005).

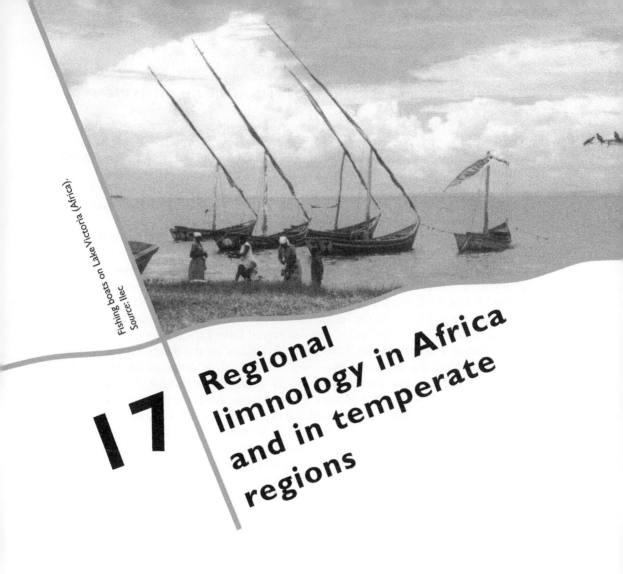

Fishing boats on Lake Victoria (Africa).
Source: Ilec

17 | Regional limnology in Africa and in temperate regions

SUMMARY

This chapter reviews and summarizes limnological studies conducted in Africa and on inland aquatic ecosystems in temperate regions. The principal lacustrine districts studied will be examined as well as the contributions from different specialists to understanding the functional mechanisms of lakes, reservoirs and rivers.

Case studies in lacustrine districts were selected that have advanced regional limnology and contributed to the development of global limnology.

Particular emphasis is given to studies on very old lakes, which present important particularities of biodiversity and functioning.

17.1   LAKES AND RESERVOIRS ON THE AFRICAN CONTINENT

Beadle (1981) published an extensive review of limnological research and studies on African rivers and lakes. From the 15th–18th century, Portuguese, French, English and German explorations and expeditions took place in the interior of Africa starting from the Senegal and **Gambia** Rivers.

The history of European exploration in Africa, according to Fage (1978) and Beadle (1981), focused on the search for the sources of the Nile, a river that played a major role in the Mediterranean civilizations.

European expeditions in Africa uncovered the existence of the **Niger River**, the origins of the Nile (Moorhead, 1962), Lake Tanganyika, Lakes Albert and Victoria, the **Zambezi** River, and lakes **Nyasa** (now Malawi), **Mweru** and **Bangweulu** (these last two were discovered by the famous British explorer David Livingstone).

Emil Pasha (or Eduard Schnitzer) was another important explorer who collected species of flora and fauna in the region of the Nile and confirmed some of the origins of this river, as identified by Speke, and discovered **Lake Edward**.

For almost a century, from 1796 to 1889, studies on the geography of the great African rivers and the lakes associated with them in their drainage basins (the Nile, Niger, Zaire and Zambezi Rivers) were mostly published in Europe. The Niger River was discovered in 1796 by the Scottish explorer Mungo Park; the discovery was considered an important landmark in geographical exploration of African rivers and lakes.

After 1890, according to Beadle (1981), scientific interest kept growing, and projects continued in the areas of geology, botany and zoology (terrestrial and aquatic). Moore's expeditions to Lake Tanganyika (Moore, 1903) in 1894 and 1897 can be considered a milestone in the limnology of African lakes and rivers. Since 1920, limnological studies continued on African lakes, with a growing number of expeditions and more consistent technically evolved studies (Worthington and Worthington, 1933). Limnological research work on African lakes and rivers intensified even more after 1945, focusing in particular on the commercial development of fisheries and fish farming, as well as the need for further study of the aquatic biology and limnology of aquatic ecosystems in Africa.

The construction of large dams/reservoirs in various countries has driven efforts in limnology. The installation of laboratories in many African universities, especially since 1950, has accelerated local-based research.

Figures 17.1 and 17.2 show the distribution of lakes and reservoirs in Africa at different latitudes, and the characteristics and major lakes in the fault valleys of the African continent ('**Rift Valley lakes**').

Dumont (1992) and Talling (1992) published extensive reviews of the factors that regulate and control the functioning of shallow lakes and wetland areas in Africa. Talling (1969) compiled annual variations in the water's surface temperature and later (1992) those in shallow lakes in general in Africa. The annual cycle of surface temperature generally follows the annual solar radiation cycle. Near the equator (on all continents) small seasonal variations occur at lower temperature in higher-altitude lakes (Löffler, 1968).

Long-term stratification in shallow lakes is rare because of wind action (Talling, 1992, defines shallow lakes as those with mean depth <5 m). However, diurnal variations are quite common (diel cycles); steep temperature gradients can occur in a lake's

Figure 17.1   Distribution of lakes and reservoirs in Africa.
Source: Talling (1993).

upper layers (1–2 m). In shallow parts of lakes near the shoreline, higher temperatues can be more marked, with differences of up to 2°C in surface water temperature during the day (Talling, 1992). Tundisi (unpublished results) made the same observation in Barra Bonita (São Paulo, Brazil). With increasing salinity, thermal heating can be greater, as shown by Melack and Kilham (1972) in lakes in East Africa.

Diurnal cycles have been extensively studied in Lake George (Talling, 1992). The interactions of temperature cycles of the water, wind regime, and depth determine the type of thermal stratification patterns and density that occur. Surface temperatures of up to 35°C were measured in some very shallow waters (~<2 m) in African lakes. Tundisi (unpublished results) measured temperatures of 36°C in shallow lakes in the Amazon. The diurnal temperature cycles limit and control the cycles of dissolved oxygen and pH. Thermal and chemical stratifications disappear with strong winds during the night linked to thermal cooling. This pattern of vertical nocturnal mixing

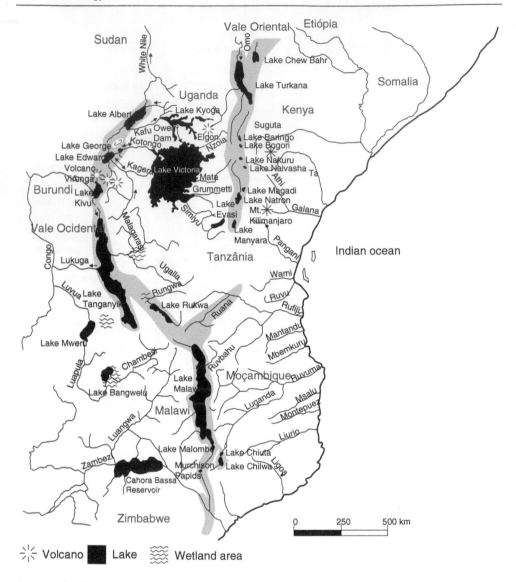

Figure 17.2   Principal features and lakes of the valleys and faults on the African continent.
Source: Beadle (1981).

significantly redistributes the dissolved oxygen and chemical elements and substances in these shallow lakes (Beadle, 1932).

Seasonal cycles in lakes in Africa are controlled by rainy and dry periods, the topography of the lakes and the precipitation/evaporation ratio. In turn, the chemical composition of these lakes depends on the regional hydrogeochemistry and precipitation/solar-radiation ratio. Table 17.1 presents the chemical composition of numerous shallow African lakes. According to Talling (1992), factors controlling the chemical composition and the limnological functioning of these shallow lakes include thermal

Table 17.1   Chemical composition of the water in shallow African lakes (most <5m mean depth).

Lake	Country	Date	K_{20} ($\mu S \cdot cm^{-1}$)	Σ Cations	Σ Anions	Na^+	K^+	Ca^{2+}	Mg^{2+} ($meq \cdot L^{-1}$)	Alk.	Cl^-	SO_4^{2-}	Total P	PO_4 P ($\mu g \cdot L^{-1}$)	Si ($mg \cdot L^{-1}$)	Total Fe ($\mu g \cdot L^{-1}$)	pH	References
Nabugabo	Uganda	Jun. 1967	25	0.198	0.199	0.090	0.028	0.060	0.020	0.140	0.040	0.019	–	–	–	–	7.0–8.2	Beadle (1981)
Tumba	Zaire	1955	24–32	–	–	–	–	0.03	0.02	0	–	–	–	–	–	–	4.5–5.0	Dubois (1959)
Bangweulu	Zambia	1960	–	0.285	0.293	0.113	0.033	0.075	0.066	0.260	0.08	0.02	–	–	–	–	–	Talling and Talling (1965)
Opi A	Nigeria	Jan. to Feb. 1980	–	0.315	–	0.113	0.049	0.100	0.053	–	–	–	–	–	–	–	–	Hare and Carter (1984)
		May 1980	15.3	–	–	–	–	–	–	0.1	–	–	–	15	–	100	6.5	
Mweru	Zambia–Zaire	Jul. 1961	76	1.03	1.05	0.20	0.032	0.375	0.418	0.83	0.141	<0.1	–	–	4.9	–	–	Talling and Talling (1965)
Tana	Ethiopia	Mar. 1964	137	1.68	1.62	0.24	0.040	0.945	0.45	1.52	0.044	0.052	–	30	6.8	–	8.4	Wood and Talling (1988)
Ras Amer	Sudan	Jan. 1956	178	–	–	–	–	1.20	–	0.81	–	–	–	200	11	–	9.1	Talling (unpublished)
George	Uganda	Jun. 1961	201	2.37	2.39	0.59	0.11	1.01	0.66	1.91	0.25	0.23	412	<18	8.5	250	9.6	Talling and Talling (1965)
Kabara	Mali	Feb. 1976	(199)	2.63	2.55	0.40	0.37	1.30	0.56	1.70	0.48	–	–	–	–	–	–	Dumont et al. (1981)
Mulehe	Uganda	Jun. 1961	260	2.94	3.09	0.470	0.246	1.085	0.63	2.18	0.34	0.65	272	220	15.9	49	8.0	Talling and Talling (1965)
Naivasha	Kenya	Jun. 1961	330	3.92	3.97	1.96	0.58	0.76	0.63	3.31	0.41	0.25	122	–	15.2	500	–	Talling and Talling (1965)
Zwei	Ethiopia	Mar. 1964	322	3.72	3.80	2.11	0.30	0.70	0.615	3.32	0.24	0.22	–	–	21.1	–	8.0	Wood and Talling (1988)
Baringo	Kenya	Dec. 1979	530	6.3	6.11	4.85	0.33	0.70	0.35	4.93	0.82	0.36	70	–	14.0	5,410	–	Talling and Rigg (unpublished)
Chad-N	Chad–	Jul. 1976	(565)	6.66	–	1.87	0.76	2.22	1.81	6.27	–	–	–	–	11.8	–	8.7	Carmouze et al. (1983)
Chad-SE	Nigeria	Aug. 1976	(45)	0.55	–	0.12	0.07	0.20	0.16	0.46	–	–	–	–	4.5	–	7.7	
Mohasi	Uganda	May 1952	–	7.47	8.60	9.1	9.1	7.10	0.41	0.76	0.22	7.41	–	–	4.1	–	–	Damas (1954)
Kitangiri	Tanzania	Jul. 1964	785	8.60	9.15	6.74	0.123	1.205	0.55	6.65	1.80	0.10–0.71	1,020	–	16.1	–	–	Talling and Talling (1965)
Abaya	Ethiopia	Feb. 1964	623	9.1	9.1	7.70	0.41	0.76	0.22	7.41	1.10	0.60	128	–	18.7	–	–	Wood and Talling (1988)
Tete pan	South Africa	Mar. 1976	(187)	3.18	–	1.70	0.03	0.30	1.15	–	–	–	–	4	–	–	–	Rogers and Breen (1980)
		Oct. 1976	(720)	11.04	–	6.70	0.05	1.26	3.03	–	–	–	–	34	–	–	–	

(continued)

Table 17.1 Chemical composition of the water in shallow African lakes (most <5 m mean depth) (continued).

Lake	Country	Date	K_{20} ($\mu S \cdot cm^{-1}$)	Σ Cations	Σ Anions	Na⁺	K⁺	Ca²⁺	Mg²⁺ ($meq \cdot L^{-1}$)	Alk.	Cl⁻	SO₄²⁻	Total P	PO₄ P ($\mu g \cdot L^{-1}$)	Si ($mg \cdot L^{-1}$)	Total Fe ($\mu g \cdot L^{-1}$)	pH	References
Hippo Pool	Uganda	Nov. 1969	978	8.58	8.25	0.65	2.61	2.60	2.72	5.27	2.82	0.15	—	1,120	—	—	6.4	Kilham (1982)
Chamo	Ethiopia	Jul. 1966	—	10.8	11.7	9.1	0.36	0.70	0.64	9.4	1.66	0.62	—	14	18	—	8.9	Wood and Talling (1988)
Chilwa	Malawi	Jan. 1970	1,000	12.85	—	11.3	0.35	0.60	0.60	6.7	7.89	—	—	5,100	—	—	8.5	McLachlan (1979)
Sonachi	Kenya	Dec. 1970	2,500	35.85	—	33.9	0.59	0.66	0.70	19.0	14.51	—	—	5,200	—	—	8.8	Talling and Rigg (unpublished)
	Kenya	Dec. 1979	4,770	58.6	59.9	53.4	4.41	0.33	0.44	52.6	4.41	2.91	450	—	32	530	—	
Marlut. Sta.1	Egypt	1966	—	59.1	59.3	45.03	1.45	2.80	9.83	5.23	45.15	8.88	—	—	—	—	—	El-Wakeel et al. (1970a, 1970b)
Rukwa-N	Tanzania	1961	5,120	51.7	67.7	49.6	2.17	<0.05	<0.08	53.3	10.79	3.44	4,500	—	54	—	—	Talling and Talling (1965)
Kilotes	Ethiopia	Apr. 1963	—	75.7	77.4	70.5	4.5	0.7	<0.6	63.4	13.6	0.4	—	5,500	15.0	—	9.6	Wood and Talling (1988)
Nakuru	Kenya	Dec. 1979	10,500	139.0	139.0	136	29.6	0.05	0.01	107	25.3	6.7	650	—	66	620	—	Talling and Rigg (unpublished)
Elmenteita	Kenya	Jul. 1969	11,700	172	182	165	7.3	<0.1	<0.1	107	55.5	2.8	—	9,200	83	—	9.4	Hecky and Kilham (1973)
Abiata	Ethiopia	Mar. 1964	15,800	228.5	240.5	222	6.5	<0.1	<0.1	166.5	51.5	22.5	—	50	60	—	10.3	Wood and Talling (1988)
Eyasi	Tanzania	Aug. 1969	23,500	301	324	300	0.24	6.1	23.7	116.4	186.5	17.3	—	86,000	8.4	9.5	9.5	Hecky and Kilham (1973)
Qarun	Egypt	Jun. 1978	—	616	532	493	6.1	23.7	93.3	3.6	181	347	191	—	—	—	—	Talling and Rigg (unpublished)
Bogoria (Hammington)	Kenya	Jan. 1970	56,400	1,275	1,205	1,235	9.9	<0.05	0.18	965	180	4.5	—	—	122	—	10.6	Hecky and Kilham (1973)
Pretoria Salt Pan	South Africa	1978–1980	(52,000)	1,264	1,249	1,260	3.3	<0.05	<0.1	400	845	5.0	9,000	7,000	120	—	10.4	Ashton and Schoeman (1983)
Metahara	Ethiopia	May 1961	72,500	784	831	774	10.4	<0.15	<0.6	580	154.6	97.5	11,000	—	—	500	9.9	Wood and Talling (1988)
Manyara	Kenya	Jun. 1961	94,000	937	1,097	935	2.4	<0.5	<2.5	806	244	47.5	65,000	—	8.9	—	—	Talling and Talling (1965)
Magadi	Kenya	Feb. 1961	160,000	1,666	1,867	1,652	13.7	<0.5	<2.5	1,180	627	50	11,000	—	117	—	—	Talling and Talling (1965)
Mahega	Uganda	May 1971	(111,300)	2,879	2,870	2,565	302	0.76	11.0	150	1,450	1,270	—	9,600	13	—	10.1	Melack and Kilham (1972)
Gaar (Wadi Natrun)	Egypt	Aug. 1976	—	—	5,620	5,959	34.8	—	—	220	4,900	500	—	4,120	—	—	10.9	Imhoff et al. (1979)

Source: Beadle (1981).

balance, variations in dissolved oxygen levels, diel variations in stratification and des-tratification, the effects of wind on the vertical circulation, and density stratification due to salinity. The interactions of sediment with water are extremely important in shallow lakes. Sediments accumulate in drainage basins, and the shape and topography of lakes can promote salinization.

Shallow African lakes often present high levels of plant and animal biomass (aquatic macrophytes, phytoplankton, zooplankton, and in some lakes, elevated biomass of fish). This biomass influences the physical and chemical functioning of the system, altering the levels of dissolved oxygen and CO_2 in the water.

Periods of precipitation and intense evaporation produce significant expansions and contractions of the masses of water and wetland areas, with important adaptations of the aquatic flora and fauna and uses of the lakes by human populations (Talling, 1992).

Dumont (1992) presented an extensive analysis of the factors regulating and controlling plant and animal species and communities in shallow lakes in Africa. The author examined fluctuations in water level (Figure 17.3a shows data from Lake Chad), water temperature, salinity, acidity, dissolved gases, turbidity, wind, substratum type, climate, and geological and human actions as basic factors in controlling fauna and flora of these shallow lakes. He also presented the case study of Lake Chad (Carmouze et al., 1983) and that of Lake George (Burgis et al., 1973).

One of the key adaptive mechanisms discussed by Dumont (1992) is the adaption to desiccation and excessive salinity (2000–$3000\,\mu S \cdot cm^{-1}$) of flora and fauna that promotes rapid growth and repopulation after periods of precipitation, dilution, and increased volume (see Figure 17.3b).

Another widely studied African lake is Tanganyika (Coulter, 1991). Extensive studies have been conducted on the lake's geography, hydrodynamics, composition of fauna and flora, zoogeographic evolution, ichthyology and benthic fauna. According to Coulter (1991), the diversity and abundance of the aquatic fauna in lakes Tanganyika, Victoria and Malawi are parallel to terrestrial fauna in diversity and specificity. Lake Tanganyika is a classic example of a biologically diverse lake, equivalent to an ancient temperate-region lake (such as Baikal in Russia).

Studies on the fish fauna of Africa's man-made lakes and reservoirs were summarized by Lévêque et al. (1988) and Lowe-McConnell (1987), who published extensive reviews of the fish faunas, their evolutionary features and comparisons with the fish fauna of South America, showing the abundance of cichlids in lakes Victoria, Tanganyika and Malawi. Figure 17.4 shows the relationship between the area of the drainage basin in systems in Africa and the number of fish species (according to Welcomme and de Merona, 1988), who calculated the quantity of biomass of the different communities of fish.

As previously mentioned, Talling (1965b) compared the seasonal cycle of Lake Victoria with that of a temperate-region lake (Windermere). Table 17.2 summarizes data on primary productivity in lakes in Africa (from Beadle, 1981).

The repercussions of diel variations on the dynamics of aquatic communities, with examples from lakes in Africa, were discussed in Chapter 7. Limnological studies were conducted on Lake Victoria (Talling and Lemoalle, 1998), Lake Chilwa (Kalk, McLachlan and Howard-Williams, 1979), Lake Chad (Carmouze, Durand and Lévêque, 1983) and Lake George (Ganf 1975; Ganf and Horne, 1975; Ganf and Viner, 1973).

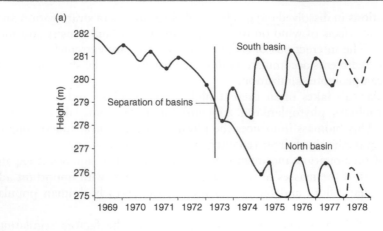

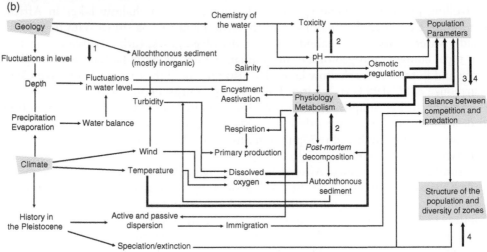

Figure 17.3   (a) Fluctuations of the water level in Lake Chad and separation into two basins during the 1970s; (b) Limiting and controlling factors in the distribution and structure of communities in shallow lakes of Africa.
Source: Dumont (1992).

Talling and Talling (1965) discussed the possible ecological effects of the **ionic composition of African lakes,** especially in terms of the presence or absence of phytoplanktonic species. Kilham (1971b) found a correlation between the level of dissolved silicate and the composition of the diatom flora.

Talling (1992) also correlated the electrical conductivity of shallow lakes and ionic levels with the levels of *Spirulina* sp. and *Aulacoseira* spp. (formerly *Melosira* spp.). Lakes with electrical conductivity (K_{20}) above $10^4 \, \mu S \cdot cm^{-1}$ and predominance of $HCO_3^- + CO_3^{2-}$, Cl^- and SO_4^{2-} present high levels of *Spirulina*. Lakes with electric conductivity (K_{20}) between 100 and $10^3 \, \mu S \cdot cm^{-1}$ and predominance of Ca^{2+}, Mg^{2+}, K^+ and Na^+ present high levels of *Aulacoseira*.

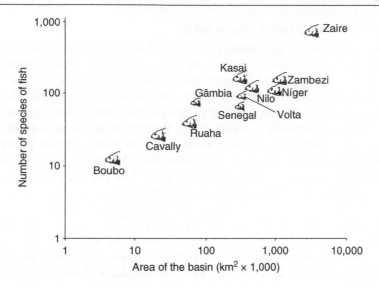

Figure 17.4 Number of fish species present in several rivers in Africa and their relationship with the area of the drainage basin.
Source: Modified from Welcomme and De Merona (1988).

Table 17.2 Gross primary production in African lakes (with inclusion of other data).

Lake	Approximate latitude	Approximate altitude (m)	Depth of collection stations (m)	Depth of euphotic zone (m)	Temperature of water in euphotic zone (°C)	Chlorophyll-a in the euphotic zone (°C) (mg.m⁻²)	Gross photosynthetic production (mg C·m⁻²·day⁻¹)	Annual production (g C.m⁻²·year⁻¹)	References
Victoria (pelagic zone)	1°S	1,230	79	13–14	24–26	35–100	1.08–4.20	950	Talling (1965b)
Tanganyika (pelagic zone)	7°S	773	500	20–25	25–27	–	0.8–1.1	–	Melack (1976)
Bunyoni	1°16′S	1,970	40	4	20	–	1.80	–	Talling (1965a)
Kivu	2°S	1,500	480	–	22–24	–	1.44	–	Degens et al. (1971b)
George (Uganda)	Equator	913	4.5	0.7	24–25	–	5.4	1980	Ganf (1969, 1975)
Chad	13°N	283	12	–	23–29	–	0.7–2.7	–	Lemoalle (1965, 1975)
Nakuru	0.2°S	1,758	3.3	–	–	–	2.3–3.2	–	Melack and Kilham (1974)
Araguandi	9°N	1,910	28.3	0.14	19–21	221–235	13–22	–	Baxter et al. (1965); Talling et al. (1973)

Source: Various sources cited in Talling and Lemoalle (1998).

Lake Nakuru – An Alkaline lake of Africa

Lake Nakuru is a shallow, highly alkaline, permanent lake located in the fault region of East Africa. The lake lies at equatorial latitude 00°24′S and 36°05′E, 1,750 m above sea level in a drainage area of 1,800 km², of which about 3,300 hectares form the lake. The main primary-producing species in the lake is the cyanobacterium *Spirulina fusiformis*, which, occurring in large blooms, supports a large fauna of the flamingo *Phoeniconaias minor*.

The lake's high alkalinity is in part due to the high evaporation rate and drainage of alkaline rocks. Some 450 species of birds are found there, of which 70 are aquatic birds. The flamingo population can reach 1 million individuals and is one of the sustaining sources of the high regional biodiversity (see Figure 17.5). The wetland areas surrounding the lake support high densities of mammals. Variations in the flamingo populations over the last 29 years are shown in Figure 17.6 for three alkaline lakes.

Lake Nakuru attracts 300,000 visitors annually, with a total annual income of US$ 24 million. The lake and its wetlands areas comprise a region of international conservation, and one of the local wetland areas protected under the Ramsar convention. It lies within the Kenyan National Park and the major threats to the ecological integrity are:

- increased suspended material as a result of agricultural activities;
- extreme fluctuations in water level caused by mis-use and over-use of groundwater;
- alterations in water quality because of excess nutrients (from agricultural waste and domestic sewage);
- effects of pesticides and herbicides due to agricultural activities in the catchment area.

Principal references on Lake Nakuru: Mavuti (1975); Rairo (1991); Vareschi (1978, 1979, 1982).

Figure 17.5 Flamingos (*Phoeniconaias minor*) on Lake Nakuru. Photo credit to J. G. Tundisi – October, 2005 (see also color plate section, plate 22).

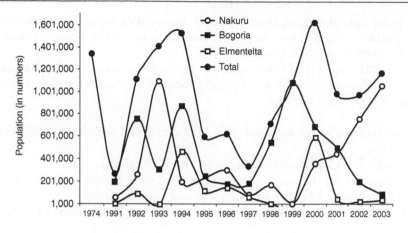

Figure 17.6 Tendencies in the fluctuations of flamingo populations in three alkaline lakes of Africa. *Source:* Vareschi (1978).

Water quality in lake Nakuru

Water temperature .... 20.5°C–27.2°C
pH 10.0–10.6
Conductivity (mS · cm^{-1}) 36–50
Salinity (g · kg^{-1}) 23–35
Dissolved oxygen (mg · L^{-1}) 5.7–23.8
BOD (mg · L^{-1}) 240–640
COD (mg · L^{-1}).. 650–1,000
TSS (mg · L^{-1}) 140–810
N total (mg · L^{-1}) 26–88
P total (mg · L^{-1}).. 8.0–12.0

BOD – Biological oxygen demand;
COD – Chemical oxygen demand;
TSS – Total suspended solids

17.1.1 Studies on African lakes: contribution to tropical and world limnology

Limnological, geographical, biological and evolutionary studies on shallow and deep lakes in Africa have made classic contributions to tropical limnology and for a long time they were cited as examples of the functioning of inland aquatic ecosystems (Margalef, 1983).

The diversity of tropical inland ecosystems, differences from continent to continent, mechanisms of variation and climatic fluctuations, the regional hydrogeochemistry, and circulation all show that there are big differences between the different inland aquatic ecosystems in the tropics. For example, Tundisi and Barbosa (1980) conducted studies on Lake Carioca (Parque Florestal do Rio Doce – MG), showing that the daily stratification of shallow lakes is one mechanism of limnological functioning, which includes partial mixing over depth. In most lakes seasonal stratification is equally important, which is not the case in shallow lakes in the tropics, where only wide diel fluctuations are significant.

Limnological studies and other scientific research on lakes, wetland areas and rivers in Africa showed many biogeophysical processes important for global scientific knowledge on limnology, such as:

▶ the interrelationships of seasonal cycle, diel cycle, and the effect of biomass on these cycles;
▶ mechanisms for adaptation to desiccation and high salinity;
▶ interactions among the components of the food web, including invertebrates as well as vertebrates, from fish to crocodiles and aquatic birds;
▶ evolutionary processes in **isolated systems**, particularly evolutionary radiation in fish.

Figure 17.7 shows the major international rivers, lakes and drainage basins in Africa, in particular the international basin of the Nile River, which is shared by ten countries.

The major environmental threats to conservation of aquatic systems on the continent of Africa, with reflections on the economy and development (UNESCO, UNEP, 2005) are:

▶ Physiography: climatic change and variability. The African continent is extremely susceptible to these factors.
▶ Threats to ecosystems:
   – Water pollution and contamination;
   – Excessive uses of water;
   – Deforestation;
   – Introduction of exotic species;
   – Access to good quality water and basic sanitation;
   – Conflicts arising from multiple uses of water (domestic and international);
   – Drop in aquifer recharge;
   – Difficulties in monitoring surface water and groundwater.

Climatic variability and availability and quality of water impact the productivity of aquatic ecosystems and their diversity. Wetland areas in Africa present wide diversity and productivity, and have regional socio-economic importance.

17.2 LIMNOLOGICAL STUDIES ON LAKES IN ENGLAND

The Lake District is near the northwest coast of England in a region known as Cumbria. Figure 17.8 presents a map of the district, with an inset showing more than a dozen lakes, all of glacial origin. Major studies on these lakes began with Pearsall (cited by Macan, 1970) and substantial contributions to basic limnological research have been made since.

Bathymetric measurements by Mill (1895, in Macan, 1970) and the development of Pearsall's ideas led to the initial hypothesis that the lakes' geographical locations

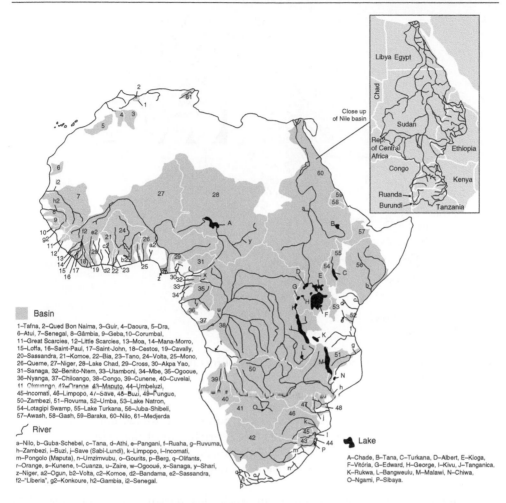

Basin

1–Tafna, 2–Qued Bon Naima, 3–Guir, 4–Daoura, 5–Dra,
6–Atui, 7–Senegal, 8–Gâmbia, 9–Geba,10–Corumbal,
11–Great Scarcies, 12–Little Scarcies, 13–Moa, 14–Mana-Morro,
15–Loffa, 16–Saint-Paul, 17–Saint-John, 18–Cestos, 19–Cavally,
20–Sassandra, 21–Komoe, 22–Bia, 23–Tano, 24–Volta, 25–Mono,
26–Queme, 27–Niger, 28–Lake Chad, 29–Cross, 30–Akpa Yao,
31–Sanaga, 32–Benito-Ntem, 33–Utamboni, 34–Mbe, 35–Ogooue,
36–Nyanga, 37–Chiloango, 38–Congo, 39–Cunene, 40–Cuvelai,
11_Okpupngo, 42–Orange, 43–Maputo, 44–Umbeluzi,
45–Incomati, 46–Limpopo, 47–Save, 48–Buzi, 49–Pungue,
50–Zambezi, 51–Rovuma, 52–Umba, 53–Lake Natron,
54–Lotagipi Swamp, 55–Lake Turkana, 56–Juba-Shibeli,
57–Awash, 58–Gash, 59–Baraka, 60–Nilo, 61–Medjerda

River

a–Nilo, b–Guba-Schebel, c–Tana, d–Athi, e–Pangani, f–Ruaha, g–Ruvuma,
h–Zambezi, i–Buzi, j–Save (Sabi-Lundi), k–Limpopo, l–Incomati,
m–Pongolo (Maputa), n–Umzimvubu, o–Gourits, p–Berg, q–Olifants,
r–Orange, s–Kunene, t–Cuanza, u–Zaire, w–Ogooué, x–Sanaga, y–Shari,
z–Niger, a2–Ogun, b2–Volta, c2–Komoe, d2–Bandama, e2–Sassandra,
f2–"Liberia", g2–Konkoure, h2–Gambia, i2–Senegal.

Lake

A–Chade, B–Tana, C–Turkana, D–Albert, E–Kioga,
F–Vitória, G–Edward, H–George, I–Kivu, J–Tanganica,
K–Rukwa, L–Bangweulu, M–Malawi, N–Chiwa,
O–Ngami, P–Sibaya.

Figure 17.7   Principal rivers, lakes and international drainage basins of Africa.
Source: Human Development Report (2006).

formed a series. The subsequent founding of the Freshwater Biological Association enabled progress in limnological research and substantially changed the purely botanical and zoological approach that had previously existed.

The lakes are usually monomictic and in occasional years, dimictic. Only the shallowest are polymictic. Basic studies on thermal stratification, thermocline development and effect of the wind on the thermal structure were conducted by Mortimer (e.g., 1951) on several lakes, including Windermere, where the author showed the effect of wind in the production of internal waves and the responses to periodic impulses. He earlier examined the relationship between thermal stratification and de-oxygenation in the hypolimnion, the redox potential and release of nutrients from the bottom (Mortimer, 1941, 1942).

The sequential aspects of thermal heating and the process of destratification were examined (Mortimer, 1955), as well as the complexation process resulting from

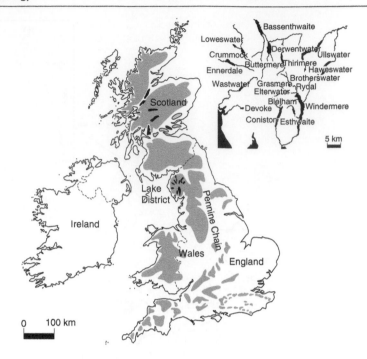

Figure 17.8   The British Isles and location of the Lake District (inset).
Source: Macan (1970).

higher oxygen levels during the oxygen circulation period. Reduction of Fe^{3+} to Fe^{2+} in the anoxic hypolimnion, which results in an enrichment of the sediment-overlying water, was also studied by Mortimer (1942). Research included identifying the steps in the iron complexation process in detail, including circulation and release during the anoxia period.

Differences in light penetration (from Secchi disc measurements) provided data allowing the placement of the lakes in a series. The ionic composition is 99% – composed of sodium, calcium, magnesium, potassium, bicarbonate, sulphate, chloride, and nitrate. Table 17.3 shows the Secchi disc readings for several lakes (Macan, 1970). Table 17.4 presents the average levels of major ions in 12 lakes (Macan, 1970). Seasonal cycles in water temperature, inorganic nutrients, the phytoplanktonic cycle, rainfall and water level in the lakes were studied over many years.

Comparison of the levels of calcium plus bicarbonate in various lakes showed there was an increase in low-productive lakes relative to productive lakes, with a tendency to increased sulphate in the productive lakes.

Differences were observed in the zooplankton as well as in the phytoplankton. The periodicity of **Asterionella formosa** Lund et al. (1964) and other diatomaceous species of phytoplankton is related to fluctuations in levels of silica and the cycles of stratification and de-stratification. Likewise, the seasonal cycle of *Aulacoseira italica* sub-species *subarctica*, studied by Lund, is related to the effects of wind on the water masses and the survival capacity of this diatom under conditions of low O_2 levels or even anoxia. According to Lund (1961), the number of filaments of *Aulocoseira (Melosira) italica*

Table 17.3 Depth of visibility of the Secchi disc in the lakes of England's Lake District.

Lake	M
Wastwater	9
Ennerdale	8.3
Buttermere	8.0
Crummock	8.0
Haweswater	5.8
Derwentwater	5.5
Bassenthwaite	2.2
Coniston	5.4
Windermere	5.5
Ullswater	5.4
Esthwaite	3.1

Source: Pearsall (1921).

Table 17.4 Average concentrations (mg · L^{-1}) of principal ions in the lakes of the Lake District.

	Ca^{2+}	Mg^{2+}	Na^+	K^+	HCO_3^-	Cl^-	SO_4^{2-}	NO_3-N
Esthwaite	8.3	3.5	4.7	0.90	18.3	7.6	9.9	0.78
Windermere S.	6.2	0.70	3.8	0.59	11.0	6.7	7.6	1.2
Windermere N.	5.7	0.61	3.5	0.51	9.7	6.6	6.9	1.2
Coniston	6.1	0.89	4.4	0.66	10.8	7.8	8.0	1.1
Ullswater	5.7	0.89	3.3	0.35	12.7	5.5	6.8	0.75
Bassenthwaite	5.3	1.2	5.0	0.66	10.0	9.1	7.4	1.1
Derwentwater	4.5	0.46	4.8	0.39	5.4	10.1	4.8	0.44
Crummock	2.1	0.78	3.7	0.31	2.9	6.8	4.5	0.35
Buttermere	2.1	0.72	3.5	0.27	2.6	6.9	4.1	0.48
Ennerdale	2.2	0.79	3.8	0.39	3.5	6.7	4.5	0.62
Wastwater	2.4	0.68	3.6	0.35	3.2	5.9	4.8	0.62
Thirlmere	3.3	0.67	3.1	0.31	4.1	5.4	6.0	0.62

Source: Macan (1970).

per liter of water depends almost exclusively on physical factors – in this case, wind and turbulence. This diatom was the subject of a classic study of its seasonal cycle as a function of thermal stratification and de-stratification (Lund, 1954, 1955).

The lakes can also be classified in terms of the composition of the phyto- and zoo-plankton. For example, in Wastwater and Ennerdale Lakes, *Staurastrum* is common in the net plankton, while in Lakes Windermere and **Esthwaite**, *Asterionella* is more common. Differences in composition of zooplankton are related to the relative presence of copepods, rotifers or cladocerans. For example, of the copepods in the zoo-plankton, *Cyclops* and *Mesocyclops* can be found in three lakes. All the lakes contain *Diaptomus gracilis* in the zooplankton (Macan, 1970), and in most lakes *Daphnia hyalina* also occurs (Smyly, 1968). Research on zooplankton has also shown predation of Cyclopoida on crustaceans, chironomids and oligochaetes (Fryer, 1957).

Many studies on the benthic fauna have been done. Macan (1950) compared the species composition of gastropods in different lakes as well as studying indicator species of the *Corixidae* (Macan, 1955), revealing a sequence from oligotrophic to eutrophic lakes. For example, *S. dorsalis* or *stuniata* is common in oligotrophic lakes and *S. fossorum* in eutrophic lakes, including equally eutrophic lakes in Denmark used for comparison (lakes Esrom and Fiuresø). Interrelationships of Oligochaeta and Protozoa (ciliated) with organic matter in the sediment and the degree of eutrophication were also described.

Comparative studies were conducted on the fish fauna (growth and feeding) from several lakes (Le Cren, 1965). Sediments in the lakes showed differences in the composition of the adjacent terrestrial flora as a function of pollen, involving differences in the carbon content in the vertical profile of the sediment. The similarities in distribution of organic carbon found in the vertical profiles of sediments in lakes in the Lake District and some lakes in the United States are probably related to the climatic factors that operated simultaneously throughout the Northern Hemisphere in the post-glacial era (Mackereth, 1966). This conclusion came from an important study on regional limnology, which compared systems from different latitudes, enabling deeper analysis of the unifying principles in limnology.

Talling (1965b) presented an interesting and important comparison between Lake Victoria, in Africa, and Windermere, in the Lake District of England. These lakes have different origins: Windermere is glacial and Victoria tectonic. Differences in the seasonal cycle of incident solar radiation, water temperature, euphotic depth and chlorophyll levels were remarkable. Talling suggested that differences in water temperature (10–12°C higher in Lake Victoria) and photosynthetic capacity may be the most important causes of the higher primary production estimate in Lake Victoria ($950\,g\,C\cdot m^{-2}\cdot year^{-1}$) than in Windermere. This type of comparative study of lake systems or lakes in different latitudes is essential for understanding the seasonal processes and factors affecting nutrient recycling, primary production of phytoplankon, and structure of the trophic network (see Figure 17.9).

17.3   OTHER STUDIES IN EUROPE

Elsewhere in Europe, important limnological studies have been conducted in Italy on **Lake Maggiore,** by the *Istituto Italiano di Idrobiologia,* in Pallanza. Limnological studies (physical and chemical) were conducted on this lake and other alpine lakes as well as on the population dynamics of planktonic species (Tonolli 1961; di Bernardi *et al.,* 1990, 1993), and are examples of regional studies that have contributed enormously to global limnology.

Many regional studies on rivers and lakes in France have helped to advance global limnology (Dussart, 1966). French researchers have also made significant contributions with their studies on tropical lakes, such as Lake Chad (Carmouze, Durand and Lévêque, 1983), and on aquatic primary productivity (Lemoalle, 1979).

Regional studies on lakes in Germany were very important for the progress of global limnology, particularly Ohle's work (1956) on Lake **Plussee,** and later work by Overbeck *et al.* (1984). Overbeck in particular examined aquatic microbiology and the importance of heterotrophic producers and biogeochemical cycles (Overbeck, 1994). Elsewhere classical regional studies on limnology include extensive studies on Lake Balaton, Hungary, located in the Carpathian basin ($>300,000\,km^2$). Over $1,000\,km$ long, the lake is a classic shallow lake formed between 12,500–10,000 years ago.

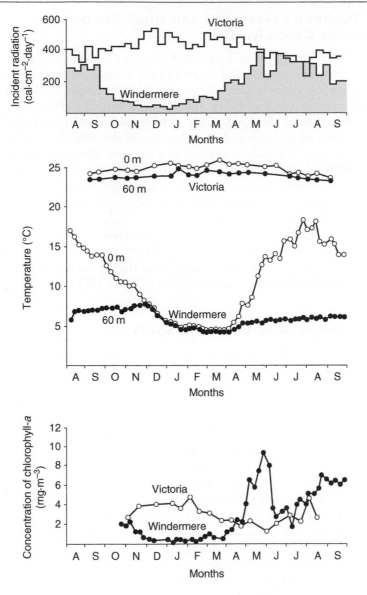

Figure 17.9   Classic comparison between a temperate-region lake (Windermere, in the Lake District of England) and an African lake (Victoria).
Source: Beadle (1981); Talling (1965).

Lake Balaton is shallow (average depth >3.0 m) and has a surface area of 593 km². The total volume of the lake's water is only 1.8 km³, and the lake has undergone severe eutrophication. Classic studies on succession in phytoplankton were conducted by Padisák (1980, 1981) and Padisák *et al.* (1984). Important studies on mechanisms controlling the growth of blooms in shallow lakes in summer were conducted by Padisák *et al.* (1988). Biró (1995a, 1995b, 1997) studied control of fishing and aquaculture production.

17.3.1   Studies on reservoirs and small fish ponds in the Czech Republic

Many limnological studies have been conducted in Czechoslovakia, including studies on management techniques and control of the ecosystems – especially ponds and reservoirs on river systems. In particular, a long series of studies examined fluctuations in levels and retention times and their effects on planktonic and benthic communities. Other studies looked at the equilibrium of biomasses in relation to inflow, outflux and sedimentation of nitrogen and phosphorus and the differences in horizontal and vertical distributions of dissolved oxygen, as a function of surface currents and vertical circulation. Similarly, other studies examined the seasonal cycle of zooplankton in reservoirs, morphometric studies on small aquaculture ponds and the interrelationships with the biomass of benthic organisms and planktonic organisms (Straškrába and Hrbáček, 1966; Hrbáček, 1966).

Regional studies conducted in Bohemia clearly show the contribution that can be made from limnological studies on pond districts or on artificial systems. A comprehensive approach for small-reservoir control and management was developed in this region based on these limnological studies (Straškrába, 1986).

17.3.2   Typology of reservoirs in Spain

According to Margalef (1976), Spain has become a country with hardly any lakes – only two natural lakes: **Sanabria** (glacial) and **Bañolas** (karst) – yet it has more than 700 dammed reservoirs. The initial purpose of the pioneer work 'Typology of Reservoirs in Spain' was to compare a large number of reservoirs located throughout the country and scientifically analyze the behaviour of these reservoirs, their limnology and processes of colonization. These reservoirs in Spain have created large artificial environments, varying in shape, size, retention time, and physical and chemical characteristics of the water. Such a varied group as this is ideal for analyzing the responses of aquatic biota, from the standpoint of diversity, species selection, primary productivity, nutrient cycles, and energy flow.

In addition to scientific purposes and comparison of the artificial aquatic ecosystems, this work also contributed to promotion of reservoir management. The project had four basic goals: a) compile further information on existing reservoirs in Spain; b) analyze the limnology of the reservoirs, their aquatic biology and contribute to basic limnology; c) identify conditions for predicting future responses of the reservoirs and their biota to impacts from the drainage basin; and d) promote criteria and make recommendations for the protection and restoration of reservoirs.

More than 100 reservoirs (104) from all areas of Spain were included in the study, with varying conditions of weather, geology, soil and vegetation. Reservoirs were included with varying capacities to provide drinking water, or water for irrigation or industrial use. Samples were collected four times each year from the deepest point (vertical profiles) in each reservoir.

The results showed that the reservoirs could be ordered in various categories: chemical composition of the water (silica or carbonates); oligotrophic and eutrophic reservoirs; accumulation of sulphate; increasing reductive conditions (in some reservoirs); and release of phosphorus and metals, such as manganese and cobalt.

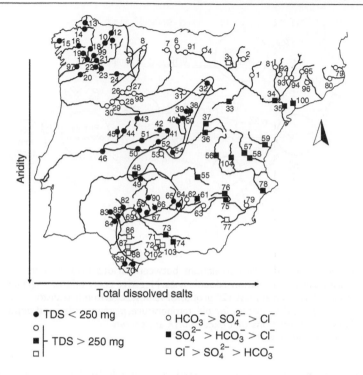

Total dissolved salts

- TDS < 250 mg
∘
■ ⎱ TDS > 250 mg
□

∘ $HCO_3^- > SO_4^{2-} > Cl^-$
■ $SO_4^{2-} > HCO_3^- > Cl^-$
□ $Cl^- > SO_4^{2-} > HCO_3^-$

Figure 17.10   Distribution of total dissolved salts in Spain and ionic composition in function of aridity. *Source:* Armengol (2008).

The study also helped to identify effects of the reservoirs on local climate and on the retention of suspended material. The nature of the rocks, soil and climate were the most important factors in determining the chemical composition of the water, especially total ionic concentration.

Figure 17.10 shows the distribution of reservoirs with basic chemical composition of sulphates, bicarbonates and chloride. Figure 17.11 shows the statistical correlation among a number of variables measured simultaneously in the same groups, in two field studies.

Studies on plankton and benthos identified 700 species of algae, 113 rotifers, 72 crustaceans and 72 chironomids, in addition to associations of organisms, for example, in reservoirs with communities of *Tabellaria* and *Aulacoseira* spp., and reservoirs with communities of *Cyclotella*, *Ceratium* and *Dinobryon* in the phytoplankton. Likewise, it was possible to classify reservoirs along the oligotrophic-eutrophic axis, based on the level of chlorophyll and inorganic nutrients (especially NO_3^{-3} PO_4).

Variations in the composition and dominance of rotifers, cladocerans, copepods, and the group of deep benthos and littoral benthos enabled a more comparative study-basis.

This project, a landmark in regional and global limnology, shows how a comparative approach, using the same method and a defined spatial-temporal axis, can enable the comparison of masses of water and distribution of organisms, and an understanding of the processes originating in the drainage basin that affect inland aquatic systems.

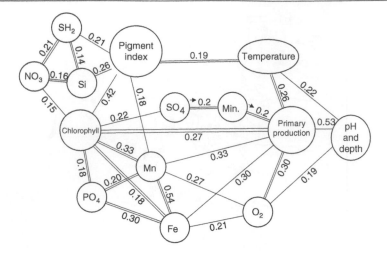

Figure 17.11   Network of statistical correlations between various measurements in reservoirs of Spain. Positive correlations: double lines. Negative correlations: single lines. Each correlation is based on at least 120 groups of values. Logarithmic transformations were made from original data, except for pH, temperature, pigment index and depth.
Only the correlations that exceed 0.14 are shown.
Source: Margalef *et al.* (1976).

A recent summary shows the major contribution made by Ramón Margalef and his collaborators to limnology on the Iberian Peninsula.

17.4 THE GREAT LAKES OF NORTH AMERICA

Together, the five lakes of the Great Lakes hold the largest single volume of fresh water on the planet. The lakes are approximately 10,000 years old, and the retention times in lakes Superior and Michigan are about 200 and 100 years, respectively (see Figure 17.12).

It is difficult to study these lakes, given their size. Limnological research and development can be done with the use of oceanographic techniques.

Table 17.5 lists some morphometric and limnological techniques used in these lakes, which occupy a total area of 245,240 km². Because the lakes are surrounded by heavily populated and industrialized areas, they are submitted to a process of continuous eutrophication, due also to the wide drainage area of the drainage basins.

Table 17.6 presents the chemical composition of these lakes. It shows a considerable increase in sulphate, chloride, calcium and dissolved solids, mainly in lakes Huron, Ontario, Erie and Michigan. However, the last three present higher values of chlorophyll-*a* (1.5 to 10–20 mg·L⁻¹) and primary production between 50–5,000 mg C·m²·day⁻¹.

The most common cyanobacteria in eutrophic lakes are *Anabaena spiroides* and *Aphanizomenon flos-aquae*. Diatoms of the genera *Aulacoseira*, *Asterionella*, and *Fragilaria* occur in Lakes Ontario, Erie and Michigan during the vertical mixing in autumn. In general, diatoms are extremely common in the Great Lakes, mainly

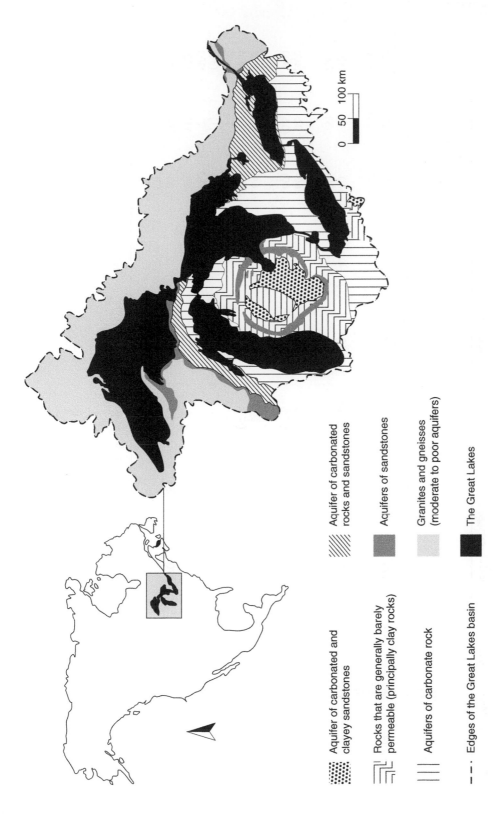

Figure 17.12 Rocky substrata of the aquifers in the basins of the North American Great Lakes.
Source: U.S. Department of the Interior; U.S. Geological Survey.

Aquifer of carbonated and clayey sandstones

Rocks that are generally barely permeable (principally clay rocks)

Aquifers of carbonate rock

– · – · Edges of the Great Lakes basin

Aquifer of carbonated rocks and sandstones

Aquifers of sandstones

Granites and gneisses (moderate to poor aquifers)

The Great Lakes

0   50   100 km

Table 17.5 Mophometric and limnological characteristics of the North American Great Lakes.

Lake	L (km) B (km) A (km²)	Z_{max} (z) (m)	Area of drainage basin (km² · 10³)	Depth of thermocline (m)	Maximum/minimum summer temperature (winter) (°C)	Retention time of water (years)	Approximate period of thermal stratification	Winter (summer) nutrients on the surface (µg · L⁻¹) NO₃–N NH₄–N	PO₄–P	SiO₂
Superior	560 256 82,000	406 (149)	125	10–30	14 (0.5)	184	Aug.–Dec.	280 (220)	0.5 (0.5)	2,200 2,000
Michigan	490 188 58,000	281 (85)	118	10–15	18–20 (<4)	104	Jul.–Dec.	300 (130)	6 (5)	1,300 (700)
Huron	330 292 60,000	228 (59)	128	15–30	18.5 (<4)	21	End of Jun.–Oct. or Nov.	260 (180)	0.5 (0.5)	1,400 800
Erie	385 85 20,000	w: 13(7.3) c: 24(18) e: 70(24)	59	w: p c: 14–20 e: 30	24 (<4)	w: 0.13 c: 1.7 e: 0.85 all: 3	Middle of Jun.–Nov.	w: 640 (80) c: 140 (20) e: 180 (20)	23 (2) 7 (1) 7 (1)	1,300 (60) 350 (30) 300 (30)
Ontario	309 85 20,000	244 (86)	70	15–20	20.5 (<4)	8	End of Jun.–Nov.	280 (40)	14 (1)	400 (100)

w – west; c – central; e – east; p – polymictic; L – maximum length; B – maximum width; A – area.
Source: Horne and Goldman (1994).

Table 17.6   Chemical composition of the waters in the North American Great Lakes.

	Superior	Michigan	Huron	Ontario	Erie
pH	7.4	8.0	8.1	8.5	8.3
Conductivity ($\mu S \cdot cm^{-1}$)	78.7	225.8	168.3	272.3	241.8
Calcium	12.4	31.5	22.6	39.3	36.7
Magnesium	2.8	10.4	6.3	9.1	8.9
Sodium	1.1	3.4	2.3	10.8	8.7
Potassium	0.6	0.9	1.0	1.2	1.4
Chlorine	1.9	6.2	7.0	23.5	21.0
Sulphate	3.2	15.5	9.7	32.4	21.1
Silica	1.4	3.1	2.4	0.3	1.5

Source: Horne and Goldman (1994).

Table 17.7   Biomass of benthic macroinvertebrates and trophic state of the North American Great Lakes.

Lakes	Regional trophic classification	Animals/m² (excluding harpacticoids and nematodes)	% Oligochaetes	% Chironomid larvae
Superior	O	392–1,720		
Huron	O – M	625–2,000	10–20	0–5
Michigan	O	660–4,265		
Erie	M – E	660–10,000	30	
Ontario	M	1,100–20,000	14–86	10–50

O – Oligotrophic; M – Mesotrophic; E – Eutrophic
Source: Margalef (1983).

Table 17.8   Maximum and minimum temperature of the North American Great Lakes.

Lake	Max. temp. in summer (°C)	Min. temp. in winter (°C)
Superior	14	0.5
Michigan	18–20	<4
Huron	18.5	<4
Erie	24	<4
Ontario	20.5	<4

Source: Horne and Goldman (1994).

of the genera *Fragilaria*, *Tabellaria*, *Asterionella*, *Synedra* and several species of *Aulacoseira*.

Table 17.7 lists the benthic macroinvertebrates in the Great Lakes (Margalef, 1983), and shows the progression of biomass of these macroinvertebrates according to the degree of trophism in each lake. Table 17.8 presents variation in temperatures in the Great Lakes.

The zooplankton in the Great Lakes is composed mainly of species of *Diaptomus* in the deepest most oligotrophic lakes. Common in shallower eutrophic lakes are cyclopoid copepods with the genus *Cyclops* and cladocerans such as *Bosmina* and *Daphnia*. Rotifers and cladocerans are common in the more productive lakes.

Many studies on eutrophication were conducted on these lakes; results were summarized in Horne and Goldman (1994).

17.5   OTHER TEMPERATE-REGION LAKES IN THE NORTHERN HEMISPHERE

In several regions and countries in the Northern Hemisphere, there are many relatively shallow lakes of glacial origin resulting from erosion. These are dimictic lakes with periods of 7–8 months of surface ice. The maximum temperature in summer is 19–20°C. Oligotrophic lakes in Scandinavia are common, with some 55,000 lakes between 60–64°N latitude. Many of these lakes have undergone a rapid eutrophication process. Some are meromictic with high levels of iron in the monimolimnion. Equally important and numerous are the small glacial lakes in North America, which include the lakes in the Experimental Lakes Area in Canada (between 49°30′ and 50°N and 93°–94°30′W) (Schindler, 1980). These lakes are shallow, with a maximum depth of 20 m, the majority dimictic and a few polymictic.

17.6   LAKES IN JAPAN

Studies on lakes in Japan provide a clear and characteristic example of regional limnology. Japan, a country made up of four main islands and many smaller ones, from the region in the north to the subtropical region in the south. Many **lakes originated from volcanoes**, depressions, fluvial activity, and movement of sand dunes (Mori *et al.*, 1984).

In addition to natural lakes, there are many flooded rice fields, small ponds and small hatcheries. Pollution and eutrophication in lakes in Japan pose a problem, since the use of aquatic organisms (fauna and flora) for food in Japan is an economically important tradition.

Through the International Biological Programme, comparative studies were conducted (Mori and Yamamoto, 1975) on various different types of lakes (oligotrophic, mesotrophic and eutrophic), rivers, and fishery ponds, examining the structure and function of communities, as well as chemical and physical aspects.

Yoshimura (1938) conducted an initial survey of lakes in Japan, based on thermal structure and dissolved oxygen. Dividing the lakes by category and percentage of saturation, the author's conclusions are presented in Table 17.9.

This was a classic comparative study by Yoshimura, who, by measuring a few variables (such as dissolved oxygen, heat profile, and transparency measured with Secchi disc), was able to categorize the lakes by type, which was extremely useful in the selection of systems for intensive study during the International Biological Programme.

Among Japan's most important lakes, Lake Biwa – a **temperate monomictic lake**, with a 674.4-km² surface and a volume of 27.8 km³ – is one of oldest on the planet. It has many endemic species, and studies of sediments conducted by Horie (1984) examined many aspects of phytoplanktonic succession, alterations in vegetation around the lake, and the chemistry of the sediment.

Lake Biwa has undergone a rapid eutrophication process (Kira, 1984), and restoration measures have been implemented: reducing the influx of nitrogen and

Table 17.9   Percentage of saturation of dissolved
oxygen in lakes of Japan.

Type of lakes	%
Brackish	77–194
Oligotrophic	86–120
Mesotrophic	68–126
Eutrophic	87–160
Dystrophic	36–112
Acidic	36–127

Source: Yoshimura (1938).

Table 17.10   Principal features of Lake Biwa.

Altitude	85 m
Length	68 km
Maximum Length	22.6 km
Perimeter	188 km
Area	674.4 km²
Shoreline development index	2.04
Maximum depth	104 m
Average depth	41.2 m
Volume	27.8 km²

Source: Horie (1984).

phosphorus, increasing wetland and flooding areas at the entrances of rivers, planting banks with grasses to prevent sedimentation, public awareness campaigns, and treatment of industrial waste.

Lake Biwa is located in the centre of Honshu Island, Japan, and has been extensively studied by limnologists, ecologists, botanists, zoologists, oceanographers and physicists. Its main features are presented in Table 17.10.

Horie (1984) published a basic summary of this work, and a 'History of Lake Biwa' (Horie, 1987). The lake has important biological characteristics, with many endemic species that have been studied extensively. Studies on primary productivity in Lake Biwa were conducted through the International Biological Programme by Mori et al. (1975), presenting the energy flow in kcal·year⁻¹.

Recent studies on Lake Biwa, focused on solving the problem of eutrophication, were initiated by the Lake Biwa Research Institute (LBRI); a publication edited by Nakamura and Nakajima (2002) describes the criteria, mechanisms and programs for restoration of the lake, heavily affected by eutrophication. Of special interest in this context was the study of the drainage basin, land use, impact on nitrogen and phosphorus levels, and responses of the communities of phytoplankton, benthos and fish to eutrophication. The impact of the restoration was clearly evident, thanks to basic research, involving climatic, physical, chemical and biological components (see Figure 17.13).

Knowledge of the role played by photo-autotrophic picophytoplankton, and the relationship with the circulation of organic matter and biogeochemical cycles caused

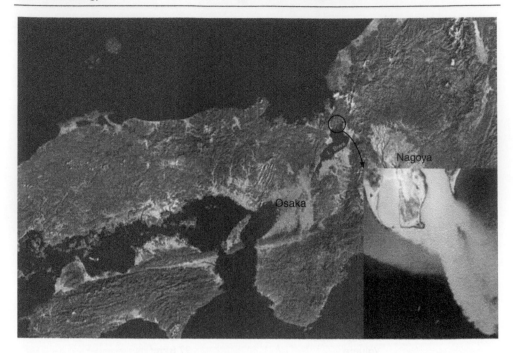

Figure 17.13   Lake Biwa, the largest lake in Japan. In the detailed inset: the intrusion of sediment in Lake Biwa, originating from the thaw (north base of the lake) (See also color plate section, plate 23).
Source: Nakamura and Nakagima (2002).

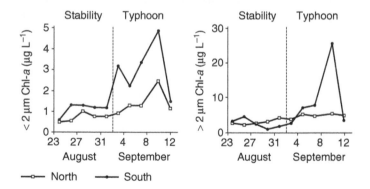

Figure 17.14   Chlorophyll-*a* of the phytoplankton <2 μm and >2 μm on the surface of the northern and southern basins of Lake Biwa before and during a typhoon.
Source: Frenette *et al.* (1996b).

by the effects of typhoons and widely fluctuating vertical circulation rates, were also of particular interest in the study of Lake Biwa and the progression of its eutrophication (see Figure 17.14) (Frenet *et al.*, 1996b).

There are two other important groups of lakes in Japan: volcanic lakes, whose features are presented in Table 17.11, and meromictic lakes (see Tables 17.12 and 17.13).

Table 17.11   Typical volcanic lakes of Japan: chemical composition of water.

Lake	District	pH	Ca^{2+}	Fe^{2+} + Fe^{3+}	SO$_4$$^{2-}$	Cl$^-$	Year of observation
				Ions in mg.L^{-1}			
Yugama	Gumma	0.6	255	320	5,349	5,010	1949
Yugama	Gumma	0.9	56	16.3	1,656	230	1968
Katanuma	Miyagi	1.8	2	5.8	1,003	3.5	1968
Okana	Miyagi	2.9	72	10.8	421.3	0.3	1968
Osoresanko	Aomori	3.1	5.2	0.4	19.9	23.8	1934
Akadoronuma	Bandai	3.2	330	69.5	2.767	7.0	1968

Source: Mori et al. (1984).

Table 17.12   Meromictic lakes in Japan.

Lake	Locale	Chlorination (%) Surface	Bottom	References
Ectogenic meromixis				
Harutori	Hokkaido	1.1	13.3 (8.5 m)	Kusuki (1937)
Mokotonuma	Hokkaido	0.17	16.8 (5 m)	Ueno (1937)
Notoro	Hokkaido	12.5	16.2 (20 m)	Kuroda et al. (1958)
Hamana	Shizuoka	11.3	14.1 (10 m)	Yoshimura (1938)
Suigetsu	Fukui	0.5–1.9	7.3–8.4 (30 m)	Matsuyama (1973)
Koyamaike	Tottori	0.01	0.15 (6 m)	Yoshimura (1973)
Kaiike	Kagoshima	6.6–9.9	18.8 (10 m)	Yoshimura (1929)
Namakoike	Kagoshima	13.6	17.0 (20 m)	Matsuyama (in prep.)
Crenogenic meromixis				
Towada	Aomori	0.010	0.014 (320 m)	Yoshimura (1934b)
Zao-okama	Miyagi	0.14	0.20 (35 m)	Yoshimura (1934b)
Shinmiyo	Tokyo	1.3	10.4 (32 m)	Yoshimura (1934a)
Biogenic meromixis				
Haruna	Gumma	0.006	0.006 (12 m)	Yoshimura (1934b)
Hangetsu	Hokkaido	0.007	0.008 (17 m)	Yoshimura (1934b)

Source: Matsuyama (1978).

In the volcanic lakes, one source of CO$_2$ is the constant supply from fumaroles at the bottom of the lake. Another source is from the decomposition of organic matter by bacteria and respiration of organisms. The principal decomposers in volcanic lakes are fungi, and not bacteria (Satake and Saijo, 1974). The photosynthetic activity measured in the lakes (**Lake Katanuma**) is the result of the presence of *Chlamydomonas avidophila*, which can survive at a pH of 1.8 to 2.0.

Meromictic lakes in Japan were studied extensively by Matsuyama (1978).

Located near coastal regions, these lakes are highly saline toward the bottom, because of the intrusion of salt water originating from the ocean. The majority of the lakes have an anoxic hypolimnion with high levels of H$_2$S in the monimolinium and also high levels of ammonia, phosphate and total CO$_2$ in this layer. High

Table 17.13   Content of sulphate in some meromictic lakes, fjords and anoxic marine waters, compared with lakes in Japan.

Water bodies	Locale	Depth zone sulphate content (m)	Maximum content (mg S · L⁻¹)	References
Big Soda Lake	USA	20–60	740	Hutchinson (1937)
Harutori Lake	Japan	4–9	630	Kusuki (1937)
Hemmelsdorfersee	German	33–43	290	Griesel (1935)
Pettaquamscutt River	USA	6–13	130	Gaines et al. (1972)
Lake Suigetsu	Japan	8–34	110	Yamamoto (1953)
Hellefjord	Norway	15–70	60	Strom (1936)
Lake Verde	USA	18–45	38	Brunskill and Ludlam (1969)
Lake Namakoike	Japan	15–21	38	Kobe Marine Observatory (1935)
Lake Ritom	Switzerland	13–45	29	Düggeli (1924)
Lake Belovod	Russia	15–25	24	Kuznetsov (1968)
Lake Kaiike	Japan	5–11	21	Matsuyama (1999)
Lake Shinmiyo	Japan	20–35	19	Yoshimura (1934a)
Lake Wakuike	Japan	3–7	18	Yoshimura (1934a)
Baía Habu	Japan	5–20	18	Ohara (1941)
Rotsee	Switzerland	10–16	15	Bachmann (1931)
Lake Sodon	Japan	8–15	15	Newcombe and Slater (1950)
Lake Hiruga	Japan	15–38	12	Yoshimura (1934a)
Lake Mokotomuma	Japan	4–6	11	Ueno (1937)
Lake Nitinat	Canada	20–200	11	Richards et al. (1965)
Black Sea		150–200	9	Sorokin (1972)
Cariaco Trench (Ocean)		400–1300	0.9	Richards and Vaccaro (1956)
Golfo Dulce	Costa Rica	150–200	0.2	Richards et al. (1971)

Source: Matsuyama (1978).

photosynthetic production rates by photosynthetic bacteria were measured in these lakes (Mori *et al.*, 1984).

Small reservoirs, wetland areas and ponds have been extensively studied. The increase in silt (due to erosion) and a large reduction in the volume of lakes (by 70–80%) are a problem in the man-made lakes in Japan.

17.7 ANCIENT LAKES

The origin of many lakes is post-glacial, which means they are 10,000 to 15,000 years old. However, about 24 lakes around the world are considered to be very ancient, or lakes with with a very long life span (Gorthner, 1994). In this section, we present the characteristics of some of these lakes. Brooks (1950) and Fryer (1995) published extensive reviews, and Martens (1997) published a detailed review of the speciation that occurs in these lakes.

Table 17.14   Age of five of the oldest lakes on Earth.

Lake	Age
Biwa (Japan)	400,000 years
Victoria (Africa)	250,000 to 750,000 years
Malawi (Africa)	3.6 to 5.6 million years
Tanganyika (Africa)	20 to 40 million years
Baikal (Russia)	25 to 30 million years

*Source:*Various sources.

Lakes that are more than 100,000 years old ('ancient lakes') have been the subject of comparative studies. They are important evolutionarily, ecologically, economically, historically and culturally, and studying them can help to explain the existence of a process of intra-lacustrine speciation that has led to groups of species ('flocks'). In some lakes, such as Tanganyika, many groups of species exist in various fish families. Since aquatic organisms share a common evolutionary history with the ecosystems in which they evolved, their evolution and speciation are the result of a long process of co-evolution and co-adaptation with the biotic and abiotic systems.

An exceptional diversity of fauna occurs in these very old extant lakes (Martens, 1997). A group with lots of species is categorized as a set only if its members are endemic to the geographically circumscribed area, as well as being closely related. The longevity of some of these lakes may explain their **evolutionary endemic radiations**. **Sympatric speciation** occurs when reproductive isolation develops within a specific continuous geographic gradient, despite the continuous gene flow.

The diversity of fauna in ancient lakes is an important biological heritage and its preservation is essential. Each of the great lakes of East Africa (Victoria, Tanganyika and Malawi) has an endemic lacustrine fauna of cichlid fish that apparently evolved from ancestral stocks of the rivers in these regions. Speciation of cichlids also occurs to a lesser extent in other lakes, such as Albert, Turkana, Edward, George and Kivu.

Table 17.14 lists the age of some of the world's oldest lakes.

With regard to endemic species of cichlids, Lake Victoria has more than 300, Malawi more than 500, and Tanganyika about 200. In Lake Titicaca, there is a genus (*Orestias*) endemic to the Andes highlands. Lake Baikal, in turn, has 56 species and subspecies of fish belonging to 14 families.

The fauna of these lakes, particularly of fish, has been threatened by the introduction of exotic species and by the excess of unregulated commercial fishing. In Lake Victoria, the introduction of Nile perch (*Lates niloticus*) began depleting the endemic cichlid fish, which occurred because of predation from the introduced species as well as the widespread use of new large-scale fishing techniques.

Traditional fishing in ancient lakes, such as those mentioned, had the characteristics of preserving fish stocks and maintaining the balance between the pressure of fishing, species diversity, and preservation.

These huge lakes influenced the culture of the populations living near them. Traditionally local populations have had important social and economic interactions with the lakes, with wide and diverse visions of the use of natural resources and the functioning of the lacustrine systems.

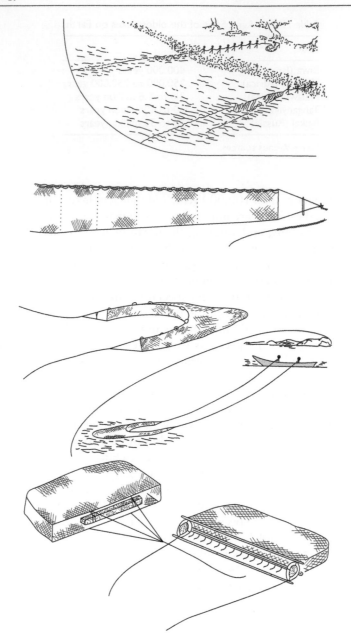

Figure 17.15   Traditional tools used for fishing on Lake Biwa.
Source: Modified from Kawanabe (1999).

Fishing systems and methods were traditionally developed by the people living around Lake Titicaca (Peru), Biwa (Japan), and the lakes of East Africa in order to exploit fishing and conserve biological resources vital for their survival (Lévêque, 1999) (see Figure 17.15).

These lakes, with a long evolutionary history, have given rise to endemic faunas of great biological and historical value. The lakes are natural laboratories and currently subject to a variety of impacts. Human impact and the physical, chemical and biological responses of the lakes vary greatly and depend on the population around the lake, their industrial and commercial activities, and ability to exploit natural resources.

Very old lakes are:

▶ **laboratories to study evolution** and biodiversity;
▶ aquatic ecosystems with large numbers of endemic species;
▶ laboratories for the study of cultural and social interactions and economic conditions of populations in these lakes over long historical periods;
▶ aquatic ecosystems in which the uses of natural resources (such as the fish fauna and flora of macrophytes in Lake Titicaca) have occurred over many generations and thousands of years.

Impacts on these lakes include:

▶ eutrophication;
▶ reduction of biodiversity;
▶ contamination and pollution from industrial activities;
▶ excessive fishing;
▶ overexploitation of natural resources (macrophytes, crustaceans and fishes).

17.7.1   Lake Baikal

Lake Baikal is estimated to be 25–30 million years old and is the oldest lake on the planet. Its maximum depth is 1620 m (fully oxygenated to the bottom) and it contains the largest volume of unfrozen water on Earth – about 1/5 of the world's total supply of surface fresh water. Currently this lake is considered to be a mega-diverse aquatic ecosystem, with the world's largest number of freshwater metazoan species.

The classic book on the biodiversity in Lake Baikal is by Kozhov (1963), entitled *The Biology of Lake Baikal*, in which many problems of development and speciation were presented and discussed. The book represents a landmark study of biodiversity.

In 1925 the first limnology laboratory on Lake Baikal was established, an initiative of the Academy of Sciences of the former Soviet Union. The lake is considered to be an aquatic ecosystem with vast **autochthonous speciation**, which is its most important process. A wide range of studies on the flora and fauna of Lake Baikal has been undertaken over time (more than 100 years).

Table 17.15 shows the morphometric characteristics of Lake Baikal and Table 17.16 presents the lake's water balance.

Tables 17.15 and 17.16 present the dimensions of the enormous size of Lake Baikal. The lake has a great diversity of species that colonized it, from the littoral zone to the pelagic zone. The conditions of life in the lake were so favourable that organisms migrating from the drainage basin of the lake and its tributaries established themselves, and a range of flora and fauna developed with a large number of genera and species.

Table 17.15   Morphometric features of Lake Baikal.

Altitude (above sea level)	455.6 m
Length	636 km
Maximum width	79.4 km
Minimum width	25 km
Average width	47 km
Area	31,500 km²
Perimeter	2,000 km
Perimeter of the islands	139.2 km
Maximum depth	1,620 m
Average depth	740 m
Volume	23,000 km³

Source: Kozhov (1963).

Table 17.16   Water balance of Lake Baikal.

	Volume (km³)	%
Water supply		
Precipitation	9.29	12.1
Superficial flux	58.75	82.7
Subterranean water flow	2.30	3.0
Total supply	71.16	98.8
Loss of water		
Drainage of surface downstream	69.39	84.8
Evaporation	10.33	14.6
	71.16	**99.4**

Source: Kozhov (1963).

Understanding the evolutionary history of how the **fauna of the tributaries** settled in Lake Baikal and multiplied into many species is one of the great challenges in the study of biogeography and the evolution of aquatic flora and fauna. More recent palaeontological studies in areas of inland centrifugal depressions in Central Asia (which in the geological past were large lakes) have shown the degree of complexity and connections existing in the drainage basin of Lake Baikal. The residual large lakes still existing in deep tectonic depressions in the vicinity of Lake Baikal still contain living relicts of the existing fauna in them and the drainage basin.

Extensive studies have been conducted on annual fluctuations in plankton, the seasonal cycles of pelagic flora and fauna, diurnal migrations, and predator-prey relationships on Lake Baikal. They have made important contributions to global scientific knowledge of aquatic biota.

Impacts on Lake Baikal

As with all other ancient lakes in the world, human activities impact Lake Baikal, especially economic activities. The paper pulp industry, large-scale use of mineral fertilizers, urbanization, intensive use of soil, increase of tourism and shipping are among the key threats to the biological integrity of Lake Baikal, and also threaten its diversity. Deforestation and increased agriculture cause other impacts.

Chemical compounds, hydrocarbons and heavy metals resulting from agricultural and industrial activities contribute to the degradation of the lake and its drainage basin. In the basin, one major source of contamination from heavy metals is the atmospheric contribution from air polluted outside the basin. The contribution from the atmosphere is greater than from the tributaries. Metals entering the Baikal from air pollution include aluminium, manganese, iron, cobalt, copper, zinc, selenium, sodium, barium, mercury and lead.

Phosphorus and nitrogen also enter the lake in the untreated sewage, and there are elevated contributions of atmospheric nitrogen-compounds. Experimental studies showed that the lake's flora and fauna are extremely sensitive to toxic substances.

Impacts on lakes and reservoirs resulting from human activities and eutrophication
Photo: J.G. Tundisi

18 Impacts on aquatic ecosystems

SUMMARY

The impact of human activities on inland aquatic ecosystems has caused a continuous and relentless deterioration of water quality and profound changes in the hydrological, biogeochemical and biodiversity cycles. This process of deterioration has had economic and social consequences, and in some cases, permanent and irreversible alterations in lakes, rivers and reservoirs. The costs of water treatment and clean-up of lakes, rivers and reservoirs are extremely high.

Another impact has been the eutrophication of inland waters. Eutrophication, increased toxicity, sedimentation in rivers and lakes, and altered hydrodynamics are some of the resulting problems commonly found in almost all countries and regions around the world. Chemical contamination of waters and effects on food webs are also consequences of human activities.

It is essential to monitor the causes and consequences of these alterations in order to diagnose the deterioration processes and restore ecosystems. Toxic substances and chemical elements such as heavy metals cause damage to inland waters and make it difficult to identify and diagnose the impacts on aquatic ecosystems and aquatic biota.

Lakes, reservoirs and rivers differ in temperate and tropical regions with the progression of eutrophication and contamination, response time of communities and concentrations of nutrients, especially nitrogen and phosphorus. There are also differences in the threshold concentrations of nitrogen and phosphorus needed to trigger the eutrophication process.

Global changes affect rivers, lakes, ponds, and wetland areas and produce synergistic effects in terms of factors that affect human health.

18.1   KEY IMPACTS AND THEIR CONSEQUENCES

All freshwater ecosystems are subject to a range of impacts resulting from human activities and multiple uses of the drainage basins in which lakes, rivers, reservoirs, wetlands and swamps are located. These impacts directly or indirectly produce alterations in estuaries and coastal waters. As the uses multiply and diversify, the impacts become more complex and it becomes more difficult to troubleshoot and resolve the resulting problems.

In addition to impacts from human activities, there are also natural impacts resulting from the very operating mechanisms of the ecosystems and the drainage basins. To a certain extent, the natural impacts are absorbed by the ecosystem, which has appropriate mechanisms, with multiple controls, to minimize these impacts. For example, the responses of rivers and lakes to fluctuations in level that occur in the great floodplains of the Amazon and Paraná rivers are part of a natural functioning process and of a cycle of responses perfectly integrated with the natural process (Junk *et al.*, 2000). However, the sum of the impacts caused by human activities is extensive and produces huge alterations in the structure and function of aquatic ecosystems.

The impacts are classified as *primary*, immediate and relevant effects (i.e., interference in the *hydrological cycle* or the influx of pollutants from point sources); *secondary,* which are much more difficult to detect or measure and equally severe (i.e., alterations in the food chain, which may have consequences that appear much later in the process); or *tertiary*, with complex long-term responses (i.e., changes in the chemical composition of sediment or modifications in species composition).

Cumulative impacts consist precisely of the interactions and **synergy** of different long-term physical, chemical or biological effects that can become irreversible over several years or decades, given the extent of accumulation of alterations occurring.

In terms of the different impacts of human activities, it is necessary to consider: a) the quantification of these impacts and their detection at an early stage in time to allow remedial actions/mitigation; b) economic assessment of the impacts and the possible effects of the degradation on local and regional socioeconomics.

There is a long history of the impacts of human activities on the water cycle and the processes of deterioration of water quality. The volume and complexity of the alterations have mostly occurred after the Industrial Revolution in the second half of the 19th century, as a result of the direct interference of human activities in the hydrological cycle and the growth of urbanization and land use for agriculture and irrigation.

Various different human activities and multiple uses create different threats and problems in terms of availability of water, causing increased risks (see Chart 18.1). Table 18.1 lists the principal modifications that occurred between 1680 and 1980, in relation to total and surface drainage in all continents.

The main difficulty in determining the qualitative and quantitative effects of an impact is that new impacts continue to occur at an accelerated rate, with multiple direct and indirect effects that call for rapid interdisciplinary action and appropriate technology to solve the problems (Somlyody, 1993).

This continuous interference has produced cumulative impacts and a large range of indirect effects (Branski *et al.*, 1989). Tundisi (1990, 2003) describes the following causes of impacts on Brazil's inland aquatic ecosystems and coastal waters from human activities:

▶ Deforestation;
▶ Irrigation;
▶ Mining;
▶ Urbanization;
▶ Road construction;
▶ Canal construction;
▶ Discharge of sewage from point and non-point sources;
▶ Discharge of industrial waste and agricultural products;
▶ Introduction of exotic species in terrestrial and aquatic systems;
▶ Removal of key species from ecosystems;
▶ Construction of dams and reservoirs;
▶ Dumping of solid waste in drainage basins;
▶ Eutrophication (cause and consequence);
▶ Construction of waterways;
▶ Impacts on water sources (deforestation, dumping of solid waste, use of drainage basins).

As a result of these impacts, many problems have resulted, producing direct and indirect effects, including:

▶ **Eutrophication**
As result of activities such as the discharge of untreated domestic sewage and industrial and agricultural run-off and waste, the eutrophication process accelerates quickly in inland aquatic ecosystems, including estuaries and coastal regions.
▶ **Increased turbidity and suspended material**
Due to the inappropriate use of drainage basins, especially deforestation, increased turbidity and suspended material are among the most serious problems affecting rivers, lakes and reservoirs. Many consequences occur, such as reduction of primary production of phytoplankton and flow capacity, damage to fish and turbines and pipes in dams/reservoirs, and alterations of thermal behaviour in rivers and reservoirs.
▶ **Loss of biological diversity**
The introduction of exotic species, deforestation, construction of dams, mining activities, and the loss of the mosaic of vegetation in the floodplain regions can cause drastic reductions in biological diversity. Reservoirs in the Amazon region also reduce the biodiversity in terrestrial ecosystems, leading to loss of native plant and animal species. Eutrophication and chemical contamination are also causes of biodiversity loss.
▶ **Alterations in the hydrological cycle and water levels**
Changes in the hydrological cycle can be attributed to the following factors: modifications in the reserve of water, construction of dams, increase or alterations in evapo-transpiration from lakes and reservoirs, modifications in the level of aquifers and alterations in recharge. Deforestation, including that in riparian forests along rivers, can cause alterations and reduce the recharge of aquifers.

Chart 18.1   Impacts on aquatic ecosystems from various human activities.

Human activity	Impact in aquatic ecosystems	Values/services at risk
Construction of dams/reservoirs	Alters the flow of rivers and the transport of nutrients and sediment and interferes with fish migration and reproduction	Alters habitats and commercial and sports fishing, as well as the deltas and their economies
Construction of dykes and channels	Destroys the connection of river with areas that previously flooded occasionally	Affects the natural fertility of floodplains and controls of flooding
Alteration of the natural channel of rivers	Ecologically damages rivers, modifies the flow of rivers	Affects the habitats, commercial and sports fishing, hydroelectricity production and transport
Drainage of wetland areas	Eliminates a key component of aquatic systems	Loss of biodiversity, natural filtering functions, recycling of nutrients, and habitats for fishes and aquatic birds
Deforestation/ use of soil	Alters drainage patterns, inhibits the natural recharge of aquifers; increases sedimentation	Alters the quality and quantity of water, commercial fishing, biodiversity and flood control
Uncontrolled pollution	Reduces water quality	Alters water supply and commercial fishing; increases treatment cost; decreases biodiversity; affects human health
Excessive removal of biomass	Reduces living resources and biodiversity	Alters commercial and sports fishing as well as the natural cycles of organisms; decreases biodiversity
Introduction of exotic species	Decreases native species; alters nutrient cycles and biological cycles	Loss of habitats, natural biodiversity and genetic stocks; alters commercial fishing
Air pollution (acid rain) and heavy metals	Alters the chemical composition of rivers and lakes	Alters commercial fishing; affects aquatic biota, recreation, human health and agriculture
Global climate change	Drastically affects the volume of water resources; alters distribution patterns of precipitation and evaporation	Affects the supply of water, its transport, the production of electrical energy, agricultural production and fishing; increases flooding and the flow of water in rivers
Population growth and general patterns of human consumption	Increases the pressure for hydroelectric construction, the pollution of water and the acidification of lakes and rivers; alters water cycles	Affects practically all economic activity that depends on services of the aquatic ecosystem

Sources: Turner *et al.* (1990a); NAS (1999); Tundisi *et al.* (2000); Tundisi (2002).

Table 18.1   Major changes in the total drainage and surface drainage on all continents between 1680 and 1980.

| Continent | Total drainage (R) 1680–1980, $m^3 \cdot year^{-1}$ | | | Surface drainage (S) $m^3 \cdot year^{-1}$ | | |
	1680	Anthropogenic alteration	1980	1680	Anthropogenic alteration	1980
Europe	3,240	−200	3,040	2,260	−410	1,850
Asia	14,550	−1,740	12,810	10,920	−1,790	9,130
Africa	4,300	−140	4,160	3,075	−1,790	2,480
N. America	6,200	−320	5,880	5,020	−1,490	3,530
S. America	10,420	−60	10,360	6,770	−320	6,450
Antarctica and Oceania	1,970	−10	1,960	1,520	−50	1,470

Source: L'vovich (1974).

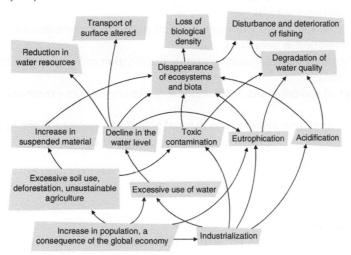

Figure 18.1   Main problems and processes related to the pollution of surface waters (lakes, rivers, reservoirs). Results of an ILEC study of 600 lakes on several different continents.
Sources: Kira (1993); Tundisi (1999).

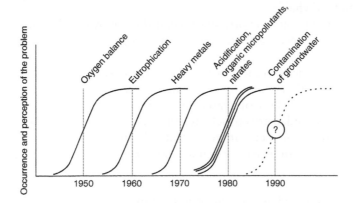

Figure 18.2   Effects of degradation observed in industrialized countries.
Source: Straškrába and Tundisi (2000).

Major impacts and problems in water quality of lakes, dams and rivers

- Pollution of organic matter from domestic sources (untreated sewage).
- Contamination by bacteria and viruses.
- Waterborne diseases.
- Eutrophication: excessive input of organic matter in aquatic systems due to influx of nutrients (especially nitrogen and phosphorus).
- Pollution from nitrate, causing public health problems.
- Hypolimnetic anoxia (in stratified lakes and reservoirs). Aggressiveness of structures. Increased levels of manganese and phosphorus. Nutrient release from sediments.
- Acidification: decrease in pH and release of metals.
- Problems of turbidity produced by suspended solids.
- Salinization resulting from excessive application of fertilizers to soil, or soil salinization in arid or semi-arid areas.
- Pollution from heavy metals.
- Impacts of toxic substances from agrochemical products. Accumulation in sediments and bioaccumulation in organisms.
- Impacts of toxic substances resulting from eutrophication and the accelerated growth of cyanobacteria.
- Discharge of oil and other chemicals into rivers, lakes, reservoirs, estuaries and coastal waters.
- Increased water temperature due to thermal pollution, imbalance in the hydrological cycles, and changes associated with changes in the global climate.

Source: Straškrába (1996).

▸ **Loss of buffering capacity**
Loss of wetland areas and reduction of vegetation in these areas and the eco-tones reduce the buffering capacity of aquatic ecosystems. Wetland areas are also regions of denitrification that affect the nitrogen cycle, functioning as natural buffering systems for the impacts, even enhancing biodiversity (see Chapters 6 and 15).

▸ **Changes in food chains**
With the introduction of exotic species and the removal of key species, multiple alterations in the feeding chains can occur. The loss of endemic fish species in large rivers of South America is an example of such alterations.

▸ **Expansion of the geographical distribution of tropical diseases**
With the rapid alterations in water quality and the construction of dams/reservoirs, an expansion of water-borne tropical diseases has occurred. For example, the expanded area of schistosomiasis can be attributed to the construction of dams/reservoirs, eutrophication, and human migration into regions with existing reservoirs or reservoirs under construction (Straškraba and Tundisi, 2000).

▸ **Toxicity**
Toxicity has increased as a result of mining operations, industrial run-off, agricultural practices, pesticides and herbicides, heavy metals and increased levels of

pollutants in food chains (Lacerda and Solomons, 1998). In addition, cyanobacteria resulting from eutrophication increase toxic substances in the water of lakes, rivers and reservoirs (Chellappa *et al.*, 2004).

▸ **Construction of dams/reservoirs**
The positive and negative impacts of the construction of dams/reservoirs were discussed in detail in Chapter 12.

Figure 18.1 presents the key impacts and their consequences, based on the analysis of 600 lakes and reservoirs around the world, a work conducted by the International Lake Environmental Committee (ILEC) (Firal, 1998). Figure 18.2 shows the sequence of degradation observed in industrialized countries.

Important quantitative and qualitative impacts include:

▸ **Urbanization**
In 1800, 29 million people lived in urban areas, which was approximately 3% of the world's population at that time. By 1986, the world's total urban population was already 2.2 billion and currently it exceeds 3 billion (approximately 50% of world population). The growth of urbanization implies an enormous alteration in the hydrological cycle due to the impermeability of urban surfaces as well as an increase in discharges of domestic waste, nitrogen and phosphorus, resulting from growth of the urban population and the volume of daily waste. Urbanization accelerates the cycle of processes, contamination and pollution (Tundisi, 2003).

▸ **Agricultural and industrial use of water resources**
The increase in the agricultural and industrial use of surface and underground water resources, as well as the resulting pollution, have caused serious alterations in the hydrological cycle and significant increases in organic and inorganic pollutants, with major effects on the aquatic biota and physical and chemical conditions of the water.

The resulting effects of these two intense impacts and discharges (agricultural and industrial) can also be attributed to technological and industrial development over the last 300 years. The introduction of new mechanisms for removing earth, constructing concrete and steel structures, drainage systems, advances in the capacity to open and operate deep wells, the increased use of pesticides and herbicides – all these developments and technical innovations have had large-scale, even global, effects on the hydrological cycle and water quality.

Per-capita water consumption increased four times between 1687 and 1987, and the consumption rate accelerated in the last 20 years of the 20th century. The exploitation of more accessible underground aquifers has been implemented since 1930 as result of **technological advances** in drilling wells and pumping water.

▸ **Sedimentation of inland ecosystems**
Sedimentation in rivers, lakes, reservoirs and flooded areas has had a large quantitative and qualitative impact because of improper use of soil and intensive agricultural practices that continually affect these ecosystems with diverse physical, chemical and biological impacts.

Alterations produced by the transport of sediments into the aquatic ecosystems include:

▶ Increased turbidity.
▶ Interference in biogeochemical cycles.
▶ Interference in the organisms: the effect on the feeding network, shading from light, and availability for phytoplankton and aquatic macrophytes.
▶ Interference in the hydrodynamics of rivers, lakes and reservoirs.
▶ Interference in hydrodynamics, in the direction and velocity of the water in rivers.
▶ Decreased availability of water in reservoirs and lakes.
▶ Accumulation of heavy metals and toxic organic substances in areas with high sedimentation.
▶ Interference in the life cycles of aquatic organisms, due to modification in substrata.
▶ Drastic decrease in levels of dissolved oxygen with increased sedimentation rates and high levels of suspended material in the water.

Figure 18.3 shows the main effects of suspended material on fish and other aquatic organisms.

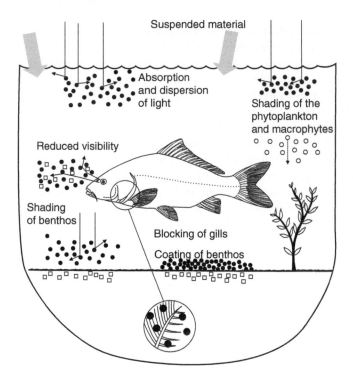

Figure 18.3   Impacts of suspended material on fish and other aquatic organisms.
   Source: Modified from Melack (1985).

The transport of sediment and suspended material carried by the drainage of water through soil and by erosion clearly depends on the type of soil (sedimentary or igneous rock), vegetation coverage on the slope and intensity of drainage. This effect is local and regional and depends on the impact produced by multiple uses: agriculture, urbanization, intensity of deforestation, soil-removing activities (construction of roads, railways, ports and channels) (Campagnoli, 2002).

Recently, Syvitski *et al.* (2005) studied the impact of human activities on the flow of the transport of sediment from continents to oceans, from the contribution of the rivers on different continents. The authors' main conclusion is that the multiple uses of soil by human activities have increased sediment transport in all rivers and soil erosion by 2.3 ± 0.6 billion metric tonnes per year. Over the last 50 years, annual sedimentation in reservoirs has been upwards of 100 billion metric tonnes of sediment, as well as 1–3 billion metric tonnes of carbon. The load of sediment flow retained in reservoirs varies from 0% on oceanic islands to a maximum of 31% in Asian reservoirs.

According to Syvitski *et al.* (2005), the intensification of human activities that alter sediment transport (for example, dam/reservoir construction) has resulting effects on the **exploitation of the aquifers,** changes in the **direction of transport of surface water,** changes in the **volume of lakes, drainage of wetland areas,** and **deforestation.**

Figure 18.4 shows the impacts caused by the construction of dams/reservoirs on all continents, and Figure 18.5 shows the sediment carried by rivers in the pre-anthropogenic period.

A recent study by Nilsjon *et al.* (2005) showed that out of 292 rivers cited, 172 have been affected by man-made reservoirs. In the areas around these reservoirs, there is increased use of water for irrigation and other economic activities per unit of water as compared to areas without reservoirs. The authors point to the risks and vulnerability

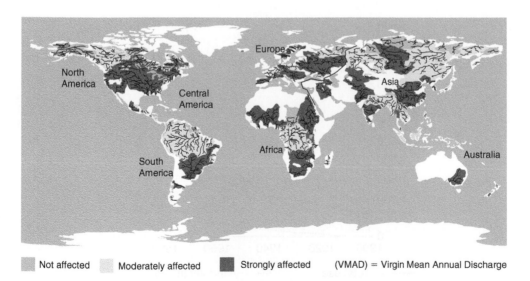

Figure 18.4   Impacts from the construction of dams and modifications of the courses of rivers in 292 drainage basins on all continents (see also color plate section, plate 24).
Source: Modified from Nilson *et al.* (2005).

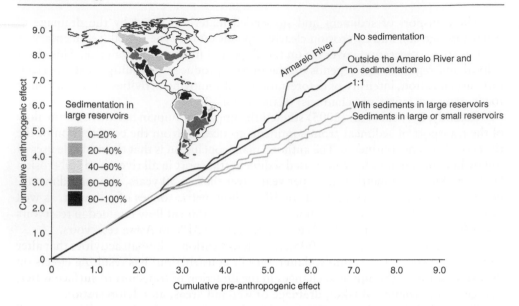

Figure 18.5   Comparison of the pre-anthropogenic loads with recent sediment loads, using data from 216 rivers, before and after the construction of dams. Data are presented as cumulative curves established by descending level of discharge (see also color plate section, plate 25). *Source*: Modified from Syvitski *et al.* (2005).

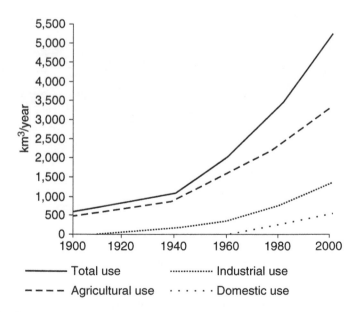

Figure 18.6   Total use of water by human activity.

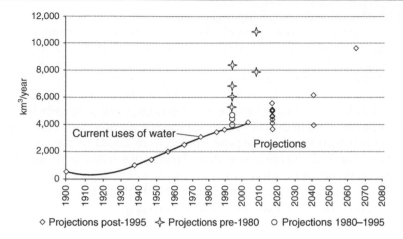

◇ Projections post-1995   ✧ Projections pre-1980   ○ Projections 1980–1995

Figure 18.7   Scenarios of current uses of water, with projections to 2080.
Source: Gleick (2000).

associated with these activities and uses of water in areas influenced by the presence of reservoirs.

In reservoir-affected areas, the aquatic biodiversity is also threatened by the loss of mechanisms and natural evolutionary processes that occur in these drainage basins and their floodplains and flooded areas. Therefore, when dams are built, considerable biodiversity and natural evolutionary processes are lost.

Figure 18.6 shows the total amounts of water usage from 1900 to 2000, and projections to 2080 are presented in Figure 18.7.

18.2   EUTROPHICATION OF INLAND WATERS: CONSEQUENCES AND QUANTIFICATION

Eutrophication is one of the most important qualitative and quantitative impacts on rivers, lakes and reservoirs, affecting virtually all freshwater ecosystems to a greater or lesser extent.

Under natural conditions, it can take several hundred years for the trophic level of a lake to increase. It depends on the lake's **inorganic load** and contributions from natural processes in drainage basins. The increase of nitrogen and phosphorus produced by human activity markedly accelerates the natural process of eutrophication, reducing the natural features of lakes and reservoirs and negatively impacting the water quality, making it unavailable for some uses and making the treatment process considerably more expensive. This eutrophication process associated with human activities has been called artificial eutrophication (Esteves and Barbosa, 1986) or cultural eutrophication (U.S. National Academy of Science, 1969; Welch, 1980; Margalef, 1983; Tundisi 1986).

The natural processes that occur in a drainage basin, and that cause a slow and steady increase in sedimentation and levels of nitrogen and phosphorus and organic matter, are related to effects of wind, erosion from rain, and addition of biological material (dead and/or decomposing organic matter, as by the addition of leaves and plant debris in riparian forests around lakes). The **eutrophication rate** in a lake basically depends on these factors, when there is a constant load of nutrients.

According to Margalef (1983), the terms oligotrophy and eutrophy were introduced by Weber (1907) to categorize wetland areas with higher or lower levels of nutrients. Naumann (1919) extended these terms to lakes. More characteristics for eutrophic lakes have been added to the initial conditions of high turbidity and levels of plankton, so that the matrix of eutrophication is currently quite complex.

The terms mesotrophic and hypereutrophic were added to the range of trophic levels. Mesotrophic indicates an intermediate status between eutrophic and oligotrophic systems, and hyperutrophic refers to a system with a high degree of eutrophication.

The rate and stages of progression of eutrophication in a lake basically depend on the following factors, given a constant load of nutrients:

18.2.1   Initial trophic state of the lake

Hypereutrophic lakes respond slowly or barely at all to addition of nitrogen and phosphorus, due to the already high rates of mobilization, self-shading and inorganic load present. Oligotrophic or mesotrophic lakes respond more quickly to the addition of nutrients from eutrophication. Trophic state also depends on the circulation, since the sediment, as discussed in the previous chapter, can serve as a source of both nutrient immobilization and internal nutrient load, due to alteration of the redox potential of the sediment based on the circulation (see Chapter 10).

18.2.2   Average depth and morphometry

Average depth is an important factor because it can increase or decrease the dilution of nutrients and, thus, increase levels per volume or area. A decrease in average depth signifies a euphotic zone closer to the sediment, as the sediment is closer to the surface of the lake. A lake's morphometry is another important factor, due to the epilimnion/hypolimnion ratios and compartmentalization that occurs in **dendritic lakes**, which can cause eutrophic segments to grow more quickly in lakes or reservoirs.

Correlations between the influx of nutrients and area of the drainage basin were demonstrated by Schindler (1971a) for lakes in Canada in conditions of little or no influence from human activity.

18.2.3   Retention time

Retention time has been studied in considerable detail. As retention time increases, so does the availability of nutrients for use. If retention time is short, the tendency for accumulation of phytoplankton is less and there is loss of cells or colonies. Tundisi and Matsumura Tundisi (1990) examined the effect of retention time on the eutrophication process in the Barra Bonita reservoir in the Médio Tietê (SP), and concluded that the rapid growth of cyanobacteria and the increase in conductivity during periods of drought were associated with increased retention time in the reservoir and thermal stability.

Water quality

According to Chapman (1992), water quality can be defined as a set of concentrations, speciation and physical partitions of inorganic or organic substances, and the composition, diversity and state of aquatic biota found in a given body of water. This quality presents temporal and spatial variations due to factors internal and external to the water body.

Human-caused pollution of the aquatic ecosystem signifies the introduction, directly or indirectly, of substances or energy resulting in:

i) negative effects on living resources;
ii) negative impacts on human health;
iii) impairment of activities in aquatic systems, such as fishing;
iv) impairment of the quality of water and its use in agricultural, economic and industrial activities;
v) reduction of amenities.

Water quality is thus used as an indicator of the conditions in an aquatic system and to assess the state of pollution, degradation or conservation of rivers, lakes, reservoirs, estuaries, coastal waters and wetland areas. This evaluation can be performed through monitoring, which is the periodic collection of information and the formation of a basic database for future actions. The limitations of the uses of water due to deterioration of its quality are shown in Chart 18.2.

Chart 18.2   Limitations in the use of water, due to degradation of its quality.

Pollutant	Uses						
	Potable water	Aquatic life	Recreation	Irrigation	Industrial uses	Energy and cooling	Transport
Pathogens	XX	N	XX	X	XX	na	na
Suspended solids	XX	XX	XX	X	X	X	XX
Organic matter	XX	X	XX	+	XX	X	na
Algae	XX	X	XX	+	XX	X	
Nitrate	XX	X	na	+	XX	na	na
Salt	XX	XX	na	XX	XX	na	na
Trace elements	XX	XX	X	X	X	na	na
Organic micro-pollutants	XX	XX	X	X	?	na	na
Acidification	X	XX	X	?	X	X	na

(xx) Increased impact, impeding use; (x) Negligible impact; (N) No impact; (na) not applicable; (+) Major impact on quality; (?) Effects not completely determined.
Source: Chapman (1992).

18.2.4   Causes of eutrophication

The main causes of cultural eutrophication (i.e., that produced by human activities) are related to the influx of domestic and industrial waste water, surface drainage, the contribution of groundwater, and the fertilizers used in agriculture. Soil erosion and excessive use of non-biodegradable detergents also cause eutrophication.

The major **sources of pollution** from agricultural ecosystems are the drainage of nitrogen and phosphorus applied to the soil and organic waste from livestock. Fertilizer can be removed by rain water and wind, increasing the levels of nitrogen and phosphorus in water.

Inorganic fertilizers are readily removed by water and drainage of soil. Recently, the use of low-soluble fertilizers, based on urea-aldehyde, has been attempted with the goal of reducing drainage. The basic combination of the large-scale application of fertilizer with nutrients in the soil involving nitrogen, phosphorus and potassium produces a stock of nutrients from which the soluble fraction is removed. Nitrogenized fertilizers contain ammonia, urea or condensations of urea-aldehyde. Solutions of ammonium nitrate and urea, with a typical content of 28–36% nitrogen, can be used with herbicides (Henderson-Sellers and Markland, 1987). They produce super-phosphated phosphorus fertilizers (P_2O_5-P), nitrophosphated in water-soluble parts or phosphated under the low-solubility granulated form, with 7–22% phosphate.

The use of **organic fertilizers** from animal excreta is also frequent.

Currently, it is widely held that phosphorus is the main cause of eutrophication caused by non-point or point sources.

18.2.5   Consequences of eutrophication and characteristics of eutrophic and oligotrophic lakes

Several qualitative and quantitative features can be identified in oligotrophic and eutrophic lakes. Eutrophication is a process of forced entry of nutrients into lakes, which consequently causes an acceleration of the nutrient cycle. Levels of phosphorus and nitrogen increase, as well as particulate carbon resulting from the accelerated production of organic matter. However, part of the cycle can be eliminated with the system's regulating mechanisms.

Urbanization also contributes to considerable quantities of household waste (often untreated). It is estimated that the contribution of nitrogen and phosphorus from human waste generally is:

Phosphorus: 2.18 g per-capita · day^{-1}
Nitrogen: 10.8 g per-capita · day^{-1}
(Henderson-Sellers and Markland, 1987)

Intense urbanization, coupled with industrial activities, has caused the discharge of nitrogen and phosphorus into lakes to increase significantly. A significant proportion of the phosphorus in household waste consists of synthetic detergents, in some countries reaching 50% of the total. Some countries have banned the use of phosphates in detergents in order to avoid eutrophication.

Industrial activities contribute significantly to eutrophication. **Contamination from chemical substances and toxic metals** resulting from various industrial processes

Chart 18.3   General characteristics of oligotrophic and eutrophic lakes.

Characteristics of lakes	Oligotrophic	Eutrophic
Physical-chemical		
Concentration of O_2 in the hypolimnion	High	Low to zero
Concentration of nutrients in the water column	Low	High
Concentration of nutrients in the sediment	Low	High
Particulate material in suspension	Low	High
Penetration of solar radiation	High	Low
Depth	Deep lake	Shallow lake
Biological		
Primary production	Low	High
Diversity of plant and animal species	High	Low
Aquatic macrophytes (density per m²)	Low	High
Biomass of phytoplankton	Low	High
Blooming of cyanophytes	Rare	Common or permanent
Characteristic groups of phytoplankton	Diatoms/ Chlorophyceae	Chlorophyceae/ cyanophytes

Source: Welch (1980); Margalef (1983).

affect many lakes. Direct discharge of untreated industrial waste can alter the pH, dissolved oxygen and temperature of natural waters.

Different types of industries contribute to the increased biochemical demand for oxygen, suspended material, phenols, sulphides, ammonia, phosphates and cyanides, and organic and inorganic substances, resulting from various types of processing (food and dairy products, refined oil, and the steel-making industry). Thermal pollution and the resulting thermal heating of lakes, reservoirs and rivers can accelerate the eutrophication process.

Eutrophication is a component in the decomposition process and often the most important. However, eutrophication leads to a wide range of other processes. Accumulation of toxic substances and heavy metals and other outcomes of human activities in drainage basins are components added to the water of rivers, lakes and reservoirs, making recuperation of eutrophicated ecosystems extremely complex and costly.

Chart 18.3 presents the classic difference between oligotrophic and eutrophic lakes. This is a concept that currently is being modified by advances in understanding the problem and the complexity of the process. However, the framework is useful for comparing extremes and establishing criteria for classification.

Eutrophic lakes, as already mentioned, present some regulatory mechanisms of eutrophication (for example, immobilization or loss of phosphorus and nitrogen), which can be summarized in the cycle shown in Figure 18.8.

18.2.6   Quantitative indicators and criteria to measure trophic state

Lakes can be grouped into categories by measuring the trophic state with various indices: oligotrophic, mesotrophic, eutrophic and hypereutrophic. These categories act as

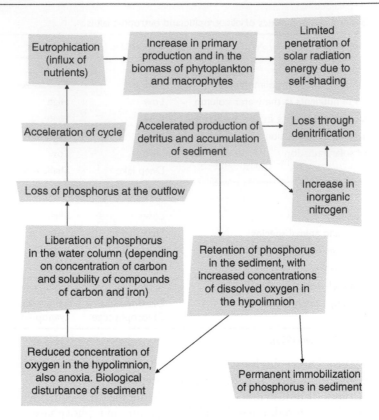

Figure 18.8   Summary of the cycle of some regulating mechanisms in eutrophication.
Source: Margalef (1983).

references and enable the monitoring of quantitative changes suffered by lakes from nutrient loads. Clearly, the dynamic characteristics of lakes and their individuality produce some deviations from this generalization, which is important in the restoration of lakes and **prevention of eutrophication.**

The trophic state is not only a measurement of the nutrient levels in lakes, but other parameters are also identified in order to prepare an **trophic state index,** based on a matrix of various indicators such as biomass of phytoplankton, zooplankton and bacterioplankton; oxygen level in the hypolimnion; transparency and total phosphorus in water. Each characteristic of the lake is assigned a numerical value, which, through an empirical formula, enables the calculation of the trophic status. Correlation between these factors is imperfect, however, and the same lake can be classified as oligotrophic, mesotrophic or eutrophic, depending on the index. For example, a reservoir in Texas (Canyon reservoir) was classified as oligotrophic by 11 indexes, mesotrophic by 4, and eutrophic by 7 (Henderson-Sellers and Markland, 1987). Therefore, the use of a single criterion is not advised for determining the trophic state index.

The criteria used to define the trophic state of a mass of water follow these variables:

▸   **Levels of nutrients** (total phosphorus, orthophosphate, total nitrogen and inorganic dissolved nitrogen – ammonia, nitrite, nitrate) – phosphorous levels below

Table 18.2   Rates of primary production indicative of trophic state.

Annual productivity rate	Oligotrophic	Eutrophic
$gC \cdot m^{-2} \cdot year^{-1}$	7–25	75–700
$mgC \cdot m^{-2} \cdot day^{-1}$	30–100	300–3,000

Source: Welch (1980).

$0.01 g \cdot m^{-3}$ (or $10 \mu g \cdot L^{-1}$) or of nitrogen below $0.3 g \cdot m^{-3}$ (or $300 \mu g \cdot L^{-1}$) can cause these nutrients to be limiting factors.

▸ **Allochthonous and autochthonous inorganic nutrient load** is measured by quantification of the uses of the drainage basin and the sediment-water interactions.

▸ **Rate of oxygen consumption in the hypolimnion** – The consumption of dissolved oxygen in the hypolimnion increases with eutrophication. This method can only be used in stratified lakes, and the various trophic levels are measured as follows:

  Oligotrophic lakes: $250 mg \cdot m^{-3} \cdot day^{-1}$
  Mesotrophic lakes: $250-550 mg \cdot m^{-3} \cdot day^{-1}$
  Eutrophic lakes: $550 mg \cdot m^{-3} \cdot day^{-1}$

(Henderson-Sellers and Markland, 1987)

▸ **Primary production of the phytoplankton** – Rodhe (1969) and Welch (1980) suggested primary production rates indicative of trophic status. Table 18.2 presents the respective values.

▸ **Chlorophyll-*a*** – For oligotrophic lakes, values of $0-4 \mu g$ chlorophyll-*a* $\cdot L^{-1}$; for mesotrophic lakes, $4-10 \mu g$ chlorophyll-*a* $\cdot L^{-1}$; and for eutrophic lakes, $10-100 \mu g$ clorophyll-*a* $\cdot L^{-1}$. However, Tundisi *et al.* (1994) found values of $150 \mu g$ chlorophyll-*a* $\cdot L^{-1}$ for Barra Bonita.

▸ **Transparency by Secchi disc** – the use of the Secchi disc to calculate (by contrast) the coefficient of vertical light attenuation, as seen in Chapter 4. This simple technique measures the total attenuation of underwater solar radiation in the lake, due to the concentration of inorganic and organic material, living or in decomposition. However, when there is an accumulation of inorganic particulate matter in the aquatic system, application of the Secchi disc should be used with caution. If the concentration of suspended matter in the water is always high, the use of Secchi disc to assess the degree of eutrophication is not recommended.

▸ **Other criteria** – criteria also used for qualitative and quantitative analysis that associate the index of trophic status with the composition of species, including the ratios of diatoms/cyanobacteria and Calanoida/Cyclopoida, as well as the biomass of benthic invertebrates. Generally, planktonic species common in eutrophic waters are *Microcystis aeruginosa*, *Microcystis flos-aquae*, *Anabaena* spp., and *Aphanizomenon flos-aquae*.

All these criteria described above should take into account the hydrological and morphometric characteristics of the lake, including volume, maximum depth, surface area, area of drainage basin, area and volume of the hypolimnion (in the case of stratified lake), and retention time.

Therefore, **geographical factors** such as **latitude** (and consequently, air temperature, annual cycle of precipitation, drainage, seasonal cycle of water temperature and wind), and **morphological, morphometric** and **hydrodynamic factors** are relevant to the understanding of eutrophication processes in lakes (Uhlmann, 1982).

Carlson (1977) presented one of the main methods for measuring trophic state. Using data from a series of lakes, he looked at the relationship between total phosphorus, chlorophyll-*a* and Secchi disc transparency, based on the following equations:

$$TSI = 10(6 - \log_2 SD)$$

$$TSI = 10\left(6 - \log_2 \frac{7.7}{0.68}\right)$$

Chl*a*

$$TSI = 10\left(6 - \log_2 \frac{64.9}{Ptotal}\right)$$

where:
TSI = Trophic State Index
SD = Secchi Depth (m)

Table 18.3 presents values for total phosphorus, chlorophyll-*a* and the Secchi disc in terms of the respective trophic state indices.

Several correlations exist between the various parameters used to measure the index of trophic status. Dillon and Rigler (1974) showed that the average chlorophyll-*a* near the surface in summer in temperate lakes and total phosphorus (*P*) is presented in the correlation:

$$\text{Chl}a = 0.0731P^{1,449}$$

The correlation presented by the OECD (1982) is:

$$\text{Chl}a = 0.28P^{0.26}$$

where:
Chl*a* – average chlorophyll-*a* level in the euphotic zone (in mg · m^{-3})
P – average annual concentration of phosphorus (also in mg · m^{-3})

Table 18.3   Indices of trophic state for phosphorus, chlorophyll-*a* and Secchi Disc.

Trophic state	Total P ($\mu g \cdot L^{-1}$)	Chlorophyll-a (mg · m^{-3})	Secchi disc (m)
20	3	0.34	16
30	6	0.94	8
40	12	2.6	4
50	24	6.4	2
60	48	20.0	1
70	96	56.0	0.5

Source: Carlson (1977).

The correlation between the depth of disappearance of the Secchi disc and clorophyll-a (in mg $\cdot$ m^{-3}) can be expressed by the following formula:

$$SD = 8.7(1 + 0.47\text{Chl}a)$$

where:
SD – Depth of Secchi in meters

Correlations are also used between the hypolimnetic oxygen demand (HOD) and the phosphorus load:

$$HOD = 1.58 + 0.37CR$$

(Welch and Perkins, 1979)
where:
HOD – hypolimnetic oxygen demand (in mg O$_2 \cdot$ m$^{-2} \cdot$ day^{-1})
C – phosphorus load per area (in mg $\cdot$ m$^{-2} \cdot$ year^{-1})
R – retention time (in years)

The limits vary widely between the various trophic state indices given by various parameters. Table 18.4 shows the values produced by the OECD (Organization for Economic Co-operation and Development) (1982) for chlorophyll-a, the Secchi disc transparency, and phosphorus load.

It should be emphasized that these limits mostly correspond to the values determined for a series of lakes in temperate regions, thus with non-tropical climatic and hydrological characteristics and point and non-point loads. For example, Tundisi and Matsumura Tundisi (1990) measured minimum values of 0.2 m for the Secchi disc in intense cyanobacterial blooms in the Barra Bonita reservoir.

However, there are difficulties in using the trophic state index when based on measurements of the Secchi disc in reservoirs and lakes in many tropical regions subject to intense discharges of suspended solids after rain and hydrological extremes.

Table 18.4 Trophic state and average values of phosphorus, chlorophyll-a and Secchi Disc.

Trophic state	Average load of phosphorus mg $\cdot$ m^{-3}	Average chlorophyll-a mg $\cdot$ m^{-3}	Maximum chlorophyll-a mg $\cdot$ m^{-3}	Maximum Secchi Disc (m) (annual avg)	Minimum Secchi Disc (m) (annual avg)
Ultra-oligotrophic	4.0	1.0	2.5	12.0	6.0
Oligotrophic	10.0	2.5	8.0	6.0	3.0
Mesotrophic	10–35	2.5–8.0	8–25	6–3	3–1.5
Eutrophic	35–100	8.0–25	25–75	3–1.5	1.5–0.7
Hyper-eutrophic	>100	>25	>75	1.5	0.7

Source: OECD (1982).

Another global index recently used by the EPA (the Environmental Protection Agency of the United States) is the lake assessment index (LAI), given by:

$$IAL = 0.25|(Chla + MAC)|2 + SD + OD + T(N,P)$$

(Henderson-Sellers and Markland, 1987)
where:
Chl*a* – chlorophyll-*a*
MAC – macrophytes
SD – Secchi disc
DO – dissolved oxygen
T (N, P) – Total nitrogen and phosphorus

The models dealing with the **balance of mass** of phosphorus in a lake correlate load, sedimentation rate, discharge, and area:

$$\frac{dP}{dT} = \frac{L}{Z} - \left(\smallint + \tau\right) \cdot P$$

(Vollenweider, 1969)
where:
L – annual load of phosphorus per area
∫ – discharge
τ – sedimentation rate
P – phosphorus level

The contributions of phosphorus from the sediment can be defined by the equation:

$$\bar{P} = \frac{L + Rs}{Z\left(\smallint + \tau\right)}$$

(Welch *et al.*, 1973)
where:
L – annual load per area
Rs – retention coefficient (estimated based on the relations between flow and sedimentation)
Z – average depth of lake
∫ and τ – values of flow and sedimentation of phosphorus, respectively.

Chart 18.4 summarizes the main criteria for definition of trophic state and the response of various parameters to the eutrophication process.

In terms of eutrophication, lakes and reservoirs in tropical regions have approximately the same symptoms and impacts that occur in lakes and reservoirs in temperate regions. However, the absence of a cold season and of a well-defined seasonal cycle in terms of temperature, as well as seasonality characterized by high rainfall in many tropical regions, produce other characteristics that make it difficult to compare the

Chart 18.4 Principal criteria for determining level of trophic state.

Physical	Chemical	Biological
Transparency (D)	Concentration of nutrients (I)	Frequency of blooms (I)
Electrical conductivity (I)	Morphometry (D)	Diversity of phytoplankton (D)
Suspensoids (I)	Deficit of oxygen in the hypolimnion (I) Supersaturation of oxygen in the epilimnion (I)	Biomass of phytoplankton (I) Chlorophyll-a (I) Zooplankton (biomass) and fish (I) Diversity of benthic fauna (D) Biomass of benthic fauna (I) Littoral vegetation (I)

I – Increase
D – Decrease
Source: Modified from Brezonick (1989), Taylor et al. (1980), and Welch (1980).

trophic state index of lakes and reservoirs in tropical and temperate regions. For example, with the heavy precipitation of summer, a huge amount of suspended material enters the tropical aquatic systems in deforested areas and areas with intense agriculture, lowering the underwater solar radiation practically to its lower limit. Low levels of nitrogen and phosphorus in tropical lakes often produce low N:P ratios, which can lead to a rapid and excessive cyanobacterial bloom, which often fixes nitrogen.

In a review of the problem, Henry et al. (1986) suggested that in natural lakes it is common for nitrogen to be a limiting factor, while in systems with advanced cultural eutrophication, phosphorus is more likely to be the limiting factor. Tropical systems seem to tolerate higher loads of phosphorus than do temperate systems, while tropical and temperate regions do not seem to differ very much in their limiting levels of nitrogen.

Trophic state indexes defined for tropical regions may differ by orders of magnitude from those for temperate regions. The individuality of lakes in response to eutrophication and initial chemical composition (which depends on regional geochemical characteristics) are also important when comparing eutrophication processes and their specific effects on lakes in temperate and tropical regions. Phosphorus and nitrogen are the principal limiting and key elements in eutrophication in tropical, temperate and sub-arctic lakes and reservoirs, but thresholds and limits differ.

Recent results from the Pan American Health Organization and the National Health Organization (1986) propose the following preliminary trophic classification for tropical lakes:

	Total phosphorus (average)
Oligotrophic	$<30 \mu g \cdot L^{-1}$
Mesotrophic	$30–50 \mu g \cdot L^{-1}$
Eutrophic	$>50 \mu g \cdot L^{-1}$

The classification focuses on the limiting level of the nutrient and on the threshold levels for eutrophication in tropical and temperate lakes. Thornton (1980) suggests,

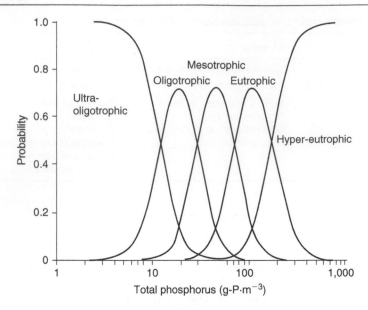

Figure 18.9   Probable distribution of trophic level in tropical lakes based on concentration of total phosphorus.
Source: Salas and Martino (1991).

for example, levels of $50–60\,\mu g\,P\cdot L^{-1}$ and $200–1000\,\mu g\,N\cdot L^{-1}$ for tropical lakes as lower limit to eutrophication.

Figures 18.9 and 18.10 show the ranges of probability proposed by Salas and Martino (1991) and Vollenweider (1968). Chart 18.5 details the basic criteria to assess eutrophication.

Table 18.5 shows the various methods to classify trophic states, the variables used and values described for different trophic states.

It is useful to compare the various trophic state indices. Carlson's index, for example, was developed for lakes in temperate regions (Carlson, 1977). Salas and Martino (1991) proposed a trophic state index for tropical lakes based on a large number of studies of tropical lakes and reservoirs. These two authors presented, in addition to a differential index, the following indices to estimate non-point sources contributing to tropical lakes, reservoirs and rivers (Table 18.6).

Table 18.7 lists the **discharge coefficients** per capita, according to Jørgensen (1989).

18.2.7   Eutrophication and cyanobacteria

Frequent cyanobacterial blooms are one of the most serious consequences of eutrophication. Many blooms have no greater repercussions other than triggering a very rapid process of increasing amounts of live particulate organic matter, which decomposes rapidly after a bloom begins to degrade. Cyanobacteria release significant amounts of dissolved organic matter (DOM) into the water, partly due to the effect from high light intensities that damage the colonies and cause mass mortality or to the decomposition of

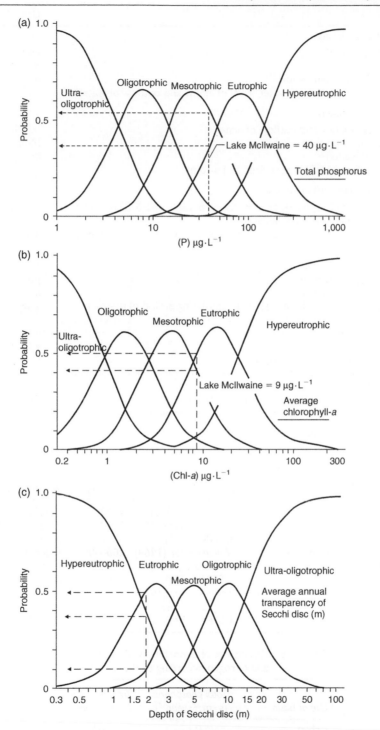

Figure 18.10 Probable distribution of trophic categories based on the total concentration of phosphorus (a), average concentration of chlorophyll-*a* (b), and visibility of the Secchi disc (c). The possible classifications are indicated for Lake McIlwaine (now Lake Chivero), Zimbabwe.

Chart 18.5 Primary criteria for assessment of eutrophication.

Parameter	Unit
Morphometric conditions	
Surface area of the lake	km^2
Volume of the lake	m^3
Minimum and maximum depths	m
Sites of influx and outflow of water (level of inflow and outflow of water)	
Hydrodynamic conditions	
Total volumes of influx and outflow of water	$m^3 \cdot day^{-1}$
Theoretical retention time	Month
Thermal stratification	Year
Conditions of flow (outflow at surface or bottom)	
Concentration of nutrients in the lake	
Reactive dissolved phosphorus; total dissolved phosphorus; total phosphorus	$mgP \cdot L^{-1}$
Nitrate; nitrite; ammonia; total nitrogen	$mgN \cdot L^{-1}$
Silicate (if diatoms constitute a significant proportion of the phytoplanktonic populations)	$mgSiO_2 \cdot L^{-1}$
Response to lake stratification	
Chlorophyll-*a*; phaeophytin	$\mu g \cdot L^{-1}$
Transparency (Secchi disc)	m
Rate of oxygen depletion in the hypolimnion (during period of thermal stratification)	$g\, O_2 \cdot m^{-2} \cdot day^{-1}$
Primary production	$g\, C \cdot m^{-3} \cdot day^{-1}$ $g\, C \cdot m^{-2} \cdot year^{-1}$
Diurnal variation in dissolved oxygen	$mg \cdot L^{-1}$
Major taxonomic groups and dominant species of phytoplankton, zooplankton and bottom fauna	Dominance and relationships
Extension of growth of periphyton in the littoral zone	Biomass ($mg \cdot m^{-2}$)
Biomass and dominance of species of aquatic macrophytes	g or $kg \cdot m^{-2}$

Source: Ryding and Rast (1989).

Table 18.5 Indices of classification of trophic state.

Variable	Source	Oligotrophic	Mesotrophic	Eutrophic
Total phosphorus ($mg \cdot m^{-3}$)	Sakamoto (1966)	2–20	10–30	10–90
	Vollenweider (1968)	5–10	10–30	30–100
	USEPA (1974)	<10	10–30	>20
Inorganic nitrogen ($mg \cdot m^{-3}$)	Vollenweider (1968)	200–400	300–650	500–1,500
Chlorophyll-*a* ($mg \cdot m^{-3}$)	Sakamoto (1966)	0.3–2.5	1–15	5–140
	USEPA (1974)	<7	7–12	>2
Biovolume of phytoplankton ($cm^3 \cdot m^{-3}$)	Vollenweider (1968)	1	3–5	10
Index of diatoms	Nygaard (1949)	0.0–0.3		0.0–1.75
Depth of Secchi (m)	USEPA (1974)	>3.7	2.0–3.7	<2.0

Table 18.6 Coefficient of exportation of phosphorus and nitrogen, according to the use of soil.

Use of basin	Total phosphorus ($g \cdot m^{-2} \cdot year^{-1}$)	Total nitrogen ($g.m^{-2} \cdot year^{-1}$)
Urban	0.1	0.5
Rural agriculture	0.05	0.5
Forest	0.01	0.5

Source: Salas and Martino (1991).

Table 18.7   Discharge coefficients per capita.

Phosphorus	Variation	800–1,800 (g·year⁻¹)
	Average	1,300 (g·year⁻¹)
Nitrogen	Variation	3.000–3,800 (g·year⁻¹)
	Average	3,400 (g·year⁻¹)

Source: Jørgensen (1989).

the bloom after extensive mortalities. This dissolved organic matter is used by bacteria, and thus, in the case of decomposing cyanobacteria, the associated bacterial flora are vast and diverse, as shown by Sandes (1998) and Panhota *et al.* (2003).

In addition to the general effects of the cyanobacterial blooms, another serious consequence, the production of toxins, threatens the aquatic biota and human health (see Figure 18.11). These toxins can cause a range of different health problems and even death when swallowed or when humans or animals come into contact with them (Carmichael and Chorus, 2001). Exposure to cyanobacteria can result in dysfunction and mortality.

The principal effects of the toxins include skin irritations, allergic responses, irritation of the mucosa, paralysis of the respiratory muscles, diarrhoea, and damage to the liver and kidneys. Epidemiological evidence showed there was increased incidence of liver cancer and rectal cancer after consumption of cyanobacteria in contaminated water (Zalewski *et al.*, 2004).

Chart 18.6 describes the main toxins caused by cyanobacteria and the effects produced. Table 18.8 shows the occurrences of cyanotoxins in several regions during the last 20 years as a result of the presence of toxic strains of cyanobacteria. Figure 18.12 illustrates the main inter-relationships of the microcystin toxin with other biological components in the aquatic system.

The removal of cyanobacteria from inland waters is complex and costly. Chart 18.7 describes how these cyanotoxins can be removed during the water treatment process through different mechanisms and methods.

18.2.8   Models of the eutrophication process

Eutrophication has occurred in many lakes, rivers and estuaries in all regions around the world. A model of the eutrophication process is therefore important as a means for resolving this problem (restoration of lake, river reservoir, minimization of effects). Introduction of a model depends on identification of the contours of the system, time scales and sub-systems. The definition of spatial and time scales and sub-systems clearly depends on a specific understanding of the limnology and ecology of the system.

With respect to eutrophication of lakes, the focus has been primarily on the growth of algae and nutrient cycles, as a basis to predict the effects of the nutrients on the process. In this case, basic knowledge of the cycles of the main phytoplanktonic species is essential. Often, cyanobacterial blooms are associated with the eutrophication process. In order to resolve the problem, the bloom periods must be identified, as well as the interrelationship with external climatic factors, influx of nutrients, and internal factors in the functioning of lakes.

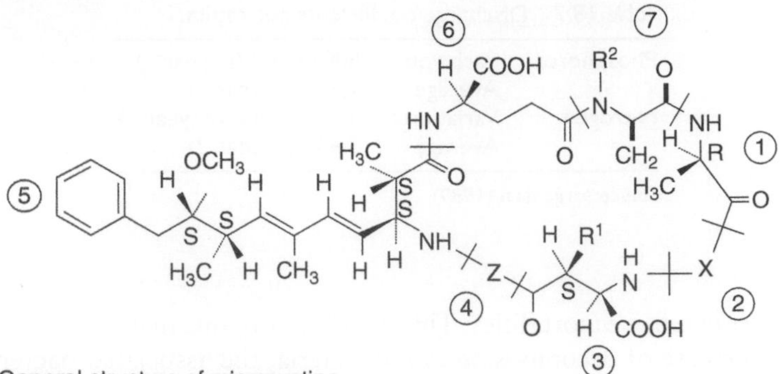

General structure of microcystins
cycle–(D–Ala^1–X^2–D–MeAsp^3–Z^4–Adda^5–D–Glu^6–Mdha^7)

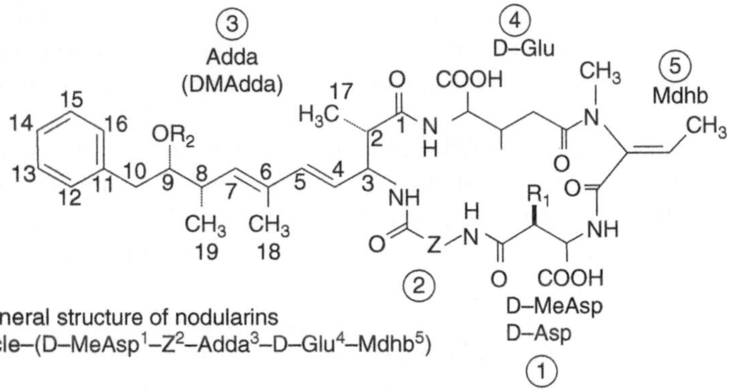

General structure of nodularins
cycle–(D–MeAsp^1–Z^2–Adda^3–D–Glu^4–Mdhb^5)

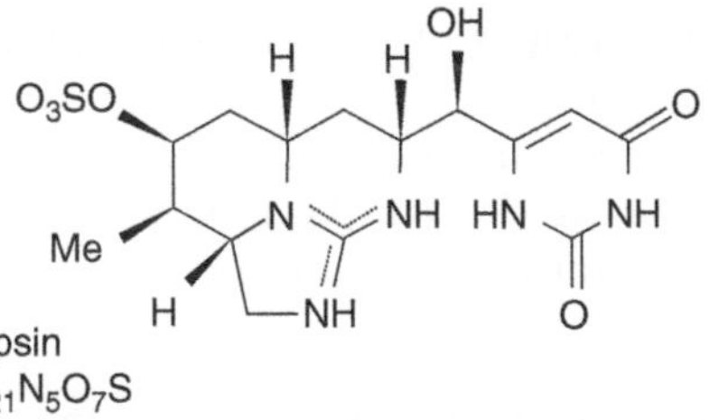

Cilindrospermopsin
MW 415; $C_{15}H_{21}N_5O_7S$

Figure 18.11 Structure of cyclic peptides and cylindrospermopsin.
Source: Chorus and Barthram (1999).

Designing a model to study and address the eutrophication process involves an extensive initial programme of sampling and collection of material. The definition of the sampling period and frequency of collection depends on the initial understanding of the main forcing functions acting on the lake or reservoir. For example, an intensive sampling programme can be conducted during periods of intense rainfall and during periods of drought. Water temperature can also be used as a basis for intensive sampling. The linking of sampling of different variables and of biological information with the identification of certain processes (such as circulation and turbulence) is essential to understanding the problem. The solution of problems related to processes and rates should be done experimentally, keeping in mind the lake or reservoir's intrinsic

Chart 18.6 Cyanotoxins of cyanobacteria

Cyanotoxins	Type of compound (molecular weight)	LD 50 $\mu g \cdot kg^{-1}$ ration	Toxic species
Hepatoxins			
Microcystin-LR	Cyclic Heptapeptide (MW 994)	50	Microcystis aeruginosa, M. wesenbergii, Oscillatoria agardhii, O. tenuis, Anabaena flos-aquae, Nostoc rivulare
Nodularin	Cyclic pentapeptide (MW 994)	50	Nodularia spumigena
Cylindrospermopsin	Hydroximethyluracil Tricyclic Guanidine (MW 415)	500	Cylindrospermopsis raciborskii, Umezakia natans
Neurotoxins			
Anatoxin-A	Secondary amine Alkaloid (MW 165)	200	Anabaena flos-aquae, O. agardhii, Aphanizomenon flos-aquae, M. aeruginosa
Anatoxin-A-S	Cyclic guanidine N-Hydroxy Methyl phosphate ester (MW 252)	20	Anabaena flos-aquae
Aphantoxin I	Purine alkaloid (Neosaxitoxin MW 315)	10	Aphanizomenon flos-aquae
Aphanatoxin II	(Saxitoxin MW299)	10	Anabaena circinalis
Cytotoxins			
Scytophycin A and B	Methylformamide (Scytophycin A MW 821) (Scytophycin B MW 819)	650	Scytonema pseudo hofmani Scytonema pseudo hofmani
Cyanobacterin	Diarylaloctone chloryde		Scytonema hofmani
Hapalindol A	Indole alkaloid		Hapalosiphon fontinalis
Acutiphycin	Macrolide		Oscillatoria acutissima
Tubercidin	Nucleotide of pyrrolopyrimidine		Tolypothrix byssoidea
Dermatoxins			
Debromoaplysiatoxin	Phenol (MW 560)		O. nigroviridis, Schizotothrix calcicola,
Oscillatoxin A	Phenol (MW 560)		O. nigroviridis, S. calicola
Lyngbiatoxin A	Alkaloideindol (MW 435)		Lyngbya majuscula

LD – Lethal dose
Source: Carmichael (1992).

operating mechanisms and the effects of the forcing functions on these processes. The use of data from the literature, with rates for different processes for lakes in tropical and temperate regions, should be done as a comparative framework.

Over the past ten years, many models of eutrophication have been developed. Table 18.9 presents the main models, with their significant features.

Table 18.8 Frequency of occurrences of masses of toxic cyanobacteria in inland aquatic ecosystems.

Country	No. of samples tested	% of samples toxic	Type of toxicity	References
Australia	231	42	Hepatotoxic	Baker and Humpage (1994)
			Neurotoxic	
Australia	31	84	Neurotoxic	Negri et al. (1997)
Brazil	16	75	Hepatotoxic	Costa and Azevedo (1994)
Canada (Alberta)	24	66	Hepatotoxic	Gorham (1962)
			Neurotoxic	
Canada (Alberta)	39	95	Hepatotoxic	Kotak et al. (1993)
			Neurotoxic	
Canada (Alberta) (3 lakes)	226	74	Hepatotoxic	Kotak et al. (1995)
Canada (Saskatchewan)	50	10	Hepatotoxic	Hammer (1968)
			Neurotoxic	
China	26	73	Hepatotoxic	Carmichael et al. (1988b)
Czech Republic	63	82	Hepatotoxic	Marsalek et al. (1996)
Denmark	296	82	Hepatotoxic	Henriksen et al. (1996b)
			SDF	
			Neurotoxic	
East Germany	10	70	Hepatotoxic	Henning and Kohl (1981)
			SDF	
Germany	533	72	Hepatotoxic	Fastner (1998)
Germany	393	22	Neurototoxic	Bumke-Vogt (1998)
Greece	18	?	Hepatotoxic	Lanaras et al. (1989)
Finland	215	44	Hepatotoxic	Sivonen (1990)
			Neurotoxic	
France (Brittany)	22	73	Hepatotoxic	Vezie et al. (1997)
Hungary	50	66	Hepatotoxic	Torokné (1991)
Japan	23	39	Hepatotoxic	Watanabe and Oishi (1980)
Holland	10	90	Hepatotoxic	Leeuwangh et al. (1983)
Norway	64	92	Hepatotoxic	Skulberg et al. (1994)
			Neurotoxic	
			SDF	
Portugal	30	60	Hepatotoxic	Vasconcelos (1994)
Scandinavia	81	60	Hepatotoxic	Berg et al. (1986)
Switzerland	331	47	Hepatotoxic	Willén and Mattsson (1997)
			Neurotoxic	
United Kingdom	50	48	Hepatotoxic	Codd and Bell (1996)
		28		
USA (Minnesota)	92	53	Not specified	Olson (1960)
			Neurotoxic	
USA (Wisconsin)	102	25	Hepatotoxic	Repavich et al. (1990)
			Neutrotoxic	
Average		59		

SDF – Slow death factors
Source: Chorus and Barthram (1992).

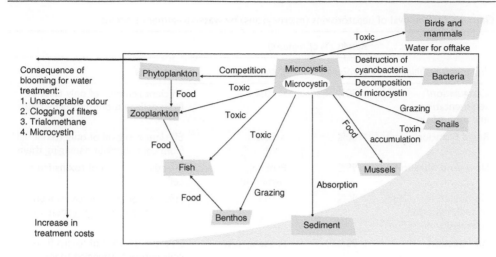

Figure 18.12   Interaction of microcystine from toxic strains of *Microcystis* spp. with components of the aquatic flora and fauna in lakes, reservoirs and rivers, and the effects on the treatment of water and its potability.
Source: Park *et al.* (2001).

Consequences of eutrophication

Eutrophication has a number of consequences that can be summarized in the following general processes:

- Anoxia (lack of oxygen in water), which causes mass mortality of fish and invertebrates and releases an odorous gas, often toxic (H_2S and CH_4).
- Blooms of algae and uncontrolled growth of aquatic plants, especially macrophytes.
- Production of toxins by some toxic species of algae.
- High levels of organic matter, which, if treated with chlorine, can produce carcinogens.
- Deterioration of recreational values of lakes and reservoirs, due to reduced transparency.
- Restricted access for fishing and recreational activities because of the accumulation of aquatic plants that can prevent locomotion and transport.
- Steep decline in biodiversity and number of plant and animal species.
- Alterations in the composition of fish species, with a decrease in their commercial value (and changes in species and loss of commercial value due to contamination).
- Decrease in dissolved oxygen concentration, especially in the deeper regions of temperate lakes in the autumn.
- Reduction in fish stocks caused by depletion of dissolved oxygen in water in the deeper regions of lakes and reservoirs.
- Chronic and acute effects on human health (Azevedo, 2001).

Chart 18.7   Removal of hepatotoxins (microcystins) by water treatment systems.

| Treatment technique | Results (% of removal) | | Comments |
	Intracellar	Extracellular	
Coagulation/ sedimentation Injection of air	>80%	<10%	Efficient removal of only toxins from cells without damaging them
Rapid filtration	>60%	<10%	Efficient removal of only toxins from cells, without damaging them
Slow sand filtration	~99%	Probably significant	Effective removal of toxins from cells
Combined techniques of sedimentation, coagulation/filtration	>90%	<10%	Efficient removal of toxins from cells without damaging them
Air injection	>90%	Probably low	Efficient removal of toxins from cells without damaging them
Adsorption – activated charcoal in powder	Negligible	>85%	Appropriate doses of activated charcoal >20 mg·L^{-1}; DOM reduces removal capacity
Adsorption – activated charcoal granules	>60%	>80%	Competition with DOM reduces the capacity and accelerates the breakdown of cells
Biologically activated charcoal granules	>60%	>90%	Biological activity in granular charcoal improves efficiency and lifetime of the system
Pre-ozonation	Very effective in efficiency of coagulation	Potential increase of extracellular concentration	Useful in low dosages to improve the coagulation of cells. Risk of release of toxins, which requires continuous monitoring
Post-clarification ozonation	Effective in coagulation	Causes lysis and release of metabolites	Effective in coagulation of cells, but increases the risk of dissolved toxic substances
Post-clarification ozonation	–	>98%	Rapid and efficient in soluble toxins
Post-filtration application of chlorine	–	>80%	Effective when the free chlorine is >0.5 mg·L^{-1} with pH < 8 and low DOM, negligible effect at pH > 8
Chloramine	–	Negligible	No effect
Chlorine dioxide	–	Negligible	No effect
Potassium permanganate	–	95%	Effective in soluble toxins
Hydrogen peroxide	–	Negligible	No effect
UV radiation	–	Negligible	Capable of degradation of L-R microcystin and anatoxin but only in high dosages, which is impractical
Membrane processes	Very high >99%	Uncertain	Depends on type of membrane. Further research is necessary

DOM – Dissolved organic material
Source: Zalewski *et al.* (2004).

Table 18.9   Principal models used for measuring the processes of eutrophication.

Type of model	Number of variables of state by layer or segment	Nutrient	Segments	Dimension (D) or Layer (L)	CS or NC	Calibrated (C) or Validated (V)	Published case studies
Vollenweider	1	P (N)	1	1L	CS	C+V	many
Imboden	2	P	1	2L, 1D	CS	C+V	3
Onelia	2	P	1	1L	CS	C	1
Larsen	3	P	1	1L	CS	C	1
Lorenzen	2	P	1	1L	CS	C+V	1
Patten	33	P, N, C	1	1L	CS	C	2
Ditoro	7	P, N	7	1L	CS	C+V	1
Canale	25	P, N, Si	1	2L, 1D	CS	C	1
Jorgensen	17	P, N, C	1	1–2L	NC	C+V	3
Cleaner	40		many	many	CS	C+V	many

P – Total phosphorus; N – Total nitrogen; C – Total carbon; Si – 'Reactive' silicon; CS – Constant; NC – Cycle of independent nutrients.
Source: Modified from Jorgensen (1980).

18.2.9   Ten-year history of the eutrophication process in Barra Bonita reservoir

The recent growth of agribusiness in the state of São Paulo, especially large-scale production of sugarcane and *Eucaliptus* spp., has resulted in large losses of biodiversity in the terrestrial system. The large biomasses of monocultures, with intensive use of fertilizers, have various different impacts on the bodies of water. The increase in export rates of nutrients from soils, associated with increased discharge of untreated domestic water, is largely responsible for the eutrophication of rivers, lakes and reservoirs (Tundisi and Matsumura Tundisi, 1992).

The Tietê basin is an example in the state of São Paulo of how the growth of agro-industrial production and rapid urbanization significantly alter aquatic ecosystems.

As the first major reservoir of waters on the Tietê River (state of São Paulo), the Barra Bonita reservoir reflects the processes of the entire catchment area, which has a population of 23 million urban dwellers, including the metropolitan region of São Paulo, Campinas, Sorocaba and the regions with extensive cultivation of sugarcane. Saggio (1992) noted that the rivers that run through the drainage basin have problems with high organic load and hypoxia, and the same problems occur in the reservoir, which is highly fertilized by nitrogen and phosphorus, leading to an intense eutrophication process.

The same water in which waste is deposited is then drained to supply drinking water. The reservoir, in addition to accumulating excess nutrients, supports activities such as fluvial transport, recreation, aquaculture and production of electricity.

To analyze the changes in water quality in the Barra Bonita reservoir over the last ten years, climatic data were obtained (on wind and precipitation, for example) as well as hydrological data (including turbine discharge, spillway discharge, total discharge, and water retention time).

Figure 18.13 shows the boundary conditions in which the reservoir functions, which are essential for understanding and controlling the eutrophication process. Precipitation and wind are two basic forcing functions, which act in different ways over time.

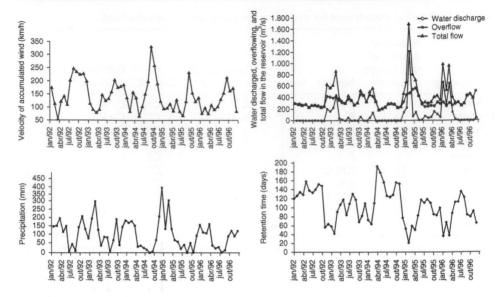

Figure 18.13   Flow conditions in the Barra Bonita reservoir.
Source: IIE/PNUMA (2001).

The **discharges of precipitation in summer** coincide with the lightest winds. In winter, the dry season, the winds are stronger, causing alterations in the vertical structure of the reservoir and more intense circulation. Retention time is also an important forcing function that increases in winter.

All of these forcing functions act on the system and produce **pulses of eutrophication,** accumulation of phytoplanktonic biomass or its loss during the opening of floodgates. The export rates of suspended material, nitrogen and phosphorus toward reservoirs downstream reflect the actions of the various forcing functions that control cycles of the Barra Bonita reservoir. The interactions of the hydrological functioning of the reservoir (as a result of the seasonal climatic cycle), the hydraulic operating systems (retention time) and the point and non-point loads of nutrients determine the cycles of eutrophication and cyanobacterial blooms (*Microcystis aeruginosa* in the summer and *Anabaena flos-aquae* in winter).

18.2.10   Denitrification in the control of eutrophication

Denitrification – a process performed by bacteria that reduces nitrate, nitrite and gaseous forms of nitrogen when dissolved oxygen becomes a limiting factor – plays a basic biochemical and ecological role in the nitrogen cycle, acting to prevent the accumulation of unwanted nitrate unassimilated by plants (Whatley and Whatley 1981). In aquatic environments with significant amounts of nitrogenated compounds from runoff of waste-water from agricultural and municipal waste sources, denitrification helps to control the degree of eutrophication, imposing a nutritional limitation on excessive growth of algae.

In normal conditions of availability of dissolved oxygen, bacteria oxidize the organic matter with the concomitant reduction of oxygen. However, with depletion of oxygen, denitrifying bacteria use the nitrate as a means to accept electrons for oxidation of the organic matter. Therefore, for denitrification to occur, the dissolved oxygen levels

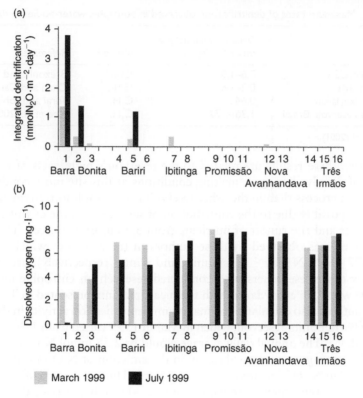

Figure 18.14   a) Integrated denitrification; b) Dissolved oxygen in six reservoirs on the Tietê River, Barra Bonita – SP.
Source: Abe *et al.* (2001).

need to be very low, less than $1\,mg \cdot L^{-1}$ (Abe *et al.*, 2000), nitrate needs to be present as an electron acceptor and organic matter as substrate in the environment.

According to Payner (1973), the linear representation can be:

$$NO_3^- \rightarrow NO_2^- \rightarrow NO \rightarrow N_2O \rightarrow N_2$$

Nitric oxide (NO) and nitrous oxide (N_2O) – intermediary gaseous products released to the atmosphere in the denitrification process – can be involved in reactions that alter global climates by reducing the ozone layer and producing the greenhouse effect (Yung *et al.*, 1976; Davidson, 1991). In fact, it is known that the N_2O level in the atmosphere has increased in recent decades at the rate of 0.25% per year (IPCC, 1990), due to indiscriminate use of nitrogen fertilizers, population growth, and progressive increase in dumping of domestic sewage.

There is a series of large cascade reservoirs in the centre portion of the middle Tietê. This system, including the Barra Bonita reservoir, shows an advanced degree of eutrophication, indicated by high primary productivity and high levels of nutrients and water conductivity (Tundisi and Matsumura Tundisi, 1990).

Abe *et al.* (2001) conducted denitrification experiments in the water column of the cascade reservoirs from the middle to lower part of the Tietê river, and verified that the highest rates were in the Barra Bonita (see Figure 18.14). According to

Table 18.10   Maximum rates of denitrification observed in eutrophic water-bodies in diverse regions.

System	Denitrification activity (mmolN · day^{-1})	Method	References
Mendota Lake, USA	0.6–1.9	$^{15}NO_3^-$	Brezonik and Lee (1968)
Lake 227, Canada	0.2–1.6	$^{15}NO_3^-$	Chan and Campbell (1980)
Lake Fukami-ike, Japan	0.64	C_2H_2	Terai and Yoh (1987)
Barra Bonita reservoir, Brazil	1.36–1.77	C_2H_2	Abe et al. (2001)

Source: Abe et al. (2001).

these authors, these results are related to the reservoir's high levels of nutrients and low levels of dissolved oxygen, making conditions at this site more conducive to the denitrification process than in the other reservoirs. The conditions in the Barra Bonita reservoir are possibly due to the contribution of sewage from the city of São Paulo in the Tietê river and the runoff of nutrients from agricultural sources. **The maximum denitrification rates** observed by these authors in the reservoir, which ranged from 1.36 to 1.77 mmol N · day^{-1} (in summer and winter respectively), are in the range observed by other researchers who conducted research on eutrophic lakes in other parts of the world, in accordance with the measurements presented in Table 18.10.

These authors also calculated the maximum denitrification activity in the water column in the Barra Bonita reservoir, which in March 1999 varied from 1.36 to 3.79 mmol $N_2O · m^{-2} · day^{-1}$. The annual daily average would be 2.58 mmol $N_2O · m^{-2} · day^{-1}$, and the annual contribution of denitrification in the emission of N_2O into the atmosphere was calculated to be 941.70 mmol $N_2O · m^{-2} · day^{-1}$. Thus, in an area of 1 km^2, the emission of N_2O into the atmosphere would be 941,000 mols per year, the equivalent of 26,348 tons of nitrogen removed from the system by denitrification processes.

In both summer and winter, denitrification occurred only in places with low levels of dissolved oxygen (< 1 mg · L^{-1}). Since the reservoirs of the middle Tietê are polymictic, thermal stratification with the formation of an anoxic layer at the bottom of the water column is a sporadic event, which is why denitrification activity was not detected in a majority of the points sampled, except in areas affected by incoming sewage, as in Station 1 and on the Piracicaba River.

Although no data exist on the sediments at the bottom of the reservoirs in the middle and lower Tietê River, it is possible that the denitrification process is occurring there more than in the water column. In the sediment, frequently anoxic, levels of organic matter and nutrients and the density of bacteria are all extremely high.

18.3   INTRODUCTION OF EXOTIC SPECIES IN LAKES, RESERVOIRS AND RIVERS

The deliberate or accidental introduction by man of exotic species (not native to a region) has caused many direct and indirect problems in aquatic ecosystems in the short-, medium- and long-term. Such introduction can cause significant modifications in the food chain. If the introduced species is a predator, for example, there are likely to be severe negative effects on the trophic structure of lakes and reservoirs.

Classic cases of exotic species being introduced into lakes and reservoirs have occurred in Lake Victoria (Africa), and the Gatún reservoir (Panama). In 1960 in Lake

Victoria, the introduction of Nile perch (*Lates niloticus*), a voracious predator, decimated the native fish populations of more than 400 species of cichlids. The cichlid species had varied feeding habits: native species fed on algae and insects, and other species ate organic matter and crustaceans. Local inhabitants currently fish only Nile perch, whose stocks are also declining due to depletion of the stocks of native cichlid species.

Another classic example of the impact of species transplanted by man is the introduction of *Cichla ocellaris* (tucunaré) in the Gatún reservoir (Panama), which due to intense predation drastically simplified the food web.

Over the past 20 years, there has been a significant increase in the incidence of introduced species around the planet due to the expansion of navigation, economic globalization and other human activities. The use of water as ballast and the occurrence of biofouling on ships and other naval structures are the main causes of the vehiculization of exotic aquatic species (Tavares and Mendonça, 2004). Ballast water in ships, a recent practice, is largely responsible for the introduction of exotic species in the oceans and in inland waters.

The latest global cases of invasive exotic species include:

▶ *Dreissena polymorpha*, a mollusc species native to the Caspian and Black Seas, arrived to Lake St. Clair in 1988 in the ballast water of an ocean-liner, and in less than ten years spread throughout all the North American Great Lakes. The species, forming massive colonies that affected structures, drastically reduced the population of native mollusc species and caused $5 billion in economic damages in the United States and Canada. This mollusc – whose popular name is 'Zebra mussel' because its shell is striped – affected fishing because it filters nanophytoplankton (<5 micrometres) and the resulting reduction in food for zooplankton produced losses of 20% in fishing the Great Lakes.

▶ The water hyacinth *Eichhornia crassipes* is native to the Amazon Basin and has spread into rivers, lakes and reservoirs throughout the tropics, affecting structures, navigation channels, and reducing light penetration and dissolved oxygen.

▶ A variety of *Vibrio cholerae* was introduced in Peru from ballast water in ships from Bangladesh, causing 10,000 deaths in three years in the country.

▶ A species of ctenophore (*Mnemiopsis leidyi*) was introduced in the Black Sea in 1980. Originally from North America, this species has completely changed the trophic network of the Black Sea because it is a voracious predator of zooplankton and fish larvae. It also invaded the **Azov Sea** and the Caspian Sea.

▶ *Pomacea caniculata*, a freshwater snail, was introduced as a food source in Southeast Asia and is now a pest in Thailand, Cambodia, Hong Kong, China and Japan. A native of the Amazons, this species was dispersed in paddies throughout Southeast Asia.

▶ In South America, recent cases of introduced species include *Corbicula fluminea* and *Limnoperna fortunei*, two species of molluscs. The latter, popularly known as the Golden Mussel, is originally from Southeast Asia, and was accidentally introduced in ballast water at the port of Buenos Aires in 1991. It spread throughout all the major rivers and lakes of the La Plata basin, and reached the lakes and rivers of the Mata-Grosso swamp and the reservoirs of the upper basin of the Paraná River. It has a high reproductive capacity, broad tolerance to environmental conditions, it attaches to substrata, causing clogging of pipes and affecting industries, power stations and water treatment stations (Mansur *et al.*, 1999, 2003, 2004a, 2004b). It altered the habitat, food networks and modified structures. Currently, a study to control this species is being conducted in Brazil (Fernandes *et al.*, 2005).

Penchaszadeh (2005) examined several issues related to exotic invertebrates in the Plata River and adjacent regions, including marine areas. He pointed out that invasive species have the following characteristics:

▸ increase in life cycle;
▸ rapid growth, sexual maturation at younger stage (for example, the course of *Limnoperna fortunei*);
▸ rapid sexual maturation, with enormous production of gametes, eggs and larvae;
▸ high fertility; ability to tolerate varied environmental conditions;
▸ association with human activities such as food (*Corbicula flumino*, *Oreochromis niloticus* – tilapia);
▸ wide genetic variability;
▸ ability to repopulate previously colonized environments.

The literature on the invasions of molluscs has increased significantly in recent years (Mansur *et al.*, 1999; Callil and Mansur, 2002, Mansur *et al.*, 2003, 2004a, 2004b, 2004c).

There are many examples of introduction of exotic fish species in inland systems of South America. These species were introduced especially with the construction of reservoirs in the Southeast and, initially, in reservoirs in the Northeast, causing substantial alterations in the habitats, feeding networks, and diversity of native species.

The introduction of *Cichla ocellaris* (tucunaré) in many reservoirs in the Southeast substantially altered the local fish fauna. Likewise, there is much evidence and studies on the introduction of exotic species in reservoirs and weirs in Brazil, with direct impact on the functioning of trophic networks and the ecology of reservoirs (Agostinho *et al.*, 1997, 1999) (see Figure 18.15).

Chart 18.8 lists the different fish species introduced into drainage basins in Brazil.

18.4   TOXIC SUBSTANCES

The concentration of toxic substances in terrestrial and aquatic ecosystems has increased substantially in recent decades. These toxic substances come from industrial and agricultural activities and the production of toxins by cyanobacteria. The entire range of toxic elements and substances – dissolved in water, accumulated in the sediment and in the food chain through the process of bioaccumulation – have chronic and acute toxic effects on aquatic organisms and ultimately, on humans.

The range of toxic elements and substances accumulated in natural waters is very large, given the variety and diversity of industrial and agricultural activities. These toxic substances are classified as:

▸ **Organic contaminants:** Thousands of organic compounds that affect the water systems have many physical, chemical and toxicological impacts. Mineral oils, petroleum products, phenols, pesticides, and polychlorinated biphenyl compounds are examples of these organic compounds, the identification of which

Figure 18.15 The golden mussel is an invasive species that since 2002 has caused major problems in the drainage basins of southern and southeastern Brazil (see also color plate section, plate 26).
Source: Furnas Centrais Electricas.

require highly specialized equipment and trained teams. Substances need to be identified, including aromatic and polychromatic hydrocarbons, different groups of pesticides, phenols, nitrosamines, derivatives of benzidine and esters, whether in particulate form or in water, after filtration in a 0.2-micrometre GFE filter. The toxins produced by cyanobacteria, as described earlier in this chapter, also play a relevant role as toxic organic substances.

▸ **Metals:** Some metals are important in maintaining the physiological processes of living tissues and organisms. These metals regulate the biochemical processes. For example, at very low concentrations, manganese, zinc and copper are essential in the physiological processes of regulation. However, at high levels, these metals can be toxic to organisms and humans. Pollution from metals currently affects many aquatic ecosystems around the planet, causing serious environmental and public health problems, especially after their bioaccumulation in the food web.

Metals can accumulate in one aquatic compartment or another, and their concentration in the sediment can be highly deleterious to water quality.

The **toxicity of metals** in water depends on the degree of oxidation of a specific metal ion and the form in which it occurs. In general, the ionic form is the most toxic; the toxicity is reduced in the case of complexation, for example, with fulvic or humic acids.

Certain conditions that favour the formation of low-molecular-weight metal-organic compounds make the compound highly toxic (for example, **methyl-mercury**).

In water, metals can occur in dissolved, colloidal or particulate form. In general, the following metals are monitored, given their ecological and toxicological importance: aluminium, cadmium, chromium, copper, iron, mercury, manganese, nickel,

Chart 18.8 Species of fish introduced to water-basins of Brazil.

Drainage basin	Alien species		Exotic species	
	Scientific name	Popular name	Scientific name	Popular name
Amazon	Prochilodus argenteus	Curimatã pacu	Oreochromis niloticus	Nile Tilapia
Araguaia/ Tocatins	Piaractus mesopotamicus Leporinus macrocephalus	Pacu Piau-açu		
Northeast	Astronatus ocellatus Plagioscion surinamensis Plagioscion squamosissimus Cichla ocellaris Cichla temensis Colossoma macropomum Piaractus mesopotamicus Arapaima gigas Colossoma brachypomum Triportheus signatus Hypophtalmus edentatus	Oscar Pacora South American silver croaker Butterfly peacock bass Pinima Tucunaré Tambaqui Pacu Pirarucu Piratitinga Sardine Mapará	Cypinus carpio Hypophthalmicthys molitrix Aristichthys nobilis Oreochromis niloticus Tilapia rendalli Clarias gariepinus	Common carp Silver carp Bighead carp Nile Tilapia Redbreast Tilapia African sharptooth catfish
São Francisco	Cichla ocellaris Astronatus ocellatus Colossoma macropomum Piaractus mesopotamicus Plagioscion squamosissimus Colossoma brachypomum Hybrid	Butterfly peacock bass Apaiari Tambaqui Pacu South America silver croaker Piratitinga Tambaqui/Pacu	Cypinus carpio Hypophthalmicthys molitrix Aristichthys nobilis Oreochromis niloticus Tilapia rendalli Hybrid Ctenopharyngodon idella Claris lazera	Common carp Silver carp Bighead Carp Tilapia of the Nile Redbreast Tilapia Red Tilapia (St. Peter) Grass Carp African catfish
Leste	Piaractus mesopotamicus Colossoma macropomum Hoplias lacerdae Prochilodus margravii Brycon lundi Lophiosiluros alexandri	Pacu Tambaqui Trairão Curimba Matrinxa Pacamã	Oreochromis niloticus Tilapia rendalli Cypinus carpio Aristichthys nobilis Ctenopharyngodon idella Clarias gariepinus	Nile Tilapia Redbreast Tilapia Common carp Bighead carp Grass carp African Catfish

Basin	Common name	Scientific name
	Surubim	*Pseudoplatistoma sp*
	Tucunaré comum	*Cichla ocellaris*
	Dorado	*Salminus maxillosus*
	Piranha	*Pygocentrus sp*
	Piau-açu	*Leporinus macrocephalus*
	Piapara	*Leporinus elongatus*
	Black Bass	*Micropterus salmoides*
	Common Carp	*Cyprinus carpio*
	Silver carp	*Aristichthys nobilis*
	Bighead carp	*Hypophthlmictys molitrix*
	Nile Tilapia	*Oreochromis niloticus*
	Redbreast Tilapia	*Tilapia rendalli*
	Zanzibar Tilapia	*Oreochromis hornorum*
	Mozambique Tilapia	*Oreochromis mossambicus*
	Aurea Tilapia	*Oreochromis aureus*
	Black Bass	*Micropterus salmoides*
	King fish	*Odontesthis bonariensis*
	Canal catfish	*Ictalurus punctatus*
	Trout	*Onchorhynchus mikss*
	African catfish	*Clarias gariepinus*
	Red Tilapia (St. Peter)	*Hybrid*
Upper Paraná	Tambaqui	*Colossoma macropomum*
	Common Tucunare	*Cichla ocellaris*
	Mapara	*Hypophthalmus edentates*
	Sardine	*Triportheus signatus*
	Piau-acu	*Leporinus macrocephalus*
	Piapara	*Leporinus elongates*
	Matrinxã	*Brycon cephalus*
	Piaui Fish	*Plagioscion squamosissimus*
	Apaiari	*Astronatus ocellatus*
	Trairão	*Hoplias lacerdae*
	Piratitinga	*Colossoma brachypomum*
	Piau/Piracajuba	*Hybrid*
	Tambaqui/Pacu	*Hybrid (Tambacu)*
	Pacu/Tambaqui	*Hybrid (Paqui)*
	Tambaqui/Pirapitinga	*Hybrid (Tambatinga)*
Paraguay	Tambaqui	*Colossoma macropomum*
	Pirapitinga	*Piaractus brachypomum*
	Common Tucunare	*Cichla ocellaris*
	Matrinxa	*Brycon cephalus*
	Common carp	*Cyprinus carpio*

Source: Rocha *et al.* (2005).

Table 18.11   Mercury concentrations ($\mu g \cdot g^{-1}$ wet weight) in fish in regions of gold exploration of Carajas (1988 to 1990).

Species of fish	Hg ($\mu g \cdot g^{-1}$ wet weight) 1988	Hg ($\mu g \cdot g^{-1}$ wet weight) 1990
Paulicea luetkeni (C)	0.80–1.46	1.25–2.30
Prochilodus nigricans (H)	0.13–0.31	0.01–0.02
Pimelodus sp. (C)	0.09–0.24	0.17–0.19
Brycon sp. (H)	0.05–0.16	0.04
Serrasalmus nattereri (C)	0.10	0.01–0.87
Leporinus sp. (H)	0.01	0.01–0.03
Hoplias malabaricus (C)	0.35–0.91	0.31

C – Carnivorous fish
H – Herbivorous fish
Source: Lacerda and Salomons (1998).

lead, and zinc. Also included are arsenic and selenium (which are not strictly metals), in addition to other toxic metals such as beryllium, vanadium, antimony and molybdenum. Measurements of the levels of iron and manganese are also considered in the analysis, since high levels of iron oxides (iron hydroxide) can occur in groundwater, and anoxic waters can have concentrations of ferrous ion (Fe^{2+}) of around $50\,mg \cdot L^{-1}$. The concentration of different metals in water varies from $0.1–0.0001\,mg \cdot L^{-1}$ and can reach very high levels from human activities.

One epidemiologically and toxicologically important element is arsenic. Arsenic contamination is a global problem, causing countless public health problems, interfering in diseases such as diabetes and generating disorders in the immune, nervous and reproductive systems. Arsenic can cause cancer, and contamination of drinking water by this element is well known (Drurphy and Guo, 2003). The exploitation of aquifers can cause the release of arsenic and heavy metals.

Another toxicologically significant metal is mercury. More than 1,400 people have died worldwide and nearly 20,000 have been affected by mercury poisoning. Mercury can quickly be transformed into methyl-mercury and become stable by the action of various microorganisms. This compound has a long retention time in aquatic biota and can cause serious harm to humans.

The classic case of mercury contamination occurred in Japan between 1956 and 1960, when more than 150 people died and nearly 1,000 were incapacitated by mercury poisoning. By December 1987, more than 17,000 people had been poisoned by methyl-mercury and 999 individuals had died (Lacerda and Salomons, 1998).

Lacerda and Salomon's book (1998) deals with the effects of mercury and the range of ecological, chemical and environmental problems arising from the distribution of the metal and its accumulation in compartments in the aquatic biota. The book also examines the contamination of people with mercury as a result of the use of this metal as an amalgam in areas where gold and silver are mined.

Tables 18.11 and 18.12 indicate the levels of mercury in fish in the gold mining region of Carajás and the levels of mercury in fish in the mining regions of Amazon, compared with other contaminated areas.

Fish in large tropical wetland areas, black-water rivers, and artificial reservoirs present higher mercury content than other aquatic systems. The **concentration of mercury in fish** does not correlate with the concentration of mercury in water,

Table 18.12   Extreme concentrations of mercury ($\mu g \cdot g^{-1}$ net weight) in various species of carnivorous fish in mineral regions of the Amazons, compared with different contaminated areas.

Species	Site	Hg ($\mu g \cdot g^{-1}$ net weight)	Researcher (S)
Paulicea luetkeni (Steindachner)	Carajás	2.19	Fernandes *et al.* (1989)
Pseudoplatystoma fasciatus L.	**Madeira River**	2.70	Pfeiffer *et al.* (1989)
Brachyplatystoma filamentosum (Lichtenstein)	**Teles Pires River**	3.82	Akagi *et al.* (1994)
Esox lucius L.	Canadian lakes	2.87	Olgivie (1991)
Mullus barbatus L.	**Tyrrhenian Sea** (Italy)	2.20	Bacci *et al.* (1990)
Esox lucius L.	Finnish lakes	1.80	Mannio *et al.* (1986)

Source: Lacerda and Salomons (1998).

because of the effect of methylation of mercury. Indeed, it is the productive capacity of methyl-mercury, determined by the biogeochemical characteristics of black-water rivers, tropical reservoirs and wetland areas, that controls the mercury levels in fish. These tropical aquatic systems are extremely rich in oxidizable organic matter, low pH (<6.0) and low conductivity ($<50 \mu S \cdot cm^{-1}$), promoting conditions for the formation of methyl-mercury. Dissolved organic material produces stable conditions and solubility of mercury in water through the process of complexation (Lacerda and Salomons, 1998).

18.5   WATER AND HUMAN HEALTH

Although water is vital for human health, it can also weaken people, produce disease through several mechanisms and increase mortality. These are consequences produced by contaminated and poor-quality water. The Water Quality and Health Council (1977) published a comprehensive list of diseases associated with water and its adverse effects on human health. The list contains 100 pathogenic organisms associated with water and nearly 100 adverse effects.

Close to 2 billion people lacked basic sanitation at the end of the 20th century and in the early 21st century. Approximately 93% of people in industrialized countries with high standards of living have access to potable water. In emerging and developing countries, the rate is only 43%.

Figure 18.16 shows the relationship between the human being, the parasites, water, and the vectors diseases of water-bourne.

Diseases associated with water can be classified into four categories:

▶ **water-borne diseases** (organisms that develop in water): cholera, typhoid and dysentery;
▶ **diseases caused by contaminated water** from organisms that do not develop in water: trachoma, leishmaniasis;
▶ **diseases related to organisms whose vectors develop in water**: malaria, filariasis, yellow fever and dengue;
▶ **diseases dispersed by water**: bacteria of various types, viruses.

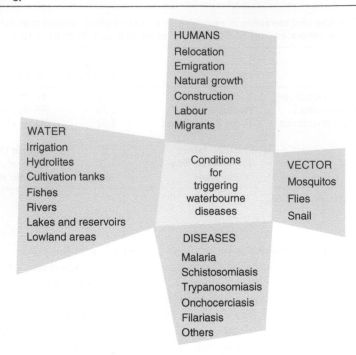

Figure 18.16   Aquatic habitats and waterborne diseases.
Source: INWEH (1992).

Chart 18.9 describes the main human health problems caused by contaminated water.

The effects of water-borne diseases can be exacerbated by climate change and in the long-term by global changes. Transmission can depend on the life cycle of vector – mosquitoes – and the microorganisms they carry. Higher temperatures and warmer climates increase the capacity of mosquitoes to infect, and increased rainfall leads to increased pools of water and ecological conditions for the development of *Anopheles*, which transmits the disease.

Dry periods can increase the number of marginal lakes in rivers and streams, resulting in favorable conditions for mosquito breeding. The *Aedes aegypti* mosquito, which carries and transmits the dengue virus, develops rapidly in hot and humid conditions. In Latin America, specifically, the increase in occurrence of malaria and dengue fever were related to events such as El Niño, due to increases in air and water temperatures. Heavy rains occurring during extreme climatic events can transport *Vibrio cholerae* and contaminate reservoirs of clean water, increasing the likelihood of the spreading and impact of the diseases.

In many regions around the world, the correlations between the incidence of malaria and the El Niño event are very significant. The increased potential for transmission of the disease occurs during greater precipitation in certain areas or lack of precipitation in others. Malaria epidemics correlated with El Niño were documented in Bolivia, Colombia, Peru, Ecuador, Venezuela, Pakistan and Sri Lanka. In South America and much of Africa, the increase of disease correlated with higher precipitation

Chart 18.9   Principal problems of human health caused by polluted and contaminated water.

Disease	Infectious agent	Type of organism causing the disease	Symptoms
Cholera	*Vibrio cholerae*	Bacteria	Severe diarrhoea; vomiting; loss of liquid.
Dysentery	*Shigella dysenteriae*	Bacteria	Inflammation of the colon that causes diarrhoea, loss of blood and intense abdominal pain.
Enteritis	*Clostridium perfringens*	Bacteria	Inflammation of the intestine; loss of appetite; diarrhoea and abdominal pains.
Typhoid fever	*Salmonella typhi*	Bacteria	Initial symptoms are headaches, loss of energy, fever; intestinal haemorrhage and spots on the skin occur in later stages of the illness.
Infectious hepatitis	Hepatitis A virus	Virus	Inflammation of the liver that causes vomiting, fever, nausea, and loss of appetite.
Poliomyelitis	Polio virus	Virus	Initial symptoms include fever, diarrhoea and muscle pains; in more advanced stages, paralysis and muscle atrophy.
Cryptosporidiosis	*Cryptosporodium* sp.	Protozoa	Infection of the colon that causes diarrhoea, loss of blood and abdominal pain.
Schistosomiasis	*Schistosoma* sp.	Worm	Tropical disease that attacks the liver, causing diarrhoea, weakness, abdominal pains.
Hookworm	*Ancylostoma* sp.	Worm	Anaemia; symptoms of bronchitis.
Malaria	*Anopheles* sp. (transmitter)	Protozoa	High fever; prostration.
Yellow fever	*Aedes* sp. (transmitter)	Virus	Anaemia.
Dengue	*Aedes* sp. (transmitter)	Virus	Anaemia.

Source: Tundisi (2003).

rates. The same is true with respect to cholera in countries such as Somalia, Congo, Kenya, Bolivia, Honduras and Nicaragua (Epstein, 1999).

18.6   GLOBAL CHANGES AND THE IMPACT ON WATER RESOURCES

Global changes currently underway that have already been identified through numerous joint studies and analyses by international organizations such as the IPCC (Intergovernmental Panel on Climate Change, 1996, 2007) clearly show the existence of significant alteration in the atmosphere increasing surface temperatures and producing other effects, such as the build-up of CO_2, CH_4 and N_2O in the atmosphere. The causes of global warming of the planet are related to the exacerbated effects of emissions. The average temperature on the Earth's surface has already risen 0.3–0.6°C over the last 100 years. Projections for the next 100 years show that the average temperature of the planet, with the scenario of higher emissions and more warming, present an average increase of 3.5°C.

There is consensus among researchers that global warming will significantly impact the water resources of the earth. The principal effects include increased drainage, changes in precipitation, increased levels in rivers, alterations in the patterns of land use, and population displacement due to alterations of local and regional climate. Higher temperatures lead to acceleration of the hydrological cycle, **frequency of floods** and droughts, and increased evapotranspiration rates, altering the filtration in the soil, soil moisture, and the distribution and cycles of aquatic organisms. Regional patterns of precipitation are changing, significantly altering the volume of lakes, rivers and reservoirs and substantially increasing the frequency of flooding in many areas around the world.

The hydrology of arid and semi-arid regions is particularly sensitive to climate change, and humid areas in the future may experience permanent desiccation, with the definitive disappearance of lakes. With the volume of water declining and acceleration of cycles, eutrophication may increase, resulting in deterioration of the water quality in springs.

Table 18.14 shows the indicators of providing water services and their historical projections from 1960 to 2010 as an example of future pressures on water use and the consequences, including global changes.

Lakes are particularly sensitive to global changes, because of their responses to climate conditions. Variation in air temperature, solar radiation and precipitation cause alterations in evaporation and heat balance in the hydrochemical regimes and biogeochemical cycles of endorheic lakes, such as the Caspian and Aral Seas (already greatly altered, as described earlier in this chapter). Lakes Titicaca, Malawi and Tanganyika may undergo many changes because of hydrological imbalances resulting from the reflows of water and increasing salinity. Changes in the water quality in these lakes are possible consequences of these climatic changes, putting available water resources at risk and promoting changes in the diversity of aquatic species due to increased salinity/conductivity.

It is also possible that the ionic balance is disrupted (for example, solutions with different balances of Na^+, F^+, Ca^{2+}, Mg^{2+}, CO_3^{2-} and SO_4^{2-}), affecting the aquatic fauna and flora. Since the chemical composition of water largely depends on the **chemical loads** resulting from the drainage basin, climatic modifications can alter chemical processes in the soil, including chemical weathering. For example, in Spain, a substantial increase of cations in the water is predicted, due to increased temperature and precipitation (Avila *et al.*, 1996). Drier and warmer conditions can cause more rapid mineralization of organic nitrogen and, thus, increased nitrogen load in rivers and lakes.

The dissolved oxygen level, which is lower at higher temperatures, is likely to undergo drastic changes, affecting aquatic life. Alterations in the flow of rivers can lead to increased eutrophication and transport of nutrients – these alterations occur with greater flow. Also, increased precipitation and drainage in other areas, with the resulting increase in flow in rivers, can increase the nutrient load into lakes, reservoirs and coastal waters.

Table 18.14   Indicators of water supply services and historical trends and projections from 1960 to 2010.

Geographic region, according to the aem	Population (millions)	Water use U_a (km³/year)	Population with access to renewable supply of water* (miles/weight/m³/year)	Use relative to renewable supply (U_a/B_a)
Asia	1960: 1,490	1960: 860	1960: 161	1960: 9
	2000: 3,230	2000: 1,553	2000: 348	2000: 17
	2010: 3,630	2010: 1,717	2010: 391	2010: 19
ex-Soviet Union	1960: 209	1960: 131	1960: 116	1960: 7
	2000: 288	2000: 337	2000: 160	2000: 19
	2010: 290	2010: 359	2010: 161	2010: 20
Latin America	1960: 215	1960: 100	1960: 25	1960: 1
	2000: 510	2000: 269	2000: 59	2000: 3
	2010: 584	2010: 312	2010: 67	2010: 4
North Africa/ Middle East	1960: 135	1960: 154	1960: 561	1960: 63
	2000: 395	2000: 284	2000: 1,650	2000: 117
	2010: 486	2010: 323	2010: 2,020	2010: 133
Sub-Saharan Africa	1960: 225	1960: 27	1960: 55	1960: <1
	2000: 650	2000: 97	2000: 163	2000: 2
	2010: 871	2010: 117	2010: 213	2010: 3
OECD region	1960: 735	1960: 552	1960: 131	1960: 10
	2000: 968	2000: 1,021	2000: 173	2000: 18
	2010: 994	2010: 1,107	2010: 178	2010: 20
World total	1960: 3,010	1960: 1,824	1960: 101	1960: 6
	2000: 6,060	2000: 3,561	2000: 204	2000: 12
	2010: 6,860	2010: 3,935	2010: 231	2010: 13

*Renewable supply calculated according to the flow of water directly from the atmosphere or evaporated from the ocean ('blue water').

Climate changes can significantly alter the pattern of thermal stratification in lakes located at latitudes between 30–45° and 65–80°. Substantial changes in ice cover of lakes were simulated based on the temperature increase produced by global changes, and in some lakes, with the decrease in the flows of rivers and tributaries and higher temperatures, the thermocline deepened.

Reduced levels of dissolved oxygen in lakes, rivers and reservoirs due to the effect of global changes and the increase in blooms due to eutrophication have been predicted as one of the most significant consequences of global changes. These changes pose a risk to aquatic life and water quality. The growing demand for water resulting from the effects of climate changes is another important result of the process. Increased demand for irrigation and domestic use is probably due to higher temperatures and the scarcity of water in some regions. So there will be changes in the availability/demand for water (Schindler *et al.*, 1996).

Figure 18.17 shows the variations in average global temperatures on the surface of the Earth, in function of the changes that cause global warming. Figure 18.18 shows the main interactions of global processes, toxication of the biosphere, soil use, and human health. Figure 18.19 presents one of the results of overuse of water.

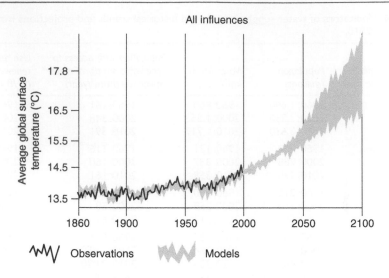

Figure 18.17   Changes in average global temperature on the Earth's surface.
Source: National Geographic Brazil (Sept./2004).

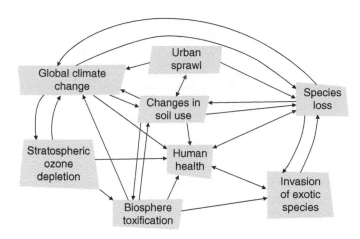

Figure 18.18   Principal interactions of global processes, toxification of biosphere, soil use and human health.
Source: Modification of Likens (2001).

Chart 18.10 summarizes some of the most critical problems resulting from the impacts on water resources and implications for management, administration and control of the impacts and for the restoration of lakes, rivers, reservoirs and wetland areas.

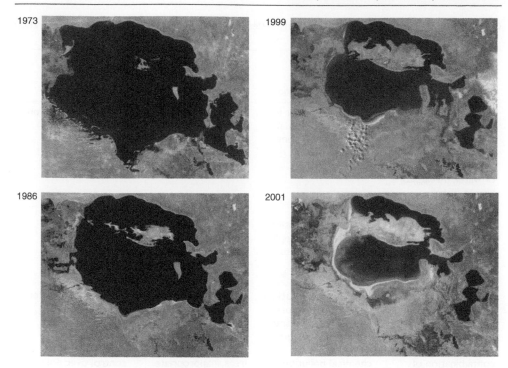

Figure 18.19   Changes in the Aral Sea as result of the overuse of tributary waters for irrigation (see also color plate section, plate 27).
Source: Millennium Ecosystem Assessment (2005).

Chart 18.10   Interactions of problems in water resources, management and administration.

Problem in water resources	Direct and indirect physical manifestations	Implications for management	Implications for organization and administration
Erosion and sedimentation: economic losses in fishing, hydroelectricity and reserve capacity	Increased sedimentation in rivers and reservoirs as a result of poor terrestrial system management	Implies the absence of sound planning and management programs; protection, restoration and technical assistance programmes	Multiple control agencies and lack of coordination in water basins
Floods: economic losses to agriculture, because of contamination by wastewater and deterioration of infrastructure	Increased flood peaks due to development of floodplains, and increased sedimentation rates; mixing of waste waters and flood waters	Lack of sound management of basins, lack of control in terrestrial systems; poor agricultural practices; lack of flood warning systems	Lack of institutional coordination and view of flooding as a broader problem calling for integrated institutional management

(continued)

Chart 18.10   Interactions of problems in water resources, management and administration (continued).

Problem in water resources	Direct and indirect physical manifestations	Implications for management	Implications for organization and administration
Irrigation: economic losses for agriculture, forestry management and domestic and industrial availability of water, threats to human health	Overuse of water for irrigation; inadequate drainage conditions; reduced flow of surface water	Deficiency or absence of management in irrigation or overuse of groundwater	Lack of coordination of institutions, especially in management of irrigation
Imbalance between supply and demand, limiting economic development	Variability of precipitation, causing uncertainty in supply and limits agricultural activities	Difficulty in managing basins; inability to predict peaks in precipitation and drought, lack of reliable database	Responsibilities spread in various agencies
Water pollution: economic losses in agriculture, fishing and industry, threats to public health; contamination of rivers, streams, lakes and reservoirs; increased costs of treating waters	**Biological pollution** caused by improper disposal of solid and liquid waste in rural and urban areas; chemical pollution from pesticides, herbicides and fertilizers; chemical pollution by industries	Absence or lack of basic sanitation programs in rural areas, lack of waste disposal systems in urban areas, inappropriate use of fertilizers and pesticides	Lack of coordination between pollution control agencies; water resource agencies with no control over pollution

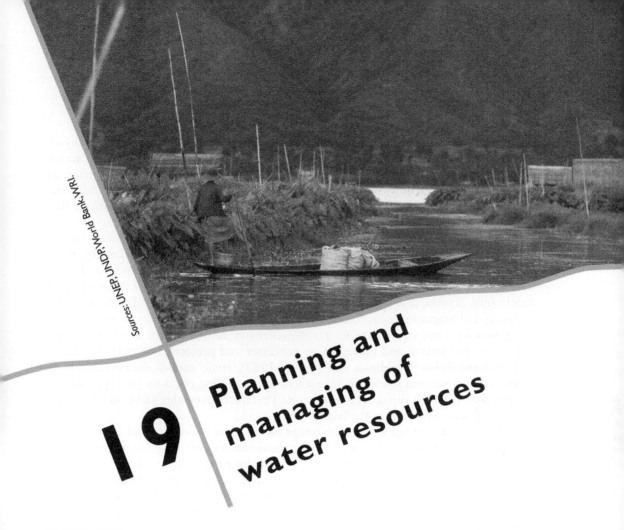

Sources: UNEP, UNDP, World Bank, WRI.

19 Planning and managing of water resources

SUMMARY

Limnological studies are essential in the implementation of measures for planning and managing water resources and tracking conservation actions and restoration of inland aquatic ecosystems.

Today, the 'economy of restoration' (conservation) is spreading quickly, promoting a new dimension in the use of basic information and database management. This 'economy of restoration' promotes revitalization of rivers, lakes and reservoirs, representing a significant conceptual advance.

For more effective management, the predictive power of limnology is used to anticipate impacts or to mitigate them. In addition, management is integrated (multiple uses considered) with a holistic vision (drainage basin as a unit). The management of rivers, lakes, reservoirs, wetland areas, and water resources needs to integrate two approaches that complement each other: for the quality and the quantity of surface water and groundwater.

Several techniques exist for the restoration and management of water basins, rivers, lakes and reservoirs. These techniques differ in costs, depending on the extent of restoration or conservation and the existing impacts. In addition to describing the various techniques, examples are presented of the implementation of actions of restoration and management in several natural and artificial systems (reservoirs). The chapter presents the basics of ecological modelling and its application, especially in restoration projects and forecasting development, useful to minimize impacts, predict effects, and promote integrated and systematic management.

19.1   LIMNOLOGY: PLANNING AND MANAGEMENT OF WATER RESOURCES

As already presented in Chapter 18, natural waters in drainage basins and virtually all inland aquatic systems are subject to a range of impacts resulting from human activities in the drainage basin and the multiple uses of water. Urban and rural drainage, as well as the discharge of waste into lakes, reservoirs and rivers significantly modify the chemical and physical characteristics of waters, producing countless alterations that render the waters unfit for human consumption and other uses.

One of the key contributions of limnology is to provide suitable conditions for the *protection, conservation* and *restoration* of inland aquatic ecosystems. By helping measure the impacts from point and non-point sources, promote implementation of databases and historic series, and develop systems of regional standards of indicators and functioning of lakes, reservoirs and rivers, limnology provides essential data for regional planning and programmes on conservation, protection and restoration of systems.

Limnology can contribute to a comprehensive evaluation of alternatives and impacts in **restoration and management processes and projects**. For example, in the case of Lake Tahoe (Goldman and Horne, 1983), basic studies showed that even extremely low nutrient levels can have a strong stimulating effect on phytoplanktonic growth and provoke rapid eutrophication. Therefore, even the addition of treated water may have an effect on the lake.

The **re-use of sewage water** in drainage basins is an important and economically viable process. In many regions, however, basic limnological studies together with studies on soil permeability, surface runoff and the basic features of natural waters showed that such re-use produces greater damage, even to groundwater, resulting in excessive costs in eutrophic lakes and rivers. Limnological studies also enable the assessment of appropriate costs, especially when dealing with large amounts of water to treat or prevent from degradation.

An important aspect in terms of multiple use is the contribution that basic limnology can provide in *monitoring* the effects of various activities on lakes, reservoirs and drainage basins. Such monitoring and tracking, well planned and executed, can be a sensitive indicator of interactions and alterations in the system. Another important aspect is that generally, when it is necessary to preserve multiple use of the lake, reservoir or river and at the same time restoration and rehabilitation is needed, **scientific limnological research** can contribute effectively through knowledge of many processes, mechanisms and interactions, and promote alternatives with various costs.

Another problem in the management of lakes, rivers and reservoirs is the use of biomass for intensive, semi-extensive or extensive cultivation of aquatic organisms. All integrated limnological studies contribute to the comprehensive understanding of communities and their interactions, the main species present and their biology, and the succession and biomass of communities, which is essential for later application to aquaculture. In addition, the species' *limits of tolerance* and the ranges of variables (such as water temperature and dissolved oxygen) are determined by limnological studies, facilitating potential application, coupled with calculations of tolerance and life cycles in the laboratory (Tavares and Rocha, 2001).

Limnology can also help to identify *biological indicators* that indicate the degree of contamination/pollution. For example, data from the Biota/FAPESP program

(Matsumura Tundisi *et al.*, 2003) showed that the calanoid copepod *Argynodiaptomus furcatus* is an indicator of low-grade eutrophication and contamination in an ecosystem, and thus its presence or absence can enable the categorization of a lake or reservoir and provide conditions for further studies to measure the principal processes.

The results obtained in already-restored lakes show that intense limnological research programmes existed or were implemented, which allowed rapid successful actions.

Therefore, establishing limnological research programmes appropriate to regional conditions provides an important source for *planning* and **regional management**. Building a scientific limnological database based is an important step in establishing ecologically based regional planning programs, enabling existing problems to be accurately identified and appropriate alternative solutions and measures to be planned, using the aquatic system as a catalyst and 'collector of events' along the drainage basins.

19.2 LIMNOLOGY AND HEALTH ASPECTS

Basic knowledge of limnology can be essential in addressing health problems. Many diseases and parasites that affect human beings are water-borne, as already shown in Chapter 18. Basic limnological studies can contribute to the following aspects:

▶ knowledge of the *life cycles* and **biology of vectors** and **parasites** and types of aquatic system in which they thrive. Such studies can indicate the vulnerability of pests or vectors to certain types of treatment or activity;
▶ establishment of **standards of tolerance** for each type of organism and distribution-limiting factors, such as temperature, dissolved oxygen (too high or too low) and other physical and chemical characteristics of the water, such as pH and conductivity;
▶ knowledge about the interactions of the various components of a community – processes such as **predator-prey interrelationships** and *parasitism* – which can be extremely useful in determining mechanisms for the biological control of certain parasites;
▶ prevention, in the search for alternative uses of natural waters, in order to avoid pollution, for example;
▶ the search for economically viable solutions to solve *health problems* and the use of more feasible and less expensive treatment systems.

19.3 LIMNOLOGY AND REGIONAL PLANNING

Limnological studies provide a comprehensive vision of the diverse and conflicting problems that occur in drainage basins. Water, as an indispensable resource for humans, has been utilized for multiple uses and it is therefore natural that the integrated use of drainage basins take into account water resources. The **uses of aquatic resources** and their development and management can provide a powerful and empirical basis for regional planning.

All regional planning needs to take into account the limnology of the different components in a drainage basin: rivers, lakes, reservoirs, and wetland areas. Alterations in the terrestrial system and the effects on rivers, reservoirs, and lakes; multiple uses of natural and artificial ecosystems; multiple uses of the water; utilization of the biomass of aquatic organisms and the disposal of industrial waste – all require basic information on the structure of inland aquatic ecosystems and their main hydrodynamic, biological, physical and chemical characteristics. In particular, the interaction of the drainage basin with the lake, reservoir or river plays an important role. Combining the ecological studies of the terrestrial system in the drainage basin with the limnological studies of lakes, reservoirs and rivers is another fundamental component.

Planning should also take into account the modifications in the drainage basin and their impacts on aquatic systems; thus, from basic limnological studies, diverse alternatives can be proposed.

Figure 19.1 presents practical applications of the scientific studies of lakes, reservoirs and rivers. Basic limnology can be used in preventive actions or for prognosis and the correction/restoration of lakes, reservoirs and rivers, or in regional planning. Therefore, one important function of basic limnology is to provide information for the sound management of aquatic ecosystems.

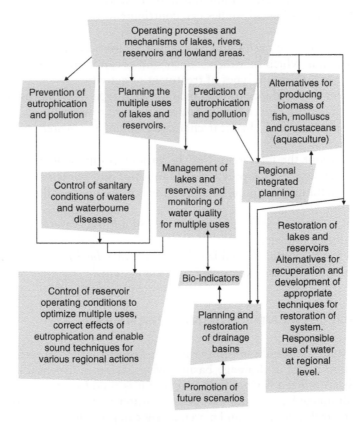

Figure 19.1   Practical applications of basic limnology.

19.4   CONCEPTUAL ADVANCES IN WATER RESOURCE MANAGEMENT

In the last decades of the 20th century, there was a conceptual advance with respect to water resource management: a shift occurred from **local, sectorial, response management** to *integrated predictive management at the ecosystem level*. This conceptual shift, still underway, is leading to significant changes in the management process, with limnology as a central science in the management of rivers, lakes and reservoirs.

Chart 19.1, extracted and adapted from Somlyody *et al.* (2001) and Tundisi (2003), shows the evolution of management systems and the existing transition phase.

Chart 19.1   Evolution of management systems and existing transition phases.

Past	Present Future (desired)
(1) General	
Local problems	Larger scale
Rapid response, reversibility	Delayed responses
Limited number of pollutants	Multiple pollutants
Limited to one medium (water)	Multiple media (solar/air/water)
Static, deterministic, predicted	Dynamic, stochastic, uncertain
Regional independence	Global independence
Point sources	Non-point sources
(2) Type of control	
Completion of process	Control of the source, closed cycle of material, control of drainage basin
Discharge patterns	Use and adaption
Purely technical	Non-technical elements
(3) Infrastructure and treatment systems	
Traditional technology	Special treatment methods, eco-technologies, small-scale natural treatment
Landfills	Reuse and recycling
Large-scale control and exploitation	Small-scale development, integrated management, conservation
Massive urban infrastructure	Localized infrastructure, development of creative systems
(4) Monitoring	
Local determinants	Networks, remote sensing, continuous measurements
Conventional parameters	Special parameters (micropollutants, eco-toxicology)
Monitoring of water	Integration of monitoring sources and effluents
Unreliable data	Improvements in **reliability**, databases, information systems
Unavailable data	Open flow of information

Chart 19.1   Evolution of management systems and existing transition phases (continued).

(5) Modelling

Topics limited to generations and processes	GIS integration, decision-making systems
Limited number of results	Scenarios, case studies, use of multi-media
Used only by specialists	Use in administration and management

(6) Project planning and evaluation

Very diffuse definitions of objectives	Well-defined objectives
Short-term vision	Long-term vision
Assessment of costs	Assessment of global environmental impact, political and social impacts
Little concern about failures or adjustments needed	Uncertainties: adaptability, resilience, vulnerability, robustness
Separate positive and negative impacts	Positive and negative impacts

(7) Science and engineering

Actions not scientifically based	Science for actions and combination of science and engineering
Isolation of problem and engineering solutions	More efficient planning
Obstacles and inter-disciplinary problems	Integration of quality, quantity, hydrology, economy, policy, science, social and management
Barely one correct paradigm, one discipline	Many accepted paradigms within the concept of disciplines

(8) Legislation, institutions for management and development

Rules: general and rigid	Rules: specific and flexible
Rapid implementation	Critical testing and analysis of processes
Little legal enforcement	Increased legal enforcement
Confused institutional organization	Clear structures and responsibilities, fewer barriers, more communication*
Decisions by politicians and administrators	**International policies**
National policies	Sustainable development (how to proceed)

*Participation of the public and NGOs, as well as specialists, managers and administrators.
Source: Somlyody *et al.* (2001).

Another important conceptual and comprehensive advance currently underway in many regions around the world (including several Brazilian states) is the adoption of the drainage basin as the basic unit for water resource planning and management. Essential characteristics (functional, operational, and geosystemic) of the unit include:

▸   The drainage basin is a physical unit with defined borders that can extend over various spatial scales, from small basins of 10, 20 or 100 to 200 km² to enormous drainage basins such as the La Plata basin (3 million km²) (Tundisi and Matsumura Tundisi, 1995).

▶ A drainage basin is a *hydrologically integrated* ecosystem, with interactive components and subsystems.

▶ It provides opportunities for development of partnerships and resolution of conflicts (Tundisi and Straškraba, 1995).

▶ Local people can participate in the **decision-making process** (Nakamura and Nakajima, 2000).

▶ Community participation is encouraged as well as **environmental and health education** (Tundisi *et al.*, 1997).

▶ A sound systematic vision for **training in management** and controlling eutrophication (managers, decision-makers and technicians) (Tundisi, 1994a).

▶ A rational form to organize databases.

▶ Alternative uses for water *sources* and resources.

▶ A sound approach for the development of a database including **biogeophysical, economic and social components**.

▶ As a physical unit with well-defined limits, the spring provides a basis for **institutional integration** (Hufschmidt and McCauley, 1986).

▶ The spring approach promotes the integration of scientists, managers and decision-makers with the general public, allowing all to work together on a physical unit with defined limits.

▶ Promotion of the **institutional integration** needed for the management of sustainable development (UNESCO, 2003).

Source: Tundisi *et al.* (1998); Tundisi and Schiel (2002).

Therefore, the drainage basin concept applied to the management of water resources extends traditional political barriers (city, state, country) to a physical unit for economic and social management, planning and development (Schiavetti and Camargo, 2002). The lack of a *systemic vision* for **water resource management** and the inability to incorporate/adapt the project to economic and social processes (see Figure 19.2) delay the planning and interfere with competent and healthy **public policies** (Biswas, 1976, 1983). The ability to develop a set of indicators is an important aspect of applying this unit in planning. The drainage basin is also a decentralized process of environmental conservation and protection, and a stimulus for community integration and institutional integration. The indicators of the conditions that provide a **quality index of the drainage basin** can represent an important step in the consolidation of decentralization and management. These include:

▶ water quality in rivers and streams;
▶ species of fish and wildlife (terrestrial fauna) present;
▶ rate of preservation or loss of wetland areas;
▶ rate of preservation or loss of native forests;
▶ rate of contamination of **sediments in rivers, lakes** and **reservoirs;**
▶ rate of preservation or contamination of the sources of water supply;
▶ urbanization rate and extent (% of area of drainage basin);
▶ urban/rural population ratio (Revenga *et al.* 1998; Tundisi *et al.*, 2002).

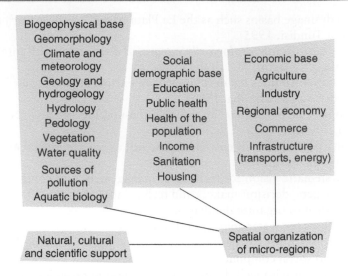

Figure 19.2   Main features of analytical projects needed in regional planning.
Source: Modified from the UNDP (1999).

Along with *quality indicators*, there should also be consideration of *indicators of the vulnerability* of the drainage basin:

▶ toxic pollutants (Pimentel and Edwards, 1982);
▶ load of pollutants;
▶ urban run-off;
▶ agricultural run-off;
▶ alterations in the population: rate of growth and/or migration/immigration;
▶ general effects of human activities (Tundisi, 1978);
▶ potential for eutrophication (Tundisi, 1986a).

For proper management of the drainage basin, it is essential to coordinate participation of the public and private sectors, consumers and the universities. Tundisi and Straškraba (1995) emphasized the following participatory aspects of these various components of the system:

University:
▶ qualitative and quantitative diagnosis of the problems;
▶ development of databases and information systems;
▶ support for implementation of *public policies*;
▶ support for methodological development and introduction of new technologies.

Public sector:
▶ implementation of public policies in the committees of the basin;
▶ implementation of projects for conservation, protection and restoration;
▶ public information and **health** and **environmental education**.

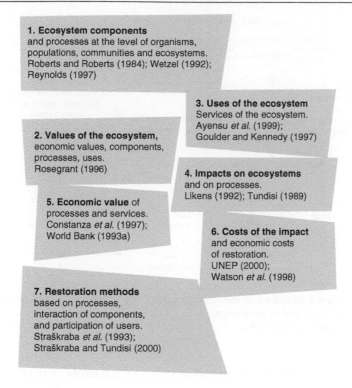

1. **Ecosystem components**
and processes at the level of organisms,
populations, communities and ecosystems.
Roberts and Roberts (1984); Wetzel (1992);
Reynolds (1997)

3. **Uses of the ecosystem**
Services of the ecosystem.
Ayensu *et al.* (1999);
Goulder and Kennedy (1997)

2. **Values of the ecosystem,**
economic values, components,
processes, uses.
Rosegrant (1996)

4. **Impacts on ecosystems**
and on processes.
Likens (1992); Tundisi (1989)

5. **Economic value** of
processes and services.
Constanza *et al.* (1997);
World Bank (1993a)

6. **Costs of the impact**
and economic costs
of restoration.
UNEP (2000);
Watson *et al.* (1998)

7. **Restoration methods**
based on processes,
interaction of components,
and participation of users.
Straškraba *et al.* (1993);
Straškraba and Tundisi (2000)

Figure 19.3   Sequence of the procedures and stages in restoration of ecosystems.
Source: Tundisi *et al.* (2003).

Private sector:
▶ support for implementation of public policies;
▶ technological development and implementation of new projects;
▶ joint financing of technologies.

Consumers and the general public:
▶ participation in *mobilization*, for *conservation* and *restoration*;
▶ information to the Public Ministry and the public sector.
▶ participation in the health education process.

A set of proposals and ideas for the management and restoration of drainage basins should be prepared, following the procedures and steps shown in Figure 19.3.

The **economy of restoration** is today a very important activity in some countries and includes a series of basic studies and functioning mechanisms to restore drainage basins, reservoirs, rivers and lakes. The restoration of inland aquatic ecosystems can create jobs, employment and income, and revitalize the declining regional economy as a result of the degradation of ecosystems, water quality and the 'services' provided by these ecosystems. The issue of 'services' should be considered a fundamental point of the restoration or conservation project.

Figure 19.4 shows the main interactions of the components of terrestrial and aquatic systems, highlighting the issue of 'services' provided by these ecosystems.

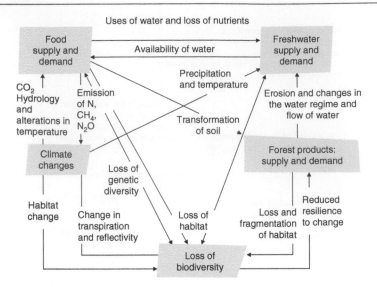

Figure 19.4　Principal interactions of the components of the water system and production system; biodiversity and global change.
Source: Ayensu *et al.* (1999).

The two approaches commonly used in water resources management are presented in Figure 19.5.

The more recent approaches that involve the existing knowledge base include the following fundamental aspects:

▶ recognition of uncertainties;
▶ recognition that the policy on planning and management will not provide 'exact' solutions but instead 'adaptive' and in stages, incorporating new ideas and methods throughout the process (Cooke and Kennedy, 1989);
▶ development of predictive capacity for interactions of clients, consumers, planners and managers;
▶ definition of precise objectives: integrated predictive and adaptive management; advancement by steps, introduction of sound eco-technologies and implementation of decision-making support systems with the participation of consumers and managers.

It is essential to build a local capacity for management based on scientific knowledge in different regional situations (IETC, 2001; Tundisi, 2007).

19.5　RECOVERY TECHNIQUES, MANAGEMENT AND CONSERVATION OF WATER RESOURCES

For a sound approach in addressing the problem, it is necessary to consider a range of attributes and characteristics of ecosystems that can be used as the basis for

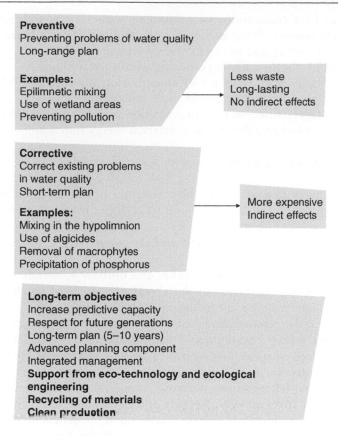

Preventive
Preventing problems of water quality
Long-range plan

Examples:
Epilimnetic mixing
Use of wetland areas
Preventing pollution

Less waste
Long-lasting
No indirect effects

Corrective
Correct existing problems
in water quality
Short-term plan

Examples:
Mixing in the hypolimnion
Use of algicides
Removal of macrophytes
Precipitation of phosphorus

More expensive
Indirect effects

Long-term objectives
Increase predictive capacity
Respect for future generations
Long-term plan (5–10 years)
Advanced planning component
Integrated management
**Support from eco-technology and ecological
engineering
Recycling of materials
Clean production**

Figure 19.5   Approaches to water resource management and long-term goals.
Source: Modified from Straškrába and Tundisi (2000).

implementing eco-technologies in the preservation and recuperation of ecosystems. Eco-technologies are understood to mean the application of technologies that take into account naturally occurring functioning mechanisms of ecosystems that promote advanced management using the knowledge acquired from basic research (Straškraba, 1985, 1986; Tundisi and Straškraba, 1995).

The functioning principles of ecosystems that can be used as a basis for adaptation of eco-technologies include:

1. Ecosystems conserve energy and matter.
2. Ecosystems preserve information.
3. Ecosystems are dissipative. The dissipation provides the mechanisms necessary to maintain order and structure. Ecosystems by dissipation (energy degradation) produce entropy.
4. Ecosystems are open to the influx of energy, matter and information. The functioning of ecosystems depends on the influx of external energy, such as solar radiation, wind and precipitation.

5. Ecosystems have components of subsystems with various self-regulating processes; often, indirect effects predominate over direct effects. There is a constant coupling of dynamic and hierarchical networks of systems.
6. Ecosystems have **self-regulating capacity**, within certain limits.
7. Ecosystems have the capacity for self-organization and adaption, which is a characteristic of organisms, populations and communities.
8. Each ecosystem is different, each organization depending on its evolutionary history (co-evolution) and the effects of external energy function.

Thus, according to these characteristics, a system should:

1. Minimize energy loss and close the circulation of matter: these two processes enable energy use to be optimized and recycling of materials to increase, which reduces external transport (downstream, in the case of reservoirs and lakes) and increases the production of biomass and the complexity of the structures in the network.
2. Improve the equilibrium between influx and outflow – considering the sensitivity of external forcing functions: the alterations produced by the effects of the forcing functions must be in equilibrium with regulation of the outflows, such as discharge, selective removal, and control of retention time. Sensitivity to the influx of toxic substances, for example, and their bio-accumulation should be considered as well as the selective capacity exerted by these substances on the biota.
3. Retention of structures of **genetic diversity** and biodiversity: conservation and increased spatial heterogeneity (wetland areas, zones of riparian forests, littoral zone) enable an increase in biodiversity and at the same time preservation of the potential for adaptation and growth contained in the genetic heritage. **Retention of biodiversity** increases the buffering capacity of the system and its capacity to recycle nitrogen and phosphorus. Wetland areas become important centres for the recolonization of aquatic ecosystems. Colonization by exotic species can be a dangerous mechanism leading to loss of biodiversity, due to the filling of unoccupied niches and the deleterious effects of predation on native species.

The implementation of any restoration project of the aquatic ecosystem requires a **diagnostic plan** to evaluate the situation of the aquatic ecosystem and its trophic level; the origin of point sources and non-point sources of nitrogen and phosphorus and contamination from pollutants; the identification of biodiversity and its responses to variations and the different impacts.

These tests include: the drainage basin and rivers, lakes or reservoirs; the development of annual or seasonal balance of nutrients; the morphometry of the system; the characteristics of the subsystems; retention time; the composition and quantification of phytoplankton and zooplankton; determination of the diversity of fish fauna and biomass, as well as planktonic and benthic biomass. These are critical diagnostic steps. It is also important to identify how the *forcing functions* work in aquatic systems, since they generally are related to climatic and hydrological factors, as already described in Chapters 4 and 12.

Chart 19.2 presents basic theoretical principles and their use in restoration of drainage basins and inland ecosystems.

Chart 19.2   Theoretical principles and their application in restoring drainage basins and inland aquatic ecosystems.

Principle	Application
Effects at the top of the food chain ('top-down effects')	Biomanipulation and control of the food chain
Effects at the base of the food chain ('bottom-up effects')	Control of chemical factors– determinants in primary production
Concept of limiting factors	Control of eutrophication through knowledge and manipulation of limiting factors
Interactions of subsystems	Interactions of the compartments in reservoirs; interactions between drainage basin and reservoirs
Negative feedback	Phytoplankton–nutrient interactions
Connectivity	Connectivity relationships between system components (predator-prey, for example)
Adaptability of ecosystem and auto-organization of ecosystem	Response of the ecosystem and anthropogenic influences
Spatial heterogeneity of ecosystem	Protection of the headwaters between the shore and littoral zone
Biological diversity and biological indicators	Reforestation, wetland areas, protection of ecotones, environmental diagnosis
Competition	Introduction of exotic species and their effects
Pulse theory	Regulation of retention time; control of pulses with the maintenance of the gallery forest
Colonization	Exploitation of the pelagic environment, monitoring the colonization and re-colonization of the reservoir, lake or river

Source: Modified from Straškrába et al. (1993b), Bozelli et al. (2000).

19.5.1 Technologies

A range of technologies needs to be applied based on the characteristics of the ecosystem and the existing conceptual basis of its functioning. These technologies begin in the drainage basin and are control measures for aquatic ecosystems. Initially, a set of information on the drainage basins should be gathered and the following issues addressed:

1. What is the area of the aquatic ecosystem, and the area of the drainage basin, and the relationship between the two?
2. What is the existing hydrographic network in the drainage basin?
3. What are the main sources of pollution existing in the drainage basin?
4. How is the mosaic in the drainage basin organized: floodplains, different types of forests, vegetation, agriculture, industry and human settlements? What is the relationship of areas among these various components?
5. What are the types and slopes of soil/land that comprise the drainage basin, considering soil erosion and its effects on the composition of the water?

6. What are the predominant types of soil use?
7. What are the consequences of these types of use? (Consider erosion, transport of suspended material, transport of pollutants, and contamination of groundwater.)
8. What are the possible consequences of deforestation for the rivers and reservoirs and lakes?
9. What is the influx (load) of nutrients (N, P) in the reservoir, river or lake?
10. What is the retention time of the water in the reservoir or lake?
11. What is the composition of the sediments in the reservoir or lake, and what are the levels of nitrogen and phosphorus?
12. Are there contaminants in the sediments? If yes, what is the level (internal load)?
13. At what rate are herbicides and pesticides applied in the area of the drainage basin?
14. How does the public use the reservoir, lake, or river, and drainage basin? (Possible uses: fishing, recreation, irrigation, transportation, generation of electricity, drinking water supply, agriculture in the drainage basin and types of crops.)
15. What is the economic value of the drainage basin in terms of production, recreation or any other type of use?
16. How did the historical development occur? (Consider the current number of inhabitants in the drainage basin and estimated population growth for the future.)
17. What data are available? (Consider maps, data on water quality, climate, remote sensing, public health problems in terms of water supply, demographic data.)
18. What is the state of **vegetation coverage**? (Include considerations of the natural vegetation and crops existing in the drainage basin.)
19. What is the status of floodplains and forests in the drainage basin? Do they need restoration or protection?
20. What is the sedimentation rate in the reservoir, lake or river?
21. What legislation regulates the drainage basin, water uses, and water management policies?
22. What are the main existing factors of impact? Consider industries (type, production, waste), mining (type, production, conservation), agriculture, and other factors.
23. Analyze the position and the distance of the foci of pollution in relation to rivers, floodplains and reservoir.

In addition to these issues, integrated planning and management refer to:

▶ planning unit – drainage basin;
▶ water as an economic factor;
▶ planning linked to social and economic projects;
▶ participation of community, consumers, organizations;
▶ community health and environmental education; technical training;
▶ ongoing monitoring, with community participation;
▶ integration of engineering, operation, and management of aquatic ecosystems;
▶ ongoing exploration and evaluation of impacts and trends;
▶ implementation of support systems for decision-making.

For an effective control of drainage basins, the following procedures (Tundisi and Straškraba, 1994; Tundisi *et al.*, 1999) should be implemented:

i) **Control of erosion** – using various methods, including reforestation. The principal consequence is to reduce the influx of suspended material and reduce eutrophication and the runoff of toxic substances into aquatic systems.

ii) **Reforestation with native species** – using various reforestation techniques, primarily riparian vegetation on slopes, which reduces the transport of suspended material, increases spatial heterogeneity, decreases the influx of nitrogen and phosphorus, and improves the recharge of the aquifer (see Chapter 11).

iii) **Restoration of rivers** – the rivers of the drainage basin transport material into the aquatic systems. The restoration of rivers by various techniques includes: restoration of banks, increase in spatial heterogeneity with diversification of the substratum, increase of reoxygenation capacity of the system with the introduction of artificial turbulence. This process reduces the eutrophication of rivers, increases the levels of dissolved oxygen, increases the diversity of the substratum, and reduces the load of nitrogen and phosphorus in the reservoir.

iv) **Conservation and restoration of wetland areas** – Wetland areas near lakes and reservoirs act as efficient buffer systems, since they reduce eutrophication and contamination (Whitaker, 1993). They also promote increased species diversity, since they can be areas of establishment for different native species, and they effectively increase the spatial heterogeneity of the reservoir. They also function well as a source of shelter, food and breeding ground for fish. The association of these wetland areas with aquatic systems is therefore critical in the projects of integrated management of the system.

v) **Construction of pre-dams ('pre-impoundments')**
An alternative to the control of sedimentation and external load in lakes and rivers is the use of **pre-dams**, which retain suspended material and reduce the influx of nitrogen and phosphorus. The construction of a series of small pre-dams can reduce soil erosion and control the flow of water and external load. Generally, these structures are built in narrow valleys, with rudimentary material, and, after they fill with sediment, the land can be reforested and erosion substantially reduced. These pre-dams can be an efficient method for controlling contamination and sedimentation and reducing point loads. Their effectiveness was demonstrated in arid and semiarid rural areas (Biswas, 1990). These pre-dams need ongoing inspection and maintenance, as they can become sites of water-borne diseases.

vi) **Re-introduction of native fish species** – Large numbers of reservoirs and lakes lose native fauna in rivers due to limitations resulting from dams, diverse impacts and introductions. The re-introduction of native fish species and the effective development of technologies for their adaptation and reproduction in inland aquatic ecosystems are important techniques of recuperation, since they enable increases in biomass and exploitation of ecological niches in the rivers, increasing the capacity of economic development in the drainage basin and aquatic systems (Agostinho *et al.*, 2001, 2004).

vii) **Maintenance of conservation areas as buffer systems** – Preserved areas can be extremely effective in the management and optimization of drainage basins.

They help reduce eutrophication, suspended material, and external load of contaminants. They also provide a substratum for invertebrates and food for fish; and they increase spatial heterogeneity and facilitate the maintenance of gene banks of native species (Tundisi *et al.*, 2003).

viii) **Management of littoral zones and shorelines** – The management of shorelines and littoral zones with reforestation enables increase in substratum for invertebrates and fish, reduction in the load of contaminants and increasing spatial heterogeneity. Management of the littoral zone involves the introduction and maintenance of filters suitable for guarding the process of sedimentation, the influx of toxic substances and pollutants, and is an effective mechanism for reducing eutrophication and degradation (Bozelli *et al.*, 2000). For example, Whitaker *et al.* (1995) showed that the wetland area found upstream of the UHE Carlos Botelho reservoir (Lobo/Broa) removes 30% of the nitrogen and phosphorus loads reaching the reservoir.

ix) **Management of the littoral zone (maintenance of banks of macrophytes)** – The maintenance and control of banks of macrophytes provide breeding areas for fish, supply of organic matter for invertebrates, and their effective performance as 'filters', recycling dissolved organic matter and contaminants. Banks of macrophytes also provide feeding areas for aquatic birds, and are important in enhancing the biodiversity of aquatic vertebrates and invertebrates (Mitsch and Gosselink, 1986).

Actions in aquatic ecosystems

1. **Reduction of light penetration** – Occurs in the case of increased turbulence and circulation of phytoplankton in the aphotic zone. Various techniques can be used to produce turbulence and reduce the euphotic zone, in order to control eutrophication.

2. **Biomanipulation** – The removal of predators of herbivorous zooplankton can increase the efficiency of removal of phytoplankton by zooplankton (see Figure 19.6).

3. **Removal and chemical isolation of sediment** – There are many techniques to remove and isolate chemicals from the sediment. Aeration, for example, accelerates the precipitation of phosphorus to the bottom of the reservoir. Many reservoirs with bottom discharge can reduce sediment within a few days (bottom discharge can lead to oxygen depletion downstream). The isolation of chemical sediment can be done with successive layers of aluminum sulphate.

4. **Aeration of lakes, reservoirs or rivers** – Various different techniques are used, which produce phosphorus accumulation in sediment, decrease the euphotic zone, causing several effects that partially reduce the organic load of the reservoir (see Figure 19.7) (Cooke and Kennedy, 1989, Cooke *et al.*, 1993).

5. **Control of retention time** – The retention time of small reservoirs and lakes can be controlled by periodically opening the different gates, coupled with the various uses of water upstream and downstream, considering the possible effects downstream. The existing theory shows that it is possible to control phytoplanktonic blooms by reducing retention time with subsequent loss of biomass downstream (Reynolds, 1997). In a simplified version, consider that in cases of short retention

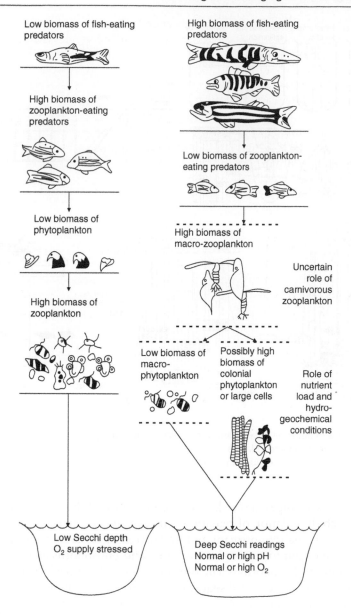

Figure 19.6 Schematic representation of biomanipulation. A low biomass of fish-eating fish implies a low phytoplanktonic biomass due to the grazing zooplankton (left side of the figure). A high biomass of fish-eating fish indicates high biomass of zooplankton and high biomass of colonial phytoplankton or larger-cell phytoplankton ($>50\,\mu$m or $>100\,\mu$m).
Source: Straškraba and Tundisi (2000).

time, phytoflagellates, Chlorophyceae and small species ($<20\,\mu$m) predominate; in cases of longer retention time, larger-sized species or colonies predominate (>50 microns) with longer reproduction time. In these conditions, blooms of *Microcystis aeruginosa*, *Anabaena* spp., and *Anabaenopsis* spp. can occur,

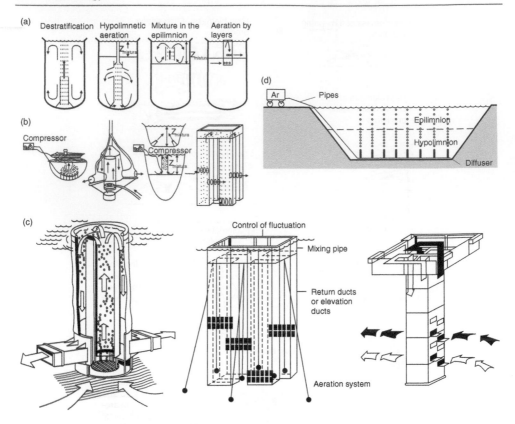

Figure 19.7 **Types of aeration in lakes and reservoirs: a – mixing types; b, c and d – types of hypolim-netic aeration and in layers.**
Source: Straškrába and Tundisi (2000).

causing major changes in the chemical composition of the water and the structure of food web. Regulation of retention time can thus be an essential factor in controlling water quality upstream and downstream. Reservoirs or lakes with long retention time tend to accumulate nitrogen and phosphorus as well as biomass.

6. **Inactivation of phosphorus** – The precipitation and removal of phosphorus from the water column and its subsequent deposition and immobilization in the sediment is an important practice in the restoration of lakes, reservoirs and rivers, along with other measures of course. To inactivate phosphorus, chemical coagulants are used; aluminium sulphate is one of the most common. The success of aluminium sulphate in the process of removing phosphorus from waste-water explains the increase in its production and commercialization in the early 1970s.

The sources of aluminium used are analyticalgrade bauxite (with low levels of iron and heavy metals) and clay with high levels of aluminium or aluminium trihydrate. One of the frequently used formulations is $Al_2(SO_4)_3 \cdot 14H_2O$ called 'anhydrous aluminum'. There are other groups of coagulants all categorized as poly-aluminate hydroxides with the addition of sulphate or calcium and polyelectrolytes.

Ferric sulphate is another widely used coagulant (with iron levels between 10–12%), but currently the most commonly used coagulant is ferric chloride.

Coagulants that contain aluminium are rarely used in public water supplies, because of the possible effects that metal can have on human health.

The inactivation of phosphorus is done by simple precipitation as a metal phosphate. Calcium hydroxide is also used, although it has the drawback of raising the pH level above 10 and other effects resulting from the high pH. The equations that represent the effect of these coagulants are:

▶ Aluminum sulphate:

$$Al_2(SO_4)_3 + 2H_3PO_4 \rightarrow 2AlPO_4 + 3H_2SO_4$$

▶ Ferric sulphate:

$$Fe_2(SO_4)_3 + 2H_3PO_4 \rightarrow 2FePO_4 + 3H_2SO_4$$

▶ Iron sulphate:

$$2FeSO_4 + 2H_3PO_4 \rightarrow 2FePO_4 + 2H_2SO_4 + 2H^+$$

The application of these chemical coagulants requires specialized techniques and knowledge of the required doses for the products chosen.

The inactivation of phosphorus from the internal loads, especially in shallow lakes, is an essential technique for restoration. Shallow lakes or deeper, stratified lakes (from 10–30 m deep) have a high internal load of phosphorus, and its removal depends on the application of layers of clay or aluminum sulphate, to prevent sediment-water exchanges.

In the inactivation of phosphorus, one main issue is identifying the cause or causes of the phosphorus load: external, point, from the drainage basin, or diffuse and internal from the sediment. For example, in 17 lakes in the state of Washington (the United States), Welsh and Jacoby (1997) showed that from 26–97% of the phosphorus load was internal.

As examples of application of technology, Chart 19.3 presents the main external control mechanisms used to restore lakes and reservoirs, applied in various different regions to reservoirs, lakes and rivers.

The joint application of these methods in aquatic ecosystems and drainage basins tends to reduce internal load and eutrophication, control undesired phytoplanktonic blooms and improve water quality, reducing the degree of organic load and toxicity. This can be highly efficient and enable sound integrated management.

In all these cases, it was essential to compile limnological information before undertaking management actions. Examples of management systems and conservation in Brazil include the Mimirauá Sustainable Development Reserve in the Amazons (see Figure 19.8), the conservation plan and management of the UHE Carlos Botelho reservoir (Lobo/Broa) (Tundisi et al., 2002), the recovery and management plan of Batata Lake in the Amazons (Bozelli, Esteves and Roland, 2000), the integrated management plan of the UHE Luis Eduardo Magalhaes reservoir (Paved/Tocantins)

Chart 19.3   External control measures for restoration of lakes and reservoirs.

1) Control of hydrological cycle and erosion

Reforestation around springs – Lake Dianchi *(China)*

Stabilization of slopes – Lake Biwa *(Japan)*

Installation/maintenance of buffer zones between agricultural areas and shoreline zone of lake or reservoir – UHE Carlos Botelho reservoir *(Lobo/Broa, Brazil)*; UHE Luis Eduardo Magalhaes reservoir *(Lajeado/Tocantins)*

Appropriate techniques for soil treatment and contour lines – Lake Biwa *(Japan)*

Treatment of affected soil with organic material – Lake Batata *(Amazonia)*

2) Legislation and control of soil use and waste discharge

Regulation restricting soil use – Lake Tahoe *(USA)*

Removal of pollutants from the drainage basin – Feitsui reservoir *(Taiwan)*

Landfills prohibited – Lake Constance *(Germany, Austria, Switzerland)*

Closing factories that emit pollutants – Lake Baikal *(Russia)*

3) Treatment of sewage and wastewaters

Construction of treatment plants on a large scale – Lake Maggiore *(Italy)*

Construction of oxidation ponds – Lake Ya-er *(China)*

Construction of small treatment plants – Lake Naka-Uni *(Japan)*

Use of septic tanks – Lake Biwa *(Japan)*

Diversion of sewage outlets – Lake McIllwaine *(Zimbabwe)*

Treatment of animal waste – Lake Furen *(Japan)*

Fermentation of water with domestic waste – Lake Chao-Chu *(China)*

Regulation of agrochemical use – Lake Kinneret *(Sea of Galilee, Israel)*

Dumping of toxic substances prohibited – Lake Orta *(Italy)*

Restoration of pH – Lake Orta *(Italy)*

4) Control of nutrients in tributaries

Use of wetland areas as filtration system and containment of heavy metals and residues – *(Hungary)*

UHE Carlos Botelho reservoir *(Lobo/Broa, Brazil)*; Billings and Guarapiranga reservoirs *(São Paulo, Brazil)*; Paraná River *(Brazil)*

Establishment of coordination and management units for national and international drainage basin – the Reno River basin (various countries); basins of the Piracicaba and Jacaré Pepira (Brazil); Plata basin; management of water resource units.

Construction of pre-dams – Lake Rorotua *(New Zealand)*

Controlled use of agricultural fertilizers – Lake Dota *(Colombia)*

Recycling of treated water through reforested areas – San Roque reservoir *(Argentina)*

Treatment of soils destined for domestic residues – several lakes in *Japan*

Re-vegetation with *Oryza* sp. and arboreal species – Lake Batata *(Amazonia)*

Shifts in water levels in reservoirs and ponds to **control macrophytes** (Cooke *et al.*, 1943)

Sources: IETC (2000); Tundisi and Straškraba (1999); Bozelli *et al.* (2000).

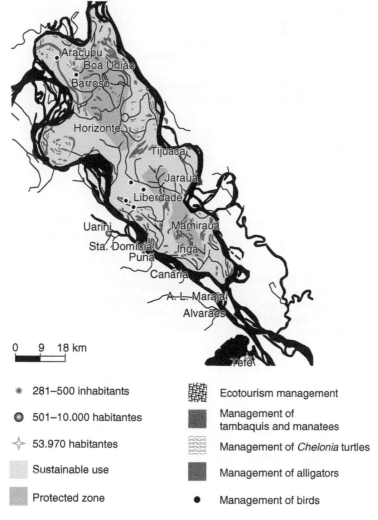

Figure 19.8   Zoning in the Mamirauá Sustainable Development Reserve (AM).
Source: Mamirauá Civil Society (1999) in Tundisi (2003).

(see Figure 19.9), and the conservation and management plan of wetlands and flood-plain areas upstream in the Paraná River (Agostinho *et al.*, 2004).

Chart 19.4 presents several measures of internal controls implemented to restore lakes and reservoirs.

In the restoration processes of all these systems, it is important to consider the costs of the technologies and their sustainability, and keep in mind that every lake, reservoir or river presents a unique situation, which requires application of restoration or conservation measures in each specific case after basic studies. In addition, indirect effects can result from application of technologies that may damage the aquatic biota and humans. For example, application of aluminum sulphate is not recommended for

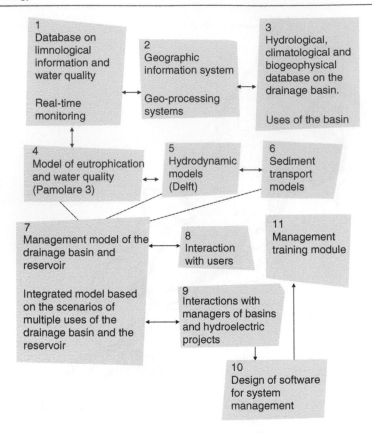

Figure 19.9   Summary of the principal models for use in the management project for the UHE Luis
Eduardo Magalhães reservoir, Tocantins.
Source: Investco /IIE; Tundisi et al. (2002, 2003).

control of phosphorus in public-supply reservoirs. Application of copper sulphate to control cyanobacteria or molluscs can be very effective for short periods of time and to control blooms in lakes and reservoirs that are not used for public supply. High levels of neurotoxins were reported after application of 0.5 mg $CuSO_4 \cdot L^{-1}$ in blooms of *Microcystis aeruginosa*, which indicates that copper should not be used to treat cyanobacterial blooms, particularly in drinking water (Cooke *et al.*, 1993), due to the effects on the cells, resulting in the release of neurotoxins.

19.6   INTEGRATED MANAGEMENT: CONSEQUENCES AND PERSPECTIVES

Integrated management can be defined as a series of preventive, corrective, mitigative and restorative efforts that maintain close to optimal conditions and that enable

Chart 19.4   Measures of internal control applied in the recuperation of various different lakes and reservoirs.

1. Physical measures	Thermal de-stratification and increased vertical mixing – **Lake Sagami** (*Japan*)
	Acceleration of rate of recycling – **Lake Bled** (*Slovenia*)
	Introduction of water with low contamination – **Lake Igsell** (*Holland*)
	Removal of sediment and the deepest layers – **Lake Baldegger** (*Switzerland*)
	Removal of sediment by dragging – **Lake Trummer** (*Switzerland*)
	Isolation of sediment with sand – *Lake Biwa* (*Japan*)
	Aeration for artificial destratification – *Pao-Cachinche Reservoir* (*Venezuela*)
2. Chemical measures	Destruction of algae – *Lake Mendota* (*USA*)
	Destruction of water hyacinth – **Kariba reservoir** (*Zambia–Zimbabwe*)
	Addition of carbonate – *Lake Orto* (*Italy*)
	Inactivation of phosphorus – *several lakes* (*USA*)
3. Measures with respect to the organisms (biological)	Removal and collection of macrophytes – *Lake Léman* (*Switzerland–France*)
	Removal of cyanophytes – *Lake Kasumigaura* (*Japan*)
	Manipulation of the food chain – biomanipulation – *Lake Paranoá* – (*Brasilia*)
	Control of macrophytes with herbivorous fish – **Lake Bong-hu** (*China*)
	Control of water level to protect vegetation – **Lake Chao-Lu** (*China*)
	Protection of river vegetation – *Lake Neusiedlersee* (*Austria – Hungary*)
	Navigation prohibited in order to prevent increase in discharge of combustible oils and lubricants – **Lake Taupo** (*New Zealand*)
	Use of locusts to control aquatic macrophytes –*Kariba reservoir* (*Zambia-Zimbabwe*) Planting wetland species (igapó) – *Lake Batata* (*Amazonia*)

Source: Straškrába *et al.* (1993); Straškrába and Tundisi (2000); Starling (2006); Bozelli *et al.* (2000); Nakamura and Nakajima (2002); Gonzalez *et al.* (2002).

rational exploitation and self-sustaining development. The principal benefits from integrated management of ecosystems include:

▶ protection and **rational use of resources of the ecosystem;**
▶ increased potential for multiple uses;
▶ reduced costs and conflicts, with better application of resources in environmental programmes;
▶ more rapid and effective restoration of ecosystems in damaged areas and better utilization of the ecosystem's available 'services'.

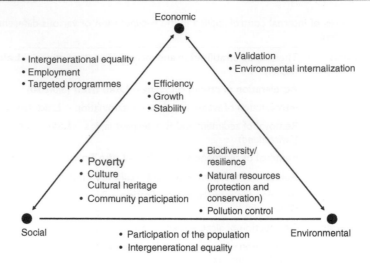

Figure 19.10 Coordination of environmental, economic, and social processes in integrated and sustainable management.

Any integrated management plan should undoubtedly be based on a broad database, which enables a deep knowledge of the ecosystem, its basic processes, and rates of transference among its components.

The basic principles of an integrated management include:

▶ identification of the system's boundaries and diagnosis;
▶ geographic, morphological, and topographic boundaries, and physiographic data;
▶ history and current situation of resource use;
▶ identification of:
  a) main users of the resource;
  b) impacts of the use of the system;
  c) the institutional organization and the institutions using the resource.
▶ information for the general public about the various uses of the resources and the institutions that use them;
▶ request for a list of **potential impacts** produced by the use of resources from each institution;
▶ establishment of an appropriate programme that enables ongoing monitoring of environmental quality, the magnitude of impacts and the degree of recovery and **persistence of impacts** in the ecosystem;
▶ integration of biogeophysical, economic and social processes (see Figure 19.10).

The integration of economic, social and environmental processes in the sustainable development programme enables the drainage basins, lakes and reservoirs to have long-term planning and an ongoing program of preservation, restoration and pollution control, represented by control of point and non-point sources (see Chart 19.5).

Chart 19.5   Principal elements in the dynamics of ecosystems of continental waters, important for long-term integrated management and planning.

Drainage basin and human activities	Probable adverse environmental effects	Inclusion of environmental considerations in development
Degree of urbanization Domestic waste	Environmental load resulting from population growth	Ordering of spaces; public health; sanitation
Mining and manufacturing industries	Localized pollution (point)	**Control of industrial pollution**
Generation of hydroelectricity	General effects on the ecosystems	Control of effluents, waste and processes; control of environmental accidents
Tourism, commerce, transportation	Effects of navigation	Control of environmental accidents
Civil construction	**Irreversible effects** on the environment	Soil use; problems of environmental engineering; operation of industrial plants
Agriculture Forestry Fishing Ecological reserves Natural parks	**Diffuse pollution** Degradation of natural resources Disruption of functioning systems	Sustainable aquaculture; regulation of fishing; use of pesticides; soil management; irrigation; control of rivers and floodplains
Natural functioning base of systems		Conservation of nature; protection of wildlife; control of droughts and flooding

Source: UNEP (1987).

19.7   ECOLOGICAL MODELS AND THEIR USE IN MANAGEMENT

Among the many resources available for conservation and restoration of aquatic ecosystems, the use of ecological models is particularly relevant, since the models can function as guiding (or reorienting) systems for basic research, due to the questions arising during implementation.

Ecological models can be defined as the formal expression of the essential elements of a problem posed in physical and mathematical terms. They represent a simplification of reality and are basically used to solve problems and as a stimulus for basic research. The intensive application of ecological models in limnology is relatively recent (last two decades) and in part is the result of limnological research developed during the International Biological Programme. Upon completion of this project, it was clear that there was a need for comprehensive studies of ecosystems to quantify processes and integrate the fundamental concepts with basic research and application. The model, of course, does not have all the details of the real ecosystem, but it does contain the essential features of the functioning of the ecosystem.

There are many ways to categorize and define models, but it is important that they present the criteria of *generality*, *realism* and *accuracy* (Vollenweider, 1987). Generality refers to the overall conception of the ecosystem; realism, to the development of hypotheses and theories; and accuracy, to specific data obtained in the real system that delimits the validity of the model.

The field of ecological modeling developed rapidly during the last decade, due to two essential factors:

▶ the development of computer technology, enabling the manipulation of complex mathematical systems;
▶ a deeper understanding of the problems of pollution, which showed the impossibility of achieving 'zero pollution'; however it can possibly be controlled with limited economic resources and the use of ecological models, for the promotion of scenarios and optimizations.

Urbanization, population growth and technological development have had a significant impact on ecosystems. Ecological models help to identify the principal characteristics of the ecosystem that are being affected; consequently, they enable the correction of the process and selection of the most appropriate environmental technology to solve the problem. Applied to lakes, reservoirs and rivers, these models allow the development of future scenarios.

According to Jørgensen (1981), ecological models also enable the understanding of the properties of the ecosystem. With the following criteria, models can:

▶ indicate the basic characteristics of complex systems;
▶ reveal flaws in the study of ecosystem and thus help establish research priorities;
▶ be used to test scientific hypotheses in simulations that (later or simultaneously) can be compared with observations.

The main focus in terms of applying ecological models in limnology is to define the nature of the problem in its spatial and temporal boundaries. Naturally, the definition and delimitation of the problem depend on the knowledge of the existing level of scientific information on the ecosystem.

Figure 19.11 (from Jørgensen, 1981) shows the main steps in ecological modelling.

19.7.1 Key concepts of models

An ecological model basically consists of the mathematical formulation of five components:

▶ *Forcing functions* or *external variables* – These are external to the ecosystem and can influence its state. Forcing functions include, for example, precipitation, solar radiation and wind, which act on the ecosystem producing changes and introducing *external energy*. Variations in these forcing functions over the course of time definitely produce modifications in the temporal functioning of the ecosystem.
▶ *Variables in state* – These variables show the *state* of the ecosystem, that is, they are important in the delimitation and assembling of the structure of the model. Variables in state include, for example, levels of inorganic nutrients, levels of phytoplankton (in biomass per m³), biomass of benthic organisms (per m²), levels

Quality indices for drainage basins

Applying quality indices to drainage basins enables advanced diagnostics and identification of conditions that promote adoption of technologies and application of alternative scenarios. Quality indices for drainage basins have four main objectives (EPA, 1998):

i)   To identify a drainage basin's conditions and vulnerability to pollution.
ii) To promote a basis for dialogue between scientists and water resource managers.
iii) To promote an integrated and systemic vision for the public and stimulate their capacity for interaction with the administrators of water basins, increasing the public's awareness of the major problems.
iv) To develop appropriate systems and methods for evaluation, facilitating the monitoring of effects from applying protective and recuperative technologies and measures.

To implement these indices, two categories need to be taken into consideration: condition and vulnerability.

In terms of condition, the following indicators can be used:

1) Indices of biotic integrity based on parasites in/on fish.
2) Indicators of the quality of water in public supply sources.
3) The index of contamination in the sediment.
4) The quality of surface waters as a function of toxic pollutants (five toxic pollutants).
5) Water quality in terms of conventional pollutants.
6) The index of loss of wetland areas.

In terms of vulnerability, the following indicators can be used:

1) Aquatic species at risk.
2) Loads of toxic pollutants.
3) Loads of conventional pollutants.
4) Potential urban drainage.
5) Potential drainage in agricultural regions.
6) Changes in population.
7) Hydrologic modifications produced by dams.

Source: Matsumura Tundisi (2006).

of zooplankton (per m^3 or m^2), and quantity of carbon, nitrogen and phosphorus in the particulate matter. Naturally, the quantity of these variables in state also depends on the initial questions and the initial problem addressed by the model applied.

▶ The chemical, physical and biological processes in the ecosystem are represented by the use of *mathematical equations*, which characterize the magnitude of the interactions between the forcing functions and variables in state. For example, some processes are relatively similar in different lakes. The mathematical representation of these processes can be extremely complex, because of the very complexity of the processes involved.

▶ The mathematical representation of processes contains *coefficients* or *parameters*, which can be considered constant for a specific ecosystem. However, it should

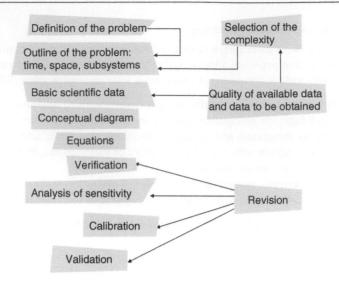

Figure 19.11   Stages of development in ecological models.
Source: Jørgensen (1981).

be emphasized that many of these coefficients are known only in their limits. Different species of plants or animals have different parameters. There is also a shortage of these coefficients for lakes and processes in tropical regions, so that the application of coefficients that exist in the literature (Jørgensen *et al.*, 1979) is not always appropriate (Tundisi, 1992). Growth rates of phytoplankton, zooplankton and fish vary greatly with the temperature of water and availability of substrates (nutrients, food). Therefore, in the application of models, these coefficients need to be calculated for individual ecosystems.

▶   *Calibration* of the model means the best correlation between the computed and observed variables of state, using the variation in a large number of parameters. Calibration can be calculated by trial and error.

Figure 19.12 shows a representation of an aquatic ecosystem in the schematic form (a) and in the form of models and flows (b).

Figure 19.13 shows a conceptual model of the nitrogen cycle in a lake. The variables of state are nitrate and ammonia, as well as nitrogen in the phytoplankton, zooplankton, fish, sediment and detritus. The main forcing functions are: the influx of nitrogen in tributaries, outflow of nitrogen through effluents, solar radiation and temperature, winds and precipitation. This conceptual model needs to be corroborated with measurements, determinations and experiments in the laboratory and field.

19.7.2   Main types of models used in limnology

Stochastic models contain influxes, stochastic disturbances, and measurements with error, at random. When the influx and outflow are equal to zero, the model will be

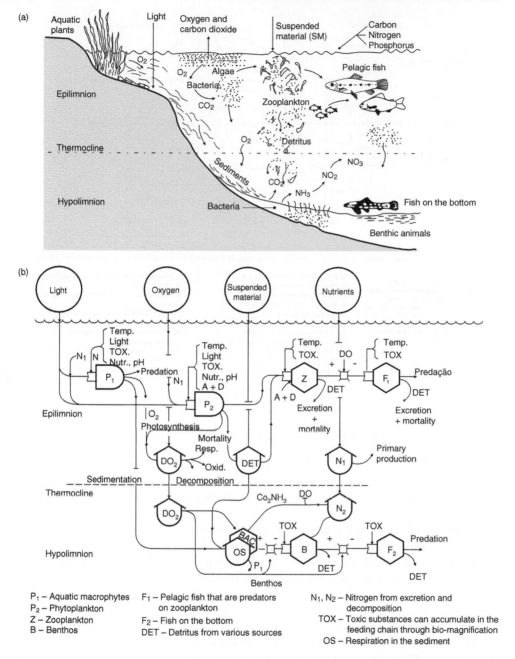

Figure 19.12   Representation of an aquatic ecosystem in the form of a schematic diagram (a) and in the form of a model and flows (b). The model and flows help determine the functions of transformation between the components of the system and present quantitative response scenarios.

Source: Jørgenson (1982).

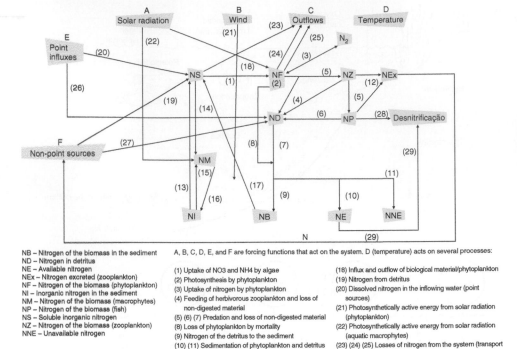

NB – Nitrogen of the biomass in the sediment
ND – Nitrogen in detritus
NE – Available nitrogen
NEx – Nitrogen excreted (zooplankton)
NF – Nitrogen of the biomass (phytoplankton)
NI – Inorganic nitrogen in the sediment
NM – Nitrogen of the biomass (macrophytes)
NP – Nitrogen of the biomass (fish)
NS – Soluble inorganic nitrogen
NZ – Nitrogen of the biomass (zooplankton)
NNE – Unavailable nitrogen

A, B, C, D, E, and F are forcing functions that act on the system. D (temperature) acts on several processes:

(1) Uptake of NO3 and NH4 by algae
(2) Photosynthesis by phytoplankton
(3) Uptake of nitrogen by phytoplankton
(4) Feeding of herbivorous zooplankton and loss of non-digested material
(5) (6) (7) Predation and loss of non-digested material
(8) Loss of phytoplankton by mortality
(9) Nitrogen of the detritus to the sediment
(10) (11) Sedimentation of phytoplankton and detritus
(12) Excretion of nitrogen by zooplankton
(13) Liberation of nutrients from the sediment
(14) Nitrification
(15) (16) Influx and outflow of nitrogen
(17) Contribution of nitrogen in the sediment to nitrogen in the water

(18) Influx and outflow of biological material/phytoplankton
(19) Nitrogen from detritus
(20) Dissolved nitrogen in the inflowing water (point sources)
(21) Photosynthetically active energy from solar radiation (phytoplankton)
(22) Photosynthetically active energy from solar radiation (aquatic macrophytes)
(23) (24) (25) Losses of nitrogen from the system (transport downstream)
(26) Contribution of soluble nitrogen from non-point sources
(27) Contribution of nitrogen in the detritus from non-point sources
(28) Loss of nitrogen in fish (through fishing or predation)
(29) Loss of nitrogen through denitrification

Figure 19.13 Model of the nitrogen cycle in a lake or reservoir.

deterministic, which implies well-defined parameters and not estimates in statistical terms.

Most models used in limnology are **deterministic,** which implies the existence of only one output for a given number of input variables.

Reductionist models incorporate, as much as possible, the details of the system. **Holistic models,** in turn, use the general principles of the system and incorporate theories on operations in the ecosystem as a **system.** The reductionist models interpret the system as the sum of its parts, while holistic models are interpreted as a single functional unit, above parts.

Dynamic models describe the response of the system to external factors and take into account differences in state over time. A **static model** assumes that all the variables of the system are independent of time.

The main dynamic models used in limnology are:

▸ Hydrodynamic models;
▸ Hydrochemical models;
▸ Ecological models.

Chart 19.6   Principal types of models generally used in limnology.

Type of model	Features
Deterministic	The predictable values are computed with precision
Stochastic	The predictable values depend on the probability of distribution
Static	The variables that define the system do not depend on time
Dynamic	The defining variables of the system are quantified by differential equations that depend on time (or space)
Linear	Uses first-degree equations
Non-linear	One or more equations are not first degree

Chart 19.7   Problems in aquatic systems and springs, the causes and consequences.

| Problem areas | Caused by or indirectly dependent on | | | |
	Discharge of nutrients	Domestic or industrial waste	Acid rain	Turbidity
Alterations in water quality – eutrophication, toxicity Increased treatment costs	XXX	XXX	XX	X
Alterations in recreational quality Increased risk (health)	XX	XX	XX	XXX
Alterations in fishing Higher mortality of fish	XX	XXX	XXX	XXX
Reduced volume Reduced flow	XX	XX	X	XXX

Frequency (importance): XXX – Very high; XX – High; X – Occasional
Source: Vollenweider (1987).

Various types of models currently used in limnology are presented in Chart 19.6.

19.7.3   Use of models for management, and prognosis of restoration of lakes, reservoirs and rivers

Minimization, correction of impacts or restoration of aquatic ecosystems can be undertaken with the use of ecological models, hence the significant value of models in solving applied problems.

Chart 19.7 presents the problems associated with lakes and reservoirs that, in general, demand a series of preventive or corrective measures in which ecological models can be used.

Irrigation, salinization, navigation and other water-borne diseases are the problem areas considered. Factors such as thermal discharges into the lakes, saline intrusion, and an excess of humic substances can also interfere.

The use of models can be applied in the following areas of management, enabling preparation of scenarios and alternatives:

1. Quantitative estimates of the **distribution of pollutants**;
2. evaluation of criteria for **contribution** of substances and elements;
3. prognosis and prediction of responses and types of responses to treatment and restoration programs;
4. optimization of manipulations in the lake;
5. management of drainage basins;
6. evaluation of social benefits for management of drainage basins (choice of alternatives);
7. resolving problems of eutrophication (partially addressed above in 2);
8. distribution of pollutants and contaminants in function of hydrodynamic modeling, considering the total circulation and circulation in various different layers;
9. integrated regional planning for drainage basins.

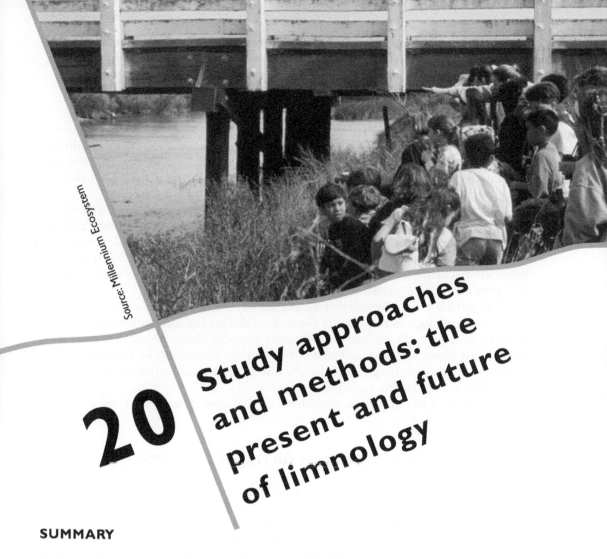

Source: Millennium Ecosystem

20 Study approaches and methods: the present and future of limnology

SUMMARY

In this chapter, we present the various different approaches to the development of the study of limnology, which include the **descriptive approach** or natural history, the **experimental approach**, plus mathematical and ecological modelling, mass balance and **predictive limnology**.

The technology of monitoring lakes, rivers, reservoirs and wetland areas will be discussed, as well as techniques to interpret limnological results. Human resource training programmes and concepts in the area of limnology will also be presented.

The authors present proposals for new advances in scientific research on limnology. Despite advances in the knowledge already developed in the area, the need still exists, especially in Brazil, for continuous progress based on descriptive limnology of a variety of aquatic ecosystems, and investment in several lines of basic research, such as aquatic microbiology, energy flows, hydrodynamic studies and further research on biological indicators.

This chapter concludes with the presentation of basic ideas on long-term ecological research. The latest scientific developments in eco-hydrology and eco-technologies are discussed here, as well as the need to promote international cooperation in South America and the sharing of knowledge and management experiences.

Studies on eutrophication and its impacts, phosphorus loads and design of scenarios to understand the effects of environmental variables in intensive cyanobacterial blooms are challenges in this area, as well as the implementation of research and management projects in drainage basins, pilot projects and demonstrative projects. Further research is also needed on the microbial loop and its impact on the functioning of food webs, biogeochemical cycles, and indirect effects on ecosystems.

20.1   THE COMPLEXITY OF INLAND AQUATIC ECOSYSTEMS

The complexity of inland aquatic ecosystems calls for the development of a range of study methods and approaches enabling the construction of scientific knowledge that represents these complexities. The task is extremely difficult, considering the diverse environments in which lakes, reservoirs, rivers and wetland areas are found, and also the importance of the origin of these systems. Often, this origin establishes the patterns of spatial and temporal complexities, as discussed in Chapter 3.

The intrinsic variability of these systems, the diversity of organisms, the processes of evolution and geographical distribution, as well as the responses of these organisms to the physical and chemical conditions of the drainage basin and aquatic ecosystems, in addition to the impacts of human activities – all these are factors that need to be considered in defining how to approach, study and sample systems so varied, complex and with great interdependence of the various components.

The publication by Bicudo and Bicudo (2004) exhaustively treated the sampling process in limnology. In this chapter the authors discuss the approach of the scientific work in limnology; the method of monitoring physical, chemical and biological variables (with recent advances in technology), and the interaction of various types of analyses and methods for greater scientific understanding (of course not complete) of these complex and greatly varied ecosystems.

The study of inland aquatic ecosystems can have multiple approaches, as the following topics will show.

20.2   DESCRIPTIVE APPROACH OR NATURAL HISTORY

This approach, used in many studies of lakes, rivers, reservoirs and wetland areas, consists of describing the system and its components, using observation and periodic measurements of physical, chemical and biological variables, thus seeking to interpret the functioning and interaction of the components. This descriptive approach, with strong emphasis on the biological component of the system, gave rise to an extensive body of information that has helped to broaden and deepen the understanding of aquatic biology, ecology, physics and water chemistry. This approach, developed over a long period in certain lakes, reservoirs or rivers, can lead to a broad and detailed database that shows the main trends of the system.

Thus, combining the *descriptive approach* with *long-term studies* of a system can be highly relevant and informative. Even if analysis is reduced to a few variables (such as water temperature, transparency by the Secchi disc, dissolved oxygen and samples of plankton and/or benthos), the data can be informative if collected over long periods. In this descriptive approach, climatic and hydrological data are also used, which, in many regions, are available on the Internet, enabling the long-term gathering of relevant information.

20.3   EXPERIMENTAL APPROACH

If the laboratory is located near a lake, reservoir, river, wetland area or coastal waters, there is a possibility of combining observations and the descriptive approach with controlled experiments in the laboratory. Goldman and Horne (1994) detailed experimental methods: experimental methods in the field, **cultures of organisms, experiments in microcosms** or **mesocosms** (see Figure 20.1). Laboratory trials and studies with pure cultures under controlled conditions offer the advantage of promoting a better assessment of certain processes (such as growth of organisms) and impacts of temperature and salinity on survival, reproduction, and feeding.

Experimental laboratory methods allow development of limited predictive capacity (such as biological indicators, for example), but, if combined with field analysis and observation, such methods can be important predictive tools (see Figure 20.2).

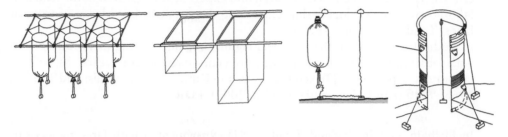

Figure 20.1   Several types of experimental systems for studies on enrichment and manipulation.

Figure 20.2   Examples of experiments of different scales and complexities to study the effects of enrichment with nitrogen and phosphorus and biomanipulation in microcosms and macrocosms. (a) Luis Eduardo Magalhães hydroelectric dam/reservoir, Tocantins; (b) Lake Suwa, Japan; (c) and (d) UHE Carlos Botelho reservoir (Lobo/Broa), Brotas–SP (see also color plate section, plate 28).

On three occasions, the authors of this book could exercise this activity and approach:

▶ in studies on the Carlos Botelho reservoir (Lobo/Broa), where intensive field collections (Tundisi and Matsumura Tundisi, 1995) were coupled with laboratory studies (Rocha *et al.*, 1978; Rietzler *et al.*, 2002), which resulted in knowledge of various processes, giving rise to duly confirmed hypotheses (Tundisi *et al.*, 1978, 2003);

▶ in the studies conducted on the lakes in the Parque Florestal do Rio Doce in eastern Minas Gerais, where field research and experimental research in the field and laboratory were developed, thus expanding the scientific knowledge of tropical lakes (Tundisi and Saijo, 1997);

▶ in studies conducted on the Cananéia lagunar region, starting in 1960, together with observations and periodic measurements over several years, as well as laboratory experiments (Teixeira, Tundisi and Kutner, 1965; Tundisi and Tundisi, 1968; Tundisi and Matsumura Tundisi, 2002).

In Blelham Tarn in England, Lund (1981) experimented with large mesocosms for 11 years. There, the control of phytoplanktonic succession and experiments with nitrogen and phosphorus enrichment made it possible to efficiently study the processes of seasonal succession and interactions of phytoplankton, zooplankton and parasites on plankton.

Large cylindrical experimental tanks each holding 18,000 m³ of water were installed in Blelham Tarn. After four years, an equivalent tube was placed in the lake. Experiments in the tubes on fertilization with phosphorus and nitrogen provided an assessment of the effects on plankton of enrichment, the effect of added phosphorus on cyanobacterial blooms and of the dynamics of phytoplanktonic succession in lakes and comparatively, in the tubes (see Figure 20.3).

Figure 20.3   A large-scale experiment to study the effects of nutrients on lakes.
Source: LTER (1998).

Both the descriptive approach as well as the experimental approach can be used in a single lake (a single system – intensively studied) or in many lakes or reservoirs, producing an intensive comparative study in regions or lake districts. For example, Margalef *et al.* (1976) developed a typology of reservoirs in Spain, studying 100 reservoirs in the country and developing a comparative study that established important foundations for understanding functioning mechanisms in reservoirs.

Tundisi *et al.* (1978) and Tundisi (1981) developed a typology of reservoirs for the state of São Paulo, comparing 50 reservoirs (with FAPESP support), which gave rise to many further studies and also helped in selecting which reservoirs to study further in-depth. Such comparative limnology also developed in the lakes of the Amazon River floodplain studied by Sioli (1984) and Junk *et al.* (2001), and the lakes of the Paraná River floodplain studied by Agostinho *et al.* (2004).

The same approach, combining experimental work with field work, was used by Rocha, Esteves and Carani (2004) in the study of coastal lagoons in the state of Rio de Janeiro, and by Bicudo *et al.* (2002) in a study of lakes in the Fontes do Ipiranga State Park (PEFI) in the city of São Paulo.

20.4 MODELING AND ECOLOGICAL MATHEMATICS

The use of ecological and mathematical modelling is an innovative and important approach because it allows quantification of key processes and elucidation of the dynamic components of aquatic ecosystems. To apply ecological models, however, it is necessary to have a large amount of basic scientific information that can serve as a foundation on which to assemble the conceptual model (its calibration and validation). Application of ecological models to lakes, reservoirs and rivers without a large volume of basic scientific information on seasonal cycles, diurnal variations, species composition, and biogeochemical cycles is not very effective and functions more as a theoretical exercise than an effective model of the system.

Two examples of this approach can be cited: the work of Jørgensen (1980a), on the river Nile (see Figure 20.4) and of Angelini and Petrere (1996) on the UHE Carlos Botelho reservoir (Lobo/Broa) (see Figure 20.5), using Ecopath software, which combines the estimates of the biomass and composition of components in an ecosystem with the theories of Ulanowitz (1986). The two models – the River Nile and the UHE Carlos Botelho reservoir (Lobo/Broa) – were able to be implemented because of the volume and quality of existing information on the structure and function of the aquatic ecosystems modelled.

According to Marani (1988), the evolution of computer technology, data-processing capacities, and steady advances in artificial intelligence have promoted a solid expansion in interactive modelling applied to natural ecosystems. The models are influenced by the choices of spatial and temporal scales, which define the parameters and determine the appropriate algorithms. Conceptual models based on the reality of the experimental and observed data also play an important role in the definition of experimental protocols and in identifying gaps in the basic research. Likewise, they are useful in organizing monitoring norms for environmental control and in the identification of forcing functions among system components.

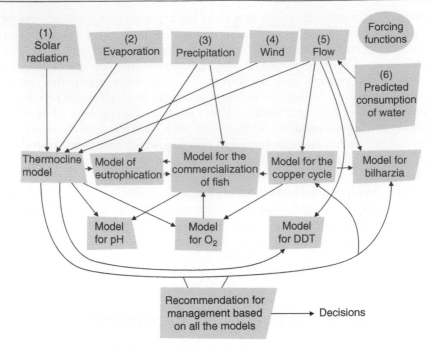

DDT – dichlorodiphenyltrichloroethane

Figure 20.4   A set of models used in studies on the Nile River.
Source: Jørgensen (1986).

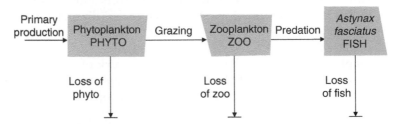

Figure 20.5   Diagram of the Ecopath model.
Source: Angelini and Pretere (2000).

Predictive limnology applied to water supply reservoirs in the São Paulo metropolitan region

In a recent study, Tundisi *et al.* (2004) showed that cold fronts act on relatively shallow reservoirs (≤30 m) in Southeast Brazil, promoting turbulence and a vertical re-organization process during the passing of the cold front. When a cold front occurs, the air temperature drops, the wind forces increase (from 3–4 to 8–12 km/hour), cloud cover increases, reducing direct solar radiation, and, in some cases, increasing indirect solar radiation.

All supply reservoirs in the São Paulo metropolitan area are shallow (from 10–30 m depth) and are therefore subjected to a set of external forces that promote vertical circulation. As a result, nutrients (such as NO_3 and PO_4) increase in the water column; there is removal of sediment from the bottom; toxic substances increase, as well as the level of suspended material – through turbulence and drainage of basins (Campagnoli, 2002). As a result, water treatment costs increase during this period.

After the passing of a cold front, the water column stabilizes and diurnal stratification occurs, promoting conditions for cyanobacterial blooms of *Microcystis* spp.

Therefore, the passage of cold fronts plays an important role, both quantitatively and qualitatively, on the functioning of the supply reservoirs in metropolitan São Paulo.

Considering that cold fronts move at a speed of 500 km.day^{-1}, it is possible to predict the impact on reservoirs and, to some extent, the impacts on the treatability of water and on treatment costs.

20.5 PREDICTIVE LIMNOLOGY

As a result of countless processes of degradation occurring in the structure and functioning of inland aquatic ecosystems, there is a growing need for predictive tools in experiments with ecological and mathematical ecological modelling.

Predictive limnology is a new approach in the field of limnology. Through interpretation of the existing information (on lakes, rivers and reservoirs), it promotes predictive models that can provide diversified scenarios of impacts on aquatic systems and the response of components: biota and biogeochemical cycles, for example. This prediction has great theoretical and applied value.

Typical issues in predictive limnology, according to Håkanson and Peters (1995), are the quality of empirical data (sampling, representative accuracy of data, and compatibility) and the capacity to establish structural and dynamic hierarchies of structural and dynamic factors, which constitute the main forcing functions acting on the system, whose alteration or stability should indicate the responses of the biological community or the abiotic factors. Predictive limnology can be applied, for example, to studies on the impacts of use of drainage basins in lakes, reservoirs or rivers; to predict cyanobacterial blooms between periods of turbulence and stratification, and in the response of organisms to the effects of heavy metals, pesticides and herbicides. It can also be applied to studies on the responses of lakes, reservoirs, rivers and wetland areas to global changes, such as the behaviour of organisms under continuous stress.

20.6 MASS BALANCE

Another approach traditionally used in limnology is that of mass balance, which considers the lake as a 'reaction tank' (Håkanson and Peters, 1995) that can undergo complete mixing during a given period of time, or can remain completely stratified for long periods. Thus, for any substance or element that enters a lake or reservoir,

Long-term ecological research

Long-term data are essential for understanding environmental changes and future management. Historically, many countries have had difficulty maintaining long-term programmes,
due to the inconsistency of available funding. Long-term ecological research projects enable
the establishment of databases on processes and phenomena that occur over extended periods of time. Barbosa and Padisák (2004) defined long-term periods for limnological studies
as five years or more, and studies lasting five to ten years as 'inter-annual variations'.

One problem is the challenge of maintaining the appropriate methods for long-
term studies throughout the project period. The methodology can present changes, thus
causing difficulties in comparing results. Barbosa and Padisák (2004) list 25 lakes and
reservoirs in 13 states where they conducted long-term research on phytoplankton.

In Brazil, long-term studies on aquatic and terrestrial ecosystems were initiated in 1998
by the CNP along with other institutions (CAPES and FINEP) as partners. Today there are
14 sites where long-term studies are being conducted on aquatic and terrestrial ecosystems.

The purpose of long-term research on aquatic ecosystems is to monitor changes
in ecosystems over time to generate databases, scientifically understand alterations in
ecosystems and processes in communities, the physical and chemical factors related to
global changes, and compare the structure and function of aquatic ecosystems over time.

the rates of influx, outflow, and sedimentation can be calculated, according to the
following formula:

$$Cdc/dt = Q \cdot C_{in} - Q \cdot C_s - KT \cdot vC$$

where:
Cdc/dt – rate of change with time t in concentration C of substance or element in
the lake
$Q \cdot C_{in}$ – influx mass of substance or element Q=inflow rate and C=concentration
$Q \cdot C_s$ – outflow mass of the substance or element C=concentration
$KT \cdot vC$ – sedimentation rate as mass flux
K – sedimentation rate (depth interval/time)
v – velocity of sedimentation
V – volume of lake
T – retention time, which is volume/discharge=V/Q

Mass balance is a widely used approach in controlling eutrophication and in the
technologies for restoration of lakes and reservoirs. It includes a series of components:

▶   point sources of substances and elements;
▶   non-point sources of substances and elements;
▶   sediment-water interactions;
▶   retention time;
▶   accumulation of substances or elements in the hypolimnion (when stratification
    occurs);
▶   rate of internal nutrient recycling.

20.7 TECHNOLOGIES OF MONITORING LAKES, RIVERS AND RESERVOIRS

Monitoring is an important step in assessment of the functioning of inland aquatic ecosystems, coastal waters, and oceanic waters. It provides an important support for future research, since it helps detect problems such as point source contamination and pollution, biological alterations (in the plankton, benthos or nekton) that may occur due to impacts from alterations in the drainage basins, and, if performed continuously for many years, it provides key information about global impacts on lakes, reservoirs, rivers, coastal waters and wetland areas.

This monitoring has two main components: **guidance monitoring**, which consists of the gathering of information on a large scale to evaluate the 'state of the system', and the **systematic monitoring** of fixed points over long periods, which provides a significant amount of basic data and enables interpretations based on correlations with forcing functions, such as the effect of winds, precipitation, solar radiation, and the impact of human activities.

If several different aquatic ecosystems are all monitored, the ecosystems and their operating mechanisms and structures can be effectively compared (Straškraba, 1993). Figure 20.6 shows how monitoring water quality comparatively can offer the conditions for selecting appropriate management strategies.

The following considerations are relevant in the monitoring process:

▶ Selection of the necessary data and information, which should be related to the research and assessment objectives.
▶ The measures of each variable must include levels of **sensitivity, detectability,** and **accuracy.**
▶ The cost/benefit analysis of monitoring should be examined. For example, a few variables can be placed at many sampling sites, or the number of variables can be increased at a few strategically selected sampling points.
▶ The level of information provided by the samples and monitoring must be considered; this level depends on the rigour of the selection of the best methods for sampling and evaluation.

In the monitoring process, the following considerations still need proper assessment when designing a monitoring project:

▶ How quickly can data be obtained?
▶ Low operating costs of monitoring.
▶ Maximum coverage for incorporation of all critical and problem areas.
▶ **Minimum error** in sampling.
▶ Absence of pre-determined ideas.
▶ Identification of the users of the information.

In terms of the problem of the costs of monitoring and obtaining information, Figure 20.7 presents a schematic illustration of the problem.

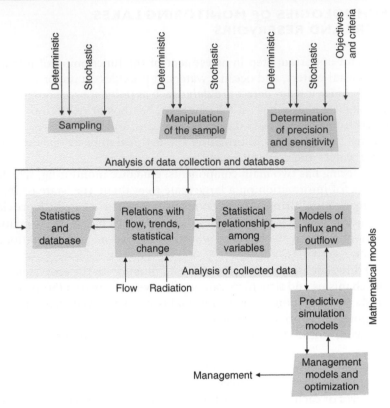

Figure 20.6 Systemic approach to determine water quality, leading to the selection of appropriate management strategies.
Source: Straškrába and Tundisi (2000).

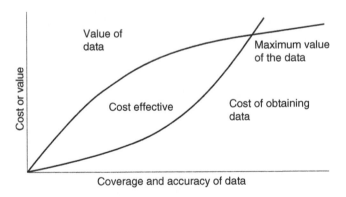

Figure 20.7 Relationship between the cost of monitoring, coverage and accuracy of information.
Source: Biswas (1990).

20.7.1 Monitoring: variables and their assessment

Chart 20.1 shows the range of determinants for systematics and orientation, for the purpose of aquatic life, water supply, recreation and irrigation.

Chart 20.1 Extent of the systematic and orientation measurements for aquatic life, water supply, recreation and irrigation

Rate of inflow and outflow of water ($m^3 \cdot s^{-1}$) E S	Type of measurement	AL	PW	RH	I
Water temperature	O and M	X	X	X	X
Dissolved oxygen	O and M	X	X	X	X
pH	O and M	X	X	X	X
Electrical conductivity	O and M	X	X	X	X
Suspended solids	O and M	X	X	X	X
Turbidity	O and M	X	X	X	X
Transparency	O and M	X	X	X	X
Chlorophyll-*a*	O and M	X	X	X	X
Phytoplankton	O and M	X	X	X	–
Zooplankton	O and M	–	X	X	–
Fish stocks	M	X	–	–	–
Aquatic macrophytes	O and M	X	X	X	–
Nitrate	M	X	X	X	–
Nitrite	M	X	X	X	–
Ammonia	O and M	X	X	X	X
Total N	O and M	X	X	X	X
Dissolved inorganic phosphate	M	X	–	–	–
Organic phosphate	M	X	X	X	–
Total phosphate	O and M	X	X	X	–
Biochemical oxygen demand	O and M	X	X	X	X
Chemical oxygen demand	M	–	–	–	–
Total organic carbon	M	X	X	X	–
Dissolved organic carbon	M	X	X	–	–
Particulate organic carbon	O and M	X	X	–	–
Iron	O and M	X	X	X	–
Magnesium	M	X	X	X	–
Chloride	M	X	X	–	–
Sulphate	M	X	X	–	–
Sodium	M	X	X	–	–
Potassium	M	X	–	–	–
Calcium	M	X	–	–	–
Magnesium	M	X	–	–	–
Fluoride	M	X	X	–	–
Heavy metals	O and M	X	X	X	X
Organic solvents	M	X	X	X	X
Iron	O and M	X	X	X	X
Pesticides	O and M	X	X	X	X
Oils and hydrocarbons	M	X	X	X	X
Colour and odour	O and M	X	X	X	–
Microbiological indicators		X	X	X	–
Faecal coliforms	O and M	X	X	X	–
Total coliforms	O and M	X	X	X	–
Pathogens	O and M	X	X	X	X

AL – Aquatic life; PW – Public water supply; RH – Recreation and health; I – Irrigation; E – Inflow into reservoir or lake; S – Outflow from reservoir or lake; O – Monitoring, orientation; M – Monitoring, systematic.
Source: Modified from Chapman *et al.* (1992); Straškraba *et al.*; Straškraba and Tundisi (2000).

The relationships of the variables to determine the water quality and evaluation of the aquatic ecosystem depend on the objectives of the monitoring – the initial questions formulated by managers and administrators. In this case, researchers, ecologists, and limnologists play an essential role in the evaluation and choice of parameters for determination.

Monitoring for guidance uses broader horizontal-spatial data and more limited samples in time. Systematic monitoring uses more limited horizontal-spatial data and detailed vertical sampling, in strategic locations in the aquatic ecosystem.

Other elements that can be included in the monitoring are: arsenic, selenium and boron, depending on the condition of the aquatic ecosystem, the proximity of chemical industries, or possible contamination detected.

20.8   MONITORING AND PREDICTIVE LIMNOLOGY

Monitoring is crucial for the establishment of predictive limnology programmes. This should enable the researcher, the water-resource manager and the administrator to have means and methods to anticipate critical situations and design scenarios that provide appropriate management conditions. *Integrated ecosystemic predictive* management must, without a doubt, be based on the limnologists' capacity of prediction and anticipation. Chart 20.2 presents comparison of the approaches used to assess the water quality with scientifically based monitoring (using biological information, with advantages and disadvantages).

Monitoring of rivers in South America using diatoms was presented by Lobo *et al.* (2004). This method was used in Brazil and Argentina to assess organic pollution and eutrophication.

20.9   INTERPRETATION OF RESULTS IN LIMNOLOGY

The range of information obtained from limnological studies need to be subjected to analyses that make it possible to improve the capacity to interpret physical, chemical and biological phenomena. Valentin (2000) published an introduction to multivariate analysis of ecological data (entitled 'Numerical Ecology'), in which he presented many technologies for treatment of results and temporal and spatial structural patterns.

The ordering of information begins, in fact, with the initial plan of approach and sampling to obtain the information. After obtaining results, statistical analyses and studies comparing samples, correlation coefficients and multiple regressions, analysis of clusters and sorting methods are techniques commonly used in ecological and limnological studies, according to Valentin (2000).

The interpretation of results also depends on the number of samples collected and their degree of representative accuracy. In horizontally homogeneous but vertically heterogeneous lakes, the number of horizontal samples can be much smaller than the number of samples on the vertical axis. In rivers, which are generally horizontally heterogeneous systems, it is necessary to take a large number of samples on the horizontal axis that can represent this heterogeneity. A vertical profile in a deeper lake, reservoir or river can be described simply in a summarized form or, if worked further,

Chart 20.2 Comparative critical analysis of various ecological methods for biological quality of the water

Ecological methods

	1. Indicator species	2. Studies of communities	3. Microbiological methods	4. Physiological and biochemical methods	5. Bioassays and toxicity, tests invertebrates and fish	6. Chemical analysis of the biota	7. Histological and morphological studies
Principal organisms used	Plant and algae invertebrates	Invertebrates	Bacteria	Invertebrates, algae and fish	Invertebrates and fish	Fish, plants and molluscs	Fish and invertebrates
Principal evaluations applied	Basic surveys; impact surveys	Impact surveys; monitoring trends	Surveys and evaluations; survey impacts	Survey of impacts; methods of preservation	Operational surveys; prevention of impacts; monitoring	Impact survey; monitoring	Survey of impacts; monitoring; evaluation of impacts
Source or effects of pollution	Pollution from organic matter; enrichment from nutrients; acidification	Pollution from organic matter or toxic waste; enrichment from nutrients	Risks to human health; pollution from organic material	Pollution from organic matter; enrichment of nutrients; toxic substances	Toxic waste; pollution from pesticides or organic matter	Toxic waste; pollution pesticides; human health risks (toxic contaminants)	Toxic residues; pollution from organic material or pesticides
Advantages	Simple to run; inexpensive; low-cost equipment	Simple to run; low cost; minimum biological knowledge needed	Relevant to human health; simple to run; relatively low-cost equipment	Rapid responses; relatively low cost; allows continuous monitoring	Usually very sensitive; rapid results; low- or high-cost options	Relevant to human health; requires less advanced equipment than that used to analyse water samples	Some methods are sensitive; simple to complex methods; low- or high-cost options
Disadvantages	Localized use; taxonomic knowledge; subject susceptible to changes in environment	The relevance of some methods does not always apply to all aquatic; ecosystems long historical series needed to monitor long-term changes; susceptible to natural environmental changes	Easily transported organisms can give false positive results far away from sources	Some methods require specialized knowledge and techniques	Laboratory tests do not always indicate situations in the field	Scientific equipment and specialized personnel needed	Specialized knowledge and personnel needed

Source: Chapman et al. (1992).

Monitoring in real time

One of the most recent trends in advanced monitoring of aquatic ecosystems is real-time monitoring. Such monitoring involves the following technology:

- Use of high-quality sensors for physical, chemical and biological measurements in the water.
- Databases.
- Data transmission via telephone, satellite or Internet.
- Coupling of water-quality sensors with climatologic measurements.

A typical real-time monitoring station produces data on the vertical profiles of the water column at certain time intervals, coupled with continuous measurements of climatic parameters – typically, solar radiation, air temperature, wind (strength and direction), relative humidity, and precipitation.

Figure 20.8 shows the advanced real-time monitoring station in the HPP Carlos Botelho reservoir (Lobo/Broa), with capacity for continuous transmission of meteorological data and execution of vertical profiles in the reservoir (12 m deep). The vertical profiles are conducted every half hour, so that all oscillations in the aquatic system and responses to climatic processes are recorded and sent in real (present) time.

This technology will enable the assessment of limnological processes along with climatologic processes; surface thermal heating and turbulence promoted by the wind; the impact of cold fronts on lakes and reservoirs; the vertical distribution of physical, chemical and biological parameters, such as thermal stratification or circulation; the vertical distribution of dissolved oxygen and conductivity; the vertical distribution of chlorophyll and redox potential. This technology also enables the detection of unusual phenomena, the effect of pulses on the vertical circulation of systems and the interactions of climate and limnology.

A more synthetic version of this technology involves the application of real-time monitoring for the evaluation of the physical, chemical and biological characteristics of the water quality of rivers. A network of monitoring stations in water basins, based on the tributaries of the main rivers, allows the integration of data on the concentration of nutrients, pH, dissolved oxygen, conductivity, turbidity and water temperature with the flow, and thus, the measurement of the load that reaches these tributaries and rivers as a function of the precipitation, drainage, impacts of erosion and increase in the BOD and chemical substances.

Real-time monitoring technology consistently advances the management of drainage basins, allowing prediction of and rapid responses to natural phenomena and the impacts of human activities, such as eutrophication, thermal pollution and discharges of toxic substances and suspended solids. Figure 20.9 lists the results obtained when a series of measurements were taken at the UHE Luis Eduardo Magalhães reservoir (Lajeado/Tocantins) (Tundisi et al., 2004).

can provide a broader range of information. Tundisi and Overbeck (2000) presented a range of information that can be obtained with a vertical profile in the deepest part of a lake or reservoir, from which it is possible to perform further analysis of theoretical and applied importance.

Figure 20.8 Advanced monitoring station in real time. SMATER® model (see also color plate section, plate 29).

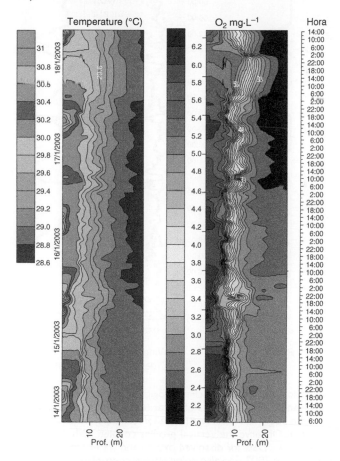

Figure 20.9 The chronological evolution of climatologic data (continued).

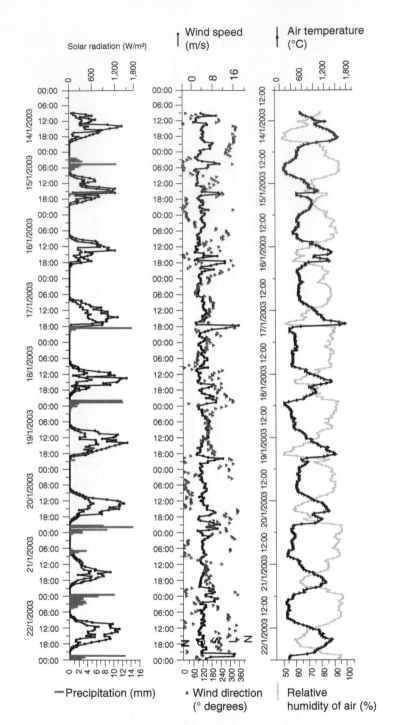

Figure 20.9 The chronological evolution of climatologic data – air temperature, wind speed and solar radiation – and limnological data of water temperature and dissolved oxygen in the UHE Luis Eduardo Magalhães reservoir (Lajeado/Tocantins), obtained with the real-time monitoring station. Observe the vertical profiles coupled with the effects of radiation, wind and temperature. Data for dissolved oxygen and water temperature are presented from 14 to 18 January 2003 (see also color plate section, plate 30).
Source: International Ecology Institute, FAPESP/Investco/FINEP (2001).

20.9.1 Basic profile data of the deeper vertical part of the lake or dam

- ▶ water temperature;
- ▶ dissolved oxygen;
- ▶ penetration of light (climate of underwater radiation);
- ▶ electrical conductivity;
- ▶ pH;
- ▶ vertical distribution of phytoplankton;
- ▶ vertical distribution of zooplankton;
- ▶ primary production by phytoplankton;
- ▶ vertical distribution of nutrients (nitrate, nitrite, ammonium, silicate, total dissolved phosphorus, total phosphorus and total nitrogen);
- ▶ vertical distribution of ions in solution (including heavy metals).

These basic data are strongly influenced by climatic forcing functions, including:

- ▶ solar radiation;
- ▶ wind (strength and direction);
- ▶ air temperature;
- ▶ precipitation.

Thus, with the determinations of the main physical, chemical and biological variables, the installation and use of a climatologic station is necessary to provide essential information on the forcing functions.

20.9.2 Analysis of sediments

- ▶ Sampling of sediment for analysis of chemical composition, including heavy metals.
- ▶ Collection of samples of interstitial water.

20.9.3 Information on drainage basins

- ▶ Use of soil
- ▶ Flow of rivers
- ▶ Nutrient load in rivers
- ▶ Type of drainage system: the extent and characteristics of drainage networks
- ▶ Relationship between area of lake and area of drainage basin.

20.9.4 Morphometric data of the lake or reservoir

- ▶ Maximum depth (Z_{max})
- ▶ Average depth (Z)
- ▶ Minimum depth (Z_{min})
- ▶ Morphology of the lake basin
- ▶ Shoreline development index

20.9.5 Information derived from measurements

▶ Z_{eu}/Z_{max} ratios (euphotic zone/max depth);
▶ Z_{eu}/Z_{af} ratios (euphotic zone/aphotic zone);
▶ Z_{eu}/Z_{mix} ratios (euphotic zone/mixing zone)
▶ Trophic state of the aquatic system;
▶ Size of the internal load of the lake (through analysis of the sediments);
▶ Influence of drainage basin on the aquatic system;
▶ Frequency of cyanobacterial blooms and potential risks;
▶ Chemical characteristics of the lakes;
▶ Degree of thermal stability or instability of the system;
▶ Degree of contamination of the water and the sediment.

20.10 HUMAN RESOURCE TRAINING IN LIMNOLOGY

Limnology is an integrative and inclusive science. The goal is to understand the scientific variations of the spatial and temporal variations and the physical, chemical and biological variations of inland aquatic systems, the interactions of these components and their variability in space and time. The **components of aquatic systems** are integrated into an interactive system in which processes extend beyond the limits of the aquatic ecosystem and are dependent on the drainage basins and their characteristics (Likens, 1984; Wetzel, 1990). Inland aquatic systems present biogeochemical cycles that depend on the biological components and drainage basins. Physical, chemical, biological, hydrological, geological and biogeochemical processes are examined and studied in large-scale spatial and temporal scales. Experiments are conducted with organisms in the laboratory, in mesocosms, and in lakes or reservoirs that serve as large experimental laboratories. However, the application of a holistic interdisciplinary approach leads to some fundamental questions: traditionally there has been a large number of studies and projects on the physics and chemistry in lakes, when biotic considerations are included, they are focused on lower trophic levels.

For many decades, ichthyological research was not treated as part of limnological research. Only recently, in the past 20 years, has research on aquatic microbiology and the microbial loop produced more evidence and gained more space. These disparities in the study of limnology and the different approaches has made it difficult to establish an interdisciplinary group for human resource training in limnology, but it is important to continue working in this direction, to promote innovation in the training of researchers with this systemic and interdisciplinary vision.

The economy of any country or region depends on water resources (quantity and quality). Therefore, regional knowledge of lakes, rivers, reservoirs, swamps and wetland areas is the first step toward proper management of these resources and their use. Limnology is an eminently interdisciplinary science that usually exceeds the boundaries of a traditional university department, and therefore it is with these basic premises that human resources should be trained: **interdisciplinary and systemic training** incorporating graduate courses in: geomorphology, hydrology, organic and inorganic chemistry of water, aquatic biology (all biological components), biogeo-chemistry and biostatistics. The task of training human resources should include theo-retical and laboratory work, field trips and collection in the field, technical visits to regions of impacts or systems and water treatment plants. As complementary training for graduate courses use and practice of the **Geographic Information System** (GIS) and hydrodynamics are recommended.

In the case of post-graduate courses, the following curriculum should be considered:

▸ Mathematics and statistics: calculus, differential equations, statistical analysis;
▸ Physics: thermodynamics, energy, solar radiation;
▸ Organic and inorganic chemistry: water chemistry;
▸ Geology and geomorphology.
▸ Climatology and hydrology;
▸ Aquatic biology (including all components of aquatic systems, from bacteria to fish): taxonomy of organisms;
▸ Aquatic biogeochemical cycles;
▸ Biostatistics;
▸ Physical limnology (hydrodynamics, circulation);
▸ Microbiology;
▸ Water analysis (practices);
▸ Theoretical ecology;
▸ Environmental law;
▸ Public communication.

Post-graduate studies in limnology need to include fieldwork, technical visits and experimental work. A complete case study (for example, a comprehensive study of a drainage basin) should be considered as a pilot program during the course. The Master's and PhD theses, of course, form part of the curriculum and the work of the advisors, but it is essential that during this thesis work the limnologist not lose the systemic and integrated vision, which will help enormously in development of limnology.

20.11 LIMNOLOGY: THEORY AND PRACTICE

Limnology is the science that points to the future of humanity. In fact, conservation and restoration of aquatic ecosystems, their operating mechanisms and biota are criti-cal to the survival of the human species and the biodiversity of the planet. Greater scientific information is vital for the promotion of mechanisms and technologies to restore and conserve lakes, rivers, swamps, reservoirs and other aquatic systems in the interior of continents. As recommended by Margalef (1980), the point of departure must always consist of observation of well-defined situations in the field and control-led experiments, which however unsatisfactory (given the inherent limitations to field work and experimentation), provide information that can be used to expand a refer-ence system and establish principles and unified indicators that integrate biological and evolutionary processes, physics (thermodynamics), and biogeochemical spatial/temporal scales.

From this point of view, limnology in Brazil has exceptional conditions to offer the world community a new and innovative vision of *systems*, *components* and *interactions*, given the superlative characteristics of biological, physical and functional diversity of inland aquatic systems in the country. It should also be considered that the presence of '**humanized systems**' such as large urban areas, reservoirs, agricultural areas, along with natural systems of large spatial scales – all these represent excellent opportunities for research and application, transforming the academic experience into a range of applications in real case studies, and not just exercises limited to experimental systems. The 'experimental systems' in Brazil, which correspond to urbanized areas and their suburbs and roads, represent a large-scale set of processes involving at least 150 million people and their reliance on inland water systems for supply, recreation, transportation, biomass production, energy and leisure. On the other hand, large wetland and swampy areas, floodplain systems in the interior of the country, present many opportunities to test and try new mechanisms of distribution, evolution, interactions of species and operations of physical, biological, chemical and biogeochemical processes.

20.12   THE FUTURE OF LIMNOLOGY: SEARCH FOR BASIC APPLICATION

As shown here, lakes, rivers reservoirs, wetland areas and other inland aquatic systems have varied responses that manifest themselves in their different physical chemical and biological components. Understanding the individuality of each aquatic system is essential. Limnology progressed considerably in the 20th century from a science that attempted to understand, with intensive study, a large number of components, to an inclusive science that seeks to understand the aquatic ecosystem as a functional set of components, some of which have very few obvious variables (Håkanson and Peters, 1995).

This was the approach taken by Vollenweider (1968, 1976, 1990) for the study to simplify, to a certain degree, the understanding of the solution to the problem of eutrophication. Studies on individual ecosystems, their responses to external forcing functions and the interactions of the components are still needed, and need to extend well into the 21st century, because these studies will help provide the wide body of scientific information necessary for the development of theories and to simplify concepts on aquatic ecosystems, using comparative data.

20.12.1   Descriptive limnology

Descriptive limnology, in a country as large as Brazil, and in many regions around the world, will continue as a priority activity and as part of the assessment of operating mechanisms. Descriptive limnology is also useful in the establishment of units and programs for conservation and restoration of ecosystems, be it rivers, lakes, wetland areas or temporary lakes. These limnological studies will help form the wide base of information needed to better understand the major operating mechanisms at the regional or continental level as well as **climatic-hydrological-limnological** interactions (Tundisi and Barbosa, 1995).

From descriptive limnology, a set of comparative studies should emerge, such as those in the lake systems of England (Macan, 1970), the reservoirs of Spain (Margalef, 1976) and the reservoirs in the state of São Paulo (Tundisi, 1981).

There is still a large intellectual gap that needs to be filled with more comparative studies, which can yield basic information over a short period of time, considering the technological advances in sampling (real-time monitoring, reliable sensors, satellite imagery and advanced geo-processing programs).

It has been clearly shown earlier in this chapter that the interactions of field and laboratory techniques, coupled with advanced analysis technologies with the latest-generation instrumentation, are leading to important advances in the description of processes in lakes.

20.12.2   Process studies

In addition to this descriptive limnology, organized and performed with new techniques, the study of *processes* needs to advance in terms of *organisms, populations, communities* and *ecosystems*. Summaries of the last ten years of research in Brazil show that the dynamic nature of aquatic ecosystems, their **transient phases** and continual organization and reorganization have been understood and incorporated in studies and research (Tundisi and Straškraba, 1999, Junk *et al.*, 2000; Junk, 2006; Bozelli *et al.*, 2000; Agostinho *et al.*, 2004). Studies of communities and their processes include the responses of communities to alterations in environmental factors (Matsumura Tundisi and Tundisi, 2003; Rietzler *et al.*, 2002), predator-prey interactions (Arcifa *et al.*, 1993, 1997) as well as the effects on benthic communities (Callisto *et al.*, 1998b, 2000).

One of the major challenges for the future of limnology is to collect information on energy flow in lakes, rivers, reservoirs and wetland areas. Advances are needed in this area, through conducting field research, experimental studies and design of models for coupling energy flow to the effects of external forcing functions (Margalef, 1968; Angelini Petrere, 1996) and alterations in biodiversity.

There is, therefore, an immediate need to expand these advances and research processes. Within this perspective, research with biological indicators becomes extremely relevant given the wide biodiversity in Brazil and the variety of situations and processes in ecosystems and communities. Limnology in Brazil can contribute significantly to global limnology because of the extent and variety of the country's inland aquatic systems, the uniqueness of the physical, chemical and biological factors (extensive floodplains, isolated lakes in the Medio Rio Doce, significant sets of cascading reservoirs in the major river drainage basins); extensive wetland areas, such as the Pantanal; and high average annual temperatures. In addition, the C:N:P ratios, the contributions from terrestrial systems into the lake systems and the responses to eutrophication can be significantly different than those usually found in temperate-zone systems. Salas and Martino (1991) addressed this possibility by describing new mechanisms for response to eutrophication in tropical lakes, especially the response of lakes and phytoplanktonic communities to phosphorus loads.

The flora and fauna in Brazil's inland aquatic ecosystems are also characteristic, and in some groups, there is a high degree of endemicity. Developmental mechanisms in undisturbed areas are dynamic. For example, Margalef (1983, 2002 and personal communication, 2003) considered large internal deltas of floodplain lakes and permanent lakes and the dynamic changes resulting from fluctuations in level, transport of sediment, and sedimentation as 'active centres of development' where there is permanent genetic flow and interactions of subpopulations, promoting speciation.

It is important to broaden the spatial-temporal limnological knowledge of this large group of ecosystems and their hydrological, physical, chemical and biological components. Salo et al. (1986) contributed to knowledge on the dynamic system in the Amazon. An important contribution that can be promoted is the physiology of these tropical and subtropical aquatic organisms in the spatial and temporal scales with varying dynamic characteristics and unique characteristics of hydrodynamic flow (Val, 1991, Val et al., 1993; Cáceres and Vieira, 1988).

20.12.3   Predictive limnology

These problems of basic research should no doubt provide the foundations for the development of predictive limnology, an important component in management and restoration processes of inland aquatic ecosystems. Predictive limnology can propose alternatives for conservation and restoration, through the implementation of models and the analysis of cost/benefit scenarios, which will support *flexible*, *adaptive*, *accessible* and *practical* **support and decision-making systems**, according to Håkanson and Peters (1995). The integration of various approaches in science and technology (such as limnology, engineering, mathematics, computer science, biology, chemistry and physics) should be considered in predictive ecology, as well as the ecosystem approach (Tundisi et al., 1995).

A study of the basic processes in terms of predictive limnology is urgently needed in Brazil to advance **hydrodynamic knowledge** and calibrate and validate hydrodynamic models, especially for large rivers, internal deltas and large and small reservoirs located at various latitudes, which are subject to multiple uses, especially in urban areas, where they are used intensively for water supply. The coupling of hydrodynamic models with hydrological cycles and hydrological and biogeochemical models is an important method to be developed further in order to understand the latitudinal distribution of biodiversity in aquatic systems and its responses to impacts.

The identification of the drainage basin as the unit of study and management and the projection of future impacts and responses in drainage basins constitute other fundamental processes in the development of predictive limnology. There is a need for more experimental studies in pilot basins; description of responses to loads of nutrients, heavy metals and toxic chemicals, as well as advanced studies on the effects of atmospheric deposition in aquatic systems (Lara et al., 2001; Martinelli et al., 1999; Moraes et al., 1998).

Figure 20.10 summarizes the different steps in the integration of scientific knowledge and the proposed systems management based on these steps.

Predictive and adaptive integrated pest management depends on reliable and consistent limnological bases. Example of systems management in Brazil include the UHE Carlos Botelho reservoir (Lobo/Broa) (Tundisi et al., 2003); Paranapanema River

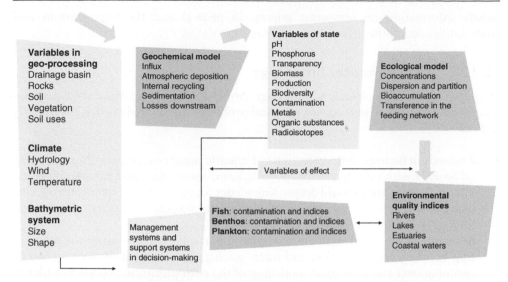

Figure 20.10   Conceptual model of evaluation of the responses of aquatic systems to the impacts in drainage basins, as a base for the implementation of environmental quality indices of ecosystem integrity and biota, as well as decision-making support systems. With the support of models and responses, scenarios can be designed that promote the necessary actions for conservation and restoration of inland ecosystems, estuaries and coastal systems.
Source: Based on Håkanson and Peters (1995); Straškrába and Tundisi (2000).

(Nogueira *et al.*, 2004); studies and management of the Paraná river (Agostinho *et al.*, 2004), in the lakes in Parque Florestal Rio Doce, in rivers of eastern Minas Gerais (Barbosa, 1994) and reservoirs on the São Francisco River (Godinho and Godinho, 2003), fishing management (Petrere, 1996, Freitas *et al.*, 2002), management and restoration of Lake Batata (Bozelli *et al.*, 2000) and the results of Bicudo *et al.* (2002). All these examples show how the limnological base and its biological, physical and chemicals components have contributed to the development of management models, with excellent practical results or potential optimal outcomes. More examples of limnological research and its application should be stimulated in aquatic ecosystems in various regions of Brazil.

20.12.4   Ecological and mathematical models as management tools

Ecological and mathematical models need to be implemented, but conceptual models should be corroborated, calibrated and validated with data from the field and laboratory. Limnology practiced in Brazil can be very effective worldwide if it can contribute knowledge of the rates of various processes at the biological, chemical and physical levels, since these rates are all based on the international literature, which contains mostly information on temperate regions in general and the organizations and communities found there (Jørgensen, 1981, 1996).

20.12.5   Experimental limnology

Two special areas of research in limnology should be encourage, since their development would contribute to resolving practical problems with global impact, particularly in Brazil:

▶   Research in biology and interactions of aquatic organisms, especially *experimental biology* with phytoplankton, zooplankton, zoobenthos and fish (Arcifa *et al.*, 1995). These studies should deepen knowledge about life cycles, inter/intra-specific relationships, tolerance levels of species, and biological indicators of stress and responses to toxic substances.
▶   Research on the problem of *eutrophication*, particularly focusing on the relationship between eutrophication and *water quality*, development of toxic strains of *cyanobacteria* and ecological modelling of the eutrophication process in order to evaluate and control it (Azevedo, 1998; Azevedo *et al.*, 1994).

For greater understanding of these areas of basic biology and eutrophication, research in aquatic microbiology and its various interactions should be encouraged. Also with regard to eutrophication, the challenge of implementing more consistent indices for tropical and subtropical regions must be faced, as well as clarifying the role of concentrations of phosphorus in accelerating eutrophication and in the doubling time of eutrophication in inland aquatic ecosystems, without sewage treatment measures and **agricultural drainage** being adopted. The existing scientific information (Straškraba and Tundisi, 2000) shows the role of phosphorus in eutrophication, but the minimum concentrations needed to trigger and accelerate the process need to be determined experimentally and with intensive comparative field studies, including the contribution from diffuse sources and the responses from the communities of phytoplankton, aquatic macrophytes, and periphyton.

20.12.6   Limnology and global changes

In the context of the advanced role of limnology in managing global processes, it is important to keep in mind the importance of limnological studies in understanding the impacts of global changes and their effects on ecosystems and communities. For example, it is solid scientific knowledge that alterations in the Amazon can provoke global changes in terrestrial and inland aquatic systems around the world, especially on the South American continent. What will happen with the responses of floodplain lakes, large reservoirs, rivers and small streams of the Amazon and the Cerrado? What is the impact of these changes on cold fronts in aquatic ecosystems of southeast Brazil? Studies of the problems and local versus global perspectives should be considered as a basic proposal for permanent action (Kumagai and Vincent, 2003).

There is also another consideration, which is the impact of **global changes** on the structure and functioning of communities, the dispersal and distribution of invasive species, the structure of the food web, as well as on the spread and incidence of water-borne diseases. These two topics – invasive species and water-borne diseases – require special studies and major investment in basic research.

Such questions must be answered by researchers, engineers and skilled technicians, with a systemic view of the problems and sound theoretical and practical training in actual case studies. Integration between research and management in *water resources* is essential, in order to prepare managers with this systemic and integrated vision, with sophisticated perception of limnological, ecological, economic and social processes (Tundisi and Matsumura Tundisi, 2003).

One big challenge of global limnology in the 21st century will be the incorporation and integration of *biogeophysical*, social and economic processes as a basis for anticipating events, and upon anticipating them, promotion of sound sustainable alternatives. Thus, the long-term studies on representative ecosystems undertaken in Brazil starting in 1998 play a key role in constructing a systemic vision. In this book, especially Chapter 16, the authors have emphasized that long-term, intensive and comparative studies in many countries are the basic structure for advancing limnology, its application, and resolution of relevant problems in conservation and restoration of aquatic ecosystems (Barbosa, 1994, 1995; Barbosa *et al.*, 1995).

20.12.7 Integration of limnology with other science and technologies

In the not-too-distant future, even in the first half of this century, the trajectories of limnology and engineering will tend to merge, due to the fact that the needs for basic research must supply the technology for management of aquatic ecosystems, especially those for immediate use, promoting services relevant to humans: water supply, production of biomass, transportation and recreation. It is undeniable that water management, treatment systems and **protection of springs** should have their costs increased as the deterioration of surface and groundwater sources accelerates, and the expansion of 'humanized systems' (Margalef, 2002) eliminates the natural processes of conservation and restoration.

Therefore, limnology will promote the scientific and conceptual bases necessary for the interventions for management. This movement, which has already begun in many countries, is expected to accelerate in Brazil and provide new opportunities for management and innovation, given the particularities of the inland aquatic ecosystems in Brazil and the expansion of the economy and urbanization (Tundisi, 1990, 2004).

The integration of limnology with oceanography is another significant challenge, which in Brazil has a geographic point of integration in the estuaries and coastal waters, where exploitation could be expanded, especially with regard to aquaculture, recreation and fishing. Joint technologies of exploitation and research will serve to promote better management capacity and the understanding of the continuum represented by inland ecosystems to estuaries, coastal lagoons and coastal waters. This *continuum* concept includes dynamic spatial/temporal processes that should be known in their natural bases, as well as the impacts and responses.

The demand for 'unifying principles' in the functioning of ecosystems, particularly inland aquatic ecosystems, should be a constant concern for academic researchers in limnology, which will greatly facilitate application and management (Margalef, 1968, 1974, 1978; Reynolds, 1997). This should be carried out with periodic synthesis and the promotion of databases that enable the interconnection of knowledge and networks. Since Brazil is a country the size of a continent with an immense variety of natural and artificial aquatic systems, the progression of research in these regions will depend on a continuous local and regional process that points to a strategy of knowledge and the necessary steps toward development. Therefore, an essential connection needs to be made between **basic research** and **application in systems** that can be used as an example situation for authorities and administrators, especially for technical and scientific purposes. For example, in two regional systems, Pantanal Mato Grosso and ponds of the Middle Rio Doce, a breakthrough is still needed in limnological research. The next stages of work on these two systems must see research and development expertise in a greater variety of lakes, as well as investigate processes involved in the microbial loop and its role in the decomposition of organic matter and the food network.

20.12.8 International cooperation

The two principal drainage basins in Brazil, the Amazon basin and the Plata basin, are ecosystems that reach into almost all countries of South America. They also have relatively similar social, economic and cultural characteristics. In view of this, international scientific cooperation in limnology and joint studies of all the major rivers, floodplain lakes and internal deltas are essential for the future progress in the region and also for the use and appropriate management of water resources. The high degree of urbanization, insufficient sanitation, deforestation, overuse of contaminated water and water-borne diseases are common problems that must be treated with joint strategies in these countries, with basic and applied scientific research and qualified and competent managerial training.

20.13 FUTURE DEVELOPMENTS

The future of limnology will include the introduction and application of the relatively recent concepts of **ecohydrology and ecotechnologies**, which incorporate theory and practical knowledge of ecosystems into management processes, producing new low-cost alternatives, avoiding the introduction of expensive heavy technology and promoting creative actions for management (Figure 20.11).

The advances in the management of drainage basins and the integration between research and management depend on these new approaches and methods. The conceptual and theoretical base was already prepared, and examples include the works of Reynolds (1997), Tundisi and Straškraba (1999) and Tundisi (2007).

Figure 20.12 summarizes the entire range of variables (chemical, biotic factors, flow regime, sources of energy, and habitat structure) that maintain the integrity of the aquatic ecosystem. These components should be the basis of scientific work in limnology, enabling a systemic approach to the dynamic processes for the purpose of conservation or restoration. Figure 20.12 represents the evolution of knowledge in limnology starting with Rawson (see Figure 1.3).

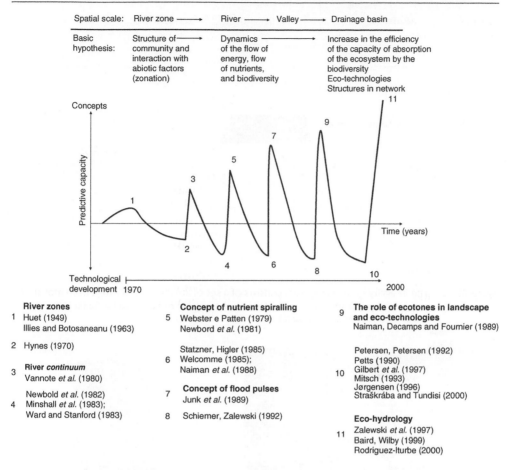

Spatial scale: River zone ⟶   River ⟶ Valley ⟶ Drainage basin

Basic   Structure of ⟶   Dynamics ⟶   Increase in the efficiency
hypothesis: community and   of the flow of   of the capacity of absorption
                interaction with   energy, flow   of the ecosystem by the
                abiotic factors   of nutrients,   biodiversity
                (zonation)   and biodiversity   Eco-technologies
                                                            Structures in network

River zones

1 Huet (1949)
   Illies and Botosaneanu (1963)

2 Hynes (1970)

3 **River *continuum***
   Vannote *et al.* (1980)

4 Newbold *et al.* (1982)
   Minshall *et al.* (1983);
   Ward and Stanford (1983)

Concept of nutrient spiralling

5 Webster e Patten (1979)
   Newbord *et al.* (1981)

   Statzner, Higler (1985)
6 Welcomme (1985);
   Naiman *et al.* (1988)

   Concept of flood pulses
7 Junk *et al.* (1989)

8 Schiemer, Zalewski (1992)

9 **The role of ecotones in landscape
   and eco-technologies**
   Naiman, Decamps and Fournier (1989)

   Petersen, Petersen (1992)
   Petts (1990)
10 Gilbert *et al.* (1997)
   Mitsch (1993)
   Jørgensen (1996)
   Straškrába and Tundisi (2000)

   Eco-hydrology
11 Zalewski *et al.* (1997)
   Baird, Wilby (1999)
   Rodriguez-Iturbe (2000)

Figure 20.11   Evolution of the limnological and ecological body of knowledge and integrated proposals, up to the most recent concept of eco-hydrology, which increases the predictive capacity and anticipation of impacts, promoting the introduction of low-cost eco-technologies that incorporate the scientific operating mechanisms of drainage basins as a basis for the conservation and restoration of aquatic ecosystems.
Source: Modified from Zalewski (2002).

20.14 TOOLS AND TECHNOLOGY

The following photos (Figure 20.13) show a range of traditional and advanced limnological instruments and technology, showing the evolution of the systems to determinations and measurements needed in order to understand the physical, chemical and biological functioning of lakes, rivers, reservoirs, wetland areas, estuaries and coastal lagoons. This set of techniques and technologies with tools used in the field, coupled with sophisticated laboratory analysis and mathematical and statistical analysis techniques, promotes an approach that seeks to embrace the complexity of aquatic ecosystems in space and time.

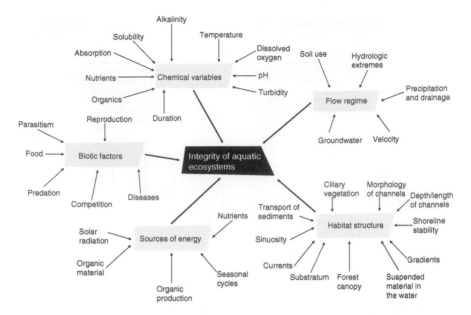

Figure 20.12   The integrity of the ecosystem consists of a set of attributes that range from morphometry and physical and chemical factors to the regime of flow and biotic factors.
Source: Somlydoy *et al.* (2001, 2006).

Figure 20.13   Traditional and advanced equipment and systems for calculating physical, chemical and biological measurements in aquatic ecosystems. (1) Collector of sampling by gravity (UWITEC, Austria), used to quantify gases in sediment; (2) Portable fluorometer to measure chlorophyll in the field; (3) Diffusion chambers (a) miniatures used to measure diffusion of gas through the air-water interface and (b) buoys of the diffusion chambers.

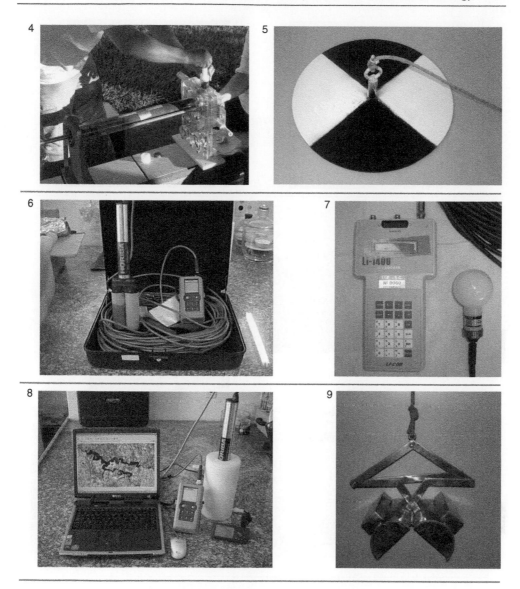

(4) Adams-Niederreiter Gas Sampler; UWITEC, Austria; (5) Secchi disc to measure water transparency; (6) Multi-parametric probe for physical and chemical estimations in water; (7) Underwater radiation meter; (8) Flow cells for continuous use and horizontal profiles in lakes, reservoirs and rivers; (9) Petersen dredge to collect benthic organisms and sediments.

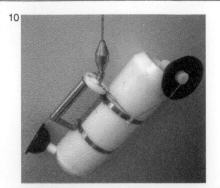

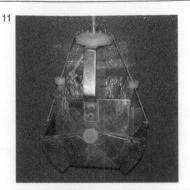

(10) Van Dorn sample bottle to collect water; (11) Birge-Ekman dredge to collect benthic organisms and sediments; (12) Net plankton; (13) Fluoroprobe sensor for selective determination of phytoplankton pigments; (14) Fluoroprobe sensor reader; (15) Multi-parameter probe reader with GPS (see also colour plate section, plate 31).

Photos: Fernando Blanco Nestor F. Mazini.

Annex I
Fish species in the São Francisco river

Superorder Clupeomorpha
    Order Clupeiformes
        Family Engraulidae
            Anchoviella vaillanti (Steindachner, 1908)
Superorder Ostariophysi
    Order Characiformes
        Family Characidae
            Subfamily Tetragonopterinae
                Astyanax bimaculatus lacustris (Reinhardt, 1874)
                Astyanax eigenmanniorum (Cope, 1894)
                Astyanax fasciatus (Cuvier, 1819)
                Astyanax scabripinnis intermedius (Eigenmann, 1908)
                Astyanax scabripinnis rivularis (Lutken, 1874)
                Astyanax taeniatus (Jenyns, 1842)
                Bryconamericus stramineus (Eigenmann, 1908)
                Creatochanes affinis (Gunther, 1864)
                Hasemania nana (Reinhardt, 1874)
                Hemigrammus brevis (Ellis, 1911)
                Hemigrammus marginatus (Ellis, 1911)
                Hemigrammus nanus (Reinhardt, 1874)
                Hyphessobrycon gr. bentosi (Durbin, 1908)
                Hyphessobrycon gracilis (Reinhardt, 1874)
                Hyphessobrycon santae (Eigenmann, 1907)
                Moenkhausia costae (Steindachner, 1907)
                Moenkhausia sanctae-filomenae (Steindachner, 1907)
                Phenacogaster franciscoensis (Eigenmann, 1911)
                Piabina argentea (Reinhardt, 1866)
                Psellogrammus kennedyi (Eigenmann, 1903)
                Tetragonopterus chalceus (Agassiz, 1829)
            Subfamily Acestrorhynchinae
                Acestrorhynchus britskii (Menezes, 1969)
                Acestrorhynchus lacustris (Reinhardt, 1874)
                Oligosarcus jenynsii (Gunther, 1891)
                Oligosarcus meadi (Menezes, 1969)
            Subfamily Cynopotaminae
                Galeocharax gulo (Cope, 1870)
            Subfamily Characinae
                Roeboides francisci (Steindachner, 1908)
                Roeboides xenodon (Reinhardt, 1849)
            Subfamily Stethaprioninae
                Brachychalcinus franciscoensis (Eigenmann, 1929)
            Subfamily Glandulocaudinae
                Hysteronotus megalostomus (Eigenmann, 1911)
            Subfamily Cheirodontinae
                Cheirodon piaba (Lutken, 1874)
                Compsura heterura (Eigenmann, 1917)
                Holoshestes heterodon (Eigenmann, 1915)

Megalamphodus micropterus (Eigenmann, 1915)
Odontostilbe sp.
Subfamily Characidiinae
Characidium fasciatum (Reinhardt, 1866)
Jobertina sp.
Subfamily Triportheinae
Triportheus guentheri (Garman, 1890)
Subfamily Bryconinae
Brycon hilarii (Valenciennes, 1849)
Brycon lundii (Reinhardt, 1874)
Brycon reinhardti (Lutken, 1874)
Subfamily Salmininae
Salminus brasiliensis (Cuvier, 1817)
Salminus hilarii (Valenciennes, 1849)
Subfamily Serrasalminae
Serrasalmus brandtii (Reinhardt, 1874)
Serrasalmus piraya (Cuvier, 1820)
Subfamily Myleinae
Myleus altipinnis (Valenciennes, 1849)
Myleus micans (Reinhardt, 1874)
Family Parodontidae
Apareiodon hasemani (Eigenmann, 1916)
Apareiodon sp. "A"
Apareiodon sp. "B"
Parodon hilarii (Reinhardt, 1866)
Family Hemiodontidae
Hemiodopsis gracilis (Gunther, 1864)
Hemiodopsis sp.
Family Anostomidae
Leporellus cartledgei (Fowler, 1941)
Leporellus vittatus (Valenciennes, 1849)
Leporinus elongatus (Valenciennes, 1849)
Leporinus marggravii (Reinhardt, 1875)
Leporinus melanopleura (Gunther, 1864)
Leporinus piau (Fowler, 1941)
Leporinus reinhardti (Lutken, 1874)
Leporinus taeniatus (Lutken, 1874)
Schizodon knerii (Steindachner, 1875)
Family Curimatidae
Steindachnerina elegans (Steindachner, 1875)
Cyphocharax gilberti (Quoy and Gaimard, 1824)
Curimatella lepidura (Eigenmann and Eigenmann, 1889)
Family Prochilodontidae
Prochilodus affinis (Reinhardt, 1874)
Prochilodus marggravii (Walbaum, 1792)
Prochilodus vimboides (Kner, 1859)
Family Erythrinidae
Hoplias aff. lacerdae (Ribeiro, 1908)
Hoplias aff. malabaricus (Bloch, 1794)
Order Siluriformes
Suborder Gymnotoidei
Family Gymnotidae
Gymnotus carapo (Linnaeus, 1758)
Family Sternopygidae
Eigenmannia virescens (Valenciennes, 1847)
Eigenmannia sp. "A"

Sternopygus macrurus (Block and Schneider, 1801)

Family Hypopomidae

Hypopomus sp.

Family Sternachidae

Apteronotus brasiliensis (Reinhardt, 1852)

Sternachella schotti (Steindachner, 1868)

Suborder Siluroidei

Family Doradidae

Franciscodoras marmoratus (Reinhardt, 1874)

Family Auchenipteridae

Glanidium albescens (Reinhardt, 1874)

Parauchenipterus galeatus (Linnaeus, 1777)

Parauchenipterus leopardinus (Borodin, 1927)

Pseudauchenipterus flavescens (Eigenmann and Eigenmann, 1888)

Pseudauchenipterus nodosus (Bloch, 1794)

Pseudotatia parva (Gunther, 1942)

Family Pimelodidae

Bagropsis reinhardti (Lutken, 1875)

Bergiaria westermanni (Reinhardt, 1874)

Cetopsorhamdia sp. (aff. C. Ihering)

Conorhynchus conirostris (Valenciennes, 1840)

Duopalatinus emarginatus (Valenciennes, 1840)

Heptapterus sp.

Imparfinis microcephalus (Reinhardt, 1875)

Imparfinis minutus (Lutken, 1875)

Lophiosilurus alexandri (Steindachner, 1876)

Microglanis sp.

Pimelodella lateristriga (Muller and Troschel, 1849)

Pimelodella laurenti (Fowler, 1941)

Pimelodella vittata (Kroyer, 1874)

Pimelodella sp.

Pimelodus fur (Reinhardt, 1874)

Pimelodus maculatus (Lacépede, 1803)

Pimelodus sp. (aff. P. blochii)

Pseudopimelodus fowleri (Haseman, 1911)

Pseudopimelodus zungaro (Humboldt, 1833)

Pseudoplatystoma coruscans (Agassiz, 1829)

Rhamdella minuta (Lutken, 1875)

Rhamdia hilarii (Valenciennes, 1840)

Rhamdia quelen (Quoy and Gaimard, 1824)

Family Trichomycteridae

Stegophilus insidiosus (Reinhardt, 1858)

Trichomycterus brasiliensis (Reinhardt, 1873)

Trichomycterus reinhardti (Eigenmann, 1917)

Family Bunocephalidae

Bunocephalus sp. "A"

Bunocephalus sp. "B"

Family Cetopsidae

Pseudocetopsis chalmersi (Norman, 1926)

Family Callichthyidae

Callichthys callichthys (Linnaeus, 1758)

Corydoras aeneus (Gill, 1861)

Corydoras garbei (R.V. Ihering, 1910)

Corydoras multimaculatus (Steindachner, 1907)

Corydoras polystictus (Regan, 1912)

Family Loricariidae
Subfamily Locariinae
Harttia sp.
Loricaria nudiventris (Valenciennes, 1840)
Rhinelephis aspera (Agassiz, 1829)
Rineloricaria lima (Kner, 1854)
Rineloricaria steindachneri (Regan, 1904)
Rineloricaria sp.
Subfamily Hypoptomatinae
Microlepidogaster sp.
Otocinclus sp.
Subfamily Hypostominae
Hypostomus alatus (Castelnau, 1885)
Hypostomus auroguttatus (Natterer and Heckel, 1853)
Hypostomus commersonii (Valenciennes, 1840)
Hypostomus francisci (Lutken, 1873)
Hypostomus garmani (Regan, 1904)
Hypostomus macrops (Eigenmann and Eigenmann, 1888)
Hypostomus cf. *margaritifer* (Regan, 1908)
Hypostomus wuchereri (Gunther, 1864)
Hypostomus sp. "A"
Hypostomus sp. "B"
Hypostomus sp. "C"
Pterygoplichthys etentaculatus (Spix, 1829)
Pterygoplichthys lituratus (Kner, 1854)
Pterygoplichthys multiradiatus (Hancock, 1828)
Superorder Acanthopterygii
Order Ciprinodontiformes
Family Poeciliidae
Poecilia hollandi (Henn, 1916)
Poecilia vivipara (Scheneider, 1801)
Order Perciformes
Family Sciaenidae
Plagioscion auratus (Castelnau, 1855)
Plagioscion squamosissimus (Gill, 1861)
Pachyurus francisci (Cuvier, 1830)
Pachyurus squamipinnis (Agassiz, 1829)
Family Cichlidae
Cichlasoma facetum (Jenyns, 1842)
Cichlasoma sanctifranciscence (Kullander, 1983)
Crenicichla lepidota (Heckel, 1840)
Geophagus brasiliensis (Quoy and Gaimard, 1824)
Order Simbranquiformes
Family Synbranchidae
Synbranchus marmoratus (Block, 1795)

Source: Godinho and Godinho (2003).

Annex 2
Species of catfish in Amazonia

Species	Common names	Distribution
Brachyplatystoma flavicans, Pimelodidae	Dourada (Brazil); zúngaro dorado (Peru); dorado or plateado (Colombia)	Wide distribution in the Amazon Basin. Similar (if not the same) to species in the Orinoco Basin. Migrates through corridors, and is found in the headwaters of many tributaries, such as those of the Negro and Madeira Rivers. It is very common in fresh water and in low salinity in the mouth of the Amazon.
Brachyplatystoma vaillantii, Pimelodidae	Piramutaba, pira-botão or mulher-ingrata (Brazil); pirabutón (Colombia); manitoa (Peru)	Occurs mainly along the Solimões/Amazon River and in the white-water tributaries. Rarely found past the first rapids, except in the Madeira River. Similar (but not the same) to the species in the Orinoco River basin.
Branchyplatystoma filamentosum, Pimelodidae	Piraíba or filhote (Brazil); zúngaro saltón (Peru); pirahiba, lechero or valentón (Colombia)	Wide distribution in the Amazon. Similar (if not the same) to the species in the basin of the Orinoco River.
Brachyplatystoma juruense, Pimelodidae	Zebra or flamengo (Brazil); siete babas (Colombia); zúngaro alianza (Peru)	Despite showing a wide distribution in the Amazon basin, it is relatively rare; not identified as an important fish for consumption.
Pseudoplatystoma fasciatum, Pimelodidae	Surubim or surubim-lenha (Brazil); pintado, rayado or pintadillo (Colombia); zúngaro (Peru)	Wide distribution in the Amazon basin, but rare or absent at the mouth. Occurs in the headwaters of all types of rivers, though different species may be involved.
Pseudoplatystoma tigrinum, Pimelodidae	Caparari or surubim-tigre (Brazil); bagre tigre (Colombia); zúngaro tigre (Peru)	Wide distribution in the Amazon basin, but rare or absent at the mouth. Seems to be rarer than the headwater catfish.
Goslinia platynema, Pilelodidae	Babão, xeréu ou barba-chata (Brazil); baboso or saliboro (Colombia)	Occurs along the major rivers of the Amazon, including in the estuary. Not yet found in tributaries of black and clear waters.
Sorubimichthys planiceps, Pimelodidae	Pirauaca or peixe-lenha (Brazil); cabo de hacha or peje leña (Colombia); acha cubo (Peru)	Wide distribution in lowland rivers of the Amazon basin to the west of the River Tapajos. Not yet recorded in black- or clear-water tributaries.
Phractocephalus hemiliopterus, Pimelodidae	Pirarara, bigorilo ou guacamaio (Brazil); pirarara or guacamayo (Colombia); pez torre (Peru)	Occurs in the basin of the Amazon and Orinoco rivers, including black- and clear-water tributaries. Reaches the headwaters and marine regions by tidal estuaries.

(continued)

Species	Common names	Distribution
Paulicea lutkeni, Pimelodidae	Jaú ou pacamão (Brazil); peje negro, chontaduro or pacamu (Colombia); cunchi mama (Peru)	Occurs in several basins of South America from northern Argentina to Venezuela, in all types of water, often in the headwaters near rapids.
Platynematichthys notatus, Pimelodidae	Cara de gato or coroatá (Brazil); capaz (Colombia)	Occurs the length of the Amazon River and in all types of rivers.
Merodontodus tigrinus, Pimelodidae	Dourada zebra (Brazil)	Only by a few specimens have been collected in the Upper Madeira River and the Caquetá River region.

Source: Barthem and Goulding (1997).

Annex 3
Species of the upper Paraná classified by reproductive strategies

1. *External fertilization, large migrations, no parental care*
Brycon orbignyanus
Piaractus mesopotamicus
Prochilodus lineatus
Pterodoras granulosus
Salminus hilarii

Leporinus elongatus
Pinirampus pirinampu
Pseudoplatystoma corruscans
Rhinelepis aspera
Salminus maxillosus

2. *External fertilization, non-migratory, no parental care*
Acestrorhynchus falcatus
Apareiodon affinis
Astyanax scabripinnis
Astyanax eignmaniorum
Astyanax schubarti
Cheirodon piaba
Steindachnerina insculpta
Hypophthalmus edentatus
Leporinus friderici
Leporinus piau
Moenkhausia intermedia
Parodon tortuosus
Plagioscion squamosissimus
Rhamdia hilarii
Schizodon knerii
Trachydoras paraguayensis

Acestrorhyncus lacustris
Aphyocharax difficilis
Astyanax bimaculatus
Astyanax fasciatus
Bryconamericus stramineus
Curimata gilberti
Gymnotus carapo
Iheringichthys labrosus
Leporinus octofasciatus
Leporellus vittatus
Oxidoras knerii
Pimelodus maculatus
Rhaphiodon vulpinus
Schizodon borellii
Schizodon nasutus

3. *External fertilization, non-migratory, parental care*
Cichla monoculus
Hoplias lacerdae
Hypostomus albopunctatus
Hypostomus comersonii
Serrasalmus marginatus
Serrasalmus spilopleura

Geophagus brasiliensis
Hoplias malabaricus
Hypostomus ancistroides
Hypostomus hermanni
Serrasalmus nattereri

4. *Internal fertilization, non-migratory, parental care*
Ageneiosus brevifilis
Ageneiosus valenciennesi
Parauchenipterus galeatus

Ageneiosus ucayalensis
Auchenipterus nuchalis

Source: Vazzoler and Menezes (1992).

Annex 4

Taxonomic groups of aquatic invertebrates found in Brazil and the State of São Paulo

Taxonomic groups	Number of species identified in the State of São Paulo	Number of species known in Brazil	% of species identified/ estimated in the State of São Paulo
Sponges	6	44	?
Cnidaria	6	7	?
Turbellarian flatworms	81	84	81
Nemertines	1	2	?
Gastrotricha	42	63	?
Gordioid nematomorphs	1	10	?
Rotifera	236	467	50
Bryozoa	6	10	60
Tardigrada	58	61	?
Bivalve molluscs	44	115	88
Gastropod molluscs	70	193	50
Polychaete annelids	3	4	30
Oligochaete annelids	46	70	?
Mites	20	332	6.6
Planktonic crustacean copepods	26	76	80
Nonplanktonic copepod crustaceans	46	120	?
Branchiopod crustaceans	84	?	70
Syncarida crustaceans	3	10	?
Decapod crustaceans	33	116	68
Ephemeroptera	8	150	?
Chironomid diptera	31	188	?
Odonata	?	641–670	?
Plecoptera	40	110	?

Source: Ismael et al. (1999).

Annex 5

Total species diversity of the major groups of freshwater animals, by zoogeographical regions

	Palearctic	Nearctic	Afrotropical	Neotropical	Oriental	Australasia	Pacific and oceanic islands	Antarctica	World
Other phyla	3,675	1,672	1,188	1,337	1,205	950	181	113	6,109
Annelids	870	350	186	338	242	210	10	10	1,761
Molluscs	1,848	936	483	759	756	557	171	0	4,998
Crustaceans	4,449	1,755	1,536	1,925	1,968	1,225	125	33	11,990
Arachnids	1,703	1,069	801	1,330	569	708	5	2	6,149
Springtails	338	49	6	28	34	6	3	1	414
Insects[a]	1,519	9,410	8,594	14,428	13,912	7,510	577	14	75,874
Vertebrates[b]	2,193	1,831	3,995	6,041	3,674	694	8	1	18,235
Total	30,316	17,072	16,789	26,186	22,360	11,860	1,080	174	125,530

[a] The zoogeographical distribution of species from several families of Diptera is incomplete, so the regional total number of species is less than the number of genera known worldwide.
[b] Only strictly freshwater fish species are included (there is also approximately 2,300 species of brackish water).
Source: Balian; Segers; Martens (2008).

Diversity of species in the major groups of freshwater vertebrates, by zoogeographical regions.

	Palaearctic	Nearctic	Afrotropical	Neotropical	Oriental	Australasia	Pacific and oceanic islands	Antarctic	World
Amphibians	160	203	828	1,698	1,062	301	0	0	4,294
Crocodiles	3	2	3	9	8	4	0	0	24
Lacertidae (lizards)	0	0	9	22	28	14	2	0	73
Snakes	6	22	19	39	64	7			153
Turtles	8	55	25	65	73	34			260
Freshwater fish	1,844	1,411	2,938	4,035	2,345	261			12,740
Mammals	18	22	35	28	18	11	0	0	124
Birds	154	116	138	145	76	62	6	1	567
Total	2,193	1,831	3,995	6,041	3,674	694	8	1	18,235

Source: Balian; Segers; Martens (2008).

Annex 6

Processes of sampling and programme for water quality management in reservoirs (Rast *et al.*, 1986)

This programme presents some important aspects on reservoir management and determination of water quality. The sampling programme and measurements must contain the following information, which is essential for determining management options and making sound decisions:

I) Water-quality management
▶ Identification of water uses in the reservoir
▶ Relationship between quality and quantity of water
▶ Water quality management objectives

II) Identify water quality in drainage basins
Factors to consider: geology and physiography, soil chemistry, land use patterns, number and location of tributaries, flow in tributaries.

▶ Primary sources of pollution in the drainage basin: point source and non-point source.
▶ Non-point sources: drainage from agricultural soils, transport of air pollution
▶ Water quality and flow of tributaries in relation to precipitation
▶ Chemical and biological transformations in river tributaries
▶ Water quality measurement in rivers

III) Reservoir limnology
▶ Uses and methods of construction
▶ Classification: geographical, morphometric, hydrological and trophic
▶ Horizontal differentiation and gradients
▶ Conditions of stratification and hydrodynamics
▶ Retention time, circulation patterns
▶ Chemical and biological processes
▶ Microbiology of water and sediment
▶ Primary and secondary production, effect of phytoplankton on water quality
▶ Fish populations: diversity, biomass, food; inter-relationship of fish industry and water quality management
▶ Effects downstream

IV) Cascading reservoirs on rivers
▶ Cascading reservoirs and alterations in water quality
▶ Reservoirs in water systems

V) Types of pollution in reservoirs:
Differentiation of diverse types of pollution

▶ Organic pollution
▶ Eutrophication of reservoirs
▶ Heavy metals
▶ Accumulation of nitrate
▶ Dissolved organic substances
▶ Acid rain and its effects
▶ Other types of pollutants
▶ Interactions of pollutants

VI) Decision-making processes
General processes involved in the effects of pollutants in reservoirs:

▶ Physical (sedimentation, light penetration, flow, turbulence)
▶ Chemical (redox potential, pH)
▶ Biological (biological decomposition, assimilation by phytoplankton, concentration in other trophic levels)

VII) Evaluation of the impact of pollution
Tests to evaluate impacts: biological testing, tests in experimental systems (micro-organisms or large tanks).

VIII) Methods to evaluate water quality
▶ Research before construction
▶ Systematic research and monitoring
▶ Frequency of sampling
▶ Sampling methods
▶ Water quality factors
▶ Methods that identify the stratification, organic pollution, acidification, toxicity, nitrate, and eutrophication
▶ Evaluation of trends by statistical analysis, classification of water quality, norms and criteria of water quality

IX) Application of mathematical models to evaluate water quality
▶ Model types and use
▶ Empirical models of water quality
▶ Eutrophication models
▶ Models to measure eutrophication and dissolved oxygen
▶ Models of acid rain
▶ Models of nitrogen and nitrate cycle
▶ Hydrodynamic models
▶ Hierarchy of models and selection of appropriate models

X) Eco-technological techniques
▶ Eco-technology: Classification of techniques and uses of appropriate technology: measures of water basin and reservoir; inactivation of pollutants, dilution, increased flow, selective removal, bio-manipulation, chemical control
▶ Combination of approaches and optimization models.

Appendix 1
Conversion table for units

Mass
$1\,kg = 1{,}000\,g = 0.001$ metric ton $= 2.20462$ pound mass (lb)
1 pound mass $= 453.593\,g$

Mol
$1\,mol = 1{,}000\,mmol = 10^6\,\mu mol = 10^9\,nmol$
$\quad = 6 \times 10^{23}$ atoms or molecules

Length
$1\,km = 1{,}000\,m = 10^5\,cm = 10^6\,mm$
$1\,m = 100\,cm = 1000\,mm = 10^6\,microns\,(\mu) = 10^{10}$ Angstroms (Å)
$\quad = 39.37$ inches (in) $= 3.2808$ feet (ft)
$\quad = 1.0936$ yards $= 0.0006214$ miles
1 foot $= 12$ inches $= 30.48\,cm$
1 yard $= 3$ feet
1 inch $= 2.54\,cm$

Volume
$1\,km^3 = 10^9\,m^3 = 10^{15}\,cm^3$
$1\,m^3 = 1{,}000\,L = 10^6\,cm^3 = 10^6\,mL = 35.3145$ cubic feet $= 246.17$ gallons
1 cubic foot $= 1{,}728$ cubic inches $= 7.48$ gallons $= 0.028317\,m^3 = 28.317\,L$
1 gallon $= 3.785\,L$

Force
1 Newton (N) $= 1\,kg \cdot m/s^2 = 10^5\,dynes = 10^5\,g.cm/s^2 = 0.22481$ pound force (lbF)
1 pound force $= 32.174$ pound mass $\cdot$ foot/s$^2 = 4.4482$ N $= 4.4482 \times 10^5$ dynes
1 kilogram pressure (kgf) $= 9.8$ N

Pressure
1 atmosphere (atm) $= 1.01325 \times 10^5\,N/m^2$ (Pa) $= 1.01325$ bars
$\quad = 1.01325 \times 10^6\,dynes/cm^2$
$\quad = 760\,mm$ Hg at $0°C$ (torr) $= 10.333$ m H_2O at $4°C$
$\quad = 14.696$ pound pressure/inch^2 (psi) $= 33.9$ feet H_2O at $4°C$
$\quad = 29.921$ inches Hg at $0°C$

Energy
1 Joule (J) $= 1$ N$\cdot$m $= 10^7$ ergs $= 10^7$ dyne$\cdot$cm $= 2.778 \times 10^7\,kW \cdot h$
$\quad = 0.23901$ calories (cal) $= 0.7376$ foot $\cdot$ pound force
$\quad = 9.486 \times 10^4$ BTU (British Thermal Unit)

Potential
1 Watt (W) $= 1$ J/s $= 0.23901$ cal/s $= 0.7376$ foot $\cdot$ pound force/s
$\quad = 9.486 \times 10^{-4}$ BTU/s $= 1.341 \times 10^{-3}$ HP (horse power)

Temperature

$T\,(K) = T\,(°C) + 273.15$	$\Delta T\,(°C) = \Delta T\,(K)$
$T\,(°R) = T\,(°F) + 459.67$	$\Delta T\,(°R) = \Delta T\,(°F)$
$T\,(°R) = 1.8\,T\,(K)$	$\Delta T\,(K) = 1.8\,\Delta T\,(°R)$
$T\,(°F) = 1.8\,T\,(°C) + 32$	$\Delta T\,(°C) = 1.8\,\Delta T\,(°F)$

K – Kelvin
R – Rankine
C – Celsius
F – Fahrenheit

Newton's Law $\quad g_c = \dfrac{1\,kg \cdot m/s^2}{N} = \dfrac{1\,g \cdot cm/s^2}{dyne} = 32.174\,\dfrac{pound\ mass \cdot foot}{s^2 \cdot pound\ pressure}$

Universal Gas Constant

$8.314\,m^3 \cdot Pa/mol \cdot K$ $\qquad\qquad 10.73\,foot^3 \cdot psia/pound \cdot mol \cdot {}^{\circ}R$

$0.08314\,L \cdot bar/mol \cdot K$ $\qquad\qquad 8.314\,J/mol \cdot K$

$0.08206\,L \cdot atm/mol \cdot K$ $\qquad\qquad 1.987\,cal/mol \cdot K$

$62.36\,L \cdot mmHg/mol \cdot K$ $\qquad\qquad 1.987\,BTU/pound \cdot mol \cdot {}^{\circ}R$

$0.7302\,foot^3 \cdot atm/pound \cdot mol \cdot {}^{\circ}R$

Conversion: units of concentration

1 umol of P = 1 at-μg of P = 31 μg of $P\text{-}PO_4^{3-}$ = 95 μg of PO_4^{3-}

1 umol of Si = 1 at-μg of Si = 28.2 μg of Si = 96.1 μg of $Si(OH)_4$

1 umol of N = 1 at-μg of N = 14 μg of $N\text{-}NO_3^-$ = 62 μg of NO_3

1 mmol of O_2 = 2 at-mg of O_2 = 32 mg of O_2 = 22.4 mL of O_2

1 mmol of HS^- = 34.1 at-mg of H_2S = 32.1 mg of S^{2-} = 22.4 mL of H_2S

10 μg of $N\text{-}NO_3^-$ = 0.71 at-μg of $N\text{-}NO_3^-$ = 0.71 μmol of $N\text{-}NO_3^-$ = 0.71 μmol of NO_3^- = 44.3 μg of NO_3^-

10 μg of $N\text{-}N_2O$ = 0.71 at-μg of $N\text{-}N_2O$ = 0.71 μmol of $N\text{-}N_2O$ = 0.357 μmol of N_2O = 15.7 μg of N_2O

Appendix 2
Time scale of geological periods

EON	ERA	System (Period)	Series (Epoch)	Years ago (millions)
Phanerozoic	Cenozoic	Quaternary	Holocene	(0.01)
			Pleistocene	1.6
		Neogene	Pliocene	5.3 (4.8)
			Miocene	23
		Palaeogene	Oligocene	(36.5)
			Eocene	53
			Palaeocene	65 (64.4)
	Mesozoic	Cretaceous	Upper	95
			Lower	135 (140)
		Jurassic	Upper	152
			Middle	180
			Lower	205
		Triassic	Upper	230
			Middle	240
			Lower	250
	Palaeozoic	Permian	Upper	260
			Lower	290
		Carboniferous	Upper	325
			Lower	355
		Devonian	Upper	375
			Middle	390
			Lower	410
		Silurian	Upper	428
			Lower	438
		Ordovician	Upper	455 (473)
			Lower	510
		Cambrian	Upper	(525)
			Lower	570 (540)
Proterozoic	Neoproterozoic		Upper	1,000
	Mesoproterozoic		Middle	1,600
	Palaeoproterozoic		Lower	2,500
Archean				

Source: Salgado-Laboriau (1994).

References

Ab'Saber, N.A. (1975) *Projeto Brasileiro para o Ensino da Geografia: Formas do Relevo*. São Paulo, Instituto de Geografia da USP (Texto Básico).

———. (1977) *Geomorfologia: Os Domínios Morfoclimáticos na América do Sul*. São Paulo, Instituto de Geografia da USP.

———. (1988) O Pantanal Matogrossense e a Teoria dos Refúgios. *Revista Brasileira de Geografia*, Rio de Janeiro, 50 (2), 9–57.

———. (1997) Espaços ocupados pela expansão dos climas secos da América do Sul, por ocasião dos períodos glaciais quaternários. *Paleoclimas*, 3, 1–19.

———. (2001) *Litoral Brasileiro*. São Paulo, Metalivros.

Abe, D.S. (1998) *Desnitrificação e caracterização filogenética das bactérias de vida livre e bactérias aderidas às partículas no hipolimnio do Lago Kizaki, Japão*. [Tese de Doutorado]. São Carlos, Escola de Engenharia de São Carlos, USP.

Abe, D.S. & Kato, K. (2000) Microbial nitrogen metabolism. In: Saijo, Y. & Hayashi, H. (Org.) *Lake Kizaki: Limnology and Ecology of a Japanese Lake*, vol. 1. Leiden, Holanda: Backhuys Publishers. pp. 195–198.

Abe, D.S., Kato, K., Terai, H., Adams, D.D. & Tundisi, J.G. (2000) Contribution of free-living and attached bacteria to denitrification in the hypolimnion of a mesotrophic japanese lake. *Microbes and Environments*, Tóquio, 15, 93–101.

Abe, D.S., Rocha, O., Matsumura-Tundisi, T. & Tundisi, J.G. (2002) Nitrification and denitrification in a series of reservoirs in the Tietê River, southeastern Brazil. *Verh. Internat. Verein. Limnol.*, Stuttgart, 28, 877–880.

Abe, D.S., Matsumura-Tundisi, T., Rocha, O. & Tundisi, J.G. (2003) Denitrification and bacterial community structure in the cascade of six reservoirs on a tropical river in Brazil. *Hydrobiologia*, Baarn, Holanda, 504, 67–76.

Abe, D.S., Tundisi, J.G., Vannucci, D. & Sidagis-Galli, C. (2006) Avaliação da capacidade de remoção de nitrogênio em uma várzea da cabeceira do reservatório de Guarapiranga, região metropolitana de São Paulo. In: Tundisi, J.G., Matsumura-Tundisi, T. & Galli, C.S. (eds.) *Eutrofização na América do Sul: causas, conseqüências e tecnologias para gerenciamento e controle*. São Carlos, IIE/ABC/IAP-São Carlos.

Åberg, B. & Rodhe, W. (1942) Über die Milieufaktoren in einigen Südschwedischen Seen. *Symb. Bot. Upsalienses*, Uppsala, Suécia, 5 (3), 383, 392, 402, 408, 413, 602, 686, 706, 711, 879.

Absy, M.L. (1979) *A palynological study of holocene sediments in the Amazon Basins*. [Tese]. Amsterdã, University of Amsterdan.

Adámoli, J. (1986a) A dinâmica das inundações no Pantanal. In: *Anais do 1o Simpósio Sobre Recursos Naturais e Sócio-Econômicos do Pantanal*. Brasília, EMBRAPA-CPAP. pp. 71–76.

———. (1986b) Fitogeografia do Pantanal. In: *Anais do 1o Simpósio Sobre Recursos Naturais e Sócio-Econômicos do Pantanal*. Brasília, EMBRAPA-CPAP, pp. 105–107.

Adams, D.D. (1996) Introduction and overview aquatic cycling and hydrosphere to troposphere transport of reduced trace gases- a review. *Mitt. Internat. Verein. Limnol.*, 25, 1–13.

Adamson, D. & Williams, E. (1980) Structural geology tectonics and the control of drainage in the Nile Basin. In: Williams, M.A.J. & Faure, H. (eds.) *The Sahara and The Nile*. Roterdã, A.A. Balkema. pp. 225–252.

Agassiz, L. (1850) *Lake Superior*: Physical Character, Vegetation, and Animals Compared With Those of Other Similar Regions. Boston, Gould, Kendal and Lincoln.

Agostinho, A.A. & Gomes, L.C. (1997) Manejo e Monitoramento de Recursos Pesqueiros: Perspectivas para o Reservatório de Segredo. In: Agostinho, A.A. & Gomes, L.C. (eds.) *Reservatório de Segredo: bases ecológicas para o manejo*. Maringá, EDUEM, pp. 275–292.

Agostinho, A.A. & Julio Jr., H.F. (1999) Peixes da bacia do alto Paraná. In: Lowe-McConnel, R.H. *Estudos ecológicos de comunidades de peixes tropicais*, cap. 16. São Paulo, EDUSP, pp. 374–399.

Agostinho, A.A., Vazzoler, A.E.A.M., Gomes, L.C. & Okada, E.K. (1993) Estratificación Espacial y Comportamiento de *Prochilodus scrofa* en Distintas Fases del Ciclo de Vida, en la Planície de Inundación del Alto Rio Paraná y Embalses de Itaipu, Paraná, Brazil. *Rev. Hydrobiol. Trop.*, Paris, 26, 79–90.

Agostinho, A.A., Borghetti, J.R., Vazzoler, A.E.A.M. & Gomes, L.C. (1994) Impacts on the Ichthyofauna on biological bases for its management. *Enviromental and Social Dimensions of Reservoir Development and Management in the La Plata Basin*. Nagoya, Japão, UNCRD. pp. 135–148.

Agostinho, A.A., Vazzoler, A.E.A.M. & Thomaz, S.M. (1995) The high River Paraná Basin: Limnological and ichthyological aspects. In: Tundisi, J.G., Bicudo, C.E.M., Matsumura-Tundisi, T. (eds.) *Limnology in Brazil*, n. 59. Rio de Janeiro, ABC/SBL. p. 103.

Agostinho, A.A., Ferretti, C.L.M., Gomes, L.C., Hahn, N.S., Suzuki, H.I., Fugi, R. & Abujanra, F. (1997a) Ictiofauna de Dois Reservatórios do Rio Iguaçu em Diferentes Fases de Colonização: Segredo e Foz de Areia. In: Agostinho, A.A. & Gomes, L.C. (eds.) *Reservatório de Segredo: bases ecológicas para o manejo*. Maringá, EDUEM. pp. 319–364.

Agostinho, A.A., Júlio, Jr., H.F., Gomes, L.C., Bini, L.M. & Agostinho, C.S. (1997b) Composição, abundância e distribuição espaço-temporal da ictiofauna. In: Vazzoler, A.E.A.M., Agostinho, A.A. & Hahn, N.S. (eds.) *A planície de inundação do alto rio Paraná: aspectos físicos, biológicos e socioeconômicos*. Maringá, EDUEM. pp. 179–208.

Agostinho, A.A., Miranda, L.E., Bini, L.M., Gomes, L.C., Thomaz, S.M. & Suzuki, H.I. (1999) Patterns of colonization in neotropical reservoirs, and prognoses on aging. In: Tundisi, J.G. & Straškraba, M. (eds.) *Theoretical Reservoir Ecology and its Applications*. São Carlos, Academia Brasileira de Ciências & Backhuys Publishers. pp. 227–265.

Agostinho, A.A., Gomes, L.C. & Zalewski, M. (2001) The importance of the floodplain for the dynamics of fish communities of the Upper River Paraná. *Ecohydrology & Hydrobiology*, Varsóvia, 1–2, 209–217.

Agostinho, A.A., Gomes, L.C., Thomaz, S.M. & Hahn, S.N. (2004) The Upper Paraná River and its floodplain: Main characteristics and perspectives for management and conservation. In: Thomaz, S.M., Agostinho, A.A. & Hahn, N.S. (eds.) *The Upper Paraná River and its Floodplain*: Physical Aspects, Ecology and Conservation. Leiden, Holanda, Backhuys Publishers. pp. 381–393.

Aidar Aragão, E. (1980) *Alguns aspectos de autoecologia de* Skeletonema costatum *(Greville) de Cananéia (25°S-48°W), com especial referência ao fator salinidade*. [Tese]. São Paulo, USP.

Allan, D. (1995) *Stream Ecology*: Structure and Functioning of Running Waters. Oxford, Chapman and Hall.

Allen, P.H. (1956) *The Rain Forests of Golfo Dulce*. Jacksonville, University of Florida Press.

Almeida-Val, V.M.F., Val, A.L. & Hochachka, P.W. (1993) Hypoxia tolerance in Amazon fishes: Status of an under-explored biological "Goldmine". In: Hochachka, P.W., Lutz, P.L., Sick, T., Rosenthal, M. & Van Den Thillart, G. (eds.) *Surviving Hypoxia*: Mechanisms of Control and Adaptation. Boca Ratón, EUA, CRC Press. pp. 436–445.

Alonso, M. (1985) *Las lagunas de la España peninsular: taxonomia, ecología y distribución de los cladoceros*. [Tese de doutorado]. Barcelona, Dep. de Ecologia, Facultad de Biologia.

American Public Health Association. (1989) *Standard Methods for Examination of Water and Wastewater*, 16th ed. Washington, APHA, AWWA, WPCF.

Angelini, R. & Pretere, Jr., M. (2000) A model for the plankton system of the Broa reservoir. *Ecological Modeling*, S. Carlos, Brazil, 26, 131–137.

Ara, K. (1998) Variabilidade temporal e produção dos copepodos no complexo estuarino Lagunar de Cananéia, São Paulo, Brasil. PhD thesis University of São Paulo, Brazil.

Aranha, J.M.R., Caramaschi, E.P. & Caramaschi, U. (1993) Ocupação Espacial, Alimentação e Época Reprodutiva de Duas Espécies de *Corydoras lacépéde* (Siluroide, Callichthyidae) Coexistentes no Rio Alambari (Botucatu, SP). *Revista Brasileira de Zoologia*, Curitiba, 10 (3), 453–466.

Araujo, D.S.D. (1992) Vegetation types of sandy coastal plains of tropical Brazil: A first approximation. In: Seeliger, U. (ed.) *Coastal Plant Communities of Latin America*. San Diego: Academic Press. pp. 337–347.

Araujo, D.S.D., Scarano, F.R., As, C.F.C., Kurtz, B.C., Zaluar, H.L.T., Montezuma, R.C.M. & Oliveira, R.C. (1998) Comunidades vegetais do Parque Nacional da Restinga de Jurubatiba. In: Esteves, F.A. (Ed.) *Ecologia das Lagoas Costeiras do Parque Nacional de Jurubatiba e do municipio de Macaé*, RJ. Macaé, NUPEM. pp. 39–62.

Araujo-Lima, C.A.R.M., Fosberg, B.R., Victoria, R. & Martinelli, L. (1986) Energy Sources for Detritivores Fishes in the Amazon. *Science*, Washington, 234, 1256–1258.

Araujo-Lima, C.A.R.M., Agostinho, A.A. & Febré, N.N. (1995) Trophic aspects of fish communities in Brazilian Rivers and Reservoirs. In: Tundisi, J.G., Bicudo, C.E.M. & Matsumura-Tundisi, T. (eds.) *Limnology in Brazil*. São Paulo, ABC/SBL. pp. 105–136.

Araujo-Lima, C.A.R.M. & Goulding, M. (1998) *Os Frutos do Tambaqui*: Ecologia, Conservação e Cultivo na Anatomia. Tefé, Sociedade Civil Mamirauá. MCT-KNPG.

Arcifa, M.S. (2000) Feeding habits of *Chaoboridae larvae* in a tropical Brazilian reservoir. *Rev. Bras. Biol.*, São Carlos, 60, 591–597.

Arcifa, M.S., Froelich, C.G., Giasenella Galvão, S.M.F. (1981) Circulation patterns and their influence on physico-chemical and biological conditions in eight reservoirs in Southern Brazil. *Verh. Internat. Ver. Limnol.*, Stuttgart, 21, 1054–1059.

Arcifa, M.S., Gomes, E.A.T. & Meschiatti, A.J. (1992) Composition and fluctuations of the zooplankton of a Tropical Brazilian Reservoir. *Arch. Hydrobiol.*, Stuttgart, 123, 479–495.

Arcifa, M.S., Silva, L.H.S. & Silva, M.H.L. (1998) The Planktonic Community in a Tropical Brazilian Reservoir: Composition, Fluctuations and Interactions. *Rev. Bras. Biol.*, São Carlos, 58, 241–254.

Armengol, J., Garcia, J.C., Comerma, M., Romero, M., Dolz, J., Roura, M., Han, B.H., Vidal, A. & Simek, K. (1999) Longitudinal processes in canyon type reservoir: The case of Sau (N.E. Spain). In: Tundisi, J.G. & Straškraba, M. (eds.) *Theoretical Reservoir Ecology and Its Applications*. Academia Brasileira de Ciências & Backhuys Publishers. pp. 313–345.

Attayde, J.L. (2000) *Direct and Indirect Effects of Fish Predation and Excretion in Aquatic Food Webs*. [Tese de Doutorado]. Lund, Suécia, Department of Ecology & Limnology, Lund University.

Ayensu, E. et al. (1999) International ecosystem assessment. *Science*, Washington, 286, 685–686.

Azam, F., Fenchel, T., Field, J.G., Graf, J.S., Meyer-Reii, L.A. & Thingstad, F. (1983) The ecological role of water-column microbes in the sea. *Marine Ecology: Progress Series*, 10, 257–263. Luhe, Alemanha, Inter Research.

Azevedo, S.M.F.O., Evans, W.R., Carmichael, W.W. & Namikoshi, M. (1994) First report of microcystins from a Brazilian isolate of the *Cyanobacterium Microcystis aeruginosa*. *Journal of Applied Phycology*, Dordrecht, Holanda, 6, 261–265.

Bahamonde, N. & Cabrea, R.A. (eds.) (1984) *Embalses, Fotosintesis y Productividad Primaria*. Santiago, Chile, Univ. Chile.

Baker, P.D. & Humpage, A.R. (1994) Toxicity associated with commonly occurring cyanobacteria in surface waters of Murray-Darling Basin, Australia. *Aust. Jour. Marine Freshwat. Res.*, Melbourne, Austrália, 45, 773–786.

Baldi, E. (1949) La Situation Actuelle de la Recherche Limnologique, Après le Congrès de Zurich. *Schweiz. Z. Hydrol.*, Zurique, 11, 637–649.

Ball, E. & Gluscksman, J. (1978) Limnological studies of Lake Wisdom, a large new Guinea Caldera with a simple fauna. *Freshwat. Biol.*, Victoria, Austrália, 8, 455–468.

Balon, E.K. & Coche, A.G. (eds.) (1974) *Lake Kariba: A Man-made Tropical Ecosystem in Central Africa*. Haia, Holanda, Springer. [*Monographiae Biologicae*, 24].

Banner, M.L. & Phillips, O.M. (1974) On the incipient breaking of small scale waves. *J. Fluid. Mech.*, 65 (part 4), 647–656.

Bannerman, M. (2001) *Mamirauá*: A Guide to the Natural History of the Amazon Flooded Forest. Tefé, IDSM.

Banarescu, P. (1995) Distribution and dispersal of freshwater animals in Africa, Pacific areas and South America. *Zoogeography of Freshwaters*, vol. 3. Wiesbaden, Alemanha, AULA-Verlag.

Barbosa, F.A.R. (1981) *Variações Diurnas (24 horas) de Parâmetros Limnológicos Básicos e da Produtividade Primária do Fitoplâncton na Lagoa Carioca – Parque Florestal do Rio Doce – M.G. Brasil*. [Tese de Doutorado]. São Carlos, UFSCAR.

Barbosa, F.A.R. (ed.) (1994) Workshop: Brazilian Programme on Conservation and Management of Inland Waters. *Acta Limnol. Brasil.*, Campinas, 5.

Barbosa, F.A.R. & Padisák, J. (2002) The Forgothen Lake Stratification Pattern: Atelomixis and its Ecological Importance. *Verh. Internat. Verein. Limnol.*, Stuttgart, 28, 1385–1395.

Barbosa, F.A.R. & Tundisi, J.G. (1980) Primary production of phytoplankton and environmental characteristics of a Shallow Quaternary Lake at Eastern Brazil. *Arch. Hydrobiol.*, Stuttgart, 90 (2), 139–161.

Barbosa, F.A.R., Padisák, J., Espíndola, E.L.G., Borics, G. & Rocha, O. (1999) The Cascading Reservoir Continuum Concept (CRCC) and its application to the River Tietê-basin, São Paulo State, Brazil. In: Tundisi, J.G. & Straškraba, M. (eds.) *Theoretical Reservoir Ecology and its Applications*. São Carlos, Rio de Janeiro, Leiden, Holanda, IIE.BAS/Backhuys Publishers. pp. 425–438.

Barthem, R.B. (1999) Varzea Fisheries in the Middle Rio Solimões: Introduction. In: Padoch, C., Ayres, J.M., Pinedo-Vasquez, M., Henderson, A. (Eds.) *Varzea*: Diversity, Development, and Conservation of Amazonia's Whitewater Floodplains. Nova York, The New York Botanical Garden Press. pp. 7–28.

Barthem, R.B. & Goulding, M. (1997) *Os Bagres Balizadores*: Ecologia, Migração e Conservação de Peixes Amazônicos. Tefé, Sociedade Civil Mamirauá, MCT, CNPq, IPAAM.

Barthem, R. & Goulding, M. (2007) *Um ecossistema inesperado*: a Amazônia revelada pela pesca. Peru, Amazon Conservation Association (ACA, Sociedade Civil Mamirauá).

Barthem, R. et al. (2003) *Aquatic Ecology of the Rio Madre de Dios: Scientific Bass for Andes-Amazon Headwaters Conservation*. Peru, ACA.

Basterrechea, M. (1986) Limnological characteristics of Lake Amatitlan, Guatemala. *Revista Brasileira de Biologia*, São Carlos, 46 (2), pp. 461–468.

Baxter, R.M. (1977) Environmental effects of dams and impoudments. *Annual Review of Ecology and Systematics*, Palo Alto, EUA, 8, 255–283.

Baxter, R.M., Prosser, M.V., Talling, J.F. & Wood R.B. (1965) Stratification in tropical African lakes at moderate altitudes (1,500 to 2,000 m). *Limnol. Oceanogr.*, Milwaukee, EUA, 10, 510–520.

Bayley, P.B. & Petrere, Jr., M. (1989) Amazon fisheries: Assessment methods, current status and management options. In: DODGE, D.P. (Ed.) Proceedings of the International Large River Symposium. *Can. Spec. Publ. Fish. Aquat. Sci.*, Ottawa, vol. 106, pp. 385–398.

Bayly, I.A.E. (1964) The Ocurrence of Calanoid Copepods in Athalassic Saline Waters in Relation to Sandy and Anionic Proportions. *Verh. Internat. Verein. Limnol.*, Stuttgart, 17, 449–455.

———. (1967) The general biological classification of aquatic environments with special reference to those of Australia. In: Weatherley, A.H. (Ed.) *Australian Inland Waters and their Fauna*: Eleven Studies. Canberra, Aust. Nat. Univ. Press.

———. (1972) Salinity tolerance and osmotic behaviour of animals in athallassic saline and marine hypersaline waters. *Ann. Rev. Ecol. Systematics*, Palo Alto, 3, 233–268.

Bayly, I.A.E. & Williams, W.D. (1973) *Inland Water and Their Ecology*. Melbourne, Longman.

———. (1966) Chemical and biological studies on some saline lakes of South-East Australia. *Aust. J. Mar. Freshwat. Res.*, Sydney, 177–228.

Beadle, L.C. (1932) Scientific results of the Cambridge expedition to the East African Lakes. 1930–1, v. 3. Observation on the bionomics of some East African Swamps. *J. Linn. Soc.*, Londres, 38, 135–155.

———. (1943) An ecological survey of some inland saline waters of Algeria. *J. Linn. Soc. (Zool.)*, Londres, 41, 218–242.

———. (1959) Osmotic and ionic regulation in relation to the classification of brackish and inland saline waters. *Arch. Oceanogr. Limnol.* (Suppl.), Waco, EUA, 143–151.

———. (1974) *The Inland Waters of Tropical Africa: An Introduction to Tropical Limnology*. Londres, Longman.

———. (1981) *The Inland Waters of Tropical Africa: An Introduction to Tropical Limnology*. 2nd edition. Londres, Longman.

Bemvenuti, C.E. (1997) Benthic invertebrates. In: Seeliger, U., Odebrecht, C. & Castello, J. (Eds.) *Subtropical Convergence Marine Ecosystem: The Coast and the Sea in the Warm Temperate Southwestern Atlantic*. Heidelberg, Springer-Verlag. pp. 43–46.

Berg, K. (1948) Biological studies on the River Susaa. *Folia Limnol. Scand.*, Copenhaguem, 4, 1–318.

Beyruth, Z. (2000) Periodic disturbances, trophic gradient and phytoplankton characteristics related to cyanobacterial growth in Guarapiranga reservoir, São Paulo, Brazil. *Hidrobiologia*, 424, 51–65.

Bianchini, Jr., I. (1982) *Aspectos do Processo de Decomposição de Plantas Aquáticas*. [Tese de Mestrado]. São Carlos, UFSCar.

———. (1985) *Estudos dos Processos de Humificação de* Nymphoides indica (L.) O. Kuntze. [Tese de Doutorado]. São Carlos, UFSCar.

———. (2003) Modelos de Crescimento e Decomposição de Macrófitas Aquáticas. In: (Thomaz, S.M., Bini, L.M.) (Org.) *Ecologia e Manejo de Macrófitas Aquáticas*, Maringá, 341 (4), 85–126.

Bianchini, Jr., I. & Toledo, A.A.P. (1996) Estudo da Mineralização de Eleocharis mutata. *Anais, 7o Seminário Regional de Ecologia, São Carlos*. Anais, vol. 3. São Carlos, UFSCar, PPG-ERN. pp. 57–72.

———. (1998) Estudo da Mineralização de *Nymphoides indica (L.) O. Kuntze. Anais, 7o Seminário Regional de Ecologia: São Carlos*, vol. 3. São Carlos, UFSCar, PPG-ERN. pp. 1315–1329.

Bianchini, Jr., I., Toledo, A.P.P. & Toledo, S.H.P.P. (1984) Influência do Tempo na Variedade e Quantidade de Polifenóis Dissolvidos, Originados da Decomposição de Plantas Aquáticas. *Anais, 4o Seminário Regional de Ecologia*. São Carlos, DCB-UFSCar. pp. 167–181.

Bicudo, C.E.M. & Bicudo, D.C. (2004) *Amostragem em Limnologia*. São Carlos, RIMA.

Bicudo, D.C., Necchi Junior, O. & Chamixaes, B.C.B. (1995) Periphyton studies in Brazil: Present status and perspectives. In: Tundisi, J.G., Bicudo, C.E.M., Matsumura-Tundisi, T. (Orgs.) *Limnology in Brazil*. Rio de Janeiro, Academia Brasileira de Ciências. pp. 37–58.

Bicudo, C.E.M., Carmo, C.F., Bicudo, D.C., Henry, R., Pião, A.C.S., Santos, C.M.S. & Lopes, M.R.M. (2002) Morfologia de Três Reservatórios do PEFI. In: Bicudo, D.C., Forti, M.C. &

Bicudo, C.E.M. (eds.) *Parque Estadal das Fontes do Ipiranga (PEFI)*: Unidade de Conservação que Resiste à Urbanização de São Paulo. São Paulo, Secretaria do Meio Ambiente do Estado de São Paulo.

Bicudo, D.C., Forti, M.C. & Bicudo, C.E.M. (eds.) (2002) *Parque Estadual das Fontes do Ipiranga (PEFI)*: Unidade de Conservação que Resiste à Urbanização de São Paulo. São Paulo, Secretaria do Meio Ambiente de Estado de São Paulo.

Bigelow, H.B. (1940) Medusae of the Templeton Crocker and Eastern Pacific ZACA Expeditions, 1936–1937. *Zoologica*, Nova York, 25, 281–321.

Bini, L.M., Tundisi, J.G., Matsumura-Tundisi, T. & Matheus, C.E. (1997) Spatial variation of zooplankton groups in a tropical reservoir (Broa Reservoir, São Paulo State-Brazil). *Hydrobiologia*, Baarn, Holanda, 357, 89–98.

Bini, L.M., Coelho, A.S., Diniz-Filho, J.A.F. (2001) Is the relationship between population density and body size consistent across independent studies? A meta-analytical approach. *Rev. Bras. Biol.*, São Carlos, 61, 1–6.

Birge, E.A. & Juday, C. (1911a) The Inland of Lakes of Wiscosin: The dissolved gases in the water and their biological significance. *Wisconsin Geological and Natural History Survey Bulletin*, Mount Vernon, EUA, 22 (2).

———. (1911b) The work of the wind in warming a lake. *Trans. Wis. Acad. Sci. Arts Lett.*, 18 (Pt. 2), 341–391. [Madison, EUA, *Wis. Acad. Sci. Arts Lett.*].

———. (1921) Further limnological observation on the finger lakes of New York. *Bull. U.S. Bur. Fish.*, vol. 37. Washington, Government Printing Office.

———. (1922) The inland lakes of Wiscosin: The plankton, its quality and chemical composition. *Wisconsin Geological and Natural History Survey Bulletin*, Mount Vernon, EUA, 64 (3).

———. (1929) Transmission of solar radiation by the waters of inland lakes. *Trans. Wis. Acad. Sci. Arts Lett.*, 24–27. [Madison, EUA, *Wis. Acad. Sci. Arts Lett.*].

———. (1934) Particulate and dissolved organic matter in Wiscosin lakes. *Ecological Monographs*, vol. 4. Washington, Ecological Society of America.

Biswas, A.K. (1976) *Systems Approach to Water Management*. Nova York: McGraw Hill, 1976.

———. (1983) Major water problems facing the world. *Water Resources Development*, vol. 1. Nova York, Routledge. pp. 1–14.

———. (1990) Objects and concepts of environmentally sound water management. In: Thant, N.C., Biswas, A.K. (eds.) *Environmentally Sound Water Management*. New Delhi: Oxford University Press. pp. 30–58.

Bloch, M.E. (1794) *Naturgeschichte der Ausländischen Fische*.

Bloss, S. & Harleman, D.F.R. (1979) Effect of wind mixing on the thermocline formation lakes and reservoirs. Report 249, MIT. Cambridge, Massassuchets, 147p.

Bo Ping Han, Virtanem, M., Koponen, J., Straškraba, M. (1999) Predictors of light limited growth and competition of phytoplancton in a web mixed water column. *J. Theor. Biol.*, Orlando, 197, 439–450.

Bond-Buckup, G. & Buckup, L. (1994) A família Aeglidae (Crustacea, Decapoda, Anomura). *Arquivos de Zoologia*, São Paulo, 34, 43–49.

Bonetto, A.A. (1970) Principal limnological features of northeastern Argentina. *Boletín de la Sociedad Argentina de Botánica*, Córdoba, Argentina, 11 (suppl.), 185–209.

———. (1975) Hydrologic regime of the Paraná River and influence on ecosystems. In: Hasler, D. (ed.) *Coupling of Land and Water Systems*. Berlim: Springer Verlag. pp. 175–197.

———. (1976) *Calidad de las aguas del Rio Paraná: introducción a su estudio ecologico*. Buenos Aires, Dirección Nacional de Construcciones Portuarias y Vias Navegables.

———. (1986a) The Paraná River System. In: Davies, B.F. & Walter, F.F. (eds.) *The Ecology of River Systems*. Dordrecht, Holanda, Dr. W. Junk. pp. 541–555.

———. (1986b) Fish of the Paraná River system. In: Davies B.F. & Walter F.F. (eds.) *The Ecology of River Systems*. Dordrecht, Holanda, Dr. W. Junk. pp. 573–588.

————. (1993) Structure and functioning of large river floodplains of Neotropical America: The Parana–Paraguay System. In: Gopal, B., Hillbricht-Ilkowska, A. & Wetzel, R.G. (eds.) *Wetlands and Ecotones*. New Delhi, National Institute of Ecology. pp. 123–138.

————. (1994). Austral Rivers of South America. In Margalef, R. (ed.) *Limnology Now*: A Paradigm of Planetary Problems. Amsterdã: Elsevier Science. p. 425–472.

Bonetto, A.A., Pignalberi, C. & Cordiviola, E. (1968) Las "palometas" o "piranas" de las aguas del Parana medio. *Acta Zool. Lilloana*, San Miguel de Tucumán, Argentina, 23, 45–65.

Bonetto, A.A., Pignalberi, C.Y., Oliveros, O. (1969) Ciclos hidrológicos del río Paraná y las poblaciones de peces contenidas en las cuencas temporarias de su valle de inundación. *Phycis.*, 29 (78), 213–223.

Bonetto, C.A. (1983) Fitoplancton y producción primaria del Paraná Medio. *Ecosur*, Rosario, Argentina, 10 (19–20), 79–102.

————. (1982) *Producción primaria del fitoplancton, concentración de pigmentos, materia orgánica y nutrientes, en la caracterización limnológica de los cuerpos de agua regionales del noreste argentino*. [Tese de Doutorado]. Buenos Aires, Universidad Nacional de Buenos Aires.

Bonetto, C.A., Zalocar, Y., Caro, P. & Vallejos, E. (1979) Producción primaria del fitoplancton del río Paraná en el área de su confluencia con el río Paraguay. *Ecosur*, Rosario, Argentina, 6 (12), 207–227.

Bonetto, C.A., Bonetto A.A. & Zalocar, Y. (1981) Contribución al conocimiento limnológico del río Paraguay en su tramo inferior. *Ecosur*, Rosario, Argentina, 8 (16), 55–88.

Borman, F.H. & Likens, G.E. (1967) Nutrient cycling. *Science*, Washington, 155, 424–429.

————. (1979) *Pattern and Processes in a Forested Ecosystem*. Nova York, Springer.

Borne, V.D.M. (1877) *Wie kann man unsere Gewässer nach den Ihnen vorkommenden Arten klassifizieren?* Cirk, Dt, Ver. 4

Borowitzka, L.J. (1981) The Microflora. *Hydrobiologia*, Baarn, Holanda, 81, 33–46.

Boyd, C.E. (1969) Production, mineral nutrient absorption, and biochemical assimilation by *Justicia americana* and *Alternanthera philoxeroides*. *Arch. Hydrobiol.*, Stuttgart, 66 (2), 139–160.

Bozelli, R.L. (1992) Composition of the zooplankton community of Batata and Mussura Lakes and of the Trombetas River, state of Para, Brazil. *Amazoniana*, Kiel, Alcmanha, 12 (2), 239–261.

Bozelli, R.L., Esteves, F.A. & Roland, F. (eds.) (2000) *Lago Batata*: Impacto e Recuperação de um Ecossistema Amazônico. Rio de Janeiro, UFRJ, SBL.

Braga, B., Rocha, O. & Tundisi, J.G. (1998) Dams and the environment: The Brazilian experience. *Water Resources Development*, Nova York, 14 (2), 126–139.

Branco, C.W.C. (1998) *Comunidades zooplanctônicas e aspectos limnológicos de três lagoas costeiras da Região Fluminense (Macaé, RJ)*. [Tese de Doutorado]. Rio de Janeiro, CCS, Instituto de Biofisica Carlos Chagas Filho, UFRJ.

Branco, C.W.C. & Senna, A.C. (1994) Factors influencing the development of *Cylindrospermopsis raciborskii* and *Microcystis aeruginosa* in the Paranoa reservoir, Brasilia, Brazil. *Algolog. Stud.*, Stuttgart, 75, 85–96.

————. (1996) Relations among heterotrophic bacteria, chlorophylla, total phytoplankton, total zooplankton and physical and chemical features in Paranoa Reservoir, Brasilia, Brazil. *Hydrobiologia*, Baarn, Holanda, 337, 171–181.

————. (1996) Phytoplankton composition, community structure and seasonal changes in a reservoir (Paranoá Reservoir, Brazil). Algological Studies 81. *Arch. Hydrobiol. Suppl.*, Stuttgart, 114, 69.

Branco, L.H.Z. (1991) Cyanophyceae *de comunidades bentônicas do Manguezal da Ilha do Cardoso, Município de Cananéia, SP*. [Tese de Mestrado]. Rio Claro, Universidade Estadual Paulista.

Branco, S.M. (1999) Água, Meio Ambiente e Saúde. In: Rebouças, A.C., Braga, B. & Tundisi, J.G. (eds.) *Águas Doces no Brasil*: Capital Ecológico, Uso e Conservação. São Paulo, Escrituras. pp. 227–247.

Brandini, F.P. (1982) Variação nictemeral de alguns fatores ecológicos na região de Cananéia (SP). *Arquivos de Biologia e Tecnologia*, vol. 25. Curitiba, Instituto de Tecnologia do Paraná. pp. 313–327.

Brandorff, G., Koste, W. & Smirnov, N. The composition and structure of rotiferan and crustacean communities of the lower Rio Nhamunda, Amazonas, Brazil. *Studies on Neotropical Fauna and Environment*, vol. 17. Londres Taylor & Francis. pp. 69–121.

Branski, J.M., Avill, J.P. de, Lopes, R.L.L., Gonçalves, A.J.T. & Tundisi, J.G. (1989) Environmental impact assessment for the porteira hydroeletric project. In: *River/Lake Basin Management: Proceedings of a Workshop*. Bangkok, Tailândia, UNCRD/UNEP/ILEC.

Brezonik, P.L. & Lee, G.F. (1968) Denitrification as a nitrogen sink in Lake Mendota, Wisconsin. *Environmental Science and Technology*, Washington, 2 (2), 120–125.

Brezonik, P.L. & Harper, C.L. (1969) Nitrogen fixation in some aquatic lacustrine environments. *Science*, Washington, 164, 1277–1279.

Brigante, J. & Espindola, E.L.G. (2003) A Bacia Hidrográfica: Aspectos Conceituais e Caracterização Geral da Bacia do Rio Mogi-Guaçu. In: *Limnologia Fluvial: um Estudo no Rio Mogi-Guaçu*. São Carlos, RIMA. pp. 1–13.

Brinson, M.M. & Nordlie, F.G. (1975) Lake Izabal, Guatemala. *Verh. Internat. Verein. Limnol.*, Stuttgart, 19, 1468–1479.

Britski, H.A., Sato, Y. & Rosa, A.B.S. (1984). *Manual de identificação de peixes da Bacia do São Francisco*. Coord. de Publicações, CODEVASF, Divisão de Piscicultura e Pesca.

Brooks, J.L. & Dodson, S.I. (1965) Predation, body size and composition of plankton. *Science*, Washington, 150, 28–35.

Buck, S. & Sazima, I. (1995) An assemblage of mailed catfishes (Loricariidae) in Southeastern Brazil: Distribution, activity, and feeding. *Ichthyological Explorations of Freshwaters*, Munique, Alemanha, 6 (4), 325–332.

Buckup, P.A. (1998) Relationships of the Characidiinae and Phylogeny of Characiform Fishes (Teleoestei: Ostariophysi). In: Malabarba, L.R., Reis, R.E., Vari, R.P., Lucena, Z.M. & Lucena C.A.S. (eds.) *Phylogeny and Classification of Neotropical Fishes*. Porto Alegre, Edipucrs. pp. 123–144.

Bukata, R.P., Jerome, J.H., Kondzafyek, K.Y. & Pozdnyakov, V.D. (1995) *Optical Properties and Remote Sensing of Inland and Coastal Waters*. Boca Ratón, CRC Press.

Burgis, I.M. & Morris, P. (1987). The Natural History of lakes. Cambridge University Press 218p.

Burgis, M.J. (1971) Lake George, Uganda: Studies in a Tropical Fresh Water Ecosystem. In: Hillbricht-Ilkowska, K.A. (ed.) *IBP UNESCO Symposium on Productivity Problems of Freshwaters*. pp. 301–309. Varsóvia/Cracóvia, Polônia.

———. (1974) Revised estimates for the biomass and production of zooplankton in Lake George, Uganda. *Freshwat. Biol.*, Victoria, Austrália, 4, 535–541.

———. (1987) Polar and Mountain Lakes. In: *The Natural History of Lakes*. Cambridge: Cambridge University Press. pp. 90–103.

Burton, G.W. & Odum, E.P. (1945) The distribution of stream fish in the vicinity of Mountain Lake, Virginia. *Ecology*, Washington, 26, 182–194.

Butcher, R.W. (1933) Studies on the ecology of rivers. I. On the distribution of macrophitic vegetation in the rivers of Britain. *J. Ecol.*, Victoria, Austrália, 21, 58–91.

Caceres, O. & Vieira, A.A.H. (1988) Some Features on the Growth of *Microcystis aeruginosa* Kuetz Emend. Elenkin in ASM – 1 Medium. *Rev. Microb.*, São Paulo, 19 (3), 223–228.

Cadernos De Estudos Avançados. (2006) Rio de Janeiro, Instituto Oswaldo Cruz, 3 (1).

Caleffi, S.A. (1994) *Represa de Guarapiranga: estudo da comunidade zooplanctônica e aspectos da eutrofização*. [Tese de Mestrado]. São Paulo, FSP/USP.

Calijuri, M.C., Tundisi, J.G. & Saggio, A.A. Um Modelo de Avaliação do Comportamento Fotossintético para Populações de Fitoplâncton Natural. *Rev. Brasil. Biol.*, São Carlos, 49 (4), 969–977.

Calliari, L.J. (1997). Geological setting. In: Seeliger, U., Odebrecht, C. & Castello, J. (eds.) *Subtropical Convergence Environments*: The Coast and Sea in the Southwestern Atlantic. Berlim, Springer-Verlag. p. 13–18.

Callil, C.T. (2003) *Base de dados direcionada à elaboração de um programa de monitoramento de águas continentais utilizando moluscos bivalves*. [Tese de Doutorado]. Porto Alegre, Pontifícia Universidade Católica do Rio Grande do Sul.

Callisto, M., Barbosa, F.A.R. & Viana, J.A. (1998) Qual a Importância de uma Coleção Científica de Organismos Aquáticos em um Projeto de Biodiversidade? *Anais do IV Simpósio de Ecossistemas Brasileiros*: Praia, Represa e Mata, 11 (4), 432–439. São Paulo, Aciesp.

Callisto, M., Marques, M.M., Barbosa, F.A.R. (2000) Deformities in larval Chironomus (Diptera, Insecta) from the Piracicaba river, Southeastern Brazil. *Verh. Internat. Verein. Limnol.*, Stuttgart, 27 (5), 2699–2702.

Callisto, M., Moretti, M., Goulart, M. (2001) Macroinvertebrados bentônicos como ferramenta para avaliar a saúde de riachos. *Revista Brasileira de Recursos Hídricos*, Porto Alegre, 6 (1), 71–82.

Callisto, M., Goulart, M., Barbosa, F.A.R. & Rocha, O. (2004a) Biodiversity assessment of Benthic macroinvertebates along a reservoir cascade in the lower São Francisco River (Northeastern Brazil). *Braz. Journ. Biol.*, 2, 29–240. São Carlos, IIE.

Callisto, M. et al. (2004b) Diversity assessment of benthic macroinvertebrates, yeasts, and microbiological indicators along longitudinal gradient in Serra do Cipó. *Braz. Journ. Biol.* 64 (4), 743–755.

Callisto, M. et al. (2004c) Malacological assessment and natural infestation of *Biomphalaria straminea* (Dunker, 1848) by *Schistosoma mansoni* (Sambon, 1907) and *Chaetgaster limnaei* (K. Von Baer, 1827) in an urban, eutrophic watershed. *Braz. Journ. Biol.*, 65 (2), 217–228.

Camargo, A.F.M. (1991) *Dinâmica do nitrogênio e do fosfato em uma lagoa marginal do rio Mogi-Guaçu (Lagoa do Mato, SP)*. [Tese de Doutorado]. São Carlos, UFSCar.

Camargo, A.F.M. & Esteves, F.A. (1995) Influence of water level variation on fertilization of an oxbow lake of Rio Mogi-Guacu, State of Sao Paulo, Brazil. *Hydrobiologia*, Baarn, Holanda, 299, 185–193.

Camargo, A.F.M., Pezzato, M.M. & Henry-Silva, G.G. (2003) Fatores limitantes à produção primária de macrófitas aquáticas. In: Thomaz, S.M. & Bini, L.M. (Orgs.) *Ecologia e manejo de macrófitas aquáticas*. Maringá, PR: Eduem. pp. 59–83.

Campagnoli, F. (2002) *A aplicação do assoreamento na definição de geoindicadores ambientais em áreas urbanas: exemplo na bacia do Alto Tietê, SP*. [Tese de Doutorado]. São Paulo, Escola Politécnica, USP.

Campos, H. (1984) Limnological study of Auraucanian lakes (Chile). *Verh. Internat. Verein. Limnol.*, 22, 1319–1327.

Campos, H. et al. (1978) Physical and chemical limnology of Lake Riñihue (Vladivia, Chile). *Arch. Hydrobiol.*, 84, 405–429.

———. (1983) Limnological studies in Lake Villarrica: Morphometry, physical, chemical, planktonic factors and primary productivity. *Arch. Hydrobiol. Suppl.*, 65 (4), 371–406.

———. (1987) Limnology of Lake Riñihue. *Limnologica.*, 18, 339–357.

———. (1988) Limnological study of Lake Biñihue (Chile): Morphometry, physics, chemistry, plankton and primary productivity. *Arch. Hydrobiol. Suppl.*, 81 (1), 37–67.

———. (1993) Limnological studies of Lake Ranco (Chile). Morphometry, physics, chemistry, plankton and primary productivity. *Arch. Hydrobiol. Suppl.*, 90 (1), 84–113.

———. (1998) *Determinación de la capacidad de carga y balance de fósforo y nitrógeneo del lago Riñihue*: Convenio Fondo de Investigaciones Pesqueras, Subcretaria de Pesca, Ministério de Economia, Fomento y Reconstrución de Chile y Universidad Austral de Chile.

Caramaschi, E.P., Mazzoni, R. & Peres-Neto, P.R. (1999) *Ecologia de peixes de riachos*. Rio de Janeiro, UFRJ. (Série Oecologia brasiliensis, 6).

Carlson, R.E. (1977) A Tropic State Index for Lakes. *Limnol. Oceanogr.*, Milwaukee, EUA, 19, 767–773.

Carmichael, W.W. (1992) Cyanobacteria secondary metabolites: The cyanotoxins. *J. Appl. Bacteriol.*, Victoria, Austrália, 72, 445–459.

Carmouze, J.P., Durand, J.R. & Lévêque, C. (eds.) (1983) *Lake Chad*: Ecology and productivity of a shallow tropical ecosystem. Haia, Holanda: Dr. W. Junk Publishers. pp. 1–575. (Série Monographiae Biologicae, 53).

Carpenter, K.E. (1928) *Life in Inland Waters*. Londres, Sidgwick & Jackson. pp. 135–177.

Carpenter, S.R. (2003) *Regime Shifts in Lake Ecosystems*. Luhe, Alemanha, Ecology Institute. (Excellence in ecology, 15).

Castelnau, F. (1855) *Animaux nouveaux ou rares recueillis pendant l'expédition dans les parties centrales de l'Amérique du Sud, de Rio de Janeiro a Lima et de Lima au Para*. 2. Poissons, introduction. I–XII, Paris. pp. 1–112, 50 pls.

Caumette, P. (1986) Phototrophic sulphur bacteria and sulphate reducing bacteria causing red waters in a shallow brackish coastal lagoon (Prevost Lagoon, France). *FEMS Microbiol. Ecol.*, Delft, Holanda, 38, 113–124.

Chamixaes, C.B.C.B. (1991) *Variação Temporal e Espacial da Biomassa, Composição de Espécies e Produtividade das Algas Perifíticas Relacionadas com as Condições Ambientais de Pequenos Rios da Bacia Hidrográfica do Ribeirão do Lobo (Itirapina, SP)*. [Tese de Doutorado]. São Paulo, USP.

Chapman, D. (ed.) (1992) *Water Quality Assessments: UNESCO, UNEP, WHO*. Londres, Chapman & Hall.

Chorus, I. & Barthram, J. (eds.) (1999) *Toxic Cyanobacteria in Water: A Guide to their Public Health Consequences, Monitoring and Management*. London/Nova York, WHO, E. & F.N. Spon.

Cintra, R. & Yamashita, C. (1990) Habitats, Abundância e Ocorrência das Espécies de Aves do Pantanal de Paconé, Mato Grosso, Brasil. *Pap. Avul. Zool.*, São Paulo, 37 (1), 1–21.

Clements, F.E. (1916) *Plant Succession: An Analysis of the Development of Vegetation*. Washington, Carnegie Institution of Washington. (Publ. 242).

Cohen, R.S., Blomberg, F., Berzins, K. & Siekevitz, P. (1977) The structure of postsynaptic densities isolated from dog cerebral cortex: I. Overall morphology and protein composition. *J. Cell Biol.*, Nova York, 74, 181–203.

Cole, G.A. (1983) *Textbook of Limnology*. 3rd edition. St. Louis, Toronto, London, The C.V. Mosby Company.

Cole, G.M. (1994) *Assessment and Remediation of Petroleum-Contaminated Sites*. Boca Ratón, EUA: Lewis Publishers.

Conde, C. & Sommaruga, R. (1999) A review of the state of limnology in Uruguay. In: Wetzel, R.G., Gopal, B. (eds.) *Limnology in Developing Countries*, vol. 2. Nova Deli, International Association for Limnology. pp. 1–31.

Constanza, R. et al. (1997) The value of the world's ecosystem services and natural capital. *Nature*, Nova York, 387, 253–260.

Conteras-Ramos, A. (1999) List of species of neotropical Megaloptera (Neuropterida). *Proc. Entomol. Soc. Washington*, Washington, 101, 274–284.

Cooke, G.D., Welch, E.B., Peterson, S.A. & Newroth, P.R. (1993) *Restoration and Management of Lakes and Reservoirs*. Boca Ratón, EUA, Lewis Publisher.

Cope, E.D. (1894) On the fishes obtained by the naturalist expedition in Rio Grande do Sul. *Proc. Am. Philos. Soc.*, Philadelphia, EUA, 33, 144, 84–108.

Copeland, B.J. (1967) Environmental characteristics of hypersaline lagoons. *Publ. Univ. Texas Inst. Marine Science*, Port Aransas, EUA, 12, 207–218.

Costa, S.M. & Azevedo, S.M.F.O. (1994) Implantação de um banco de culturas de cianofíceas tóxicas. Iheringia. *Série Botân.*, Porto Alegre, 45, 69–74.

Costa, W.J.E.M. (1987) Feeding habits of a fish community in a tropical coastal stream, Rio Mato Grosso, Brazil. *Stud. Neotrop. Fauna Environ.*, Londres, 3, 145–153.

Coveney, P. & Highfield, R. (1995) *Frontiers of Complexity: The Search for Order in a Chaotic World.* London, Faber and Faber.

Covich, A.P. (1988) Geographical and historical comparisons of neotropical streams: Biotic diversity and detrital processing in highly variable habitats. *J. N. Am. Benthol. Soc.*, Salt Lake City, EUA, 7, 361–386.

Cowgill, U.M. (1977a) Actividades Humanas y sus Efectos en Lagos Permanentes, In: Organización de los Estados Científicos y Tecnológicos. *Seminario Sobre Medio Ambiente y Represas*, Montevidéo, Tomo I. pp. 281–312, 368.

———. (1977b) The Hydrogeochemistry of Linsley Pond, North Brandorf, Zones of the Lake. *Arch. Hydrobiol. Suppl.*, Stuttgart, 53, 1–47. In: Organización de los Estados Científicos y Tecnológicos. *Seminario Sobre Medio Ambiente y Represas*, Montevidéo, Tomo I. p. 368.

———. (1977c) The hydrogeochemistry of Linsley Pond, North Branford, Connecticut. The mineralogy of the rocks and soils of the basin and the surface muds of the lake. *Arch. Hydrobiol.*, Stuttgart, 79 (1), 36–61.

Cowgill, U.M. & Hutchinson, G.E. (1966) El Lago de Santa Fé. *Trans. Amer. Philos. Soc. (N.S.)*, Philadelphia, EUA, 53, 7.

———. (1970) Chemistry and mineralogy of the sediments and their source minerals. *Trans. Amer. Philos. Soc.*, Philadelphia, EUA.

Crul, C.M. (1997) *Limnology and Hydrology of Lakes Tanganyka and Malawi: Studies and Reports in Hydrology*, vol. 54. UNESCO.

Cruz, O.G. (1893) *Un Nouvel Appareil pour la Recolte des eaux à Differentes Profondeurs pour L'analyse des Microbes.* Rio de Janeiro, Imprensa Brasileira.

Cunha, M.B. & Bianchini, Jr., I. (1998) Cinéticas de mineralização aeróbia de celulose e lignina durante a degradação de *Cabomba piauhyensis* e *Scirpus cubensis*. *Acta Limnol. Brasil.*, Campinas, 10 (2), 59–69.

Cuvier, G. (1817) *Le règne animal distribué d'après son organisation.* Tome II, contenant les reptiles, les poissons, les mollusques et les annélidés. Paris, Deterville.

Cuvier, G.L.C.F.D. (1830) *Le Règne Animal distribué d'après son Organisation.* 2nd edition. vol. 3. Paris, Déterville. pp. 162–170.

Da Silva C.J. (1990) *Influência da Variação do Nível da Água sobre a Estrutura e Funcionamento de uma Área Alagável do Pantanal Matogrossense (Pantanal de Barão de Malgaço, Município de Santo Antônio de Leverger e Barão de Melgaço – MT).* [Tese de Doutorado]. São Carlos, UFSCar, Departamento de Ciências Biológicas.

Da Silva, C.J. & Esteves, F.A. (1993) Biomass of three macrophytes in the Pantanal of the Mato Grosso, Brazil. *Int. J. Ecol. Environ. Sci.*, Leiden, Holanda, 19, 11–23.

———. (1995) Dinâmica das Características Limnológicas das Baías Porto de Fora e Acurizal em Função da Variação de Nível da Água (Pantanal de Mato Grosso). I. *Estrutura, funcionamento e manejo de ecossistemas brasileiros.* Rio de Janeiro, UFRJ. pp. 47–60. (Série Oecologia brasiliensis).

Dahl, F. (1894) Die Copepodenfauna des Unteren Amazonas. *Ber. Naturf. Ges.*, Freiburg, 8, 10–23.

Darrigran, G. (2005) Prevención y Control de Bivalvos de Agua Dulce: Caso Mejilón Dorado en la Región Neotropical. In: Nogueira, M.G., Henry, R., Jorcin, A. (Orgs.) *Ecologia de Reservatórios*: Impactos Potenciais, Ações de Manejo e Sistema em Cascata, Avaré, vol. 9. pp. 235–250.

Davies B.R. & Walker, K.F. (eds.) (1986) *The Ecology of River Systems.* Dordrecht, Holanda, Dr. W. Junk Publishers.

Dawson, A. (1992) *Ice Age Earth: Late Quaternary Geology and Climate.* Londres, Routledge.

De Bernardi, R. & Giussiani, G. (eds.) (2001) *Diretrizes para o Gerenciamento de Lagos*. Vol. 7: Biomanipulação para o Gerenciamento de Lagos e Reservatórios. Trad. Dino Vannuccci. São Carlos, ILEC/IIE.

De Meis, M.R. & Tundisi, J.G. (1997) Geomorphological and limnological processes as a basis for lake typology: The Middle Rio Doce Lake System. In: Tundisi, J.G. & Saijo, Y. (eds.) *Limnological Studies on the Rio Doce Valley Lakes*, Brazil. São Paulo, USP, EESC, CRHEA. pp. 25–48.

Decamps, H. (1996) The renewal of floodplain forests along rivers: A landscape perspective. *Verh. Internat. Verein. Limnol.*, 26 (1), 35–59.

Degens, E.T., Von Herzen, R.P., Wong, H.K., Deuser, W.G. & Jannasch, H.W. (1973) Lake Kivu, structure and chemistry of East African Rift Lake. *Geol. Rdsch.*, Berlim, 62, 245–377.

Degens, E.T., Kempe, S. & Ritchie, J.E. (eds.) (1991) *Biogeochemistry of the Major World Rivers*. Chichester, Inglaterra, SCOPE, John Wiley & Sons. (SCOPE 42).

Depetris, P.J. (2007) The geochemistry of the Parana River: An overview. In: Iriondo, M.H. et al. (eds.) *The Middle Parana River: Limnology of a Subtropical Wetland*. Germany, Springer Heidelberg. pp. 143–174.

Depetris, P.J. & Paolini, J.E. (1991) Biogeochemical aspects of South American Rivers: The Paraná and the Orinoco. In: Degens, E.T., Kempe, S. & Ritchie, J.E. (eds.) *Biogeochemistry of the Major World Rivers*. Chichester, Inglaterra, John Wiley & Sons. pp. 105–122. (SCOPE 42).

Di Persia, D.H. & Olazarri, J. (1986) Zoobenthos of the Uruguay system. In: Davies, B.R. & Walker K.F. (eds.) *The Ecology of River Systems*, Dordrecht, Holanda, Dr. W. Junk Publishers. pp. 623–629.

Dillon, T.M. (1982) Vertical overturns: A comparison of Thorpe and Ozmidov length scales. *J. Geophys. Res.*, Washington, 87, 9601–9613.

Dillon, P.J. & Rigler, F.H. (1974) The phosphorus–chlorophyll relationship in lakes. *Limnol. Oceanogr.*, Milwaukee, EUA, 19, 767–773.

Dodson, S.I. (2005) *Introduction to Limnology*. Nova York, McGraw-Hill.

Dubois, T. (1955) Epidemiological aspects of pollution of indoor swimming pools and problems of their purification. *Arch. Belg. Med. Soc.*, Bruxelas, 13 (1), 1–16.

Dumont, H.J. (1986a) Zooplankton of the Niger system. In: Davies, B.R. & Walker, K.F. (eds.) *The Ecology of River Systems*. Dordrecht, Holanda, Dr. W. Junk Publishers.

———. (1986b) Zooplankton of the Nile system. In: Davies, B.R. & Walker, K.F. (eds.) *The Ecology of River Systems*. Dordrecht, Holanda, Dr. W. Junk Publishers.

———. (1992) The regulation of plant and animal species and communities in African shallow lakes and wetlands. *Rev. Hydrobiol. Trop.*, Paris, 4, 303–346.

———. (2005) Biodiversity: A resource with a monetary value? *Hydrobiologia*, Baarn, Holanda, 542, 11–14.

Dumont, H.J., Miron, I., Dall'asta, U., Decraemer, W., Claus, C. & Somers, D. (1973) Limnological aspects of some Moroccan Atlas Lakes, with reference to some physical and chemical variables, the nature and distribution of the phyto- and zooplankton, including a note on possibilities for the development of an inland fishery. *Int. Rev. Ges. Hydrobiol.*, Weinheim, Alemanha, 58, 33–60.

Dumont, H.J., Tundisi, J.G. & Roche, K. (eds.) (1990) *Intrazooplankton Predation*. Dordrecht, Holanda, Kluwer.

Dunlap, W.J. & MacNabb, J.F. (1973) *Subsurface biological activity in relation to ground water pollution*. Washington, U.S. Environmental Protection Agency. (Environmental protection technology series, EPA-660/2-73-014).

Dussart, B. (1966) *Limnologie: L'Etude des Eaux Continentales*. Paris, Gauthier Villars.

———. (1984) Some crustacea. Copepoda from Venezuela. *Hydrobiologia*, Baarn, Holanda, 113, 15–23.

Dussart, B. & D. Defaye. (2002) *World Directory of Crustacea*. Copepoda of inland waters. I. Calaniformes. Leiden, Holanda, Backhuys Publishers.

Dussart, B.H., Fernando, C.H., Matsumura-Tundisi, T., Shiel, R.J. (1984) A review of systematics, distribution and ecology of tropical freshwater zooplankton. *Hydrobiologia*, Baarn, Holanda, 113, 77–91.

Dyrssen, D. & Wedborg, M. In: Goldberg, E.D. (ed.) (1974) *The Sea: Ideas and Observations on the Nature of Seawater*. Nova York, Wiley-Interscience. pp. 181–195.

Eberly, W.R. (1975) The use of oxygen deficit measurements as an index of eutrophication in temperate dimictic lakes. *Verh. Internat. Verein. Limnol.*, Stuttgart, 19, 439–441.

Edmondson, W.T. (1960) Reproductive rates of rotifers in natural populations. *Mem. Ist. Ital. Idrobiol.*, Verbiana-Pallanza, Itália, 12, 21–77.

Edmondson, W.T. & Wimberg, G.C. (1971) *A Manual on Methods for the Assessment of Secondary Productivity in Freshwaters*. 1st edition. Vol. 17. Oxford, Blackwell.

Elouard J.-M. & Gibon F.-M. (2001) *Biodiversité et Biotypologie des eaux Continentales de Madagascar*. Montpellier, França, IRD, CNRE, LRSAE.

Elster, H.J. (1954) Über die Populationsdynamik von *Eudiaptomus gracilis* Sars und *Heterocope borealis* Fischer im Bodensee-Obersee. *Arch. Hydrobiol. Suppl.*, Stuttgart, 20, 546–614.

———. (1974) History of limnology. *Verh. Internat. Verein Limnol.*, Stuttgart, 20, 7–30.

Elster, H.J. & Ohle, W. (1978) *Ergebnisse der Limnologie*. Stuttgart.

Esteves, F.A. (1979) Die Bedeutung der aquatischen Macrophyten für den Stoffhaushaalt des Schonsees. II. Die organischen Hauptstandteile und Energiegehalt der aquatischen Macrophyten. *Arch. Hydrobiol. Suppl.*, Stuttgart, 57, 144–187.

———. (1983) Levels of phosphate, calcium, magnesium and organic matter in the sediments of some Brazilian reservoirs and implications for the metabolism of the ecosystems. *Arch. Hydrobiol.*, 96, 129–138.

———. (1988) *Fundamentos de Limnologia*. Rio de Janeiro, Interciência/FINEP.

———. (1998) *Ecologia das Lagoas Costeiras do Parque Nacional da Restinga de Jurubatiba e do Município de Macaé*. Rio de Janeiro, Nupem, Dep. Ecologia, UFRJ.

Esteves, K.E. & Aranha, J.M. (1999) Ecologia trófica de peixes de riachos. In: Caramaschi, E.P., Mazzoni, R. & Peres-Neto, P.R. (eds.) *Ecologia de Peixes de Riachos*. Rio de Janeiro, UFRJ. pp. 157–182. (Série Oecologia brasiliensis).

Esteves, F.A., Ferreira, J.R., Pessenda, L.C.R. & Mortati, J. (1981) Análises Preliminaries Sobre o Teor e a Distribuição de Metais em Sedimento do Estado de São Paulo. *Anais do II Simpósio Regional de Ecologia*. São Carlos, UFSCar. pp. 323–345.

Esteves, F.A., Ishii, I.H. & Camargo, A.F.M. (1984) Pesquisas limnológicas em 14 lagoas do estado do Rio de Janeiro. In: Lacerda, L.D., Araujo, D.S.D., Cerqueira, R. & Turcq, B. (Orgs.) *Restingas*: Origem, Estrutura e Processos. Niterói, CEUFF. pp. 443–454.

Esteves, F.A., Amorim, J.C., Cardoso, E.L. & Barbosa, F.A.R. (1985) Caracterização limnológica preliminar da Represa de Três Marias (MG) com base em alguns parâmetros ambientais básicos. *Cie. e Cult.*, Campinas, 37 (4), 608–617.

Esteves, F.A., Barbosa, F. & Bicudo, C.E.M. (1995) Limnology in Brazil: Origin, development and perspectives. In: Tundisi, J.G., Bicudo, C.M. & Matsumura-Tundisi, T. (eds.) *Limnology in Brazil*. Rio de Janeiro, ABC/SBL.

Ezcurra De Drago, I. (1974) Las espécies sudamericanas de Corvomeyenia Weltner (Porifera, Spongillidae). *Physis*, Buenos Aires, B, 33 (87), 233–240.

Fairbridge, R.W. (ed.) (1968a) *Encyclopedia of Geomorphology*. Nova York, Reynold.

Fairbridge, R.W. (1968b) Lake Maracaibo. In: Fairbridge, R.W. (ed.) *Encyclopedia of Geomorphology*. Nova York, Reynold. pp. 614–616.

Falkenmark, M. (1999) Lakes in a global perspective: Pearls in a river string. Keynote paper. *Sustainable Lake Management Conference*. Copenhagen.

Falkenmark, M., Anderson, L., Castensson, R., Sundblad, K., Batchelor, C., Gardiner, J., Lyle, C., Peters, N., Pettersen, B., Quinn, P., Rockstrom, J. & Yapijakis, C. (1999) *Water: A Reflection of Land Use. Options for Countering Land and Water Mismanagement.* Estocolmo, Conselho Suéco de Pesquisa em Ciências Naturais.

Falkowski, P.G., Miriam, E.K., Andrew, H.K., Antonietta, Q., Raven, J.A., Schofield, O. & Taylor, F.J.R. (2004) The evolution of modern eukaryotic phytoplankton. *Science*, Washington, 305.

Falótico, M.H.B. (1993) *Características Limnológicas e Aspectos da Composição e Distribuição da Comunidade Zooplanctônica em um Reservatório da Região Amazônica em sua Fase de Enchimento (Reservatório de Samuel, Rondônia).* [Tese de Mestrado]. São Carlos, USP, Escola de Engenharia de São Carlos.

Fernandes, V.O. (1993) *Estudos limnológicos na Lagoa de Jacarepaguá (RJ): variáveis abióticas e mudanças na estrutura e dinâmica da comunidade perifítica em* Typha domingensis Pers. [Tese de Mestrado]. São Carlos, UFSCar.

———. (1997) Variations in dry weight, organic matter and chlorophyll a of the periphytic community in Imboassica lagoon, Rio de Janeiro, RJ, Brazil. *Verh. Internat. Verein. Limnol.*, Stuttgart, 26, 1445–1447.

Fernando, C.H. & Holcik, J. (1991) Fish in reservoirs. *Int. Rev. Ges. Hydrobiol.*, Weinheim, Alemanha, 76, 149–167.

Fernando, C.H., Tudorancea, C. & Mengestou, S. (1990) Invertebrate zooplankton predator composition and diversity in tropical lentic waters. *Hydrobiologia*, Baarn, Holanda, 198, 13–31.

Fisher, H.B., List, E.J., Koh R.C.Y., Imberger J. & Brooks N.H. (1979) *Mixing in Inland and Coastal Waters.* Nova York, Academic Press.

Fitkau, E.J., Irmler, U., Junk, W.J., Reiss, F. & Schmidt, G.W. (1975) Productivity, biomass and population dynamics in Amazonian water bodies. In: Golley, F.B. & Me Dina, E. (eds.) *Tropical Ecological Systems: Trends in Terrestrial and Aquatic Research.* Nova York, Berlim, Springer-Verlag.

Fogg, G.E. (1965) *Algal Cultures and Phytoplankton Ecology.* Madison, University of Wisconsin Press.

Forbes, S.T. (1887) The Lake as a microcosm. *Bull. Peoria (Ill.) Sci. Assoc.*, Peoria, EUA, Peoria Sci. Ass. pp. 77–87.

Forel, F.A. (1892) *Le Léman: monografie limnologique.* Tome 1. Géographie, Hidrographie, Géologie, Climatologie, Hydrologie. Lausanne, F. Rouge, xiii.

Forel, F.A. (1895) Le *Léman: monografie limnologique.* Tome 2. Mécanique Chimie, Thermique, Optique, Acoustique. Lausanne, F. Rouge.

Forel, F.A. (1901) *The Study of Lakes.* Handbuch der Seenkunde. Stuttgart, Verlag Von Engelhorn.

Forel, F.A. (1904) *Le Léman: monografie limnologique.* Tome 1. Biologie, Histoire, Navegation, Peche. Lausanne, F. Rouge.

Forte Pontes, M.C. (1980) *Produção Primária, Fitoplâncton e Fatores Ambientais do Lago D. Helvécio, Parque Florestal do Rio Doce, MG.* [Tese de Mestrado]. São Carlos, UFSCar, Programa de Pós-Graduação em Ecologia Rec. Nat.

Fowler, H.W. (1941) A collection of freshwater fishes obtained in Eastern Brazil by Dr. Rodolph Von Ihering. *Proc. Acad. Nat. Sci.*, Philadelphia, 93, 123–199.

Franco, G.M.M. (1982) *Ciclo sazonal da produção primária, "standing-stock" do fitoplâncton e fatores ambientais na Represa de Jacaré-Pepira (Brotas, SP).* [Tese de Doutorado]. São Carlos, UFSCar.

Frank, L.A. (1990) *The Big Splash.* Secaucus, EUA, Carol Publishing.

Frey, D.G. (ed.) (1969) Symposium on Paleo-Limnology. *Mitt. Internat. Verein. Limnol.*, Stuttgart, 17.

Funk, J.L. & Campbell, R.S. (1953) The population of larger fishes in Black River, Missouri. *Univ. Mo. Stud.*, Columbia, EUA, 26, 69–82.

Furch, K. (1984) Water chemistry of the Amazon Basin: The distribution of chemical elements among freshwaters. In: Sioli, H. (ed.) *The Amazon: Limnology and Landscape Ecology of a Mighty Tropical River and its Basin*. Dordrecht, Holanda, Dr. W. Junk Publishers. pp. 167–200.

Furnas. (2005) *Mexilhão-dourado: Cartilha*. Rio de Janeiro.

Gaarder, T. & Gran, H.H. (1927) Investigations of the Production of Plankton in the Oslo Fjord. *Rapp. Proc. Verb. Cons. Int. Explor. Mer*, Copenhaguem, 42, 3.

Ganf, G.G. (1972) The regulation of net primary production in Lake George, Uganda, East Africa. In: Kajak, Z. & Hillbricht-Ilkowska, A. (eds.) *Productivity Problems of Freswaters*. Cracóvia, Polish Scientific Publishers. pp. 693–708.

———. (1974) Phytoplankton biomass and distribution in a shallow eutrophic lake (Lake George, Uganda). *Oecologia*, Berlim, 16, 9–29.

———. (1975) Photosynthetic production and irradiance-photosynthesis relationships of the phytoplankton from a shallow equatorial lake (Lake George, Uganda). *Oecologia*, Berlim, 18, 165–183.

Ganf, G.G. & Horne, A.J. (1975) Diurnal stratification, photosynthesis and nitrogen fixation in a shallow, equatorial lake (Lake George, Uganda). *Freshwat. Biol.*, Victoria, Austrália, 5, 13–39.

Ganf, G.G. & Viner, A.B. (1973) Ecological stability in a shallow equatorial lake (Lake George, Uganda). *Trans. R. Soc. B.*, Londres, 84, 321–346.

Garcia, C.A.E. (1997) Hydrographic characteristics. In: Seeliger, U., Odebrecht, C. & Castello, J. (eds.) *Subtropical Convergence Marine Ecosystem. The Coast and the Sea in the Warm Temperate Southwestern Atlantic*. Heidelberg, Alemanha, Springer Verlag, Heidelberg. pp. 18–20.

Garman, S. (1980) On the species of the genus *Anostomus*. *Bull. Essex Inst.*, Essex, 22, 15–23.

Garrels, R.M. & Thompson, M.E. (1962) A chemical model for sea water at 25 degrees C and one atmosphere total pressure. *Am. J. Sci.*, New Haven, EUA, 260, 57–66.

Garutti, V. (1988) Distribuição longitudinal da ictiofauna de um córrego na região noroeste do Estado de São Paulo, Bacia do Rio Parana. *Rev. Brasil. Biol.*, São Carlos, 48, 747–759.

Gessner, F. (1955) *Hydrobotanik, Die physiologisken Grundlagen der Pflanzenverbreitung im Wasser*. I. Energiehaus-halt. Berlim, VEB Deutscher Verlag der Wissenschaften.

———. (1959) *Hydrobotanik, Die physiologisken Grundlagen der Pflanzenverbreitung im Wasser*. II. Stoffhaushalt. Berlim, VEB Deutscher Verlag der Wissenschaften.

Gibbs, R.J. (1970) *Mechanisms Controlling World Water Chemistry*. New York, Science. pp. 1088–1090.

———. (1972) Water chemistry of the Amazon River. *Geochim. Cosmoch. Acta*, St. Louis, EUA, 36, 1061–1066.

Gill, T. (1861) Notes on some genera of fishes of the western coast of North America. *Proc. Acad. Nat. Sci. Philad.*, Philadelphia, 13, 164–168.

Giller P.S. & Malmqvist B. (1998) *The Biology of Streams and Rivers*. Oxford, Oxford University Press.

Godinho, A.L. & Godinho, H.P. (2003) Breve visão do São Francisco. In: Godinho, H.P. & Godinho, A.L. (eds) *Águas, Peixes e Pescas no São Francisco, Minas Gerais*. Belo Horizonte, CNPq/PADCT, Editora PUC Minas. pp. 15–24.

Goldman, C.R. & Horne, A.J. (1983) *Limnology*. Nova York, McGraw-Hill.

Golterman, H.L. (1975) *Physiological Limnology*. Amsterdã, Elsevier.

———. (1988) The calcium- and ironbound phosphate phase diagram. *Hydrobiologia*, Baarn, Holanda, 159, 149–151.

Golterman, H.L., Clymo, R.S. & Ohmstad, M.A. (1978) *Methods for Chemical, Physical and Chemical Analysis of Freshwaters*. Oxford, Blackwell Scientific Publications.

Gomes, J.H.C. (1994) *Distribuição, alimentação e período reprodutivo de duas espécies de Tetragonopterinae (Osteichthyes) sintópicas no rio Ubatiba (Maricá, RJ)*. [Tese de Mestrado]. Rio de Janeiro, UFRJ, Museu Nacional.

Gonçalves, J.F., Callisto, M.F.P. & Fonseca, J.J.L. (1998) Relação entre a composição granulométrica do sedimento e as comunidades de macroinvertebrados bentônicos nas lagoas Imboassica, Cabiúnas e Comprida. In: Esteves, F.A. (ed.) *Ecologia das Lagoas Costeiras do Parque Nacional da Restinga de Jurubatiba e do Município de Macaé (RJ)*, cap. 3.2. Rio de Janeiro, NUPEM/UFRJ.

Gonzales, E., Ortaz, M., Peñaherrera, C., Matos, M. & Mardoza, J. (2003) Fitoplancton de Cinco Embalses de Venezuela con Diferentes Estados Tróficos. *Limnética*, Barcelona, 22 (1–2), 15–35.

Gonzalez, E.J. (1998) Natural diet of zooplankton in a tropical reservoir (Embalse El Andino, Venezuela). *Verh. Internat. Verein. Limnol.*, Stuttgart, 26, 1930–1934.

Gonzalez, E.J., Matsumura-Tundisi, T. & Tundisi, J.G. (2006) Dieta natural del zooplancton en dos embalses tropicales (Broa y Barra Bonita, Brasil) con diferentes estados tróficos. In: Tundisi, J.G., Matsumura-Tundisi, T. & Galli, S.C. (eds.) *Eutrofização na América do Sul*. São Carlos, IIE. pp. 457–472.

Gonzales-Rio, T. & Revilla M.A. (1991) Evaluation of parameters affecting yield viability and cell division of Actinidiaceae protoplasts. 8th International protoplast symposium. *Physiol. Plant.*, Copenhaguem, 82 (1), A85.

Gopal, B. (1973) A survey of the Indian studies on ecology and production of wetland and shallow water communities. *Pol. Arch. Hydrobiol.*, Varsóvia, 20 (1), 21–29.

Goulder, L.H. & Kennedy, D. (1997) Valuing ecosystem services: Philosophical bases and empirical methods. In: Daily, G.C. (ed.) *Nature's Services: Societal Dependence on Natural Ecosystems*. Washington, Island Press.

Goulding, M. (1979) *Ecologia da Pesca do Rio Madeira*. Manaus, Conselho Nacional de Desenvolvimento Científico e Tecnológico (INPA).

———. (1980) *The Fishes and the Forest: Explorations in Amazonian Natural History*. Berkeley, Los Angeles, Londres, University of California Press.

———. (1981) *Man and Fisheries on an Amazon Frontier*. Haia, Holanda, Dr. W. Junk Publishers.

———. (1999) Section I: Fish and Fisheries, Introduction. In: Padoch, C., Ayres, J.M., Pinedo-Vasquez, M. & Henderson, A. (eds.) *Varzea: Diversity, Development, and Conservation of Amazonia's Whitewater Floodplains*. Nova York, Botanical Garden Press. pp. 3–6.

Goulding, M., Carvalho, M.L. & Ferreira, E.G. (1988) *Rio Negro: Rich Life in Poor Water: Amazonian Diversity and Foodchain Ecology as seen Through Fish Communities*. Haia, Holanda, SPB Academic Publishing.

Gran, H.H. & Braarud, T. (1935) A quantitative study of the phytoplankton in the Bay of Fundy and the Gulf of Maine (including observations of hydrography, chemistry, and turbidity). *J. Biol. Bd. Can.*, Ottawa, Canadá, 1, 279–467.

Grannemann, G.G. et al. (2000) *The Importance of Ground Water in the Great Lakes Region*. Lansing, EUA, U.S. Geological Survey. (Water-Resources Investigations Report 00–4008).

Guerrero, R., Pedrós-Alió, C., Esteve, I. & Mas, J. (1987) Communities of phototrophic sulfur bacteria in lakes of the Spanish Mediterranean Region. *Acta Acad. Aboen.*, Abo, Finlândia, 47 (2), 125–151.

Guerrero, R.A., Acha, E.M., Framinan, M.B. & Lasta, C.A. (1997) Physical oceanography of the Rio de la Plata estuary, Argentina. *Cont. Shelf Res.*, Amsterdã, 17, 727–742.

Gunatilaka, A. (1988) *Nutrient Cycling in Tropical Asian Reservoirs: Some Important Aspects with Special Reference to Parakrama Samudra, Sri Lanka*. Ottawa, Canadá, IDRC.

Hakanson, L. (1981) *A Manual of Lake Morphometry*. Berlim, Springers Verlag.

Hakanson, L. & Peters, R.H. (1995) *Predictive Limnology: Methods for Predictive Modelling*. Amsterdã, SPB Academic Publishing.

Hall, H.J. (1972) *Evaluation of Measurement Methods and Instrumentation of Odorous Compounds in Stationary Sources*, Volume I: State of the Art. Athens, EUA, U.S. Environment Protection Agency. (U.S. EPA Publ. No. APTD–1180).

Hallam, J.C. (1959) Habitat and associated fauna of four species of fish in Ontario streams. *J. Fish. Res. Bd. Can.*, Ottawa, Canadá, 16, 147–173.

Hammer, U.T. (1986) *Saline Lake Ecosystems of the World*. Dordrecht, Holanda, Boston, EUA, Lancaster, Inglaterra, Dr. W. Junk Publishers, vol. 59. (Monographiae Biologicae, 59).

Hanor, J.S. (1969) Barite saturation in sea water. *Geochim. Cosmochim. Acta*, St. Louis, 33 (7), 894–898.

Harbott, B.J. (1982) Studies on algal dynamics and primary productivity in Lake Turkana. In: Hopson, A.J. (ed.) *Lake Turkana. A Report on the Findings of the Lake Turkana Project, 1972–1975*, vol. 1. Londres, Overseas Development Administration. pp. 109–161.

Hardy, E.R. (1980) Composição do zooplâncton em cinco lagos da Amazônia Central. *Acta Amazon.*, Manaus, 10 (3), 577–609.

Hardy, E.R., Robertson, B. & Koste, W. (1984) About the relationship between the zooplankton and fluctuating water levels of Lago Camaleão, a Central Amazonian varzea lake. *Amazoniana*, Kiel, Alemanha, 9 (1), 43–52.

Hare, L. & Carter, J.C.H. (1984) Diel and seasonal physico-chemical fluctuations in a small natural West African Lake. *Freshwat. Biol.*, Victoria, Austrália, 14 (6), 597–610.

Harris G.P. (1978) *Photosynthesis, Productivity and Growth: The Physiological Ecology of Phytoplankton*. Stuttgart, Schweizerbart'sche Verlagsbuchhandlung. (Ergebnisse der Limnologie, vol. 10).

———. (1984) Phytoplankton productivity and growth measurements: Past, present and future. *J. Plankton Res.*, Oxford, 6, 219–237.

———. (1986) *Phytoplankton Ecology: Structure, Function and Fluctuation*. Londres, Chapman and Hall.

———. (1987) Time series analysis of water quality data from Lake Ontario: Implications for the measurement of water quality in large and small lakes. *Freshwat. Biol.*, Victoria, Austrália, 18, 389–403.

Harrison, A.D. & Elsworth, J.F. (1958) Hydrobiological studies on the Great Berg River Western Cape Province. Part I. General description, chemical studies and main features on the flora and fauna. *Trans. R. Soc. Afr.*, Cidade do Cabo, África do Sul, 35, 125–226.

Hart, R.C. (1987) Population dynamics and production of five crustacean zooplankters in a subtropical reservoir during years of contrasting turbidity. *Freshwat. Biol.*, Victoria, Austrália, 18, 287–318.

Hassan, R., Scholes. R. & Ash, N. (2005) *Ecosystems and Human Well-being: Current State and Trends: Findings of the Condition and Trends Working Group*. Washington, Island Press. (Millennium ecosystem assessment, 1).

Hawkes, H.A. (1975) River zonation and classification. In: Whitton, B.A. (ed.) *River Ecology*. Berkely, EUA, University of California Press. pp. 313–374.

Heath, R. (1988) Ground water. In: Speidel, D.H., Ruesdisili, L.C. & Agnew, A.F. (eds.) *Perspectives on Water: Uses and Abuses*. Oxford, Oxford University Press., pp. 73–89.

Heckel, J.J. (1840) Johann Natterer's neue Flussfische Brasilien nach den Beobachtungen und Mitteilungen des Entdeckers beschrieben (Erste Abteilung, Die Labroiden). *Ann. Wien Mus. Naturgesell.*, Vienna, 2, 325–471.

Hecky, R.E. & Fee, E.J. (1981) Primary production rates of algal growth in Lake Tanganyika. *Limnol. Oceanogr.*, Milwaukee, EUA, 26, 532–544.

Hecky, R.E. & Kilham, P. (1973) Diatoms in alkaline lakes: ecology and geochemical implications. *Limnol. Oceanogr.*, Milwaukee, EUA, 18, 53–71.

Heide, J. Van der (1982) *Lake Brokopondo. Filling Phase Limnology of a Man Made Lake in the Humid Tropics.* [Tese de Doutorado.] Amsterdã, University of Amsterdã.

Henderson, H.F. & Welcomme, R.L. (1974) *The Relationship of Yield to Morphoedaphic Index and Numbers of Fishermen in African Inland Fisheries.* Roma, FAO/CIFA. (Occasional Paper I).

Henderson, H.F. & Sellers, B. (1984) *Engineering Limnology.* Londres, Pitman Publish. Inc.

Henderson, H.F., Sellers, B. & Markland, H.R. (1987) *Decaying Lakes: The Origins and Control of Cultural Eutrophication.* Chichester, Inglaterra, John Wiley & Sons.

Henning, M. & Kohl, J.G. (1981) Toxic blue-green algae water blooms found in some lakes in the German Democratic Republic. *Int. Rev. Ges. Hydrobiol.*, Weinheim, Alemanha, 66 (4), 553–561.

Henry, R. (1990) Amônia ou fosfato como agente estimulador do crescimento do fitoplâncton na Represa de Jurumirim (Rio Paranapanema, SP). *Rev. Brasil. Biol.*, São Carlos, 50 (4), 883–892.

————. (1992) The oxygen deficit in Jurumirim Reservoir (Paranapanema River, Sao Paulo, Brazil). *Jpn. J. Limnol.*, Osaka, Japão, 53 (4), 379–384.

————. (1993a) Thermal regime and stability of Jurumirim Reservoir (Paranapamena River, Sao Paulo, Brazil). *Int. Rev. Ges. Hydrobiol.*, Weinheim, Alemanha, 78, 501–511.

————. (1993b) Primary production by phytoplankton and its controlling factors in Jurumirim Reservoir (Sao Paulo, Brazil). *Rev. Brasil. Biol.*, São Carlos, 53 (3), pp. 489–499.

————. (1999a) Heat budgets, thermal structure and dissolved oxygen in Brazilian Reservoirs. In: Tundisi, J.G. & Straskraba, M. (eds.) *Theoretical Reservoir Ecology and its Applications.* Leiden, Holanda, IIE, BAS, Backhuys Publishers. pp. 125–152.

————. (1999b) *Ecologia de Reservatórios: Estrutura, Função e Aspectos Sociais.* Botucatu, Fapesp-Fundrio.

————. (Org.) (2003) *Ecótonos nas Interfaces dos Ecossistemas Aquáticos.* São Carlos, RIMA.

Henry, R. & Barbosa, F.A.R. (1989) Thermal structure, heat content and stability of two lakes in the National Park of Rio Doce Valley (Minas Gerais, Brazil). *Hydrobiologia*, Baarn, Holanda, 171, 189–199.

Henry, R. & Tundisi, J.G. (1982a) Efeitos do enriquecimento artificial por nitrato e fosfato no crescimento da comunidade fitoplanctônica da Represa do Lobo (Broa, Brotas, Itirapina, SP). *Ciência e Cultura*, Campinas, 34 (4), 518–524.

————. (1982b) Evidence of limitation by molybdenum and nitrogen on the growth of the phytoplankton community of the Lobo Reservoir (São Paulo, Brazil). *Rev. Hydrobiol. Trop.*, Paris, 15 (3), 201–208.

————. (1983) Responses of the phytoplankton community of a tropical reservoir (São Paulo, Brazil) to the enrichment with nitrate, phosphate and EDTA. *Int. Rev. Ges. Hydrobiol.*, Weinheim, Alemanha, 68 (6), 853–862.

————. (1986) Artificial enrichment and its effects on the surface phytoplankton of Lake Dom Helvecio (River Doce, MG, Brazil) during the isothermy period. *Anais do Simpósio Internacional de Algas: A Energia do Amanhã.* São Paulo, Instituto Oceanográfico, USP.

————. (1988) O Conteúdo de Calor e a Estabilidade em Dois Reservatórios com Diferentes Tempos de Residência. In: Tundisi, J.G. (ed.) *Limnologia e Manejo de Represas*, vol. I, Tomo 1. São Carlos, ACIESP/EESC/USP. pp. 299–322.

Henry, R., Hino, K., Gentil, J.G. & Tundisi, J.G. (1985a) Primary production and effects of enrichment with nitrate and phosphate on phytoplankton in the Barra Bonita Reservoir (State of São Paulo, Brazil). *Int. Rev. Ges. Hydrobiol.*, Weinheim, Alemanha, 70, 561–273.

Henry, R., Hino, K., Tundisi, J.G. & Ribeiro, J.S.B. (1985b) Responses of phytoplankton in Lake Jacaretinga to enrichment with nitrogen and phosphorus in concentrations similar to those in the River Solimões (Amazon, Brazil). *Arch. Hydrobiol.*, Stuttgart, 103 (4), 453–477.

Henry, R., Tundisi, J.G. & Curi, P.R. (1983) Fertilidade potencial em ecossistemas aquáticos: Estimativa através de experimentos de eutrofização artificial. *Ciência e Cultura*, Campinas, 35 (6), 789–804.

———. (1984) Effects of phosphorus and nitrogen enrichment on the phytoplankton in a tropical reservoir (Lobo Reservoir, Brazil). *Hydrobiologia*, Baarn, Holanda, 118, 25–41.

Henry, R., Pontes, C.F. & Tundisi, J.G. (1989) O Déficit de Oxigênio no Lago Dom Helvécio (Parque Florestal do Rio Doce, Minas Gerais). *Rev. Brasil Biol.*, São Carlos, 49 (1), 251–260.

Hill, G. & Rai, H. (1982) A preliminary characterization of the tropical lakes of the Central Amazon by comparison with polar and temperate systems. *Arch. Hydrobiol.*, 96, 97–111.

Hino, K., Tundisi, J.G. & Reynolds, C.S. (1986) Vertical distribution of phytoplankton in a stratified lake (Lago Dom Helvécio, Southeastern Brazil). *J. Limnol.*, Verbiana-Pallanza, Itália, 47 (3), 239–246.

Horie, S. (ed.) (1984) *Lake Biwa*. Dordrecht, Holanda, Dr. W. Junk Publishers.

———. (1987) *History of Lake Biwa*. Kyoto, Universidade de Kyoto.

Horie, S. (1984) Advanced report on the Pleistocene glaciation in the Japanese islands in connection with Lake Biwa stratigraphic study. *Proc. Jpn. Acad. Ser. B Phys. Biol. Sci.*, Tóquio.

Horne, A.J. & Goldman, C.R. (1994) *Limnology*. 2nd edition. Nova York, McGraw-Hill.

Hoskin, C.M. (1959) Studies of oxygen metabolism of streams of North Carolina. *Publ. Inst. Mar. Sci. Univ. Tex.*, Austin, EUA, 6, 186–192.

Howard-Williams, C. & Lenton, G.M. (1975) The role of the littoral zone in the functioning of a shallow lake ecosystem. *Freshwat. Biol.*, Victoria, Austrália, 5, 445–459.

Howard-Williams, C. & Junk, W.J. (1977) The chemical composition of central Amazonian aquatic macrophytes with special reference to their role in the Ecosystem. *Arch. Hydrobiol.*, 79 (4), 446–464.

Huet, M. (1949) Aperçu des relations entre la pente et les populations piscicoles des eaux courantes. *Schweiz. Z. Hydrol.*, Zurique, 11, 332–351.

Huet, M. (1954) Biologie profils en long et en travers des eaux courantes. *Bull. Fr. Piscic.*, Paris, 178, 41–53.

Hufschmidt, M.M. & McCauley, D. (1986) *Strategies for Integrated Water Resources Management in a River/Lake Basin Context*. Otsu/Nagoya, UNEP, UNCRD, ILEC.

Hulbert, S.H. (1978) The measurement of niche overlap and some relatives. *Ecology*, Washington, 59, 67–77.

———. (1981) *Andean Lake and Flamingo Investigations*. Tech. Report 2. Results of Three Flamingo Censuses conducted between December 1978 and July 1980. San Diego, University of San Diego.

———. (1982) Limnological studies of flamingo diets and distributions. *Nat. Geogr. Soc. Res. Rep.*, Washington, 14, 351–356.

Hurlbert, S.H., Lopez, M. & Keith, J.O. (1984) Wilson's phalarope in the Central Andes and its interaction with the Chilean flamingo. *Rev. Chile. Hist. Nat.*, Santiago, Chile, 57, 47–57.

Huszar, V.L.M. & Reynolds, C.S. (1997) Phytoplankton periodicity and sequences of dominance in an Amazonian flood-plain lake (Lago Batata, Pará, Brazil): Responses to gradual environmental change. *Hydrobiologia*, Baarn, Holanda, 346, 169–181.

Huszar, V.L.M., Caraco, N.F., Roland, F. & Cole, J. (2006) Nutrient chlorophyll relationships in tropical-subtropical lakes: Do temperate models fit? *Biogeochemistry*, Spring.

Hutchinson, G.E. (1957) *A Treatise on Limnology*. Vol. I. Geography Physics and Chemistry. Nova York, John Wiley & Sons.

———. (1958) Concluding Remarks. *Cold Spring Harbour Symposium of Quantitative Biology*, 22, 415–427.

————. (1967) *A Treatise on Limnology. Vol. II. Introduction to Lake Biology and the Limnoplankton.* Nova York, John Wiley & Sons.

————. (1975) *A Treatise on Limnology. Vol. III. Limnological Botany.* Nova York, John Wiley & Sons.

————. (1993) *A Treatise on Limnology. Vol. IV. The Zoobenthos.* Nova York, John Wiley & Sons.

Hynes, H.B.N. (1961) The invertebrate fauna of a Welsh mountain stream. *Arch. Hydrobiol.*, Stuttgart, 57, 344–388.

————. (1970) *The Ecology of Running Waters.* Toronto, University of Toronto Press.

————. (1975) Downstream drift of invertebrates in a river in southern Ghana. *Freshwat. Biol.*, Victoria, Austrália, 5, 515–532.

Illies, J. (1953) Die Besiedlung der Fulda (inbesondere das Benthos der Salmoniden region) nach dem jetzigen Stand der Untersuchung. *Ber. Limnol. Flusstn. Freudenthal*, Munique, 5, 1–28.

————. (1961a) Versuch einer allgemein biozönotischen Gleiderung der Fliessgewässer. *Int. Rev. Ges. Hydrobiol.*, Weinheim, Alemanha, 46, 205–213.

————. (1961b) Gebirgsbäch in Europa und in Südamerika – ein limnologischer Vergleich. *Verh. Int. Verein. Theor. Angew. Limnol.*, Stuttgart, 14, 517–523.

Illies, J. & Botosaneanu, L. (1963) Problèmes et méthodes de la classification et de la zonation écologique des eaux courantes, considérées surtout du point de vue faunistique. *Mitt. Internat. Verein. Theor. Angew. Limnol.*, 12, 1–57.

Imberger, J. (1994) Transport process in lakes: A review. In: Margalef, R. (ed.) *Limnology Now: A Paradigm of Planetary Problems.* Nova York, Elsevier Science. pp. 99–194.

Imberger, J. & Patterson, J.C. (1981) A dynamic reservoir simulation model: Dyresm 5. In: Fisher, H.B. (ed.) *Transport Models for Inland Coastal Waters.* New York, Academic Press. pp. 310–316.

Imboden, D., Weiss, R.F., Craing, H., Michel, R.L. & Goldman, C.R. (1977) Lake chemistry and tritium mixing study, *Limnol. Oceanogr.*, Milwaukee, EUA, 22, 1039–1051.

Infante, A. (1978) Natural food of herbivorous zooplankton of Lake Valencia (Venezuela). *Arch. Hydrobiol.*, Stuttgart, 82, 347–358.

————. (1988) *El Plankton de las Águas Continentales.* Washington, Secretaria General de la Organización de los Estados Americanos. (Série de Biologia, Monografia 33).

Infante, A. & Riehl, W. (1984) The effect of cyanophyta upon zooplankton in an eutrophic tropical lake (Lake Valencia, Venezuela). In: Dumont, H.J. (ed.) *Tropical Zooplankton.* Haia, Holanda, Dr. W. Junk. pp. 293–298.

Infante, A., Infante, O. & Vegas, T. (1992) *Caracterización limnológica de los embalses Camatagua, Guanapito y Lagartijo, Venezuela.* Proyecto Multinacional de Medio Ambiente y Recursos Naturales. Washington, Caracas, Organización de los Estados Americanos, Universidad Central de Venezuela.

International *Scientific Workshop on Ecosystem Dynamics in Freshwater Wetlands and Shallow Water Bodies* (1981) Minsk-Pinsk-Tskaltubo. Proceedings. Moscou, Centre of International Projects GKNT, 1982.

IPCC WGI. (1990) Climate change: The IPCC scientific assessment. In: Houghton, R.A. et al. (eds.) Cambridge University Press, Cambridge, UK. Japanese lakes and its dependence on lake depth. *Arch. Hydrobiol.* 62, 1–28.

Japanese lakes and its significance in the photosynthetic production of phytoplankton communities. *Bot. Mag.*, Tokyo, 79, 77–88.

Ismael, D. et al. (eds.) (1999) *Biodiversidade do Estado de São Paulo, Brasil.* vol. 4: Invertebrados de Água doce. São Paulo, Programa Biota/Fapesp. pp. 141–148.

Istvánovics, V. (1999) Mechanisms and efficiency of phosphorus retention as influenced by the external and internal loads in the Kis-Balaton Reservoir, and the effect of nutrient load reduction on the water quality of Lake Balaton. In: UNESCO-UNEP (eds.) *Advanced Study Course in Ecohydrology.* Lodz, Hungria, University of Lodz.

Jackson, M.L. (1960) *Soil Chemical Analysis*. Londres, Constable and Co Ltd.

Jenkin, P.M. (1957) The filter-feeding and food of flamingos (Phoenicopteri). *Phil. Trans. R. Soc. B*, Londres, 240, 401–493.

Joly, C.A. & Bicudo, C.E.M. (Orgs.) (1999) *Biodiversidade do Estado de São Paulo, Brasil: síntese do conhecimento ao final do século XX*, vol. 7. São Paulo, Fapesp.

Jónasson, P.M. (1978) Zoobenthos of lakes. Proceedings: 20th Congress, Internationale Vereinigung für Theoretische und Angewandte Limnologie. *Internat. Verein. Theor. Angew. Limnol.*, Copenhaguem, 20 (1), 13–37.

Jorcin, A. (1997) *Distribuição espacial (vertical e horizontal) do macrozoobenthos na região extramarina de Cananéia (SP) e suas relações com algumas variáveis físicas e químicas.* [Tese de Doutorado]. São Paulo, USP.

Jorgensen, S.E. (1976) A eutrophication model for a lake. *Ecol. Model.*, 2, 147–165.

———. (1980a) Water quality and environmental impact model of the Upper Nile Basin. *Water Suppl. Manage.*, 4 (3), 147–153.

———. (1980b) *Lake Management*. Oxford, Londres, Pergamon Press.

———. (1981) *Application of Ecological Modelling in Environmental Management*. Amsterdã, Elsevier Science.

———. (1990) Application of Models in Limnological Research. *Internat. Verein. Theor. Angew. Limnol.*, Stuttgart, 24 (1), 61–67.

———. (1992) *Integration of Ecosystem Theories: A Pattern*. Copenhaguem, Royal Danish School of Pharmacy, Department of Environmental Chemistry.

———. (1995) State of the art of ecological modelling in limnology. *Ecol. Model.*, Amsterdã, 78, 101–115.

———. (1996) The application of ecosystem theory in limnology. *Verh. Internat. Verein. Limnol.*, Stuttgart, 26, 181–192.

Jorgensen, S.E. & Löffler, H. (1995) *Gerenciamento de litorais lacustres*. Shiga, Japão, Ilec, UNEP. (Diretrizes para o gerenciamento de lagos, 3).

Jorgensen, S.E. & De Bernardi, R. (1998) The use of structural dynamic models to explain successes and failures of biomanipulation. *Hydrobiologia*, Baarn, Holanda, 379, 147–158.

Juday, C. (1911) Limnological studies on some lakes in Central America. *Trans. Wis. Acad. Sci. Arts Lett.*, Madison, EUA, 18, 214–250.

Junk, W.J. (1970) Investigation on the Ecology and Production-biology of the "Floating Meadows" (Paspalo-Echinochloetum) on the Middle Amazon. Part I: The Floating Vegetation and its Ecology. *Amazoniana*, Kiel, Alemanha, 2, 449–495.

———. (1973) Investigation on the Ecology and Production-biology of the "Floating Meadows" (Paspalo-Echinochloetum) on the Middle Amazon. Part II. The Aquatic Fauna in the Root Zone of Floating Vegetation. *Amazoniana*, Kiel, Alemanha, 4, 9–102.

———. (1982) Amazonian floodplains: Their ecology, present and potential use. *Rev. Hydro. Trop.*, Paris, 15 (4), 285–301.

———. (1983a) Ecology of swamps on the Middle Amazon. In: GORE, A.L.P. (ed.) *Ecosystems of the World. Mires, Swamps, Bog, Fen and Moor. 3. Regional Studies.* Amsterdã, Elsevier. pp. 269–294.

———. (1983b) As Águas da Região Amazônica. In: Salati, E., Schubart, H., Junk, W.J. & Oliveira, A.R. (eds.) *Amazônia: Desenvolvimento, Integração e Ecologia*. Brasília, ANPq, Editora Brasilliense. pp. 45–100.

———. (1986) Aquatic plants of the Amazon system. In: Davies, B.R. & Walker, K.F. (eds.) *The Ecology of River Systems*. Dordrecht, Holanda, Dr. W. Junk. pp. 319–337.

Junk, W.J. (ed) (1997) *The Central Amazon Floodplain: Ecology of a Pulsing System (Ecological Studies)*. Berlim, Heidelberg, Springer-Verlag.

Junk, W.J. (2000) The Central Amazon River Floodplain: Concepts for the sustainable use of its resources. In: Junk, W.J., Ohly, J.J., Piedade, M.T.F. & Soares, M.G.M. (eds.) *The Central*

Amazon Floodplain: Actual Use and Options for a Sustainable Management. Leiden, Holanda, Backhuys Publishers B.V. pp. 75–94.

————. (2006) Flood pulsing and the linkages between terrestrial, aquatic and wetland systems. *Verh. Internat. Verein. Limnol.*, Stuttgart, 29, 11–38.

Junk, W.J. & Howard-Williams, C. (1984) Ecology of aquatic macrophytes in Amazonia. In: Sioli, H. (ed.) *The Amazon: Limnology and Landscape Ecology of a Mighty Tropical River and its Basin.* Dordrecht, Dr. W. Junk. (Monographiae Biologicae, 56). pp. 269–293.

Junk, W.J. & Piedade, M.T.F. (1993) Biomass and primary production of herbaceous plant communities in the Amazon floodplain. *Hydrobiologia*, Baarn, Holanda, 263, 155–162.

Junk, W.J. & Da Silva, C.J. (1995) Neotropical floodplains: A comparison between the Pantanal of Mato Grosso and the Large Amazonian River Floodplains. In: Tundisi, J.G., Bicudo, C.E.M. & Tundisi, T.M. (eds.) *Limnology in Brazil.* Rio de Janeiro, Academia Brasileira de Ciências, Sociedade Brasileira de Limnologia. pp. 195–217.

Junk, W.J., Bayley, P.B. & Sparks, R.E. (1989) The flood pulse concept in river-floodplain systems. In: Dodge, D.P. (ed.) Proceedings of the International Large River Symposium (LARS). *Can. Spec. Publ. Fish. Aquat. Sci.*, Ottawa, 106, 110–127.

Junk, W.J., Ohly, M.T.F., Piedade, M.T.F. & Soares, M.G.M. (2000) *The Central Amazon Floodplain: Actual use and Options for a Sustainable Management.* Leiden, Holanda, Backhuys Publishers.

Kajak, Z. & Hilbricht-Illkovska, A. (1972) *Productivity Problems of Freshwaters: Proceedings of the IBPENESCO Symposium on Productivity.* Varsóvia, Polish Scientific Publishers.

Kalff, J. (2002) *Limnology: Inland Water Ecosystems.* Nova Jersey, EUA, Prentice Hall.

Kalk, M., McLachlan, A.J. & Williams, H.C. (eds.) (1979) *Lake Chilwa: Studies of Change in a Tropical Ecosystem.* Haia, Boston, Londres, Dr. W. Junk Publisher.

Kalwasinska, A. & Donderski, W. (2005) Neustonic versus planktonic bacteria in eutrophic lake. *Pol. J. Ecol.*, Lomianki, Polônia, 53 (4), 571–577.

Kennedy, R.H. (1999) Reservoir design and operation: Limnological implication and management opportunities. In: Tundisi, J.G. & Straskraba, M. (eds.) *Theoretical Reservoir Ecology and Its Application.* São Carlos, Academia Brasileira de Ciências, International Institute of Ecology, Backuys Publishers. pp. 1–28.

Kenya Wildlife Service. *Lake Nakuru: Ramsar site*, Kenya. [S.l], s.d.

Kester, D.R. & Pytkowicz, R.M. (1969) Sodium, magnesium, and calcium sulfate ion-pairs in seawater at 25°C. *Limnol. Oceanogr.*, Milwaukee, EUA, 14 (5), 686–692.

King, R.D. & Tyler, P.A. (1983) Sulphide pool and Lake Morrison, meromictic lakes of Southwest Tasmania. *Arch. Hydrobiol.*, Stuttgart, 96 (2), 139–163.

Kirk, J.T.O. (1983) *Light and Photosynthesis in Aquatic Ecosystems.* Cambridge, Cambridge University Press.

————. (1985) Effects of suspensoids (turbidity) on penetration of solar radiation in aquatic ecosystems. *Hydrobiologia*, Baarn, Holanda, 125, 195–208.

Kjerfve, B. (ed.) (1994) *Coastal Lagoon Processes.* Amsterdã, Elsevier.

Kleerekoper, H. (1944) *Introdução ao Estudo da Limnologia*, 2a edition. Porto Alegre, Editora UFRGS.

Kner, R. (1854) Die Hypostomiden. Zweite Hauptgruppe der Familie der Panzerfische. (Loricata vel Goniodontes). *Denkschriften der Mathematisch–Naturwissenschaftliche Classe der Kaiserlichen Akademie der Wissenschaften in Wien*, Vienna, 7, 251–286.

Knoppel, H.A. (1970) Food of Central Amazonian fishes. *Amazoniana*, Kiel, Alamanha, 2, 257–352.

Knoppers, B. (1994) Aquatic primary production in coastal lagoons. In: Kjerfve, B. (ed.) *Coastal Lagoon Processes.* Amsterdã, Sleiver. pp. 243–276.

Kolkwitz, R. & Marsson, K. (1909) Okologie der tierischen Saprobien. *Int. Rev. Ges. Hydrobiol. Hydrogr.*, Weinheim, Alemanha, 2, 126–152.

Komárková, J., Marvan, P. (1978) Primary production and functioning of algae in the fish-pond littoral. In: Dykyjová, D., Květ, J. (eds.) *Pond Littoral Ecosystems: Structure and Functioning. Ecol. Stud.*, Berlin, Springer-Verlag, 28, 321–337.

Kotak, B.G., Kenefick, S.L., Fritz, D.L., Rousseaux, C.G., Prepas, E.E. & Hrudey, S.E. (1993) Occurrence and toxicological evaluation of cyanobacterial toxins in Alberta lakes and farm dugouts. *Wat. Res.*, Washington, 27, 495–506.

Kotak, B.G., Lam, A.K.Y., Prepas, E.E., Kenefik, S.L. & Hrudey, S.E. (1995) Variability of the hepatotoxin, microcystin-LR, in hypereutrophic drinking water lakes. *J. Phycol.*, San Marcos, EUA, 31, 128–263.

Kottelat, M. & Whitten, T. (1996) Freshwater biodiversity in Asia with special reference to fish. *World Bank Technical Paper*, 343, Washington.

Kullander, S.O. (1983) *A Revision of the South American Cichlid Genus* Cichlasoma *(Teleostei: Cichlidae)*. Estocolmo, Naturhistoriska Riksmuseet.

Kumagai, M. & Vincent, W.F. (2003) *Freshwater Management: Global Versus Local Perspectives*. Tóquio, Springer-Verlag.

L'vovich, M.I. (1979) *World Water Resource and Their Future*. Washington, American Geophysical Union. (World Population Data Sheet).

Lacerda, L.D. & Salomons, W. (1998) *Mercury From Gold and Silver Mining: A Chemical Time Bomb?* Nova York, Springer-Verlag. (Environmental Science Series).

Lallana, V.H. (1980) Productividad de Eichhornia crassipes (Mart.) Solms en una laguna isleña de la cuenca del río Paraná Medio. II: Biomasa y dinámica de población. *Ecología*, Buenos Aires, 5, 1–16.

Lamotte, M. & Boulière, F. (1983) Energy flow and nutrient cycling in tropical savannas. In: Boulière, F. (ed.) *Ecosystems of the World, vol. 13: Tropical Savannas*. Amsterdã, Elsevier. pp. 583–603.

Lampert, W. (1997) Zooplankton research: The contribution of limnology to general ecological paradigms. *Aquat. Ecol.*, Eist, Holanda, 31, 19–27.

Lampert, W. & Sommer, U. (1997a) *Limnoecology: The Ecology of Lakes and Streams*. James F. Haney (trans.) Nova York, Oxford, Oxford University Press.

———. (1997b) Ecosystem perspectives. In: Lampert, W. & Sommer, U. (eds.) *Limnoecology: The Ecology of Lakes and Streams*. James F. Haney (trans.) Nova York, Oxford, Oxford University Press. Cap. 8, pp. 188–337.

———. (1997c) Interaction. In: Lampert, W. & Sommer, U. (eds.) *Limnoecology: The Ecology of Lakes and Streams*. James F. Haney (trans.) Nova York, Oxford, Oxford University Press. Cap. 6, pp. 160–253.

Lampert, W. & Wolf, H.G. (1986) Cyclomorphosis in *Daphnia cucullata*: Morphometric and population genetic analyses. *J. Plankton Res.*, Oxford, 8 (2), 289–303.

Lanaras, T., Tsitsamis, S., Chlichlia, C. & Cook, C.M. (1989) Toxic cyanobacteria in Greek freshwaters. *J. Appl. Phycol.*, Dordrecht, Holanda, 1, 67–73.

Lansac-Toha, F.A., Velho, L.F.M. & Bonecker, C.C. (1999) Estrutura da comunidade zoo-planctônica antes e após a formação do reservatório de Corumbá (GO). In: Henry, R. (ed.) *Ecologia de Reservatórios: Estrutura, Função e Aspectos Sociais*. São Paulo, Fundibio/FAPESP. pp. 347–374.

Lara, L.B.S.L., Artaxo, P., Martinelli, L.A., Victoria, R., Camargo, P.B., Krusche, A., Aynes, G.P., Ferraz, E.S.B. & Ballester, M.V. (2001) Chemical composition of rainwater and anthropogenic influences in the Piracicaba River Basin, Southeast Brazil. *Atmos. Environ.*, Amsterdã, 37, 4937–4945.

Lasenby, D.C. (1975) Development of oxygen deficits in 14 Southern Ontario Lakes. *Limnol. Oceanogr.*, Milwaukee, EUA, 20, 993–999.

Lavenu, A. (1991) Formación geológica y evolución. In: Dejoux, C.Y. & Iltis, A. (eds.) *El Lago Titicaca*. La Paz, Orstom, Hisbol. pp. 19–30.

Lazaro, X. (1987) A review of planktivorous fish: Their evolution, feeding behaviours, selectivities and impacts. *Hydrobiologia*, Baarn, Holanda, 146, 97–167.

Le Cren, E.D. & Lowe McConnell, R.H. (eds.) (1980) *The Functioning of Freshwater Ecosystems*. Cambridge, Londres, Nova york, Melbourne, Cambridge University Press. (Int. Biol. Progr. 22).

Leeuwangh, P., Kappers, F.I. & Dekker, M. (1983) Toxicity of cyanobacteria in Dutch lakes and reservoirs. *J. Aquat. Toxicol.*, Nova York, 4 (1), 63–72.

Legendre, L. & Demers, S. (1984) Towards dynamic biological oceanography and limnology. *Can. J. Fish. Aquat. Sci.*, Ottawa, 41, 2–19.

Lemoalle, J. (1981) Photosynthetic production and phytoplankton in the euphotic zone of some African and temperate lakes. *Rev. Hydrobiol. Trop.*, Paris, 11, 31–37.

Leslie, J. (1838) *Treatises on Various Subjects of Natural and Chemical Philosophy*. Edimburgo, The Encyclopaedia Britannica. Cited in Murray and Pullar (1910). pp. 91–92.

Lévêque, C. (1997) *Biodiversity Dynamics and Conservation: The Freshwater Fish of Tropical Africa*. Cambridge, Cambridge University Press.

Lévêque, C., Bruston, M.N. & Ssentongo, G.M. (1988) *Biologie et écologie des Poissons d'Eau Douce Africains*. Paris, Editions de L'ORSTOM. (Colletion Travaux et Documents 216).

Lévêque, C., Balian, E.V. & Martens, K. (2005) An assessment of animal species diversity in continental waters. *Hydrobiologia*, Baarn, Holanda, 542 (1), 39–67.

Lewis, W.M. (1974) Primary production in the plankton community of a tropical lake. *Ecol. Monogr.*, Washington, 44, 377–379.

———. (1979) *Zooplankton Community Analysis: Studies on a Tropical System*. Nova York, Springer-Verlag.

Lewis, J.R.W.M. (1983) A revised classification of lakes based on mixing. *Can. J. Fish. Aquat. Sci.*, 40, 1779–1787.

Likens, G.E. (1983) A priority for ecological research. *Bull. Ecol. Soc. Am.*, Washington, 64, 234–243.

———. (1984) Beyond the shoreline: A watershed-ecosystem approach. *Verh. Internat. Verein. Limnol.*, 22, 1–22.

———. (1988) *Long Term Studies in Ecology: Approaches and Alternatives*. Nova York, Springer-Verlag.

———. (1992) *The Ecosystem Approach: Its Use and Abuse*. Luhe, Alemanha, Ecology Institute. (Excellence in Ecology 3).

Likens, G.E. & Borman, F.H. (1972) Nutrient cycling in ecosystems. In: Wiens, J.A. (ed.) *Ecosystem Structure and Function*. Corvallis, EUA, Oregon State University Press. pp. 25–67.

Likens, G.E. & Bormann, F.H. (1974) Linkages between terrestrial and aquatic ecosystems. *BioScience*, 24 (8), 447–456.

Likens, G.E., Bormann, F.H. & Johnson, N.M. (1972) Acid rain. *Environment*, 14 (2), 33–40.

Lima, C.A. & Goulding, M. (1998) *Os frutos do tambaqui: ecologia, conservação e cultivo na Amazônia*. Tefé, Sociedade Civil Mamirauá, Brasília, CNPq. (Estudos do Mamirauá, 4).

Lima, W.C., Marins, M.A. & Tundisi, J.G. (1978) Influence of wind on the standing stock of *Melosira italica* (Ehr.) Kutz. *Rev. Bras. Biol.*, São Carlos, 43, 317–320.

Lind, O.T. (1979) *Handbook of Common Methods in Limnology*. 2nd edition. St. Louis, EUA, The C.V. Mosby Co.

Linnaeus, C. (1758) *Systema Naturae*, 10th edition. vol. 1. Estocolmo, Salvi.

Livingstone, D.A. & Melack, J.M. (1984) *Some Lakes and Reservoirs of subsaharan Africa*. Amsterdã, Elsevier. (Ecosystems of the World, 23).

Lockwood, A.P.M. (1963) *Animal Body Fluids and Their Regulation*. Portsmouth, EUA, Heinemann Educational Books Ltd.

Löffler, H. (1956) Ergebnisse des Österreichischen Iran expedition 1940/50, Limnologische Beobachtungen an Iranischen Binnengewässern: I. Teil. *Hydrobiologia*, Baarn, Holanda, 8, 201–238.

———. (1961) Beitrage zur kenntnis der Iranischen Binnengewässern: II. Regional-limnologische Studie mit besonderer Berucksichtigung Crustaceenfauna. *Int. Rev. Ges. Hydrobiol.*, Weinheim, Alemanha, 46, 309–406.

———. (1968) Tropical mountain lakes. Their distribution, ecology and zoogeographical importance. *Colloq. Geogr.*, Bonn, Alemanha, 9, 57–76.

———. (1972) Contribution to the limnology of high mountain lakes in Central America. *Int. Rev. Hydrobiol.*, Weinheim, Alemanha, 57 (3), 357–408.

———. (1982) Shallow lakes. *Proceedings of the International Scientific Workshop on Ecosystem Dynamics in Freshwater Wetlands and Shallow Water Bodies.* Moscou, SCOPE, UNEP. pp. 1–22.

Lombardo, C.P. & Gregg, M.C. (1989) Similarity scaling of and in a convecting surface boundary layer. *J. Geophys. Res.*, Washington, 94, 6273–6284.

Lorenzen, C.J. (1966) A method for the continuous measurement of in vivo chlorophyll concentration. *Deep Sea Res.*, Nova York, 13, 223–227.

Lowe-McConnell, R.H. (1975) *Fish Communities in Tropical Freshwaters.* Nova York, Longman.

———. (1984) The status of studies on Southern America freshwater food fish. In: Zaret, T.M. (ed.) *Evolutionary Ecology of Neotropical Freshwater Fishes.* Haia, Holanda, Dr. W. Junk.

———. (1999) *Estudos Ecológicos das Comunidades de Peixes Tropicais.* Vazzoler, A.E.A., Agostinho, A.A. & Cunninghan, P. (trans.) São Paulo, EDUSP. (Base, 3).

Lund, J.W.G. (1950) Studies on *Asterionella formosa* Hass. II. Nutrient depletion and the spring maximum. *J. Ecol.*, Londres, 38, 1–35.

———. (1954) The seasonal cycle of the plankton diatom *Melosira italica* (Ehr.) Kütz. subsp. *subarctica* O. Müll. *J. Ecol.*, Londres, 42, 151–179.

———. (1955) Further observations on the seasonal cycle of *Melosira italica* (Ehr.) Kütz. subsp. *subarctica* O. Müll. *J. Ecol.*, Londres, 43, 91–102.

———. (1961) The periodicity of micro-algae in three English lakes. *Verh. Internat. Verein. Limnol.*, Stuttgart, 14, 147–154.

———. (1964) Primary production and periodicity of phytoplankton. *Verh. Internat. Verein. Limnol.*, Stuttgart, 15, 37–56.

———. (1965) The ecology of the freshwater phytoplankton. *Biol. Rev.*, Cambridge, 40, 231–293.

———. (1981) *Investigations on Phytoplankton with Special Reference to Water Usage.* Cumbria, Inglaterra, Freshwater Biological Association. (Occasional Publication 13).

Lundberg, P., Ranta, E., Ripa, J. & Kaitala, V. (2000) Population variability in space and time. *Trends Ecol. Evol.*, Nova York, 15, 460–464.

Lütken, C.F. (1873) On spontaneous division in the Echinodermata and other Radiata. *Ann. Mag. Nat. Hist. (series 4)*, 12, 323–337, 391–399. [Traduzido e ampliado por Dallas, W.S., de Lütken, C.F. (1872) *Ophiuridarum novarum vel minus cognitarum descriptiones nonnullae.* Oversigt over det Kongelige Danske Videnskabernes Selskabs Forhandlinger, 1872].

———. (1874) *Ichthyographike Bidrag*, II, Nyeeller mindre vel kjendte Malleformer fra forskjellige Verdensdele, Vidensk. Medd. Copenhaguem, Naturh. Foren. pp. 190–220.

———. (1875) Velhas Flodens Fiske. Danske Vidensk. *Selsk. Skrifter.*, Copenhaguem, 12, 122–254.

Macan, T.T. (1961) A review of running water studies. *Verh. Internat. Verein. Theor. Angew. Limnol.*, Stuttgart, 14, 587–602.

———. (1970) *Biological Studies of the English Lakes.* Londres, Longman.

———. (1973) *Ponds and Lakes.* Londres, George Allen & Unwin Ltd.

Madsen, J.D. & Adams, M.S. (1988) The seasonal biomass and productivity of the submerged macrophytes in a polluted Wisconsin Stream. *Freshwat. Biol.*, Victoria, Austrália, 20, 41–50.

Maltchik, L., Rolon, A.S. & Groth, C. (2004) The effects of flood pulse on the macrophyte community in a shallow lake of Southern Brazil. *Acta Limnol. Bras.* 16, 2. Botucatu, SP, Brazil.

Mann, K.H. (1975) Patterns of energy flow. In: Whitton, B.A. (ed.) *River Ecology*. Londres, Blackwell Scientific Publications.

Mannio, J., Verta, M., Kortelainen, P. & Rekolainen, S. (1986) The effect of water quality on the mercury concentration of Northern Pike (*Esox lucius*) in the Finnish forest lakes and reservoirs. *Publ. Water Res. Inst. (Vesihallitus)*, Helsinque, Finlândia, 65, 32–43.

Mansur, M.C.D., Richinitti, L. & Dos Santos, C.P. (1999) *Limnoperna fortunei* (Dunker, 1857), molusco bivalve invasor, na bacia do Guaíba, Rio Grande do Sul, Brasil. *Biociências*, Porto Alegre, 7, 147–150.

Mansur, M.C.D., Pinheiro Dos Santos, C., Darrigran, G., Hydrich, I., Calli, C. & Rossoni Cardoso, F. (2003a) Primeiros dados quali-quantitativos do mexilhão-dourado, *Limnoperna fortunei* (Dunker), no Delta do Jacuí, no Lago Guaíba e na Laguna dos Patos, Río Grande do Sul, Brasil e alguns aspectos de sua invasão no novo ambiente. *Rev. Brasil Zool.*, São Carlos, 20, 75–84.

Mansur, M.C.D., Heydrich, I., Pereira, D., Richinitti, L.M.Z., Tarasconi, J.C. & Rios, E.C. (2003b) Moluscos. In: Fontana, C.S., Bencke, G.A. & Reis, R.E. (eds.) *Livro vermelho da fauna ameaçada de extinção no Rio Grande do Sul*. Porto Alegre, EDIPUCRS. pp. 49–71.

Mansur, M.C.D., Quevedo, C.B., Santos, C.P. & Callil, C.T. (2004a) Prováveis vias da introdução de *Limnoperna fortunei* (Dunker, 1857) (Mollusca, Bivalvia, Mytilidae) na bacia da Laguna dos Patos, Rio Grande do Sul e novos registros de invasão no Brasil pelas bacias do Paraná e Paraguai. In: Silva, J.S.V. & Souza, R.C.C.L. (eds.) *Água de lastro e bioinvasão*. Rio de Janeiro, Interciência. pp. 33–38.

Mansur, M.C.D., Callil, C.T., Cardoso, F.R. & Ibarra, J.A.A. (2004b) Uma retrospectiva e mapeamento da invasão de espécies de Corbicula (Mollusca, Bivalvia, Veneroida, Corbiculidae), oriundas do sudeste asiático, na América do Sul. In: Silva, J.S.V. & Souza, R.C.C.L. (eds.) *água de lastro e bioinvasão*. Rio de Janeiro, Interciência. pp. 39–58.

Mansur, M.C.D., Cardoso, F.R., Ribeiro, L.A., Santos, C.P., Thormann, B.M., Fernandes, F.C. & Richinitti, L.M.Z. (2004c) Distribuição e conseqüências após cinco anos da invasão do mexilhão-dourado, *Limnoperna fortunei*, no estado do Rio Grande do Sul, Brasil (Mollusca, Bivalvia, Mytilidae). *Biociências*, Porto Alegre, 12 (2), 165–172.

Marchese, M. & Ezcurra de Drago, I. (2006) Bentos como Indicadores de Condiciones Tróficas del Rio Paraná Medio. In: Tundisi, J.G., Matsumura-Tundisi, T. & Galli, C.S. (eds.) *Eutrofização na América do Sul: Causas, Conseqüências e Tecnologias de Gerenciamento e Controle*. São Carlos, IIE. pp. 339–362.

Margalef, R. (1962) Adaptación, ecología y evolución: nuevas formas de plantear antiguos problemas. *Bol. R. Soc. Esp. Hist. Nat. (B)*, Madri, 60, 231–246.

———. (1967) *Perspectives in Ecological Theory*. Chicago, The University of Chicago Press.

———. (1969) Diversity and stability: A practical proposal and a model of interdependence. *Brookhaven Symp. Biol.*, Brookhaven, EUA, 22, 25–37.

———. (1974) *Ecologia*. Barcelona, Ediciones Omega.

———. (1975) Typology of reservoirs. *Verh. Internat. Verein. Limnol.*, Stuttgart, 19, 1811–1816.

———. (1978) Diversity. *Monogr. Oceanogr. Methodol.*, Paris, 6, 251–260.

———. (1978) Life forms of phytoplankton as survival alternatives in an unstable environment. *Oceanol. Acta*, Amsterdã, 14, 493–509.

———. (1980) *La Biosfera: Entre la Termodinâmica y el Juego*. Barcelona, Ediciones Omega.

———. (1983) *Limnologia*. Barcelona, Ediciones Omega.

————. (1991) *Teoria de los Sistemas Ecológicos*. Barcelona, Publicaciones de la Universidad de Barcelona.

————. (1994) *Limnology Now: A Paradigm Of Planetary Problems*. Amsterdã, Elsevier Science.

————. (1997) Our biosphere. In: Kinne, O. (ed.) *Excellence in Ecology*. Luhe, Alemanha, Ecology Institute.

————. (2002) La Superfície del Planeta y la Organización de la Biosfera: Reacción a los Nuevos Mecanismos Añadidos por el Poder Creciente de los Humanos. *MUNIBE Ciencias Naturales*, Donostia-San Sebastián, 53, 7–14.

Margalef, R., Planas, D.M., Armengol, J.B., Celma, A.V., Fornells, P.N., Serra, A.G., Santillana, T.J. & Miyanes, M.E. (1976) *Limnologia de los embalses españoles*. Dirección General de Obras Hidráulicas. Madri, Ministerio de Obras Públicas.

Margulis, L. & Schwartz, K.V. (1998) *Five Kingdoms: An Illustrated Guide to the Phyla of Life on EARTH*. 3rd edition. Nova York, W.H. Freeman.

Marlier, G. (1951) La biologie d'un ruisseau de plaine Le Smohain. *Mém. Inst. R. Sci. Nat. Belg.*, Bruxelas, 114, 1–98.

————. (1967) Ecological studies on some lakes of the Amazon valley. *Amazoniana*, Kiel, Alemanha, 1, 91–115.

Marsalek, J., Dutka, B.J., McCorquodale, A.J. & Tsanis, I.K. (1996) Microbiological pollution in the Canadian upper great lakes connecting channels. *Wat. Sci. Technol.*, Londres, 33, 349–356.

Marshall, S.M. & Orr, A.P. (1972) *The Biology of a Marine Copepod*. Berlim, Springer-Verlag.

Martens, K. & Behen, F.A.A. (1994) A checklist of the recent non-marine ostracods (Crustacea, Ostracoda) from the inland waters of South America and adjacent islands. *Travaux Scientifiques du Musee National D'Histoire Naturelle de Luxembourg*, Luxemburgo, 22, 1–84.

Martin, D.B. & Arneson, R.D. (1978) Comparative limnology of a deep discharge reservoir and a surface discharge lake on the Madison River Montana. *Freshwat. Biol.*, Victoria, Austrália, 8, 33–42.

Martineli, L.A. et al. (1999) Land-cover changes and the 13-C composition of riverine particulate organic matter in the Piracicaba River Basin (Southeast Region of Brazil). *Limnol. Oceanogr.*, Milwaukee, EUA, 44, 1826–1833.

Martinelli, L.A., Krusche, A.V. & Victoria, R.L. (1999) Effects of sewage on the chemical composition of the Piracicaba River, Brazil. *Wat. Air Soil Pollut.*, Baarn, Holanda, 110, 67–79.

Matheus, C.E. & Tundisi, J.G. (1988) Estudo físico-químico e ecológico dos rios da bacia hidrográfica do Ribeirão e Represa do Lobo. In: Tundisi, J.G. (ed.) *Limnologia e manejo de represas*, vol. 1, tomos 1–2. São Paulo, USP, ACIESP, FAPESP, UNEP. (Série Monografias em Limnologia). pp. 419–472.

Matsumura-Tundisi, T. (1972) *Aspectos ecológicos do zooplâncton da região lagunar de Cananeia com especial referência aos Copepoda (Crustacea)*. [Tese de Doutorado]. São Paulo, USP.

————. (1986) Latitudinal distribution of Calanoida copepods in freshwater aquatic systems of Brazil. *Rev. Bras. Biol.*, São Carlos, 43 (3), 527–553.

————. (1987) Processos de Eutrofização da Represa de Barra Bonita. *Barra Bonita, Relatório Científico* FAPESP. (Processo 85/0653-2).

Matsumura-Tundisi, T. & Okano, W.Y. (1983) Seasonal fluctuations of copepod populations in Lake Dom Helvecio (Parque Florestal, Rio Doce, Minas Gerais, Brazil). *Rev. Hydrobiol. Trop.*, Paris, 16, 35–39.

Matsumura-Tundisi, T. & Rocha, O. (1983) Occurrence of copepods (Calanoida, Cyclopoida and Harpacticoida) from Broa reservoir (São Carlos, SP, Brazil). *Rev. Bras. Biol.*, São Carlos, 13 (1), 1–17.

Matsumura-Tundisi, T. & Tundisi, J.G. (1976) Plankton studies in a lacustrine environment. I. Preliminary investigations on the seasonal cycle of zooplankton. *Oecologia*, Berlin, 25, 265–270.

———. (2003) Calanoida (Copepoda) species composition changes in the reservoirs of São Paulo State (Brazil) in the last twenty years. *Hydrobiologia*, Baarn, Holanda, 504, 215–222.

———. (2005) Plankton richness in a eutrophic reservoir (Barra Bonita Reservoir, SP, Brazil). *Hydrobiologia*, Baarn, Holanda, 542 (1), 367–378.

Matsumura-Tundisi, T., Cintra, R.H., Tundisi, J.G., Bernardes, S. & Martins, C.R. (2005) Projeto PIPE/FAPESP Hotel Ambiental: Diagnóstico, Adequação e Inovação Ambiental no Setor Hoteleiro. In: Unidade Temática de Ciência, Tecnologia e Capacitação da Rede de Mercocidades, 2005, São Carlos. *Anais da I Mostra de Ciência e Tecnologia em Políticas Públicas Municipais.* São Carlos, Suprema Gráfica e Editora, 1, 60–68.

Matsumura-Tundisi, T., Rietzler, A.C., Espíndola, E.L.G., Tundisi, J.G. & Rocha, O. (1990) Predation on *Ceriodaphnia cornuta* and *Brachionus calyciflorus* by two *Mesocyclops* species coexisting in Barra Bonita Reservoir (SP, Brazil). *Hydrobiologia*, Baarn, Holanda, 198, 141–151.

Matsumura-Tundisi, T., Tundisi, J.G. & Tavares, L.H.S. (1984) Diel migration and vertical distribution of Cladocera in lake D. Helvécio (Minas Gerais, Brazil). Tropical Zooplankton: Series Development in Hydrobiology. *Hydrobiologia*, Baarn, Holanda, 113, 299–306.

Matsumura-Tundisi, T., Tundisi, J.G., Abe, D.S., Rocha, O. & Starling, F. (2006) Limnologia de águas interiores: impactos, conservação e recuperação de ecossistemas aquáticos. In: Tundisi, J.G., Rebouças, A. & Braga, B. (eds.) *águas Doces no Brasil: capital ecológico, uso e conservação*, 3rd edition. São Paulo, Escrituras Editora. pp. 203–240.

Matsumura-Tundisi, T., Tundisi, J.G., Azevedo, C.O., Luzia, A.P. & Silva, W.M. (2003) *Cartilha da água, Série 2:* Os organismos que vivem na água. Contribuição à Educação Ambiental e à Ciência. São Carlos, São Paulo, IIE/BIOTA-FAPESP.

Matsumura-Tundisi, T., Tundisi, J.G., Rocha, O. & Calijuri, M.C. (1997) The ecological significance of the metalimnion in lakes of Middle Rio Doce Valley. In: Saijo, Y. & Tundisi, J.G. (eds.) *Limnological Studies on the Rio Doce Valley Lakes, Brazil.* Rio de Janeiro, São Paulo, Academia Brasileira de Ciências, USP. pp. 373–390.

Matsumura-Tundisi, T., Tundisi, J.G., Saggio, A., Oliveira Neto, A. & Espindola, E.G. (1991) Limnology of Samuel Reservoir (Brazil, Rondônia) in the Filling Plase. *Verh. Internat. Verein. Limnol.*, Stuttgart, 24, 1482–1488.

Matsuyama, M. (1978) Limnological aspects of meromictic Lake Suigetsu: Its environmental conditions and biological metabolism. *Nagasaki University Bull. Fac. Fisher.*, Nagasaki.

Matsuyama, M. & Moom, S.W. (1999) Dissolved H_2S around *Chromatium* sp. blooming in lake Kaike protects against feeding pressure by a *Hypotrich ciliate. Jap. J. Limnol.*, 60, 87–96.

Matvienko, B. & Tundisi, J.G. (1996) Biogenic gases and decay of organic matter. In: international *Workshop on Greenhouse Gas Emissions from Hydroeletric Reservoirs*, Anais. . . . Rio de Janeiro, Eletrobrás. pp. 1–6.

May, R.M. (1973) *Stability and Complexity in Model Ecosystems.* Princeton, EUA, Princeton University Press.

Medri, M.E., Bianchini, E., Pimenta, J.A., Colli, S. & Muller, C. (2002) Estudos Sobre Tolerância ao Alagamento em Espécies Arbóreas Nativas da Bacia do Rio Tibagi. In: Medri, M.E., Bianchini, E., Shibatta, O.A. & Pimenta, J.A. (eds.) *A Bacia do Rio Tibagi*, Londrina. Londrina, Eduel. pp. 133–172.

Meibeck, M. (1995) Global distribution of lakes. In: Her, A., Imboden, D. & Gat, J. (eds.) *Physics and Chemistry of Lakes.* 2nd edition. Nova York, Springer-Verlag. pp. 1–32.

Melack, J.M. (1976) Primary productivity and fish yields in tropical lakes. *Trans. Am. Fish Soc.*, Bethesda, EUA, 105, 575–580.

————. (1978) Morphometric physical and chemical features of volcanic crater lakes of Western Uganda. *Arch. Hydrobiol.*, Baarn, Holanda, 84, 430–453.

————. (1979) Photosynthetic rates in four tropical African fresh waters. *Freshwat. Biol.*, Victoria, Austrália, 9, 555–571.

————. (1984) Amazon floodplain lakes: Shape, fetch and stratification. *Verh. Internat. Verein. Limnol.*, Stuttgart, 22, 1278–1282.

————. (1985) Interactions of detrital particulates and plankton. In: Davies, B.R., Walmsley, R.D. (eds.) *Perspectives in Southern Hemisphere Limnology. Developments in Hydrobiology*, 28. Dordrecht, W. Junk. pp. 209–220.

Melack, J.M. & Fisher, T.R. (1983) Diel oxygen variations and their ecological implications in Amazon floodplain lakes. *Arch. Hydrobiol.*, Stuttgart, 98 (4), 422–442.

Melack, J.M. & Kilham, P. (1974) Photosynthetic rates of phytoplankton in East African alkaline, saline lakes. *Limnol. Oceanogr.*, Milwaukee, EUA, 19, 743–455.

Melack, J.M., Kilham, P. & Fisher, T.R. (1982) Responses of phytoplankton to experimental fertilization with ammonium and phosphate in a African soda lake. *Oecologia*, Berlim, 52, 321–326.

Melão, M.G.G. & Rocha, O. (1996) Macrofauna associada a *Metania spinata* (Carter, 1881), Porífera, Metaniidae. *Acta Limnol. Brasil.*, Campinas, 8, 59–64.

Melão, M.G.G. & Rocha, O. (2004) Life history, biomass and production of two planktonic cyclopoid copepods in a shallow subtropical reservoir. *J. Plankton Res.*, Oxford, 26 (8), 909–923.

Melo, G.A.S. (2003) *Manual de Identificação dos Crustáceos Decapoda de Água Doce do Brasil*. São Paulo, Edições Loyola.

Menezes, N.A. (1969) Systematics and evolution of the tribe Acestrorhynchini (Pisces, Characidae). *Arq. Zool.*, São Paulo, 18, 1–150.

Menezes, N.A. (1994) Importância da ictiofauna dos ecossitemas aquáticos brasileiros. In: COMASE. *Seminários sobre fauna aquática e o setor elétrico brasileiro: reuniões temáticas preparatórias*. Caderno 3, Conservação. Rio de Janeiro, Eletrobras. pp. 7–13.

Menezes, N.A. & Vazzoler, A.E. (1992) Reproductive characteristics of Characiformes. In: HAMLETT, W.C. (ed.) *Reproductive Biology of South American Vertebrates*. Nova York, Springer. pp. 101–114.

Menezes, N.A. et al. (2007) Peixes de Água Doce da Mata Atlântica. In: MUSEU de Zoologia. *Lista Preliminar das Espécies e Comentários sobre Conservação de Peixes de água Doce Neotropicais*. São Paulo, Museu de Zoologia, USP. p. 407.

Mereschkowsky, C. (1905) Über Natur und Ursprung der Chromatophoren in Pflanzenreiche. *Biol. Zbl.* 25, 593–604.

Mesquita, H.S.L. (1983) Suspended particulate organic carbon and phytoplankton in the Cananéia estuary (25S, 48W), Brazil. *Oceanogr. Trop.*, Montpellier, França, 18 (1), 55–68.

Mianzan, H. et al. (2001) The Rio de la Plata estuary, Argentina-Uruguay. In: Seeliger, U. & Kjerfve, B. (eds.) *Coastal Marine Ecosystems of Latin America*. Berlim, Springer, 2001. (Ecological studies, 144). Cap. 13, pp. 185–204.

Mieczan, T. (2005) Periphytic ciliates in littoral zone of three lakes of different trophic status. *Pol. J. Ecol.*, Lomianki, Polônia, 53, 105–111.

Migoto, A.E. & Tiago, C.G. (1999) In: *Biodiversidade do Estado de São Paulo, Brasil: síntese do conhecimento ao final do século XX, 3: Invertebrados Marinhos. (44) Síntese*. São Paulo, FAPESP.

Millero, F.J. (1975a) The physical chemistry of estuaries. In: Church, Tm.M. (ed.) *Marine Chemistry in the Coastal Environment*. Philadelphia, American Chemical Society.

————. (1975b) The state of metal ions in seawater. *Reimpresso de*, Thalassia Jugoslavica, 11 (1), 2, 53–84.

Minshall, G.W., Petersen, R.C., Cummins, K.W., Bott, T.L., Dedell, J.R., Cushing, C.E. & Vannote, R.L. (1983) Interbiome comparison of stream ecosystem dynamics. *Ecol. Monogr.*, Washington, 53, 1–25.

Miranda, L.B., Castro, B.M. & Kjerfve, B. (2002) *Princípios de Oceanografia Física de Estuários*. São Paulo, Edusp.

Mitsch, W.J. (1996) Managing the world's wetlands: Preserving and enhancing their ecological functions. *Verh. Internat. Verein. Limnol.*, Stuttgart, 26, I, 139–148.

Mitsch, W.J. & Gosselink, J.G. (1986) *Wetlands*. Nova York, Van Nostrand Reinhold.

Mitsch, W.J., Straškraba, M. & Jorgensen, S. (1988) *Wetland Modelling*. Amsterdã, Elsevier.

Montenegro Guillén, S. (2003) Lake Cocibolca/Nicaragua. In: Lakenet (ed.) *World Lake Basin Management Initiative*. Annapolis, USA, LakeNet.

Moore, J.E. (1939) *A Limnological Study of Certain Lakes of Southern Saskatchewan with Special Reference to Salinity*. [Tese de Mestrado]. Saskatoon, Canadá, University of Saskatchewan.

Mori, S. & Yamamoto, G. (1975) Productivity of communities in Japanese inland waters. *JIBP Synthesis*, v. 10. Tokyo, University of Tokyo Press. 436 pp.

Morren, A. & Morren, C. (1841) *Recherches sur la Rubéfaction des Eaux et leur Oxygénation par les Animalcules et les Algues*. Bruxelas, Imp. de L'Académie Royale.

Mortimer, C.H. (1941) The exchange of dissolved substances between mud and water in lakes (Parts I and II). *J. Ecol.* 29, 280–329.

———. (1951) *Water Movements in Stratified Lakes, Deduced from Observations in Windermere and Model Experiments*, tomo 3. Bruxelas, Union Geod. Geophy. Intern. pp. 335–348.

———. (1974) Lake hydrodynamics. *Mitt. Internat. Verein. Limnol.*, Stuttgart, 20, 124–197.

Moss, B. (1988) *Ecology of Fresh Waters: Man and Medium*. 2nd edition. Liverpool, Blackwell Scientific Publications.

Moura, J.R.S., Oliveira, P.T.T.M. & Meis, M.R.M. (1978) Insight to the morphometry of the drowned valleys. *Brazil. Geogr. Studies (UGI)*, Belo Horizonte, 1, 78–100.

Mourão, G.M. (1989) *Limnologia Comparativa de três Lagoas (duas "Baias" e uma "Salina") do Pantanal de Nhecolândia, MS*. São Carlos, UFSCAR.

Müller, J. & Troschel, F.H. (1849) *Beschreibung und Abbildung neuer Fische*. Berlim, Verlag von Veit und Comp.

Müller, K. (1951) Fische und Fischeregionen der Fulda. *Ber. Limnol. Flusst. Freudenthal*, Munique, 2, 18–23.

Mullin J.B. & Riley J.P. The colorimetric determination of silicate with special reference to sea and natural waters. *Anal. Chim. Acta*, Nova York, 12, 162–176.

Munawar, M., Lawrence, S.G., Munawar, I.F. & Malley, D.F. (eds.) (2000) *Aquatic Ecosystems of Mexico*. Leiden, Holanda, Backhuys Publishers.

Munk, W.H. & Riley, G.A. (1952) Absorption of nutrients by aquatic plants. *J. Mar. Res.*, Londres, 11, 215–240.

Nakamoto, N., Marins, M.A. & Tundisi, J.G. (1976) Synchronous growth of a freshwater diatom *Melosira italica* under natural environment. *Oecologia*, Berlim, 23, 179–184.

Nakamura, M. & Nakajima, T. (eds.) (2002) *Lake Biwa and Its Watersheds: A Review of LBRI Research Notes*. Otsu, Japão, Lake Biwa Research Institute.

National Geographic Brasil. (2004) São Paulo, Abril, ano 5, 53, set.

Naumann, E. (1919) Nagra synpunkter angaende limnoplanktons okologi med sarskild hansyn till fytoplankton. *Svensk Botanisk Tidskrift*, Uppsala, Suécia, 13, 129–163.

———. (1921) Einige Grundlinien der regionalen Limnologie. Lunds Universit. *Årsskrift N.F. Avd.*, Lund, Suécia, 2 (17), 1–22.

———. (1926) Die Bodenablagerungen de Süsswassers. *Arch. Hydrobiol.*, Stuttgart, 18.

———. (1931) Limnologische Terminologie. In: Abderhalden, E. (ed.) *Handbuch der Biologischen Arbeitsmethoden*. Abt. IX, Teil 8. Berlin/Viena, Urban & Schwarzenberg.

———. (1932) Grundzuege der regionalen Limnologie. *Die Binnengewässer*, Stuttgart, 11, 176.

Necchi, Jr., O. (1992) Macroalgae dynamics in a spring in Sao Paulo State, southeastern Brazil. *Arch. Hydrobiol.*, Stuttgart, 124, 489–499.

Necchi, Jr., O. & Pascoaloto, D. (1993) Seasonal dynamics of macroalgal communities in the Preto River Basin, São Paulo, Southern Brazil. *Arch. Hydrobiol.*, Stuttgart, 129 (4), 231–252.

Neiff, J.J. (1978) Fluctuaciones de la Vegetación Acuatica en Ambientes del Valle de Inundación del Parana Medio. *Physis* (B), Buenos Aires, 38, 41–53.

———. (1986) Aquatic plants of Parana System. In: Davies, B.R. & Walker, K.F. (eds.) *The Ecology of River Systems*. Dordrecht, Holanda, Dr. W. Junk. pp. 557–571.

———. (1990) Ideas para la Interpretación Ecológica del Paraná. *Interciencia*, Caracas, 15, 424–441.

———. (1996) Large rivers of South America: Toward the new approach. *Verh. Internat. Verein. Limnol.*, 26, 167–180.

Neiff, J.J. et al. (2000) Prediction of colonization by macrophytes in the Yaciretá reservoir of the Paraná River (Argentina and Paraguay). *Rev. Bras. Biol.*, 60 (4), 615–626.

Neiff, J.J. & Poi de Neiff, A. (1978) Estadios sucesionales en los camalotes chaqueños y su fauna asociada. Etapa geral *Pistia stratioides* – Eichornia crassipes. *Phycis*. Sección B, Buenos Aires, 38 (95), 29–39.

———. (1984) Cambios estacionales en la biomasa de *Eichhornia crassipes* (Mart.) Solms y su fauna asociada en una laguna del Chaco (Argentina). *Ecosur*, Rosario, Argentina, 11 (21–22), 51–60.

———. (2005) Connectivity as a basis for the management of aquatic plants. In: Thomaz, S.M. & Bini, L.M. (eds.) *Ecologia e Manejo de Macrófitas Aquáticas*. Maringá, Editora da UEM. pp. 39–58.

Nelson, J.S. (1994) *Fishes of the World*. 3rd edition. Chichester, Inglaterra, John Wiley & Sons.

Nelson, S.J. (2006) *Fishes of the World*. 4th edition, xv. Canada, John Wiley & Sons.

Nelson, W.S. (1979) Experimental studies of selective predation on amphipods. Consequences for amphipod distribution and abundance. *J. Exp. Mar. Biol. Ecol.*, Amsterdã, 38, 225–245.

Neumann-LeitãO, S., Paranhos, J.D.N. & De Souza, F.B.V.A. (1989) Zooplâncton do Açude de Apipucos, Recife/PE (Brasil). *Arquivos de Biologia e Tecnologia*, Curitiba, 32 (4), 803–821.

Niencheski, L.F. & Baumgarten, M.G. (1997) Environmental chemistry. In: Seeliger, U., Odebrecht, C. & Castello, J. (eds.) *Subtropical Convergence Marine Ecosystem. The Coast and the Sea in the Warm Temperate Southwestern Atlantic*. Heidelberg, Alemanha, Springer-Verlag, Heidelberg. pp. 20–23.

Nilsso, C. et al. (2005) Fragmentation and flow regulation of the world's large river systems. *Science*, 308 (5.720), 405–408. New York, Washington, DC, EUA.

Nogueira, C.R., Bonecker, A.C.T. & Bonecker, S.L.C. (1988) Zooplâncton da Baía de Guanabara (RJ-Brasil): Composição específica e variações espaço-temporais. In: Brandini, F.P. (ed.) *Memórias do III Encontro Brasileiro de Plâncton, Caiobá, Paraná*. Caiobá, Sociedade Brasileira de Plâncton. pp. 150–156.

Nogueira, M.G. (2001) Zooplankton composition, dominance and abundance as indicators of environmental compartmentalization in Jurumirim Reservoir (Paranapanema River), Sao Paulo, Brazil. *Hydrobiologia*, Baarn, Holanda, 455 (1), 1–18.

Nogueira, M.G. & Matsumura-Tundisi, T. (1994) Limnologia de um Sistema Artificial Raso (Represa do Monjolinho, São Carlos, SP). I. Dinâmica das Variáveis Físicas e Químicas. *Rev. Bras. Biol.*, São Carlos, 54 (1), 147–159.

Nowicki, M. (1889) *Fishes of River Systems of Wisla, Styr, Dniestr and Prut in Galicja* (em polonês). Cracóvia/Wydz/Krajowy, Polônia, [s.n].

Nygaard, G. (1949) Hydrobiological studies on some Danish ponds and lakes, part II. *Kong. Dansk. Vidensk. Selsk. Biol.*, Copenhaguem, 7, 1–293.

Odum, E.P. (1969) The strategy of ecosystem development. *Science*, Washington, 164, 262–270.

Odum, H.T. (1956) Primary production in flowing waters. *Limnol. Oceanogr.*, Milwaukee, EUA, 1, 102–117.

———. (1957) Trophic structure and productivity of Silver Springs, Florida. *Ecol. Monogr.*, Washington, 27, 55–112.

Odum, H.T. & Hoskin, C.M. (1957) Metabolism of a laboratory stream microcosm. *Publs. Inst. Mar. Sci. Univ. Tex.*, Port Aransas, EUA, 4, 115–133.

OECD. (1982) *Eutrophication of Waters: Monitoring, Assessment and Control.* Paris, OECD.

Ohle, W. (1952) Die hypolimnische Kohlendioxyd Akkumulation als produktionsbiologischer Indikator. *Arch. Hydrobiol.*, Stuttgart, 46, 153–285.

Okano, W.Y. (1980) *Padrão de migração vertical e flutuação sazonal das principais espécies de Copepoda (Crustacea) do lago Dom Helvécio, Parque Florestal do Rio Doce, MG. Lago Dom Helvécio, Minas Gerais, Brasil.* [Tese de Mestrado]. São Carlos, UFSCar.

Oldani, N.O. (1990) Variaciones de la abundancia de peces del valle del río Paraná (Argentina). *Rev. Hydrobiol. Trop.*, 23 (1), 67–76.

Oliff, W.D. (1960) Hydrobiological studies on the Tugela River System. Part I. The main Tugela River. *Hydrobiologia*, Baarn, Holanda, 14, 281–385.

Oliveira, E.C. (1996) *Introdução à Biologia Vegetal.* São Paulo, Edusp.

Oliveira, M.A. (1996) *Ecologia do perifíton de susbstrato artificial em cursos d'água do trecho medio do Arroio Sampaio, Mato Leitão, RS, Brasil.* [Tese de Mestrado]. Porto Alegre, Universidade Federal do Rio Grande do Sul.

Olivier, S.R. (1953) Contribution to the limnological knowledge of the Salada Grande Lagoon. *Verh. Internat. Verein. Limnol.*, Stuttgart, 12, 296–301.

Ortmann, A. (1897) Carcinologische Studien. *Zool. Jahrb. System.*, Alemanha, 10, 258–372.

Oullet, M. & PAGE, P. (1990) Some limnological aspects of crater lake, Quebec, Ungava. *Verh. Internat. Vererin. Limnol.*, Stuttgart, 24, 348.

Padisák, J., Toth, L.G. & Rajev, M. (1988) The role of storms in the summer succession of the phytoplankton community in a shallow lake (Lake Balaton, Hungry). *J. Plankton Res.*, Oxford, 10 (2), 249–265.

Padisák, J., Reynolds, C.S. & Sommer, U. (1993) *Intermediatic Disturbance Hypothesis in Phytoplankton Ecology.* Dordrecht, Holanda, Kluwer Academic Publishers. (Developments in Hydrobiology 81) (Reimpresso de Hydrobiologia 249.)

Paggi, J.C. (1978) Revisión de las especies argentinas del género Diaphanosoma Fischer (Crustacea, Cladocera). *Acta Zool. Lilloana*, San Miguel de Tucumán, Argentina, 33, 43–65.

———. (1980) Aportes al conocimiento de la fauna Argentina de cladoceros. III: *Euryalona fasciculata* Daday 1905 y *Euryalona occidentalis* Sars 1901. *Rev. Asoc. Cienc. Nat. Litoral*, Santo Tomé, Argentina, 11, 145–160.

———. (1990) Ecological and biogeographical remarks on the rotifer fauna of Argentina. *Rev. Hydrobiol. Trop.*, Paris, 23, 297–311.

Panitz, C.M.N. (1980) *Estudo Comparativo do Perifíton em Diferentes Substratos Artificiais na Represa do Lobo ("Broa"), São Carlos, SP.* [Tese de Mestrado]. São Carlos, UFSCar.

Panosso, R.F., Attayde, J.L. & Muehe, D. (1998) Morfometria das lagoas Imboassica, Cabiúnas, Comprida e Carapebus: Implicações para seu funcionamento e manejo. In: Esteves, F.A. (ed.) *Ecologia das lagoas costeiras do Parque Nacional da Restinga de Jurubatiba e do Município de Macaé (RJ).* Rio de Janeiro, Nupem-UFRJ. Cap. 3.1, pp. 91–108.

Pantulu, V.R. (1986) Fish of the Lower Mekong Basin. In: Davies, B.R. & Walker, K.F. (eds.) *The Ecology of River Systems.* Dordrecht, Holanda, Dr. W. Junk Publishers.

Paolini, J., Herrera, R. & Nemeth, A. (1983) Hydrochemistry of the Orinoco and Caroni Rivers. In: Degens, E.T., Kempe, S. & Soliman, H. (eds.) *Transport of Carbon and Minerals in Major World Rivers.* Hamburgo: Im Selbst. Geol-Paläont. Inst. Univ. Hamburg,

UNEP/SCOPE. (Mitt. Geol-Paläont. Inst. Univ. Hamburg, SCOPE/UNEP Sonderh. 55). Part 2, pp. 223–236.

Paredes, J.F., Paim, A.J., Da Costa-Doria, E.M. & Rocha, W.L. (1983) São Francisco River: Hydrological studies in the dammed lake of Sobradinho. In: Degens, E.T., Kempe, S. & Soliman, H. (eds.) *Transport of Carbon and Minerals in Major World Rivers*. Hamburgo: Im Selbst. Geol-Paläont. Inst. Univ. Hamburg, UNEP/SCOPE. (Mitt. Geol-Paläont. Inst. Univ. Hamburg, SCOPE/UNEP Sonderb. 55). Part 2, pp. 193–202.

Patten, B.C. (ed.) (1992a) *Wetlands and Shallow Continental Water Bodies:* Vol 1. Natural and Human Relationships. Haia, Holanda, SPB Academia Publishing.

———. (1992b) *Wetlands and Shallow Continental Water Bodies:* Vol. 2. Case Studies. Haia, Holanda, SPB Academia Publishing.

Payne, A.I. (1986) *The Ecology of Tropical Lakes and Rivers*. Nova York, John Wiley & Sons.

Pearsall, W.H. (1959) Production ecology. *Sci. Prog. Twent. Cent.*, Londres, 47, 106–111.

Pelaez-Rodríguez, M. & Matsumura-Tundisi, T. (2002) Rotifer production in a shallow artificial lake (Lobo-Broa reservoir, SP, Brazil). *Brasil. J. Biol.*, 62 (3), 509–516.

Penchaszadeh, P.E. (Coord.) (2005) *Invasores: Invertebrados Exóticos en el Rio de La Plata y Region Marina Aledaña*. Buenos Aires, Eudeba.

Penha, J.M.F., Da-Silva, C.J. & Bianchini, Jr., I. (1999) Productivity of the aquatic macrophyte *Pontederia lanceolata* Nutt. (Pontederiaceae) on floodplains of the Pantanal Mato-Grossense, Brazil. *Wetl. Ecol. Mgmt*, Baarn, Holanda, 7, 155–163.

Pennak, R.W. (1989) *Fresh-water Invertebrates of the United States: Protozoa to Mollusca*. 3rd edition. Nova York, John Wiley & Sons.

Perotti de Jorda, N.M. (1980) Campaña limnológica "Keratella I" en el río Paraná Medio: pigmentos y productividad primaria de ambientes leníticos. *Ecología*, Buenos Aires, 4, 63–68.

———. (1984) Estudios limnológicos en una sección transversal del tramo medio del río Paraná. IX: Biomasa y productividad. *Rev. Asoc. Cienc. Nat. Litoral*, Santo Tomé, Argentina, 15 (2), 117–133.

Petracco, P. (1995) *Determinação da Biomassa e Estoque de Nitrogênio e Fósforo de Polygonum spectabile Mart. e Paspalum repens Berg. da Represa de Barra Bonita (SP)*. [Tese de Mestrado]. São Carlos, USP.

Petrere, M. (1996) Fisheries in large tropical reservoirs in South America. *Lakes Reserv. Res. Mgmt*, Kusatsu-shi, Japão, 2, 111–133.

Petrere, M. & Agostinho, A.A. (1993) La pesca en el tramo brasileño del rio Paraná. *FAO, Informe de Pesca*, Roma, 490, 52–72.

Petts, G.E. (1984) *Impounded Rivers: Perspectives for Ecological Management*. Nova York, Wiley-Intersience Publication.

Petts, G.E. & Amoros, C. (ed.) (1996) *Fluvial Hydrosystems*. Londres, Chapman & Hall.

Petts, G.E. & Calow, P. (eds.) (1996) *River Restoration*. Liverpool, Blackwell Science.

Pfeiffer, W.C., Lacerda, L.D., Malm, O., Souza, C.M.M., Silveira, E.G. & Bastos, W.R. (1989) Mercury concentrations in inland waters of gold-mining areas in Rondonia, Brazil. *Sci. Total Environ.*, Amsterdã, 87/88, 233–240.

Pflug, R. (1969a) Quaternary lakes of Eastern Brazil. *Photogrametria*, Amsterdã, 24, 29–35.

———. (1969b) Das Überschüttungsrelief des Rio Doce, Brasilien. *Zeitschrift für Geomorphologie*, Stuttgart, 13, 141–162.

Pieczynska, E. (1990) Littoral habitats and communities. In: Jørgensen, S.E., Löffler, H. (eds.) *Guidelines for Lake Management: Lake Shore Management*, 3. Japan, Ilec. pp. 39–71.

Pielou, E. (1988) *Freshwater*. Chicago, The University of Chicago Press.

Pimentel, D. & Edwards, C.A. (1982) Pesticides and ecosystems. *BioScience*, Washington, 32, 7.

Pinto-Coelho, R.M. (1987) Flutuações sazonais e de curta duração na comunidade zooplanctônica do Lago Paranoá, Brasília, DF, Brasil. *Rev. Brasil. Biol.*, São Carlos, 47, 17–29.

Pinto Coelho, R., Giani, A. & Von Sperling, E. (eds.) (1994) *Ecology and Human Impact on Lakes and Reservoirs in Minas Gerais*. Belo Horizonte, ICB/UFMG.

Platt, T. & Li, W.K. (1986) Photosynthetic picoplankton. *Can. J. Fish. Aquat. Sci.*, Toronto, 214, 159–204.

Pomeroy, L.R. (1974) The ocean's food web, a changing paradigm. *BioScience*, Washington, 24, 499–504.

Pompeo, M.L.M. (1996) *Ecologia de* Echinochloa polystachya *(H.B.K.) Hitchcock na Represa de Jurumirim (Zona de Desembocadura do rio Pararapanema, SP)*. [Tese de Doutorado]. São Carlos, Escola de Engenharia de São Carlos, USP.

Pompeo, M.L.M. & Moschini-Carlos, V. (2003) *Macrófitas aquáticas e perifíton: aspectos ecológicos e metodológicos*. São Carlos, Rima Editora.

Por, F.D. (1986) Stream type diversity in the Atlantic lowland of the Jureia area (subtropical Brazil). *Hydrobiologia*, Baarn, Holanda, 131, 39–45.

Por, F.D. (1995) *The Pantanal of Mato Grosso, (Brazil): World's Largest Wetlands*. Dordrecht, Kluuer Academic Publishers.

Por, F.D., Shimizu, G.Y., Almeida Prado-Por, M.S., Lansac, F.A. & Rocha Oliveira, I. (1984) The Blackwater River Estuary of Rio Una do Prelado (São Paulo, Brazil): Preliminary hydrobiological data. *Rev. Hydrobiol. Trop.*, Paris, 17 (3), 245–258.

Por Dov, F. (1994) *Guia ilustrado de manguezal brasileiro*. São Paulo, Instituto de Biociências da USP.

Porter, K.G. (1973) Selective grazing and differential digestion of algae by zooplankton. *Nature*, Londres, 244, 179–180.

Porto, L.M.S. (1994) Dieta e Ciclo de Atividades Diuturno de *Pimelodella lateristringa* (Müller & Troschel, 1849) (Siluroidei, Pimelodidade) no Rio Ubatiba, Maricá, Rio de Janeiro. *Rev. Brasil. Biol.*, São Carlos, 54 (3), 459–468.

Post, W. (1981) Biology of the yellow-shouldered Blackbird *Agelaius xanthomus* on a tropical island. *Bull. Florida St. Mus. Biol. Sci.*, Gainesville, EUA, 26, 125–202.

Pott, V.J. & Pott, A. (eds.) (2000) *Plantas Aquáticas do Pantanal*. Brasília, EMBRAPA. (Comunicação para Transferência de Tecnologia).

Pytkowicz, R.M. & Hawley, J.E. (1974) Bicarbonate and carbonate ion-pairs and a model of seawater at 25C. *Limnol. Oceanogr.*, Milwaukee, EUA, 19 (2), 223–234.

Queiroz, H.L. & Crampton, W.G. (eds.) (1999) *Estratégias para manejo de recursos pesqueiros em Mamirauá*. Brasília, Sociedade Civil Mamirauá, CNPq.

Quirós, R. (2002) The nitrogen to phosphorus ratio for lakes: A cause or a consequence of aquatic biology? In: Fernandez-Cirelli, A. & Chalar, G. (eds.) *El Agua em Iberoamérica*: *de la Limnologia a la Gestión en Sudamérica*. Buenos Aires, CYTED XVII. pp. 123–141.

Quoy J.R.C. & Gaimard J.R. Zoologie. In: Freycinet, L. De (eds.) (1824) *Voyage autour du Monde entrepris par ordre du Roi, exécuté sur les corvettes de S.M.* L'Uranie et La Physicienne, pendant les années 1817, 1818, 1819 et 1820. Paris, Pillet Ainé.

Rai, H. & Hill, G. (1981) Observations on heterotrophic activity in lake Janauari a ria/várzea of the Central Amazon. *Verh. Internat. Verein Limnol.*, Stuttgart, 21, 683–688.

———. (1984) Primary production in the Amazonian aquatic ecosystem. In: Sioli, H. (ed.) *Amazon: Limnology of a Mighty Tropical River and its Basin*. Stuttgart, Dr. W. Junk Publ. pp. 311–335.

Rast, W. & Lee, G.F. (1978) *Summary Analysis of the North American (U.S. portion) OECD Eutrophication Project: Nutrient Loading-Lake Response Relationships and Trophic Indices*. Corvallis, EUA, U.S. EPA, 1978. (EPA-600/3-78-008).

Rast, W., Holland, M. & Olof, R.S. (1989) *Eutrophication: Management Framework for the Policy Maker*. Paris, UNESCO. (MAB Digest).

Rawson, D.S. (1939) *Some Physical and Chemical Factors in the Metabolism of Lakes*. Washington, American Association for the Advancement of Science. (AAAS Publication, 10).

———. (1951) The total mineral content of lake waters. *Ecology*, Washington, 32 (4), 669–672.

———. (1955) Morphometry as a dominant factor in the productivity of large lakes. *Verh. Internat. Verein. Limnol.*, Stuttgart, 12, 164–175.

———. (1956) Algal indications of trophic lake types. *Limnol. Oceangr.*, Milwaukee, EUA, 1, 18–25.

Raymont, J.E.G. (1963) *Plankton and Productivity in the Oceans*. Oxford, Londres, Pergamon Press.

Rebouças, A. (1994) Water crisis: Facts and myths. *An. Acad. Bras. Ciências*, Rio de Janeiro, 66 (suppl. 1), 136–147.

Rebouças, A., Braga, B. & Tundisi, J.G. (eds.) (2006) *Águas Doces no Brasil: Capital Ecológico, Uso e Conservação*. Rio de Janeiro, São Paulo, Academia Brasileira de Ciências, IEA (USP), Escrituras.

Redfield, A.C. (1958) The biological control of chemical factors in the environment. *Amer. Sci.*, Research Triangle Park, NC, EUA, 46, 205–221.

Regali, Seleghim, M. & Godinho, M.J.L. (2004) Peritrich epibiont protozoans in the zooplankton of a subtropical shallow aquatic ecosystem (Monjolinho reservoir, São Carlos, Brazil). *J. Plankton Res.*, 26 (5), 501–508.

Reid, J.W. & Turner, P.N. (1988) Planktonic Rotifera, Copepoda and Cladocera from Lagos Açu and Viana, State of Maranhão, Brazil. *Rev. Brasil. Biol.*, São Carlos, 48 (3), 485–495.

Reiss, F. (1977) Qualitative and quantitative investigations on the macrobenthic fauna of Central Amazon lakes. I. Lago Tupé, a black water lake on the lower Rio Negro. *Amazoniana*, Kiel, Alemanha, 6, 203–235.

Repavich, W.M., Sonzogni, W.C., Standridge, J.H., Wedepohl, R.E. & Meisner, L.F. (1990) Cyanobacteria (blue-green algae) in Wisconsin waters: Acute and chronic toxicity. *Wat. Res.*, Londres, 24, 225–231.

Revenga, C., Murray, S., Abramovitz, J. & Hammond, A. (1998) *Watershed of World: Ecological Value and Vulnerability*. Washington, World Resources Institute.

Revue D'hydrobiologie Tropicale. (1992) Paris, Orstom, 25, 2.

Revue D'hydrobiologie Tropicale. (1992) Paris, Orstom, 25, 4.

Reynolds, C.S. (1973a) The phytoplankton of Crose Mere, Shropshire. *Eur. J. Phycol.*, Abigdon, Inglaterra, 8 (2), 153–162.

———. (1973b) Growth and buoyancy of *Microcystis aeruginosa* Kutz. emend. Elenkin in a shallow eutrophic lake. *Proc. R. Soc. Lond.*, Londres, 184, 1074, 29–50. Series B, Biological Sciences).

———. (1978) Notes on the phytoplankton periodicity of Rostherne Mere, Cheshire, 1967–1977. *Eur. J. Phycol.*, Abidgon, Inglaterra, 13 (4), 329–335.

———. (1980) Phytoplankton assemblages and their periodicity in stratifying lake systems. *Holarc. Ecol.*, Lund, Suécia, 3 (3), 141–159.

———. (1984) *The Ecology of Freshwater Phytoplankton*. Cambridge, Cambridge University Press. (Cambridge Studies in Ecology).

———. (1986) Experimental manipulations of the phytoplankton periodicity in large limnetic enclosures in Belham Tarn, English Lake District. *Hydrobiologia*, Baarn, Holanda, 138, 43–64.

———. (1992a) The role of fluid motion in the dynamics of phytoplankton in lakes and rivers. In: Giller, P.S., Hildrew, A.G. & Raffaelli, D.G. (eds.) *Aquatic Ecology Scale, Patterns and Process*. Londres, Blackwell Scientific Publications. pp. 141–187.

———. (1992b) Eutrophication and management of planktonic algae: What Vollenweider couldn't tell us. In: Sutcliffe, D.W. & Jones, J.G. (eds.) *Eutrophication: Research and Application to Water Supply*. Roma, FBA. pp. 4–29.

———. (1994) The long, the short and the stalled: On the attributes of the phytoplankton selected by physical mixing in lakes and rivers. *Hydrobiologia*, Baarn, Holanda, 289, 9–21.

————. (1995) Successional change in the planktonic vegetations: Species, structures, scales. In: Joint, I. (ed.) *The Molecular Ecology of Aquatic Microbes*. Berlim, Springer-Verlag. pp. 115–132.

————. (1997a) On the vertical distribution of phytoplankton in the middle Rio Doce Vale lakes. In: Tundisi, J.G. & Saijo, Y. (eds.) *Limnological Studies on the Rio Doce Valley Lakes, Brazil*. Rio de Janeiro, Academia Brasileira de Ciências, São Paulo, USP. pp. 227–242.

————. (1997b) *Vegetation Processes in the Pelagic: A Model for Ecosystem Theory*. Luhe, Alemanha, Inter-Research Science Center, Ecology Institute. (Excellence in Ecology, 9).

————. (1999) Modelling phytoplankton dynamics and its application to lake management. *Hydrobiologia*, Baarn, Holanda, 395–396, 123–131.

Reynolds, C.S. & Walsby, A.E. (1975) Water blooms. *Biol. Rev.*, Cambridge, 50, 437–481.

Reynolds, C.S., Wiseman, S.W., Godfrey, B.M. & Butterwick, C. (1983a) Some effects of artificial mixing on the dynamics of phytoplankton populations in large limnetic enclosures. *J. Plankton Res.*, Oxford, 5 (2), 203–234.

Reynolds, W.D., Elrick, D.E. & Topp, G.C.A. (1983b) Reexamination of the constant head well permeameter for measuring saturated hydraulic conductivity above the water table. *Soil Sci.*, Riverwood, EUA, 136, 250–268.

Ribeiro, A.M. (1908) *Peixes da Ribeira*. Kosmos, Rio de Janeiro, 5, 1–5.

Richardson, J.S. (1992) Food, microhabitat or both? Macroinvertebrate use of leaf accumulations in a montane stream. *Freshwat. Biol.*, Victoria, Austrália, 27, 169–176.

Rietzler, A.C., Rocha, O. & Espindola, E.L.G. (2004) Produção Secundária das Espécies do Zooplâncton do Reservatório de Salto Grande, Estado de São Paulo. In: Espíndola, E.L.G., Leite, M.A. & Dornfeld, C.B. (Orgs.) *Reservatório de Salto Grande (Americana, SP)*, Caracterização, Impactos e Propostas de Manejo. São Carlos, RIMA. pp. 199–219.

Righi, G. (1984) *Manual de identificação de invertebrados Límnicos do Brasil*. Brasília, CNPq.

Rigler, H.F. & Peters, R.H. (1995) *Science and Limnology*. Luhe, Alemanha, Inter-Research Science Center, Ecology Institute. (Excellence in Ecology 6).

Riley, G.A. (1942) The relationship of vertical turbulence and spring diatom flowerings. *J. Mar. Res.*, Londres, 6, 67–87.

Ringelberg, J. (1991) The relation between ultimate and proximate aspects of diel vertical migration in *Daphnia hyalina*. *Verh. Internat. Verein. Limnol.*, Stuttgart, 24, 2804–2807.

Roberts, R.D. & Roberts, T.M. (eds.) (1984) *Planning and Ecology*. Boca Ratón, EUA, Chapman and Hall.

Robertson, B.A., Hardy, E.R. () Zooplankton of Amazonian lakes and rivers. In: Sioli, H. (ed.) *The Amazon. Limnology and Landscape Ecology of a Mighty Tropical River and Its Basin*. Dordrecht, W. Junk. pp. 337–352.

Rocha, O. et al. (2005) *Espécies invasoras em águas doces: estudos de caso e propostas de manejo*. São Carlos, Editora Universidade Federal de São Carlos.

Rocha, O., Matsumura-Tundisi, T. & Tundisi, J.G. (1982) Seasonal fluctuation of *Argyrodiaptomus furcatus* populations in Lobo Reservoir (Sao Carlos, SP, Brazil). *Trop. Ecol.*, Nova Déli, 23 (1), 134–150.

Rocha, O. & Matsumura-Tundisi, T. (1984) Biomass and production of *Argyrodiaptomus furcatus*, a tropical calanoid copepod in Broa Reservoir, Southern Brazil. *Hydrobiologia*, Baarn, Holanda, 113, 307–311.

————. (1997) Respiration rates of epilimnetic and metalimnetic zooplankton in a stratified tropical lake. In: Tundisi, J.G. & Saijo, Y.S. (eds.) *Limnological Studies in the Rio Doce Valley Lakes*. São Paulo, Crhea (USP), Rio de Janeiro, Academia Brasileira de Ciências. pp. 113–120.

Rocha, O., Matsumura-Tundisi, T., Tundisi, J.G. & Fonseca, C.P. (1990) Predation on and by pelagic turbellaria in some lakes in Brazil. *Hydrobiologia*, Baarn, Holanda, 198, 91–101.

Rocha, O., Sendacz, S. & Matsumura-Tundisi, T. (1995) Composition, biomass and productivity of zooplankton in natural lakes and reservoirs in Brazil. In: Tundisi, J.G., Bicudo, C.E.M. & Matsumura-Tundisi, T. (eds.) *Limnology in Brazil*. Rio de Janeiro, ABC/SBL. pp. 151–166.

Rocha, O., Matsumura-Tundisi T., Espíndola, E.L.G., Roche, K.F. & Rietzler, A.C. (1999) Ecological theory applied to reservoir zooplankton. In: Tundisi, J.G. & Straškraba, M. (eds.) *Theoretical Reservoir Ecology and Its Applications*. Leiden, Holanda, Backhuys Publishers. pp. 29–51.

Rodhe, W. (1969) Crystallisation of eutrophication concepts in Northern Europe. In: Eutrophication: Causes, consequences, correctives. *Proceedings of an International Symposium on Eutrophication*. Washington, National Academy of Sciences. pp. 50–64.

Rodhe, W. (1974) Plankton, Planktic, planktonic. *Limn. Oceanogr.*, Milwaukee, EUA, 19, 360.

Rodrigues, L., Thomaz, S.M., Agostinho, A.A. & Gomes, L.C. (2005) *Biocenoses em Reservatórios:* Padrões Espaciais e Temporais. São Carlos, RIMA.

Rodrigues, M.T. (2003) Herpetofauna da Caatinga. In: Leal, I.R., Tabarelli, M. & Da Silva, J.M.C. (eds.) *Ecologia e Conservação da Caatinga*. Recife, Ed. Universitária da UFPE. pp. 181–236.

Rodrigues, R.R. & Leitão Filho, H.F. (eds.) (2001) *Matas ciliares: conservação e recuperação*. São Paulo, Edusp, Fapesp.

Rodrigues, S.L. (2003) *Comparação da Atenuação da Radiação Solar e dos Fatores que Determinam o Clima de Radiação Solar Subaquatica em Lagos e Reservatórios do Brasil*. [Tese de Doutorado]. São Carlos, EESC/USP.

Rodrigues-Filho, S. & Muller, G. (1999) *A Holocene Sedimentary Record from Lake Silvana, SE Brazil*. Nova York, Springer-Verlag. (Lecture Notes in Earth Sciences).

Rodriguez, G. (1975) Some aspects of the ecology of tropical estuaries. In: Golley, F.B. & Me Dina, E. (eds.) *Tropical Ecological Systems*. Nova York, Springer. pp. 313–333.

Roland, F. (1998) Produção fitoplanctonica em diferentes classes de tamanho nas lagoas Imboassica e Cabiunas. In: Esteves, F.A. (ed.) *Ecologia das lagoas costeiras do Parque Nacional da Restinga de Jurubatiba e do Municipio de Macaé (RJ)*. Macaé, NUPEM. pp. 159–175.

Roldán, G. (1988) *Guía para el estudio de los macroinvertebrados acuáticos del Departamento de Antioquia*. Bogotá, Presencia.

———. (1992) Fundamentos de Limnologia neotropical. *Editorial Universidad de Antiorquia*, Modelling. 529 pp.

———. (1999) Los macroinvertebrados y su valor. *Rev. Acad. Col. Cienc. Exac. Fis. y Natur.*, 23 (88), 375–387.

———. (2001) Los macroinvertebrados de agua dulce en los Andes tropicales. In: Primack, R.P., Feisinger, R.D., Mazardo, F. (2001) *Fundamentos de conservación biológica, perspectivas latinoamericanas*. Mexico, Fondo de Cultura Economica. pp. 122–125.

Roldán, G., Posada, J.A. & Gutiérrez, J.C. (2001) Estúdio limnológico de los recursos hídricos del parque de Piedras Blancas. *Acad. Col. Cienc. Exat. Fis. Nat. 9*. (Colección Jorge Alvarez Lleras).

Romero, J.R. & Imberger, I. (1999) Seasonal horizontal gradient of dissolved oxygen in a temperate Austral reservoir. In: Tundisi, J.G. & Strakraba, M. (eds.) *Theoretical Reservoir Ecology and its Applications*. Rio de Janeiro, BAS, São Carlos, IIE, Leiden, Holanda, Backhuys Publishers. pp. 211–226.

Rosa, C.A., Rezende, M.A., Barbosa, F.A.R., Morais, P.B. & Franzot, S.P. (1995) Yeast diversity in a mesotrophic lake on the karstic plateau of Lagoa Santa, MG-Brazil. *Hydrobiologia*, Baarn, Holanda, 308, 103–108.

Round, F.E. (1981) *The Ecology of Algae*. Cambridge, Cambridge University Press.

Ruppert, E.E. & Barnes, R.D. (1996) *Zoologia dos invertebrados*, 6th edition. São Paulo, Rocca.

Ryding, S.O. & Rast, W. (1989) *The Control of Eutrophication of Lakes and Reservoirs*. Lancaster, Inglaterra, UNESCO, The Parthenon Publishing Group. (Man and Biosphere Series 1).

Ryther, J.H. (1969) Photosynthesis and fish production in the sea. The production of organic matter and its conversion to higher forms of life vary throughout the world ocean. *Science*, Washington, 166, 72–76.

Sabino, J. & Castro, R.M.C. (1990) Alimentação, Período de Atividade e Distribuição Espacial dos Peixes de um Riacho da Floresta Atlântica. *Rev. Brasil. Biol.*, São Carlos, 50 (1), 23–36.

Saggio, A. & Imberger, J. (1998) Internal wave weather in a stratified lake. *Limnol. Oceanogr.* 43 (8), 1780–1795.

Saijo, Y. & Tundisi, J.G. (eds.) (1985) *Limnological Studies in Central Brazil*. Nagoya, Instituto de Pesquisa da Água, Universidade de Nagoya.

Sakamoto, M. (1966) Primary production by phytoplankton community in some Japanese lakes and its dependence on lake depth. *Arch. Hydrobiol.*, Stuttgart, 62, 1–28.

Sakamoto, M., Tilzer, M.M., Gachter, R., Rai, H., Collos, Y., Tscumi, P., Berner, P., Zbaren, D., Zbaren, J., Dokulil, M., Bossard, P., Uehlinger, U. & Nusch, E.A. (1984) Joint field experiments for comparisons of measuring methods of phothosynthetic production. *J. Plankton Res.*, Oxford, 6, 2.

Salas, H. & Martino, P. (1991) A simplified phosphorus trophic state model for warm-water tropical lakes. *Wat. Res.*, Londres, 25 (3), 341–350.

Sampaio, E.V., Rocha, O., Matsumura-Tundisi, T. & Tundisi, J.G. (2002) Composition and abundance of zooplankton in the limnetic zone of seven reservoirs of the Paranapanema river, Brazil. *Braz. J. Biol.*, São Carlos, 62, 525–545.

Sanders, H.L. (1969) Benthic marine diversity and the stability-time hypothesis. In: Woodwell, G.M. & Smith, H.H. (eds.) *Diversity and Stability in Ecological Systems*, vol. 22. Nova York, Brookhaven National Laboratory. pp. 71–81.

Sandes, M.A.L. (1998) *Estudos Ecológicos em Florescimento de Microcystis (Cyanobacteria – Cyanophycea) e Interações com a Flora Bacteriana na Represa de Barra Bonita – Médio Tietê/SP*. [Tese de Doutorado]. São Paulo, USP.

Santos, G.M. & Ferreira, E.J.G. (1999) Peixes da Bacia Amazonica. In: Lowe-McConnell, R.H. (ed.) *Estudos Ecologicos de Comunidades de Peixes Tropicais*. São Paulo, Edusp. pp. 345–373.

Santos, J.E. & Pires Salatiel, J. (2000) *Estudos Integrados em Ecossitsemas*. São Carlos, RIMA.

Sars, G.O. (1901) Contribution to the knowledge of the freshwater Entomostraca of South America. Part. I. Cladocera. *Archiv fur Matematik og Naturvidenskab*, Christiania (Oslo), Noruega, 22 (6), 3–27.

Sathyendranath, S & Platt, T. (1990) The light field in the ocean: Its modification and exploration by the pelagic biota. In: Herring, P., Campbell, A.K., Whitfield, M. & Maddock, L. (eds.) *Light and Life in the Sea*. Cambridge, Cambridge University Press. pp. 3–18.

Sato, Y. & Godinho, H.P. (1999) Peixes da bacia do rio São Francisco. In: Lowe-McConnell, R.H. (ed.) *Estudos Ecológicos de Comunidades de Peixes Tropicais*. São Paulo, Edusp. pp. 401–413.

Scarabino, V., Maytia, S. & Caches, M. (1975) Carta binomica litoral del Departamento de Montevideo. I. Niveles superiores del sistema litoral. *Comunicaciones de la Sociedad Malacológica del Uruguay*, Montevidéo, 4 (29), 117–129.

Schafer, A. (1985) *Fundamentos de Ecologia e Biogeografia das Águas Continentais*. Porto Alegre, Editora da Universidade UFRGS.

Schaeffer Novelli, Y., Mesquita, H. & Cintron, G. (1990) The Cananéia lagoon estuarine system, SP, Brazil. *Estuaries*, Port Republic, EUA, 13, 193–203.

Scheffer, M. (1998) *Ecology of Shallow Lakes*. Boca Ratón, EUA, Chapman & Hall.

Schiavetti, A. & Camargo, A.F.M. (2002) *Conceitos de bacias hidrográficas*. Ilhéus, Editora da UESC.

Schiemer, F. & Boland, K.T. (1996) *Perspectives in Tropical Limnology*. Amsterdã, SPB Academic Publishing.

Schmidt, G.W. (1973) Primary production of phytoplankton in three types of Amazonian waters II. The limnology of a tropical floodplain lake in Central Amazônia (Lago do Castanho). *Amazoniana*, Kiel, Alemanha, 4 (2), 139–203.

———. (1976) Primary production of phytoplankton in three types of Amazonian waters. IV: The primary productivity of phytoplankton in a bay on the lower Rio Negro (Amazonas, Brasil). *Amazoniana*, Kiel, Alemanha, 5 (4), 517–528.

Schmidt, W. (1928) Über die Temperatur- und Stabilitäts-verhältnisse von Seen. *Geogr. Ann.*, Estocolmo, 10, 145–177.

Schindler, D.W., Bayley, S.E., Parker, B.R., Beaty, K.G., Cruikshank, D.R., Fee, E.J., Schindler, E.U. & Stainton, M.P. (1996) The effects of climatic warming on the properties of boreal lakes and streams at the Experimental Lakes Area, Northwestern Ontario. *Limnol. Oceanogr.*, Milwaukee, EUA, 41, 1004–1017.

Schulz, R.W., Vischer, H.F., Cavaco, J.E.B., Santos, E.M., Tyler, C.R., Goos, H.J.T.H. & Bogerd, F. (2001) Gonadotropins, their receptors, and the regulation of testicular functions in fish. *Comp. Biochem. Physiol. Part B*, Nova York, 129, 407–417.

Schwarzbold, A. (1992) *Efeitos do regime de inundação do rio Mogi-Guaçu (SP) sobre a estrutura, diversidade, produção e estoques do perifiton de* Eichornia azurea *(Sw) Kunth da Lagoa do Infernão*. [Tese de Doutorado]. São Carlos, UFSCar.

Schwoerbel, J. (1987) *Handbook of Limnology*. Chichester, Inglaterra, Ellis Horwood.

Science Progress. (1974) Oxford, Blackwell Scientific Publications, 61, 241.

Seeliger, U. & Kjerfve, B. (eds.) (2001) *Coastal Marine Ecosystems of Latin America*. Berlim, Springer. (Ecological studies, 144).

Sendacz, S. (1993) *Estudo da comunidade zooplanctonica de Lagoas Marginais do Rio Paraná Superior* [Tese de Doutorado]. São Paulo, USP.

Sendacz, S. & Kubo, E. (1982) Copepoda (Calanoida e Cyclopoida) of reservoirs in São Paulo State. *Bol. Inst. Pesca*, São Paulo, 9, 51–89.

Sendacz, S. & Melo-Costa, S.S. (1991) Caracterização do Zooplâncton do Rio Acre e Lagos Lua Nova, Novo Andirá e Amapá (Amazônia, Brasil). *Rev. Brasil. Biol.*, São Carlos, 51, 463–470.

Sendacz, S., Kubo, E. & Cestarolli, M.A. (1985) Limnologia de reservatórios do Sudeste do Estado de São Paulo. *Bol. Inst. Pesca*, São Paulo, 12, 187–207.

Shejlkvalc, B.L. et al. (2001) Heavy metal surveys in Nordic lakes: Levels patterns and relation to critical limits. *Ambio*, 30, 1–10.

Shiklomanov, I. (1998) *World Water Resources: A New Appraisal and Assessment for the 21st Century*. Paris, IHP, UNESCO.

Sidagis Galli, C.V. (2003) *Análise da função de uma várzea na ciclagem de nitrógeno*. [Tese de Doutorado]. São Carlos, EESC/USP.

Silva, L.H.S. (1999) Fitoplâncton de um reservatório eutrófico (lago Monte Alegre). Ribeirão Preto, São Paulo, Brasil. *Rev. Braz. J. Biol.*, São Carlos, 59 (2), 281–303.

Silva, W.M. & Matsumura-Tundisi, T. (2005) Taxonomy, ecology and geographical distribution of the species of the genus *Thermocyclops* Kiefer, 1927 (Copepoda, Cyclopoida) in São Paulo State, Brazil, with description of a new species. *Braz. J. Biol.*, São Carlos, 65 (3), 521–531.

Simpson, E.H. (1949) Measurement of diversity. *Nature*, Washington, 163, 688.

Sioli, H. (1975) Tropical river: The Amazon. In: Whitton, B. (ed.) *River Ecology*. Liverpool, Blackwell. (Studies in Ecology, 2). pp. 461–486.

———. (1984) *The Amazon: Limnology and Landscape Ecology of a Mighty Tropical River and Its Basin*. Dordrecht, Holanda, Dr. W. Junk Publishers.

Sivonen, K. (1990) Effects of light, temperature, nitrate, orthophosphate and bacteria on growth of and hepatoxin production by *Oscillatoria agardhii* strains. *Appl. Environ. Microbiol.*, Washington, 56 (9), 2658–2666.

Sipaúba Tavares, L.H., Rocha, O. (2001) *Produção de plâncton (fitoplâncton e zooplâncton) para alimentação de organismos aquáticos*. São Carlos, Rima Editora.

Skulberg, O.M., Underdal, B. & Utkilen, H. (1994) Toxic waterblooms with cyanophytes in Norway: Current knowledge. *Algol. Stud.*, Stuttgart, 75, 279–289.

Smayda, T.J. (1980) The phytoplankton species succession. In: Morris, I. (ed.) *The Physiological Ecology of Phytoplankton*. Berkeley, EUA, University of California Press. pp. 497–570.

Somlyody, L. & Varis, O. (2006) Freshwater under pressure. In: International Review For Environmental Strategies. Institute for Global Environmental Strategies, 6, 2.

Somlyody, L., Yates, D. & Varis, O. (2001) Challenges to freshwater management. *Ecohydrol. Hydrol.*, Varsóvia, 1 (1–2), 65–95.

Sondergaard, M. (1997) Bacteria and organic carbon in lakes. In: Jersen, S. & Pédersen, O. (eds.) *Freshwater Biology: Priorities and Development in Danish Research*. Copenhaguem, Freshwater Biological Laboratory. pp. 138–161.

Sorokin, Y.I. (1999a) *Aquatic Microbial Ecology: A Textbook for Students in Environmental Sciences*. Leiden, Holanda, Brackhuys.

———. (1999b) Aquatic microbial communities: The phytoplankton. In: Sorokin, Y.I. (ed.) *Aquatic Microbial Ecology: A Textbook for Students in Environmental Sciences*. Leiden, Brackhuys. Cap. 1, pp. 5–47.

Speidel, D.H., Ruedisili, L.C. & Agnew, A.F. (eds.) (1988) *Perspectives on Water: Uses and Abuses*. Oxford, Oxford University Press.

Spies, M. & Reiss, F. (1996) Catalog and bibliography of neotropical and Mexican Chironomidae (Insecta, Diptera). *Spixiana Suppl.*, Munique, 22, 61–119.

Spix, J.B. & Agassiz, L. (1829) *Selecta Genera et Species Piscium*. Munique, C. Wolf. Springer Netherlands. Vol. 159, 2/March. (Biomedical and Life Sciences).

Stankovic, S. (1960) *The Balkan Lake Ohrid and its Living World*. Dordrecht, Holanda, Dr. W. Junk.

Stanley, D.J. & Wetzel, F.C. (eds.) (1985) *Geological Evolution of the Mediterranean Basin*. Nova York, Springer.

Starling, F.L.R.M. (1998) *Development of Biomanipulation Strategies for the Remediation of Eutrophication Problems in an Urban Reservoir, Lago Paranoá, Brazil*. [Tese de Doutorado]. Stirling, Escócia, University of Stirling, Institute of Aquaculture.

Steemann Nielsen, E. (1951) *The Marine Vegetation of Isefjord*. Copenhaguem, Meddelelser. (Série Plankton, v. 5, n. 4).

———. (1952) The use of radioactive Carbon (C14) for measuring organic production in the sea. *J. Cons. Int. Explor. Mer.*, Copenhaguem, 18 (2), 117–140.

———. (1975) *Marine Photosynthesis with Special Emphasis on the Ecological Aspects*. Amsterdã, Elsevier.

Straskraba, M. (1973) Limnological basis for modelling reservoir ecosystems. Ackerman, W.C., Wite, G.F. & Worthington, E.B. (eds.) *Man Made Lakes: Their Problems and Environmental Effects*. Washington, AGU. (Geophysical Monograph Series, 17). pp. 517–535.

———. (1985) Managing of eutrophication by means of ecotechnology and mathematical modeling. In: LAKE Pollution and Recovery, *Proc. Internat. Congress*, Rome 15–18th April 1985. Roma, European Water Pollution Control Association. pp. 17–27.

———. (1986) Ecotechnological measuring against eutrophication. *Limnologica*, Berlim, 17, 239–249.

Strašskraba, M. (1994) Ecotechnological models of reservoir water quality management. *Ecol. Model.* 74, 1–38.

———. (1996a) Ecotechnological methods for managing non point source pollution in watersheds, lakes and reservoirs. *Wat. Sci. Tech.*, Londres, 33 (4–5), 73–80.

———. (1996b) Lake and reservoir management. *Verh. Internat. Verein. Limnol.*, Stuttgart, 26, 193–209.

————. (1998) Coupling of hydrobiology and hydrodynamic: Lakes and reservoirs. In: Imberger, J. (ed.) *Physical Processes in Lakes and Oceans*. Washington, American Geophysical Union. (Coastal and Estuarine Studies, 54). pp. 623–644.

————. (1999) Retention time as a key variable of reservoir limnology. In: Tundisi, J.G. & Straškraba, M. (eds.) *Theoretical Reservoir Ecology and its Applications*. Leiden, Holanda, Backhuys Publishers, São Carlos, IIE. pp. 385–410.

Straškraba, M. & Tundisi, J.G. (1999) Reservoir ecosystem functioning: Theory and applications. In: Straškraba, M. & Tundisi, J.G. (eds.) *Theoretical Reservoir Ecology and its Applications*. Leiden, Holanda, Backhuys. pp. 565–597.

————. (2000) *Gerenciamento da qualidade da água de represas*. Tradução Dino Vannucci. São Carlos, ILEC, IIE. (Diretrizes para o gerenciamento de lagos, 9).

Straškraba, M., Blazka, P., Brandl, Z., Hejzlar, P., Komárková, J., Kubecka, J., Nesmerák, I., Procházková, L., Straskrabová, V. & Vyhnálek, V. (1993a) Framework for investigation and evaluation of water quality in Czechoslovakia. In: Straškraba, M., Tundisi, J.G. & Duncan, A. (eds.) *Comparative Reservoir Limnology and Water Quality Management*. Dordrecht, Holanda, Kluwer Academic Publishers. pp. 169–212.

Straškraba, M., Tundisi, J.G. & Duncan, A. (eds.) (1993b) *Comparative Reservoir Limnology and Water Quality Management*. Dordrecht, Holanda, Kluwer Academic Publishers.

————. (1993c) State of the art of reservoir limnology and water quality management. In: Straškraba, M., Tundisi, J.G. & Duncan, A. (eds.) *Comparative Reservoir Limnology and Water Quality Management*. Dordrecht, Holanda, Kluwer Academic Publisher. pp. 213–288.

Strixino, G. & Trivinho-Strixino, S. (1998) Povoamentos de Chironomidae (Diptera) em lagos artificiais. In: Nessimian, J.L. & Carvalho, A.L. (eds.) *Ecologia de Insetos Aquáticos*. Rio de Janeiro, PPG-UFRJ. (Série Oecologia brasiliensis, 5). pp. 141–154.

Stumm, W. (ed.) (1985) *Chemical Processes in Lakes*. Nova York, John Wiley & Sons.

Sverdrup, H.U., Johnson, M.W. & Fleming, R.H. (1942) *The Oceans: Their Physics, Chemistry and General Biology*. Nova Jersey, EUA, Prentice Hall.

Swanson, F.J. (1980) Geomorphology and ecosystems. . In: Warning, R.W. (ed.) *Forests: Fresh Perspectives from Ecosystem Analysis*. *Proc. Ann. Biol. Coll. (1979)*. Corvallis, EUA, Oregon State University Press. pp. 159–170.

Syvitski, J.P.M., Vörösmarty, C.J., Kettner, A.J. & Green, P.A. (2005) Impact of humans on the flux of terrestrial sediment to the global coastal ocean. *Science*, Washington, 308 (5720), 376–380.

Talling, J.F. (1957a) Diurnal changes of stratification and photosynthesis in some tropical African waters. *Proc. Roy. Soc. London. Ser. B, Biol. Sci.*, Londres, 147 (926), 57–83.

————. (1957b) Some observations on the stratification of Lake Victoria. *Limnology and Oceanography*, Milwaukee, EUA, 2 (3), 213–221.

————. (1957c) The phytoplankton population as a compound photosynthetic system. *New Phytologist*, Lancaster, Inglaterra, 56 (2), 133–149.

————. (1957d) Photosynthetic characteristics of some freshwater plankton diatoms in relation to underwater radiation. *New Phytologist*, Lancaster, Inglaterra, 56 (1), 29–50.

————. (1965a) The photosynthetic activity of phytoplankton in East African lakes. *Int. Rev. Hydrobiol.*, Weinheim, Alemanha, 50, 1–32.

————. (1965b) Comparative problems of phytoplankton production and photosynthetic activity in a tropical and a temperate lake. *Mem. Ist. Ital. Idrobiol.*, Roma, supl. 18, 399–424.

————. (1966) The annual cycle of stratification and phytoplankton growth in Lake Victoria (East Africa). *Int. Revue Hydrobiol.*, Weinheim, Alemanha, 51 (4), 545–621.

————. (1969) The incidence of vertical mixing and some biological and chemical consequences in tropical lakes. *Verh. Internat. Verein. Limnol.*, Stuttgart, 17, 998–1012.

————. (1971) The underwater light climate as a controlling factor in the production ecology of freshwater phytoplankton. *Mitt. Int. Verein. Limnol.*, 19, 214–243.

———. (1976) The depletion of carbon dioxide from lake water by phytoplankton. *J. Ecol.*, Londres, 64, 79–121.

———. (1996) Tropical limnology and the Sunda Expedition. In: Schiemer, F. & Boland, K.T. (Eds). *Perspectives in Tropical Limnology*. Haia, Holanda, SPB Academic Publishing. pp. 19–26, 347.

Talling, J.F. (1999). Some English lakes as diverse and active ecosystems: a factual summary and source book. Freshwater Biological Association, Ambleside.

Talling, J.F. (2006). A brief history of the scientific study of tropical African inland waters. Freshwater Forum 26, 3–37.

Talling, J.F. & Lemoalle, J.L. (1998) *Ecological Dynamics of Tropical Inland Waters*. Cambridge, Cambridge University Press.

Talling, J.F. & Talling, I.B. (1965) The chemical composition of African lake waters. *Int. Revue Ges. Hydrob.*, Weinheim, Alemanha, 50 (3), 421–463.

Talling, J.F., Wood, R.B., Prosser, M.V. & Baxter, R.M. (1973) The upper limit of photosynthetic productivity by phytoplankton: Evidence from Ethiopian Soda Lakes. *Freshwat. Biol.*, Victoria, Austrália, 3, 53–76.

Tavares, M. & Mendonça, Jr., J.B. (2004) Introdução de Crustáceos Exóticos no Brasil: Uma Roleta Ecológica. In: Silva, J & Souza, R. (eds.) *Água de Lastro e Bioinvasão*. Rio de Janeiro, Interciência, 59–76.

Teal, J. (1962) Energy flow in the salt marsh ecosystem of Georgia. *Ecology*, Washington, 43 (4), 614–624.

Teixeira, C. & Tundisi, J.G. (1967) Primary production and phytoplankton in equatorial waters. *Bull. Mar. Sci.*, Miami, 17 (4), 884–891.

———. (1981) The effects of nitrogen and phosphorus enrichments on the phytoplankton in the region of Ubatuba, Brazil. *Bol. Inst. Oceanogr. USP*, São Paulo, 30, 77–86.

Teixeira, C., Tundisi, J.G. & Kutner, M.B. (1965) Plankton studies in a mangrove environment. II. The standing stock and some ecological factors. *Bol. Inst. Oceanogr. USP*, São Paulo, 14, 13–42.

Teixeira, R.L. (1989) Aspectos da ecologia de alguns peixes do arroio Bom Jardim, Triunfo, RS. *Rev. Brasil. Biol.*, São Carlos, 49, 183–192.

The Japanese Journal of Limnology. (1996) *Osaka, The Japanese Society of Limnology*, 57, 4 (2), dez. Edição especial.

The Japanese Journal of Limnology. (1999) *Osaka, The Japanese Society of Limnology*, 60 (1).

The Japanese Journal of Limnology. (2001) *Osaka, The Japanese Society of Limnology*, 62 (3), out.

Thienemann, A. (1925) *Die Binnengewässer Mitteleuropas: eine limnologische Einführung*. Stuttgart, Verlag. (Die Binnengewässer 1).

———. (1931) Der Produktionsbegriff in der Biologie. *Arch. Hydrobiol.*, Stuttgart, 22, 616–622.

Thomaz, S.M. & Bini, L.M. (1998) Ecologia e manejo de macrófitas aquáticas em reservatórios. *Acta Limnol. Brasil.*, Campinas, 10, 103–116.

———. (1999) A expansão das macrófitas aquáticas e implicações para o manejo de reservatórios: um estudo na represa de Itaipu. In: Henry, R. (ed.) *Ecologia de reservatórios: estrutura, função e aspectos sociais*. Botucatu, FAPESP/FUNDIBIO. p. 597–625.

———. (2003) *Ecologia e Manejo de Macrófitas Aquáticas*. Maringá, EDUEM.

Thomaz, S.M. & Bini, L.M. (eds.) (2005) *Ecologia e Manejo de Macrófitas Aquáticas*. Maringá, Eduem. pp. 39–58.

Thomaz, S.M., Agostinho, A.A. & Hahn, N.S. (2004) *The Upper Paraná River and its Floodplain: Physical Aspects, Ecology and Conservation*. Leiden, Holanda, Backhuys Publishers.

Thomaz, S.M., Bini, L.M. & Alberte, S.M. (1997) Limnologia do reservatório de Segredo: padrães de variação espacial e temporal. In: Agostinho, A.A. & Gomes, L.C. (eds.) *Reservatório de Segredo: bases ecológicas para o manejo*. Maringá, EDUEM. p. 19–37.

Thomaz, S.M., Roberto, M.C. & Bini, L.M. (1997) Caracterização limnológica dos ambientes aquáticos e influência dos níveis fluviométricos. In: Vazzoler, A.E.M.A., Agostinho, A.A. & Hahn, N.S. *A planície de inundação do alto Rio Paraná:* aspectos físicos, biológicos e sócio-econômicos. Maringá, Eduem. pp. 73–102.

Thompson, K. (1976) The primary productivity of African Wetlands with particular reference to the Okavango Delta. *Proceedings of the Symposium on the Okavango Delta and its Future Utilization.* Gaborone, Botsuana, Nat. Mus., Botswana Soc. pp. 67–79.

Thornton, I. (1965) Nutrient content of rainwater in the Gambia. *Nature,* Londres, 205, 1025.

Thorp, J.H. & Covich, A.P. (1991) *Ecology and classification of North American freshwater invertebrates.* San Diego, Academic Press.

Thorpe, S.A. (1977) Turbulence and mixing in a Scottish Loch. *Philos. Trans. R. Soc. Lond. A,* Londres, 286, 125–181.

Thorton, K.W., Kimmel, B.L. & Paine, F.E. (eds.) (1990). *Reservoir Limnology: Ecological Perspectives.* Nova York, Wiley Interscience.

Tiercelin, J.J. & Mondeguer, A. (1991) The geology of the Tanganyika Trough. In: Coulter, G. (ed.) *Lake Tanganyika and Its Life.* Oxford, Oxford Univ. Press. pp. 7–48.

Timms, B.V. (1993) *Lake Geomorphology.* Adelaide, Gleneagles.

Tundisi, J.G. (1965) *Some Aspects of the Production of Extracellular Substances by Marine Algae.* [Tese de Mestrado]. Southampton, Inglaterra, University of Southampton.

———. (1969a) *Produção Primária, "Standing-Stock" e Fracionamento de Fitoplâncton na Região Lagunar de Cananéia.* [Tese de Doutorado]. São Paulo, USP.

———. (1969b) Plankton studies in a mangrove environment: Its biology and primary production. In: Lagunas Costeras. *Mem. Simp. Int. Lagunas Costeras,* vol. 28–30. Cidade do México, UNAM, UNESCO. pp. 485–494.

———. (1970) O plâncton estuarino. *Contribuições Inst. Oceanográfico,* USP, São Paulo, 19, 1–22. (Série Oceanogr. Biol.).

———. (1977a) *Produção Primária, "Standing-Stock", Fracionamento do Fitoplâncton e Fatores Ecológicos em Ecossistema Lacustre Artificial (Represa do Broa, São Carlos).* [Tese de Livre Docência]. Ribeirão Preto, Faculdade de Filosofia, Ciências e Letras de Ribeirão Preto, USP.

———. (1977b) As obras hidráulicas e seu impacto sobre os ecossistemas. Simp. Bacias Hidrográficas e seu planejamento regional. In: *IX Congr. Bras. de Engenharia Sanitária.* pp. 74–81.

———. (1981a) Limnology in Brazil, present status and perspectives for future research development. In: Mori, S. & Ikusima, I. (eds.) *Proceedings* Kyoto, Japão: Workshop for the Promotion of Limnology in Developing Countries. p. 113–117., 1981a.

———. (1981b) Ecological studies at the lagunar region of Cananéia. *Proceedings of a workshop.* Roma, FAO, UNESCO, SCOR. pp. 298–304.

———. (1981c) Typology of reservoirs in Southern Brazil. *Verh. Int. Verein. Limnol.,* Stuttgart, 21, 1031–1039.

———. (1983a) Tipologia de Represas do Estado de São Paulo. *Anais do Encontro sobre Ecologia Aplicada.* São Paulo, USP. pp. 15–27.

———. (1983b) A review of basic ecological processes interacting with production and standing stock of phytoplankton in lakes and reservoirs in Brazil. *Hydrobiologia,* Baarn, Holanda, 100, 223–243.

———. (1984) Estratificação hidráulica em reservatórios e suas conseqüências ecológicas. *Ciência e Cultura,* Campinas, 36 (9), 1498–1504.

———. (1985) Represas Artificiais: Perspectivas para o Controle e Manejo da Qualidade da Água para usos Múltiplos. *Anais do VI Simpósio Brasileiro de Hidrologia e Recursos Hídricos,* vol. 4. Rio de Janeiro, ABRH. pp. 1–22.

————. (1986) Limnologia de Represas Artificiais. *Boletim de Hidráulica e Saneamento EESC/USP*, São Paulo, 11, 1–46.

————. (1988a) Impactos Ecológicos da Construção de Represas: Aspectos Específicos e Problemas de Manejo. In: Tundisi, J.G. (ed.) *Limnologia e Manejo de Represas*. São Paulo, USP. pp. 1–75. (Série Monografias em Limnologia, vol. I, tomo 1).

————. (1988b) *Limnologia e Manejo de Represas*. São Paulo, USP. (Série Monografias em Limnologia, Vol. I, Tomo 1).

————. (1989) Environmental impact of the Porto Primavera Dam in the wetlands of the Feio River. *Report to the Public Ministry*, S. Paulo State, Environmental Section. São Paulo. pp. 1–18.

————. (1990a) Distribuição espacial, seqüência temporal e ciclo sazonal do fitoplâncton em represa: fatores limitantes e controladores. *Rev. Bras. Biol.*, São Carlos, 50, 937–955.

————. (1990b) Perspectives for ecological modelling of tropical and subtropical reservoirs in South America. *Ecol. Modelling*, Amsterdã, 52, 7–20.

————. (1990c) Ecology and development: Perspective for a better society. *Physiol. Ecol. Jpn*, Kyoto, 27, 93–130.

————. (1990d) Key factors of reservoir functioning and geogarphic aspects of reservoir limnology. Chairman's Overview. *Arch. Hydrobiol./Beith Ergeih. Limnol.*, Stuttgart, 33, 645–646.

————. (1990e) Distribuição espacial, sequência temporal e ciclo sazonal do fitoplâncton em represas: fatores limitantes e controladores. *Rev. Brasil. Biol.*, São Carlos, 50 (4), 937–955.

————. (1993a) The environmental impact assessment of lakes and reservoirs. In: Salánski, J. & Itsvánovics, V. (eds.) *Limnological Bases of Lake Management: Proceedings of the ILEC/UNEP Int. Training Course*. Tihany, Hungria, ILEC/UNEP. pp. 38–50.

————. (1993b) Man-made lakes: Theoretical basis for reservoir management. *Verh. Internat. Verein. Limnol.*, Stuttgart, 25, 1153–1156.

————. (1993c) Represas do Paraná Superior: limnologia e bases científicas para o gerenciamento. In: Boltovskoy, A. & Lopez, H.L. (eds.) *Conferências de Limnologia*. La Plata, Unlp/Conycet. pp. 41–52.

————. (1994) Tropical South América: Present and perspectives. In: Margalef, R. (ed.) *Limnology Now: A Paradigm of Planetary Problems*. Amsterdã, Elsevier Science. pp. 353–424.

————. (1995) Reservatórios na Amazônia. In: Hashimoto, M. & Banett, F.D. (eds.) *Diretrizes para o gerenciamento de lagos*, vol. 2. São Carlos, ILEC/UNEP. p. 177–181.

————. (1997) Climate. In: Tundisi, J.G. & Saijo, Y. (eds.) *Limnological Studies on the Rio Doce Valley Lakes*, Brazil. São Paulo, USP, ABC. pp. 7–13.

————. (1999) Limnologia no século XXI: Perspectivas e Desafios. *7o Congresso Brasileiro de Limnologia*. São Carlos, IIE. pp. 1–24.

————. (2000) Limnologia e Gerenciamento de Recursos Hídricos: Avanços Conceituais e Metodológicos. *Ciência e Ambiente: Gestão das Águas*, vol. 21. Santa Maria, Universidade Federal de Santa Maria. pp. 10–21.

————. (2001) Gerenciamento da Qualidade da Água: Interação entre Pesquisa, Desenvolvimento Tecnológico e Políticas Públicas. *Rev. Bras. Pesq. Desenvolvimento*, São Paulo, 3 (2), 57–68.

————. (2003) *Água no século XXI:* enfrentando a escassez. São Carlos, RIMA, IIE.

————. (2004) Barragens. In: Almanaque *Brasil Sócio-Ambiental:* Água. São Paulo, ISA. pp. 255–257.

————. (2005a) Águas Futuras. In: Moss, G. & Moss, M. (eds.) *Brasil das Águas:* Revelando o Brasil verde e amarelo. Rio de Janeiro, Supernova. pp. 148–151.

————. (2005b) Gerenciamento integrado de bacias hidrográficas e reservatórios: Estudos de caso e perspectivas. In: Nogueira, M.G., Henry, R. & Jorcin, A. (Orgs.) *Ecologia de Reservatórios:* Impactos potenciais, ações de manejo e sistemas em cascata. São Carlos, RIMA. pp. 1–21.

————. (2006) Novas perspectives para a gestão de recursos hídricos. *Revista USP*, São Paulo, 70, 24–35.

————. (2007a) Coupling surface and groundwater research: A new step forward towards water management. International Centres for innovation, research, development and capacity building in water management. In: OECD. *Integrating Science and Technology into Development Policies: An International Perspective*. Paris, OECD. pp. 163–170.

————. (2007b) Exploração do potencial hidrelétrico da Amazônia. *Revista de Estudos Avançados: USP*, São Paulo, 21 (59), 109–117.

Tundisi, J.G. et al. (2003) *Plano de gerenciamento e otimização de usos múltiplos da bacia hidrográfica e do reservatório da UHE Luiz Eduardo Magalhães*. São Carlos, IIE/IIEGA, FINEP.

Tundisi, J.G. & Barbosa, F.A.R. (1995) Conservation of aquatic ecosystems: Present status and perspectives. In: Tundisi, J.G., Bicudo, C.F.M. & Matsumura-Tundisi, T. (eds.) *Limnology in Brazil*. Rio de Janeiro, Academia Brasileira de Ciências, Sociedade Brasileira de Limnologia. pp. 365–371.

Tundisi, J.G. & Matsumura-Tundisi, T. (1968) Plankton studies in mangrove environment. V. Salinity tolerances of some planktonic crustaceans. *Bol. Inst. Oceanogr. USP*, São Paulo, 17 (1), 57–65.

————. (1972) Some aspects of the seasonal cycle of the phytoplankton in tropical inshore waters. *Ciência e Cultura*, Campinas, 24 (2), 189–192.

————. (1982) Estudos Limnológicos no Sistema de Lagos do Médio Rio Doce, Minas Gerais, Brasil. In: ANAIS *do II Sem. Regional de Ecologia*. São Carlos, UFSCar. pp. 133–258.

————. (1984) Comparative limnological studies at three lakes in Tropical Brasil. *Verh. Internat. Verein. Limnol.*, Stuttgart, 22, 1310–1314.

————. (1990) Limnology and eutrophication of Barra Bonita Reservoir, São Paulo State, Southern Brazil. *Arch. Hydrobiol. Beih. Ergebn. Limnol.*, Stuttgart, 33, 661–676.

————. (1994) Plankton diversity in a warm monomictic lake (Dom Helvécio, Minas Gerais) and a Polymictic reservoir (Barra Bonita): A comparative analysis of the intermediate disturbance hypothesis. *An. Acad. Bras. Ciênc.*, Rio de Janeiro, 66 (supl. 1), 15–28.

————. (1995) The Lobo-Broa Ecosystem research. In: Tundisi, J.G., Bicudo, C.F.M. & Matsumura-Tundisi, T. (eds.) *Limnology in Brazil*. Rio de Janeiro, Academia Brasileira de Ciências, Sociedade Brasileira de Limnologia. pp. 219–244.

————. (2001) The Lagoon Region and Estuary Ecosystem of Cananéia, Brazil. In: Seelinger, U. & Kjerfve, B. (eds.) *Coastal Marine Ecosystems of Latin America*, cap. 9. Nova York, Springer. pp. 119–130. (Ecological Studies, 144).

————. (2002) *Lagos e reservatórios*. Qualidade da água: O impacto da eutrofização (tradução), vol. 3. São Carlos, pp. 1–28.

————. (2003) Integration of research and management in optimizing uses of reservoirs: The experience in South America and Brazilian case studies. *Hydrobiologia*, Baarn, Holanda, 500, 231–242.

Tundisi, J.G. & Mussara, M.L. (1986) Morphometry of four lakes in the Rio Doce Valley system and its relationships with primary production of phytoplankton. *Rev. Bras. Biol.*, São Carlos, 46 (1), 159–171.

Tundisi, J.G. & Saijo, Y. (eds.) (1997a) *Limnological studies on the Rio Doce Valley Lakes, Brazil*. Rio de Janeiro, Academia Brasileira de Ciências.

Tundisi, J.G. & Schiel, D. (2002) A Bacia Hidrográfica como Laboratório Experimental para o Ensino de Ciências, Geografia e Educação Ambiental. In: Schiel, D. & Mascarenhas, S. (eds.) *O Estudo das Bacias Hidrográficas: uma Estratégia para a Educação Ambiental*. São Carlos, RIMA, pp. 12–17. IEA, CDCC, Ford Fundation.

Tundisi, J.G. & Straskraba, M. (1994) Ecological basis for the application of ecotechnologies to watershed/reservoir recovery and management. In: Workshop: Brazilian Programme on

Conservation and Management of Inland Waters. *Acta Limnologica Brasiliensia*, Campinas, V (especial), 49–72.

———. (1995) Strategies for building partnerships in the context of river basin management: The role of ecotechnology and ecological engineering. *Lakes Reservoirs: Res. Manage.*, Kasatsu-shi, Japão, 1, 31–38.

———. (1999) *Theoretical Reservoir Ecology and Its Applications*. Leiden, Holanda, IIE, BAS, Backhuys Publishers.

Tundisi, J.G., Matsumura-Tundisi, T., Strixino, S., Martins, M.A., Santos, E.P. & Matos, E. (1972a) *Ecological Studies in a Lacustrine Environment. I. First Assessment of the Environmental Factors*. Progress Report 1. [S.n., s.l.].

———. (1972b) *Ecological Studies in a Lacustrine Environment. II. First Assessment of the Environmental Factors*. Progress Report 2. [S.n., s.l.].

Tundisi, J.G., Matsumura-Tundisi, T., Gentil, J.G. & Nakamoto, N. (1977) Primary production, standing-stock of phytoplankton and ecological factors in a shallow tropical reservoir (Represa do Broa, São Carlos, Brazil). *Sem. Med. Amb. y Rep.*, Montevidéo, 138–172.

Tundisi, J.G., Gentil, J.G. & Dirickson, M.C. (1978a) Seazonal cycle of primary production of nanno and microphytoplankton in a shallow tropical reservoir. *Rev. Bras. de Botânica*, São Paulo, 1, 35–39.

Tundisi, J.G., Matsumura-Tundisi, T., Barbosa, F.A.R., Gentil, J.G., Rugani, C., Pontes, M.C.F., Aleixo, R.C., Okano, W.Y. & Santos, L.C. (1978b) *Estudos Limnológicos no Sistema de Lagos do Parque Florestal do Rio Doce, MG*. São Carlos, UFSCar, DCB, Convênio CETEC UFSCar.

Tundisi, J.G., Matsumura-Tundisi, T., Teixeira, C., Kutner, M.B. & Kinoshita, L. (1978c) Plankton studies in mangrove environment. IX. Comparative studies with oligotrophic coastal waters. *Rev. Bras. Biol.*, São Carlos, 38 (2), 301–320.

Tundisi, J.G., Forsberg, B.R., Devol, A.H., Zaret, T.M., Tundisi, T.M., Dos Santos, A., Ribeiro, J.S. & Hardy, E. (1984) Mixing patterns in Amazon Lakes. *Hydrobiol.*, Baarn, Holanda, 108 (3–15).

Tundisi, J.G., Matsumura-Tundisi, T., Henry, R., Rocha, O. & Hino, K. (1988a) Comparações do Estado Trófico de 23 Reservatórios do Estado de São Paulo: Eutrofização e Manejo. In: Tundisi, J.G. (ed.) *Limnologia e Manejo de Represas*. São Carlos, USP-EESC/CRHEA/ACIESP. pp. 165–204. (Série: Monografias em Limnologia, v. I, Tomo 1).

Tundisi, J.G., Matsumura-Tundisi, T. & Calijuri, M.C. (1988b) The Lobo (Broa) Ecosystem and Reservoirs in Brazil. In: Expert *Group Workshop on River/Lake Basin:* Approach to Environmentally Sound Management of Water Resources. Nagoya, Japão, Unep/ilec/UNCRD p. 28.

———. (1990a) Limnology and management of reservoirs in Brazil. In: Straškraba, M., Tundisi, J.G. & Duncan, A.M. (eds.) *Comparative Reservoir Limnology and Water Quality Management*. Dordrecht, Holanda, Kluwer Academic Publishers.

Tundisi, J.G., Matsumura-Tundisi, T., Calijuri, M.C., Novo, E.M.L.M. & Brandi, J.M. (1990b) River/reservoir approaches to water resources management: The Lobo (Broa) reservoir case study and the hydropower development in the Amazon and Middle Tietê River basins, Brazil. The conceptual training of qualified human resouces in Brazil. In: Third *Expert Group Workshop on River/Lake Basin Approaches to Environmentally Sound Mangement of Water Resources*. Otsu, Japão, Uncrd/ilec/unep. pp. 1–46.

Tundisi, J.G., Matsumura-Tundisi, T., Rocha, O., Henry, R. & Calijuri, M.C. (1992) *Estudo comparado dos mecanismos de funcionamento das Represas de Barra Bonita (Médio Tietê) e Jurumirim (Paranapanema) e dos impactos das bacias hidrográficas*. São Paulo, Fapesp. (Projeto Temático 0612/91–5).

Tundisi, J.G., Bicudo, C.E.M. & Barbosa, F.A.R. (1995a) Synthesis. In: Tundisi, J.G., Bicudo, C.F.M. & Matsumura-Tundisi, T. (eds.) *Limnology in Brazil*. Rio de Janeiro, Academia Brasileira de Ciências, Sociedade Brasileira de Limnologia. pp. 373–376.

Tundisi, J.G., Bicudo, C.E.M. & Matsumura-Tundisi, T. (eds.) (1995b) *Limnology in Brazil*. Rio de Janeiro, Academia Brasileira de Ciências, Sociedade Brasileira de Limnologia.

Tundisi, J.G., Matheus, C.E., Campos, E.G.C. & Moraes, A.J. de. (1997a) Use of the hydrographic basin and water quality in the training of school teachers and teaching of environmental science in Brazil. In: Jorgensen, S.E., Kawashima, M. & Kira, T. (eds.) *A Focus on Lakes/ Rivers in Environmental Education*. Kusatsu-shi, Japão, ILEC.

Tundisi, J.G., Matsumura-Tundisi, T., Fukuara, H., Mitamura, O., Rocha, O., Guillén, S.M., Henry, R., Calijuri, M.C., Ibañez, M.S.R., Espíndola, E.L.G. & Govoni, S. (1997b) Limnology of fifteen Lakes. In: Saijo, Y. & Tundisi, J.G. (eds.) *Limnological studies on the Rio Doce Valley Lakes*, Brazil. Rio de Janeiro, Academia Brasileira de Ciências, São Paulo, USP. pp. 409–440.

Tundisi, J.G., Matsumura-Tundisi, T., Rocha, O., Gaeta Espíndola, E.L., Rietsler, A., Ibañez, M.S., Costa Neto, P., Caliguri, M.C. & Pompeo, M. (1998a) Aquatic biodiversity as last a consequence of the diversity of habitats and functioning mechanisms. *Anais Academia Brasileira Ciência*, Rio de Janeiro, 70 (4, parte I), 767–774.

Tundisi, J.G., Rocha, O., Matsumura-Tundisi, T. & Braga, B. (1998b) Reservoir management in South America. *Water Resour. Develop.*, Nova York, 14 (2), 141–155.

Tundisi, J.G., Matsumura-Tundisi, T. & Rocha, O. (1999) Theoretical basis for reservoir management. In: Tundisi, J.G. & Straškraba, M. (eds.) *Theoretical Reservoir Ecology and Its Applications*. Dordrecht, Holanda, IIE, BAS, Backhurys Publishers. pp. 505–528.

Tundisi, J.G., Braga, B. & Rebouças, A. (2000a) Water for sustainable development: The Brazilian perspectives. In: Miranda, C.E.R. (ed.) *Transition to Sustainability*. Rio de Janeiro, Academia Brasileira de Ciências. pp. 237–246.

Tundisi, J.G., Matsumura-Tundisi, T., Rocha, O. & Espíndola, E.G. (2000b) Limnologia y Gerenciamiento Integrado de Presas en America Del Sur: Avanzos Recientes y Nuevas Perspectivas. In: Tundisi, J.G. (Org.) *Presas, Desarrollo y Medio Ambiente*. São Paulo, Workshop on Dams, Development and the Environment. pp. 17–30.

Tundisi, J.G., Matsumura-Tundisi, T. & Reis, V.L. (2002) Sustainable water resources in South America: The Amazon and La Plata basins. In: Jansky, L., Nakayama, M. & Uitto, J. (eds.) *Lakes and Reservoirs as International Water Systems, Towards World Lake Vision*. Tóquio, United Nations University Press. pp. 28–44.

Tundisi, J.G., Matsumura-Tundisi, T. & Rodriguez, S.L. (2003a) *Gerenciamento e Recuperação das Bacias Hidrográficas dos Rios Itaqueri e do Lobo e da Represa Carlos Botelho (Lobo-Broa)*. São Carlos, IIE, IIEGA, PROAQUA, ELEKTRO.

Tundisi, J.G., Matsumura-Tundisi, T., Abe, D.S. & Reis, V.L. (2003b) *The Management of Reservoirs in Brazil: New Conceptual Advances and Integration of Mathematical and Ecological Modelling*. Montevidéo, CYTED.

Tundisi, J.G., Matsumura-Tundisi, T., Arantes, Jr., J.D., Tundisi, J.E., Manzini, N.F. & Ducrot, R. (2004) The response of Carlos Botelho (Lobo-Broa) reservoir to the passage of cold fronts as reflected by physical, chemical and biological variables. *Braz. J. Biol.*, São Carlos, 64 (1), 177–186.

Tundisi, J.G., Dupas, F.A., Rohm, S.A., Poli, M.N., Matsumura-Tundisi, T., Pareschi, D.C., Souza, A.T.S. & Shibata, O.A. (2005a) Uso atual e uso potencial do solo no município de São Carlos-SP: Base do planejamento urbano e rural. *Anais da I Mostra de Ciência e Tecnologia em Políticas Públicas Municipais:* Unidade Temática de Ciência, Tecnologia e Capacitação da Rede de Mercocidades (26 a 29 agosto/2004). São Carlos, Suprema. pp. 46–53.

Tundisi, J.G., Matsumura-Tundisi, T., Pareschi, D.C., Dupas, F.A., Souza, A.T.S. & Shibata, O.A. (2005b) Diagnóstico e Prognóstico da Qualidade da água dos Rios que Compõem as Bacias Hidrográficas do Município de São Carlos-SP Relacionados ao Uso e Ocupação do Solo. *Anais da I Mostra de Ciência e Tecnologia em Políticas Públicas Municipais:*

Unidade Temática de Ciência, Tecnologia e Capacitação da Rede de Mercocidades (26 a 29 Agosto/2004). São Carlos, Suprema. pp. 69–72.

Tundisi, J.G., Abe, D.S., Matsumura-Tundisi, T., Tundisi, J.E.M. & Vanucci, D. (2006a) Reservatórios da Região Metropolitana de São Paulo: conseqüências e impactos da eutrofização e perspectivas para o gerenciamento e recuperação. In: Tundisi, J.G., Matsumura-Tundisi, T. & Galli, C.S. (eds.) *Eutrofização na América do Sul*: causas, *conseqüências e tecnologias para gerenciamento e controle*. São Carlos, Instituto Internacional de Ecologia, Associação Instituto Internacional de Ecologia e Gerenciamento Ambiental, Eutrosul. pp. 161–180.

Tundisi, J.G., Tundisi, T.M. & Galli, C.S. (2006b) *Eutrofização na América do Sul: causas, consequências e tecnologias para gerenciamento e controle*. São Carlos, IIE.

Tyler, J.E. (1968) The Secchi Disc. *Limnol. Oceanogr.*, Milwaukee, EUA, 13, 1–6.

Uhlmann, D. (1983) *Hydrobiologie*. Viena, Gustav Fisher Viena.

Uieda, V.S. (1995) *Comunidades de Peixes em Riacho Litorâneo: Composição, Habitat e Hábitos*. [Tese de Doutorado]. Campinas, Unicamp.

Ulmer, G. (1907a) Neue Trichopteren. *Notes from the Leyden Museum*, Leiden, Holanda, 29, 1–53.

———. (1907b) Monographie der Macronematinae. *Collections Zoologiques du Baron Edm. de Selys Longchamps*, Bruxelas, 6 (2), 1–121.

———. (1907c) Trichoptera. In Wytsman, P. (ed.) *Genera Insectorum*, parte 60. Bruxelas, L. Desmet Verteneuil [etc.].

UNEP. (2000) *World Resources 2000–2001: People and Ecosystems: The Fraying Web of Life*. Washington, UNDP, UNEP, World Bank, WRI.

UNEP, IIE, World Bank, UNESCO, ANA, ANEEL. (1999) *Planejamento e Gerenciamento de Lagos e Reservatórios:* Uma Abordagem Integrada do Problema da Eutrofização (Ed. em português, tradução D. Vannucci). São Carlos, UNEP, IIE, World Bank, UNESCO, ANA, ANEEL.

UNEP-IETC. (2005) *Handbook on Phytotechnology for Water Quality Improvement and Wetland Management Through Modelling Applications*. [S.l.].

UNESCO. (2003) Compartilhar a água e definir o interesse comum. In: *Água para todos, água para a vida*. Paris, UNESCO. pp. 25–26. (Informe das Nações Unidas sobre o desenvolvimento dos recursos hídricos no mundo).

United Nations University. (1993) *International Network on Water, Environment and Health (INWEH)*. [S.l.]. (Report of the International Feasibility Study Team).

US LTER Network. (1998) *The International Long Term Ecological Research Network*. Novo México.

Val, A.L. (1993) Adaptations of fish to extreme conditions in freshwaters. In: Bicudo, J.E. (ed.) *The Vertebrate Gas Transport Cascade: Adaptations to Environment and Mode of Life*. Boca Ratón, EUA, CRC Press. pp. 43–53.

Val, A.L., Almeida Val, V.M.F. & Randall, D.J. (1996) *Physiology and Biochemistry of the Fishes of the Amazon*. Manaus, INPA.

Valenciennes, A. (1840) Siluroidei. In: Cuvier, G. & Valenciennes, A. *Natural History of the Fishes*, vol. 15. Paris, Estrasburgo, Chez PITOIS. pp. 1–464 (em francês).

———. (1847) Poissons. Catalogue des principales espéces de poissons, raportées de l'Amerique Meridionale. In: D'orbigny, A. (ed.) *Voyage dans l'Amerique Meridionale*, Paris, 5 (2), 1–11.

———. (1849) Des Cynodons. In: Cuvier, G. & Valenciennes, M.A. *Histoire naturelle des poissons*, vol. 22. Paris, P. Bertrand. p. 323–336.

Van der Heide, J. (1982) *Lake Brokopondo:* Filling Phase Limnology of a Man-Made Lake in the Humid Tropics. Amsterdã, Universidade de Amsterdã.

Vannote, R.L., Minshall, G.W., Cummins, K.W., Sedell, J.R. & Cushing, C.E. (1980) The river continuum concept. *Can. J. Fish. Aquat. Sci.*, Toronto, 37, 130–137.

Vannuci, M. & Tundisi, J.G. (1962) Las Medusas presentes en los museos de La Plata y Buenos Aires. *Bol. Mus. Arg. Cienc. Nat. "Bernardino Rivadavia"*, Buenos Aires, 10 (10).

Vareschi, E. (1978) The ecology of Lake Nakuru (Kenya). I. Abundance and feeding of the lesser flamingo. *Oecologia*, Berlim, 32, 11–35.

———. (1982) The ecology of Lake Nakuru (Kenya). III. Abiotic factors and primary production. *Oecologia*, Berlim, 55, 81–101.

Vargas, I.J., Loureiro, C.G.C. & Milward de Andrade, R. (1976) *Anais do I Encontro Nacional sobre Limnologia, Piscicultura e Pesca Continental*. Belo Horizonte, Fundação João Pinheiro, Centro de Recursos Naturais.

Vaz, M.M., Petrere, Jr., M., Martinelli, L.A. & Mozeto, A.A. (1999) The dietary regime of detritivorous fish from the River Jacare-Pepira, Brasil. *Fish. Manag. Ecol.*, Victoria, Austrália, 6, 121–132.

Vazzoler, A.E.A.M. & Menezes, N.A. (1992) Síntese de conhecimentos sobre o comportamento reprodutivo dos Characiformes da América do Sul (Teleostei, Ostariophysi). *Rev. Brasil. Biol.*, São Carlos, 52, 627–640.

Vazzoler, A.E.A.M., Agostinho, A.A. & Hahn, N.S. (1997a) *A planície de inundação do alto Rio Paraná: aspectos físicos, biológicos e sócio-econômicos*. Maringá: Eduem. 460 pp.

Vazzoler, A.E.A.M., Lizama, P.H.M. & Inada, P. (1997b) Influências ambientais sobre a sazonalidade reprodutiva. In. Vazzoler, A.E.A.M., Agostinho, A.A. & Hahn, N.S. *A planície de inundação do alto Rio Paraná: aspectos físicos, biológicos e sócio-econômicos*. Maringá, Eduem. pp. 268–280.

Vezie, C., Brient, L., Sivonen, K., Bertru, G., Lefeuvre, J.C. & Alkinoja-Salonen, M. (1997) Occurence of microcystin-containing cyanobacterial blooms in freshwaters of Brittany (France). *Arch. Hydrobiol.*, Stuttgart, 139, 401–413.

Viner, A.B. (1973) Responses of mixed phytoplankton population to nutrient enrichment of ammonia and phosphate and some associated ecological implications. *Proc. R. Soc. London B.*, Londres, 183, 351–370.

———. (1977) The sediments of Lake George (Uganda): IV. Vertical distribution of chemical features in relation to ecological history and nutrient cycle. *Arch. Hydrobiol.*, Stuttgart, 80, 40–69.

Visser, S.A. (1961) Chemical composition of rainwater in Kampala, Uganda, and its relation to meteoreological and tropical conditions. *J. Geophys. Res.*, Washington, 66, 3759–3765.

Volkmer-Ribeiro, C. & Pauls, S.M. (2000) Esponjas de Agua Dulce (Porifera, Demospongiae) de Venezuela. *Acta Biologica Venezuelica*, Caracas, 20 (1), 1–28.

Vollenweider, R.A. (1965) *Scientific Fundamentals of the Eutrophication of Lakes and Flowing Waters, with Particular Reference to Nitrogen and Phosphorus as Factors in Eutrophication*. Paris, OECD. (DAS/CSI/68.27).

———. (1969) Possibilities and limits of elementary models concerning the budget of substances in lakes. *Arch. Hydrobiol.*, Stuttgart, 66, 1–36.

———. (1974) *A Manual on Methods for Measuring Primary Production in Aquatic Environments*, 2nd edition. Londres, Blackwell Scientific Publications. (IBP Handbook, 12).

———. (1976) Advances in defining critical loading levels for phosphorus in lake management. *Mem. Ist. Ital. Idrobiol.*, Verbiana-Pallanza, Itália, 33, 53–83.

———. (1987) Scientific concepts and methodologies pertinent to lake research and lake restoration. *Schweiz. Z. Hydrol.*, Zurique, 49 (2), 129–147.

Von Sperling, E. (1999) *Morfologia de Lagos e Represas*. Belo Horizonte: DESA/UFMG.

Walbaum, J.J. (1792) *Petri Arte di Sueci Genera Piscium . . .* Emendata et aucta a Iohanne Iulio Walbaum, M.D. Grypeswaldiae (Greifswald), Alemanha, Ant. Ferdin. Roese.

Walker, B.H. (1992) Biodiversity and ecological redundancy. *Cons. Biol.*, Victoria, Austrália, 6, 18–23.

Walker, I. (1995). Amazonian Streams and Small Rivers. In: Tundisi, J.G., Bicudo, C.F.M. & Matsumura-Tundisi, T. (eds.) *Limnology in Brazil*. Rio de Janeiro, Academia Brasileira de Ciências, Sociedade Brasileira de Limnologia. pp. 166–193.

Ward, H. & Whipple, G.C. (1959) *Fresh-water Biology*, 2nd edition. Nova York, John Wiley & Sons.

Ward, J.V. & Stanford, J.A. (1983) The intermediate disturbance hypothesis: An explanation for biotic diversity patterns in lotic ecosystems. In: Fontaine, T.D. & Bartell, S.M. (eds.) *Dynamic of Lotic Ecosystems*. Michigan, Ann Arbor Science Publishers. pp. 347–356.

———. The four-dimensional nature of lotic ecosystems. *J. N. Am. Benthol. Soc.*, 8, 2–8.

———. (1992) *Aquatic Insect Ecology: 1. Biology and habitat*. Nova York, John Wiley & Sons.

Watanabe, M.F. & Oishi, S. (1980) Toxicities of *Microcystis aeruginosa* collected from some lakes, reservoirs, ponds and moat in Tokyo and adjacent regions. *Jpn. J. Limnol.*, Osaka, 41, 5–9.

Watanabe, T. (1990) Perifiton: Comparação de metodologias empregadas para caracterizar o nível de poluição das águas. *Acta. Limnol. Brasil.*, Campinas, 3, 593–615.

Watson, R.T., Dixon, J.A., Harburg, S., Janetos, A.C. & Moss, R.H. (1998) *Protecting Our Planet, Securing Our Future: Linkages Among Global Environmental Issues and Human Needs*. Washington, UNEP, NASA, World Bank.

Weber, C.A. (1907) Aufbau und Vegetation der Moore Norddeutschlands. *Beibl. Bot. Jahrb.*, Stuttgart, 90, 19–34.

Welch, E.B. (1980) *Ecological Effects of Waste Water*. Cambridge, Cambridge University Press.

Welch, P.S. (1935) *Limnology*. Nova York, McGraw-Hill.

———. (1948) *Limnological Methods*. Nova York, McGraw-Hill.

Welcomme, R.L. (1979) *Fisheries Ecology of Floodplain Rivers*. Londres, Longman.

———. (1985) River Fisheries. *FAO Fish. Tech. Pap.*, Roma, 262, 330.

Wesenberg-Lund, C. (1910) Summary of our knowledge regarding various limnological problems. In: Murray, J. & Pullar, L. (eds.) *Bathymetrical Survey of the Scottish Freshwater Lochs*, vol. I. Edimburgo, Challenger Office. pp. 374–438.

Westlake, D.F. (1963) Comparations of plants productivity. *Biol. Rev.*, Cambridge, 38, 385–425.

———. (1965) Some basic data for investigations of the productivity of aquatic macrophytes. *Memorie Dell'Instituto Italiano di Idrobiologia Dott. Marco de Marchi*, Verbiana-Pallanza, 18, 229–248.

Wetzel, R.G. (1964) A comparative study of the primary productivity of higher aquatic plant, periphyton and the phytoplankton in a large shallow lake. *Int. Rev. Ges. Hydrobiol. Hydrogr.*, Leipzig, Alemanha, 49, 1–64.

———. (1975) *Limnology*. Philadelphia: W.B. Saunders Company.

Wetzel, R.G. (ed.) (1983a) *Periphyton on Freshwater Ecosystems*. Haia, Holanda, Dr. W. Junk Publishers, Kluwer Academic Publishers.

———. (1983b) *Limnology*, 2nd edition. Philadelphia, Saunders Company.

———. (1983c) Recommendation for future research on periphyton. In: Wetzel, R.G. (ed.) *Periphyton of Freshwater Ecosystems*. Haia, Holanda, Dr. W. Junk Publ. pp. 339–346.

———. (1990a) Reservoir ecosystems: Conclusion and speculations, In: Thorton, K.W., Kimmel, B. & Payne, F.E. (eds.) *Reservoir Limnology: Ecological Perspectives*. Nova York, Wiley Interscience. pp. 227–238.

———. (1990b) Land–water Interfaces: Metabolic and Limnological Regulators. *Verh. Int. Vererin. Limnol.*, Stuttgart, 24, 6–24.

———. (1992) Clean water: A fading resource. *Hydrobiologia*, Baarn, Holanda, 243/244, 21–30.

———. (2001) *Limnology: Lake and River Ecosystems*, 3rd edition. Nova York, Elsevier Academic Press.

Wetzel, R.G. & Likens, G.E. (1991) *Limnological Analysis*, 2nd edition. Nova York, W.B. Saunders Company.

Wetzel, R.G. & Richard, D. (1996) Application of secondary production methods to estimates of net aboveground primary production of emergent aquatic macrophytes. *Aquatic Botany*, Nova York, 53, 109–120.

Whatley, J.M. & Whatley, F.R. (1981) Chloroplast evolution. *New Phytol.*, Lancaster, Inglaterra, 87, 233–247.

Whitaker, V. & Matvienko, B. (1998) The denitrification potential and hydrological conditions in the wetlands of the Lobo reservoir. *Verh. Internat. Verein. Limnol.*, Stuttgart, 26, 1377–1380.

Whitaker, V.A., Matvienko, B. & Tundisi, J.G. (1995) Spatial heterogeneity of physical and chemical conditions in a tropical reservoir wetland. *Lakes & Reservoirs*, Kusatsu-shi, Japão, 1 (3), 169–176.

Whitton, B.A. (ed.) (1975) *River Ecology: Studies in Ecology*, vol. 5. Liverpool, Blackwell Scientific Publications.

Wilhemy, H. (1958) Umlaufseen und Dammuferseen tropischer Tieflandflusse. *Zeitschrift für Geomorphologie*, Stuttgart, 2, 27–54.

Willen, T. & Mattsson, R. (1997) Water-blooming and toxin-producing cyanobacteria in Swedish fresh and brackish waters, 1981–1995. *Hydrobiologia*, Baarn, Holanda, 353, 181–192.

Williams, T., Gier, K.M. & Thouveny, N. (1993) Preliminary 50m palaeomagnetic records from Lac du Bouchet, Haute Loire, France. In: Negendank, J.F.W. & Zolitschka, B. (eds.) *Palaeolimnology of European Maar Lakes*. Berlim, Springer. pp. 367–376.

Williams, W.D. (1964) A contribution of lake typology in Victoria, Australia. *Verh. Internat. Verein. Linnol.*, Stuttgart, 15, 158–168.

———. (1986) Conductivity and salinity of Australian Salt Lakes. *Aust. J. Mar. Freshwat. Res.*, Victoria, Austrália, 37, 177–182.

———. (1972) The uniqueness of salt lake ecosystems. In: Kajak, S. & Hillbricht-Illkowska, A. (eds.) *Productivity Problems of Freshwaters*. Varsóvia, Academia Polonesa de Ciência. pp. 349–361.

———. (1996) The Largest, the Biggest and Lowest Lakes in the World: Saline Lakes. *Ver. Internat. Verein Limnol.*, Stuttgart, 26, 61–79.

World Bank. (1993) *Water Resources Management: A World Bank Policy Paper*. Washington, World Bank, 1993.

World Resources Institute. (1991) World Resource: 1990–1991. Oxford, WRI, UNEP, UNDP, Oxford University Press.

———. (2000) *A Guide to World Resources 2000–2001: People and Ecosystems: The Fraying Web of Life*. Washington.

———. (2005) *Ecosystems and Human Well-Being: Wetlands and Water: Synthesis*. Washington.

Worthington, E.B. (ed.) (1975) *The Evolution of the IBP*. Cambridge: Cambridge University Press.

Wright, S.T. (1927) A revision of the South American species of Diaptomus. *Trans. Am. Microsc. Soc.*, Malden, EUA, 46, 73–121.

———. (1935) Three new species of Diaptomus from Northeast Brazil. *Ann. Acad. Brasil. Sci.*, Rio de Janeiro, 7 (3), 213–233.

———. (1937a) A review of some species of Diaptomus from São Paulo. *Ann. Acad. Brasil. Sci.*, Rio de Janeiro, 9 (1), 65–82.

———. (1937b) Chemical conditions in some waters of northeast Brazil. *Annaes Academia Bras. Sciencias*, 9 (4), 278.

Yoshimura, S. (1938) Dissolved oxygen of the lake waters of Japan. *Sci. Rep. T.B.D. Se. C.*, Tóquio, 2 (8), 64–215.

Zalewski, M. (2002) Ecohydrology: The use of ecological and hydrological processes for sustainable management of water resouces. *Hydrol. Sci. J.*, Wallingford, 47, 825–834.

Zalewski, M., Janauer, G.A. & Jolankai, G. (1997) *Ecohydrology: A new Paradigm for the Sustainable Use of Aquatic Resources*. Paris, UNESCO. (UNESCO IHP-V Projects 2.3/2.4: Technical Document in Hydrobiology, 7).

Zalewski, M., Harper, D.M. & Robarts, R.D. (2004a) Ecohydrology from theory to action. Proceedings of the International Conference, Wierzba, Poland, 18–21 May 2003. *Int. J. Ecohydrol. Hydrobiol.*, Varsóvia, 4 (3), 229–355.

Zalewski, M. & Loktowska, I.W. (eds.) (2004b) *Integrated Watershed Management: Ecohydrology & Phytotechnology Manual*. Otsu/Nagoya, Japão, Unesco, IHP, Unep, TETC.

Zalocar, Y., Bonetto, C.A. & Lancelle, H. (1982) Algunos aspectos limnológicos de la laguna La Herradura (Formosa, Argentina). *Ecosur*, Rosario, Argentina, 9 (18), 171–188.

Zalocar, Y., Poi de Neiff, S.G.A. Casco, S.L. (2007) Abundância e diversidade do fitoplâncton do rio Paraná (Argentina) 220 km a jusante do reservatório de Yacyretá. *Braz. J. Biol.*, 67 (1), 53–63.

Zani Teixeira, M.L. (1983) *Contribuição ao conhecimento da ictiofauna da baía do Trapandé, complexo estuarino-lagunar de Cananéia*. [Tese de Mestrado]. São Paulo, USP.

Zaret, T.M. (1984) Central American Limnology and Gatun Lake, Panamá. In: Taub, F.B. (ed.) *Ecosystems of the World: Lakes and Reservoirs*. Amsterdã, Elsevier. pp. 447–465.

Zaret, T.M. & Paine, R.T. (1973) Species introduction in a tropical lake. *Science*, Washington, 182 (2), 449–455.

Watershed and water body index

General index

Color plates

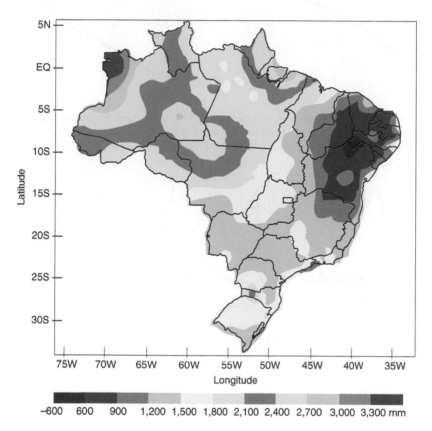

−600 600 900 1,200 1,500 1,800 2,100 2,400 2,700 3,000 3,300 mm

Plate 1 Characteristics of **average annual precipitation** (in mm) in Brazil (CPTEC/Inpe) (see also Figure 2.7, page 35).
Source: Rebouças *et al.* (2002).

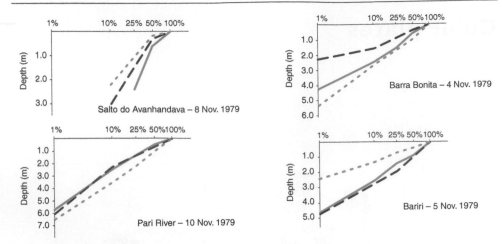

Plate 2   Relative penetration of solar radiation of various wavelengths (logarithmic scale) in various
reservoirs in the state of São Paulo (see also Figure 4.3, page 66).
Source: Reservoir Typology project of the state of São Paulo – Fapesp).

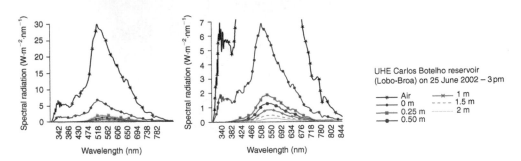

Plate 3   Spectral distribution of underwater solar radiation in the UHE Carlos Botelho (Lobo/Broa)
reservoir 25 June 2002 (3 pm) (see also Figure 4.4, page 66).
Source: Rodrigues (2003).

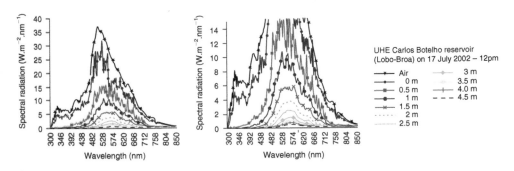

Plate 4   Spectral distribution of underwater solar radiation in the Barra Bonita (SP) reservoir of São
Paulo, on 17 July 2002 (12 pm) (see also Figure 4.5, page 67).
Source: Rodrígues (2003).

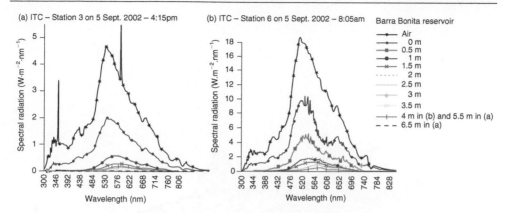

(a) ITC – Station 3 on 5 Sept. 2002 – 4:15pm   (b) ITC – Station 6 on 5 Sept. 2002 – 8:05am   Barra Bonita reservoir

Air
0 m
0.5 m
1 m
1.5 m
2 m
2.5 m
3 m
3.5 m
4 m in (b) and 5.5 m in (a)
6.5 m in (a)

Plate 5   Spectral distribution of underwater solar radiation in the Barra Bonita (SP) reservoir (see also Figure 4.6, page 67).
Source: Rodrigues (2003).

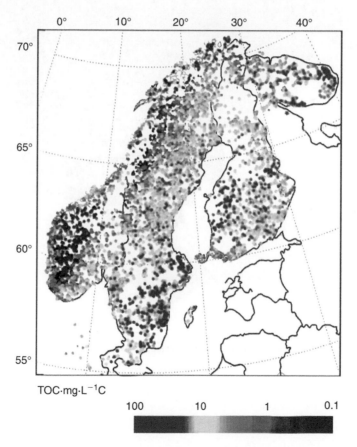

TOC·mg·L^{-1}C

Plate 6   Distribution of total organic carbon in waters in Scandinavia (see also Figure 5.3, page 106).
Source: Skjelvale *et al.* (2001) in Eloranta (2004).

Plate 7   Concentration of total carbon in natural surface water of Brazil (Brazil Water Project) (see also Figure 5.4, page 107).

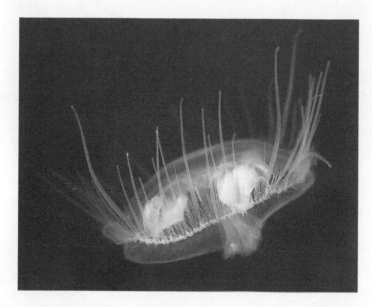

Plate 8   Craspedacusta sowerbii (see also Figure 6.9, page 153).

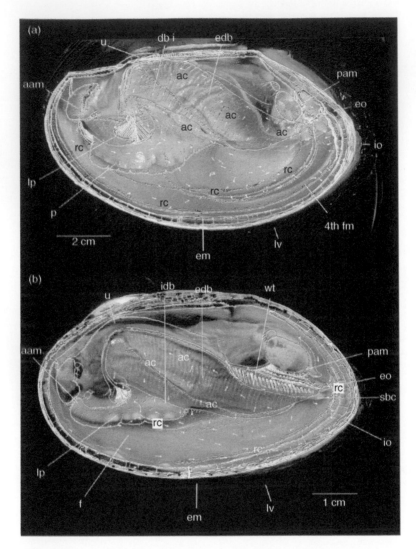

Plate 9   a) *Anodontites trapesialis*; b) *A. elongantus*. Schematic representation of the ciliary currents (acceptance and rejection). Abbreviations: eo – exhalant opening, io – inhalant opening, em – edge of mantle, ac – acceptance current, rc – rejection current, sbc – supra branchial chamber, 4th fm – 4th fold of the mantle, edb – external demibranch, idb – internal demibranch, aam – anterior adductor muscle, pam – posterior adductor muscle, f – foot, lp – labial palps, wt – water tubules, lv – left valve, u – umbo (see also Figure 6.10, page 155).
Sources: a) Lamarck (1819); b) Swaison (1823) in Callil (2003).

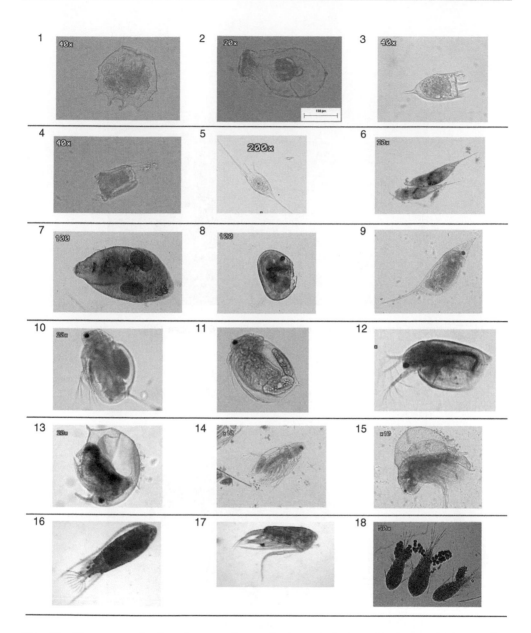

Plate 10 **Common planktonic organisms.** Rotifers: *1–6* (*1* – *Brachionus dolabratus; 2 –
Asplanchna sieboldi; 3 – Keratella cochlearis; 4 – Polyarthra vulgaris; 5 – Kellicotia bostoniensis;
6 – Trichocerca cylindrica chattoni*); *7 – Turbellaria; 8 – Ostracoda;* Cladocerans: *9–15*
(*9 – Daphnia gessneri; 10 – Moina minuta; 11 – Ceriodaphnia cornuta; 12 – Simocephalus sp.;
13 – Bosmina hagmanni; 14 – Diaphanosoma birge; 15 – Holopedium amazonicum*); Copepods:
16–18 (*16 – Notodiaptomus iheringi, female; 17 – Notodiaptomus iheringi, female; 18 – Three
genera of Cyclopoida: Acanthocyclops, Mesocyclops and Thermocyclops*) (see also Figure 6.14,
page 161).

Plate 11   Tambaqui (*Colossoma macropomus*) (see also Figure 6.18, page 170).

Plate 12   Capybara (*Hydrochoerus hydrochaeris*), the biggest rodent in the world: 1 m long, 50 cm tall and weighing 60 kg (see also Figure 6.19, page 172).

Plate 13   Amazonian manatee (*Trichechus inunguis*) (see also Figure 6.20, page 172).

Plate 14   White-faced Whistling Duck (*Dendrocygna viduata*) (see also Figure 6.21, page 173).

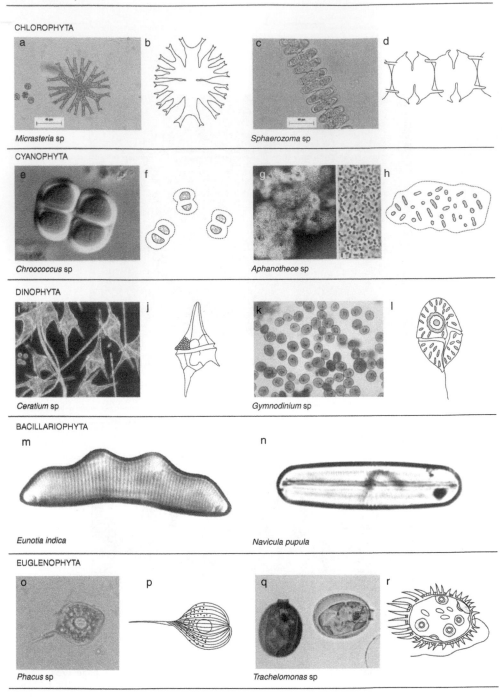

CHLOROPHYTA

a   b

Micrasteria sp

c   d

Sphaerozoma sp

CYANOPHYTA

e   f

Chroococcus sp

g   h

Aphanothece sp

DINOPHYTA

i   j

Ceratium sp

k   l

Gymnodinium sp

BACILLARIOPHYTA

m

Eunotia indica

n

Navicula pupula

EUGLENOPHYTA

o   p

Phacus sp

q   r

Trachelomonas sp

Plate 15   Representative examples of the different divisions of phytoplankton (see also Figure 7.3, page 202).
Sources: Prescott, 1978 (b, d), Canter-Lund and Lund, 1995 (e, g, i, k, q); Hino and Tundisi, 1984 (m, n); Mizuno, 1968 (p); Bicudo and Menezes, 2005 (j, l, r); Silva, 1999 (f, h); Thais Ferreira Isabel (a, c); Ana Paula Lucia (o).

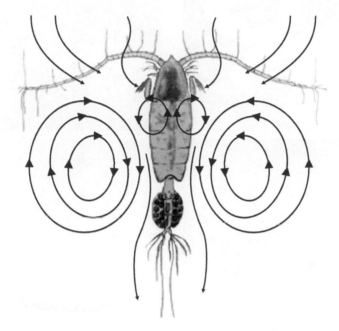

Plate 16   Filtering mechanisms of herbivorous copepods, represented by an original scheme for
Calanus finmarchicus (Marshall, 1972) but valid for herbivorous copepods of inland waters in
general (see also Figure 8.13, page 279).
Source: Modified from Marshall and Orr (1972).

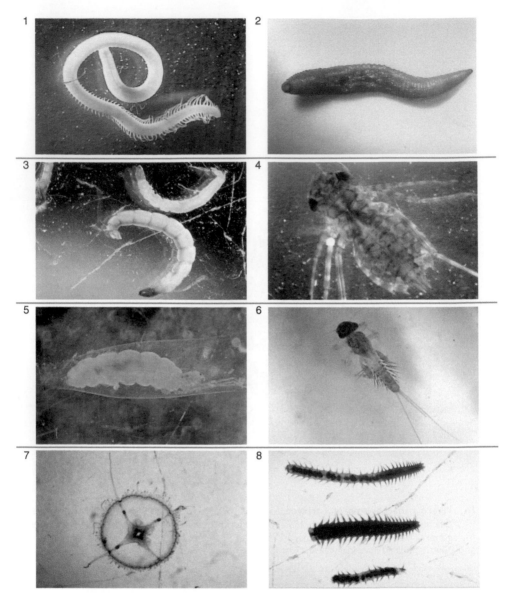

Plate 17   Organisms resistant to pollution: (1) *Branchiura sowerbyi* (Oligochaeta, Tubificidae, collected in the Ibitinga reservoir, middle Tietê, SP), (2) Hirudinea, Glossiphonidae (collected in the Xingu River, AM), (3) *Coelotanypus* sp (Chironomidae larvae, Tanypodinae); organisms tolerant to average pollution – (4) Libellulidae (Odonata larva, collected in the Ibitinga reservoir, middle Tietê, SP); organisms sensitive to pollution – (5) Trichoptera (larva inside the house, collected in the Xingu River, AM), (6) Ephemeroptera, Leptophlebiidae (larva, collected on the Xingu River, AM), (7) *Craspedacusta sowerbyi* (Cnidaria, rare, collected in the Tocantins river, TO), (8) Freshwater Polychaeta (often collected in the Xingu River, MA) (see also Figure 8.21, page 292).
Photos: Daniela Pareschi Cambeses.

Plate 18 Experimental systems for study and response of limiting nutrients (see also Figure 10.12, page 366).

Plate 19   Terrestrial vegetation flooded and in decomposition in the UHE **Luis Eduardo Magalhães** Lageado reservoir, Tocantins River (see also Figure 12.15, page 432).

Plate 20   Man-made urban reservoir – Paranoá Lake, Brasilia (see Annex 6) (see also Figure 12.17, page 439).

Plate 21   UHE Carlos Botelho reservoir (Lobo/Broa), where limnological studies began with methods that led to the project 'Typology of reservoirs in the state of São Paulo' and the other projects in Brazil, especially in the Southeast region (see also Figure 16.20, page 604).

Plate 22   Flamingos (Phoeniconaias minor) on Lake Nakuru. Photo credit to J. G. Tundisi – October, 2005 (see also Figure 17.5, page 616).

Plate 23   Lake Biwa, the largest lake in Japan. In the detailed inset: the intrusion of sediment in Lake Biwa, originating from the thaw (north base of the lake) (see also Figure 17.13, page 632).
Source: Nakamura and Nakagima (2002).

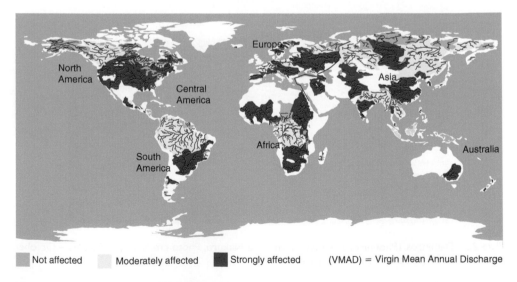

Plate 24   Impacts from the construction of dams and modifications of the courses of rivers in 292 drainage basins on all continents (see also Figure 18.4, page 649).
Source: Modified from Nilson *et al.* (2005).

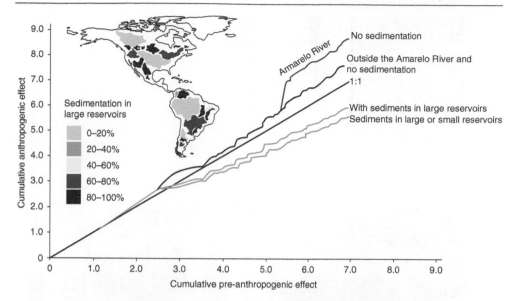

Plate 25 Comparison of the pre-anthropogenic loads with recent sediment loads, using data from 216 rivers, before and after the construction of dams. Data are presented as cumulative curves established by descending level of discharge (see also Figure 18.5, page 650).
Source: Modified from Syvitski *et al.* (2005).

Plate 26 The golden mussel is an invasive species that since 2002 has caused major problems in the drainage basins of southern and southeastern Brazil (see also Figure 18.15, page 677).
Source: Furnas Centrais Electricas.

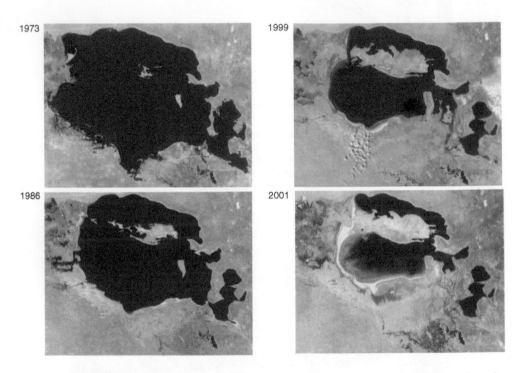

Plate 27   Changes in the Aral Sea as result of the overuse of tributary waters for irrigation (see also
Figure 18.19, page 687).
Source: Millennium Ecosystem Assessment (2005).

Plate 28   Examples of experiments of different scales and complexities to study the effects of enrichment with nitrogen and phosphorus and biomanipulation in microcosms and macrocosms. (a) Luis Eduardo Magalhães hydroelectric dam/reservoir, Tocantins; (b) Lake Suwa, Japan; (c) and (d) UHE Carlos Botelho reservoir (Lobo/Broa), Brotas–SP (see also Figure 20.2, page 723).

Plate 29   Advanced monitoring station in real time. SMATER® model (see also Figure 20.8, page 735).

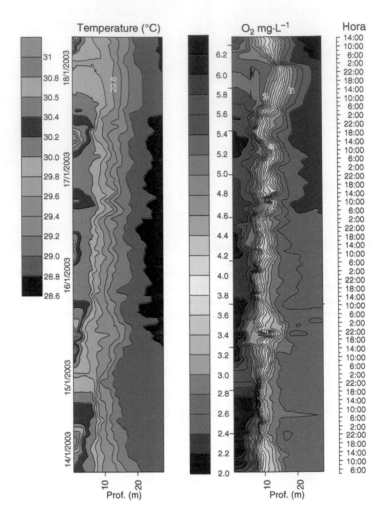

Plate 30   The chronological evolution of climatologic data – air temperature, wind speed and solar radiation – and limnological data of water temperature and dissolved oxygen in the UHE Luis Eduardo Magalhães reservoir (Lajeado/Tocantins), obtained with the real-time monitoring station. Observe the vertical profiles coupled with the effects of radiation, wind and temperature. Data for dissolved oxygen and water temperature are presented from 14 to 18 January 2003 (see also Figure 20.9, page 735–736).
Source: International Ecology Institute, FAPESP/Investco/FINEP (2001).

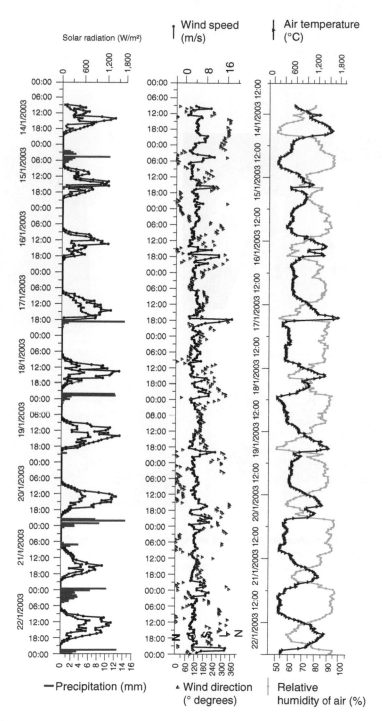

Plate 30   Continued

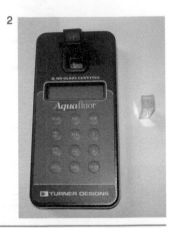

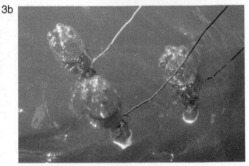

Plate 31   Traditional and advanced equipment and systems for calculating physical, chemical and biological measurements in aquatic ecosystems. (1) Collector of sampling by gravity (UWITEC, Austria), used to quantify gases in sediment; (2) Portable fluorometer to measure chlorophyll in the field; (3) Diffusion chambers (a) miniatures used to measure diffusion of gas through the air-water interface and (b) buoys of the diffusion chambers

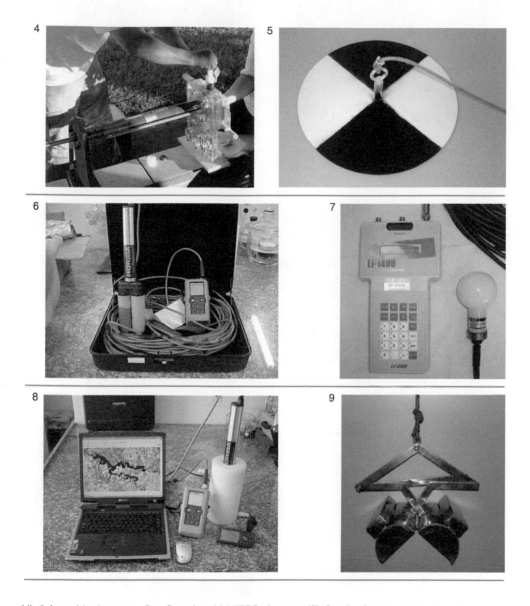

(4) Adams-Niederreiter Gas Sampler; UWITEC, Austria; (5) Secchi disc to measure water transparency; (6) Multi-parametric probe for physical and chemical estimations in water; (7) Underwater radiation meter; (8) Flow cells for continuous use and horizontal profiles in lakes, reservoirs and rivers; (9) Petersen dredge to collect benthic organisms and sediments.

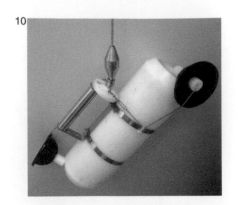

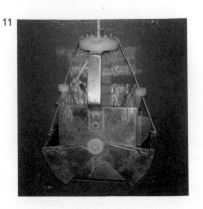

(10) Van Dorn sample bottle to collect water; (11) Birge-Ekman dredge to collect benthic organisms and sediments; (12) Net plankton; (13) Fluoroprobe sensor for selective determination of phytoplankton pigments; (14) Fluoroprobe sensor reader; (15) Multi-parameter probe reader with GPS (see also Figure 20.13, page 748–750)..

Photos: Fernando Blanco Nestor F. Mazini.

About the authors

José Galizia Tundisi is president of the International Institute of Ecology São Carlos-SP. He holds a doctorate in Biological Sciences from the USP and the University of Southampton in England. He has extensively researched the ecology of estuaries, reservoirs and lakes. He has advised Doctorate and Masters theses in limnology, oceanography, ecology and natural resources. He has published many scientific articles and books.

As a consultant, he has advised and participated in management projects and restoration of water basins, reservoirs and lakes in 38 nations around the world, including countries in the Western Hemisphere and Africa and the Far East (China, Japan, Thailand), with support from the World Bank, the Organization of American States, and UNEP (The United Nations Environmental Program).

He is a title member of the Brazilian Academy of Sciences, the Institute of Ecology (Germany), the International Network on Water and Human Health (The United Nations, Canada), and the Third World Academy of Sciences (Triestre, Italy).

He received the Augusto Ruschi Medal from the Brazilian Academy of Sciences; the Bouthors Galli Award from the United Nations; the Anisio Teixeira Award from Brazil's Ministry of Education and the Conrad Wessel Foundation award for applied science of water, among other things. He is at present full professor of Environmental Quality at Universidade Feevale, Novo Hamburgo, Rio Grande do Sul, Brazil.

Takako Matsumura is professor emeritus of the Federal University of São Carlos-SP and director of research at the International Institute of Ecology. She has authored four books and more than 100 articles in national and international journals. She has coordinated major limnological research projects funded by Fapesp, CNPq and CTHIDRO, including the Biota/Faspesp project. She has been an advisor for countless Masters and Doctorate theses. Her principal lines of research include limnology, management and restoration of water basins, water quality and eutrophication. She has contributed significantly to studies on distribution of zooplankton in inland waters and phytoplanktonic and zooplanktonic biodiversity in reservoirs.

Printed and bound by CPI Group (UK) Ltd, Croydon, CR0 4YY

18/10/2024

01776249-0006